Mechanics of Mechanisms and Machines

Mechanics of Mechanisms and Machines

Ilie Talpasanu and
Alexandru Talpasanu

CRC Press
Taylor & Francis Group
Boca Raton London New York

CRC Press is an imprint of the
Taylor & Francis Group, an **informa** business

eResource material is available for this title at https://www.crcpress.com/Mechanics-of-Mechanisms-and-Machines/Talpasanu-Talpasanu/p/book/9781498735476

CRC Press
Taylor & Francis Group
6000 Broken Sound Parkway NW, Suite 300
Boca Raton, FL 33487-2742

Printed on acid-free paper

International Standard Book Number-13: 978-1-4987-3547-6 (Hardback)

Library of Congress Cataloging-in-Publication Data

Names: Talpasanu, Ilie, author. | Talpasanu, Alexandru, author.
Title: Mechanics of mechanisms and machines / Ilie Talpasanu and Alexandru Talpasanu.
Description: Boca Raton : Taylor & Francis, a CRC title, part of the Taylor & Francis imprint, a member of the Taylor & Francis Group, the academic division of T&F Informa, plc, 2018.
Identifiers: LCCN 2018036250 | ISBN 9781498735476 (hardback : acid-free paper)
Subjects: LCSH: Mechanical movements—Mathematical models. | Mechanics.
Classification: LCC TJ181 .T35 2018 | DDC 621.8—dc23
LC record available at https://lccn.loc.gov/2018036250

Visit the Taylor & Francis Web site at
http://www.taylorandfrancis.com

and the CRC Press Web site at
http://www.crcpress.com

Printed and bound by CPI Group (UK) Ltd, Croydon, CR0 4YY

In memory of our parents Sofia and Nicolae Talpasanu

In memory of Professor Mihail Atanasiu and Professor Radu Voinea

Contents

Preface

This book presents novel analytical methods for mechanisms with respect to kinematic, dynamic, and static aspects. What is presented is a set of analytic equations for positions, velocities, accelerations, and forces, which can be used in the control algorithms of industrial and commercial mechanisms and robots. These equations are presented for various types of mechanisms, and a dynamic method of computation can be derived for complex mechanisms whereby the method shown in this book can be applied to almost any mechanism. In addition, equations for actuating forces/torques and reaction forces/torques are presented as they apply to a variety of planar and spatial open cycle, single- and multiple-cycle linkages, and gear trains and cams.

The authors started the development of these equations 35 years ago while searching for efficient computer algorithms and how they can be applied to real-world mechanisms. These equations presented in this book are an alternative to the graphical vector polygon or free-body diagram methods. In kinematics, a novel matroid method is presented, which introduces an approach based on an oriented edge graph, also known as a digraph, and how this can be applied to a mechanism. To apply the matroid method, vector equations are defined, which correspond to the number of rows in a cycle basis matrix; consequently, the number of columns in the cycle basis matrix corresponds to the number of joints in the mechanism. The equations developed for dynamic analysis, also known as the IT equations, have a standard form, namely that all mechanisms with the same digraph share the same equations.

The text requires a basic knowledge of college-level algebra as well as dynamics courses. The topics presented are of interest for undergraduate students in mechanical engineering, mechanical engineering technology, aerospace engineering, biomechanics, and robotics. Students working on their final school projects, researchers, design engineers involved in mechatronics, robotics, vehicle dynamics, and biomechanics will find this text as a useful guide, which can be used during the design and analysis of direct and inverse kinematics and dynamics of various mechanisms.

The results presented in this book for the various analysis of a large variety of mechanisms were performed using an Engineering Equation Solver: Klein, S.A., EES-Engineering Equation Solver, F-Chart Software, http://fchart.com.

The results are compared and correlated with implementations of the same mechanisms in the SolidWorks Motion simulation software. SolidWorks is a registered trademark of Dassault Systemes SolidWorks Corporation.

Animations

Readers will find the animation clips referenced in the text at https://www.crcpress.com/Mechanics-of-Mechanisms-and-Machines/Talpasanu-Talpasanu/p/book/9781498735476. The animations may be found under the "Downloads & Updates" tab on that webpage.

Acknowledgments

Sincere thanks to Leslie Talpasanu for reviewing the manuscript.

The authors would like to thank Prof. Stephen Chomyszak from Wentworth Institute of Technology who made useful suggestions. We thank the following students for testing the simulation files:

Allison Sirois, Richard McNabb, Ryan Shanley, Ismail Gomaa, Kelley Coates, Michael Cushera, Zeke Fowler, Nicholas Timm, Faisal Alfalastini, Aous Ymani, Joshua Martin, and Collin Ritter.

Our gratitude is expressed to Jonathan W. Plant from CRC Press.

Authors

Ilie Talpasanu is a professor in the Department of Mechanical Engineering at the Wentworth Institute of Technology, Boston. He earned a PhD at the University of Texas at Arlington and Doctor Engineer at the Polytechnic University of Bucharest. He was a topic co-organizer for American Society of Mechanical Engineers (ASME) International Mechanical Engineering Congress and Exposition, ASME International Design and Engineering Technical Conferences and Computers and Information in Engineering Conference, and a member in the executive committee of the American Society of Mechanical Engineers—West Texas Section. He is the author and co-author for several books and book chapters. His publications are at peer-reviewed journals such as *ASME Journal of Mechanical Design, ASME Journal of Mechanisms and Robotics, International Journal of Mechanisms and Robotic Systems, SAE International Journal of Passenger Cars—Mechanical Systems, Mechanism and Machine Theory,* and ASME, IEEE, IFTOMM conferences. Ilie has been happily married to Dalia Talpasanu for 40 years.

Alexandru Talpasanu graduated from the Georgia Institute of Technology with a BS in computer engineering as well as an MS in electrical and computer engineering. Alex currently works as a senior software engineer. From his young age, he has been interested in robotics and its applications. Fascinated with how computers work, he started his quest to understand the magic beyond the keyboard. Alex enjoys programming and algorithm development as well as developing enterprise level computer applications. He finds robotics to be a fascinating field and has volunteered as a judge for various robotic competitions on several occasions. He is married to his lovely wife, Leslie Talpasanu.

1

Background

1.1 Links and Links Numbering

Links are considered as rigid members of a mechanism.

For any planar or spatial, open or closed cycle mechanism with n links, their labeling begins with 0 assigned to the fixed link (ground), then sequentially from 1 to m mobile links. For any mechanism with n total links, the total number of mobile links is therefore

$$m = n - 1 \tag{1.1}$$

where

m = total number of mobile links

n = total number of links.

When a link only requires two connections or less, it is defined as simple, as shown in Figure 1.1. However, when more connections are required in the mechanism's design, the link is defined as complex (Figure 1.2). This text treats links as rigid, keeping their length constant when acted on by forces. Other nonrigid mechanism members such as belts, cables, and chains are also modeled as rigid links of constant length. Elastic parts such as springs are not considered as links. The elastic forces in springs will be considered in statics and dynamics of mechanisms, although they are not considered in kinematics since, in the study of motion, the forces are omitted.

1.2 Joints and Joints Labeling

A joint is a connection between two links.

The joints are labeled with CAPITAL LETTERS, and in case of multiple jointed links, like in Figure 1.4, each joint is labeled with a different letter.

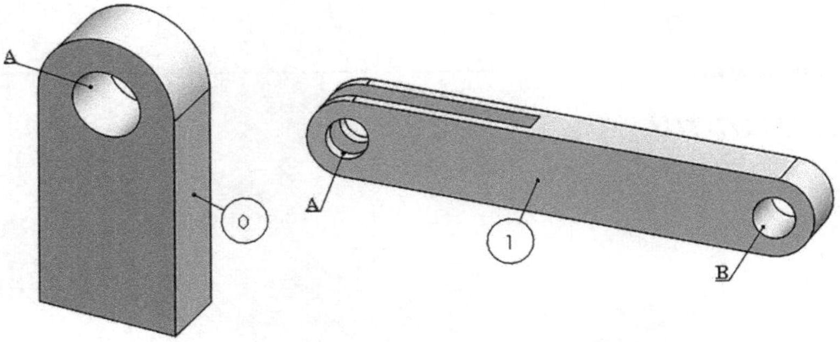

FIGURE 1.1
Simple links. Numbers indicate links and capital letter alphabets indicate joints.

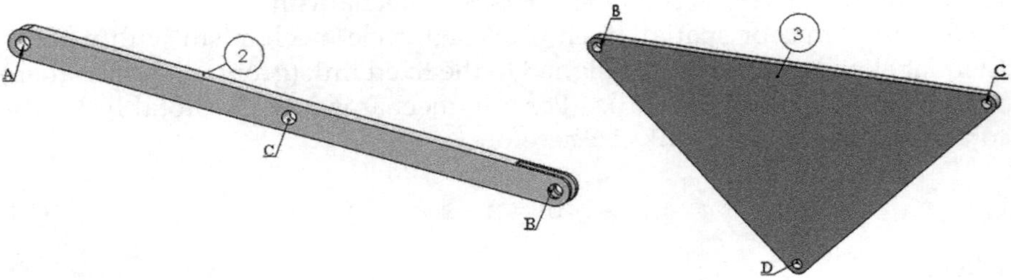

FIGURE 1.2
Complex links.

Any joint allows from one to five relative independent motions between the links. This movement can be either linear (sliding) or rotational. In Figure 1.3, an arrow represents a linear motion, and a curved arrow represents a rotation. The number of relative motions allowed by the joint is the number f of the joint's degrees of freedom (DOF). We use f to denote the joint's number of DOF, $f = 1, 2, \ldots, 5$. For the entire mechanism, j_f is the total number of joints with DOF = f.

- Figure 1.3a: The prismatic joint C, also called a sliding pair, allows one relative linear motion between links 0 and 5, and has $f = 1$ DOF.
- Figure 1.3b: The revolute joint A, which is called a turning pair, pin, or hinge, allows one relative rotation between links 1 and 2, and has $f = 1$ DOF.
- Figure 1.3c: The screw joint A allows two relative motions between nut 1 and screw 2. However, since one complete revolution of the screw causes a linear advance (of a distance known as the screw's lead), only one motion is independent. Therefore, the screw joint has $f = 1$ DOF.

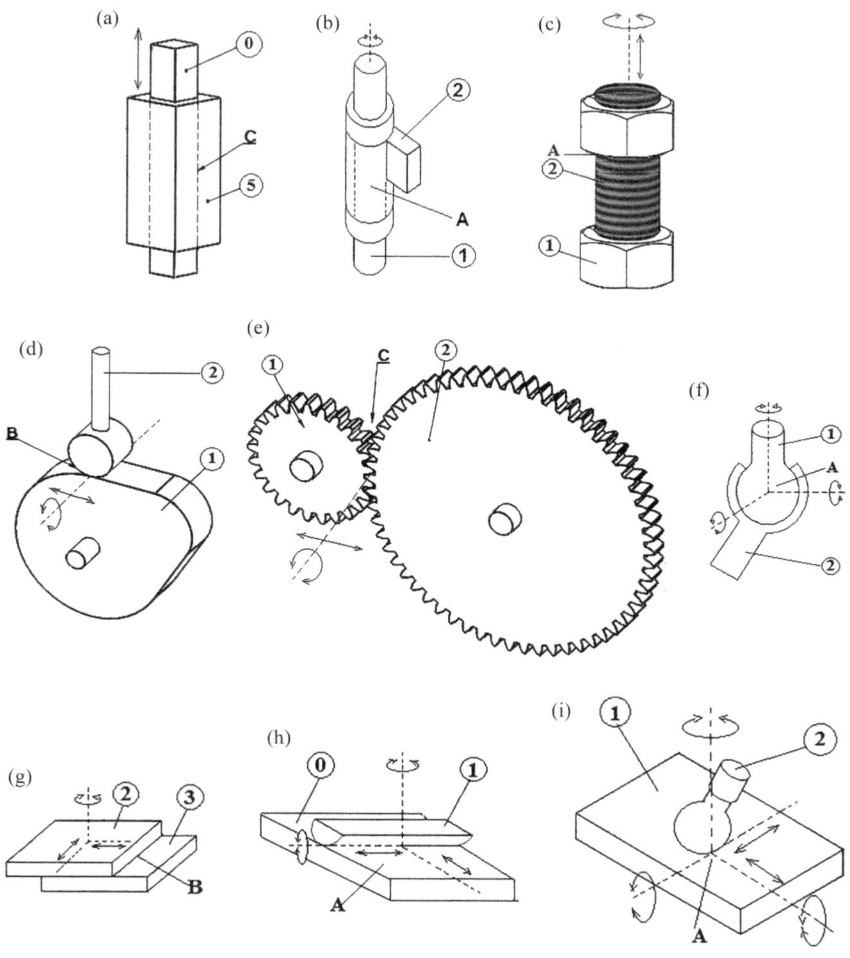

FIGURE 1.3
(a) Prismatic joint; (b) revolute joint; (c) screw joint; (d) cam joint; (e) gear joint; (f) spherical joint; (g) plane–plane joint; (h) cylinder–plane joint; (i) sphere–plane joint.

- Figure 1.3d: The cam joint B allows one linear and one revolute motion between links 1 and 2, so it has f = 2 DOF.
- Figure 1.3e: The gear joint C allows one linear and one revolute motion between links 1 and 2, so it has f = 2 DOF.
- Figure 1.3f: The spherical joint A allows three relative rotations between links 1 and 2, so it has f = 3 DOF.
- Figure 1.3g: The plane–plane joint B allows two linear and one revolute motion between the two flat surfaces of links 2 and 3, so it has f = 3 DOF.

- Figure 1.3h: The cylinder–plane joint A allows two linear and two revolute motions between the two surfaces of links 0 and 1, so it has f = 4 DOF.
- Figure 1.3i: The sphere–plane joint A allows three linear and two revolute motions between the two surfaces of links 1 and 2, so it has f = 5 DOF.

Some mechanisms have multiple links that are connected at the same geometric point. When this happens, multiple joints are created at the point. The joints are labeled with different letters, that is, A and B, where A is the joint between link 1 and link 2, and B is the joint between link 1 and link 3, as shown in Figure 1.4.

1.3 Graph Representation of a Mechanism

1.3.1 Graph

In this text, graph theory [1] is adopted along with matrix algebra as a model-building tool for mechanisms, although the vector equations are presented as reference.

A graph drawing is a visual model of the links and joints of a mechanism, like in Figure 1.5a. The nodes represent the mechanism's links, whereas the edges represent the mechanism's joints connecting the nodes.

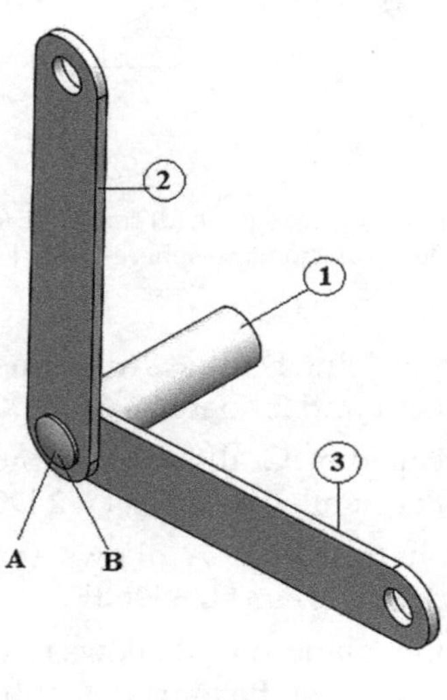

FIGURE 1.4
Multiple joined links.

Each edge in a graph is a two-node incident, like how a joint connects two links.

A simple graph, Figure 1.5b, has a set of n = 7 nodes, N(n) = {0, 1, 2, 3, 4, 5, 6}, representing the links of a mechanism. The nodes are connected by the set of j = 9 edges, E(j) = {A, B, C, D, E, F, G, H, I}, representing the joints between the links.

When examining a mechanism, the degree of a link (represented by a node in the graph) is defined as the number of joints (represented by edges in the graph) connected to that link. Graph theory also includes isolated nodes; however, isolated nodes with zero incident edges are not considered in this text since all links (or nodes in the graph) of a mechanism are connected to transmit motion or power.

In Figure 1.5, the degree of node 0 is 3, the degree of node 2 is 2, and the degree of node 5 is 4.

1.3.2 Labeling of Nodes and Edges

Node labeling in a graph begins with node 0 assigned to the ground, then from 1 to m assigned to the mobile links. The total number n of nodes is therefore

$$n = m + 1 \tag{1.2}$$

In Figure 1.5, the graph depicts a mechanism with n = 7 links in the set N(n) = {0, 1, 2, 3, 4, 5, 6} and m = 7 − 1 = 6 mobile links in the set N(m) = {1, 2, 3, 4, 5, 6}.

The j edges in a graph are labeled in alphabetical order with CAPITAL LETTERS. E(j) = {A, B, C, D, E, F, G, H, I}.

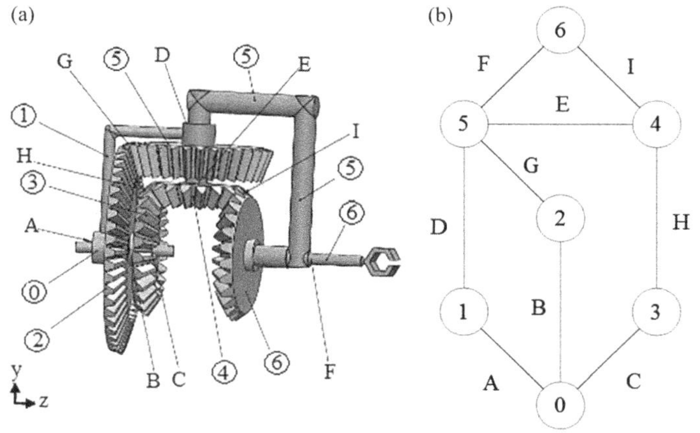

FIGURE 1.5
(a) Mechanism; (b) Graph.

1.3.3 Digraph

A *digraph* is a graph with oriented edges. As will be shown later in the text when kinematics is further explored, each oriented edge will be assigned a relative velocity and relative acceleration. For the mechanism's statics and dynamics, each oriented edge will be assigned a reaction force and reaction moment [2].

Although the edges could be arbitrarily oriented, in this text, all edges are oriented from the lower node's number toward the higher node's number. As an example, edge A is oriented from node 0 to node 1, edge B is oriented from node 0 to node 2, etc., as shown in Figure 1.6.

Each arrow has a tail and a head. For a directed edge j, a pair of two numbers $[n_{tail}, n_{head}]$ is assigned and is represented as an arrow oriented from node n_{tail} to node n_{head} (i.e., edge A is assigned the pair [0, 1], etc.). For the digraph in Figure 1.6, the set of j = 9 directed edges, E(j) = {A, B, C, D, E, F, G, H, I}, is shown in Equation (1.3):

$$\begin{aligned} E(j) &= \{A, B, C, D, E, F, G, H, I\} \\ &= \{[0,1], [0,2], [0,3], [1,5], [4,5], [5,6], [2,5], [3,4], [4,6]\} \end{aligned} \tag{1.3}$$

1.3.4 Paths

A path is a sequence of alternating nodes and edges, starting with a node and ending with a node.

In a path, the edges are distinct, meaning no edge can be traversed more than once, and each edge has a direction: **positive** if the edge is traversed from n_{tail} to n_{head} and **negative** if the edge is traversed from n_{head} to n_{tail}.

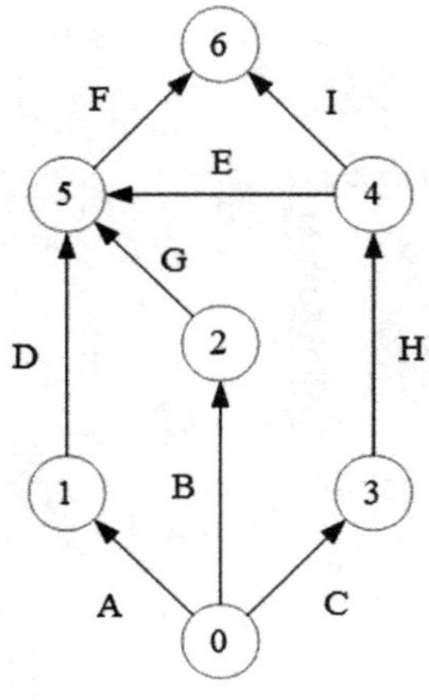

FIGURE 1.6
Digraph.

A connected digraph is a digraph where a path exists between every pair of nodes.

Two types of paths are defined for a digraph: open and closed.

1.3.5 Open Paths

An open path is a path where the first and last nodes are separate. An open path's direction is oriented along the route of the start node toward the end node.

In Figure 1.6, a path can be observed that is initiated at node 0 and that ends at node 4. The path's direction is oriented from 0 to 4. The following sequence describes the open path's direction: {0, A^+, 1, D^+, 5, E^-, 4}. The superscripts represent which direction the edge was traversed in order to make the connection to the next node.

1.3.6 Closed Paths (Cycles)

A *cycle* is a closed path where each node is used once, except for the start and end nodes being the same.

Cycles can be observed as the connected regions (areas) within the planar graph.

The number c of independent cycles is the number of disjoint internal regions in the graph, as shown in Figure 1.7. How these independent cycles were chosen will be discussed further in Section 1.3.7.

The borders (perimeters) of disjoint regions constitute c = 3 independent cycles.

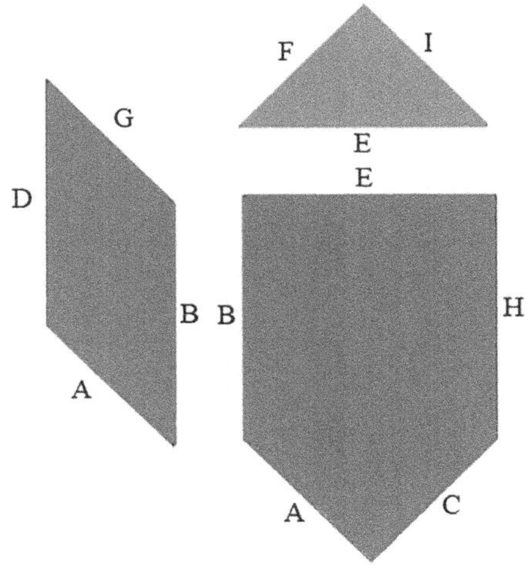

FIGURE 1.7
Disjoint internal regions.

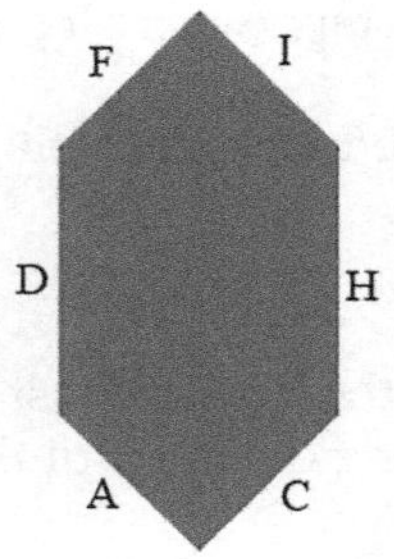

FIGURE 1.8
External region.

The region external to the graph is called the peripheral cycle (Figure 1.8).

The internal region delimited by the peripheral cycle is the union of the disjoint internal regions; therefore, the peripheral cycle is not independent.

Then, the total number c_{tot} of cycles is provided by Equation (1.4):

$$c_{tot} = c + 1 \tag{1.4}$$

The total number of cycles includes the peripheral cycle. In this case, $c_{tot} = 3 + 1 = 4$.

Theorem 1.1: Euler's Formula

In a graph, the relation between the number n of nodes, the number j of edges, and the total number c of cycles is provided by Equation (1.5):

$$c_{tot} = j - n + 2 \tag{1.5}$$

Considering Equation (1.4), the number c of independent cycles is provided by Equation (1.6):

$$c = j - n + 1 \tag{1.6}$$

Statement 1.1

The number of independent cycles in the mechanism is the difference between the number j joints and the number m of mobile links, as shown in Equation (1.7):

$$c = j - m \tag{1.7}$$

Proof: Considering Equations (1.1) and (1.6), then Equation (1.7) holds.

1.3.7 Tree and Spanning Tree

A tree is a connected digraph with no cycles.

A spanning tree is a tree that contains the fewest possible numbers of edges, while still containing all the original n nodes. Additionally, when examining a spanning tree, an edge is instead defined as arc a.

The spanning tree in Figure 1.9 has a = 6 arcs in the set: E(a) = {A, B, C, D, E, F}, and n = 7 nodes in the set: N(n) = {0, 1, 2, 3, 4, 5, 6}.

A chord, c, is defined as an edge that was in the original digraph, but was removed in creating the spanning tree. Chords can be arbitrarily chosen so long as there are no closed cycles or isolated nodes remaining. Therefore,

$$c = j - a \tag{1.8}$$

A cut-set is the representation of all existing chords, Q(c) = E(j) – E(a).

The chords are labeled with the last letters in the edge set.

The spanning tree in Figure 1.9 with a = 6 arcs in the set, E(a) = {A, B, C, D, E, F}, was extracted from the digraph with j = 9 edges in the set, E(j) = {A, B, C, D, E, F, G, H, I}, by cutting c = 9 – 6 = 3 chords in the cut-set, Q(c) = {G, H, I}.

Statement 1.2

The number of chords is the same as the number of independent cycles.

Proof: Adding, one at a time, a chord from the cut-set to a given spanning tree generates c independent cycles, which is called a cycle basis.

Each cycle is oriented by the inverse direction of the chord, starting from the chord's tail.

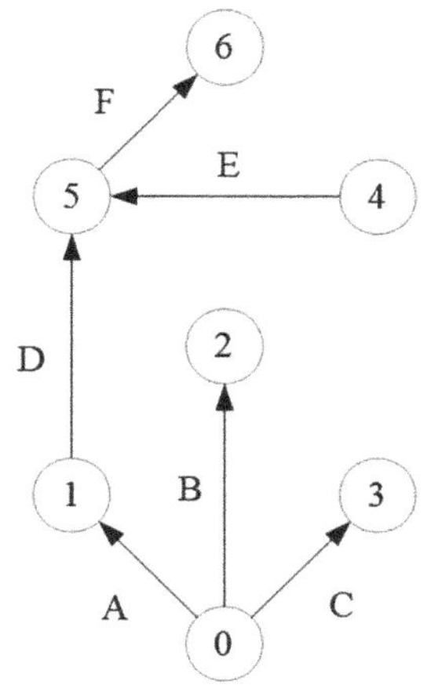

FIGURE 1.9
Spanning tree.

A superscript "minus" is assigned to an edge which is contrary to the cycle's direction, and "plus" if the edge is in the cycle's direction.

For the tree in Figure 1.9, adding c = 3 chords in the cut-set {G, H, I}, one at a time, generates c = 3 independent cycles, which is expressed as sets of edges:

$C_G = \{A^+, B^-, D^+, G^-\}$, as shown in Figure 1.10a
$C_H = \{A^+, C^-, D^+, E^-, H^-\}$, as shown in Figure 1.10b
$C_I = \{E^+, F^+, I^-\}$, as shown in Figure 1.10c.

Statement 1.3

The spanning tree contains (n – 1) arcs (Equation (1.9)):

$$a = n - 1 \tag{1.9}$$

Proof: Considering Equation (1.6): $c = j - n + 1$ and Equation (1.8): $c = j - a$, then Equation (1.9) holds.

Statement 1.4

In a mechanism, the number of joints in the spanning tree is the same as the number of mobile links, which is given in Equation (1.10):

$$a = m \tag{1.10}$$

Proof: Considering Equation (1.1): $m = n - 1$ and Equation (1.9): $a = n - 1$, then Equation (1.10) holds.

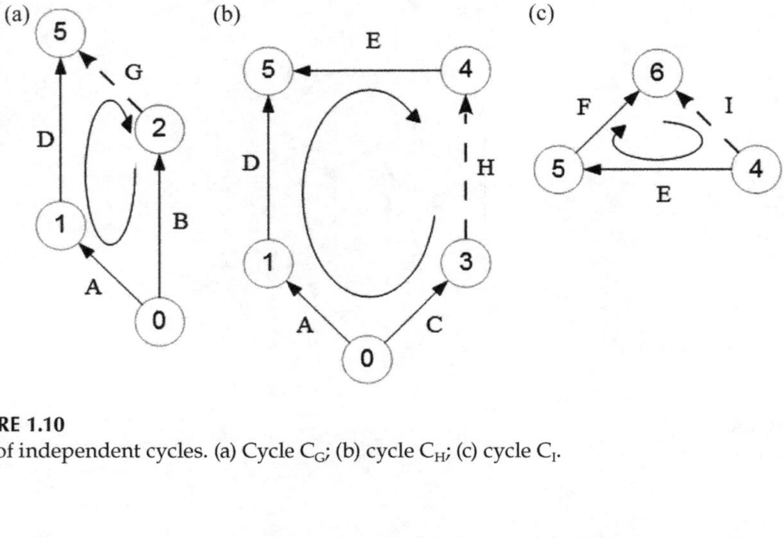

FIGURE 1.10
Base of independent cycles. (a) Cycle C_G; (b) cycle C_H; (c) cycle C_I.

1.3.8 Matrix Description of a Digraph

Several matrices are associated with any digraph that shows the relationship between its nodes and edges. For a directed edge j, a pair of two numbers $[n_{tail}, n_{head}]$ is assigned, with the convention $n_{tail} < n_{head}$.

Each edge is represented as an arrow, which starts at n_{tail} and ends at n_{head}. Furthermore, +1 is assigned for all n_{head} nodes and –1 is assigned for all n_{tail} nodes, shown as superscript.

For the digraph in Figure 1.6, with n = 7 nodes and j = 9 edges in the set E(j) = {A, B, C, D, E, F, G, H, I}, its edges are oriented as shown in Equation (1.3) reproduced as:

$$E(j) = \{[0^-,1^+], [0^-,2^+], [0^-,3^+], [1^-,5^+], [4^-,5^+], [5^-,6^+],$$
$$[2^-,5^+], [3^-,4^+], [4^-,6^+]\}$$

1.3.9 Incidence Nodes–Edges Matrix, $\underline{G}$

The incidence nodes–edges matrix, $\underline{G}$, is defined as a n × j matrix in which each row is assigned to a node, and each column is assigned to an edge:

$$\underline{G} = \overset{\text{Edges}}{\begin{bmatrix} * & \cdots & * \\ \vdots & \underline{g}_{n,j} & \vdots \\ * & \cdots & * \end{bmatrix}} \text{Nodes,} \quad \text{where } \underline{g}_{n,j} = \begin{Bmatrix} -1 & \text{if node n is } n_{tail} \\ +1 & \text{if node n is } n_{head} \\ 0 & \text{otherwise} \end{Bmatrix} \qquad (1.11)$$

For example, the digraph in Figure 1.6 has the incidence nodes–edges matrix, which is given by

$$\underline{G} = \overset{\begin{matrix}\text{Edges}\\ \text{A} \quad \text{B} \quad \text{C} \quad \text{D} \quad \text{E} \quad \text{F} \quad \text{G} \quad \text{H} \quad \text{I}\end{matrix}}{\begin{bmatrix} -1 & -1 & -1 & 0 & 0 & 0 & 0 & 0 & 0 \\ +1 & 0 & 0 & -1 & 0 & 0 & 0 & 0 & 0 \\ 0 & +1 & 0 & 0 & 0 & 0 & -1 & 0 & 0 \\ 0 & 0 & +1 & 0 & 0 & 0 & 0 & -1 & 0 \\ 0 & 0 & 0 & 0 & -1 & 0 & 0 & +1 & -1 \\ 0 & 0 & 0 & +1 & +1 & -1 & +1 & 0 & 0 \\ 0 & 0 & 0 & 0 & 0 & +1 & 0 & 0 & +1 \end{bmatrix}} \begin{matrix} 0 \\ 1 \\ 2 \\ 3 \\ 4 \\ 5 \\ 6 \end{matrix} \text{ Nodes} \qquad (1.12)$$

Each column has two equal and opposite entries, since each edge is assigned a pair of two numbers [–1, +1]. Hence, the sum of the entries in each column is zero.

For a given matrix with n rows, a simple procedure of adding the entries of its rows could conclude the row's independence. If all entries are not zero in the new row generated after the row addition, then the initial matrix has n independent rows. The rank of the matrix is the number of its independent rows.

The rows in matrix $\underline{G}$ are not independent. This is proven by adding to the first row of all the other rows, thus obtaining a row with all zero entries. Hence, its rank is less than n, it can be at most: $n - 1 = m$.

Theorem 1.2: Kirchhoff's Matrix Tree Theorem

A spanning tree is not unique in a digraph. The number of distinct spanning trees of a digraph is determined as the value of the cofactor, det $[\mathbf{K}_{11}]$, of matrix $\mathbf{K} = \underline{G} * \underline{G}^T$. Matrix $\mathbf{K}_{11}$ results from removing the first row and column from the matrix $\mathbf{K}$, which is the product of a $\underline{G}$ matrix and its transpose, $\underline{G}^T$.

1.3.10 Reduced Incidence Nodes–Edges Matrix, G

The rows in matrix $\underline{G}$ are not independent. By deleting one row (the first row, which corresponds to the fixed link), one can assure the independence of the other rows.

The reduced incidence nodes–edges matrix, **G**, is defined as a $m \times j$ matrix obtained by deleting the first row representing the node 0 (the fixed link in mechanism) from the $\underline{G}$ incidence node–edge matrix.

$$\begin{array}{c} \text{Edges} \\ \begin{array}{ccccccccc} A & B & C & D & E & F & G & H & I \end{array} \end{array}$$

$$\mathbf{G} = \left[\begin{array}{cccccc:ccc} +1 & 0 & 0 & -1 & 0 & 0 & 0 & 0 & 0 \\ 0 & +1 & 0 & 0 & 0 & 0 & -1 & 0 & 0 \\ 0 & 0 & +1 & 0 & 0 & 0 & 0 & -1 & 0 \\ 0 & 0 & 0 & 0 & -1 & 0 & 0 & +1 & -1 \\ 0 & 0 & 0 & +1 & +1 & -1 & +1 & 0 & 0 \\ 0 & 0 & 0 & 0 & 0 & +1 & 0 & 0 & +1 \end{array}\right] \begin{array}{c} 1 \\ 2 \\ 3 \\ 4 \\ 5 \\ 6 \end{array} \text{Nodes} \qquad (1.13)$$

If $j = a + c$ edges are partitioned into $a = m$ arcs and c chords, the columns in the $m \times j$ matrix G can be partitioned as shown in Equation (1.14):

$$\mathbf{G} = \left[\begin{array}{c:c} \mathbf{G}_a & \mathbf{G}_c \end{array}\right] \qquad (1.14)$$

where

- $-m \times m$ is a square matrix G_a and is called the reduced incidence nodes–arcs tree and it contains the arcs of the tree.
- $-m \times c$ matrix G_c is called *the reduced incidence nodes–chords* matrix and contains only the chords of the graph.

The two matrices are

$$
\begin{array}{c}
\text{Arcs} \\
\begin{array}{cc}
 & \begin{array}{cccccc} A & B & C & D & E & F \end{array} \\
\mathbf{G}_a = & \left[\begin{array}{cccccc}
+1 & 0 & 0 & -1 & 0 & 0 \\
0 & +1 & 0 & 0 & 0 & 0 \\
0 & 0 & +1 & 0 & 0 & 0 \\
0 & 0 & 0 & 0 & -1 & 0 \\
0 & 0 & 0 & +1 & +1 & -1 \\
0 & 0 & 0 & 0 & 0 & +1
\end{array}\right]
\begin{array}{c} 1 \\ 2 \\ 3 \\ 4 \\ 5 \\ 6 \end{array}
\text{ Nodes;}
\end{array} \\
\text{Chords} \\
\begin{array}{cc}
 & \begin{array}{ccc} G & H & I \end{array} \\
\mathbf{G}_c = & \left[\begin{array}{ccc}
0 & 0 & 0 \\
-1 & 0 & 0 \\
0 & -1 & 0 \\
0 & +1 & -1 \\
+1 & 0 & 0 \\
0 & 0 & +1
\end{array}\right]
\begin{array}{c} 1 \\ 2 \\ 3 \\ 4 \\ 5 \\ 6 \end{array}
\text{ Nodes}
\end{array}
\end{array}
\tag{1.15}
$$

From Equation (1.15), the transposed matrix of G_c is

$$
\mathbf{G}_c^T = \left[\begin{array}{cccccc}
0 & -1 & 0 & 0 & +1 & 0 \\
0 & 0 & -1 & +1 & 0 & 0 \\
0 & 0 & 0 & -1 & 0 & +1
\end{array}\right]
\begin{array}{c} G \\ H \\ I \end{array}
\tag{1.16}
$$

1.3.11 The Path Matrix, Z

Referring to Figure 1.8, let us consider the paths that start from node 0 in the tree toward any node m.

A superscript "–" sign is assigned to an arc that has an orientation opposite to the path's direction, and "+" if the arc is in the path's direction. All the paths for the tree in Figure 1.7 are as follows: $Z_1 = \{0, A^+, 1\}$, $Z_2 = \{0, B^+, 2\}$, $Z_3 = \{0, C^+, 3\}$, $Z_4 = \{0, A^+, 1, D^+, 5, E^-, 4\}$, $Z_5 = \{0, A^+, 1, D^+, 5\}$, $Z_6 = \{0, A^+, 1, D^+, 5, F^+, 6\}$.

The path matrix, Z, is defined as an a × m square matrix in which each row is assigned to an arc, and each column is assigned to an open path toward each node m.

$$\mathbf{Z} = \overset{\text{Nodes}}{\begin{bmatrix} * & \cdots & * \\ \vdots & z_{a,m} & \vdots \\ * & \cdots & * \end{bmatrix}} \text{Chords, } z_{a,m}$$

$$= \left\{ \begin{array}{ll} -1 & \text{if arc a is on the path and opposite to the path} \\ +1 & \text{if arc a is on to the path and same direction as the path} \\ 0 & \text{if arc a is not on the path} \end{array} \right\} \quad (1.17)$$

The transpose matrix $\mathbf{Z}^T$ is used later in the text for the determination of links' absolute angular velocities.

For example, the spanning tree in Figure 1.9 has the path matrix and the transposed shown below:

$$\mathbf{Z} = \overset{\text{Nodes}}{\overset{\begin{matrix} 1 & 2 & 3 & 4 & 5 & 6 \end{matrix}}{\begin{bmatrix} 1 & 0 & 0 & 1 & 1 & 1 \\ 0 & 1 & 0 & 0 & 0 & 0 \\ 0 & 0 & 1 & 0 & 0 & 0 \\ 0 & 0 & 0 & 1 & 1 & 1 \\ 0 & 0 & 0 & -1 & 0 & 0 \\ 0 & 0 & 0 & 0 & 0 & 1 \end{bmatrix}}} \begin{matrix} A \\ B \\ C \\ D \\ E \\ F \end{matrix} \quad \text{Arcs;}$$

$$\mathbf{Z}^T = \overset{\text{Arcs}}{\overset{\begin{matrix} A & B & C & D & E & F \end{matrix}}{\begin{bmatrix} 1 & 0 & 0 & 0 & 0 & 0 \\ 0 & 1 & 0 & 0 & 0 & 0 \\ 0 & 0 & 1 & 0 & 0 & 0 \\ 1 & 0 & 0 & 1 & -1 & 0 \\ 1 & 0 & 0 & 1 & 0 & 0 \\ 1 & 0 & 0 & 1 & 0 & 1 \end{bmatrix}}} \begin{matrix} 1 \\ 2 \\ 3 \\ 4 \\ 5 \\ 6 \end{matrix} \quad \text{Nodes} \quad (1.18)$$

1.3.12 Spanning Tree Matrix, T

Starting or ending from each chord c, a path is defined.

For the tree in Figure 1.9, the following edge sets belonging to the three paths: $T_G = \{A^+, B^-, D^+\}$, $T_H = \{A^+, C^-, D^+, E^-\}$, and $T_I = \{E^+, F^+\}$ will be generated.

The c × a matrix T has a row for each chord c, which represents the cut-edge, and the columns represent the arcs a that belong to the tree.

Its entries $c_{c,a}$ are defined as follows:

$$c_{c,a} = \left\{ \begin{array}{ll} -1 & \text{if arc a is on the path and opposite to the path} \\ +1 & \text{if arc a is on to the path and same direction as the path} \\ 0 & \text{if arc a is not on the path} \end{array} \right\} \tag{1.19}$$

The path matrix for the spanning tree in Figure 1.9 is

$$\mathbf{T} = \text{Paths:} \begin{array}{r} T_G \rightarrow \\ T_H \rightarrow \\ T_I \rightarrow \end{array} \overset{\begin{array}{cccccc} A & B & C & D & E & F \end{array}}{\begin{bmatrix} 1 & -1 & 0 & 1 & 0 & 0 \\ 1 & 0 & -1 & 1 & -1 & 0 \\ 0 & 0 & 0 & 0 & 1 & 1 \end{bmatrix}}$$

It was shown in [3] that the automatic entries of matrix T are calculated from the multiplication of two digraph matrices with Equation (1.20)

$$\mathbf{T} = \mathbf{G}_c^T * \mathbf{Z}^T \tag{1.20}$$

The two factors are $\mathbf{G}_c^T$ from Equation (1.16) and $\mathbf{Z}^T$ from Equation (1.18). The superscript T designates the transpose operation.

Example: Equations (1.16) and (1.18) are used in Equation (1.20) to obtain the spanning tree matrix **T**:

$$\mathbf{T} = \begin{bmatrix} 1 & -1 & 0 & 1 & 0 & 0 \\ 1 & 0 & -1 & 1 & -1 & 0 \\ 0 & 0 & 0 & 0 & 1 & 1 \end{bmatrix}$$

1.3.13 Cycle Basis Incidence Matrix, C

Closed paths (cycles) are generated when c chords are added, one at a time, to a given spanning tree.

The direction of each cycle is assigned by the inverse chord's direction, that is, clockwise or counterclockwise.

A superscript '–' sign is assigned to an edge which is contrary to the cycle's direction, and '+' if the edge is in the path's direction.

In Figure 1.10, the set of three cycles is a cycle basis, shown as sets of edges. To create the incidence matrix, start at the tail of each chord (dotted edge) and traverse according to the cycle direction (cycle arrow), noting the edge letter and adding a superscript of '+' or '–' if the edge arrow opposes the cycle direction arrow '–' or not '+', as shown below:

$$C_G = \{B^-, A^+, D^+, G^-\},\ C_H = \{C^-, A^+, D^+, E^-, H^-\}, C_I = \{E^+, F^+, I^-\} \quad (1.21)$$

The cycle basis incidence matrix, C, is defined as a $c \times j$ matrix in which each row is assigned to a cycle, and each column is assigned to an edge.

$$\begin{matrix} & \text{Edges} & \\ \mathbf{C} = & \begin{bmatrix} * & \cdots & * \\ \vdots & c_{c,j} & \vdots \\ * & \cdots & * \end{bmatrix} & \text{Cycles} \end{matrix} \quad (1.22)$$

where

$$c_{c,j} = \begin{Bmatrix} -1 & \text{if edge j is on the cycle and opposite to the cycle's direction} \\ +1 & \text{if edge j is onto the cycle and same direction as the cycle's direction} \\ 0 & \text{if the edge is not on the cycle} \end{Bmatrix} \quad (1.23)$$

The set of arcs in the tree is augmented with one chord to generate one cycle.
Therefore, the tree matrix T is augmented with the $c \times c$ diagonal matrix U_c to generate the cycle basis matrix C.

$$\mathbf{C} = \left[\ \mathbf{T} \ \vdots\ \mathbf{U}_c\ \right] \quad (1.24)$$

where

$$\mathbf{U}_c = \begin{bmatrix} -1 & 0 & 0 \\ 0 & -1 & 0 \\ 0 & 0 & -1 \end{bmatrix}, \text{ opposite of a } 3 \times 3 \text{ unit matrix}$$

In Figure 1.9, the cycle basis, in matrix form, is

$$\begin{matrix} & \text{Edges} & \\ & \begin{matrix} A & B & C & D & E & F & G & H & I \end{matrix} & \\ \mathbf{C} = & \left[\begin{array}{cccccc:ccc} 1 & -1 & 0 & 1 & 0 & 0 & -1 & 0 & 0 \\ 1 & 0 & -1 & 1 & -1 & 0 & 0 & -1 & 0 \\ 0 & 0 & 0 & 0 & 1 & 1 & 0 & 0 & -1 \end{array}\right] & \begin{matrix} C_G \\ C_H \\ C_I \end{matrix} \ \text{Cycles} \end{matrix} \quad (1.25)$$

The matrix C shown in Equation (1.25) with three independent rows is a cycle matroid with rank $c = 3$.

Statement 1.5

The peripheral cycle is not independent.

Proof: A peripheral cycle, defined in Figure 1.8 by the addition of two chords H and I to the tree, is $C_{Per} = \{A^+, C^-, D^+, F^+, H^-, I^-\}$. Let us consider matrix $\underline{C}$, with the addition of an extra row to matrix

$$\underline{C} = \begin{array}{c} \\ \left[\begin{array}{ccccccccc} A & B & C & D & E & F & G & H & I \\ 1 & -1 & 0 & 1 & 0 & 0 & -1 & 0 & 0 \\ 1 & 0 & -1 & 1 & -1 & 0 & 0 & -1 & 0 \\ 0 & 0 & 0 & 0 & 1 & 1 & 0 & 0 & -1 \\ 1 & 0 & -1 & 1 & 0 & 1 & 0 & -1 & -1 \end{array}\right] \end{array} \begin{array}{l} \\ C_G \\ C_H \\ C_I \\ C_{Per} \end{array} \quad \text{Total Cycles} \qquad (1.26)$$

The addition of the entries in the second row C_H, to the entries from the third row C_I, could generate the entries of the fourth row C_{Per}, and therefore, this cycle is clearly not independent.

Statement 1.6

A cycle basis is a set of independent cycles extracted from a base of cycles.

Proof: Since the set of four rows in Equation (1.26) is not independent, a set of at most three rows is independent.

A base of cycles is a set of three rows taken from a set of four rows: $\{C_G, C_H, C_I\}$, $\{C_G, C_H, C_{Per}\}$, $\{C_H, C_I, C_{Per}\}$, and $\{C_G, C_I, C_{Per}\}$. Sets of three disjoint regions are attached for each of the above four cases, and the independent cycles are the disjoint regions' perimeters.

A cycle basis is any of the four sets in the base. For example, the tree in Figure 1.9 generates the set $\{C_G, C_H, C_I\}$, with matrix C of rank $c = 3$.

1.3.14 Cycle Matroid Fundamentals

Matroids are generalizations of graphs and matrices. Matroid theory [4,5], which combines the properties of graphs and matrices, appears to be little known to the mechanism community. However, it is used to define the rigidity of trusses and has also been used in the study of electrical systems and structural analysis [6,7].

Let $\underline{C}$ be the set of all cycles of a digraph, and let I be the collection of independent subsets of cycles.

A *cycle* matroid M is defined as the pair $M = (\underline{C}, I)$.

It is shown that for the collection $\underline{C}$ of all cycles, the following conditions hold:

1. The empty set is not in $\underline{C}$.
2. No member is a proper subset of another member of $\underline{C}$.
3. If C_X and C_Y are members of $\underline{C}$ and $j \in (C_X \cap C_Y)$, then $(C_X \cup C_Y)$ -{j} contains a member of $\underline{C}$.

The bases are maximum independent sets of cycles; their independence is evidenced by the independent rows in the incidence cycles–edges matrix $\underline{C}$.

A cycle basis is a set in the base.

The rank r of the matroid is the number of elements in the cycle basis. All bases have the same rank.

For the digraph in Figure 1.6, the set of all cycles is $\underline{C} = \{\{C_G\}, \{C_H\}, \{C_I\}, \{C_{Per}\}\}$, and the set of independent cycles is $I = \{\{C_G\}, \{C_H\}, \{C_I\}, \{C_{Per}\}, \{C_G, C_H\}, \{C_G, C_I\}, \{C_G, C_{Per}\}, \{C_H, C_I\}, \{C_H, C_{Per}\}, \{C_I, C_{Per}\}, \{C_G, C_H, C_I\}, \{C_G, C_H, C_{Per}\}, \{C_H, C_I, C_{Per}\}$, and $\{C_G, C_I, C_{Per}\}\}$. One could notice that the set $\underline{C}$ is not included in set I, since $\underline{C}$ is not independent.

A base with rank $r = 3$ for the *cycle* matroid is $\{C_G, C_H, C_I\}, \{C_G, C_H, C_{Per}\}, \{C_H, C_I, C_{Per}\}, \{C_G, C_I, C_{Per}\}$.

A cycle basis is any set of three cycles in the base. The cycle basis $\{C_G, C_H, C_I\}$ is generated from the tree in Figure 1.9. Each cycle basis is a maximal independent set of cycles, and their independence is evidenced by the independent rows in the cycles–edges matrix C, as shown in Equation (1.25), which has the rank $r = 3$.

Statement 1.7

A set of independent kinematic equations are written along a set of independent cycles, called cycle basis, which are a base of the cycle matroid. The kinematic equations generated using any of these cycle bases will lead to the same results. Mechanisms with multiple cycles, where the cycle basis in a cycle matroid accounts for the generation of independent kinematics equations, are reported in [2].

1.4 Number of Independent Cycles in a Mechanism for Planar and Spatial Mechanisms

The total number j of joints is the sum of all joints with f = 1, 2, 3, 4, and 5 DOF.

$$j = j_1 + j_2 + j_3 + j_4 + j_5 \tag{1.27}$$

The number c of independent cycles is determined by using Euler formula (Equation (1.7)):

$$c = j - m$$

with

m = number of mobile links

j = number of joints

1.5 Mobility of Planar Mechanisms

For planar mechanisms, the mobility M can be predicted with Gruebler formula (Equation (1.28)):

$$M = 3 \cdot m - 2 \cdot j_1 - j_2 \tag{1.28}$$

where

m = number of mobile links

j_1 = number of f = 1 DOF joints

j_2 = number of f = 2 DOF joints

1.6 Mobility of Spatial Mechanisms

For spatial mechanisms, the mobility M can be predicted with Kutzbach formula (Equation (1.29)):

$$M = 6 \cdot m - 5 \cdot j_1 - 4 \cdot j_2 - 3 \cdot j_3 - 2 \cdot j_4 - j_5 \tag{1.29}$$

where

m = number of mobile links

j_f = number of f DOF joints (f = 1, …, 5)

References

1. Bondy, J. A., and Murty, U. S. R., *Graph Theory with Applications*, Macmillan: London, North-Holland, New York, 1976.

2. Talpasanu, I., Optimisation in kinematic and kinetostatic analysis of rigid systems with applications in machine design, *PhD Dissertation*, University Politehnica, Bucharest, 1991.
3. Talpasanu, I., A general method for kinematic analysis of robotic wrist mechanisms, *ASME Journal of Mechanisms and Robotics*, 7(3), 2015.
4. Whitney, H., On the abstract properties of linear dependence, *American Journal of Mathematics*, 57, 1935.
5. Oxley, J., *Matroid Theory*, Oxford University Press: New York, 1992.
6. Recski, A., *Matroid Theory and Its Applications in Electric Network Theory and Statics*, Springer: Berlin, Heidelberg, 1989.
7. Kaveh, A., *Optimal Structural Analysis*, Research Studies Press (John Wiley): Exeter, UK, 1997.

2

Kinematics of Open Cycle Mechanisms

Kinematics comes from the Greek word "kinetikos," which means motion. Kinematics is the section of classical mechanics dealing with position, displacement, velocity, and acceleration of rigid bodies and points located on rigid bodies; it disregards the forces that produce the motion. An open chain system, usually used in manipulation processes and robotics, is a set of m mobile links connected by j joints.

Since actuating forces and torques at joints are neglected, kinematics provides a preliminary analysis of the design of the system of rigid bodies. Not all systems of rigid bodies are mechanisms; the characteristic of mechanism will be discussed further in this section. The "performance" of a mechanism under actuating loads is presented further in Chapter 4, where the designer determines the dynamic reactions in joints to validate the design choice.

Examples from robotics manipulation

The spatial TRRT mechanism with four joints: translation (prismatic) T, revolute R, revolute R, and translation (prismatic) T is illustrated in Figure 2.1 (Animation 2.1). Applications for this mechanism include manipulation of parts for manufacturing.

2.1 Link and Joint Labeling, Frames, Home Position of Mechanism, Mechanism's Digraph, and Mechanism's Mobility

2.1.1 Link and Joint Labeling

In Figure 2.1, the links are labeled with numbers, starting with 0 which is assigned to the fixed link (base). There are four mobile links, m = 4, which have been assigned the numbers 1, 2, 3, and 4, as well as four joints labeled with capital letters: A (prismatic), B (revolute), C (revolute), and D (prismatic). Each joint has been assigned a pair of two geometric points, located on two different joined surfaces. In Figure 2.1, A_0 is located on the top surface of link 0, whereas A is located on the horizontal inner surface of a slot on link 1. B_1 (not shown if Figure 2.1) is located on a hollow-interior surface on link 1, whereas B is located on the pin surface on link 2. C_2 (not shown if Figure 2.1) is located on a hollow-interior surface on link 2, whereas C is located on the

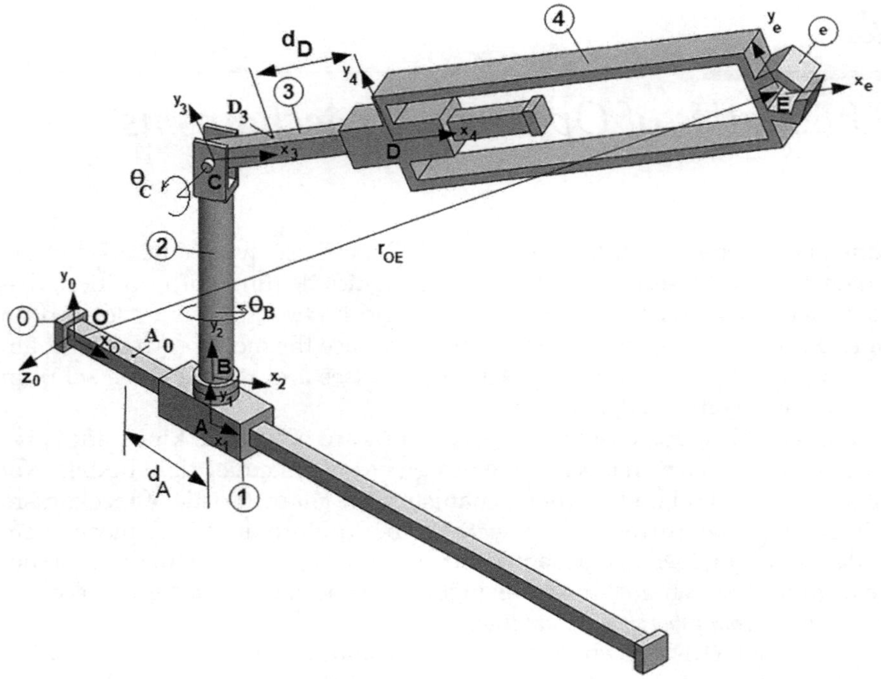

FIGURE 2.1
Open cycle spatial mechanism with revolute and prismatic joints.

pin's surface in link 3. D_3 is located on the top surface of link 3, whereas D is located on the inner surface of a slot on link 4.

A coordinate system named frame is attached to each link and is located as follows: at O for link 0 (fixed), at A for link 1, at B for link 2, at C for link 3, and at D for link 4. For this mechanism, the end-effector e (grasping device, wrist) is rigidly attached to link 4; therefore, it can be modeled as link 4&e. The point of interest labeled E is the end-effector's frame origin, locates the part to be manipulated. The two frames with origins at E and D, both located on the link 4, are geometrically distinct from each other. The E's coordinates (x_e, y_e, z_e) are identified relative to D's frame (D, x_4, y_4, z_4). In Figure 2.1, there is no relative motion between frames D and E. Although there are mechanisms where there is a relative motion between frames D and E, these mechanisms have a *flange* between links 4 and e, which will be demonstrated further in Example 2.

Each of the joints in the mechanism is 1 degree of freedom (DOF): two prismatic (description "T" at A and C) and two revolute (description "R" at B and C). The mechanism is labeled as TRRT and has $j_1 = 4$. There are no joints with 2, 3, 4, or 5 DOF, thus $j_2 = j_3 = j_4 = j_5 = 0$. The total number of joints from Equation (1.27) is

$$j = j_1 + j_2 + j_3 + j_4 + j_5 = 4 \text{ Joints} \tag{2.1}$$

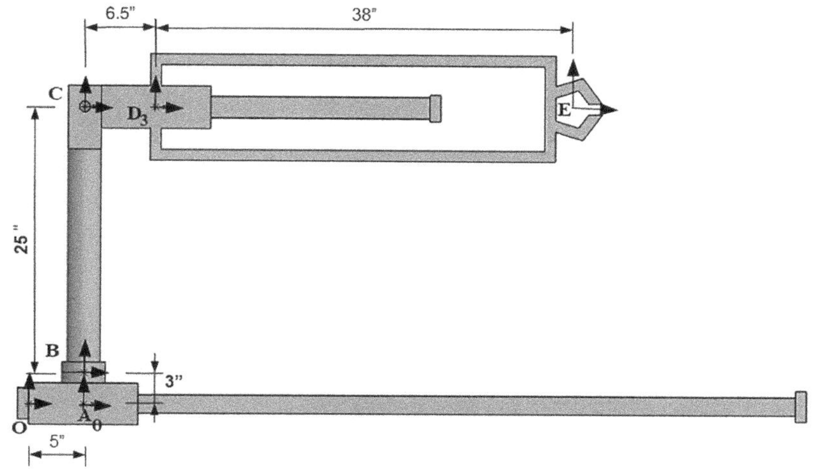

FIGURE 2.2
Home position.

2.1.2 Home Position of Mechanism

The home position is defined as the initial configuration of the mechanism prior to the start of motion. The link's displacement, linear or angular, is measured from the home position. Point A_0 on link 0 and point D_3 on link 3 define the initial positions of sliders A and D's. The initial angular position at joint B is for axis x_2 parallel and same direction as x_1, whereas the initial angular position at joint joint C is for axis y_3 parallel and same direction as y_2 (Figure 2.2).

2.1.3 Mechanism's Digraph

The number of cycles c is determined from Equation (2.2):

$$c = j - m = 4 - 4 = 0 \text{ (No cycles)} \tag{2.2}$$

Since there are no cycles (closed paths), the digraph is a spanning tree which is a characteristic of open cycle mechanisms.

Statement 2.1

Open cycle mechanism's digraph is a spanning tree

The digraph attached to the mechanism is illustrated in Figure 2.3a, with $n = m + 1 = 5$ nodes corresponding to the mechanism's $m = 4$ mobile links and the node 0 for the fixed link. The $j = 4$ are corresponding to the mechanism's joints.

The digraph's edges are oriented from the lower node number to the higher node number.

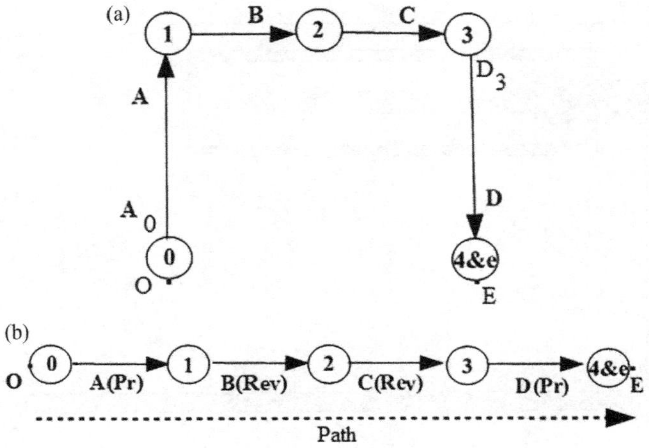

FIGURE 2.3
(a) Spanning tree; (b) path.

A *path* is defined as a sequence of nodes and edges, see Figure 2.3b. The path starts at node 0 and ends at end-effector e. In the digraph, the home position A_0 is illustrated on node 0, whereas the home position D_3 on node 3.

2.1.4 Mechanism's Mobility

For *spatial* mechanisms, the mobility M can be predicted by the Kutzbach formula, Equation (2.3), introduced in Chapter 1 as Equation (1.29):

$$M = 6\,m - 5\,j_1 - 4\,j_2 - 3\,j_3 - 2\,j_4 - j_5 = 4 \tag{2.3}$$

Obtaining M = 4 means the mechanism has 4 DOF; therefore, four motors are required for actuation.

2.2 Direct and Inverse Analysis: Frame Orientation and Position for the Spatial Open Cycle Mechanisms

There are two formulations for the orientation and positional analysis of mechanisms:

- In the *Direct Analysis,* the actuation parameters are given, and they are used to determine the orientations and positions of the joints and end-effector.

- In the *Inverse Analysis*, the orientations and positions of the joints and end-effector are given, and they are used to determine the required actuation parameters.

2.3 Incidental and Transfer (IT) Notation

In this text, a joint displacement vector has the notation of an edge in the tree, whereas the coordinate system attached to a mobile link has the notation of the node in the tree.

2.3.1 Notation for Joint Displacement Based on Incidental Digraph's Edge

For the linear angular displacements $d_{Z_{m-1,m}}$ or angular displacements $\theta_{Z_{m-1,m}}$, a subscript is assigned—a letter corresponding to the tree's edge and a set of two subscript numbers showing that the relative displacement is between link m and link m−1.

The TRRT mechanism has the tree shown in Figure 2.3, and its joint displacements in time t are

- Joint A: $d_{A_{0,1}}(t)$—linear displacement at A between link 0 and link 1
- Joint B: $\theta_{B_{1,2}}(t)$—angular displacement at B between link 1 and link 2
- Joint C: $\theta_{C_{2,3}}(t)$—angular displacement at C between link 2 and link 3
- Joint D: $d_{D_{3,4}}(t)$—linear displacement at D between link 3 and link 4

2.3.2 Notation for Frames Based on Digraph's Nodes

For a local frame (Z, x_m, y_m, z_m) attached to each mobile link, a subscript m is assigned which corresponds to the tree's node. The frame origin Z is located at joint Z. The fixed coordinate system (O, x_0, y_0, z_0), named base, is attached to the tree's node 0, which has its origin at the fixed point O.

The joint displacements defined above are vectors with their components expressed with respect to local frames. Vector addition is possible if their components are expressed with respect to the same frame. The common frame is typically the base. Therefore, a frame transformation is required when transferring the vector's components from the local frames to the base frame to perform vector addition.

The IT notation is further described using the three distinct types of rotations about x_m, y_m, z_m, or translation between the two frames.

2.4 Relative Frames Orientation and Relative Rotation Matrix

2.4.1 Relative Rotation Matrix about x_m-Axis: IT Notation

Let us consider two frames ($Yx_{m-1}, y_{m-1}, z_{m-1}$) attached to link m − 1 and (Zx_m, y_m, z_m) attached to link m.

The unit vectors of frame (Z, x_m, y_m, z_m) are shown in Figure 2.4:

- $\mathbf{i}_m$ is the unit vector for the axis of rotation x_m, noted $\mathbf{u}_{Z_{m-1,m}}$
- $\mathbf{j}_m$ is the unit vector for axis y_m in the plane perpendicular to the axis of rotation x_m
- $\mathbf{k}_m$ is the unit vector for the z_m-axis, defined by Equation (2.4)

$$\mathbf{i}_m \times \mathbf{j}_m = \mathbf{k}_m \tag{2.4}$$

At Z, within this perpendicular plane, there is a *reference axis* ($||y_{m-1}$), parallel to axis y_{m-1} with the unit vector $\mathbf{j}_{m-1}$. Angular displacement $\theta_{Z_{m-1,m}}$ measures the angle between the *two y-axes.* The positive direction is considered from the $||y_{m-1}$ axis rotating counterclockwise toward y_m-axis, which is given in Equation (2.5):

$$\mathbf{j}_{m-1} \times \mathbf{j}_m = \mathrm{Sin}\,\theta_{Z_{m-1,m}} \cdot \mathbf{u}_{Z_{m-1,m}} \tag{2.5}$$

Therefore, the first unit vector $\mathbf{j}_{m-1}$ in the cross product is rotated by the angle $\theta_{Z_{m-1,m}}$ to locate the direction of the unit vector $\mathbf{j}_m$, according to the right-hand rule.

The proof of Equation (2.5) is shown in Appendix 2.1.

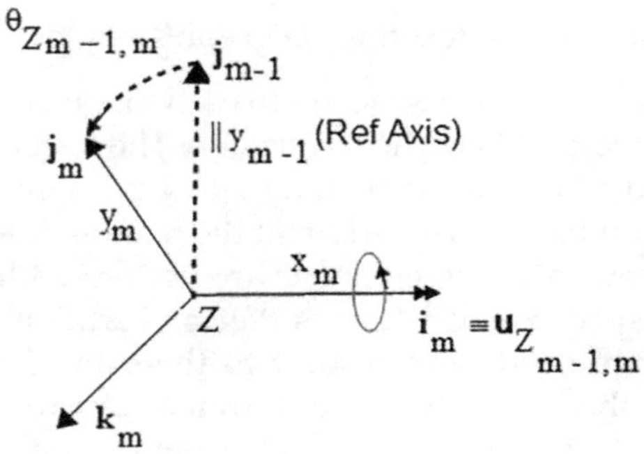

FIGURE 2.4
Rotation about x_m-axis. The reference axis ($||y_{m-1}$) is rotated by the angle $\theta_{Z_{m-1,m}}$ to locate the direction of axis y_m.

2.4.1.1 Relative Rotation Matrix about x_m

The *directional angles* $\alpha_{m,m-1}$, $\beta_{m,m-1}$, and $\gamma_{m,m-1}$ define the frame m's orientation relative to frame m−1. These angles are between 0° and 180° and are measured between the positive directions of x_{m-1}-, y_{m-1}-, z_{m-1}-axes and the Cartesian unit vectors $\mathbf{i}_m$, $\mathbf{j}_m$, and $\mathbf{k}_m$, respectively.

There are nine directional cosines, and Appendix 2.2 shows the six relations between them. Therefore, only three angles are independent.

The *directional cosines* noted $l_{m,m-1}$, $m_{m,m-1}$, and $n_{m,m-1}$ are shown in Equation (2.6). They are the unit vectors' components expressed as a function of relative angular displacement, $\theta_{Z_{m-1,m}}$. The upper scripts I, II, and III are assigned for the components of i_m, j_m, and k_m, respectively. Figure 2.4 is used to determine the directional cosines:

$$i_m = \begin{Bmatrix} \text{Cos}\,\alpha^{I}_{m-1,m} \\ \text{Cos}\,\beta^{I}_{m-1,m} \\ \text{Cos}\,\gamma^{I}_{m-1,m} \end{Bmatrix} = \begin{Bmatrix} l^{I}_{m-1,m} \\ m^{I}_{m-1,m} \\ n^{I}_{m-1,m} \end{Bmatrix} = \begin{Bmatrix} \text{Cos}\,0^\circ \\ \text{Cos}\,90^\circ \\ \text{Cos}\,90^\circ \end{Bmatrix} = \begin{Bmatrix} 1 \\ 0 \\ 0 \end{Bmatrix} = \mathbf{u}_{Z_{m-1,m}}$$

$$j_m = \begin{Bmatrix} \text{Cos}\,\alpha^{II}_{m-1,m} \\ \text{Cos}\,\beta^{II}_{m-1,m} \\ \text{Cos}\,\gamma^{II}_{m-1,m} \end{Bmatrix} = \begin{Bmatrix} l^{II}_{m-1,m} \\ m^{II}_{m-1,m} \\ n^{II}_{m-1,m} \end{Bmatrix} = \begin{Bmatrix} \text{Cos}\,90^\circ \\ \text{Cos}\,\theta_{Z_{m-1,m}} \\ \text{Cos}\left(90^\circ - \theta_{Z_{m-1,m}}\right) \end{Bmatrix} = \begin{Bmatrix} 0 \\ \text{Cos}\,\theta_{Z_{m-1,m}} \\ \text{Sin}\,\theta_{Z_{m-1,m}} \end{Bmatrix}$$

$$k_m = \begin{Bmatrix} \text{Cos}\,\alpha^{III}_{m-1,m} \\ \text{Cos}\,\beta^{III}_{m-1,m} \\ \text{Cos}\,\gamma^{III}_{m-1,m} \end{Bmatrix} = \begin{Bmatrix} l^{III}_{m-1,m} \\ m^{III}_{m-1,m} \\ n^{III}_{m-1,m} \end{Bmatrix} = \begin{Bmatrix} \text{Cos}\,90^\circ \\ \text{Cos}\left(90^\circ + \theta_{Z_{m-1,m}}\right) \\ \text{Cos}\,\theta_{Z_{m-1,m}} \end{Bmatrix} = \begin{Bmatrix} 0 \\ -\text{Sin}\,\theta_{Z_{m-1,m}} \\ \text{Cos}\,\theta_{Z_{m-1,m}} \end{Bmatrix} \tag{2.6}$$

The relative rotation matrix about x_m is defined as a 3 × 3 matrix whose columns I, II, and III are the unit vectors from Equation (2.6), as shown in Equation (2.7):

$$\mathbf{D}_{Z_{m-1,m}} = \overset{\begin{matrix}\text{I} & \text{II} & \text{III}\end{matrix}}{\left[\begin{array}{c|c|c} \mathbf{i}_m & \mathbf{j}_m & \mathbf{k}_m \end{array}\right]} = \begin{matrix} x_{m-1} \\ y_{m-1} \\ z_{m-1} \end{matrix} \overset{\begin{matrix} i_m & j_m & k_m \end{matrix}}{\left[\begin{array}{c|c|c} 1 & 0 & 0 \\ 0 & \text{Cos}\,\theta_{Z_{m-1,m}} & -\text{Sin}\,\theta_{Z_{m-1,m}} \\ 0 & \text{Sin}\,\theta_{Z_{m-1,m}} & \text{Cos}\theta_{Z_{m-1,m}} \end{array}\right]} \tag{2.7}$$

2.4.2 Relative Rotation Matrix about y_m-Axis: IT Notation

The frame's (Z, x_m, y_m, z_m) unit vectors are shown in Figure 2.5:

- j_m is the unit vector for the axis of rotation y_m, noted $\mathbf{u}_{Z_{m-1,m}}$

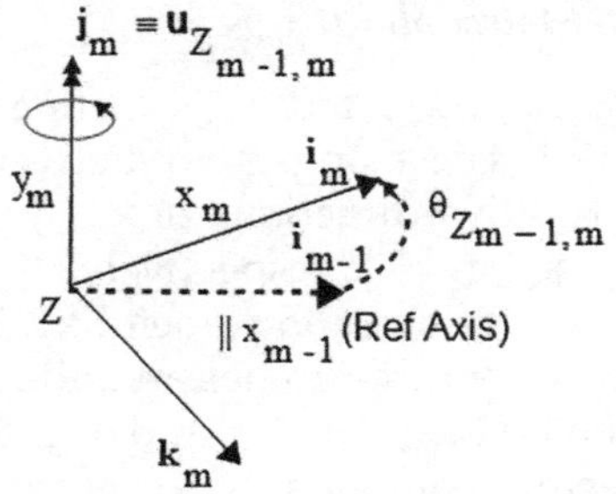

FIGURE 2.5
Rotation about axis y_m. The reference axis $(||x_{m-1})$ is rotated by the angle $\theta_{Z_{m-1,m}}$ to locate the direction of axis x_m.

- i_m is the unit vector for axis x_m, located at Z, in the plane perpendicular to the axis of rotation y_m.

  At Z, within this perpendicular plane, there is a *reference axis* $(||x_{m-1})$, parallel to axis x_{m-1} with the unit vector i_{m-1}. Angular displacement $\theta_{Z_{m-1,m}}$ measures the angle between these *two x-axes.* The positive direction is considered from the $||x_{m-1}$-axis rotating counterclockwise toward x_m-axis, which is given in Equation (2.8):

$$\mathbf{i}_{m-1} \times \mathbf{i}_m = \operatorname{Sin}\theta_{Z_{m-1,m}} \cdot \mathbf{u}_{Z_{m-1,m}} \tag{2.8}$$

Therefore, the first unit vector $\mathbf{i}_{m-1}$ in the cross product is rotated by the angle $\theta_{Z_{m-1,m}}$ to locate the direction of the unit vector $\mathbf{i}_m$, according to the right-hand rule.

The proof of Equation (2.8) is shown in Appendix 2.1:

- $\mathbf{k}_m$ is the unit vector for axis z_m defined by Equation (2.9).

$$\mathbf{i}_m \times \mathbf{j}_m = \mathbf{k}_m \tag{2.9}$$

2.4.2.1 Relative Rotation Matrix about y_m-Axis

From Figure 2.5, the components (directional cosines) of unit vectors $\mathbf{i}_m$, $\mathbf{j}_m$, and $\mathbf{k}_m$, expressed as functions of angular displacement $\theta_{Z_{m-1,m}}$, are shown in Equation (2.10):

$$\mathbf{i}_m = \begin{Bmatrix} \operatorname{Cos}\alpha^I_{m-1,m} \\ \operatorname{Cos}\beta^I_{m-1,m} \\ \operatorname{Cos}\gamma^I_{m-1,m} \end{Bmatrix} = \begin{Bmatrix} l^I_{m-1,m} \\ m^I_{m-1,m} \\ n^I_{m-1,m} \end{Bmatrix} = \begin{Bmatrix} \operatorname{Cos}\theta_{Z_{m-1,m}} \\ 0 \\ \operatorname{Cos}\left(90^\circ + \theta_{Z_{m-1,m}}\right) \end{Bmatrix} = \begin{Bmatrix} \operatorname{Cos}\theta_{Z_{m-1,m}} \\ 0 \\ -\operatorname{Sin}\theta_{Z_{m-1,m}} \end{Bmatrix}$$

$$\mathbf{j}_m = \begin{Bmatrix} \cos\alpha^{II}_{m-1,m} \\ \cos\beta^{II}_{m-1,m} \\ \cos\gamma^{II}_{m-1,m} \end{Bmatrix} = \begin{Bmatrix} l^{II}_{m-1,m} \\ m^{II}_{m-1,m} \\ n^{II}_{m-1,m} \end{Bmatrix} = \begin{Bmatrix} \cos 90^\circ \\ \cos 0^\circ \\ \cos 90^\circ \end{Bmatrix} = \begin{Bmatrix} 0 \\ 1 \\ 0 \end{Bmatrix} = \mathbf{u}_{Z_{m-1,m}}$$

$$\mathbf{k}_m = \begin{Bmatrix} \cos\alpha^{III}_{m-1,m} \\ \cos\beta^{III}_{m-1,m} \\ \cos\gamma^{III}_{m-1,m} \end{Bmatrix} = \begin{Bmatrix} l^{III}_{m-1,m} \\ m^{III}_{m-1,m} \\ n^{III}_{m-1,m} \end{Bmatrix} = \begin{Bmatrix} \cos(90^\circ - \theta_{Z_{m-1,m}}) \\ 0 \\ \cos\theta_{Z_{m-1,m}} \end{Bmatrix} = \begin{Bmatrix} \sin\theta_{Z_{m-1,m}} \\ 0 \\ \cos\theta_{Z_{m-1,m}} \end{Bmatrix} \tag{2.10}$$

The relative rotation matrix about y_m is defined as a 3 × 3 matrix whose columns (I, II, and III) are the unit vectors from Equation (2.10), as shown in Equation (2.11):

$$\mathbf{D}_{Z_{m-1,m}} = \overset{\quad\text{I}\qquad\text{II}\qquad\text{III}}{\left[\begin{array}{c:c:c} \mathbf{i}_m & \mathbf{j}_m & \mathbf{k}_m \end{array}\right]} = \begin{matrix} x_{m-1} \\ y_{m-1} \\ z_{m-1} \end{matrix} \overset{\qquad\mathbf{i}_m\qquad\qquad\mathbf{j}_m\qquad\qquad\mathbf{k}_m}{\left[\begin{array}{c:c:c} \cos\theta_{Z_{m-1,m}} & 0 & \sin\theta_{Z_{m-1,m}} \\ 0 & 1 & 0 \\ -\sin\theta_{Z_{m-1,m}} & 0 & \cos\theta_{Z_{m-1,m}} \end{array}\right]} \tag{2.11}$$

2.4.3 Relative Rotation Matrix about z_m-Axis: IT Notation

The unit vectors for frame (Z, x_m, y_m, z_m) are shown in Figure 2.6:

- $\mathbf{k}_m$ is the unit vector for the axis of rotation z_m, noted $\mathbf{u}_{Z_{m-1,m}}$
- $\mathbf{i}_m$ is the unit vector for axis x_m, located, at Z, in the plane perpendicular to axis of rotation z_m.

At Z, within this perpendicular plane there is a *reference axis* ($||x_{m-1}$), parallel to the x_{m-1} axis with the unit vector i_{m-1}. Angular displacement $\theta_{Z_{m-1,m}}$

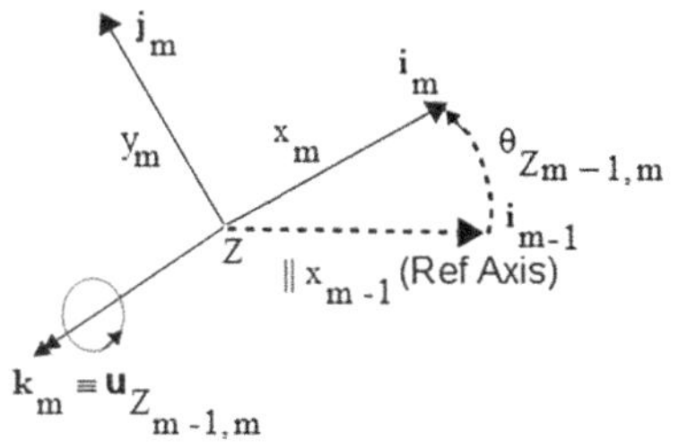

FIGURE 2.6
Rotation about axis z_m. The reference axis ($||x_{m-1}$) is rotated by the angle $\theta_{Z_{m-1,m}}$ to locate the direction of axis x_m.

measures the angle between the *two x-axes*. The positive direction is considered from the $||x_{m-1}$ axis rotating counterclockwise toward x_m-axis, which is given in Equation (2.12):

$$\mathbf{i}_{m-1} \times \mathbf{i}_m = \operatorname{Sin}\theta_{Z_{m-1,m}} \cdot \mathbf{u}_{Z_{m-1,m}} \tag{2.12}$$

Therefore, the first unit vector $\mathbf{i}_{m-1}$ in the cross product is rotated by the angle $\theta_{Z_{m-1,m}}$ to locate the direction of the unit vector $\mathbf{i}_m$, according to the right-hand rule.

The proof of Equation (2.12) is shown in Appendix 2.1.

- $\mathbf{j}_m$ is the unit vector for axis y_m defined by Equation (2.13):

$$\mathbf{k}_m \times \mathbf{i}_m = \mathbf{j}_m \tag{2.13}$$

2.4.3.1 Relative Rotation Matrix about z_m-Axis

From Figure 2.6, the components (directional cosines) of unit vectors $\mathbf{i}_m$, $\mathbf{j}_m$, and $\mathbf{k}_m$, expressed as functions of angular displacement $\theta_{Z_{m-1,m}}$, are shown in Equation (2.14):

$$\mathbf{i}_m = \begin{Bmatrix} \operatorname{Cos}\alpha^{I}_{m-1,m} \\ \operatorname{Cos}\beta^{I}_{m-1,m} \\ \operatorname{Cos}\gamma^{I}_{m-1,m} \end{Bmatrix} = \begin{Bmatrix} l^{I}_{m-1,m} \\ m^{I}_{m-1,m} \\ n^{I}_{m-1,m} \end{Bmatrix} = \begin{Bmatrix} \operatorname{Cos}\theta_{Z_{m-1,m}} \\ \operatorname{Cos}(90^\circ - \theta_{Z_{m-1,m}}) \\ \operatorname{Cos}90^\circ \end{Bmatrix} = \begin{Bmatrix} \operatorname{Cos}\theta_{Z_{m-1,m}} \\ \operatorname{Sin}\theta_{Z_{m-1,m}} \\ 0 \end{Bmatrix}$$

$$\mathbf{j}_m = \begin{Bmatrix} \operatorname{Cos}\alpha^{II}_{m-1,m} \\ \operatorname{Cos}\beta^{II}_{m-1,m} \\ \operatorname{Cos}\gamma^{II}_{m-1,m} \end{Bmatrix} = \begin{Bmatrix} l^{II}_{m-1,m} \\ m^{II}_{m-1,m} \\ n^{II}_{m-1,m} \end{Bmatrix} = \begin{Bmatrix} \operatorname{Cos}(90^\circ - \theta_{Z_{m-1,m}}) \\ \operatorname{Cos}\theta_{Z_{m-1,m}} \\ \operatorname{Cos}90^\circ \end{Bmatrix} = \begin{Bmatrix} -\operatorname{Sin}\theta_{Z_{m-1,m}} \\ \operatorname{Cos}\theta_{Z_{m-1,m}} \\ 0 \end{Bmatrix}$$

$$\mathbf{k}_m = \begin{Bmatrix} \operatorname{Cos}\alpha^{III}_{m-1,m} \\ \operatorname{Cos}\beta^{III}_{m-1,m} \\ \operatorname{Cos}\gamma^{III}_{m-1,m} \end{Bmatrix} = \begin{Bmatrix} l^{III}_{m-1,m} \\ m^{III}_{m-1,m} \\ n^{III}_{m-1,m} \end{Bmatrix} = \begin{Bmatrix} \operatorname{Cos}90^\circ \\ \operatorname{Cos}90^\circ \\ \operatorname{Cos}0^\circ \end{Bmatrix} = \begin{Bmatrix} 0 \\ 0 \\ 1 \end{Bmatrix} = \mathbf{u}_{Z_{m-1,m}}$$

$$\tag{2.14}$$

The relative rotation matrix about z_m is defined as a 3×3 matrix whose columns (I, II, and III) are the unit vectors from Equation (2.14), as shown in Equation (2.15):

$$\mathbf{D}_{Z_{m-1,m}} = \begin{array}{c} \\ \\ \end{array}\overset{\begin{array}{ccc} \text{I} & \text{II} & \text{III} \end{array}}{\left[\begin{array}{c|c|c} \mathbf{i}_m & \mathbf{j}_m & \mathbf{k}_m \end{array}\right]} = \begin{array}{c} x_{m-1} \\ y_{m-1} \\ z_{m-1} \end{array} \overset{\begin{array}{ccc} \mathbf{i}_m & \mathbf{j}_m & \mathbf{k}_m \end{array}}{\left[\begin{array}{c|c|c} \text{Cos}\,\theta_{Z_{m-1,m}} & -\text{Sin}\,\theta_{Z_{m-1,m}} & 0 \\ \text{Sin}\,\theta_{Z_{m-1,m}} & \text{Cos}\,\theta_{Z_{m-1,m}} & 0 \\ 0 & 0 & 1 \end{array}\right]} \tag{2.15}$$

Notation: For relative entities, the two subscripts m − 1 and m can be dropped, resulting the notation $\mathbf{D}_Z$.

2.4.4 Properties of a Relative Rotation Matrix

Appendix 2.2 shows the proofs for the following properties for the relative matrices $\mathbf{D}_Z$, which are valid for the rotation about the x_m, or y_m, or z_m axes.

a. The matrix **D** is an *orthonormal* matrix because the entries obey two properties:

$$\begin{array}{l} \text{The sum of the square of the entries of each column is 1} \\ \text{The unit vectors are perpendicular (orthogonal) to each other.} \end{array} \tag{2.16a}$$

b. The determinant of a relative matrix is 1

$$\det \mathbf{D}_Z = 1 \tag{2.16b}$$

c. The inverse, $\mathbf{D}_Z^{-1}$, which corresponds to the negative rotation $-\theta_Z$, is the same as its transposed matrix

$$\mathbf{D}_Z^{-1} = \mathbf{D}_Z^{T} \tag{2.16c}$$

From this property, notice that there is no need to calculate the inverse; instead, the inverse matrix is replaced by its transposed matrix which simplifies the calculations.

d. The product of the orthogonal matrix and its transposed maxtrix is a 3×3 unit matrix

$$\mathbf{D}_Z \mathbf{D}_Z^{T} = \mathbf{U}_{3\times 3} \tag{2.16d}$$

e. The equation of constrained motion for a prismatic joint. If two frames are in *translation* relative to each other, then their x, y, and z axes remain parallel, therefore $\theta_Z = 0$, a *null rotation*, which means there is no rotation. Since there is no relative rotation between frames, the matrix becomes the 3×3 unit matrix:

$$\mathbf{D}_Z = \begin{matrix} \\ x_{m-1} \\ y_{m-1} \\ z_{m-1} \end{matrix} \begin{matrix} i_m & j_m & k_m \\ \left[\begin{matrix} 1 \\ 0 \\ 0 \end{matrix}\right. & \begin{matrix} 0 \\ 1 \\ 0 \end{matrix} & \left.\begin{matrix} 0 \\ 0 \\ 1 \end{matrix}\right] \end{matrix} = \mathbf{U}_{3\times 3} \tag{2.16e}$$

2.5 Open Cycle Mechanisms: The Relative Rotation Matrices along the Tree

Figure 2.7 illustrates the sequence (path) of nodes and edges in the tree, which starts at node 0 and ends at the last node m. A relative matrix $\mathbf{D}_Z$ is assigned to each edge in the digraph and represents the frame m's orientation relative to frame $m-1$.

2.5.1 Direct and Inverse Relative Rotation Matrix

The direction of an oriented edge in the digraph could be coincident or opposite with the path's sense.

a. *The path's sense coincides with the edge's sense.* In this case, the edge has assigned a direct relative matrix, noted $\mathbf{D}_z$. For example, in Figure 2.7, all edges have their sense coincident to the path; therefore, their assigned relative matrices are direct.
b. *The path sense is opposite to the edge's sense.* In this case, the inverse of a relative matrix (its transposed) is associated with the edge, noted $\mathbf{D}_Z^{-1} = \mathbf{D}_Z^T$. Equation (2.16c) shows that the inverse of an orthogonal matrix is its transposed matrix.

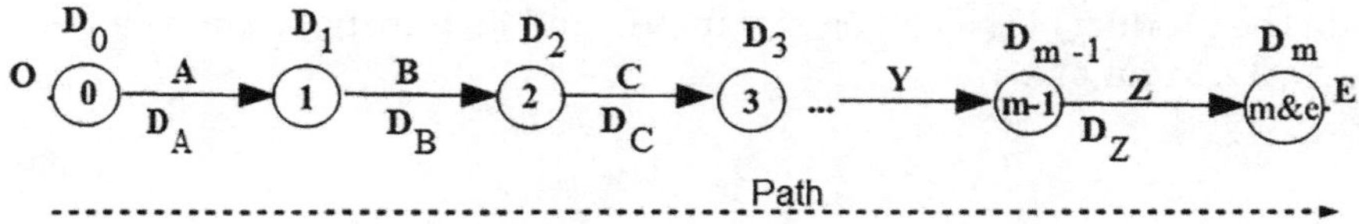

FIGURE 2.7
Relation between absolute and relative matrices along the tree.

2.6 Additional Frames on the Same Link

At the end-effector e, an additional frame is introduced (E, x_e, y_e, z_e). Since points Z and E are located on the same link m, the frame at E is translated (without rotation) relative to the last frame (Z, x_m, y_m, z_m). Therefore, the rotation matrix is the 3×3 unit matrix: $\mathbf{D}_E = \mathbf{U}$. In robotic applications, the part or tool to be grasped is referenced with respect to the end-effector's frame.

An additional required frame is frame g, located at the center of mass (COM) G on link m. The frame at Z is translated parallel to itself (null rotation) into the new frame (G, x_g, y_g, z_g) with origin at G_m. Therefore, the rotation matrix is the unit matrix: $\mathbf{D}_G = \mathbf{U}$. This frame will be used later in the text to describe the velocity of link m, acceleration, and moments of inertia in the dynamic analysis.

2.7 The Absolute Rotation Matrix for Links' Orientation

In the direct orientation analysis, the absolute orientations (relative to the fixed frame) of frame m and end-effector e are defined by the *direction angles* α_m, β_m, and γ_m. Each angle is between 0° and 180°. The angles are measured between the positive x_0, y_0, z_0 axes and the Cartesian unit vectors $\mathbf{i}_m$, $\mathbf{j}_m$, and $\mathbf{k}_m$, respectively.

Analog to the previously defined relative rotation matrices, the components for the unit vectors in the absolute rotation matrices are the *directional cosines* l_m, m_m, and n_m.

A superscript I, II, III is assigned for the directional cosines of $\mathbf{i}_m$, $\mathbf{j}_m$, and $\mathbf{k}_m$ (see Equation (2.17)):

$$\mathbf{i}_m = \begin{Bmatrix} \text{Cos}\,\alpha_m^{I} \\ \text{Cos}\,\beta_m^{I} \\ \text{Cos}\,\gamma_m^{I} \end{Bmatrix} = \begin{Bmatrix} l_m^{I} \\ m_m^{I} \\ n_m^{I} \end{Bmatrix}; \quad \mathbf{j}_m = \begin{Bmatrix} \text{Cos}\,\alpha_m^{II} \\ \text{Cos}\,\beta_m^{II} \\ \text{Cos}\,\gamma_m^{II} \end{Bmatrix} = \begin{Bmatrix} l_m^{II} \\ m_m^{II} \\ n_m^{II} \end{Bmatrix};$$

$$\mathbf{k}_m = \begin{Bmatrix} \text{Cos}\,\alpha_m^{III} \\ \text{Cos}\,\beta_m^{III} \\ \text{Cos}\,\gamma_m^{III} \end{Bmatrix} = \begin{Bmatrix} l_m^{III} \\ m_m^{III} \\ n_m^{III} \end{Bmatrix} \tag{2.17}$$

Appendix 2.2 shows the properties in Equation (2.16a), which are equivalent to six relations between the nine directional cosines. Therefore, only three angles are independent.

The orientation of the frame of link m is modeled by a 3 × 3 matrix $\mathbf{D}_m$ whose columns are the components of the $\mathbf{i}_m$, $\mathbf{j}_m$, and $\mathbf{k}_m$ unit vectors, as shown in Equation (2.18):

$$\mathbf{D}_m = \overset{\begin{matrix} \text{I} & \text{II} & \text{III} \end{matrix}}{\left[\begin{array}{c|c|c} \mathbf{i}_m & \mathbf{j}_m & \mathbf{k}_m \end{array}\right]} = \begin{matrix} x_0 \\ y_0 \\ z_0 \end{matrix} \overset{\begin{matrix} \mathbf{i}_m & \mathbf{j}_m & \mathbf{k}_m \end{matrix}}{\left[\begin{array}{c|c|c} \mathrm{Cos}\,\alpha_m^{I} & \mathrm{Cos}\,\alpha_m^{II} & \mathrm{Cos}\,\alpha_m^{III} \\ \mathrm{Cos}\,\beta_m^{I} & \mathrm{Cos}\,\beta_m^{II} & \mathrm{Cos}\,\beta_m^{III} \\ \mathrm{Cos}\,\gamma_m^{I} & \mathrm{Cos}\,\gamma_m^{II} & \mathrm{Cos}\,\gamma_m^{III} \end{array}\right]} = \left[\begin{array}{c|c|c} l_m^{I} & l_m^{II} & l_m^{III} \\ m_m^{I} & m_m^{II} & m_m^{III} \\ n_m^{I} & n_m^{II} & n_m^{III} \end{array}\right] \tag{2.18}$$

2.8 Direct Links' Orientation Analysis for Open Cycle Mechanisms

Let us consider a path along the tree in Figure 2.7. The path originates at node 0 and continues from left to right. Node 0 is assigned the matrix $\mathbf{D}_0$. This matrix is the unit matrix since the orientation of frame 0 relative to itself is null rotation, Equation (2.16e).

The absolute matrix $\mathbf{D}_m$ represents the orientation of mobile link m relative to the fixed frame.

It is the result of post multiplication of the absolute matrix of a previous node m−1 and the relative matrix at the edge Z between the adjacent nodes m−1 and m, as shown in recursion Equation (2.19):

$$\mathbf{D}_m = \mathbf{D}_{m-1} * \mathbf{D}_Z \tag{2.19}$$

Equation (2.19) is written for each edge A, B, C,…, along the path. Then, Equation (2.20) holds.

Statement 2.2

For direct orientation analysis, the absolute matrix at node m is computed from the post multiplication of all relative matrices of the edges located before the node m, as shown in Equation (2.20):

$$\mathbf{D}_m = \mathbf{D}_A\big|_A * \mathbf{D}_B\big|_B * \mathbf{D}_C\big|_C * \ldots * \mathbf{D}_Z\big|_Z = \mathbf{D}_{m-1} * \mathbf{D}_Z \tag{2.20}$$

Thus, to determine $\mathbf{D}_1$, consider only the terms on the left side of bracket $|_A$. To determine $\mathbf{D}_2$, consider only the terms on the left side of bracket $|_B$.

The recursive procedure along the path in Figure 2.7 is developed in Equations (2.21).

- *Node 0*: The link 0 is fixed, so its absolute rotation matrix is null rotation Therefore, the unit 3 × 3 matrix:

$$\mathbf{D}_0 = \mathbf{U} \tag{2.21a}$$

$$\text{Edge A is assigned relative matrix: } \mathbf{D}_A \tag{2.21b}$$

- *Node m = 1*: The first link's first orientation matrix from Equation (2.20):

$$\mathbf{D}_1 = \mathbf{D}_0 * \mathbf{D}_A = \mathbf{D}_A \tag{2.21c}$$

$$\text{Edge B is assigned relative matrix: } \mathbf{D}_B \tag{2.21d}$$

- *Node m = 2*: The second link's orientation matrix from Equation (2.20):

$$\mathbf{D}_2 = \mathbf{D}_A * \mathbf{D}_B = \mathbf{D}_1 * \mathbf{D}_B \tag{2.21e}$$

$$\text{Edge C is assigned relative matrix: } \mathbf{D}_C \tag{2.21f}$$

- *Node m = 3*: The third link's orientation matrix from Equation (2.20):

$$\mathbf{D}_3 = \mathbf{D}_A * \mathbf{D}_B * \mathbf{D}_C = \mathbf{D}_2 * \mathbf{D}_C \tag{2.21g}$$

$$\ldots\text{Edge Z is assigned relative matrix: } \mathbf{D}_Z \tag{2.21h}$$

- *Node m*: The m link's orientation matrix from Equation (2.20):

$$\mathbf{D}_m = \mathbf{D}_A * \mathbf{D}_B * \mathbf{D}_C * \cdots * \mathbf{D}_Z = \mathbf{D}_{m-1} * \mathbf{D}_Z \tag{2.21i}$$

End-Effector's Orientation: The Mechanism's Absolute Rotation Matrix, $\mathbf{D}_e^{mech}$

The end-effector e (grasping device, wrist) is rigidly attached to link m, represented in graph by node m&e. The points E and Z are located on the same link, m&e. Point E is assigned the frame (E, x_e, y_e, z_e). The frame at E is translated (null rotation) relative to the frame (Z, x_m, y_m, z_m). Therefore, the relative rotation matrix is the unit matrix:

$$\mathbf{D}_E = \mathbf{U} \tag{2.21j}$$

Equation (2.22) shows that frame E has the same orientation as the frame of link m.

Input parameters are the actuated angular displacements θ_Z. The matrix multiplications from Equation (2.20) determine the entries of matrix $\mathbf{D}_e$ as functions of the actuated angular displacements.

- *The end-effector's frame's orientation* assists in the determination of direction angles for the absolute orientation of frame E, as shown in Equation (2.22):

$$\mathbf{D}_e = \mathbf{D}_m * \mathbf{D}_E = \mathbf{D}_m \tag{2.22}$$

  In robotics literature [1–3], the columns in matrix $\mathbf{D}_e$ are components of three vectors. Thus, the first vector labeled with superscript I is the *approach vector,* **a**; the second vector labeled II is the *orientation vector,* **o**; the third one is the *normal vector,* **n**.
- *The mechanism absolute rotation matrix,* $\mathbf{D}_e^{mech}$, is assigned to the end-effector's frame orientation.

  The superscript "mech" suggests that the orientation of the end-effector is determined from the manipulator's point of view (links and joints):

$$D_e^{mech} = \left[\begin{array}{c|c|c} \text{Cos}\,\alpha_e^{I} & \text{Cos}\,\alpha_e^{II} & \text{Cos}\,\alpha_e^{III} \\ \text{Cos}\,\beta_e^{I} & \text{Cos}\,\beta_e^{II} & \text{Cos}\,\beta_e^{III} \\ \text{Cos}\,\gamma_e^{I} & \text{Cos}\,\gamma_e^{II} & \text{Cos}\,\gamma_e^{III} \end{array}\right] \tag{2.23}$$

- *The governing equations for direct orientation analysis* are determined from Equation (2.24) by equating the entries from the left-side matrix to the entries from the right-side matrix:

$$\mathbf{D}_e^{mech} = \mathbf{D}_e \tag{2.24}$$

Thus, the absolute orientation of frame E, as a function of angular actuating parameters at the joint, is determined:

$$\begin{Bmatrix} \alpha_e^{I} \\ \beta_e^{I} \\ \gamma_e^{I} \end{Bmatrix} = \begin{Bmatrix} \text{Arccos}(l_a) \\ \text{Arccos}(m_a) \\ \text{Arccos}(n_a) \end{Bmatrix}; \quad \begin{Bmatrix} \alpha_e^{II} \\ \beta_e^{II} \\ \gamma_e^{II} \end{Bmatrix} = \begin{Bmatrix} \text{Arccos}(l_o) \\ \text{Arccos}(m_o) \\ \text{Arccos}(n_o) \end{Bmatrix};$$
$$\begin{Bmatrix} \alpha_e^{III} \\ \beta_e^{III} \\ \gamma_e^{III} \end{Bmatrix} = \begin{Bmatrix} \text{Arccos}(l_n) \\ \text{Arccos}(m_n) \\ \text{Arccos}(n_n) \end{Bmatrix} \tag{2.25}$$

Note: The nine entries in orthogonal matrix $\mathbf{D}_e$ are related by six nonlinear equations: three for unit vector magnitude and three for perpendicular unit vectors, as shown next:

$$(l_a)^2+(m_a)^2+(n_a)^2=1;\quad (l_o)^2+(m_o)^2+(n_o)^2=1;\quad (l_n)^2+(m_n)^2+(n_n)^2=1$$

$$l_a\cdot l_0+m_a\cdot m_0+n_a\cdot n_0=0;\quad l_a\cdot l_n+m_a\cdot m_n+n_a\cdot n_n=0;$$

$$l_o\cdot l_n+m_o\cdot m_n+n_o\cdot n_n=0$$

Figure 2.8 shows the three direction angles between unit vectors $\mathbf{i}_e$, $\mathbf{j}_e$, $\mathbf{k}_e$ and the fixed axes x_0, y_0, and z_0, respectively.

A common practice is to express the nine entries as functions of three independent angles known as Euler angles. A different approach is to select three angles of rotation about the three axes, known as yaw–pitch–roll angles. In this text, we introduce three angles α_e^I, β_e^{II}, and γ_e^{III}, named *IT direction angles.* and they are different than the Euler's or yaw–pitch–roll angles. The equations for IT direction angles are useful in the inverse orientation analysis. These angles are functions of angular displacements in the joints from the above IT notation and are derived from the cosine entries located on the main diagonal in matrix $\mathbf{D}_e$.

2.8.1 Example of Direct Links' Orientation for a Spatial Mechanism with 4 DOF

The TRRT mechanism, Figure 2.1, has m = 4 links, j = 4 joints, and mobility M = 4.

A direct orientation analysis is presented, where the following geometric parameters are provided

Data: The M = 4 actuating parameters in matrix $\mathbf{q}_M = \{d_A, \theta_B, \theta_C, d_D\}^T$

Constraint equations for angular displacement: There are two prismatic joints at A and D; therefore, there are two constraint equations for the relative rotations: $\theta_A = \theta_D = 0°$

Home position: It is the initial position of mechanism, t = 0 s, where all actuating positional parameters are zero (see Figure 2.2):

$$d_A = 0\text{ in.};\quad \theta_B = 0°;\quad \theta_C = 0°;\quad d_D(t) = 0\text{ in.}$$

Objective: Analyze frame orientation

Procedure for analyzing frame orientation:

1. Relative and absolute rotation matrices
2. The mechanism's absolute rotation matrix at any time
3. The governing equations for direct orientation analysis and the IT direction angles at any time

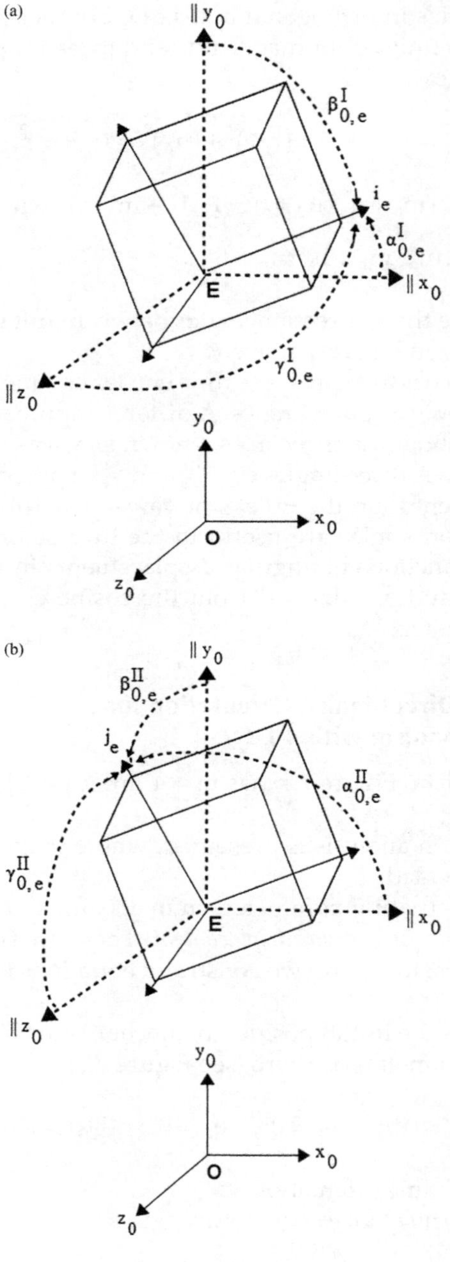

FIGURE 2.8
End-effector's orientation angles with (a) x_0-axis, (b) y_0-axis, and (c) z_0-axis.

(Continued)

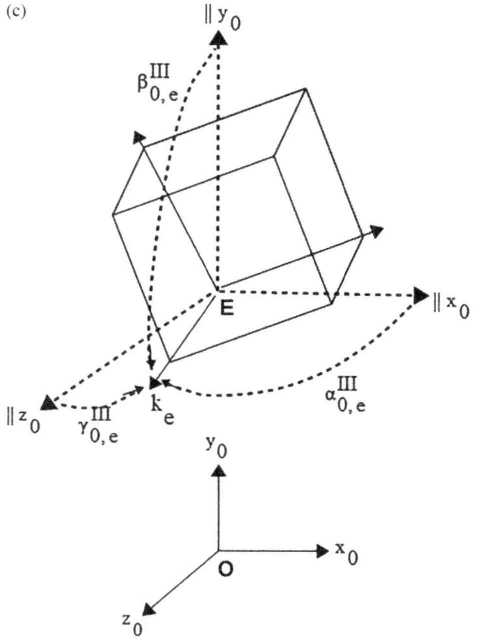

FIGURE 2.8 (CONTINUED)
End-effector's orientation angles with (a) x_0-axis, (b) y_0-axis, and (c) z_0-axis.

4. Direction angles for the links' frames at the home position
5. Direction angles for the end-effector's frame at the home position

Solution: We consider the path in the tree, Figure 2.9, where the edges are assigned the relative matrices, and the nodes are assigned the absolute rotation matrices.

1. Relative and absolute rotation matrices
   - *Node 0*: The link 0 is fixed, its absolute rotation matrix is null rotation; therefore, the unit 3×3 matrix is

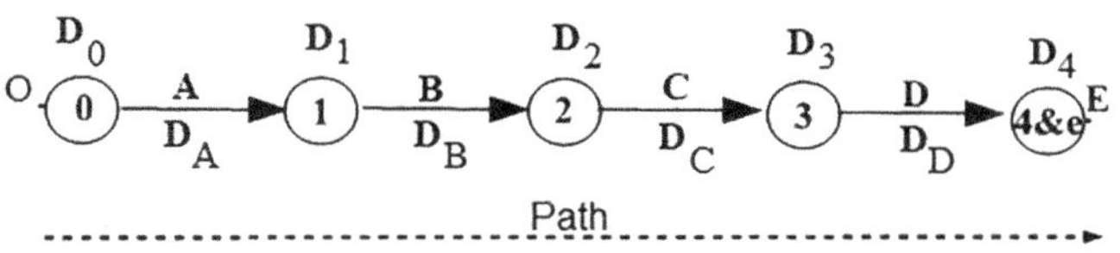

FIGURE 2.9
Relative and absolute matrices along the tree for TRRT mechanism.

$$\mathbf{D}_0 = \begin{matrix} & i_0 & j_0 & k_0 \\ i_0 & 1 & 0 & 0 \\ j_0 & 0 & 1 & 0 \\ k_0 & 0 & 0 & 1 \end{matrix} = \mathbf{U}_{3\times 3} \quad (2.26a)$$

- *Edge A (prismatic joint)*: There is a translation of the frame A, x_1, y_1, z_1 with respect to frame O, x_0, y_0, z_0, therefore a constraint rotation, $\theta_A = 0$:

$$\mathbf{D}_A = \begin{bmatrix} 1 & 0 & 0 \\ 0 & 1 & 0 \\ 0 & 0 & 1 \end{bmatrix} = \mathbf{U}_{3\times 3} \quad (2.26b)$$

- *Node m = 1*: The orientation of link 1's frame (A, x_1, y_1, z_1) with respect to the fixed frame results from its absolute matrix:

$$\mathbf{D}_1 = \mathbf{D}_0 * \mathbf{D}_A = \mathbf{D}_A = \begin{matrix} & i_1 & j_1 & k_1 \\ i_0 & 1 & 0 & 0 \\ j_0 & 0 & 1 & 0 \\ k_0 & 0 & 0 & 1 \end{matrix} \quad (2.26c)$$

- *Edge B (revolute joint)*: There is a relative rotation of the frame (B, x_2, y_2, z_2) about y_2 with respect to the frame (A, x_1, y_1, z_1) (Equation (2.11)):

$$\mathbf{D}_B = \begin{bmatrix} C\theta_B & 0 & S\theta_B \\ 0 & 1 & 0 \\ -S\theta_B & 0 & C\theta_B \end{bmatrix} \quad (2.26d)$$

where

$C\ \theta_B$ and $S\ \theta_B$ are the abbreviations for Cos θ_B and Sin θ_B, respectively.

- *Node m = 2*: The orientation of the second link's frame (B, x_2, y_2, z_2) with respect to the fixed frame results from its absolute matrix:

$$\mathbf{D}_2 = \mathbf{D}_A * \mathbf{D}_B = \mathbf{D}_1 * \mathbf{D}_B = \begin{matrix} & i_2 & j_2 & k_2 \\ i_0 & C\theta_B & 0 & S\theta_B \\ j_0 & 0 & 1 & 0 \\ k_0 & -S\theta_B & 0 & C\theta_B \end{matrix} \quad (2.26e)$$

- *Edge C (revolute joint)*: There is a relative rotation about z_3 of the frame (C, x_3, y_3, z_3) with respect to frame (B, x_2, y_2, z_2) (Equation (2.15)):

$$\mathbf{D}_C = \begin{bmatrix} C\theta_C & -S\theta_C & 0 \\ S\theta_C & C\theta_C & 0 \\ 0 & 0 & 1 \end{bmatrix} \tag{2.26f}$$

- *Node m = 3*: The orientation of link 3's frame (C, x_3, y_3, z_3) with respect to the fixed frame results from its absolute matrix:

$$\mathbf{D}_3 = \mathbf{D}_A * \mathbf{D}_B * \mathbf{D}_C = \mathbf{D}_2 * \mathbf{D}_C = \begin{matrix} & i_3 & j_3 & k_3 \\ i_0 & C\theta_B \cdot C\theta_C & -C\theta_B \cdot S\theta_C & S\theta_B \\ j_0 & S\theta_C & C\theta_C & 0 \\ k_0 & -S\theta_B \cdot C\theta_C & S\theta_B \cdot S\theta_C & C\theta_B \end{matrix} \tag{2.26g}$$

- *Edge D (prismatic joint)*: There is a translation of the frame (D, x_4, y_4, z_4) with respect to frame C, x_3, y_3, z_3, therefore a constraint rotation, $\theta_D = 0$:

$$\mathbf{D}_D = \begin{bmatrix} 1 & 0 & 0 \\ 0 & 1 & 0 \\ 0 & 0 & 1 \end{bmatrix} \tag{2.26h}$$

- *Node m = 4*: The orientation of link 4's frame (D, x_4, y_4, z_4) with respect to the fixed frame results from its absolute matrix:

$$\mathbf{D}_4 = \mathbf{D}_A * \mathbf{D}_B * \mathbf{D}_C * \mathbf{D}_D = \mathbf{D}_3 * \mathbf{D}_D = \begin{matrix} & i_4 & j_4 & k_4 \\ i_0 & C\theta_B \cdot C\theta_C & -C\theta_B \cdot S\theta_C & S\theta_B \\ j_0 & S\theta_C & C\theta_C & 0 \\ k_0 & -S\theta_B \cdot C\theta_C & S\theta_B \cdot S\theta_C & C\theta_B \end{matrix} \tag{2.26i}$$

2. The mechanism's absolute rotation matrix at any time, $\mathbf{D}_e^{mech}$
   - *End-effector E*: There is a translation of the frame (E, x_e, y_e, z_e) relative to frame (D, x_4, y_4, z_4), that is, no relative rotation, $\theta_D = 0$:

$$\mathbf{D}_E = \begin{matrix} & i_e & j_e & k_e \\ i_4 & 1 & 0 & 0 \\ j_4 & 0 & 1 & 0 \\ k_4 & 0 & 0 & 1 \end{matrix} \tag{2.26j}$$

The orientation of the end-effector's frame (E, x_e, y_e, z_e) with respect to the fixed frame (O, x_0, y_0, z_0) is determined from Equation (2.26k):

$$\mathbf{D}_e = \mathbf{D}_A * \mathbf{D}_B * \mathbf{D}_C * \mathbf{D}_D = \mathbf{D}_4 * \mathbf{D}_E = \mathbf{D}_4 = \begin{matrix} \\ i_0 \\ j_0 \\ k_0 \end{matrix} \begin{matrix} i_e & j_e & k_e \\ \left[\begin{matrix} C\theta_B \cdot C\theta_C \\ S\theta_C \\ -S\theta_B \cdot C\theta_C \end{matrix}\right. & \begin{matrix} -C\theta_B \cdot S\theta_C \\ C\theta_C \\ S\theta_B \cdot S\theta_C \end{matrix} & \left.\begin{matrix} S\theta_B \\ 0 \\ C\theta_B \end{matrix}\right] \end{matrix} \tag{2.26k}$$

3. The governing equations for direct orientation analysis. The IT direction angles at any time.

   These equations and angles are obtained from

$$\mathbf{D}_e^{mech} = \mathbf{D}_e \tag{2.26l}$$

or

$$\begin{bmatrix} \text{Cos}\,\alpha_e^{I} & \text{Cos}\,\alpha_e^{II} & \text{Cos}\,\alpha_e^{III} \\ \text{Cos}\,\beta_e^{I} & \text{Cos}\,\beta_e^{II} & \text{Cos}\,\beta_e^{III} \\ \text{Cos}\,\gamma_e^{I} & \text{Cos}\,\gamma_e^{II} & \text{Cos}\,\gamma_e^{III} \end{bmatrix} = \begin{bmatrix} C\theta_B \cdot C\theta_C & -C\theta_B \cdot S\theta_C & S\theta_B \\ S\theta_C & C\theta_C & 0 \\ -S\theta_B \cdot C\theta_C & S\theta_B \cdot S\theta_C & C\theta_B \end{bmatrix}$$

By equating the entries within the matrices in Equation (2.26l), the direction angles between the end-effector's frame and the fixed frame can be found; see Equation (2.26m):

$$\begin{Bmatrix} \alpha_e^{I} \\ \beta_e^{I} \\ \gamma_e^{I} \end{Bmatrix} = \begin{Bmatrix} \text{Arccos}(C\theta_B \cdot C\theta_C) \\ \text{Arccos}(S\theta_C) \\ \text{Arccos}(-S\theta_B \cdot C\theta_C) \end{Bmatrix}; \quad \begin{Bmatrix} \alpha_e^{II} \\ \beta_e^{II} \\ \gamma_e^{II} \end{Bmatrix} = \begin{Bmatrix} \text{Arccos}(-C\theta_B \cdot S\theta_C) \\ \theta_C \\ \text{Arccos}(S\theta_B \cdot S\theta_C) \end{Bmatrix};$$

$$\begin{Bmatrix} \alpha_e^{III} \\ \beta_e^{III} \\ \gamma_e^{III} \end{Bmatrix} = \begin{Bmatrix} \text{Arccos}(S\theta_B) \\ 90^\circ \\ \theta_B \end{Bmatrix} \tag{2.26m}$$

where the inverse trigonometric functions "Arccos" have values within the interval 0°–180°.

The IT direction angles are the entries on the diagonal:

$$
\begin{aligned}
\alpha_e^{I} &= \text{Arccos}(C\theta_B \cdot C\theta_C) \\
\beta_e^{II} &= \theta_C \\
\gamma_e^{III} &= \theta_B
\end{aligned}
\tag{2.26n}
$$

4. Direction angles for the links' frames at the home position.
   In Figure (2.2), at the home position, the angular displacements are zero:

$$
\theta_B = 0^\circ; \quad \theta_C = 0^\circ
$$

which are considered in Equations (2.26c), (2.26e), (2.26g), and (2.26i).
   Using this information, the form of the absolute rotation matricies can be deduced (Equation (2.26o)):

$$
\mathbf{D}_1 = \mathbf{D}_2 = \mathbf{D}_3 = \mathbf{D}_4 = \begin{bmatrix} 1 & 0 & 0 \\ 0 & 1 & 0 \\ 0 & 0 & 1 \end{bmatrix} \tag{2.26o}
$$

The absolute matrices in Equation (2.26o) are all unit matrices. Therefore, the mechanism at the home position has its frames 1, 2, 3, and 4 with axes parallel to frame 0's axis.

5. Direction angles for the end-effector's frame at the home position.
   If we consider $\theta_B = 0^\circ$ and $\theta_C = 0^\circ$ in Equation (2.26l), then

$$
\begin{bmatrix} \text{Cos}\,\alpha_e^{I} & \text{Cos}\,\alpha_e^{II} & \text{Cos}\,\alpha_e^{III} \\ \text{Cos}\,\beta_e^{I} & \text{Cos}\,\beta_e^{II} & \text{Cos}\,\beta_e^{III} \\ \text{Cos}\,\gamma_e^{I} & \text{Cos}\,\gamma_e^{II} & \text{Cos}\,\gamma_e^{III} \end{bmatrix} = \begin{bmatrix} 1 & 0 & 0 \\ 0 & 1 & 0 \\ 0 & 0 & 1 \end{bmatrix} \tag{2.26p}
$$

Equating the entries within the left and right matrices from Equation (2.26p), the direction angles between the end-effector and the fixed frame's axes are

$$
\begin{Bmatrix} \alpha_e^{I} \\ \beta_e^{I} \\ \gamma_e^{I} \end{Bmatrix} = \begin{Bmatrix} 0^\circ \\ 90^\circ \\ 90^\circ \end{Bmatrix}; \quad \begin{Bmatrix} \alpha_e^{II} \\ \beta_e^{II} \\ \gamma_e^{II} \end{Bmatrix} = \begin{Bmatrix} 90^\circ \\ 0^\circ \\ 90^\circ \end{Bmatrix}; \quad \begin{Bmatrix} \alpha_e^{III} \\ \beta_e^{III} \\ \gamma_e^{III} \end{Bmatrix} = \begin{Bmatrix} 90^\circ \\ 90^\circ \\ 0^\circ \end{Bmatrix} \tag{2.26q}
$$

Thus, when the mechanism is at home position, the two frames e and 0 have parallel axes (Figure 2.2).

The IT direction angles are the entries on the diagonal:

$$\alpha_e^{I} = 0^\circ$$
$$\beta_e^{II} = 0^\circ \qquad (2.26r)$$
$$\gamma_e^{III} = 0^\circ$$

2.9 Inverse Links' Orientation Analysis along a Closed Path (Cycle): The Matroid Method

The direct analysis described earlier finds the end-effector's orientation based on angular displacements at the joints. The inverse analysis finds the actuating angular displacements to put the end-effector in a desired orientation, thereby stating the manipulator's control requirements.

The open path starts at node 0 and ends at node m&e. If we imagine the path to end at the initial node 0, then a closed path, called a *cycle*, is generated. For open cycle mechanisms, this closed path is a fictitious cycle. Within a closed path, a fictitious edge, called a *chord*, is considered; see Figure 2.10a,b. In the digraph, the chord is oriented by the same rules as all the other edges in the digraph (i.e., from lower node number 0 to higher number m). For open cycle spatial mechanisms, the chord is associated with a 6 DOF *fictitious joint* at E between node 0 and node m. But for planar mechanisms, there is a 3 DOF fictitious joint.

From Equation (2.20), the absolute matrix of the last node in the closed path, node 0, is the unit matrix, and it is determined by post multiplications

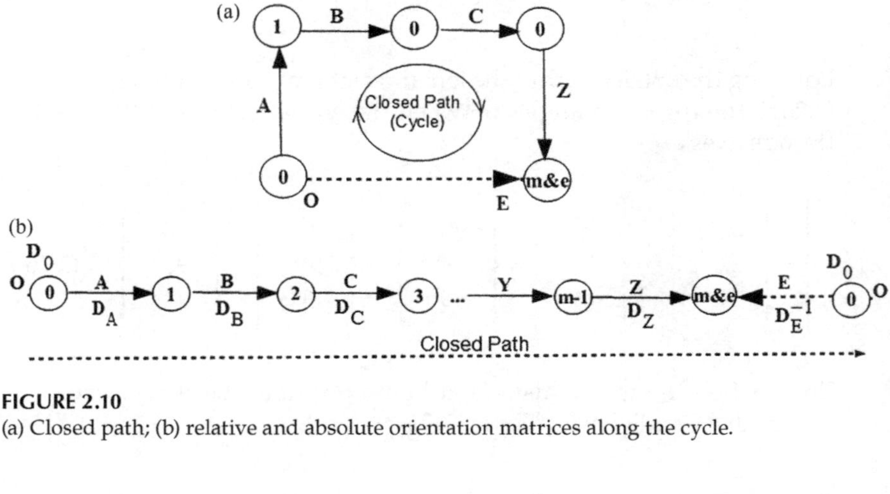

FIGURE 2.10
(a) Closed path; (b) relative and absolute orientation matrices along the cycle.

of all relative matrices of the edges located before node 0 (Equation (2.27)). Since the sense of chord E is opposite to the closed path's sense, then the inverse relative matrix attached to the chord will show in the matrix product:

$$\mathbf{D}_A * \mathbf{D}_B * \mathbf{D}_C * \cdots * \mathbf{D}_Z * \mathbf{D}_E^{-1} = \mathbf{U} \tag{2.27}$$

However, there is no need to determine the inverse of a matrix, since in Equation (2.16c), the inverse of an orthogonal matrix is the same as its transposed matrix is shown.

$$\mathbf{D}_E^{-1} = \mathbf{D}_E^{T} \tag{2.28}$$

2.9.1 Independent Equations Generated from Entries in Cycle Basis Matrix (Cycle Matroid)

In Section 1.3.13, a procedure is developed, based on digraph matrices, that generates closed paths. A closed path (cycle) is generated when the c = 1 chord is added to a given set of edges, called arcs, in the spanning tree. The direction of the cycle is assigned by the inverse chord's direction (i.e., clockwise in Figure 2.10a). A superscript "–" sign is assigned to an edge that is contrary to the cycle's direction and "+" if the edge is in the path's direction. The set of edges (arcs and chord) is $C_E = \{A^+, B^+, C^+, \ldots, Y^+, Z^+, E^-\}$.The cycle basis incidence matrix, C, is defined as a $c \times j$ matrix in which one row is assigned to a single cycle and each column is assigned to an edge.

The single row matrix C, shown in Equation (2.29), is named a *cycle matroid* and has rank c = 1:

$$\begin{array}{c} \text{Edges} \\ \begin{array}{ccccccccc} & A & B & C & & Y & Z & E & \\ C = [& +1 & +1 & +1 & \ldots & +1 & +1 & -1 &] C_E \text{ Cycle} \end{array} \end{array} \tag{2.29}$$

Statement 2.3

For inverse orientation, the result of post multiplications of relative matrices from all arcs and a chord along a cycle is the unit matrix. The entries in the cycle basis matrix C are the superscripts for relative matrices in the matrix product. Thus, "+1" denotes a direct matrix, and a "–1" denotes an inverse in the matrix product.

Equation (2.27) is written along the closed path, as shown in Equation (2.30):

$$\mathbf{D}_A^{+1} * \mathbf{D}_B^{+1} * \mathbf{D}_C^{+1} * \cdots * \mathbf{D}_Z^{+1} * \mathbf{D}_E^{-1} = \mathbf{U} \tag{2.30}$$

Note:

- Equation (2.30) is generally applicable for the orientation of mechanisms with single or multiple cycles. In the case of multiple-cycle mechanisms, one equation is written for each cycle, as shown in Chapter 3.
- Equation (2.30) generates independent equations for orientation.

This property of generating independent equations comes from the one-to-one relation that exists between the sign of each factor in Equation (2.30) and the entries +1 or –1 within the cycle basis matrix C. Section 1.3.13 shows the matrix C, named a *cycle matroid*, which has only independent rows; therefore, the Equation (2.30) is also independent. Later in this chapter, based on the row entries in the cycle matroid, independent equations are generated for velocities and accelerations (as explained in Chapter 4).

2.9.2 The Task Orientation Matrix

The matrix D_E^{task}, called the *task orientation* matrix, has the desired orientation for the axes of frame e.

Its entries are provided from the design specifications, computer-aided design (CAD) drawings, or a picture (in the case where there is machine vision implemented):

$$D_E^{task} = \begin{bmatrix} \text{Cos}\,\alpha_e^{I} & \text{Cos}\,\alpha_e^{II} & \text{Cos}\,\alpha_e^{III} \\ \text{Cos}\,\beta_e^{I} & \text{Cos}\,\beta_e^{II} & \text{Cos}\,\beta_e^{III} \\ \text{Cos}\,\gamma_e^{I} & \text{Cos}\,\gamma_e^{II} & \text{Cos}\,\gamma_e^{III} \end{bmatrix} \tag{2.31}$$

It is viewed in the digraph as edge's E matrix, when a path is initiated at node 0 from right to left (inverse path) (Figure 2.10b).

A post multiplication by D_E^{task} on both sides of Equation (2.30) results in Equation (2.32):

$$\mathbf{D}_A^{+1} * \mathbf{D}_B^{+1} * \mathbf{D}_C^{+1} * \ldots * \mathbf{D}_Z^{+1} = \mathbf{D}_E^{task} \tag{2.32}$$

Equation (2.33) is the inverse orientation equation for open cycle mechanisms.

Since the product of the matrices on the left results in D_e^{mech}, then Equation (2.32) for inverse orientation analysis reduces to

$$D_e^{mech} = D_E^{task} \tag{2.33}$$

In Equation (2.33), only three entries in the left-hand-side matrix are independent. The angular actuation parameters are determined by equating the entries within the left (function of actuating angular displacements) and right (task-provided) matrices. In general, this leads to a system of nonlinear equations involving trigonometric functions. In the following three examples, the equations and solutions are shown for actuating angular displacements to provide the task orientation of end-effector.

2.9.3 Example of Inverse Orientation Analysis for an Open Cycle Spatial Mechanism with 4 DOF

For the TRRT mechanism inverse orientation analysis, the task matrix is provided, which is shown in the right side of Equation (2.35a). We determine a set of three equations to calculate the actuated angular displacements at the joints (Figure 2.11).

$$\mathbf{D}_A^{+1} * \mathbf{D}_B^{+1} * \mathbf{D}_C^{+1} * \mathbf{D}_D^{+1} = \mathbf{D}_E^{task} \tag{2.34}$$

The left side of Equation (2.34) was determined in Equation (2.26i). When inverse positional analysis is applied, Equation (2.34) becomes

$$\begin{bmatrix} C\theta_B \cdot C\theta_C & -C\theta_B \cdot S\theta_C & S\theta_B \\ S\theta_C & C\theta_C & 0 \\ -S\theta_B \cdot C\theta_C & S\theta_B \cdot S\theta_C & C\theta_B \end{bmatrix} = \begin{bmatrix} \text{Cos}\,\alpha_e^I & \text{Cos}\,\alpha_e^{II} & \text{Cos}\,\alpha_e^{III} \\ \text{Cos}\,\beta_e^I & \text{Cos}\,\beta_e^{II} & \text{Cos}\,\beta_e^{III} \\ \text{Cos}\,\gamma_e^I & \text{Cos}\,\gamma_e^{II} & \text{Cos}\,\gamma_e^{III} \end{bmatrix} \tag{2.35a}$$

The IT direction angles are located on the diagonal of the right-side matrix:

$$\begin{bmatrix} C\theta_B \cdot C\theta_C & * & * \\ * & C\theta_C & * \\ * & * & C\theta_B \end{bmatrix} = \begin{matrix} \\ i_0 \\ j_0 \\ k_0 \end{matrix} \begin{matrix} \begin{matrix} i_e & \quad j_e & \quad k_e \end{matrix} \\ \begin{bmatrix} \text{Cos}\,\alpha_e^I & * & * \\ * & \text{Cos}\,\beta_e^{II} & * \\ * & * & \text{Cos}\,\gamma_e^{III} \end{bmatrix} \end{matrix} \tag{2.35b}$$

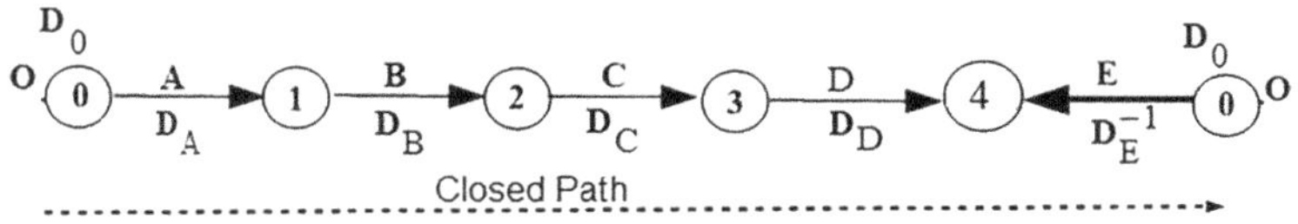

FIGURE 2.11
Relative and absolute orientation matrices along the closed cycle for TRRT mechanism.

Three equations for inverse orientation are obtained by identifying the entries from the left and right diagonals.
Equation 1:

$$C\theta_B \cdot C\theta_C = \text{Cos}\,\alpha_e^I, \text{ where } \alpha_e^I \text{ is the angle between } x_e \text{ and } x_0 \quad (2.35c)$$

Equation 2:

$$\theta_C = \beta_e^{II}, \text{ where } \beta_e^I \text{ is the angle between } y_e \text{ and } y_0 \quad (2.35d)$$

Equation 3:

$$\theta_B = \gamma_e^{III}, \text{ where } \gamma_e^{III} \text{ is the angle between } z_e \text{ and } z_0 \quad (2.35e)$$

Notice that the second and third equations provide actuating solutions θ_C and θ_B for any task (desired) values of β_e^{II} and γ_e^{III}. The first equation is not independent; therefore, the task orientation α_e^I is dependent on the values of β_e^{II} and γ_e^{III}.

2.9.4 Example of Inverse Orientation Analysis for an Open Cycle Spatial Mechanism with 5 DOF

The TRRTR mechanism with five joints: translation T, revolute R, revolute R, translation T, and revolute R, Figure 2.12a, is obtained from the TRRT mechanism in Figure 2.1 with the addition of link 5 and a flange at E. The flange allows a relative rotation about axis x_E between links 5 and 4. The mechanism has $m = 5$ links, $j = 5$ joints (A, B, C, D, and E), and mobility $M = 5$ which can be determined from Equation (2.3):

$$M = 6 \cdot m - 5 \cdot j_1 - 4j_2 - 3 \cdot j_3 - 2j_4 - j_5 = 6 \cdot 5 - 5 \cdot 5 - 4 \cdot 0 - 3 \cdot 0 - 2 \cdot 0 - 0 = 5\,\text{DOF}$$

The five actuating parameters are shown in matrix $q_M = \left\{ d_A \mid \theta_B \mid \theta_C \mid d_D \mid \theta_E \right\}^T$.

Determine the absolute orientation matrix and the IT angles, used for the inverse orientation analysis, for the TRRTR manipulator.

Solution: In the digraph shown in Figure 2.12b, the additional node 5 is separated from node 0 by a chord T oriented from node 0 toward node 5. The *tool frame* (T, x_t, y_t, z_t) is translated from the frame (E, x_5, y_5, z_5), both frames are attached to link 5. The orientation of the tool frame (T, x_t, y_t, z_t) with respect to the fixed frame (O, x_0, y_0, z_0) is determined from Equation (2.22) and expanded to Equation (2.36):

$$\mathbf{D}_A^{+1} * \mathbf{D}_B^{+1} * \mathbf{D}_C^{+1} * \mathbf{D}_D^{+1} * \mathbf{D}_E^{+1} = \mathbf{D}_T^{task} \quad (2.36)$$

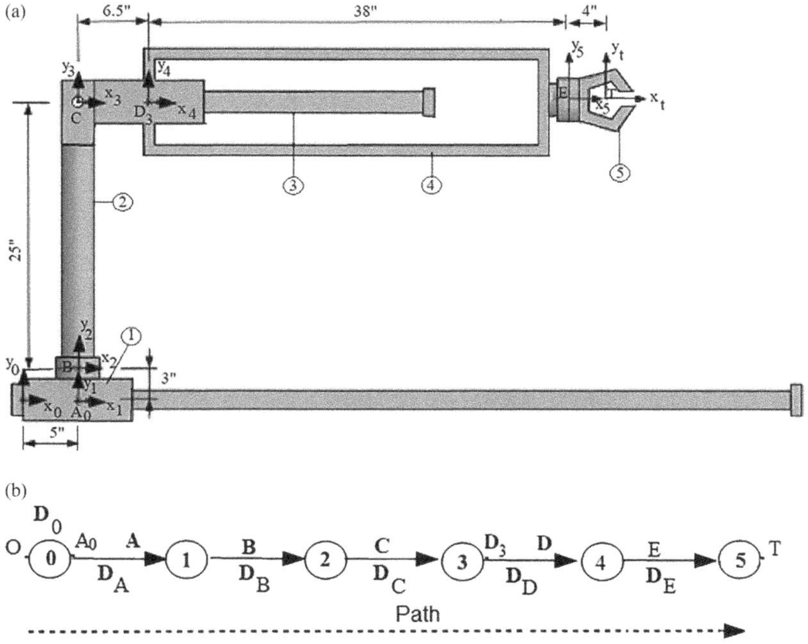

FIGURE 2.12
(a) The TRRTR spatial mechanism; (b) relative and absolute orientation matrices along the closed path.

Considering Equation (2.33), then Equation (2.36a) is obtained and expanded as follows:

$$\mathbf{D}_5 = \mathbf{D}_T^{task} \tag{2.36a}$$

$$\begin{bmatrix} C\theta_B \cdot C\theta_C & -C\theta_B \cdot S\theta_C \cdot C\theta_E + S\theta_B \cdot S\theta_E & C\theta_B \cdot S\theta_C \cdot S\theta_E + S\theta_B \cdot C\theta_E \\ S\theta_C & C\theta_C \cdot C\theta_E & -C\theta_C \cdot S\theta_E \\ -S\theta_B \cdot C\theta_C & S\theta_B \cdot S\theta_C \cdot C\theta_E + C\theta_B \cdot S\theta_E & -S\theta_B \cdot S\theta_C \cdot S\theta_E + C\theta_B \cdot C\theta_E \end{bmatrix} = \mathbf{D}_T^{task} \tag{2.36b}$$

where

$\mathbf{D}_5 = \mathbf{D}_4 * \mathbf{D}_E$

$$= \begin{matrix} \\ i_0 \\ j_0 \\ k_0 \end{matrix} \begin{matrix} \qquad i_5 \qquad\qquad\qquad j_5 \qquad\qquad\qquad k_5 \\ \begin{bmatrix} C\theta_B \cdot C\theta_C & -C\theta_B \cdot S\theta_C \cdot C\theta_E + S\theta_B \cdot S\theta_E & C\theta_B \cdot S\theta_C \cdot S\theta_E + S\theta_B \cdot C\theta_E \\ S\theta_C & C\theta_C \cdot C\theta_E & -C\theta_C \cdot S\theta_E \\ -S\theta_B \cdot C\theta_C & S\theta_B \cdot S\theta_C \cdot C\theta_E + C\theta_B \cdot S\theta_E & -S\theta_B \cdot S\theta_C \cdot S\theta_E + C\theta_B \cdot C\theta_E \end{bmatrix} \end{matrix} \tag{2.36c}$$

- $\mathbf{D}_4$ is the absolute rotation matrix of link 4, see Equation (2.26i)
- $\mathbf{D}_E$ is the relative rotation matrix about axis x_E with the angular displacement θ_E, see Equation (2.7):

$$\mathbf{D}_E = \begin{bmatrix} 1 & 0 & 0 \\ 0 & \text{Cos}\,\theta_E & -\text{Sin}\,\theta_E \\ 0 & \text{Sin}\,\theta_E & \text{Cos}\,\theta_E \end{bmatrix} \tag{2.36d}$$

The IT direction angles are located on the diagonal of the right-hand-side matrix:

$$\begin{bmatrix} C\theta_B \cdot C\theta_C & * & * \\ * & C\theta_C \cdot C\theta_E & * \\ * & * & C\theta_B \cdot C\theta_E - S\theta_B \cdot S\theta_C \cdot S\theta_E \end{bmatrix} = \begin{bmatrix} C\alpha_t^I & * & * \\ * & C\beta_t^{II} & * \\ * & * & C\gamma_t^{III} \end{bmatrix} \tag{2.36e}$$

The three equations for inverse orientation are obtained by equating the entries from the left and right diagonals.
Equation 1:

$$C\theta_B \cdot C\theta_C = C\alpha_t^I \tag{2.36f}$$

Equation 2:

$$C\theta_C \cdot C\theta_E = C\beta_t^{II} \tag{2.36g}$$

Equation 3:

$$C\theta_B \cdot C\theta_E - S\theta_B \cdot S\theta_C \cdot S\theta_E = C\gamma_t^{III} \tag{2.36h}$$

Notice that the three equations provide actuating solutions θ_C, θ_B, and θ_E for the task (desired) values of α_t^I, β_t^{II}, and γ_t^{III}. This is a nonlinear system involving trigonometric functions.

The disadvantage of solving a system of nonlinear equations is the existence of multiple solutions for the trigonometric functions. In Section 2.13, a method is developed to generate these equations without trigonometric functions.

2.9.5 Example of Inverse Orientation Analysis for an Open Cycle Spatial Mechanism with 6 DOF

Ihe TRRTRT mechanism with six joints: translation T, revolute R, revolute R, translation T, revolute R, and translation T, Figure 2.13a, is obtained from

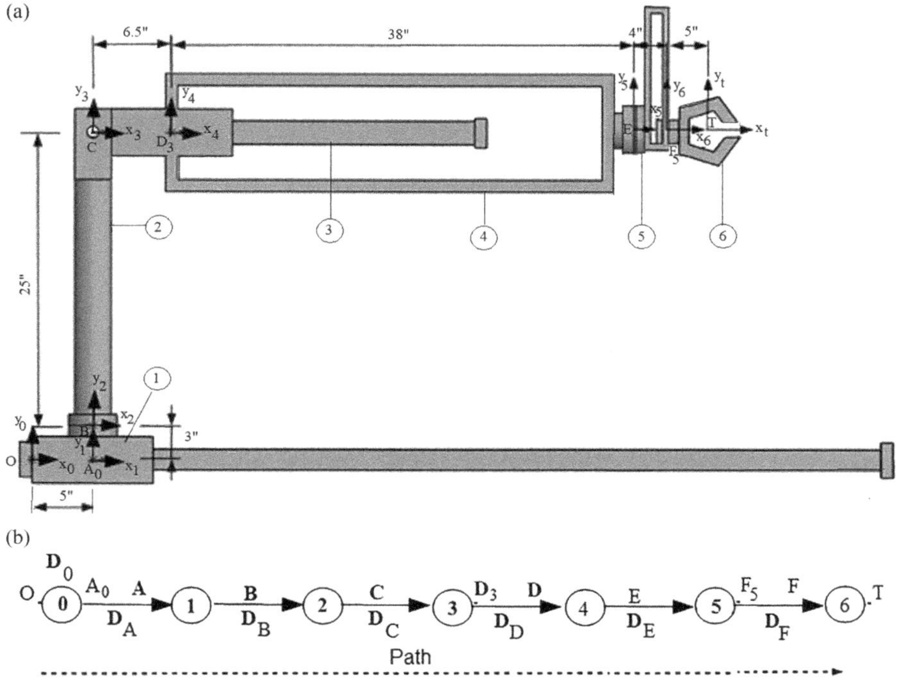

FIGURE 2.13
(a) The TRRTR spatial mechanism; (b) relative and absolute orientation matrices along the closed path.

the TRRTR mechanism in Figure 2.12a with the addition of link 6. The flange E allows for relative rotation about axis x_5 between links 5 and 4. Rigidly attached to link 5 is the guide for prismatic joint F, which allows for translation along axis y_6. The home position of slider is noted F_5. The mechanism has m = 6 links, j = 6 joints (A, B, C, D, E, and F), and mobility M = 6, which can be determined from Equation (2.3):

$$M = 6\cdot m - 5\cdot j_1 - 4j_2 - 3\cdot j_3 - 2j_4 - j_5 = 6\cdot 6 - 5\cdot 6 - 4\cdot 0 - 3\cdot 0 - 2\cdot 0 - 0 = 6\,\text{DOF}$$

The six actuating parameters are shown in matrix $\mathbf{q}_M = \left\{ d_A \mid \theta_B \mid \theta_C \mid d_D \mid \theta_E \mid d_F \right\}^T$.

Determine the absolute orientation matrix and the IT angles used for the inverse orientation analysis, for the TRRTRT manipulator.

Solution: In the digraph shown in Figure 2.13b, the additional node 6 is separated from node 0 by a chord T oriented from node 0 toward node 6. The *tool frame* (T, x_t, y_t, z_t) is translated from the frame (F, x_6, y_6, z_6); both frames are attached to link 6. The orientation of the tool frame (T, x_t, y_t, z_t) with respect to the fixed frame (O, x_0, y_0, z_0) is determined from Equation (2.22) and was expanded to Equation (2.37):

$$\mathbf{D}_A^{+1} * \mathbf{D}_B^{+1} * \mathbf{D}_C^{+1} * \mathbf{D}_D^{+1} * \mathbf{D}_E^{+1} * \mathbf{D}_F^{+1} = \mathbf{D}_T^{task} \tag{2.37}$$

Considering Equation (2.35), then Equation (2.37a) is obtained and expanded as follows:

$$\mathbf{D}_6 = \mathbf{D}_T^{task} \tag{2.37a}$$

$$\begin{bmatrix} C\theta_B \cdot C\theta_C & -C\theta_B \cdot S\theta_C \cdot C\theta_E + S\theta_B \cdot S\theta_E & C\theta_B \cdot S\theta_C \cdot S\theta_E + S\theta_B \cdot C\theta_E \\ S\theta_C & C\theta_C \cdot C\theta_E & -C\theta_C \cdot S\theta_E \\ -S\theta_B \cdot C\theta_C & S\theta_B \cdot S\theta_C \cdot C\theta_E + C\theta_B \cdot S\theta_E & -S\theta_B \cdot S\theta_C \cdot S\theta_E + C\theta_B \cdot C\theta_E \end{bmatrix} = \mathbf{D}_T^{task} \tag{2.37b}$$

where

$$\mathbf{D}_6 = \mathbf{D}_5 * \mathbf{D}_F$$

$$= \begin{matrix} & i_6 & j_6 & k_6 \\ i_0 & C\theta_B \cdot C\theta_C & -C\theta_B \cdot S\theta_C \cdot C\theta_E + S\theta_B \cdot S\theta_E & C\theta_B \cdot S\theta_C \cdot S\theta_E + S\theta_B \cdot C\theta_E \\ j_0 & S\theta_C & C\theta_C \cdot C\theta_E & -C\theta_C \cdot S\theta_E \\ k_0 & -S\theta_B \cdot C\theta_C & S\theta_B \cdot S\theta_C \cdot C\theta_E + C\theta_B \cdot S\theta_E & -S\theta_B \cdot S\theta_C \cdot S\theta_E + C\theta_B \cdot C\theta_E \end{matrix} \tag{2.37c}$$

- $\mathbf{D}_5$ is the the absolute rotation matrix of link 5 (Equation (2.36c))

$$\mathbf{D}_F = \begin{bmatrix} 1 & 0 & 0 \\ 0 & 1 & 1 \\ 0 & 0 & 1 \end{bmatrix} \text{ is the null rotation (translation) matrix along } y_6 \tag{2.37d}$$

The IT direction angles are located on the diagonal of the right-hand-side matrix:

$$\begin{bmatrix} C\theta_B \cdot C\theta_C & * & * \\ * & C\theta_C \cdot C\theta_E & * \\ * & * & C\theta_B \cdot C\theta_E - S\theta_B \cdot S\theta_C \cdot S\theta_E \end{bmatrix}$$

$$= \begin{bmatrix} C\alpha_t^I & * & * \\ * & C\beta_t^{II} & * \\ * & * & C\gamma_t^{III} \end{bmatrix} \tag{2.37e}$$

The three equations for inverse orientation are obtained by equating the entries from the left and right diagonals. The equations coincide with Equations (2.36f), (2.36g), and (2.36h).

Equation 1:

$$C\theta_B \cdot C\theta_C = C\alpha_t^I \tag{2.37f}$$

Equation 2:

$$C\theta_C \cdot C\theta_E = C\beta_t^{II} \tag{2.37g}$$

Equation 3:

$$C\theta_B \cdot C\theta_E - S\theta_B \cdot S\theta_C \cdot S\theta_E = C\gamma_t^{III} \tag{2.37h}$$

Notice that the nonlinear system involving trigonometric functions provides actuating solutions θ_C, θ_B, and θ_E for the task (desired) values of α_t^I, β_t^{II}, and γ_t^{III}.

In Section 2.13, a method is developed to generate equations without trigonometric functions.

2.10 Direct Positional Analysis: Governing Equations for Open Cycle Mechanisms

2.10.1 Transformation of Vector Components between Frames

The frame origins are shifted relative to each other. Thus, the origin Z of link m's frame is shifted relative to origin Y of link m − 1's frame by a vector $\mathbf{L}_{YZ}^{m-1}$. This vector is expressed easier with respect to the local frame m − 1, shown as a superscript. The addition of vectors requires transformation of their components from local frames to the fixed frame, as shown in Equation (2.38). These equations are shown in matrix form in Equation (2.38a):

$$\begin{aligned} x_{YZ}^0 &= l_{m-1}^I \cdot x_{YZ}^{m-1} + l_{m-1}^{II} \cdot y_{YZ}^{m-1} + l_{m-1}^{III} \cdot z_{YZ}^{m-1} \\ y_{YZ}^0 &= m_{m-1}^I \cdot x_{YZ}^{m-1} + m_{m-1}^{II} \cdot y_{YZ}^{m-1} + m_{m-1}^{III} \cdot z_{YZ}^{m-1} \\ z_{YZ}^0 &= n_{m-1}^I \cdot x_{YZ}^{m-1} + n_{m-1}^{II} \cdot y_{YZ}^{m-1} + n_{m-1}^{III} \cdot z_{YZ}^{m-1} \end{aligned} \tag{2.38}$$

$$\begin{Bmatrix} x_{YZ}^0 \\ y_{YZ}^0 \\ z_{YZ}^0 \end{Bmatrix} = \begin{bmatrix} l_{m-1}^I & l_{m-1}^{II} & l_{m-1}^{III} \\ m_{m-1}^I & m_{m-1}^{II} & m_{m-1}^{III} \\ n_{m-1}^I & n_{m-1}^{II} & n_{m-1}^{III} \end{bmatrix} \begin{Bmatrix} x_{YZ}^{m-1} \\ y_{YZ}^{m-1} \\ z_{YZ}^{m-1} \end{Bmatrix} \tag{2.38a}$$

Therefore, Equation (2.39) for transfer of components with respect to the fixed frame 0 holds:

$$\mathbf{L}_{YZ}^{0} = \mathbf{D}_{m-1} * \mathbf{L}_{YZ}^{m-1} \tag{2.39}$$

The symbol * in Equation (2.39) represents the multiplication of a matrix and a column vector.

This transfer is shown by multiplying the left side of the vector written in the local frame by its absolute rotation matrix.

2.10.2 Linear Displacement at Prismatic, Cylindrical, and Helical Joints

Each joint is assigned a pair of geometric points that are located on two different joined surfaces. The prismatic or cylindrical joint Z, between link m−1 (guide) and link m (slider), is assigned two points: Z_{m-1} (home) located on the guide's surface and Z located on the slider's surface. The linear vector displacement $\mathbf{d}_{Z_{m-1}Z}$, noted $\mathbf{d}_Z$, is the translation vector between the two points Z_{m-1} and Z. The two points are both shown on edge Z in digraph:

$$\mathbf{d}_{Z}^{m-1} = \mathbf{d}_{Z} \cdot \mathbf{u}_{Z}^{m-1} \tag{2.40}$$

where

- d_Z is a scalar used to measure the linear displacement, variable in time, between points Z_{m-1} and Z
- $\mathbf{u}_Z^{m-1}$ is the unit vector along the guide. It has the components expressed in the guide's frame m−1.

Equation (2.40) also applies for cylindrical and helical joints, as shown later in the text.

2.10.3 Linear Displacement at Revolute, Spherical, and Meshing Joints

- The *revolute joint* Z, between link m−1 and link m, is assigned a point Z_{m-1} located on the hollow surface of link m−1, although Z_m is located on the surface of a pin on link m. The two points are coincident at Z; there is no displacement along the rotation axis' unit vector $\mathbf{u}$, and therefore $d_Z = 0$. This equation also applies for a spherical joint, as shown later.
- If the *meshing joint* Z between gear m−1 and gear m is assumed to be no-slip, then $d_Z = 0$.

2.10.4 Constraint Equations for Angular and Linear Displacements

$$\begin{aligned} &\mathbf{d}_Z = \mathbf{0} \text{ is for revolute or meshing joints} \\ &\boldsymbol{\theta}_Z = \mathbf{0} \text{ is for prismatic joints} \end{aligned} \tag{2.41}$$

The first constraint in Equation (2.41) is the *constraint equation* for no linear displacement in a revolute joint or meshing joint.

The second constraint in Equation (2.41) is the *constraint equation* for no angular displacement in a prismatic joint.

Linear relative displacements are assigned to the edges in a digraph. The direction of an oriented edge in a digraph could be coincident or opposite with the path's sense:

a. If the path's sense coincides with the edge's sense, then the edge is assigned a positive relative displacement, noted $+\mathbf{d}_Z$. For example, in Figure 2.11, all edges (arrows in digraph) have the same sense as the path; therefore, their assigned relative displacements are all positive.
b. If the path sense is opposite to edge's sense, then the edge is assigned a negative relative displacement, noted $-\mathbf{d}_Z$.

2.10.5 Translation Vectors between Frame Origins and Position Vectors for Frame Origins along the Open Path

In a mechanism, the frames are located with their origins at joints (edges in a digraph). Figure 2.14 illustrates a graph for a mechanism with the following joints: A (prismatic), B (revolute), C (revolute),…, Z (prismatic). Point E is the origin of the end-effector.

Joint A (prismatic): Equation (2.39), for the transfer with respect to the fixed frame, is written as

$$\mathbf{L}_{OA}^{0} = \mathbf{D}_0 * \left(\mathbf{L}_{OA_0}^{0} + \mathbf{d}_A^{0}\right) = \mathbf{L}_{OA_0}^{0} + \mathbf{d}_A \cdot \mathbf{u}_A^{0} \tag{2.42}$$

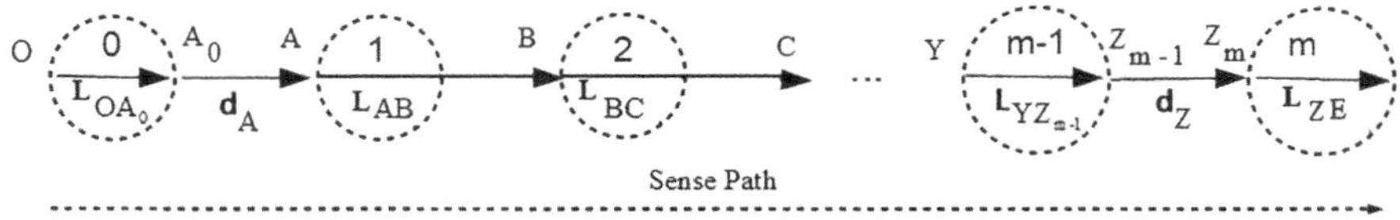

FIGURE 2.14
Translation vectors between frame origins along the tree.

Joint B (revolute): It has constraint for no linear displacement. In this case, Equation (2.42) is written as

$$\mathbf{L}_{AB}^{0} = \mathbf{D}_{1} * \left(\mathbf{L}_{AB}^{1} + 0\right) \tag{2.43}$$

where $\mathbf{d}_B = 0$, and the superscript 1 indicates that the vector is with respect to the local frame 1

Joint Z (prismatic): The point Y (origin of frame m−1) is translated at Z (origin of frame m). In the digraph, Figure (2.14), this shifting extends over node m−1 and its adjacent edge Z. The vectors are written with respect to the local frame m−1 (Equation (2.44)):

$$\mathbf{L}_{YZ}^{m-1} = \mathbf{L}_{YZ_{m-1}}^{m-1} + \mathbf{d}_{Z}^{m-1} \tag{2.44}$$

and then with respect to the fixed frame 0 from Equation (2.39):

$$\mathbf{L}_{YZ}^{0} = \mathbf{D}_{m-1} * \left(\mathbf{L}_{YZ_{m-1}}^{m-1} + \mathbf{d}_{Z}^{m-1}\right) = \left(\mathbf{L}_{YZ_{m-1}}^{0} + \mathbf{d}_{Z} \cdot \mathbf{u}_{Z}^{0}\right) \tag{2.44a}$$

where

- $\mathbf{L}_{YZ_{m-1}}$ is the constant magnitude translation vector from point Y to Z_{m-1}(home position) with both points located on link m−1. In the digraph, Figure (2.14), this vector is represented across the node m−1.
- $\mathbf{d}_Z$ is the variable magnitude, translation vector from home position Z_{m-1} to Z. This vector describes the linear displacement at joint Z and is represented in the digraph along the oriented edge.

Each vector in Equation (2.39), if expressed in terms of the fixed frame, is written as the difference between two position vectors:

$$\mathbf{L}_{YZ}^{0} = \mathbf{r}_{Z} - \mathbf{r}_{Y} \tag{2.45}$$

Considering Equations (2.44a) and (2.45), Equation (2.46) can be developed:

$$\mathbf{r}_{Z} = \mathbf{r}_{Y} + \mathbf{D}_{m-1} * \left(\mathbf{L}_{YZ}^{m-1} + \mathbf{d}_{Z}^{m-1}\right) \tag{2.46}$$

Figure 2.15 illustrates the position vectors at each frame's origins.

A recursive procedure starting from 0 is extended at A, B, C,…, Z along the path allows Equation (2.46) to be developed into Equation (2.47).

Statement 2.4

The position vector $\mathbf{r}_Z$ is the sum of shifting vectors between frame origins:

$$\mathbf{r}_{Z} = \mathbf{L}_{OA}^{0}\Big|_{A} + \mathbf{L}_{AB}^{0}\Big|_{B} + \mathbf{L}_{BC}^{0}\Big|_{C} + \cdots + \mathbf{L}_{YZ}^{0}\Big|_{Z} \tag{2.47}$$

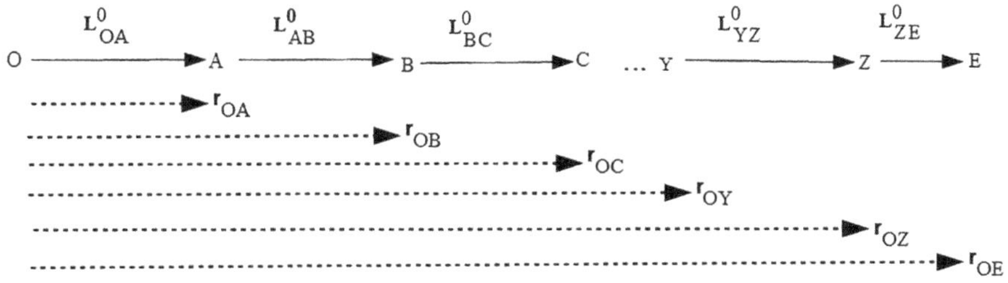

FIGURE 2.15
Position vectors for frame origins and vectors between origins.

For example, for a mechanism with all joints allowing linear displacements $\mathbf{d}_Z$, Equation (2.48) holds:

$$\mathbf{r}_Z = \mathbf{D}_0 * \left(\mathbf{L}^0_{OA_0} + \mathbf{d}^0_A\right)\Big|_A + \mathbf{D}_1 * \left(\mathbf{L}^1_{AB} + \mathbf{d}^1_B\right)\Big|_B + \mathbf{D}_2 * \left(\mathbf{L}^2_{BC} + \mathbf{d}^2_C\right)\Big|_C + \cdots + \mathbf{D}_{m-1} * \left(\mathbf{L}^{m-1}_{YZ} + \mathbf{d}^{m-1}_Z\right)\Big|_Z \quad (2.48)$$

Equation (2.48) is reduced to Equation (2.49) with vectors expressed in the fixed frame:

$$\mathbf{r}_Z = \left(\mathbf{L}^0_{OA_0} + \mathbf{d}_A \cdot \mathbf{u}^0_A\right)\Big|_A + \left(\mathbf{L}^0_{AB} + \mathbf{d}_B \cdot \mathbf{u}^0_B\right)\Big|_B + \left(\mathbf{L}^0_{BC} + \mathbf{d}_C \cdot \mathbf{u}^0_C\right)\Big|_C + \cdots + \left(\mathbf{L}^0_{YZ_{m-1}} + \mathbf{d}^0_Z \cdot \mathbf{u}^0_Z\right)\Big|_Z \quad (2.49)$$

Equation (2.48) is the governing equation for joint position vectors.

Thus, to determine $\mathbf{r}_A$, the term on the left side of bracket $|_A$ is considered. To determine $\mathbf{r}_B$, all of the terms on the left side of bracket $|_B$ are considered.

Example: An open cycle mechanism has the following joints: A (prismatic), B (revolute), C (revolute),…, Z (prismatic).The recursive procedure along the path in Equation (2.48) is developed as follows:

- *Edge A (prismatic)*: Position vector of the frame's A origin:

$$\mathbf{r}_A = \mathbf{r}_O + \mathbf{D}_0 * \left(\mathbf{L}^0_{OA_0} + \mathbf{d}^0_A\right) = \left(\mathbf{L}^0_{OA_0} + \mathrm{d}_A \cdot \mathbf{u}^0_A\right) \quad (2.50a)$$

  where $\mathbf{r}_O = 0$, and $\mathbf{D}_0 = \mathbf{U}_{3\times3}$, a unit matrix

- *Edge B (revolute)*: Position vector of the frame's B origin:

$$\mathbf{r}_B = \mathbf{r}_A + \mathbf{D}_1 * \left(\mathbf{L}^1_{AB} + 0\right) \quad (2.50b)$$

- *Edge C (revolute)*: Position vector of the frame's C origin:

$$\mathbf{r}_C = \mathbf{r}_B + \mathbf{D}_2 * \left(\mathbf{L}_{BC}^2 + 0\right) \tag{2.50c}$$

- ...*Edge Z (prismatic)*: Position vector of the frame's Z origin:

$$\mathbf{r}_Z = \mathbf{r}_Y + \mathbf{D}_{m-1} * \left(\mathbf{L}_{YZ}^{m-1} + \mathbf{d}_Z^{m-1}\right) \tag{2.50d}$$

2.10.6 The End-Effector Position Vector

Equation (2.48) is extended to point E:

$$\mathbf{r}_E = \mathbf{L}_{OA}^0 + \mathbf{L}_{AB}^0 + \mathbf{L}_{BC}^0 + \cdots + \mathbf{L}_{YZ}^0\Big|_Z + \mathbf{L}_{ZE}^0\Big|_E \tag{2.50e}$$

Since the sum of the terms from the left side of the $|_Z$ is $\mathbf{r}_Z$, then

- Position vector of the frame's E origin is

$$\mathbf{r}_E = \mathbf{r}_Z + \mathbf{D}_m * \left(\mathbf{L}_{ZE}^m + \mathbf{0}\right) \tag{2.50f}$$

where **0** is the null relative displacement, because the length along link m between Z and E remains constant over time.

2.10.7 Equations for Direct Positional Analysis

The mechanism's position vector is $\mathbf{r}_E^{mech}$. The superscript "mech" is added to the end-effector's position vector, which suggests that the position of the end-effector is determined from the manipulator's point of view (links and joints):

$$\mathbf{r}_E^{mech} = \begin{Bmatrix} x_E^{mech} \\ y_E^{mech} \\ z_E^{mech} \end{Bmatrix} \tag{2.51}$$

Equating the entries in Equations (2.50f) and (2.51) yields

$$\mathbf{r}_E^{mech} = \mathbf{r}_E \tag{2.52}$$

Thus, a system of three equations is determined (two for planar systems) for frame's E coordinates as functions of the angular actuating parameters at the joints.

2.10.8 Joint Position Matrix, r

In the direct positional analysis, the position vectors of all the joints (A, B, C,..., Y, Z) and end-effector E are determined by providing the M actuation angular and linear parameters.

Equation (2.53) shows the output matrix, **r**. Its columns are the components of the position vectors expressed in the fixed frame:

$$\mathbf{r} = \left[\begin{array}{c|c|c|c|c} \mathbf{r}_A & \mathbf{r}_B & \mathbf{r}_C & \cdots & \mathbf{r}_Z \end{array}\right] = \overset{\quad A \qquad B \qquad C \qquad \cdots \qquad Z}{\left[\begin{array}{c|c|c|c|c} x_A & x_B & x_C & \cdots & x_Z \\ y_A & y_B & y_C & \cdots & y_Z \\ z_A & z_B & z_C & \cdots & z_Z \end{array}\right]} \tag{2.53}$$

2.10.9 COM Position Vectors

The velocity, acceleration, static, and dynamic analyses are determined by the link m's COM position vector $\mathbf{r}_{G_m}$, which is given in Equation (2.54):

$$\mathbf{r}_{G_m} = \mathbf{r}_Z + \mathbf{D}_m * \mathbf{L}^m_{ZG_m} \tag{2.54}$$

where

- G_m is the COM
- Vector $\mathbf{L}^m_{ZG_m}$ from joint Z to the COM is expressed with respect to the local frame m. Its coordinates, Equation (2.55), are determined from the CAD drawings:

$$\mathbf{L}^m_{ZG_m} = \left\{ x_{ZG_m} \quad y_{ZG_m} \quad z_{ZG_m} \right\}^T \tag{2.55}$$

2.10.10 Example of Direct Positional Analysis for a Spatial Mechanism with 4 DOF

The TRRT mechanism, Figure 2.1, has m = 4 links, j = 4 joints, and mobility M = 4.

Data: The following data are provided for direct positional analysis.

- The vectors with constant magnitude (from CAD model) are written with respect to local frames:

$$\mathbf{L}^0_{OA_0} = \begin{Bmatrix} 5 \\ 0 \\ 0 \end{Bmatrix}; \quad \mathbf{L}^1_{AB} = \begin{Bmatrix} 0 \\ 3 \\ 0 \end{Bmatrix}; \quad \mathbf{L}^2_{BC} = \begin{Bmatrix} 0 \\ 25 \\ 0 \end{Bmatrix}; \quad \mathbf{L}^3_{CD_3} = \begin{Bmatrix} 6.5 \\ 0 \\ 0 \end{Bmatrix}; \quad \mathbf{L}^4_{DE} = \begin{Bmatrix} 38 \\ 0 \\ 0 \end{Bmatrix}$$

- The unit vectors for linear displacements d_A and d_D at prismatic joints A and D are as follows:

$$\mathbf{u}_A^0 = \begin{Bmatrix} 1 \\ 0 \\ 0 \end{Bmatrix}; \quad \mathbf{u}_D^3 = \begin{Bmatrix} 1 \\ 0 \\ 0 \end{Bmatrix}$$

- The relative constraint equations for revolute joints B and C and the end-effector E are as follows:

$$\mathbf{d}_B = \mathbf{d}_C = \mathbf{d}_E = \mathbf{0}$$

- The $M = 4$ actuating parameters in matrix $\mathbf{q}_M = \{\, d_A \,\vdots\, \theta_B \,\vdots\, \theta_C \,\vdots\, d_D \,\}^T$
- The home position, in Figure 2.2, is the initial position of the mechanism at $t = 0\,\text{s}$, where all actuating positional parameters are zero:

$$d_A = 0 \text{ in.}; \quad \theta_B = 0°; \quad \theta_C = 0°; \quad d_D = 0 \text{ in.}$$

Objective: To determine

1. Each joint's position vector at any time t
2. The end-effector Cartesian coordinates at any time t
3. The governing equations for direct position analysis
4. For the home position: the joints' position vector and the joints' position matrix, **r**
5. For the home position: the end-effector's position vector

Solution: We consider the path in the tree, see Figure 2.16:

1. Joint position vectors at any time t

   The recursive procedure along the path in Figure 2.13 is developed in Equations (2.56).

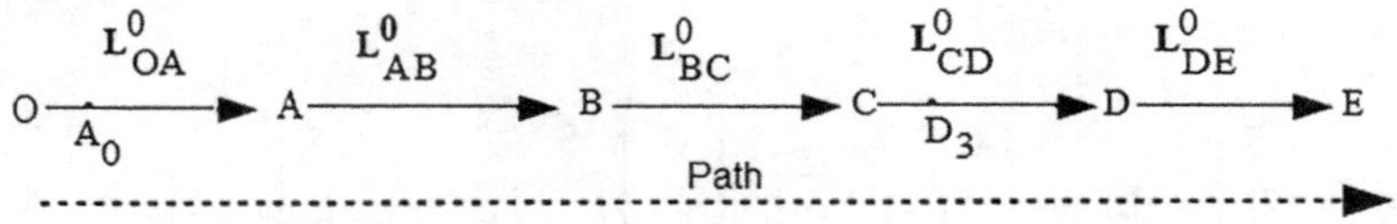

FIGURE 2.16
Translation between frame origins along the open path.

- *Edge A (prismatic)* :Position vector of frame's A origin:

$$\mathbf{r}_A = \mathbf{D}_0 * \left(\mathbf{L}^0_{OA_0} + d_A \cdot \mathbf{u}^0_A\right) = \begin{bmatrix} 1 & 0 & 0 \\ 0 & 1 & 0 \\ 0 & 0 & 1 \end{bmatrix} * \left(\begin{Bmatrix} 5 \\ 0 \\ 0 \end{Bmatrix} + d_A \cdot \begin{Bmatrix} 1 \\ 0 \\ 0 \end{Bmatrix}\right) = \begin{Bmatrix} 5+d_A \\ 0 \\ 0 \end{Bmatrix} \tag{2.56a}$$

where

- $\mathbf{D}_0 = \mathbf{U}$
- $\mathbf{L}^0_{OA_0}$ is the translation vector from origin O to reference point A_0, with respect to frame 0
- $\mathbf{d}^0_A = d_A \cdot \mathbf{u}^0_A$ is the linear displacement translation vector from point A_0 to origin A, along the unit vector $\mathbf{u}^0_A$

- *Edge B (revolute)*: Position vector of frame's B origin:

$$\mathbf{r}_B = \mathbf{r}_A + \mathbf{D}_1 * \left(\mathbf{L}^1_{AB} + \mathbf{0}\right) = \begin{Bmatrix} 5+d_A \\ 0 \\ 0 \end{Bmatrix} + \begin{bmatrix} 1 & 0 & 0 \\ 0 & 1 & 0 \\ 0 & 0 & 1 \end{bmatrix} * \begin{Bmatrix} 0 \\ 3 \\ 0 \end{Bmatrix} = \begin{Bmatrix} 5+d_A \\ 3 \\ 0 \end{Bmatrix} \tag{2.56b}$$

where

- $\mathbf{D}_1 = \mathbf{U}$ is from Equation (2.26c)
- $\mathbf{L}^1_{AB}$ is the translation vector from origin A to origin B, with respect to frame 1
- **0**, the revolute joint has zero linear displacement (null translation; therefore, it is a constraint displacement)

- *Edge C (revolute)*: Position vector of frame's C origin:

$$\mathbf{r}_C = \mathbf{r}_B + \mathbf{D}_2 * \left(\mathbf{L}^2_{BC} + \mathbf{0}\right)$$

$$= \begin{Bmatrix} 5+d_A \\ 3 \\ 0 \end{Bmatrix} + \begin{bmatrix} C\theta_B & 0 & S\theta_B \\ 0 & 1 & 0 \\ -S\theta_B & 0 & C\theta_B \end{bmatrix} * \begin{Bmatrix} 0 \\ 25 \\ 0 \end{Bmatrix} = \begin{Bmatrix} 5+d_A \\ 28 \\ 0 \end{Bmatrix} \tag{2.56c}$$

where

- $\mathbf{D}_2$ is from Equation (2.26e)
- $\mathbf{L}^2_{BC}$ is the translation vector from origin B to origin C, with respect to frame 2
- **0**, the revolute joint has zero linear displacement (null translation; therefore, it is a constraint displacement)

- *Edge D (prismatic)*: Position vector of frame's D origin:

$$\mathbf{r}_D = \mathbf{r}_C + \mathbf{D}_3 * \left(\mathbf{L}^2_{CD_3} + d_D \cdot \mathbf{u}^3_D\right)$$

$$= \begin{Bmatrix} 5+d_A \\ 28 \\ 0 \end{Bmatrix} + \begin{bmatrix} C\theta_B \cdot C\theta_C & -C\theta_B \cdot S\theta_C & S\theta_B \\ S\theta_C & C\theta_C & 0 \\ -S\theta_B \cdot C\theta_C & S\theta_B \cdot S\theta_C & C\theta_B \end{bmatrix} * \left(\begin{Bmatrix} 6.5 \\ 0 \\ 0 \end{Bmatrix} + \begin{Bmatrix} d_D \\ 0 \\ 0 \end{Bmatrix} \right)$$

$$= \begin{Bmatrix} 5+d_A+(6.5+d_D)\cdot C\theta_B \cdot C\theta_C \\ 28+(6.5+d_D)\cdot S\theta_C \\ -(6.5+d_D)\cdot S\theta_B \cdot C\theta_C \end{Bmatrix} \tag{2.56d}$$

where

- $\mathbf{D}_3$ is from Equation (2.26g)
- $\mathbf{L}^3_{CD_3}$ is the translation vector from origin C to reference point D_3; both points are located on the same node 3 (link 3)
- $\mathbf{d}^3_D = d_D \cdot \mathbf{u}^3_D$ is the linear displacement translation vector from point D_3 to origin D, along the unit vector $\mathbf{u}^3_D$

2. End-effector Cartesian coordinates at any time t
   - Position vector of frame's E origin:

$$\mathbf{r}_E = \mathbf{r}_D + \mathbf{D}_4 * \left(\mathbf{L}^4_{DE} + \mathbf{0}\right) = \begin{Bmatrix} 5+d_A+(6.5+d_D)\cdot C\theta_B \cdot C\theta_C \\ 28+(6.5+d_D)\cdot S\theta_C \\ -(6.5+d_D)\cdot S\theta_B \cdot C\theta_C \end{Bmatrix}$$

$$+ \begin{bmatrix} C\theta_B \cdot C\theta_C & -C\theta_B \cdot S\theta_C & S\theta_B \\ S\theta_C & C\theta_C & 0 \\ -S\theta_B \cdot C\theta_C & S\theta_B \cdot S\theta_C & C\theta_B \end{bmatrix} * \begin{Bmatrix} 38 \\ 0 \\ 0 \end{Bmatrix}$$

$$= \begin{Bmatrix} 5+d_A+(44.5+d_D)\cdot C\theta_B \cdot C\theta_C \\ 28+(44.5+d_D)\cdot S\theta_C \\ -(44.5+d_D)\cdot S\theta_B \cdot C\theta_C \end{Bmatrix} \tag{2.56e}$$

where

- $\mathbf{D}_4$ is from Equation (2.26i)
- $\mathbf{L}^4_{DE}$ is the translation vector from origin D to reference point E; both points are located on the same node 4 (link 4)
- $\mathbf{0}$ is the null relative displacement vector between origins D and E, and the distance between them remains constant in time.

3. The governing equations for direct position analysis
   Based on $\mathbf{r}_E^{mech} = \mathbf{r}_E$, a system of three equations is determined for the coordinates of frame E as functions of the actuating parameters at the joints:

$$\begin{Bmatrix} x_E^{mech} \\ y_E^{mech} \\ z_E^{mech} \end{Bmatrix} = \begin{Bmatrix} 5 + d_A + (44.5 + d_D) \cdot C\theta_B \cdot C\theta_C \\ 28 + (44.5 + d_D) \cdot S\theta_C \\ -(44.5 + d_D) \cdot S\theta_B \cdot C\theta_C \end{Bmatrix} \quad (2.56f)$$

4. For the home position: the joints' position vector and the vector position matrix
   At $t = 0\,s$, where all actuating positional parameters are zero and the mechanism is at the home position:

$$d_A = 0 \text{ in.}; \quad \theta_B = 0°; \quad \theta_C = 0°; \quad d_D = 0 \text{ in.}$$

   For zero actuating parameters, Equations (2.56a)–(2.56d) denotes the joint position vectors, with their coordinates:

$$\mathbf{r}_A = \begin{Bmatrix} x_A \\ y_A \\ z_A \end{Bmatrix} = \begin{Bmatrix} 5 \\ 0 \\ 0 \end{Bmatrix}; \quad \mathbf{r}_B = \begin{Bmatrix} x_B \\ y_B \\ z_B \end{Bmatrix} = \begin{Bmatrix} 5 \\ 3 \\ 0 \end{Bmatrix};$$

$$\mathbf{r}_C = \begin{Bmatrix} x_C \\ y_C \\ z_C \end{Bmatrix} = \begin{Bmatrix} 5 \\ 28 \\ 0 \end{Bmatrix}; \quad \mathbf{r}_D = \begin{Bmatrix} x_D \\ y_D \\ z_D \end{Bmatrix} = \begin{Bmatrix} 11.5 \\ 28 \\ 0 \end{Bmatrix} \quad (2.56g)$$

   Therefore, the joints' position matrix from Equation (2.53) is

$$\mathbf{r} = \left[\begin{array}{c|c|c|c} \mathbf{r}_A & \mathbf{r}_B & \mathbf{r}_C & \mathbf{r}_D \end{array}\right] = \begin{array}{c} \begin{array}{cccc} A & B & C & D \end{array} \\ \left[\begin{array}{c|c|c|c} 5 & 5 & 5 & 11.5 \\ 0 & 3 & 28 & 28 \\ 0 & 0 & 0 & 0 \end{array}\right] \end{array} \quad (2.56h)$$

5. For the home position: the end-effector's position vector
   For zero actuating parameters, Equation (2.56e) defines the end-effector's position vector:

$$\mathbf{r}_E^{mech} = \begin{Bmatrix} x_E^{mech} \\ y_E^{mech} \\ z_E^{mech} \end{Bmatrix} = \begin{Bmatrix} 49.5 \\ 28 \\ 0 \end{Bmatrix} \quad (2.56i)$$

The above calculated coordinates correspond with those of point E in Figure (2.2).

2.10.11 Simulations for a Spatial Mechanism with 4 DOF

In Animation 2.1, the TRRT manipulator starts from the *home position*, picks up a cylindrical part from a location called the *part grasped*, and places it in a hole located at the *target reached* location. Two additional positions are of interest: the *part approach* just before the part grasp position and the *target approach* just before the target reach. At these two positions, the manipulator's controller evaluates its actual orientation/position in order to compare it with the predefined ones (previously taught). The following five positions of end-effector, named *states*, occur over the time interval: $0 \le t \le 4$ s.

- Home position (h), at $t = 0$ s, Figure 2.25
- Part-approach position (pa), at $t = 1$ s, Figure 2.27
- Part-grasped position (pg), at $t = 2$ s, Figure 2.29
- Target-approach position (ta), at $t = 3$ s, Figure 2.31
- Target-reached position (tr), at $t = 4$ s, Figure 2.32

The $M = 4$ input actuation parameters called *joint coordinates* are included in a column matrix:

$$q_M(t) = \left\{ \begin{matrix} d_A & \theta_B & \theta_C & d_D \end{matrix} \right\}^T \tag{2.57a}$$

In this example, the joint coordinates chosen are stepwise linear functions in time, that is, constant velocity actuation (velocity controlled), shown in Figure 2.17 or analytically in Equation (2.57b).

Solution: For this application, two sets of results are generated. The first set of results is calculated several positions in time from Equations (2.57c) and (2.57d). The first set of results is compared to the second set of results that was determined using SolidWorks (SW) Motion simulation for the mechanism assembly built in SW.

2.10.11.1 Input SW Simulation

The assembly is built with SW. Then, SW Motion simulations are run for the $M = 4$ actuation input motors. Figure 2.17 illustrates the SW input motors with joint displacements chosen as linear in time (i.e., constant velocity actuation).

2.10.11.2 Output from SW Simulation

The coordinates of the end-effector's origin E and its path when performing the proposed task are illustrated in Figure 2.18a. Appendix 2.3 shows

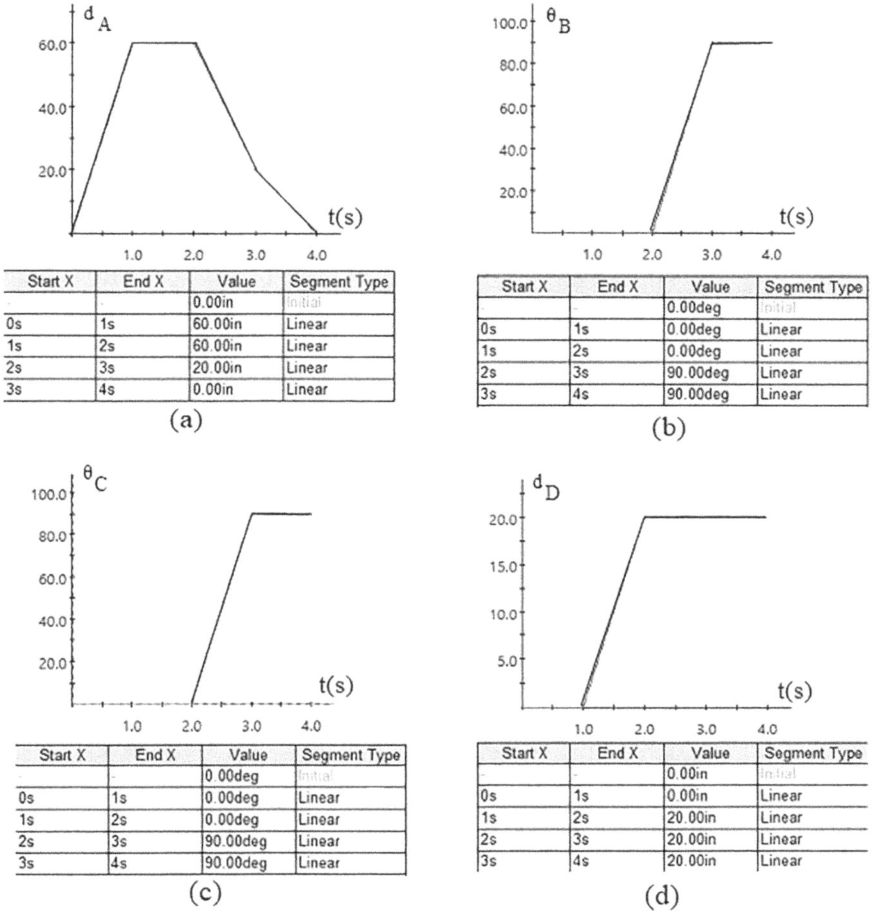

(a)

Start X	End X	Value	Segment Type
	-	0.00in	Initial
0s	1s	60.00in	Linear
1s	2s	60.00in	Linear
2s	3s	20.00in	Linear
3s	4s	0.00in	Linear

(b)

Start X	End X	Value	Segment Type
	-	0.00deg	Initial
0s	1s	0.00deg	Linear
1s	2s	0.00deg	Linear
2s	3s	90.00deg	Linear
3s	4s	90.00deg	Linear

(c)

Start X	End X	Value	Segment Type
	-	0.00deg	Initial
0s	1s	0.00deg	Linear
1s	2s	0.00deg	Linear
2s	3s	90.00deg	Linear
3s	4s	90.00deg	Linear

(d)

Start X	End X	Value	Segment Type
	-	0.00in	Initial
0s	1s	0.00in	Linear
1s	2s	20.00in	Linear
2s	3s	20.00in	Linear
3s	4s	20.00in	Linear

FIGURE 2.17
(a) Joint A—linear displacement; (b) joint B—angular displacement; (c) joint C—angular displacement; (d) joint D—linear displacement.

the tabulated values for the end-effector coordinates over time from the SW Motion Study/Results and Plots. The results are input into an Excel spreadsheet and shown in Figure 2.18b.

The joint displacements at time t are shown in Equation (2.57b):

$$d_A(t) = \begin{cases} 60 \cdot t & 0 \le t \le 1 \\ 60 & 1 < t \le 2 \\ 60 - 40 \cdot (t-2) & 2 < t \le 3 \\ 20 - 20 \cdot (t-3) & 3 < t \le 4 \end{cases} \text{[in.]}; \quad \theta_B(t) = \begin{cases} 0 & 0 \le t \le 1 \\ 0 & 1 < t \le 2 \\ 90 \cdot (t-2) & 2 < t \le 3 \\ 90 & 3 < t \le 4 \end{cases} [^\circ]$$

$$\theta_C(t)=\begin{cases}0 & 0\le t\le 1\\ 0 & 1<t\le 2\\ 90\cdot(t-2) & 2<t\le 3\\ 90 & 3<t\le 4\end{cases}\ [°];\quad d_D(t)=\begin{cases}0 & 0\le t\le 1\\ 20\cdot(t-1) & 1<t\le 2\\ 20 & 2<t\le 3\\ 20 & 3<t\le 4\end{cases}\ [\text{in.}] \tag{2.57b}$$

2.10.11.3 Output from Engineering Equation Solver (EES) Calculation

Outputs from EES:

- The orientation of the end-effector's frame (i.e., the entries in matrix D_e^{mech}).

  The direction angles between the end-effector's frame and the fixed frame are reproduced here in Equation (2.57c), from Equation (2.26m), are

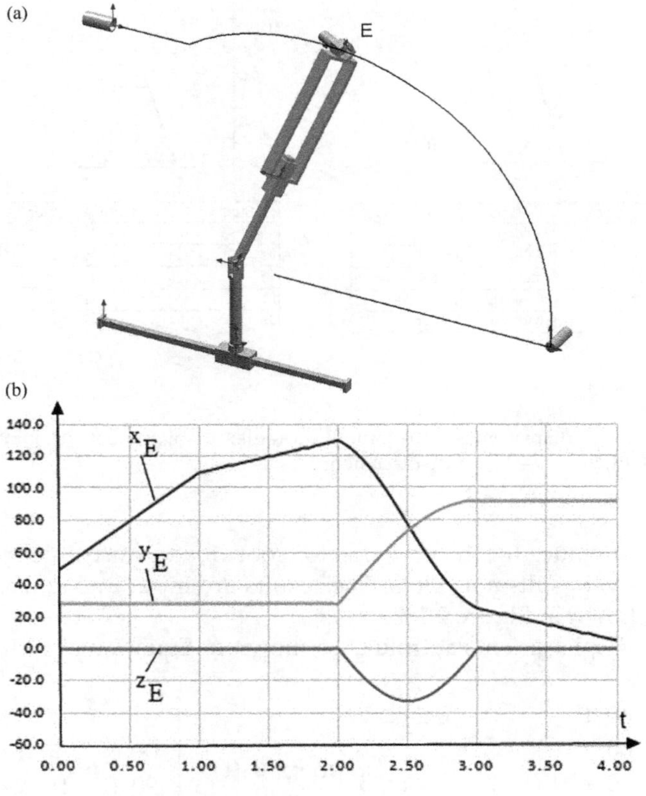

FIGURE 2.18
(a) Path generated by the effector in time; (b) end-effector's coordinates over time.

$$\begin{Bmatrix} \alpha_e^I \\ \beta_e^I \\ \gamma_e^I \end{Bmatrix} = \begin{Bmatrix} \text{Arccos}(C\theta_B \cdot C\theta_C) \\ \text{Arccos}(S\theta_C) \\ \text{Arccos}(-S\theta_B \cdot C\theta_C) \end{Bmatrix}; \begin{Bmatrix} \alpha_e^{II} \\ \beta_e^{II} \\ \gamma_e^{II} \end{Bmatrix} = \begin{Bmatrix} \text{Arccos}(-C\theta_B \cdot S\theta_C) \\ \theta_C \\ \text{Arccos}(S\theta_B \cdot S\theta_C) \end{Bmatrix};$$

$$\begin{Bmatrix} \alpha_e^{III} \\ \beta_e^{III} \\ \gamma_e^{III} \end{Bmatrix} = \begin{Bmatrix} \text{Arccos}(S\theta_B) \\ 90^\circ \\ \theta_B \end{Bmatrix} \tag{2.57c}$$

- The coordinates of the end-effector's origin E
  From Equation (2.56f), coordinates along its path when performing the proposed task are

$$\mathbf{r}_E^{mech} = \begin{Bmatrix} x_E^{mech} \\ y_E^{mech} \\ z_E^{mech} \end{Bmatrix} = \begin{Bmatrix} 5 + d_A + (44.5 + d_D) \cdot C\theta_B \cdot C\theta_C \\ 28 + (44.5 + d_D) \cdot S\theta_C \\ -(44.5 + d_D) \cdot S\theta_B \cdot C\theta_C \end{Bmatrix} \tag{2.57d}$$

- *Home position state*:
  Input: The joint coordinates at $t = 0\,s$

$$\mathbf{q}^h = \{ d_A \quad \theta_B \quad \theta_C \quad d_D \}^T = \{ 0'' \quad 0^\circ \quad 0^\circ \quad 0'' \}^T \tag{2.57e}$$

Output: The end-effector's orientation matrix, the direction angles, and its origin position:

$$\mathbf{D}_e^{mech} = \begin{bmatrix} 1 & 0 & 0 \\ 0 & 1 & 0 \\ 0 & 0 & 1 \end{bmatrix}; \begin{Bmatrix} \alpha_e^I \\ \beta_e^I \\ \gamma_e^I \end{Bmatrix} = \begin{Bmatrix} 0^\circ \\ 90^\circ \\ 90^\circ \end{Bmatrix}; \begin{Bmatrix} \alpha_e^{II} \\ \beta_e^{II} \\ \gamma_e^{II} \end{Bmatrix} = \begin{Bmatrix} 90^\circ \\ 0^\circ \\ 90^\circ \end{Bmatrix};$$

$$\begin{Bmatrix} \alpha_e^{III} \\ \beta_e^{III} \\ \gamma_e^{III} \end{Bmatrix} = \begin{Bmatrix} 90^\circ \\ 90^\circ \\ 0^\circ \end{Bmatrix}; \quad \mathbf{r}_{OE}^{mech} = \begin{Bmatrix} 49.5 \\ 28 \\ 0 \end{Bmatrix} \tag{2.57f}$$

- *Part-approach state*:
  Input: The joint coordinates state at $t = 1\,s$:

$$\mathbf{q}^{pa} = \{ d_A \quad \theta_B \quad \theta_C \quad d_D \}^T = \{ 60'' \quad 0^\circ \quad 0^\circ \quad 0'' \}^T \tag{2.57g}$$

Output: The end-effector's orientation matrix, the direction angles, and its origin position:

$$\mathbf{D}_{e}^{mech} = \left[\begin{array}{c|c|c} 1 & 0 & 0 \\ 0 & 1 & 0 \\ 0 & 0 & 1 \end{array}\right]; \quad \begin{Bmatrix} \alpha_e^{I} \\ \beta_e^{I} \\ \gamma_e^{I} \end{Bmatrix} = \begin{Bmatrix} 0^\circ \\ 90^\circ \\ 90^\circ \end{Bmatrix};$$

$$\begin{Bmatrix} \alpha_e^{II} \\ \beta_e^{II} \\ \gamma_e^{II} \end{Bmatrix} = \begin{Bmatrix} 90^\circ \\ 0^\circ \\ 90^\circ \end{Bmatrix}; \quad \begin{Bmatrix} \alpha_e^{III} \\ \beta_e^{III} \\ \gamma_e^{III} \end{Bmatrix} = \begin{Bmatrix} 90^\circ \\ 90^\circ \\ 0^\circ \end{Bmatrix}; \quad \mathbf{r}_{OE}^{mech} = \begin{Bmatrix} 109.5 \\ 28 \\ 0 \end{Bmatrix} \tag{2.57h}$$

- *Part-grasped state*:
  Input: The joint coordinates state at t = 2 s:

$$\mathbf{q}^{pg} = \{\, d_A \quad \theta_B \quad \theta_C \quad d_D \,\}^T = \{\, 60'' \quad 0^\circ \quad 0^\circ \quad 20'' \,\}^T \tag{2.57i}$$

Output: The end-effector's orientation matrix, the direction angles, and its origin position:

$$\mathbf{D}_{e}^{mech} = \left[\begin{array}{c|c|c} 1 & 0 & 0 \\ 0 & 1 & 0 \\ 0 & 0 & 1 \end{array}\right]; \quad \begin{Bmatrix} \alpha_e^{I} \\ \beta_e^{I} \\ \gamma_e^{I} \end{Bmatrix} = \begin{Bmatrix} 0^\circ \\ 90^\circ \\ 90^\circ \end{Bmatrix};$$

$$\begin{Bmatrix} \alpha_e^{II} \\ \beta_e^{II} \\ \gamma_e^{II} \end{Bmatrix} = \begin{Bmatrix} 90^\circ \\ 0^\circ \\ 90^\circ \end{Bmatrix}; \quad \begin{Bmatrix} \alpha_e^{III} \\ \beta_e^{III} \\ \gamma_e^{III} \end{Bmatrix} = \begin{Bmatrix} 90^\circ \\ 90^\circ \\ 0^\circ \end{Bmatrix}; \quad \mathbf{r}_{OE}^{mech} = \begin{Bmatrix} 129.5 \\ 28 \\ 0 \end{Bmatrix} \tag{2.57j}$$

- *Target-approach state*:
  Input: The joint coordinates state at t = 3 s:

$$\mathbf{q}^{ta} = \{\, d_A \quad \theta_B \quad \theta_C \quad d_D \,\}^T = \{\, 20'' \quad 90^\circ \quad 90^\circ \quad 20'' \,\}^T \tag{2.57k}$$

Output: The end-effector's orientation matrix, the direction angles, and its origin position:

$$\mathbf{D}_e^{mech} = \begin{bmatrix} 0 & 0 & 1 \\ 1 & 0 & 0 \\ 0 & 1 & 0 \end{bmatrix}; \quad \begin{Bmatrix} \alpha_e^I \\ \beta_e^I \\ \gamma_e^I \end{Bmatrix} = \begin{Bmatrix} 90^\circ \\ 0^\circ \\ 90^\circ \end{Bmatrix};$$

$$\begin{Bmatrix} \alpha_e^{II} \\ \beta_e^{II} \\ \gamma_e^{II} \end{Bmatrix} = \begin{Bmatrix} 90^\circ \\ 90^\circ \\ 0^\circ \end{Bmatrix}; \quad \begin{Bmatrix} \alpha_e^{III} \\ \beta_e^{III} \\ \gamma_e^{III} \end{Bmatrix} = \begin{Bmatrix} 0^\circ \\ 90^\circ \\ 90^\circ \end{Bmatrix}; \quad \mathbf{r}_{OE}^{mech} = \begin{Bmatrix} 25 \\ 92.5 \\ 0 \end{Bmatrix} \tag{2.57l}$$

- *Target-reached state:*
  Input: The joint coordinates state at t = 4 s:

$$\mathbf{q}^{tr} = \begin{Bmatrix} d_A & \theta_B & \theta_C & d_D \end{Bmatrix}^T = \begin{Bmatrix} 0'' & 90^\circ & 90^\circ & 20'' \end{Bmatrix}^T \tag{2.57m}$$

Output: The end-effector's orientation matrix, the direction angles, and its origin position:

$$\mathbf{D}_e^{mech} = \begin{bmatrix} 0 & 0 & 1 \\ 1 & 0 & 0 \\ 0 & 1 & 0 \end{bmatrix}; \quad \begin{Bmatrix} \alpha_e^I \\ \beta_e^I \\ \gamma_e^I \end{Bmatrix} = \begin{Bmatrix} 90^\circ \\ 0^\circ \\ 90^\circ \end{Bmatrix};$$

$$\begin{Bmatrix} \alpha_e^{II} \\ \beta_e^{II} \\ \gamma_e^{II} \end{Bmatrix} = \begin{Bmatrix} 90^\circ \\ 90^\circ \\ 0^\circ \end{Bmatrix}; \quad \begin{Bmatrix} \alpha_e^{III} \\ \beta_e^{III} \\ \gamma_e^{III} \end{Bmatrix} = \begin{Bmatrix} 0^\circ \\ 90^\circ \\ 90^\circ \end{Bmatrix}; \quad \mathbf{r}_{OE}^{mech} = \begin{Bmatrix} 5 \\ 92.5 \\ 0 \end{Bmatrix} \tag{2.57n}$$

2.10.12 Example of Direct Positional Analysis for a Spatial Mechanism with 5 DOF

The TRRTR mechanism, in Figure 2.12a, has the tool frame F offset from E by the vector $\mathbf{L}_{EF}^5 = \begin{Bmatrix} 4 \\ 0 \\ 0 \end{Bmatrix}$.

The position vector of the tool frame origin F is given in Equation (2.58):

$$\mathbf{r}_F = \mathbf{r}_E + \mathbf{D}_5^0 * \left(\mathbf{L}_{EF}^5 + \mathbf{0}\right) = \begin{Bmatrix} 5 + d_A + (48.5 + d_D)\cdot C\theta_B \cdot C\theta_C \\ 28 + (48.5 + d_D)\cdot S\theta_C \\ -(48.5 + d_D)\cdot S\theta_B \cdot C\theta_C \end{Bmatrix} \tag{2.58}$$

where

- $\mathbf{r}_E$ is the position vector for origin of frame 5 from Equation (2.56e)
- $\mathbf{D}_5^0$ is the link 5 rotation matrix from Equation (2.36c)
- $\mathbf{L}_{EF}^5$ is the vector offset of frame F relative to E, written with respect to the local frame 5
- **0**, indicated no relative displacement between origins E and T; the distance between them remains constant over time.

2.10.13 Example of Direct Positional Analysis for a Spatial Mechanism with 6 DOF

The TRRTRT mechanism, in Figure 2.13a, has the home position F_5 on prismatic joint F offset from E by the vector $\mathbf{L}_{EF_5}^5 = \begin{Bmatrix} 4 \\ 0 \\ 0 \end{Bmatrix}$. This vector is written with respect to the local frame 6:

$$\mathbf{r}_F = \mathbf{r}_E + \mathbf{D}_5^0 * \left(\mathbf{L}_{EF_5}^5 + \mathbf{u}_F^5 \cdot d_F\right)$$

$$= \begin{Bmatrix} 5 + d_A + (48.5 + d_D) \cdot C\theta_B \cdot C\theta_C + (-C\theta_B \cdot S\theta_C \cdot C\theta_E + S\theta_B \cdot S\theta_E) \cdot d_F \\ 28 + (48.5 + d_D) \cdot S\theta_C + (C\theta_C \cdot C\theta_E) \cdot d_F \\ -(48.5 + d_D) \cdot S\theta_B \cdot C\theta_C + (S\theta_B \cdot S\theta_C \cdot C\theta_E + C\theta_B \cdot S\theta_E) \cdot d_F \end{Bmatrix} \tag{2.59a}$$

where

- $\mathbf{r}_E$ is the position vector for the origin of frame 5 from Equation (2.56e)
- $\mathbf{D}_5^0$ is the rotation matrix of link 5 from Equation (2.36c)
- $\mathbf{u}_F^5 = \begin{Bmatrix} 0 \\ 1 \\ 0 \end{Bmatrix}$ is the tool's frame origin T offset vector from E, written with respect to the local frame 5
- **0**, indicated no relative displacement between origins E and F; the distance between them remains constant over time.

The tool frame position vector is offset from F by the vector $\mathbf{L}_{FT}^6 = \begin{Bmatrix} 5 \\ 0 \\ 0 \end{Bmatrix}$.

This vector is written with respect to the local frame 6:

$$\mathbf{r}_T = \mathbf{r}_F + \mathbf{D}_6^0 * \left(\mathbf{L}_{FT}^6 + 0\right)$$

$$= \begin{Bmatrix} 5 + d_A + (53.5 + d_D) \cdot C\theta_B \cdot C\theta_C + (-C\theta_B \cdot S\theta_C \cdot C\theta_E + S\theta_B \cdot S\theta_E) \cdot d_F \\ 28 + (53.5 + d_D) \cdot S\theta_C + (C\theta_C \cdot C\theta_E) \cdot d_F \\ -(53.5 + d_D) \cdot S\theta_B \cdot C\theta_C + (S\theta_B \cdot S\theta_C \cdot C\theta_E + C\theta_B \cdot S\theta_E) \cdot d_F \end{Bmatrix} \tag{2.59b}$$

2.11 Inverse Positional Analysis: Governing Equations for Open Cycle Mechanisms: The Task Position Vector

In general, the open path starts at origin O and ends at E. If we imagine the path to end at the initial origin O, then a closed path (cycle) is generated. This closed path is a fictitious cycle, created by adding a fictitious edge, named chord E to the open path. The chord is associated with a fictitious joint E between node 0 and node m.

Equation (2.47) is extended from point E to the starting point O:

$$\mathbf{r}_O = \mathbf{L}_{OA}^0 + \mathbf{L}_{AB}^0 + \mathbf{L}_{BC}^0 + \cdots + \mathbf{L}_{YZ}^0 + \mathbf{L}_{ZE}^0\Big|_E - \mathbf{L}_{OE}^0\Big|_O \tag{2.60}$$

In Figure 2.19 the edge E is oriented from node 0 to node m (lower number to higher number); therefore, the orientation is inverse to the direction of the path, which is depicted using a minus sign for the variable magnitude vector $\mathbf{L}_{OE}^0$. This vector is the position vector of origin E; therefore:

$$\mathbf{L}_{OE}^0 = \mathbf{r}_E \tag{2.60a}$$

Considering that the position vector of frame 0 is a zero vector: $\mathbf{r}_O = \mathbf{0}$, then the Equation (2.60) becomes

$$\mathbf{L}_{OA}^0 + \mathbf{L}_{AB}^0 + \mathbf{L}_{BC}^0 + \cdots + \mathbf{L}_{ZE}^0\Big|_E = \mathbf{L}_{OE}^0 \tag{2.60b}$$

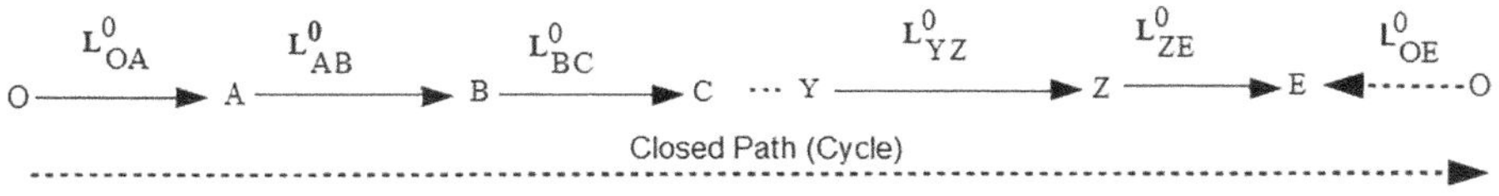

FIGURE 2.19
Translation vectors between frame origins along the closed path.

The *task position vector* is the end-effector's position vector with the components for a specific task applied:

$$\mathbf{L}_{OE}^{0} = \mathbf{r}_{E}^{task} \tag{2.61}$$

It is viewed in Figure 2.19 as edge's E position vector when a path is initiated at 0 direction from right to left (inverse path).

Equation (2.60) becomes

$$\mathbf{L}_{OA}^{0} + \mathbf{L}_{AB}^{0} + \mathbf{L}_{BC}^{0} + \cdots + \mathbf{L}_{ZE}^{0} = \mathbf{r}_{E}^{task} \tag{2.62}$$

Equation (2.62) is the inverse positional equation for open cycle mechanisms.

Since the sum on the left side of Equation (2.62) is equal to $\mathbf{r}_{E}^{mech}$, then Equation (2.63) for the inverse orientation analysis holds:

$$\mathbf{r}_{E}^{mech} = \mathbf{r}_{E}^{task} \tag{2.63}$$

The linear actuation parameters are determined by equating the entries on the left (function of actuating displacements) to the entries on the right (task-provided) matrix. In general, this leads to a system of nonlinear equations involving trigonometric functions, as shown in the next examples.

2.11.1 Example of Inverse Positional Analysis for a Spatial Mechanism with 4 DOF

For the TRRT mechanism's inverse positional analysis, the task matrix $\mathbf{r}_{E}^{task}$ is provided. Knowing the task matrix, we can determine a set of three equations to calculate the actuated linear displacements at the joints. Considering Equations (2.56f) and (2.63), then the three equations to determine the linear displacements at the joints are

$$\left\{ \begin{matrix} 5 + d_A + (44.5 + d_D) \cdot C\theta_B \cdot C\theta_C \\ 28 + (44.5 + d_D) \cdot S\theta_C \\ -(44.5 + d_D) \cdot S\theta_B \cdot C\theta_C \end{matrix} \right\} = \left\{ \begin{matrix} x_E^{task} \\ y_E^{task} \\ z_E^{task} \end{matrix} \right\} \tag{2.64}$$

The three equations extracted from Equation (2.64) represent a system of three nonlinear equations, which involve the trigonometric functions of unknown angular displacements; therefore, multiple solutions exist for the trigonometric functions. In Section 2.13, a procedure is presented to generate the equations for the inverse positional analysis without the involvement of trigonometric functions of unknown displacements.

2.11.2 Example of Inverse Positional Analysis for a Spatial Mechanism with 5 DOF

For the TRRTR mechanism's inverse positional analysis, the task matrix $\mathbf{r}_E^{task}$ is provided. Knowing the task matrix, we determine a set of three equations to calculate the actuated linear displacements at joints.

Considering Equation (2.58), then the three equations to determine the linear displacements at the joints are

$$\begin{Bmatrix} 5 + d_A + (48.5 + d_D) \cdot C\theta_B \cdot C\theta_C \\ 28 + (48.5 + d_D) \cdot S\theta_C \\ -(48.5 + d_D) \cdot S\theta_B \cdot C\theta_C \end{Bmatrix} = \begin{Bmatrix} x_F^{task} \\ y_F^{task} \\ z_F^{task} \end{Bmatrix} \tag{2.65}$$

The three equations extracted from Equation (2.65) represent a system of three nonlinear equations, which involve the trigonometric functions of unknown angular displacements. The disadvantage of solving the systems of nonlinear equations shown in Equation (2.65) is that multiple solutions exist for the trigonometric functions. In Section 2.13, a procedure is presented to generate the equations for the inverse positional analysis without the involvement of trigonometric functions of unknown displacements.

2.12 The System of Combined Equations for Inverse Orientation and Positional Analysis

The desired task orientation $\mathbf{D}_e^{task}$ and task position $\mathbf{r}_E^{task}$ are determined based on the information from a CAD model or a camera.

- For the TRRT with 4 DOF mechanism, the equations for inverse positional analysis are obtained as follows:
  - Considering Equations (2.35b)–(2.35d) from the inverse orientation:

    Equation 1:

    $$C\theta_B \cdot C\theta_C = \text{Cos}\,\alpha_e^I \tag{2.66a}$$

    Equation 2:

    $$\text{Cos}\,\theta_C = \text{Cos}\,\beta_e^{II} \tag{2.66b}$$

    Equation 3:

    $$\text{Cos}\,\theta_B = \text{Cos}\,\gamma_e^{III} \tag{2.66c}$$

- Considering Equation (2.64) for the inverse positional analysis:
  Equation 4:

$$5 + d_A + (44.5 + d_D) \cdot C\theta_B \cdot C\theta_C = x_E^{task} \tag{2.66d}$$

Equation 5:

$$28 + (44.5 + d_D) \cdot S\theta_C = y_E^{task} \tag{2.66e}$$

Equation 6:

$$-(44.5 + d_D) \cdot S\theta_B \cdot C\theta_C = z_E^{task} \tag{2.66f}$$

- For the TRRTR with 5 DOF mechanism, the equations for inverse orientation and positional analysis are obtained as follows:
  - Considering Equations (2.36b)–(2.36d) from the inverse orientation:
    Equation 1:

$$C\theta_B \cdot C\theta_C = \text{Cos}\,\alpha_e^I \tag{2.67a}$$

Equation 2:

$$C\theta_C \cdot C\theta_E = \beta_e^{II} \tag{2.67b}$$

Equation 3:

$$-S\theta_B \cdot S\theta_C \cdot S\theta_E + C\theta_B \cdot C\theta_E = \gamma_e^{III} \tag{2.67c}$$

  - Considering Equation (2.65) for the inverse positional analysis:
    Equation 4:

$$5 + d_A + (48.5 + d_D) \cdot C\theta_B \cdot C\theta_C = x_F^{task} \tag{2.67d}$$

Equation 5:

$$28 + (48.5 + d_D) \cdot S\theta_C = y_F^{task} \tag{2.67e}$$

Equation 6:

$$-(48.5 + d_D) \cdot S\theta_B \cdot C\theta_C = z_F^{task} \tag{2.67f}$$

Note:

- The disadvantage of solving the system of nonlinear Equations (2.66) and (2.67) is that multiple solutions exist for the trigonometric functions.

- In the next section, a procedure is presented to generate the equations for the inverse positional analysis without the involvement of trigonometric functions of unknown displacements. It is based on writing the equations in a matrix called a *Latin matrix*.

2.13 The Matroid Method: Equations Based on Latin Matrix and Cycle Matroid Entries

The procedure to generate the equations for inverse positional analysis without trigonometric functions of unknown angular displacements:

Consider the following difference in Equation (2.60):

$$\mathbf{L}^0_{\mathrm{OA}} - \mathbf{L}^0_{\mathrm{OE}} = \left(\mathbf{L}^0_{\mathrm{OA}_0} + \mathbf{d}_{\mathrm{A}}\right) - \mathbf{r}_{\mathrm{OE}} = \left(\mathbf{L}^0_{\mathrm{EA}_0} + \mathbf{d}_{\mathrm{A}}\right) = \mathbf{L}^0_{\mathrm{EA}} \tag{2.68}$$

Then, Equation (2.69) holds:

$$\mathbf{L}^0_{\mathrm{EA}} + \mathbf{L}^0_{\mathrm{AB}} + \mathbf{L}^0_{\mathrm{BC}} + \cdots + \mathbf{L}^0_{\mathrm{ZE}} = \mathbf{0} \tag{2.69}$$

Therefore, the sum of the translation vectors between the frame origins along a closed path is zero.

2.13.1 The Latin Matrix

The Latin matrix, Equation (2.70), is defined as a matrix whose columns are the translation vectors between the frame origins along a closed path [4]. The number of columns in this matrix is m + 1 (i.e., the number of nodes in the path):

$$\begin{array}{ccccc} 0 & 1 & 2 & \ldots & m \end{array}$$

$$\Delta\mathbf{L} = \left[\begin{array}{c:c:c:c:c} \mathbf{L}^0_{\mathrm{EA}} & \mathbf{L}^0_{\mathrm{AB}} & \mathbf{L}^0_{\mathrm{BC}} & \cdots & \mathbf{L}^0_{\mathrm{ZE}} \end{array}\right] = \left[\begin{array}{c:c:c:c:c} \mathbf{r}_{\mathrm{A}} - \mathbf{r}_{\mathrm{E}} & \mathbf{r}_{\mathrm{B}} - \mathbf{r}_{\mathrm{A}} & \mathbf{r}_{\mathrm{C}} - \mathbf{r}_{\mathrm{B}} & \cdots & \mathbf{r}_{\mathrm{E}} - \mathbf{r}_{\mathrm{Z}} \end{array}\right] \tag{2.70}$$

Statement 2.5

The sum of entries in the Latin matrix, along a closed path (cycle), is a zero vector (Equation (2.69)).

Proof: The addition of entries in the right-side matrix, Equation (2.70), results in a zero vector because each position vector appears twice and with opposite signs.

Therefore, the sum of entries from the left side of Equation (2.70) equals a zero 3×1 matrix. These translation vectors are written with components in the fixed frame after pre-multiplication by absolute matrices (Equation (2.71)):

$$\mathbf{D}_0 * \left(\mathbf{L}^0_{\mathrm{EA}_0} + \mathbf{d}_{\mathrm{A}}\right) + \mathbf{D}_1 * \left(\mathbf{L}^1_{\mathrm{AB}} + \mathbf{d}_{\mathrm{B}}\right) + \mathbf{D}_2 * \left(\mathbf{L}^2_{\mathrm{BC}} + \mathbf{d}_{\mathrm{C}}\right) + \cdots + \mathbf{D}_{\mathrm{m}} * \left(\mathbf{L}^{\mathrm{m}}_{\mathrm{ZE}} - \mathbf{d}_{\mathrm{E}}\right) = 0 \tag{2.71}$$

Statement 2.6

The Latin matrix is the sum of two matrices, one with translation vectors at the nodes and the second with displacement vectors at the edges:

$$\Delta\mathbf{L} = \Delta\mathbf{L}^{\mathrm{nodes}} + \Delta\mathbf{L}^{\mathrm{edges}} \tag{2.72}$$

Proof: From Equation (2.71) after distributing the matrix multiplications, noted "*", then Equation (2.72) holds. The two matrices are the same size; therefore, they can be added. The number of columns in the first matrix is m + 1, whereas the second matrix has j + 1 columns. Because a tree has the properties m = j (the number of mobile links is the same as the number of joints), then m + 1 = j + 1, and the two matrices have the same size.

The matrix $\Delta\mathbf{L}^{\mathrm{nodes}}$ is called the *link constraints matrix*. For inverse positional analysis, shown further in Section 2.13.3, the entries are the constraint equations for the vectors' magnitude and the constraint equations for the constant angle between two vectors located on two different links:

$$\Delta\mathbf{L}^{\mathrm{nodes}} = \begin{array}{c} \begin{array}{ccccc} 0 & 1 & 2 & \cdots & \mathrm{m} \end{array} \\ \left[\begin{array}{c|c|c|c|c} \mathbf{D}_0 * \mathbf{L}^0_{\mathrm{EA}_0} & \mathbf{D}_1 * \mathbf{L}^1_{\mathrm{AB}} & \mathbf{D}_2 * \mathbf{L}^2_{\mathrm{BC}} & \cdots & \mathbf{D}_{\mathrm{m}} * \mathbf{L}^{\mathrm{m}}_{\mathrm{ZE}} \end{array}\right] \end{array} \tag{2.73}$$

- The matrix $\Delta\mathbf{L}^{\mathrm{edges}}$ is called the joints' relative constraints matrix:

$$\Delta\mathbf{L}^{\mathrm{edges}} = \begin{array}{c} \begin{array}{ccccc} \mathrm{A} & \mathrm{B} & \mathrm{C} & \cdots & \mathrm{E} \end{array} \\ \left[\begin{array}{c|c|c|c|c} \mathbf{D}_0 * (+\mathrm{d}_{\mathrm{A}}) & \mathbf{D}_1 * (+\mathrm{d}_{\mathrm{B}}) & \mathbf{D}_2 * (+\mathrm{d}_{\mathrm{C}}) & \cdots & \mathbf{D}_{\mathrm{m}} * (-\mathrm{d}_{\mathrm{E}}) \end{array}\right] \end{array} \tag{2.74}$$

The constraint equations for relative motion:

- For all revolute joints in the mechanism, the constraint equation (null relative displacement) is shown as zero vectors in the $\Delta\mathbf{L}^{\mathrm{edges}}$ columns.

  Example: For revolute joints B and C in the TRRT mechanism, the constraints are $\mathbf{d}_{\mathrm{B}} = \mathbf{d}_{\mathrm{C}} = 0$.

- A null linear displacement (constraint) applies for translation (null rotation) between two points located on the same link.
  Example: The end-effector's origin E and point D are located on link 4; therefore, there is null linear displacement, so the constraint is $\mathbf{d}_E = 0$.

The set of edges C_E along the closed path (cycle) contains all oriented edges. The edge orientation is shown as a superscript. Thus, in the set $C_E = \{A^+, B^+, C^+, \ldots, Y^+, Z^+, E^-\}$, all edges are direct (superscript +1), except E which is inverse (superscript −1). From Equation (2.29), the set of numbers +1, −1, and 0 (in case the edge does not belong to the path) are the entries in a single row matrix *cycle matroid-C,* which has rank c = 1:

$$\begin{matrix} & & & & \text{Edges} & & & \\ & A & B & C & & Y & Z & E \\ C = [& +1 & +1 & +1 & \ldots & +1 & +1 & -1 \end{matrix} \;] C_E \text{ Cycle} \tag{2.75}$$

The joint's relative constraints matrix $\Delta\mathbf{L}^{\text{edges}}$ has its vector entries multiplied by the sign from the **C**-matroid. Therefore, the Equation (2.76) holds:

$$\begin{matrix} & A & B & C & \cdots & E \\ \Delta\mathbf{L}^{\text{edges}} = [& c_A \cdot \mathbf{d}_A & c_B \cdot \mathbf{d}_B & c_C \cdot \mathbf{d}_C & \cdots & c_E \cdot \mathbf{d}_E \;] \end{matrix} \tag{2.76}$$

Note: Mechanisms with multiple cycles are presented in Chapter 3, for which the cycle-basis incidence matrix C is defined as a c × j matrix where each row is assigned to a cycle and each column is assigned to an edge.

2.13.2 Algorithm for Automatic Generation of Latin Matrix Based on Digraph Matrices

This procedure is valid for any planar or spatial, single- or multiple-cycle mechanism, based on the digraph's matrices.

Statement 2.7

The entries in the Latin matrix are automatically generated from Equation (2.77) [4]:

$$\Delta\mathbf{L} = -\mathbf{r}_Z * \underline{\mathbf{G}}^T \tag{2.77}$$

where

$\mathbf{r}_Z$ is the digraph joint position vector matrix
$\underline{G}^T$ is the transposed of incidence links–joints matrix $\underline{G}$ from Equation (1.11)

Equation (2.77) is used to numerically or symbolically generate the entries in the Latin matrix.

Example 1: The algorithm is applied to the TRRT mechanism at the home position as follows:

- *The cycle basis matrix* (matroid) **C**, Equation (1.22), is applied to this example in Equation (2.77a):

$$\mathbf{C} = \begin{array}{c} \begin{matrix} A & B & C & D & E \end{matrix} \\ \left[\begin{array}{c:c:c:c:c} +1 & +1 & +1 & +1 & -1 \end{array} \right] \end{array} \tag{2.77a}$$

- *The digraph joint position vector matrix* with vector entries $c_Z{\cdot}\mathbf{r}_Z$ is defined by multiplication of the joint position vectors by the C entries:

$$\mathbf{r}_Z = \begin{array}{c} \begin{matrix} A & B & C & D & E \end{matrix} \\ \left[\begin{array}{c:c:c:c:c} c_A \cdot \mathbf{r}_A & c_B \cdot \mathbf{r}_B & c_C \cdot \mathbf{r}_C & c_D \cdot \mathbf{r}_D & c_E \cdot \mathbf{r}_E \end{array} \right] \end{array}$$

$$= \left[\begin{array}{c:c:c:c:c} +\mathbf{r}_A & +\mathbf{r}_B & +\mathbf{r}_C & +\mathbf{r}_D & -\mathbf{r}_E \end{array} \right] = \left[\begin{array}{c:c:c:c:c} 5 & 5 & 5 & 11.5 & -49.5 \\ 0 & 3 & 28 & 28 & -28 \\ 0 & 0 & 0 & 0 & 0 \end{array} \right] \tag{2.77b}$$

- *The incidence links–joints matrix* $\underline{G}$ and its transposed matrix are shown in Equation (2.77c):

$$\begin{array}{c} \text{Edges} \rightarrow \\ \begin{matrix} A & B & C & D & E \end{matrix} \end{array}$$

$$\underline{G} = \left[\begin{array}{c:c:c:c:c} -1 & 0 & 0 & 0 & -1 \\ +1 & -1 & 0 & 0 & 0 \\ 0 & +1 & -1 & 0 & 0 \\ 0 & 0 & +1 & -1 & 0 \\ 0 & 0 & 0 & +1 & +1 \end{array} \right] \begin{matrix} 0 \\ 1 \\ 2 \\ 3 \\ 4 \end{matrix} \quad \begin{matrix} \text{Nodes} \\ \downarrow \end{matrix} \quad ;$$

$$\begin{array}{c} \text{Nodes} \rightarrow \\ \begin{array}{ccccc} 0 & 1 & 2 & 3 & 4 \end{array} \end{array}$$

$$\underline{G}^{T} = \left[\begin{array}{c:c:c:c:c} -1 & +1 & 0 & 0 & 0 \\ 0 & -1 & +1 & 0 & 0 \\ 0 & 0 & -1 & +1 & 0 \\ 0 & 0 & 0 & -1 & +1 \\ -1 & 0 & 0 & 0 & +1 \end{array}\right] \begin{array}{c} A \\ B \\ C \\ D \\ E \end{array} \quad \begin{array}{c} \text{Edges} \\ \downarrow \end{array} \tag{2.77c}$$

then, Equation (2.77) is applied to generate the entries in the Latin matrix (Equation (2.77d)):

$$\begin{array}{ccccc} 0 & 1 & 2 & 3 & 4 \end{array}$$

$$\Delta \mathbf{L} = -\mathbf{r}_Z * \underline{G}^{T} = \left[\begin{array}{c:c:c:c:c} \mathbf{r}_A - \mathbf{r}_E & \mathbf{r}_B - \mathbf{r}_A & \mathbf{r}_C - \mathbf{r}_B & \mathbf{r}_D - \mathbf{r}_C & \mathbf{r}_E - \mathbf{r}_D \end{array}\right]$$

$$= \left[\begin{array}{c:c:c:c:c} \mathbf{L}_{EA}^{0} & \mathbf{L}_{AB}^{0} & \mathbf{L}_{BC}^{0} & \mathbf{L}_{CD}^{0} & \mathbf{L}_{DE}^{0} \end{array}\right]$$

or

$$\Delta \mathbf{L} = -\mathbf{r}_Z * \underline{G}^{T} = -\left[\begin{array}{c:c:c:c:c} 5 & 5 & 5 & 11.5 & -49.5 \\ 0 & 3 & 28 & 28 & -28 \\ 0 & 0 & 0 & 0 & 0 \end{array}\right] * \left[\begin{array}{c:c:c:c:c} -1 & +1 & 0 & 0 & 0 \\ 0 & -1 & +1 & 0 & 0 \\ 0 & 0 & -1 & +1 & 0 \\ 0 & 0 & 0 & -1 & +1 \\ -1 & 0 & 0 & 0 & +1 \end{array}\right]$$

$$= \left[\begin{array}{c:c:c:c:c} -44.5 & 0 & 0 & 6.5 & 38 \\ -28 & 3 & 25 & 0 & 0 \\ 0 & 0 & 0 & 0 & 0 \end{array}\right] \tag{2.77d}$$

In the Latin matrix, the sum of the entries of each row is zero.
Note:

- Equation (2.71) has general applicability for mechanisms with single or multiple cycles. In the case of mechanisms with multiple cycles, one equation is written for each cycle.
- The Latin matrix, based on its property that the sum of entries in each row is zero, generates the equations for the inverse positional analysis.

2.13.3 Example of Matroid Method on Inverse Positional Analysis for a Spatial Mechanism with 4 DOF

For the TRRT mechanism, determine the entries in the Latin matrix as functions of the actuating parameters in the column matrix $\mathbf{q}_M = \left\{ d_A \mid \theta_B \mid \theta_C \mid d_D \right\}^T$. This matrix is used further for the inverse positional analysis.

Solution:

Input data:

- The frame origins' position vectors (Equations (2.50)):

$$\mathbf{r}_A = \begin{Bmatrix} 5+d_A \\ 0 \\ 0 \end{Bmatrix}; \quad \mathbf{r}_B = \begin{Bmatrix} 5+d_A \\ 3 \\ 0 \end{Bmatrix}; \quad \mathbf{r}_C = \begin{Bmatrix} 5+d_A \\ 28 \\ 0 \end{Bmatrix};$$

$$\mathbf{r}_D = \begin{Bmatrix} 5+d_A+(6.5+d_D)\cdot C\theta_B \cdot C\theta_C \\ 28+(6.5+d_D)\cdot S\theta_C \\ -(6.5+d_D)\cdot S\theta_B \cdot C\theta_C \end{Bmatrix}; \tag{2.78a}$$

$$\mathbf{r}_E = \begin{Bmatrix} 5+d_A+(44.5+d_D)\cdot C\theta_B \cdot C\theta_C \\ 28+(44.5+d_D)\cdot S\theta_C \\ -(44.5+d_D)\cdot S\theta_B \cdot C\theta_C \end{Bmatrix}$$

- The Latin matrix with vector entries:

$$\begin{matrix} & 0 & 1 & 2 & 3 & 4 \end{matrix}$$

$$\Delta\mathbf{L} = \left[\begin{array}{c|c|c|c|c} \mathbf{L}_{EA}^0 & \mathbf{L}_{AB}^0 & \mathbf{L}_{BC}^0 & \mathbf{L}_{CD}^0 & \mathbf{L}_{DE}^0 \end{array} \right] \tag{2.78b}$$

$$= \left[\begin{array}{c|c|c|c|c} \mathbf{r}_A - \mathbf{r}_E & \mathbf{r}_B - \mathbf{r}_A & \mathbf{r}_C - \mathbf{r}_B & \mathbf{r}_D - \mathbf{r}_C & \mathbf{r}_E - \mathbf{r}_D \end{array} \right]$$

- The Latin matrix containing scalar entries that are determined from the difference between joint position vectors:

$$\begin{matrix} & 0 & 1 & 2 & 3 & 4 \end{matrix}$$

$$\Delta\mathbf{L} = \left[\begin{array}{c|c|c|c|c} x_{EA} & x_{AB} & x_{BC} & x_{CD} & x_{DE} \\ y_{EA} & y_{AB} & y_{BC} & y_{CD} & y_{DE} \\ z_{EA} & z_{AB} & z_{BC} & z_{CD} & z_{DE} \end{array} \right]$$

$$= \left[\begin{array}{c:c:c:c:c} -(44.5+d_D)\cdot C\theta_B \cdot C\theta_C & 0 & 0 & (6.5+d_D)\cdot C\theta_B \cdot C\theta_C & 38\cdot C\theta_B \cdot C\theta_C \\ -28-(44.5+d_D)\cdot S\theta_C & 3 & 25 & (6.5+d_D)\cdot S\theta_C & 38\cdot S\theta_C \\ (44.5+d_D)\cdot S\theta_B \cdot C\theta_C & 0 & 0 & -(6.5+d_D)\cdot S\theta_B \cdot C\theta_C & -38\cdot S\theta_B \cdot C\theta_C \end{array}\right] \tag{2.78c}$$

In the Latin matrix, the sum of the entries of each row is zero at any time.

2.13.3.1 Equations for Inverse Positional Analysis

A system of nonlinear equations *without trigonometric functions* of variables is generated as follows:

- Equations (2.35b)–(2.35d) developed before from the inverse orientation analysis are shown in Equations (2.79a)–(2.79c) with the trigonometric functions of variables (left side) substituted with the known terms from the right side.
  Equation 1:

$$C\theta_B \cdot C\theta_C = \mathrm{Cos}\,\alpha_e^{I} \tag{2.79a}$$

  Equation 2:

$$\mathrm{Cos}\,\theta_C = \mathrm{Cos}\,\beta_e^{II} \tag{2.79b}$$

  Equation 3:

$$\mathrm{Cos}\,\theta_B = \mathrm{Cos}\,\gamma_e^{III} \tag{2.79c}$$

  The rounding errors are minimized when calculations are based from entries in the Latin matrix, compared to when calculations are based on trigonometric functions. Trigonometric functions can be avoided in equations, as shown in the following method:
- The equations for inverse positional analysis are written with unknown entries in the Latin matrix only.
- A set of three equations is written because the sum of the entries in the Latin matrix equals zero.
  The components of the vector in column 0 with unknown magnitude L_{EA} are expressed as functions of the provided end-effector position and orientation.
- The trigonometric functions are eliminated by squaring the vector components (vector magnitude equations).
- Additional equations are written where there is a constraint constant angle between two vectors (dot product equations), or for three coplanar vectors, there are mixed product equations.

- The number of unknowns is the same as the number of nonlinear equations.

The components in the last two columns $\mathbf{L}^0_{CD}$ and $\mathbf{L}^0_{DE}$, shown in Equation (2.79d), are chosen to be unknowns in the equations:

$$\Delta\mathbf{L} = \left[\begin{array}{c|c|c|c|c} 5+d_A-x_E & 0 & 0 & x_{CD} & x_{DE} \\ -y_E & 3 & 25 & y_{CD} & y_{DE} \\ -z_E & 0 & 0 & z_{CD} & z_{DE} \end{array}\right] \tag{2.79d}$$

This choice of variables has the advantage of generating linear equations for the sum of the entries in each row of the Latin matrix.

Equations for sum of the entries from each row of the Latin matrix:

Equation 4:

$$(5+d_A-x_E)+x_{CD}+x_{DE}=0; \quad \text{sum of x entries is zero} \tag{2.79e}$$

Equation 5:

$$-y_E+28+y_{CD}+y_{DE}=0; \quad \text{sum of y entries is zero} \tag{2.79f}$$

Equation 6:

$$-z_E+z_{CD}+z_{DE}=0; \quad \text{sum of z entries is zero} \tag{2.79g}$$

The vector $\mathbf{L}^0_{EA}$ from Equation (2.68) is written as a function of the provided end-effector coordinates $\mathbf{r}_E$ from Equation (2.78a) and the provided orientation from Equation (2.79a):

$$\mathbf{L}^0_{EA} = \mathbf{L}^0_{OA} - \mathbf{r}_E = \left(\mathbf{L}^0_{OA_0} + d_{A_{0,1}}\right) - \mathbf{r}_E$$

And in matrix form this vector is

$$\mathbf{L}^0_{EA} = \begin{Bmatrix} x_{EA} \\ y_{EA} \\ z_{EA} \end{Bmatrix} = \begin{Bmatrix} 5+d_A-x_E \\ -y_E \\ -z_E \end{Bmatrix} = \begin{Bmatrix} -(44.5+d_D)\cdot C\theta_B\cdot C\theta_C \\ -y_E \\ -z_E \end{Bmatrix}$$
$$= \begin{Bmatrix} -(44.5+d_D)\cdot C\alpha^I_e \\ -y_E \\ -z_E \end{Bmatrix} \tag{2.79h}$$

Substituting the product of two cosines in the x_{EA} component by $C\alpha^I_e$, then:

Equation 7:

$$5 + d_A - x_E = -(44.5 + d_D) \cdot C\alpha_e^I \tag{2.79i}$$

Equations for elimination of trigonometric functions between components of vectors from columns 0, 3, and 4 (vector magnitude equations):

The advantage of choosing the variables, as shown in Equation (2.79d), is to generate equations for "vector magnitude." Such equations are written by squaring the vector's components with respect to the local frame. If expressed with respect to the fixed frame, the magnitude equation is the same as when written with respect to the local frame; it does not depend on trigonometric functions of unknown angular displacements (since the magnitude is conserved in any frame), as shown in the next three equations.

Equation 8:

$$(5 + d_A - x_E)^2 + (-y_E)^2 + (-z_E)^2 = (L_{EA})^2 \tag{2.79j}$$

Equation 9:

$$(x_{CD})^2 + (y_{CD})^2 + (z_{CD})^2 = (6.5 + d_D)^2 \tag{2.79k}$$

Equation 10:

$$(x_{DE})^2 + (y_{DE})^2 + (z_{DE})^2 = 38^2 \tag{2.79l}$$

Equations for constant angle between two vectors (dot product equation):

In the case where two links maintain a constant angle between each other (i.e., connected by a prismatic joint), then a constraint angle equation is assigned. The equation is written for the dot product between two vectors.

The vectors in column 3 and 4 keep the constant 0° angle between them:

$$\mathbf{L}_{CD}^0 \cdot \mathbf{L}_{DE}^0 = 38 \cdot (6.5 + d_D)$$

Equation 11:

$$x_{CD} \cdot x_{DE} + y_{CD} \cdot y_{DE} + z_{CD} \cdot z_{DE} = 38 \cdot (6.5 + d_D) \tag{2.79m}$$

Equations for coplanar vectors (mixed product equation):

In the case where three links remain coplanar during motion, the constraint equation is the mixed product (scalar triple product) of three vectors and equals zero. The mixed product is the scalar product of the first vector multiplied by the vector product of the remaining two vectors. Geometrically, the mixed product is the volume of a cube with edges of the three vectors. If the three vectors are coplanar, then the cube's volume is zero. Analitically, the cube's volume is the determinant which has as columns the coordinates from the three vectors:

$$\mathbf{L}_{BC}^{0} \cdot \left(\mathbf{L}_{CD}^{0} \times \mathbf{L}_{DE}^{0}\right) = 0 \Rightarrow \begin{vmatrix} 0 & x_{CD} & x_{DE} \\ 25 & y_{CD} & y_{DE} \\ 0 & z_{CD} & z_{DE} \end{vmatrix} = 0$$

Equation 12:

$$x_{CD} \cdot z_{DE} - x_{DE} \cdot z_{CD} = 0 \tag{2.79n}$$

2.13.4 Example Solution for Inverse Orientation and Positional Equations for a 4 DOF Mechanism

Since Equation (2.71a) is substituted in Equation (2.79i) in the subsection above, the system of 11 nonlinear equations, Equation (2.79e), through (2.79m), is solved for 11 unknowns.

Data: There are six orientation and position parameters from the desired task: $(\alpha_e^{I}, \beta_e^{II}, \gamma_e^{III}, x_E, y_E, z_E)$

Unknowns: The four actuating parameters in matrix $q_M = \{\, d_A \mid \theta_B \mid \theta_C \mid d_D \,\}^T$ and the seven additional unknown entries in the Latin matrix $(L_{EA}, x_{CD}, y_{CD}, z_{CD}, x_{DE}, y_{DE}, z_{DE})$ which correspond to the 11 equations discussed above.

Solving a nonlinear system involving trigonometric functions might involve the existence of multiple solutions because there could be multiple valid manipulator configurations between joints and links in which to place the end-effector in the same orientation and position.

2.13.4.1 Using EES for Inverse Orientation and Positional Analysis of the TRRT 4 DOF Robotic Mechanism

EES used in this text is a general equation-solving program that can numerically solve thousands of coupled nonlinear algebraic and differential equations. The program is also used in the text to verify unit consistency and create quality plots from parametric tables of data.

From the EES menu, the following commands are used when solving the nonlinear system of equations:

- In the Equations Window, the user types the equations as illustrated in Figure 2.20a.

  The user can copy and paste the text inside the Equations Window. EES code uses the following symbols to indicate mathematical functions: multiplication as "*", an exponent as "**" or "^", a subscript A

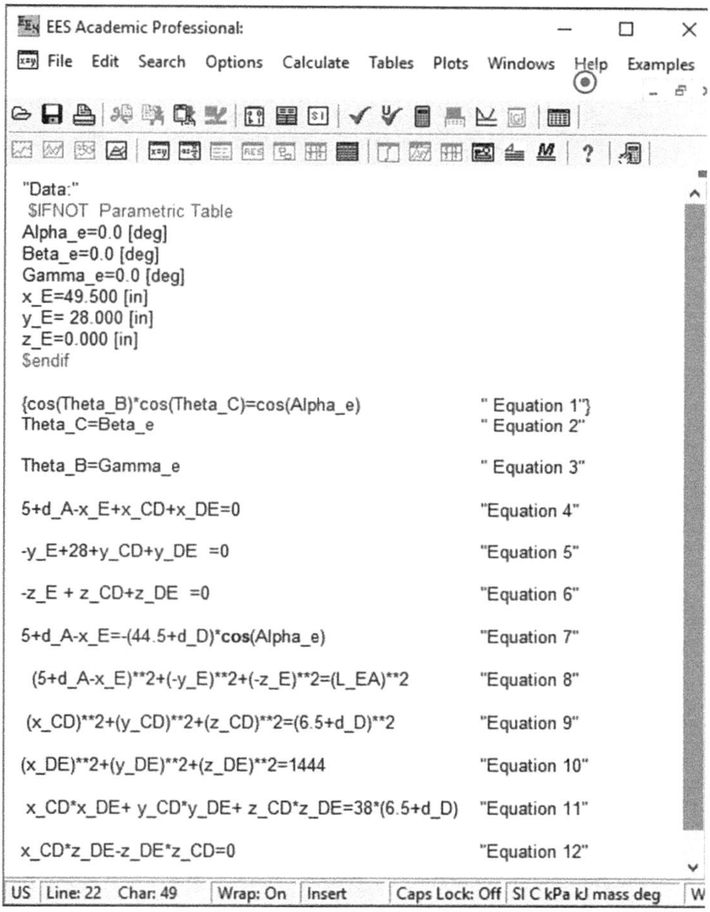

FIGURE 2.20a
Equations Window.

as in d_A, an angular argument in trigonometric functions shows between round brackets ( ).

To write a comment or description, highlight text in the Equations Window and right click to select Comment " ". To disinclude an equation from the calculations, highlight the equation in the Equations Window and right click to select Comment { }; see Equation 1 in Figure 2.20a, for example.

- From EES menu, select Windows, and then Formatted Equations. Figure 2.20b shows the formatted text, which is useful for checking eventual typing errors.

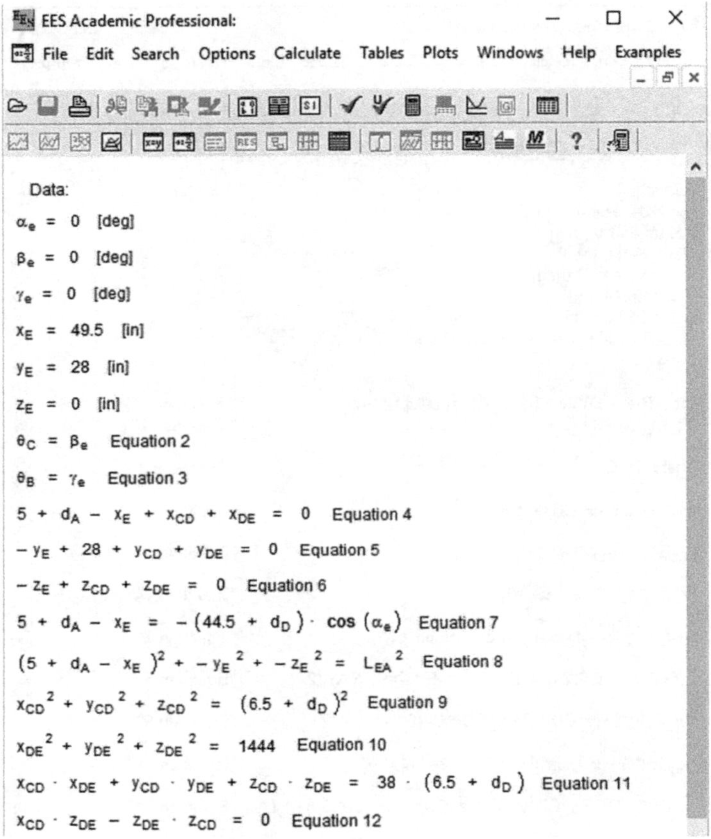

FIGURE 2.20b
Formatted Equations in Tables/New Parametric Table, the user can run the program multiple times with various inputs to solve the equations. The user inserts the number of runs and a provided time interval: set the First Value and the Last Value. Then, from Calculate select Solve Table. Parametric tables are used further in the text for the direct positional analysis. For example, Figure 2.20j shows a plot created with the data from a Parametric Table.

- The *$IFNOT* Parametric Table allows the user to solve the equations for different individual sets of input data, which is used in inverse positional analysis. Thus, the equations included between the EES instruction $IFNOT Parametric Table and $endif in Figure 2.20a allow the user to change the input orientation and position data from home position to part approach or to part grasp, target approach, and target reached.
- From the EES menu select Options/Variable Info, shown in Figure 2.20c.

- In the Variable Information window, Figure 2.20c, the user inserts the initial (guess) values and lower and upper limits for the variables. Usually only guess values are required when solving nonlinear systems of equations. Upper/lower bounds are needed if the calculations result in mathematical errors, such as square roots of negative numbers; otherwise, they can be left as the defaults.

  This is an example of lower and upper limits for linear displacements: $d_A \in [0'';60'']$ and $d_D \in [0'';20'']$. Both limits are extracted from stroke specifications for the manipulator. The variables x_{DE}, y_{DE}, z_{DE} are in the interval [−38, 38] because the components of this vector cannot exceed its magnitude $L_{DE} = 38''$, see Equation (2.10).

Note: The superscripts *I, II, III* columns in the orientation matrix of the end-effector are skipped for easiness in the EES code for the variables, α_e^I, β_e^{II}, and γ_e^{III}. Also, the subscripts "0,e" for the end-effector coordinates $x_{E0,e}$, $y_{E0,e}$, and $z_{E0,e}$ are skipped in the code.

- *Home-position state*:

  Input: Desired task parameters: $\alpha_e = 0°$, $\beta_e = 0°$, $\gamma_e = 0°$, $x_E = 49.500$, $y_E = 28$, $z_E = 0$

  Solution: Using the previously explained code, Figure 2.20d shows solution in EES. The solution is

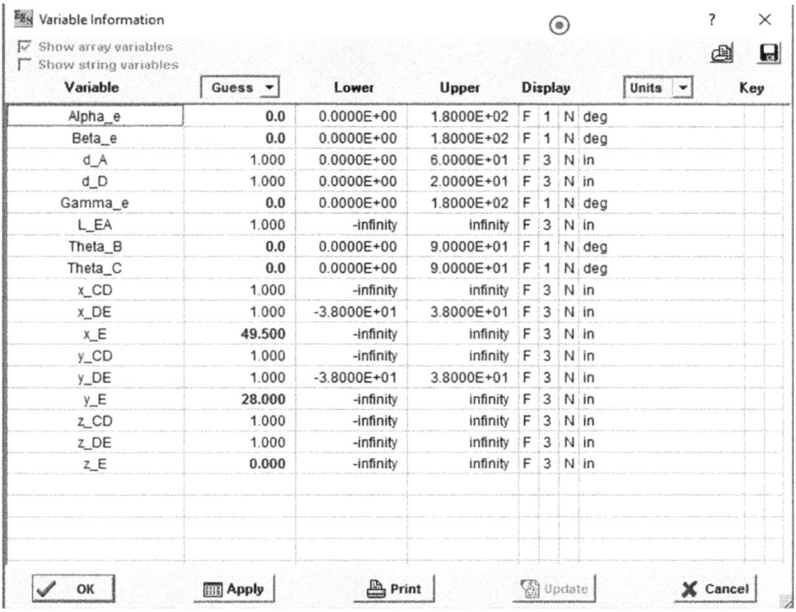

Variable	Guess	Lower	Upper	Display			Units	Key
Alpha_e	0.0	0.0000E+00	1.8000E+02	F	1	N	deg	
Beta_e	0.0	0.0000E+00	1.8000E+02	F	1	N	deg	
d_A	1.000	0.0000E+00	6.0000E+01	F	3	N	in	
d_D	1.000	0.0000E+00	2.0000E+01	F	3	N	in	
Gamma_e	0.0	0.0000E+00	1.8000E+02	F	1	N	deg	
L_EA	1.000	-infinity	infinity	F	3	N	in	
Theta_B	0.0	0.0000E+00	9.0000E+01	F	1	N	deg	
Theta_C	0.0	0.0000E+00	9.0000E+01	F	1	N	deg	
x_CD	1.000	-infinity	infinity	F	3	N	in	
x_DE	1.000	-3.8000E+01	3.8000E+01	F	3	N	in	
x_E	49.500	-infinity	infinity	F	3	N	in	
y_CD	1.000	-infinity	infinity	F	3	N	in	
y_DE	1.000	-3.8000E+01	3.8000E+01	F	3	N	in	
y_E	28.000	-infinity	infinity	F	3	N	in	
z_CD	1.000	-infinity	infinity	F	3	N	in	
z_DE	1.000	-infinity	infinity	F	3	N	in	
z_E	0.000	-infinity	infinity	F	3	N	in	

FIGURE 2.20c
Variable Information Window.

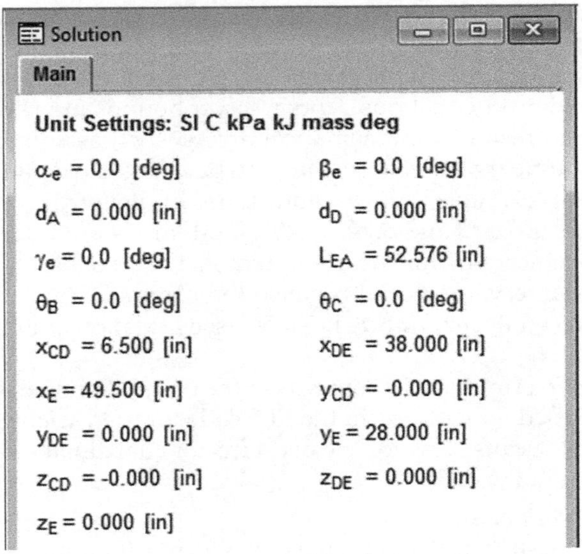

FIGURE 2.20d
Home position.

– Actuating joint coordinates:

$$\mathbf{q}^{h} = \left\{ \begin{array}{c:c:c:c} d_A & \theta_B & \theta_C & d_D \end{array} \right\}^T = \left\{ \begin{array}{cccc} 0'' & 0^\circ & 0^\circ & 0'' \end{array} \right\}^T \quad (2.79o)$$

– Latin matrix entries: (L_{EA}, x_{CD}, y_{CD}, z_{CD}, x_{DE}, y_{DE}, z_{DE}) = (52.576, 6.500, 0, 0, 38.000, 0, 0) in.

- *Part-approach state*:
Input: Desired task parameters: $\alpha_e = 0^\circ$, $\beta_e = 0^\circ$, $\gamma_e = 0^\circ$, $x_E = 109.500$, $y_E = 28.000$, $z_E = 0$
Solution: Using the previously explained code, Figure 2.20e shows solution in EES. The solution is

– Actuating joint coordinates:

$$\mathbf{q}^{pa} = \left\{ \begin{array}{cccc} d_A & \theta_B & \theta_C & d_D \end{array} \right\}^T = \left\{ \begin{array}{cccc} 60'' & 0^\circ & 0^\circ & 0'' \end{array} \right\}^T \quad (2.79p)$$

– Latin matrix entries: (L_{EA}, x_{CD}, y_{CD}, z_{CD}, x_{DE}, y_{DE}, z_{DE}) = (52.576, 6.500, 0, 0, 38.000, 0, 0) in.

- *Part-grasped state*:
Input: Desired task parameters: $\alpha_e = 0^\circ$, $\beta_e = 0^\circ$, $\gamma_e = 0^\circ$, $x_E = 129.500$, $y_E = 28.000$, $z_E = 0$

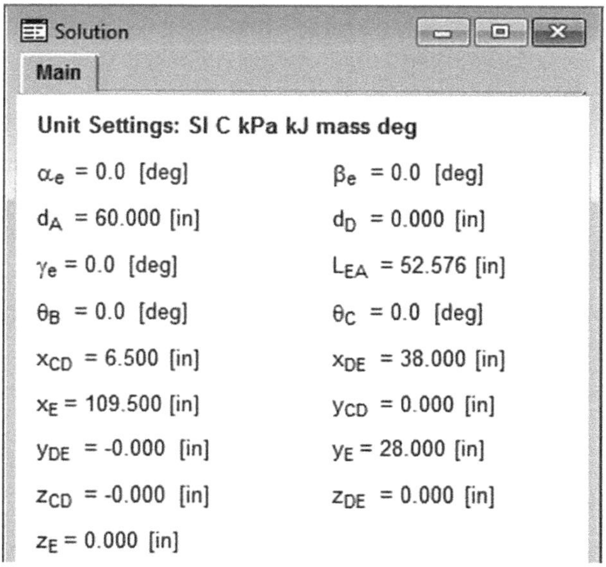

FIGURE 2.20e
Part approach.

Solution: Using the previously explained code, Figure 2.20f shows solution in EES. The solution is

– Actuating joint coordinates:

$$\mathbf{q}^{pg} = \left\{ \begin{array}{cccc} d_A, & \theta_B, & \theta_C, & d_D \end{array} \right\}^T = \left\{ \begin{array}{cccc} 60'' & 0^\circ & 0^\circ & 20'' \end{array} \right\}^T \tag{2.79q}$$

– Latin matrix entries: $(L_{EA}, x_{CD}, y_{CD}, z_{CD}, x_{DE}, y_{DE}, z_{DE}) = (70.315, 26.500, 0, 0, 38.000, 0, 0)$ in.

- *Target-approach state*:
  Input: Desired task parameters: $\alpha_e = 90^\circ$, $\beta_e = 90^\circ$, $\gamma_e = 90^\circ$, $x_E = 25.000$, $y_E = 92.500$, $z_E = 0$

  Solution: Using the previously explained code, Figure 2.20g shows solution in EES. The solution is

  – Actuating joint coordinates:

$$\mathbf{q}^{ta} = \left\{ \begin{array}{c:c:c:c} d_A & \theta_B & \theta_C & d_D \end{array} \right\}^T = \left\{ \begin{array}{cccc} 20'' & 90^\circ & 90^\circ & 20'' \end{array} \right\}^T \tag{2.79r}$$

  – Latin matrix entries: $(L_{EA}, x_{CD}, y_{CD}, z_{CD}, x_{DE}, y_{DE}, z_{DE}) = (92.500, 0, 0, 26.500, 0, 38.000, 0)$ in.

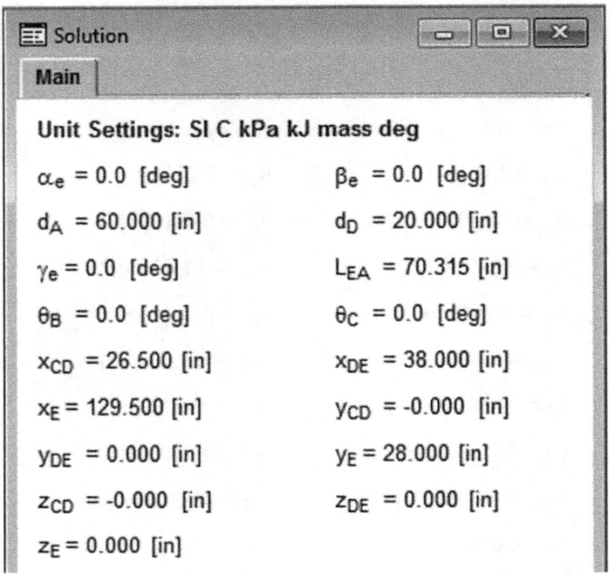

FIGURE 2.20f
Part grasp.

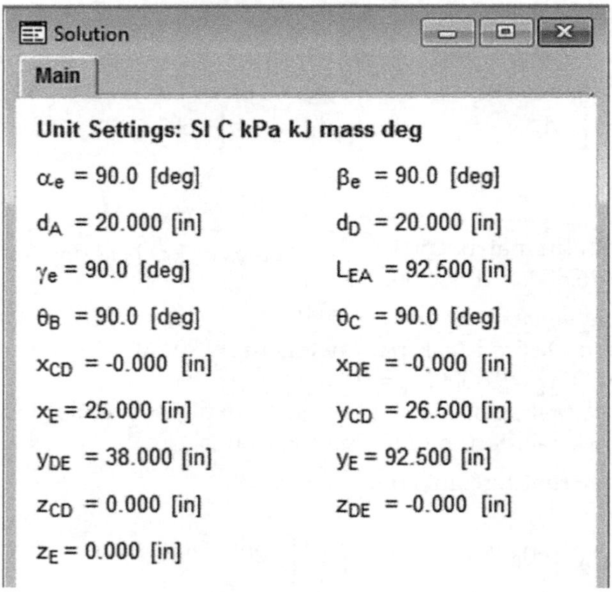

FIGURE 2.20g
Target approach.

- *Target-reached state*:
  Input: Task parameters: $\alpha_e = 90°$, $\beta_e = 90°$, $\gamma_e = 90°$, $x_E = 5.000$, $y_E = 92.500$, $z_E = 0$
  Solution: Using the previously explained code, Figure 2.20h shows solution in EES. The solution is
  - Actuating joint coordinates:

$$\mathbf{q}^{tr} = \left\{ d_A, \mid \theta_B, \mid \theta_C, \mid d_D \right\}^T = \left\{ 0'' \quad 90° \quad 90° \quad 20'' \right\}^T \quad (2.79s)$$

  - Latin matrix entries: $(L_{EA}, x_{CD}, y_{CD}, z_{CD}, x_{DE}, y_{DE}, z_{DE}) = (92.500, 0, 26.500, 0, 0, 38.000, 0)$ in.

  In the inverse analysis of the TRRT robotic mechanism with 4 DOF, the target orientation and position x_E, y_E, z_E of the robot's end-effector E are provided. Then, the angular displacements of the linear motors in joints A and D and revolute motors B and C are determined. The results for links' coordinates x_{CD}, y_{CD}, z_{CD}, x_{DE}, y_{DE}, z_{DE} are used to evaluate velocity and acceleration.

2.13.4.2 EES for Direct Positional Analysis of the TRRT 4 DOF Robotic Mechanism

The results from the above inverse analysis can be compared to the results from the direct positional analysis. For the direct positional analysis, the

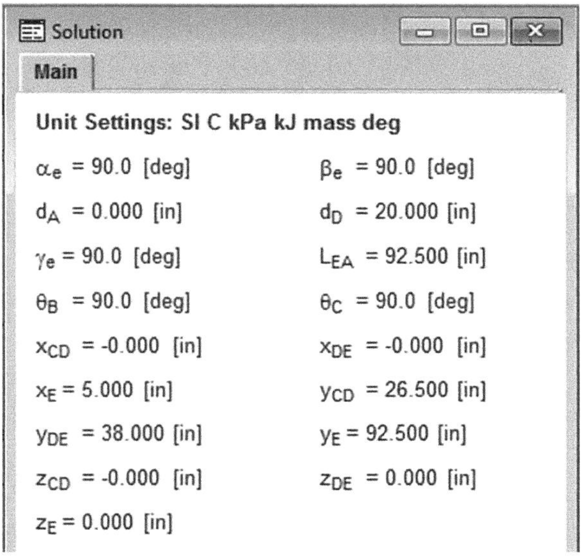

FIGURE 2.20h
Target reached.

displacements for actuated joints A, B, C, and D are shown in Equation (2.57b) as piecewise functions of time t. Then, end-effector's Cartesian coordinates x_E, y_E, z_E are determined from Equation (2.56f).

- In the Equations Window, Figure 2.20i, the user types the Equation (2.56f). Notice that for each displacement, a piecewise function of time t is applied using an *If Then Else statement*. From the Help menu, the user can open the pdf file EES Manual. In Chapter 5 of the EES Manual, there are instructions for typing the text of an *If* statement.
- In Tables/New Parametric Table, insert the number of runs and the time interval: First Value:0 and Last (linear):4. Then, from Calculate menu select Solve Table. Appendix 2.4 shows the EES Parametric Table with the end-effector's coordinates for a time interval of 4 s.
- In Plots/New Plot Window/X–Y Plot, select t for the x-Axis and x_E for the Y-Axis. Repeat this procedure for y_E and z_E. The three previously generated graphs can be displayed in the same space using Plots/Overlay Plot, illustrated in Figure 2.20j.

Equations Window

```
"Data:t=4 [sec]-simulation time for 100 runs"
Function d_a(t)
    If (t >= 0) AND (t <= 1) Then
      d_a = 60*t
    Endif
    If (t > 1) AND (t <= 2) Then
      d_a = 60
    Endif
    If (t > 2) AND (t <= 3) Then
      d_a = 60-40*(t-2)
    Endif
    If (t > 3) AND (t <= 4) Then
      d_a = 20-20*(t-3)
    Endif
End

Function d_d(t)
    If (t >= 0) AND (t <= 1) Then
      d_d = 0
    Endif
    If (t > 1) AND (t <= 2) Then
      d_d = 20*(t-1)
    Endif
    If (t > 2) AND (t <= 3) Then
      d_d = 20
    Endif
    If (t > 3) AND (t <= 4) Then
      d_d = 20
    Endif
End
```

Equations Window

```
Function theta_b(t)
    If (t >= 0) AND (t <= 1) Then
      theta_b = 0
    Endif
    If (t > 1) AND (t <= 2) Then
      theta_b = 0
    Endif
    If (t > 2) AND (t <= 3) Then
      theta_b = 90*(t-2)
    Endif
    If (t > 3) AND (t <= 4) Then
      theta_b = 90
    Endif
End

Function theta_c(t)
    If (t >= 0) AND (t <= 1) Then
      theta_c = 0
    Endif
    If (t > 1) AND (t <= 2) Then
      theta_c = 0
    Endif
    If (t > 2) AND (t <= 3) Then
      theta_c = 90*(t-2)
    Endif
    If (t > 3) AND (t <= 4) Then
      theta_c = 90
    Endif
End

"Equations for the direct positional analysis of robot's end effector:"

x_E = 5+d_a(t)+(44.5+d_d(t))*cos(theta_b(t))*cos(theta_c(t))
y_E=28+(44.5+d_d(t))*sin(theta_c(t))
z_E=-(44.5+d_d(t))*sin(theta_b(t))*cos(theta_c(t))
```

FIGURE 2.20i
Equations Window.

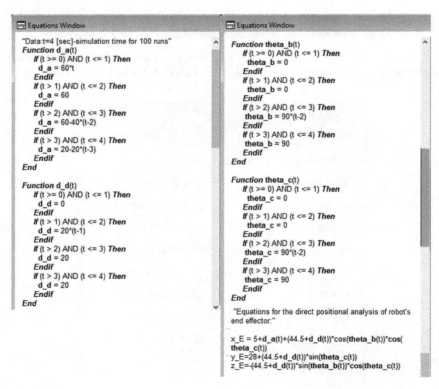

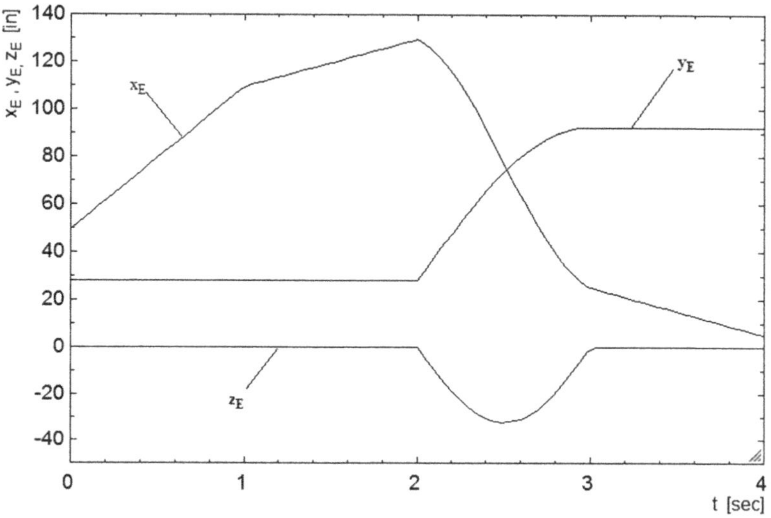

FIGURE 2.20j
End-effector's coordinates.

2.13.5 Conclusions

For the inverse analysis, a nonlinear algebraic system of equations, generated in the text from the Latin matrix entries, converges to a solution. This approach is an alternative to using a system of nonlinear equations based on trigonometric functions of unknown angular displacements.

All the calculated entries in the EES Parametric Table, Appendix 2.4, coincide with the results from SW Motion simulation, Appendix 2.3. The two sets of values are shown in Figure 2.20j (EES data) and in Figure 2.18b (SW motion data).

Therefore, the Latin matrix method validates the results from the 3D assembly, created using an SW. The validation refers to the capability of a mechanism in reaching a task position, velocity, and acceleration. The method applies for spatial and planar, open or closed multiple cycle mechanisms, shown further in Chapter 3.

2.14 The IT Method of Relative Homogeneous Matrices: Combined Equations for Direct Orientation and Positional Analysis

The previous sections presented separate equations to determine the orientation and position.

Next, an approach is presented where the equations for orientation and position can be combined into a single set of equations; this new approach can be useful in computer programming. This approach is based on the use of 4 × 4 matrices, called *relative homogeneous matrices; one of these matrices is assigned to each edge (joint).*

In this text, the equations are modeled from the digraph. The novel *absolute homogeneous matrix* is assigned to each node (link). A relation between the two types of homogeneous matrices is later introduced.

2.14.1 Absolute Homogeneous Matrix for Link (Node Digraph), $\mathbf{R}_m$

A 3 × 4 non-square matrix is by combining the following two matrices:

- $\mathbf{D}_m$ is the 3 × 3 absolute rotation matrix from Equation (2.20)
- $\mathbf{r}_{OZ}$ is the 3 × 1 joint Z's position vector

Equation (2.80) shows a 4 × 4 square matrix that consists of the previously discussed 3 × 4 matrix with a fourth row added to it. Within this row, the first three entries are zero, and the fourth entry is 1, which is called the scaling factor:

$$\mathbf{R}_m = \left[\begin{array}{c|c} \mathbf{D}_m & \mathbf{r}_Z \\ \hline 0 & 1 \end{array}\right] = \left[\begin{array}{ccc|c} C\alpha_m^{I} & C\alpha_m^{II} & C\alpha_m^{III} & x_Z \\ C\beta_m^{I} & C\beta_m^{II} & C\beta_m^{III} & y_Z \\ C\gamma_m^{I} & C\gamma_m^{II} & C\gamma_m^{III} & z_Z \\ \hline 0 & 0 & 0 & 1 \end{array}\right] \tag{2.80}$$

2.14.2 Relative IT Homogeneous Matrix for Joint (Edge Digraph), $\mathbf{IT}_{YZ}$

Another 3 × 4 non-square matrix is created from the two matrices:

- $\mathbf{D}_Z$ is the 3 × 3 relative rotation matrix from either Equation (2.7), (2.11), or (2.15).
- $\mathbf{L}_{YZ}^{m-1}$ is the 3 × 1 translation vector between the frame origins Y and Z which is an entry in the Latin matrix expressed in local frame m – 1.

Equation (2.81) shows a 4 × 4 square matrix that consists of the previously discussed 3 × 4 matrix with a fourth row added to it. Within this additional row, there are three zeros and a one, just as there was with matrix $\mathbf{R}_m$:

$$
\mathbf{IT}_{YZ} = \left[\begin{array}{c|c} \mathbf{D}_Z & \mathbf{L}_{YZ}^{m-1} \\ \hline 0 & 1 \end{array}\right] = \left[\begin{array}{c|c} \mathbf{D}_Z & \mathbf{L}_{YZ_{m-1}}^{m-1} + \mathbf{d}_Z^{m-1} \\ \hline 0 & 1 \end{array}\right]
$$

$$
= \left[\begin{array}{ccc|c} C\alpha_{m-1,m}^{I} & C\alpha_{m-1,m}^{II} & C\alpha_{m-1,m}^{III} & x_{YZ}^{m-1} \\ C\beta_{m-1,m}^{I} & C\beta_{m-1,m}^{II} & C\beta_{m-1,m}^{III} & y_{YZ}^{m-1} \\ C\gamma_{m-1,m}^{I} & C\gamma_{m-1,m}^{II} & C\gamma_{m-1,m}^{III} & z_{YZ}^{m-1} \\ \hline 0 & 0 & 0 & 1 \end{array}\right] \tag{2.81}
$$

The matrix in Equation (2.81) is partitioned into two sub-matrices, shown in Equation (2.82):

$$
\mathbf{I} = \left[\begin{array}{c} \mathbf{D}_Z \\ \hline 0 \end{array}\right] \text{ and } \mathbf{T} = \left[\begin{array}{c} \mathbf{L}_{YZ}^{m-1} \\ \hline 1 \end{array}\right] \tag{2.82}
$$

where

- The matrix on the left **I** is called the *incidental matrix* because its entries are functions of the angular displacements of the joints (tree's oriented edges).
- The matrix on the left **T** is called the *translation* matrix because its entries are the translation vector between the two frame origins.

Note: The reader that is familiar with the Hartemberg and Denavit (HD) notation might notice the difference between a homogeneous matrix based on HD notation and one based on the IT notation that is implemented in this text.

HD notation:

- The rotation axes are all chosen as the z-axes. The entries in each homogeneous matrix are functions of two angular parameters: a constant twist angle, α, and a variable rotation angle, θ.
- The linear displacement axes are all chosen as the z-axes. The entries in each homogeneous matrix are functions of two linear parameters: a constant distance a along the common perpendicular between two axes and a variable linear displacement d measured from a point located at the intersection between the sliding direction and the common perpendicular.

IT notation:

- The rotation axes are chosen as any of the x-, y-, or z-axes. The entries in each homogeneous matrix are functions of a single angular parameter, the variable rotation angle, θ, measured from either the x- or y-axes.
- The linear displacement axes are chosen as any of the x-, y-, or z-axes. The entries in each homogeneous matrix are functions of a single linear parameter: a variable *linear displacement d* measured from a reference point chosen to be the home position.

2.14.3 Relations between Absolute and Relative Homogeneous Matrices

The relative homogeneous matrix $\mathbf{IT}_{YZ}$ has entries written in local frame m−1 as functions of relative angular and linear displacements. When premultiplied by the previous absolute matrix on the path $\mathbf{R}_{m-1}$, its entries are with respect to the fixed frame.

Link m's absolute matrix $\mathbf{R}_m$ is defined as a multiplication of two homogeneous matrices (Equation (2.83)):

$$\mathbf{R}_m = \mathbf{R}_{m-1} * \mathbf{IT}_{YZ} \tag{2.83}$$

Proof: Considering Equations (2.20) and (2.46), then the matrix product from the right side of Equation (2.83) becomes its left-side matrix.

Thus,

$$\mathbf{R}_{m-1} * \mathbf{IT}_{YZ} = \left[\begin{array}{c|c} \mathbf{D}_{m-1} & \mathbf{r}_Y \\ \hline 0 & 1 \end{array}\right] * \left[\begin{array}{c|c} \mathbf{D}_Z & \mathbf{L}_{YZ}^{m-1} \\ \hline 0 & 1 \end{array}\right]$$

$$= \left[\begin{array}{c|c} \mathbf{D}_{m-1} * \mathbf{D}_Z & \mathbf{r}_Y + \mathbf{D}_{m-1} * \mathbf{L}_{YZ}^{m-1} \\ \hline 0 & 1 \end{array}\right] = \left[\begin{array}{c|c} \mathbf{D}_m & \mathbf{r}_Z \\ \hline 0 & 1 \end{array}\right] = \mathbf{R}_m$$

2.14.4 Direct Orientation and Positional Combined Equations on Open Cycle Mechanisms

Each node in the path is assigned an absolute matrix, whereas each edge is assigned a relative matrix (Figure 2.21).

Statement 2.8

The absolute homogeneous matrix at node m is determined from the right-side multiplication of all relative matrices located before the node m in the path:

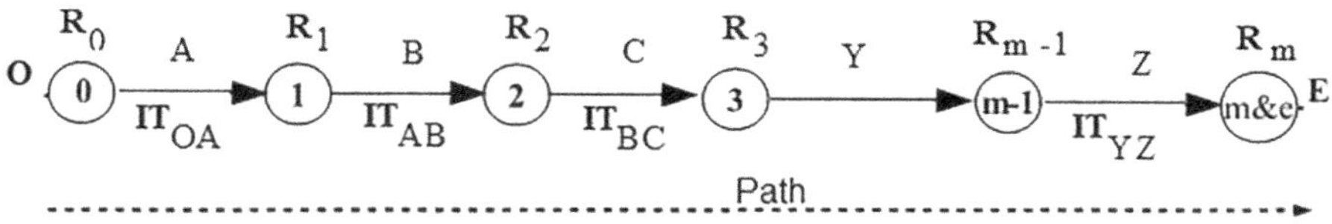

FIGURE 2.21
Relation between absolute and relative homogeneous matrices along the tree.

$$\mathbf{R}_m = \mathbf{IT}_{OA}\big|_{m=1} * \mathbf{IT}_{AB}\big|_{m=2} * \mathbf{IT}_{BC}\big|_{m=3} * \cdots * \mathbf{IT}_{YZ}\big|_{m} \tag{2.84}$$

Next, the recursion process is developed in detail.

Node 0: Absolute homogeneous matrix for the fixed link 0 is the unit 4×4 matrix because there is zero rotation of the fixed frame relative to itself and the position vector of the origin is a zero vector:

$$\mathbf{R}_0 = \left[\begin{array}{c|c} \mathbf{D}_0 & \mathbf{r}_O \\ \hline 0 & 1 \end{array}\right] = \mathbf{U}_{4\times4} = \left[\begin{array}{ccc|c} 1 & 0 & 0 & 0 \\ 0 & 1 & 0 & 0 \\ 0 & 0 & 1 & 0 \\ \hline 0 & 0 & 0 & 1 \end{array}\right] \tag{2.84a}$$

Edge A: Relative homogeneous matrix at prismatic joint A is defined as

$$\mathbf{IT}_{OA} = \left[\begin{array}{c|c} \mathbf{D}_A & \mathbf{L}^0_{OA} \\ \hline 0 & 1 \end{array}\right] = \left[\begin{array}{c|c} \mathbf{D}_A & \mathbf{L}^0_{OA_0} + \mathbf{d}^0_A \\ \hline 0 & 1 \end{array}\right] \tag{2.84b}$$

Node 1: Absolute homogeneous matrix for link 1 is determined from

$$\mathbf{R}_1 = \mathbf{R}_0 * \mathbf{IT}_{OA} = \mathbf{IT}_{OA} = \left[\begin{array}{c|c} \mathbf{D}_1 & \mathbf{r}_A \\ \hline 0 & 1 \end{array}\right] \tag{2.84c}$$

Output: From Equation (2.84c), the orientation matrix $\mathbf{D}_1$ and position vector $\mathbf{r}_A$ are identified.

Edge B: Relative homogeneous matrix at revolute joint B is defined as

$$\mathbf{IT}_{AB} = \left[\begin{array}{c|c} \mathbf{D}_B & \mathbf{L}^1_{AB} + \mathbf{0} \\ \hline 0 & 1 \end{array}\right] \tag{2.84d}$$

Node 2: Absolute homogeneous matrix for link 2 is determined from

$$\mathbf{R}_2 = \mathbf{IT}_{OA} * \mathbf{IT}_{AB} = \mathbf{R}_1 * \mathbf{IT}_{AB} = \left[\begin{array}{c|c} \mathbf{D}_2 & \mathbf{r}_B \\ \hline 0 & 1 \end{array}\right] \tag{2.84e}$$

Output: From Equation (2.84e), the orientation matrix $\mathbf{D}_2$ and position vector $\mathbf{r}_B$ are identified.

Edge C: Relative homogeneous matrix at revolute joint C is defined as

$$\mathbf{IT}_{BC} = \left[\begin{array}{c|c} \mathbf{D}_C & \mathbf{L}_{BC}^2 + \mathbf{0} \\ \hline 0 & 1 \end{array}\right] \tag{2.84f}$$

Node 3: Absolute homogeneous matrix for link 3 is determined from

$$\mathbf{R}_3 = \mathbf{IT}_{OA} * \mathbf{IT}_{AB} * \mathbf{IT}_{BC} = \mathbf{R}_2 * \mathbf{IT}_{BC} = \left[\begin{array}{c|c} \mathbf{D}_3 & \mathbf{r}_C \\ \hline 0 & 1 \end{array}\right] \tag{2.84g}$$

Output: From Equation (2.84g), the orientation matrix $\mathbf{D}_3$ and position vector $\mathbf{r}_C$ are identified.

…*Edge Z*: Relative homogeneous matrix at prismatic joint Z is defined as

$$\mathbf{IT}_{YZ} = \left[\begin{array}{c|c} \mathbf{D}_Z & \mathbf{L}_{YZ}^{m-1} \\ \hline 0 & 1 \end{array}\right] = \left[\begin{array}{c|c} \mathbf{D}_Z & \mathbf{L}_{YZ_{m-1}}^{m-1} + \mathbf{d}_Z^{m-1} \\ \hline 0 & 1 \end{array}\right] \tag{2.84h}$$

Node m: Absolute homogeneous matrix for link m is determined from

$$\mathbf{R}_m = \mathbf{R}_{m-1} * \mathbf{IT}_{YZ} = \left[\begin{array}{c|c} \mathbf{D}_m & \mathbf{r}_Z \\ \hline 0 & 1 \end{array}\right] \tag{2.84i}$$

Output: From Equation (2.84i), the orientation matrix $\mathbf{D}_m$ and position vector $\mathbf{r}_Z$ are identified.

2.14.5 End-Effector Absolute Homogeneous Matrix

Edge E: Relative homogeneous matrix at joint E is defined as

$$\mathbf{IT}_{ZE} = \left[\begin{array}{c|c} \mathbf{D}_E & \mathbf{L}_{ZE}^m + \mathbf{0} \\ \hline 0 & 1 \end{array}\right] = \left[\begin{array}{c|c} \mathbf{U} & \mathbf{L}_{ZE}^m \\ \hline 0 & 1 \end{array}\right] \tag{2.84j}$$

Equation (2.84j) defines a parallel axes translation from Z to E (both located on the same link m) by the unit matrix $\mathbf{U}$ with a null relative displacement vector (since there is a constant distance between Z and E):

$$\mathbf{D}_E = \mathbf{U}; \quad \mathbf{d}_E = 0 \tag{2.84k}$$

Node m&e: For the end-effector E located on link m, the absolute homogeneous matrix is

$$\mathbf{R}_e = \mathbf{IT}_{OA} * \mathbf{IT}_{AB} * \mathbf{IT}_{BC} * \cdots * \mathbf{IT}_{YZ} * \mathbf{IT}_{ZE} = \mathbf{R}_m * \mathbf{IT}_{ZE} = \left[\begin{array}{c|c} \mathbf{D}_e & \mathbf{r}_E \\ \hline 0 & 1 \end{array}\right] \quad (2.84l)$$

Output: From Equation (2.84l), the orientation matrix $\mathbf{D}_e$ and position vector $\mathbf{r}_E$ are identified.

Equation (2.84l), called the *mechanism's absolute homogeneous matrix*, is the mechanism's input–output relation for orientation and position which can be simplified to Equation (2.84m):

$$\mathbf{R}_e^{\text{mech}} = \mathbf{R}_e \quad (2.84m)$$

The superscript "mech" suggests that the orientation and position for the end-effector are determined from the mechanism's point of view (links and joints).

where

$$\left[\begin{array}{ccc|c} C\alpha_e^{I} & C\alpha_e^{II} & C\alpha_e^{III} & x_{OE}^{\text{mech}} \\ C\beta_e^{I} & \text{Cos}\,\beta_e^{II} & C\beta_e^{III} & y_{OE}^{\text{mech}} \\ C\gamma_e^{I} & C\gamma_e^{II} & C\gamma_e^{III} & z_{OE}^{\text{mech}} \\ \hline 0 & 0 & 0 & 1 \end{array}\right] = \left[\begin{array}{c|c} \mathbf{D}_e & \mathbf{r}_E \\ \hline 0 & 1 \end{array}\right]; \quad (2.84n)$$

The *input parameters* for direct analysis, the set of joints' displacements $q_M(t)$ shown in Equation (2.85), are the M actuating parameters which appear as the entries in the right-side matrix in Equation (2.84n):

$$q_M(t) = \left\{ \begin{array}{c} \theta_Z \\ d_Z \end{array} \right\} \quad (2.85)$$

The *output parameters*, the absolute orientation and position of the end-effector, form the set $(C\alpha_e^I, C\beta_e^I, C\gamma_e^I, \ldots, x_E, y_E, z_E)$, which appear as the entries in the left-side matrix in Equation (2.77n).

2.14.6 Example for Combined Equations on a 4 DOF Open Cycle Mechanism: Method of Relative Homogeneous Matrices

The TRRT mechanism, Figure 2.1, has m = 4 links, j = 4 joints, and mobility M = 4.

Data: The following data are provided for direct positional analysis by using homogeneous matrices.

- The vectors in the Latin matrix with respect to local frames:

$$\mathbf{L}^0_{OA_0} = \begin{Bmatrix} 5 \\ 0 \\ 0 \end{Bmatrix}; \quad \mathbf{L}^1_{AB} = \begin{Bmatrix} 0 \\ 3 \\ 0 \end{Bmatrix}; \quad \mathbf{L}^2_{BC} = \begin{Bmatrix} 0 \\ 25 \\ 0 \end{Bmatrix}; \quad \mathbf{L}^3_{CD_3} = \begin{Bmatrix} 6.5 \\ 0 \\ 0 \end{Bmatrix};$$

$$\mathbf{L}^4_{DE} = \begin{Bmatrix} 38 \\ 0 \\ 0 \end{Bmatrix}$$

- The unit vectors for linear displacements d_A and d_D at the prismatic joints A and D:

$$\mathbf{u}^0_A = \begin{Bmatrix} 1 \\ 0 \\ 0 \end{Bmatrix}; \quad \mathbf{u}^3_D = \begin{Bmatrix} 1 \\ 0 \\ 0 \end{Bmatrix}$$

- The relative constraint equations for revolute joints B and C and the end-effector E:

$$\mathbf{d}_B = \mathbf{d}_C = \mathbf{d}_E = \mathbf{0}$$

- The M = 4 actuation parameters: $d_A(t)$; $\theta_B(t)$; $d_C(t)$; $d_D(t)$
- The home position, shown Figure 2.2, is the initial position of the mechanism at t = 0 s, where all actuating positional parameters are zero:

$$d_A = 0 \text{ in.}; \quad \theta_B = 0°; \quad \theta_C = 0°; \quad d_D = 0 \text{ in.}$$

Objective: The following are determined by direct positional analysis:

1. The links' orientation matrices and frame position vectors at any time
2. The mechanism's absolute homogeneous matrix at any time
3. Equations for direct orientation and position analysis
4. The links' orientation and frame position vectors at the home position
5. The end-effector orientation and frame position vector at the home position

Solution:

1. The links' orientation matrices and frame position vectors as functions of time.

   We consider the path in the tree (Figure 2.22):

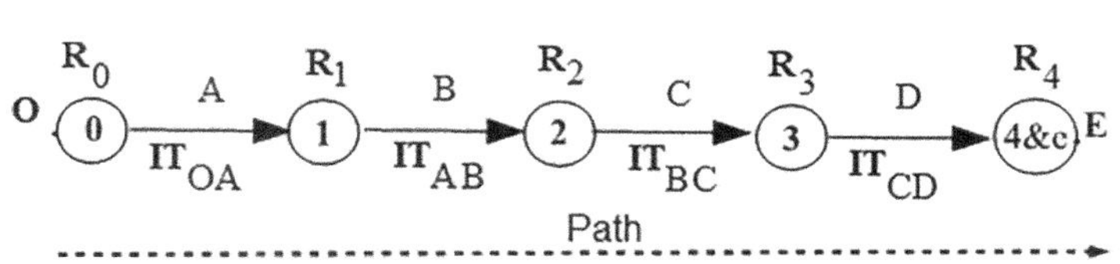

FIGURE 2.22
Absolute and relative homogeneous matrices along the tree for TRRT mechanism.

Node 0: Absolute homogeneous matrix for the fixed link 0 is the unit 4×4 matrix:

$$\mathbf{R}_0 = \left[\begin{array}{c|c} \mathbf{D}_0 & \mathbf{r}_O \\ \hline 0 & 1 \end{array}\right] = \mathbf{U}_{4\times4} = \left[\begin{array}{ccc|c} 1 & 0 & 0 & 0 \\ 0 & 1 & 0 & 0 \\ 0 & 0 & 1 & 0 \\ \hline 0 & 0 & 0 & 1 \end{array}\right] \tag{2.86a}$$

Edge A (prismatic): Relative homogeneous matrix at joint A is defined as

$$\mathbf{IT}_{OA} = \left[\begin{array}{c|c} \mathbf{D}_A & \mathbf{L}^0_{OA_0} + \mathbf{d}^0_A \\ \hline 0 & 1 \end{array}\right] = \left[\begin{array}{ccc|c} 1 & 0 & 0 & 5+d_A \\ 0 & 1 & 0 & 0 \\ 0 & 0 & 1 & 0 \\ \hline 0 & 0 & 0 & 1 \end{array}\right] \tag{2.86b}$$

Node 1: Absolute homogeneous matrix for link m = 1 is determined to be

$$\mathbf{R}_1 = \mathbf{R}_0 * \mathbf{IT}_{OA} = \left[\begin{array}{ccc|c} 1 & 0 & 0 & 5+d_A \\ 0 & 1 & 0 & 0 \\ 0 & 0 & 1 & 0 \\ \hline 0 & 0 & 0 & 1 \end{array}\right] \tag{2.86c}$$

Output: From Equation (2.86c), the orientation matrix $\mathbf{D}_1$ and position vector $\mathbf{r}_A$ are identified.
Edge B (revolute): Relative homogeneous matrix at joint B is defined as

$$\mathbf{IT}_{AB} = \left[\begin{array}{c|c} \mathbf{D}_B & \mathbf{L}^1_{AB} + \mathbf{0} \\ \hline 0 & 1 \end{array}\right] = \left[\begin{array}{ccc|c} C\theta_B & 0 & S\theta_B & 0 \\ 0 & 1 & 0 & 3 \\ -S\theta_B & 0 & C\theta_B & 0 \\ \hline 0 & 0 & 0 & 1 \end{array}\right] \tag{2.86d}$$

Node 2: Absolute homogeneous matrix for link m = 2 is determined to be

$$\mathbf{R}_2 = \mathbf{R}_1 * \mathbf{IT}_{AB} = \left[\begin{array}{ccc|c} C\theta_B & 0 & S\theta_B & 5+d_A \\ 0 & 1 & 0 & 3 \\ -S\theta_B & 0 & C\theta_B & 0 \\ \hline 0 & 0 & 0 & 1 \end{array}\right] \tag{2.86e}$$

Output: From Equation (2.86e), the orientation matrix $\mathbf{D}_2$ and position vector $\mathbf{r}_B$ are identified.

Edge C (revolute): Relative homogeneous matrix at joint C is defined as

$$\mathbf{IT}_{BC} = \left[\begin{array}{c|c} \mathbf{D}_C & \mathbf{L}_{BC}^2 + \mathbf{0} \\ \hline 0 & 1 \end{array}\right] = \left[\begin{array}{ccc|c} C\theta_C & -S\theta_C & 0 & 0 \\ S\theta_C & C\theta_C & 0 & 25 \\ 0 & 0 & 1 & 0 \\ \hline 0 & 0 & 0 & 1 \end{array}\right] \tag{2.86f}$$

Node 3: Absolute homogeneous matrix for link m = 3 is determined to be

$$\mathbf{R}_3 = \mathbf{R}_2 * \mathbf{IT}_{BC} = \left[\begin{array}{ccc|c} C\theta_B \cdot C\theta_C & -C\theta_B \cdot S\theta_C & S\theta_B & 5+d_A \\ S\theta_C & C\theta_C & 0 & 28 \\ -S\theta_B \cdot C\theta_C & S\theta_B \cdot S\theta_C & C\theta_B & 0 \\ \hline 0 & 0 & 0 & 1 \end{array}\right] \tag{2.86g}$$

Output: From Equation (2.86g), the orientation matrix $\mathbf{D}_3$ and position vector $\mathbf{r}_C$ are identified.

Edge D (prismatic): Relative homogeneous matrix at joint D is defined as

$$\mathbf{IT}_{CD} = \left[\begin{array}{c|c} \mathbf{D}_D & \mathbf{L}_{CD}^3 + \mathbf{d}_D^3 \\ \hline 0 & 1 \end{array}\right] = \left[\begin{array}{ccc|c} 1 & 0 & 0 & 6.5+d_D \\ 0 & 1 & 0 & 0 \\ 0 & 0 & 1 & 0 \\ \hline 0 & 0 & 0 & 1 \end{array}\right] \tag{2.86h}$$

Node 4: Absolute homogeneous matrix for link m = 4 is determined to be

$$\mathbf{R}_4 = \mathbf{R}_3 * \mathbf{IT}_{CD}$$

$$= \left[\begin{array}{ccc|c} C\theta_B \cdot C\theta_C & -C\theta_B \cdot S\theta_C & S\theta_B & 5+d_A+(6.5+d_D)\cdot C\theta_B \cdot C\theta_C \\ S\theta_C & C\theta_C & 0 & 28+(6.5+d_D)\cdot S\theta_C \\ -S\theta_B \cdot C\theta_C & S\theta_B \cdot S\theta_C & C\theta_B & -(6.5+d_D)\cdot S\theta_B \cdot C\theta_C \\ \hline 0 & 0 & 0 & 1 \end{array}\right] \tag{2.86i}$$

Output: From Equation (2.86i), the orientation mechanism's absolute homogeneous matrix at any time is identified.

Relative homogeneous matrix at origin E of the end-effector is defined as

$$\mathbf{IT}_{DE} = \left[\begin{array}{c|c} \mathbf{D}_E & \mathbf{L}^4_{DE} + \mathbf{0} \\ \hline 0 & 1 \end{array}\right] = \left[\begin{array}{ccc|c} 1 & 0 & 0 & 38 \\ 0 & 1 & 0 & 0 \\ 0 & 0 & 1 & 0 \\ \hline 0 & 0 & 0 & 1 \end{array}\right] \quad (2.86j)$$

where

- the orientation of frame E relative to frame D is defined by the unit matrix, $\mathbf{D}_E = \mathbf{U}$
- the translation by vector $\mathbf{L}^4_{DE}$ from origin D to E (both located on link 4) is a null displacement vector, $\mathbf{d}_E = \mathbf{0}$

End-effector E: Absolute homogeneous matrix for the end-effector is determined from

$$\mathbf{R}^{mech}_e = \mathbf{R}_4 * \mathbf{IT}_{DE}$$

$$= \left[\begin{array}{ccc|c} C\theta_B \cdot C\theta_C & -C\theta_B \cdot S\theta_C & S\theta_B & 5 + d_A + (44.5 + d_D) \cdot C\theta_B \cdot C\theta_C \\ S\theta_C & C\theta_C & 0 & 28 + (44.5 + d_D) \cdot S\theta_C \\ -S\theta_B \cdot C\theta_C & S\theta_B \cdot S\theta_C & C\theta_B & -(44.5 + d_D) \cdot S\theta_B \cdot C\theta_C \\ \hline 0 & 0 & 0 & 1 \end{array}\right]$$

(2.86k)

Output:

- Orientation matrix of the end-effector: $\mathbf{D}^{mech}_e = \mathbf{D}_e$

$$\begin{array}{c} \\ i_0 \\ j_0 \\ k_0 \end{array} \begin{array}{c} \begin{array}{ccc} \quad i_e \quad & \quad j_e \quad & \quad k_e \quad \end{array} \\ \left[\begin{array}{c|c|c} \text{Cos}\,\alpha^{I}_e & \text{Cos}\,\alpha^{II}_e & \text{Cos}\,\alpha^{III}_e \\ \text{Cos}\,\beta^{I}_e & \text{Cos}\,\beta^{II}_{0,e} & \text{Cos}\,\beta^{III}_e \\ \text{Cos}\,\gamma^{I}_e & \text{Cos}\,\gamma^{II}_e & \text{Cos}\,\gamma^{III}_e \end{array}\right] \end{array}$$

$$= \left[\begin{array}{c|c|c} C\theta_B \cdot C\theta_C & -C\theta_B \cdot S\theta_C & S\theta_B \\ S\theta_C & C\theta_C & 0 \\ -S\theta_B \cdot C\theta_C & S\theta_B \cdot S\theta_C & C\theta_B \end{array}\right] \quad (2.86l)$$

- The position vector of the end-effector's origin: $\mathbf{r}_E^{mech} = \mathbf{r}_E$

$$\begin{Bmatrix} x_E^{mech} \\ y_E^{mech} \\ z_E^{mech} \end{Bmatrix} = \begin{Bmatrix} 5 + d_A + (44.5 + d_D) \cdot C\theta_B \cdot C\theta_C \\ 28 + (44.5 + d_D) \cdot S\theta_C \\ -(44.5 + d_D) \cdot S\theta_B \cdot C\theta_C \end{Bmatrix} \tag{2.86m}$$

2. Equations for direct orientation and position analysis
   - Input actuating parameters: M = 4 angular and linear joint displacements within the matrix

$$q_M(t) = \{\ d_A \ \vdots\ \theta_B \ \vdots\ \theta_C \ \vdots\ d_D\ \}^T \tag{2.86n}$$

   - *Equations for orientation*:
     The entries in equation $\mathbf{D}_e^{mech} = \mathbf{D}_e$ are identified. The angles between the end-effector's frame and the fixed frame are shown in Equation (2.86o):

$$\begin{Bmatrix} \alpha_e^I \\ \beta_e^I \\ \gamma_e^I \end{Bmatrix} = \begin{Bmatrix} \text{Arccos}(C\theta_B \cdot C\theta_C) \\ \text{Arccos}(S\theta_C) \\ \text{Arccos}(-S\theta_B \cdot C\theta_C) \end{Bmatrix};$$

$$\begin{Bmatrix} \alpha_e^{II} \\ \beta_e^{II} \\ \gamma_e^{II} \end{Bmatrix} = \begin{Bmatrix} \text{Arccos}(-C\theta_B \cdot S\theta_C) \\ \theta_C \\ \text{Arccos}(S\theta_B \cdot S\theta_C) \end{Bmatrix}; \quad \begin{Bmatrix} \alpha_e^{III} \\ \beta_e^{III} \\ \gamma_e^{III} \end{Bmatrix} = \begin{Bmatrix} \text{Arccos}(S\theta_B) \\ 90^\circ \\ \theta_B \end{Bmatrix} \tag{2.86o}$$

     where the inverse trigonometric functions "Arccos" have values within interval 0°–180°.

   - Equations for position:
     From Equation (2.86m):

$$\begin{aligned} x_E^{mech} &= 5 + d_A + (44.5 + d_D) \cdot C\theta_B \cdot C\theta_C \\ y_E^{mech} &= 28 + (44.5 + d_D) \cdot S\theta_C \\ z_E^{mech} &= -(44.5 + d_D) \cdot S\theta_B \cdot C\theta_C \end{aligned} \tag{2.86p}$$

Note: The combined orientation and position equations obtained here from the symbolic multiplication of homogeneous matrices are equivalent to those developed independently in the orientation Equation (2.26m) and position Equation (2.56f).

3. The links' orientation and frame position vectors at the home position

   At the home position in Figure (2.2), the angular and linear displacements are zero:

$$d_A = 0 \text{ in.}; \quad \theta_B = 0°; \quad \theta_C = 0°; \quad d_D = 0 \text{ in.}$$

The above absolute $\mathbf{R}_m$ matrices, where m = 1, 2, 3, and 4, are shown in Equation (2.86q):

$$\mathbf{R}_1 = \left[\begin{array}{c|c} \mathbf{D}_1 & \mathbf{r}_A \\ \hline 0 & 1 \end{array}\right] = \left[\begin{array}{ccc|c} 1 & 0 & 0 & 5 \\ 0 & 1 & 0 & 0 \\ 0 & 0 & 1 & 0 \\ \hline 0 & 0 & 0 & 1 \end{array}\right];$$

$$\mathbf{R}_2 = \left[\begin{array}{c|c} \mathbf{D}_2 & \mathbf{r}_B \\ \hline 0 & 1 \end{array}\right] = \left[\begin{array}{ccc|c} 1 & 0 & 0 & 5 \\ 0 & 1 & 0 & 3 \\ 0 & 0 & 1 & 0 \\ \hline 0 & 0 & 0 & 1 \end{array}\right];$$

$$\mathbf{R}_3 = \left[\begin{array}{c|c} \mathbf{D}_3 & \mathbf{r}_C \\ \hline 0 & 1 \end{array}\right] = \left[\begin{array}{ccc|c} 1 & 0 & 0 & 5 \\ 0 & 1 & 0 & 28 \\ 0 & 0 & 1 & 0 \\ \hline 0 & 0 & 0 & 1 \end{array}\right]; \tag{2.86q}$$

$$\mathbf{R}_4 = \left[\begin{array}{c|c} \mathbf{D}_4 & \mathbf{r}_D \\ \hline 0 & 1 \end{array}\right] = \left[\begin{array}{ccc|c} 1 & 0 & 0 & 11.5 \\ 0 & 1 & 0 & 28 \\ 0 & 0 & 1 & 0 \\ \hline 0 & 0 & 0 & 1 \end{array}\right]$$

From Equation (2.86q), the orientation matrices are unit matrices: $\mathbf{D}_m = \mathbf{U}$, where m = 1, 2, 3, and 4. Therefore, all frames have axes parallel to the fixed frame axes, see Figure (2.2).

Identifying the entries within the last columns of the matrices in Equation (2.86q), then the position vectors are shown in Equation (2.86r):

$$\mathbf{r}_A = \begin{Bmatrix} x_A \\ y_A \\ z_A \end{Bmatrix} = \begin{Bmatrix} 5 \\ 0 \\ 0 \end{Bmatrix}; \quad \mathbf{r}_B = \begin{Bmatrix} x_B \\ y_B \\ z_B \end{Bmatrix} = \begin{Bmatrix} 5 \\ 3 \\ 0 \end{Bmatrix};$$

$$\mathbf{r}_C = \begin{Bmatrix} x_C \\ y_C \\ z_C \end{Bmatrix} = \begin{Bmatrix} 5 \\ 28 \\ 0 \end{Bmatrix}; \quad \mathbf{r}_D = \begin{Bmatrix} x_D \\ y_D \\ z_D \end{Bmatrix} = \begin{Bmatrix} 11.5 \\ 28 \\ 0 \end{Bmatrix} \tag{2.86r}$$

4. The end-effector orientation and its frame position vector at the home position

   When all angular and linear displacements are zero, Equation (2.86k) becomes

$$\mathbf{R}_e = \left[\begin{array}{c|c} \mathbf{D}_4 & \mathbf{r}_E \\ \hline 0 & 1 \end{array}\right] = \left[\begin{array}{ccc|c} 1 & 0 & 0 & 49.5 \\ 0 & 1 & 0 & 28 \\ 0 & 0 & 1 & 0 \\ \hline 0 & 0 & 0 & 1 \end{array}\right] \tag{2.86s}$$

   Therefore, $\mathbf{D}_4 = \mathbf{U}$, and the mechanism at the home position has the end-effector's frame axes parallel to the fixed frame axes, see Figure (2.2). By identifying the entries within the last columns in Equation (2.86s), then is solved for

$$\mathbf{r}_E = \begin{Bmatrix} 49.5 \\ 28 \\ 0 \end{Bmatrix} = \begin{Bmatrix} x_E \\ y_E \\ z_E \end{Bmatrix} \tag{2.86t}$$

   The end-effector coordinates correspond with those of point E in Figure (2.2).

2.15 Inverse Orientation and Positional Combined Equations along a Closed Path: The Homogeneous Matrix Method

The open path starts at node 0 and ends at node m. If we imagine the path to continue to the initial node 0, then the path is closed and called a cycle (Figure 2.23).

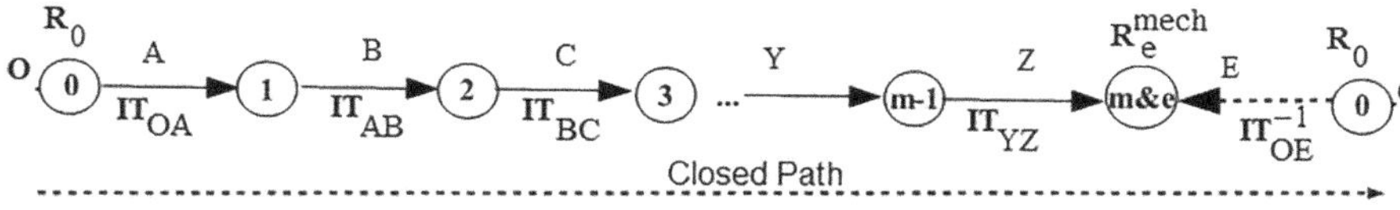

FIGURE 2.23
Relation between absolute and relative homogeneous matrices along the closed path.

2.15.1 The Direct and Inverse Sign of Relative IT Matrices

The procedure to generate closed paths is shown in Section 1.3.13. A superscript "–" sign is assigned to an edge which is contrary to the cycle's direction and a "+" if the edge is in the path's direction. The set of **D** relative orientation matrices are assigned to edges and are related by Equation (2.30). Similarly, a set of **IT** relative homogeneous matrices are assigned to the edges within a closed path and are related by their multiplication in Equation (2.87).

Statement 2.9

The result of post multiplications of relative homogeneous matrices from all edges and a chord along a closed path is the 4 × 4 unit matrix. The entries in the cycle basis matrix C are the superscripts for the relative matrices in the matrix product. Thus, "+1" denotes a direct matrix and "–1" denotes an inverse matrix:

$$\mathbf{IT}_{OA}^{+1}\Big|_A * \mathbf{IT}_{AB}^{+1}\Big|_B * \mathbf{IT}_{BC}^{+1}\Big|_C * \cdots * \mathbf{IT}_{YZ}^{+1}\Big|_Z * \mathbf{IT}_{ZE}\Big|_E * \mathbf{IT}_{OE}^{-1} = \mathbf{U}_{4\times4} \quad (2.87)$$

Note: Equation (2.87) is applicable for planar and spatial mechanisms, with single or multiple cycles. In the case of multiple cycles, one matrix equation is written for each cycle.

2.15.2 The Inverse of Homogeneous Matrix

The fact that there are edges with opposite signs in the path results in assigning an inverse homogeneous matrix to such an edge, defined in Equation (2.88):

$$\mathbf{R}_m^{-1} = \left[\begin{array}{c|c} \mathbf{D}_m^T & -\mathbf{D}_m^T * \mathbf{r}_Z \\ \hline 0 & 1 \end{array}\right]; \quad \mathbf{IT}_{YZ}^{-1} = \left[\begin{array}{c|c} \mathbf{D}_Z^T & -\mathbf{D}_Z^T * \mathbf{L}_{YZ}^{m-1} \\ \hline 0 & 1 \end{array}\right] \quad (2.88)$$

Proof: The proof is based on the product between the homogeneous matrix and its inverse:

$$\mathbf{R}_m * \mathbf{R}_m^{-1} = \mathbf{U}_{4\times4}; \quad \mathbf{IT}_{YZ} * \mathbf{IT}_{YZ}^{-1} = \mathbf{U}_{4\times4}$$

Note:

- The use of the transposed matrix to determine an inverse matrix simplifies the calculation.
- Equation (2.88) for inverse matrices is valid for homogeneous matrices, since their calculation is based on the property that the inverse of a rotation matrix **D** is the same as the transposed matrix.

Thus, the **D** matrices are orthogonal, and Equation (2.88) applies for computation of inverse matrices, although they do not apply in general for any 4×4 matrices.

2.15.3 The Task Absolute Homogeneous Matrix

In Equation (2.87), the product of the factors on the left side of the vertical line $|_E$ is the matrix $\mathbf{R}_e^{mech}(q)$. Its entries are a function of the manipulator's joints' coordinates $q_M(t)$. This homogeneous matrix is the *absolute homogeneous matrix from the mechanism's point of view.*

Equation (2.87) holds:

$$\mathbf{R}_e^{mech}(q) * \mathbf{IT}_{OE}^{-1} = \mathbf{U} \tag{2.89}$$

If we post-multiply by $\mathbf{IT}_{OE}$ on both sides, then Equation (2.89) becomes Equation (2.90):

$$\mathbf{R}_e^{mech}(q) = \mathbf{IT}_{OE}^{+1} \tag{2.90}$$

This suggests that within an inverse path, initiated at the right-side node 0 moving toward the node m&e, edge E is a direct edge. Therefore, $\mathbf{IT}_{OE}^{+1}$ is a direct matrix with the assigned superscript +1, see Figure 2.24.

This matrix, which shows how the end-effector is oriented from the fixed frame, is the *absolute homogeneous matrix* from the *task point of view.*

The task $\mathbf{R}_e^{task}$ matrix has constant numerical entries that are derived from the CAD model or image, if a camera is implemented:

$$\mathbf{R}_e^{task} = \mathbf{IT}_{OE}^{+1} \tag{2.91}$$

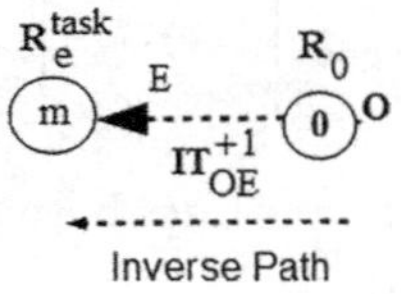

FIGURE 2.24
Task matrix for the inverse path.

Considering Equations (2.90) and (2.91), then:

$$\mathbf{R}_e^{mech}(q) = \mathbf{R}_e^{task} \tag{2.92}$$

Equation (2.92) is viewed as the *desired state* of the manipulator in reaching a *task* orientation and position.

In the case where the entries in the left-side matrix (mechanism) are the same as the right-side matrix entries (desired information obtained from CAD drawings or vision), there is no need to adjust the orientation/position of the end-effector (controller correction is not needed).

Equation (2.92) is used further in robot programming, where an overall task is achieved in steps, called *states of the joint coordinates* $q_M(t)$. Each state is modeled by solving for Equation (2.92), which is equivalent to having the end-effector reaching the desired orientations and positions.

An alternative approach to recording the entries in $\mathbf{R}_e^{mech}(q)$ is the approach of *teaching-by-doing*. When taught, the end-effector is given several points in which its orientations and positions are recorded as strings of data for joint coordinates $q_M(t)$.

2.15.4 Example for Orientation and Positional Analysis of a 4 DOF Robotic Mechanism without Vision: Introduction to Robot Programming

Previously in the text, the positional analysis for joints, links, and their connecting equations on the digraph was developed from the manipulator's point of view. The end-effector achieves a *task* when it is part of a *workstation*, which integrates additional components. An example of a task is to pick the part from a pallet, magazine, or conveyor line and place it in a hole at the target location.

Next, the interdependence between the end-effector and the other components in the workstation is modeled using a matrix which could be with or without vision (camera).

The end-effector is brought to several locations such as home position, part grasp, part approach, target reach, and target approach. IT matrices are defined for these locations, which have entries that are constant over time, as illustrated in the next example. For direct analysis, the left- and right-side entries are compared in Equation (2.92).

Example 1: Simulation for the direct positional analysis of a TRRT mechanism using homogeneous matrices. The M = 4 input actuation parameters, shown as a column matrix $\mathbf{q}_M$, with chosen stepwise linear functions in time (i.e., constant velocity actuation). The simulation time is within the interval $0 \le t \le 4$ s.

Vision is not implemented, and the joint displacements at each state are shown in Equation (2.57b), reproduced here as Equations (2.93) and (2.94):

$$d_A(t)=\begin{Bmatrix} 60\cdot t & 0\le t\le 1 \\ 60 & 1<t\le 2 \\ 60-40\cdot(t-2) & 2<t\le 3 \\ 20-20\cdot(t-3) & 3<t\le 4 \end{Bmatrix}[\text{in.}];$$

$$\theta_B(t)=\begin{Bmatrix} 0 & 0\le t\le 1 \\ 0 & 1<t\le 2 \\ 90\cdot(t-2) & 2<t\le 3 \\ 90 & 3<t\le 4 \end{Bmatrix}[°]; \tag{2.93}$$

$$\theta_C(t)=\begin{Bmatrix} 0 & 0\le t\le 1 \\ 0 & 1<t\le 2 \\ 90\cdot(t-2) & 2<t\le 3 \\ 90 & 3<t\le 4 \end{Bmatrix}[°];$$

$$d_D(t)=\begin{Bmatrix} 0 & 0\le t\le 1 \\ 20\cdot(t-1) & 1<t\le 2 \\ 20 & 2<t\le 3 \\ 20 & 3<t\le 4 \end{Bmatrix}[\text{in.}] \tag{2.94}$$

For each state:

1. Derive the absolute homogeneous matrix for the mechanism's point of view $\mathbf{R}_e^{mech}$ (Equation (2.86k)).
2. Derive the task $\mathbf{R}_e^{task}$ matrix with constant entries from the CAD model (Equation (2.91)).
3. Check the state of the manipulator when it reaches a provided task orientation and position from $\mathbf{R}_e^{mech}(q)=\mathbf{R}_e^{task}$ (Equation (2.92)).
4. Write symbolic instructions to be used for computer programming (the programming language is chosen by the reader).

Solution: The homogeneous matrices are determined for each state as follows:

- *The home-position state*
  1. Derive the absolute homogeneous matrix for mechanism's point of view $\mathbf{R}_e^{mech}$.

     The end-effector is brought to the home position E^h (Figure 2.25).

     At t = 0 s, the joint coordinate state $\mathbf{q}^h$, with all actuation displacements equal to zero, is

$$\mathbf{q}^h=\{\, d_A \mid \theta_B \mid \theta_C \mid d_D \,\}^T=\{\, 0 \quad 0 \quad 0 \quad 0 \,\}^T \tag{2.95}$$

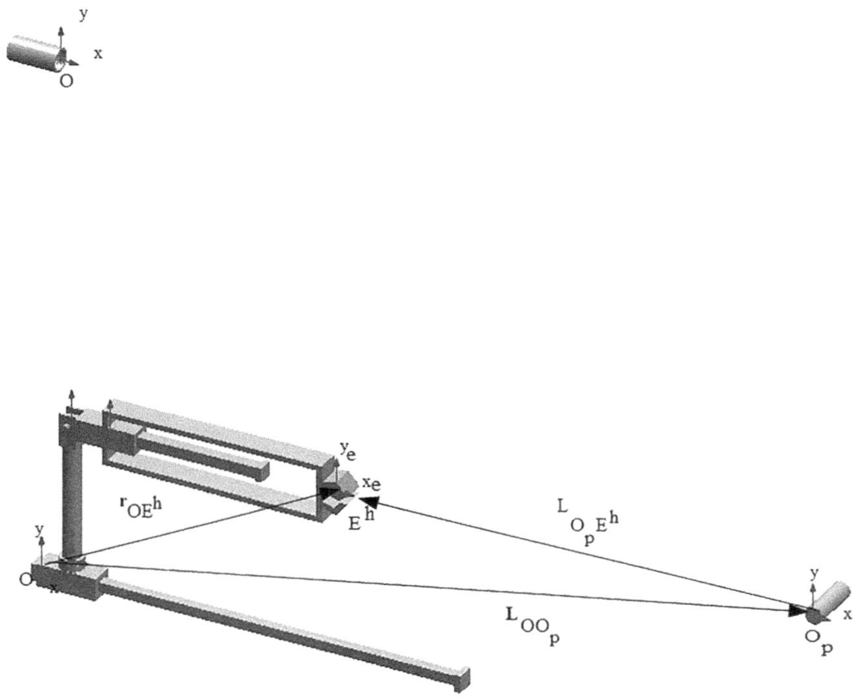

FIGURE 2.25
Location of the end-effector's frame at the home position.

From the mechanism's point of view, the desired matrix of reaching the home position $\mathbf{R}_e^{mech}(q^h)$ is obtained by assigning q^h in Equation (2.86k):

$$\mathbf{R}_e^{mech}(q^h) = \left[\begin{array}{ccc|c} 1 & 0 & 0 & 49.5 \\ 0 & 1 & 0 & 28 \\ 0 & 0 & 1 & 0 \\ \hline 0 & 0 & 0 & 1 \end{array}\right] \quad (2.96)$$

2. Derive the constant entries' task matrix $\mathbf{R}_e^{task}$ from the CAD model.

   The *part* p to be grasped (e.g., a cylindrical pin) is located on the fixed link 0 and has its own frame O_p. In the digraph, the part is represented by node 0&p. The end-effector is brought to home position E^h, and the superscript "task" is replaced by "h" (Figure 2.26).

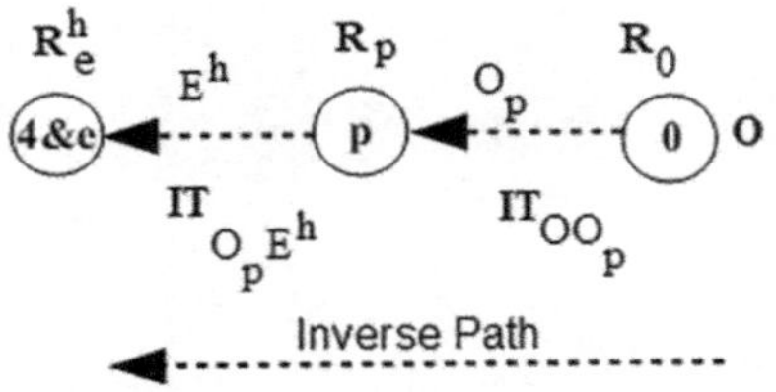

FIGURE 2.26
Digraph for home position.

Equation (2.91) is written along the inverse path as

$$\mathbf{R}_e^h = \mathbf{IT}_{OE^h} = \mathbf{IT}_{OO_p} * \mathbf{IT}_{O_pE^h} = \left[\begin{array}{ccc|c} 1 & 0 & 0 & 49.5 \\ 0 & 1 & 0 & 28 \\ 0 & 0 & 1 & 0 \\ \hline 0 & 0 & 0 & 1 \end{array}\right] \tag{2.97}$$

The two matrices in the above product are determined from the CAD model:

– The matrix $\mathbf{IT}_{OO_p}$ shows the orientation and position of the part's frame O_p as viewed in the global frame O:

$$\mathbf{IT}_{OO_p} = \left[\begin{array}{c|c} \mathbf{D}_{O_p} & \mathbf{L}^0_{OO_p} \\ \hline 0 & 1 \end{array}\right] = \left[\begin{array}{ccc|c} 1 & 0 & 0 & 129.5 \\ 0 & 1 & 0 & 28 \\ 0 & 0 & 1 & 0 \\ \hline 0 & 0 & 0 & 1 \end{array}\right] \tag{2.98}$$

This matrix has constant entries and is used for *calibration of the part position*. In the case of repeated part grasping, it should be redefined with the location of each new part.

– The matrix $\mathbf{IT}_{O_pE^h}$ shows the orientation and position of the end-effector's frame E^h as viewed in part's frame O_p (Equation (2.99)):

$$\mathbf{IT}_{O_pE^h} = \left[\begin{array}{c|c} \mathbf{D}_E & \mathbf{L}^p_{O_pE^h} \\ \hline 0 & 1 \end{array}\right] = \left[\begin{array}{ccc|c} 1 & 0 & 0 & -80 \\ 0 & 1 & 0 & 0 \\ 0 & 0 & 1 & 0 \\ \hline 0 & 0 & 0 & 1 \end{array}\right] \tag{2.99}$$

3. Check the state of the manipulator when it reaches a provided task orientation and position from $\mathbf{R}_e^{mech}(q) = \mathbf{R}_e^{task}$ (Equation (2.92)).

The two matrices in Equations (2.97) and (2.96) have identical entries; therefore, Equation (2.92) is verified.

4. The next step would be to write software program functions which would be used for controller/computer programming. The programming language is chosen by the reader. Symbols such as 'READ', 'MOVE', 'GRASP', 'RELEASE' will be provided to guide the reader. The instruction in a program when the manipulator is located at home position is symbolically written as

$$\text{READ}: \mathbf{R}^{h}_{0,e}; \quad \text{Entries from Equation (2.96) when the end-effector is at home state} \tag{2.100}$$

- *The Part-Approach State*

1. Derive the absolute homogeneous matrix for mechanism's point of view $\mathbf{R}_e^{mech}$.

   The end-effector is brought to the part-approach position E^{pa} (Figure 2.27).

   At t = 1 s, the joint coordinate state $\mathbf{q}^{pa}$ is

$$\mathbf{q}^{pa} = \left\{ \begin{array}{c:c:c:c} d_A & \theta_B & \theta_C & d_D \end{array} \right\}^T = \left\{ \begin{array}{cccc} 60'' & 0^\circ & 0^\circ & 0'' \end{array} \right\}^T \tag{2.101}$$

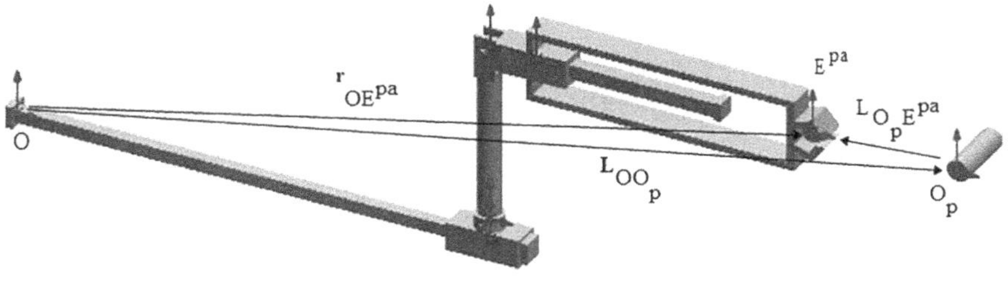

FIGURE 2.27
Location of end-effector's frame at the part-approach position.

From the manipulator's point of view, the desired matrix of reachingthe part-approach position is obtained by assigning q^{pa} in Equation (2.86k):

$$\mathbf{R}_e^{mech}\left(q^{pa}\right)=\left[\begin{array}{ccc|c} 1 & 0 & 0 & 109.5 \\ 0 & 1 & 0 & 28 \\ 0 & 0 & 1 & 0 \\ \hline 0 & 0 & 0 & 1 \end{array}\right] \quad (2.102)$$

2. Derive the task matrix $\mathbf{R}_e^{task}$ with constant entries from the CAD model (Equation (2.91)).
   The end-effector is moved to the part-approach position, noted E^{pa}, and the superscript "task" is replaced by "pa," (Figure 2.28).
   Equation (2.91) is written along the inverse path as task matrix:

$$\mathbf{R}_e^{pa}=\mathbf{IT}_{OE^{pa}}=\mathbf{IT}_{OO^p} * \mathbf{IT}_{O_pE^{pa}}=\left[\begin{array}{ccc|c} 1 & 0 & 0 & 109.5 \\ 0 & 1 & 0 & 28 \\ 0 & 0 & 1 & 0 \\ \hline 0 & 0 & 0 & 1 \end{array}\right] \quad (2.103)$$

The two matrices in the above product are determined from the CAD model:

- The matrix $\mathbf{IT}_{OO_p}$ describes the orientation and position of the part's frame O_p as viewed in the global frame O. This matrix $\mathbf{IT}_{OO_p}$ was previously derived in Equation (2.98) and can seen in Equation (2.104):

$$\mathbf{IT}_{OO_p}=\left[\begin{array}{c|c} \mathbf{D}_{O_p} & \mathbf{L}_{OO_p}^0 \\ \hline 0 & 1 \end{array}\right]=\left[\begin{array}{ccc|c} 1 & 0 & 0 & 129.5 \\ 0 & 1 & 0 & 28 \\ 0 & 0 & 1 & 0 \\ \hline 0 & 0 & 0 & 1 \end{array}\right] \quad (2.104)$$

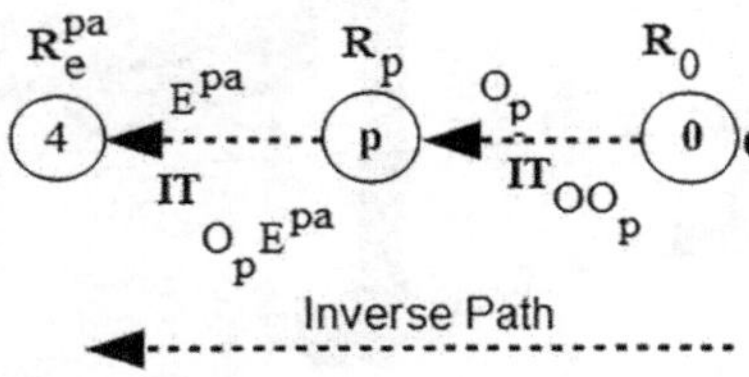

FIGURE 2.28
Part-approach position.

The matrix $\mathbf{IT}_{O_pE^{pa}}$ describes the orientation and position of the end-effector's frame E^{pa} as viewed in the part's frame O_p. This matrix $\mathbf{IT}_{O_pE^{pa}}$ can be seen in Equation (2.105):

$$\mathbf{IT}_{O_pE^{pa}} = \left[\begin{array}{c|c} \mathbf{D}_E & \mathbf{L}^p_{O_pE^{pa}} \\ \hline 0 & 1 \end{array}\right] = \left[\begin{array}{ccc|c} 1 & 0 & 0 & -20 \\ 0 & 1 & 0 & 0 \\ 0 & 0 & 1 & 0 \\ \hline 0 & 0 & 0 & 1 \end{array}\right] \quad (2.105)$$

3. Check the state of the manipulator when it reaches a provided task orientation and position from $\mathbf{R}_e^{mech}(q) = \mathbf{R}_e^{task}$ (Equation (2.92)).

   The two matrices in Equations (2.102) and (2.103) have identical entries; therefore, Equation (2.92) is verified.
4. Write a software function to be used for computer programming.

   The software function in a program for when the manipulator moves from home state to part approach is symbolically written as

   Software symbol for motion is MOVE: (2.106)

   Solve Equation (2.102) for the entries in $\mathbf{R}_e^{mech}\left(q^{pa}\right)$. The end-effector moves from the home-state with $\mathbf{R}_e^h$ (Equation (2.96)) to the part-approach state with $\mathbf{R}_e^{pa}$ (Equation (2.103)).

- *The part-grasped state*

1. Derive the absolute homogeneous matrix for the mechanism's point of view $\mathbf{R}_e^{mech}$ (Equation (2.86k)).

   The end-effector is brought to the part in the part-grasped position E^{pg} (Figure 2.29).

   At t = 2 s, the joint coordinate state $\mathbf{q}^{pg}$ is

$$\mathbf{q}^{pg} = \left\{ \begin{array}{c|c|c|c} d_A & \theta_B & \theta_C & d_D \end{array} \right\}^T = \left\{ \begin{array}{cccc} 60'' & 0^\circ & 0^\circ & 20'' \end{array} \right\}^T \quad (2.107)$$

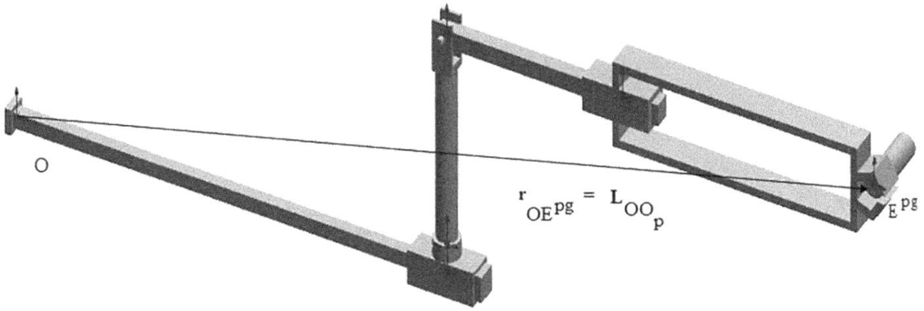

FIGURE 2.29
Location of end-effector's frame at the part-grasped position.

From the manipulator's point of view, the desired matrix of reaching the part-grasped position is obtained by assigning q^{pg} in Equation (2.86k):

$$\mathbf{R}_e^{mech}\left(q^{pg}\right)=\left[\begin{array}{ccc|c}1 & 0 & 0 & 129.5\\ 0 & 1 & 0 & 28\\ 0 & 0 & 1 & 0\\ \hline 0 & 0 & 0 & 1\end{array}\right] \tag{2.108}$$

2. Derive the task matrix $\mathbf{R}_e^{task}$ with constant entries from the CAD model (Equation (2.91)).

   The end-effector is moved to the part-grasped position, noted E^{pg}, and the superscript "task" is replaced by "pg," (Figure 2.30).

   Equation (2.91) is written along the inverse path as task matrix:

$$\mathbf{R}_e^{pg}=\mathbf{IT}_{OE^{pg}}=\mathbf{IT}_{OO_p}*\mathbf{IT}_{O_pE^{pg}}=\left[\begin{array}{ccc|c}1 & 0 & 0 & 129.5\\ 0 & 1 & 0 & 28\\ 0 & 0 & 1 & 0\\ \hline 0 & 0 & 0 & 1\end{array}\right] \tag{2.109}$$

The two matrices in the above product are determined from the CAD model:

- The matrix $\mathbf{IT}_{OO_p}$ describes the orientation and position of part's frame, O_p, as viewed in the global frame O. This matrix $\mathbf{IT}_{OO_p}$ was computed before in Equation (2.98):

$$\mathbf{IT}_{OO_p}=\left[\begin{array}{c|c}\mathbf{D}_{O_p} & \mathbf{L}_{OO_p}^0\\ \hline 0 & 1\end{array}\right]=\left[\begin{array}{ccc|c}1 & 0 & 0 & 129.5\\ 0 & 1 & 0 & 28\\ 0 & 0 & 1 & 0\\ \hline 0 & 0 & 0 & 1\end{array}\right] \tag{2.110}$$

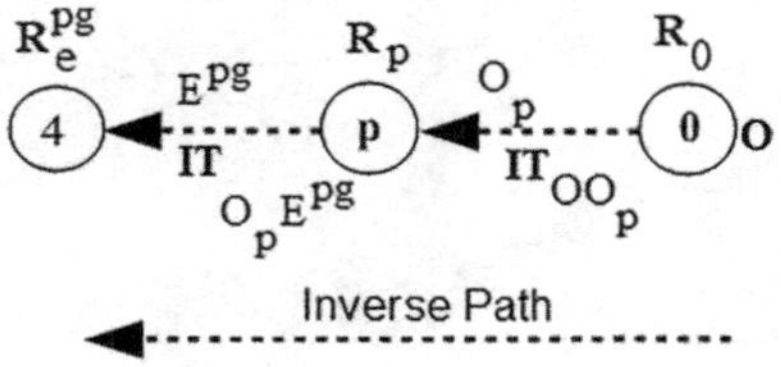

FIGURE 2.30
Part-grasped position.

– The matrix $\mathbf{IT}_{O_pE^{pg}}$ shows the orientation and position of the end-effector's frame, E^{pg}, as viewed in the part's frame O_p (Equation (2.111)):

$$\mathbf{IT}_{O_pE^{pg}} = \left[\begin{array}{c|c} \mathbf{D}_E & \mathbf{L}^p_{O_pE} \\ \hline 0 & 1 \end{array}\right] = \left[\begin{array}{ccc|c} 1 & 0 & 0 & 0 \\ 0 & 1 & 0 & 0 \\ 0 & 0 & 1 & 0 \\ \hline 0 & 0 & 0 & 1 \end{array}\right] \tag{2.111}$$

Its entries are defined as follows:

– the origin of the end-effector's frame is chosen to be coincident to the part's frame origin, $\mathbf{L}^p_{O_pE} = 0$.

– the two frame axes (x_e, y_e, z_e) and (x_p, y_p, z_p) are chosen to be oriented the same, $\mathbf{D}_E = \mathbf{U}$.

3. Check the state of the manipulator when it reaches a provided task orientation and position from $\mathbf{R}_e^{mech}(q) = \mathbf{R}_e^{task}$ (Equation (2.92)).

   The two matrices in Equations (2.108) and (2.109) have identical entries; therefore, Equation (2.92) is verified.

4. Write a software function to be used for computer programming.

   The software function in a program for when the manipulator moves from home state to part approach is symbolically written as

   Software symbol for motion is MOVE: (2.112)

   Solve Equation (2.108) for the entries in $\mathbf{R}_e^{mech}\left(q^{pg}\right)$. This command moves the end-effector from the part-approach state with $\mathbf{R}_e^{pa}$ (Equation (2.103)) to the part-grasped state with $\mathbf{R}_e^{pg}$ (Equation (2.109)).

   The next instruction in the program is to close the gripper, and as a result, the part becomes attached to the end-effector:

   Software symbol for closing the gripper is GRASP:
   Close the gripper located on the end-effector to grasp the part (2.113)

- *The target-approach state*

1. Derive the absolute homogeneous matrix for the mechanism's point of view $\mathbf{R}_e^{mech}$ (Equation (2.86k)).

   The target *t* where the part is to be placed (e.g., a cylindrical hole) is located on the fixed link 0 and has its own frame O_t.

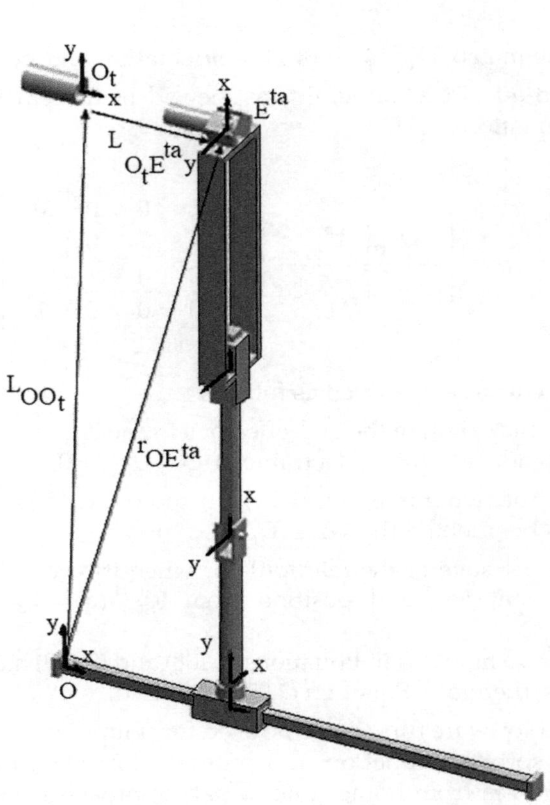

FIGURE 2.31
Location of end-effector at the target-approach position.

The end-effector is brought to the target-approach position $\mathbf{E}_e^{ta}$ (Figure 2.31).

At t = 3 s, the joint coordinate state $\mathbf{q}^{ta}$ is

$$\mathbf{q}^{ta} = \left\{ \begin{array}{c:c:c:c} d_A & \theta_B & \theta_C & d_D \end{array} \right\}^T = \left\{ \begin{array}{cccc} 20'' & 90^\circ & 90^\circ & 20'' \end{array} \right\}^T \quad (2.114)$$

From the manipulator's point of view, the desired matrix of reaching the target-approach position is obtained by assigning q^{ta} in Equation (2.86k):

$$\mathbf{R}_e^{mech}\left(q^{ta}\right) = \left[\begin{array}{ccc:c} 0 & 0 & 1 & 25 \\ 1 & 0 & 0 & 92.5 \\ 0 & 1 & 0 & 0 \\ \hdashline 0 & 0 & 0 & 1 \end{array} \right] \quad (2.115)$$

2. Derive the task matrix $\mathbf{R}_e^{task}$ with entries from the CAD model (Equation (2.92)).

   In the digraph, the target is represented by the frame O_t on link 0. The procedure for defining the relationship between the end-effector and the target is the same as before, where the symbol p is replaced by t.

   The end-effector is moved to the target-approach position, noted E^{ta}.

   Equation (2.91) is written along the inverse path as the task matrix:

$$\mathbf{R}_e^{ta} = \mathbf{IT}_{OE^{ta}} = \mathbf{IT}_{OO_t} * \mathbf{IT}_{O_tE^{ta}} = \left[\begin{array}{ccc|c} 0 & 0 & 1 & 25 \\ 1 & 0 & 0 & 92.5 \\ 0 & 1 & 0 & 0 \\ \hline 0 & 0 & 0 & 1 \end{array}\right] \quad (2.116)$$

The two matrices in the above product are determined from the CAD model:

- The matrix $\mathbf{IT}_{OO_t}$ that describes the orientation and position of the target's frame, O_t, as represented in the global frame O is

$$\mathbf{IT}_{OO_t} \left[\begin{array}{c|c} \mathbf{D}_{O_t} & \mathbf{L}^0_{OO_t} \\ \hline 0 & 1 \end{array}\right] = \left[\begin{array}{ccc|c} 1 & 0 & 0 & 3 \\ 0 & 1 & 0 & 92.5 \\ 0 & 0 & 1 & 0 \\ \hline 0 & 0 & 0 & 1 \end{array}\right] \quad (2.117)$$

  This matrix has constant entries and is used for *calibration of the target position.*

- The matrix $\mathbf{IT}_{O_tE^{ta}}$ describes the orientation and position of the end-effector's frame E^{ta}, as viewed in the target's frame O_t (Equation (2.118)):

$$\mathbf{IT}_{O_tE^{ta}} = \left[\begin{array}{c|c} \mathbf{D}_E & \mathbf{L}^t_{O_tE^{ta}} \\ \hline 0 & 1 \end{array}\right] = \begin{array}{c} \\ x_t \\ y_t \\ z_t \\ \\ \end{array} \begin{array}{c} \begin{array}{ccc} i_e & j_e & k_e \end{array} \\ \left[\begin{array}{ccc|c} C90^\circ & C90^\circ & C0^\circ & 22 \\ C0^\circ & C90^\circ & C90^\circ & 0 \\ C90^\circ & C0^\circ & C90^\circ & 0 \\ \hline 0 & 0 & 0 & 1 \end{array}\right] \end{array}$$

$$= \left[\begin{array}{ccc|c} 0 & 0 & 1 & 22 \\ 1 & 0 & 0 & 0 \\ 0 & 1 & 0 & 0 \\ \hline 0 & 0 & 0 & 1 \end{array}\right] \quad (2.118)$$

It's entries are defined as follows:

- The origin of the end-effector's frame is chosen offset from the target's frame origin:

$$x_{O_tE^{ta}} = 22''; \quad y_{O_tE^{ta}} = 0''; \quad z_{O_tE^{ta}} = 0''$$

The end-effector axes (x_e, y_e, z_e) are rotated relative to the frame (x_t, y_t, z_t), as shown below in the rotation matrix $\mathbf{D}_E$. From Figure 2.31, one could observe that

$$x_e \perp x_t; x_e \parallel y_t; x_e \perp z_t; \quad y_e \perp x_t; y_e \perp y_t; y_e \parallel z_t;$$

$$z_e \parallel x_t; z_e \perp y_t; z_e \perp z_t$$

Based on $z_e \parallel x_t$, one could notice that the z-axis of the cylindrical part (grasped in the end-effector) is parallel to the x-axis of the cylindrical hole at the target location.

3. Check the state of the manipulator when it reaches a provided task orientation and position from $\mathbf{R}_e^{mech}(q) = \mathbf{R}_e^{task}$ (Equation (2.92)).

   The two matrices in Equations (2.115) and (2.116) have identical entries; therefore, Equation (2.92) is verified.

4. Write software program functions to be used for computer programming.

   The software function in a program for when the manipulator moves from the part-grasped state to the target-approach state is symbolically written as

$$\text{Software symbol for motion is MOVE:} \tag{2.119}$$

Solve Equation (2.115) for the entries in $\mathbf{R}_e^{mech}\left(q^{ta}\right)$. The end-effector moves from the part-grasped state with $\mathbf{R}_e^{pg}$ (Equation (2.109)) to the target-approach state with $\mathbf{R}_e^{ta}$ (Equation (2.116)).

- *The target-reached state*

  In the target-reached position, noted E^{tr}, the end-effector is brought over the target (Figure 2.32).

  At t = 4 s, the joint coordinate state $\mathbf{q}^{tr}$ is

$$\mathbf{q}^{tr} = \left\{ \; d_A \;\vdots\; \theta_B \;\vdots\; \theta_C \;\vdots\; d_D \; \right\}^T = \left\{ \; 0'' \quad 90^\circ \quad 90^\circ \quad 20'' \; \right\}^T \tag{2.120}$$

From the manipulator's point of view, the desired matrix in reaching the target-reached position is obtained by assigning q^{tr} in Equation (2.86k):

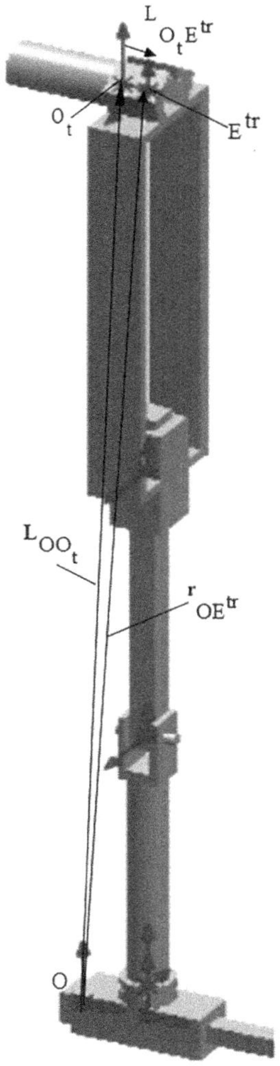

FIGURE 2.32
Location of end-effector at the target-reached position.

$$\mathbf{R}_{e}^{mech}\left(q^{tr}\right)=\left[\begin{array}{ccc|c} 0 & 0 & 1 & 5 \\ 1 & 0 & 0 & 92.5 \\ 0 & 1 & 0 & 0 \\ \hline 0 & 0 & 0 & 1 \end{array}\right] \tag{2.121}$$

Equation (2.91) is written along the inverse path as task matrix:

$$\mathbf{R}_e^{tr} = \mathbf{IT}_{OE^{tr}} = \mathbf{IT}_{OO_t} * \mathbf{IT}_{O_tE^{tr}} = \left[\begin{array}{ccc|c} 0 & 0 & 1 & 5 \\ 1 & 0 & 0 & 92.5 \\ 0 & 1 & 0 & 0 \\ \hline 0 & 0 & 0 & 1 \end{array}\right] \tag{2.122}$$

- The matrix $\mathbf{IT}_{OO_t}$ describes the orientation and position of the part's frame, O_t, as viewed in the global frame O. This matrix was computed before from Equation (2.117) and can be seen in Equation (2.123):

$$\mathbf{IT}_{OO_t} = \left[\begin{array}{c|c} \mathbf{D}_{O_t} & \mathbf{L}_{OO_t}^{0} \\ \hline 0 & 1 \end{array}\right] = \left[\begin{array}{ccc|c} 1 & 0 & 0 & 3 \\ 0 & 1 & 0 & 92.5 \\ 0 & 0 & 1 & 0 \\ \hline 0 & 0 & 0 & 1 \end{array}\right] \tag{2.123}$$

- The matrix $\mathbf{IT}_{O_tE^{tr}}$ shows the orientation and position of the end-effector's frame E^{tr} as viewed in the target's frame O_t. It is defined as follows in Equation (2.124):
- The end-effector's frame origin is offset from the target's frame origin, which is a design requirement inserting a pin into a hole:

$$x_{O_tE^{tr}} = 2''; y_{O_tE^{tr}} = 0''; z_{O_tE^{tr}} = 0''$$

- The end-effector axes (x_e, y_e, z_e) are rotated relative to frame (x_t, y_t, z_t), as shown in the rotation matrix $\mathbf{D}_E$. One could notice that the z-axis of the cylindrical part is coincident with the x-axis of the cylindrical hole:

$$\mathbf{IT}_{O_tE^{tr}} = \left[\begin{array}{c|c} \mathbf{D}_E & \mathbf{L}_{O_tE^{tr}}^{t} \\ \hline 0 & 1 \end{array}\right] = \begin{array}{c} \\ x_t \\ y_t \\ z_t \\ \\ \end{array} \begin{array}{c} \begin{array}{ccc} \quad i_e & \quad j_e & \quad k_e \end{array} \\ \left[\begin{array}{ccc|c} C90^\circ & C90^\circ & C0^\circ & 2 \\ C0^\circ & C90^\circ & C90^\circ & 0 \\ C90^\circ & C0^\circ & C90^\circ & 0 \\ \hline 0 & 0 & 0 & 1 \end{array}\right] \end{array}$$

$$= \left[\begin{array}{ccc|c} 0 & 0 & 1 & 2 \\ 1 & 0 & 0 & 0 \\ 0 & 1 & 0 & 0 \\ \hline 0 & 0 & 0 & 1 \end{array}\right] \tag{2.124}$$

1. Check the state of the manipulator when it reaches a provided task orientation and position from $\mathbf{R}_e^{mech}(q) = \mathbf{R}_e^{task}$ (Equation (2.92)).

   The two matrices in Equations (2.121) and (2.122) have identical entries; therefore, Equation (2.92) is verified.

2. Write software program functions to be used for computer programming.

   In a program, the command for when the manipulator moves from the target-approach state to the target-reached state is symbolically written as

$$\text{Software symbol for motion is MOVE:} \tag{2.125}$$

Solve Equation (2.121) for the entries in $\mathbf{R}_e^{man}\left(q^{tr}\right)$. The end-effector moves from the target-approach state with $\mathbf{R}_e^{ta}$ (Equation (2.116)) to the target-reached state with $\mathbf{R}_e^{tr}$ (Equation (2.122)).

Next, there is an instruction in the program to open the gripper, and as a result of this command, the part is left at the target location.

Software symbol for opening the gripper is:

$$\begin{array}{l}\text{RELEASE (open the gripper located on the}\\ \text{end-effector to release the part)}\end{array} \tag{2.126}$$

Note:

- When the robot approaches the part or the target, a controller compares the joint coordinates data: $q_M(t)$ calculated or stored, and desired q^{pa} or q^{ta}.
- If there is a difference between them, then the controller makes the correction(s) on the actuators to drive the end-effector to the right orientation and position.
- The mathematical modeling of each state consists of solving Equation (2.92) for the joint coordinate state $q_M(t)$ by using the entries in $\mathbf{R}_e^{mech}(q)$. The task homogeneous matrices $\mathbf{R}_e^{task}$ have constant entries, which are determined once and can be carried throughout the program.
- Two types of positional analysis are formulated: direct and inverse. Both types of analysis require solving Equation (2.92).

  The direct analysis is based on the given $q_M(t)$, which dictates the parameters in $\mathbf{R}_e^{task}$ (i.e., the pallet at the part-grasped position and the conveyor at the target-reached position).

  The inverse analysis consists in solving Equation (2.92) for $q_M(t)$. Finding the solution requires solving a nonlinear system of equations.

2.15.5 Orientation and Positional Analysis of a Robotic Mechanism with Vision

The pick-and-place application refers to picking up the part from its location (magazine, pallet, or conveyor line) and moving the part toward the target location where it is inserted into an opening or onto a pallet, etc.

When vision is implemented in the workstation, the camera c, having its own frame origin at O_c, is attached to the fixed frame 0. By taking a picture of the part, the orientation and position of the part frame is determined, and in turn, the homogeneous matrix $\mathbf{IT}_{OO_p}$ is defined. Thus, the orientation and position of the part's frame O_p is determined as viewed in the global frame O. This method uses the input from the camera instead from the design specifications (CAD model) that were shown before in Equation (2.98):

$$\mathbf{IT}_{OO_p} = \left[\begin{array}{c|c} \mathbf{D}_{O_p} & \mathbf{L}^0_{OO_p} \\ \hline 0 & 1 \end{array}\right] \tag{2.127}$$

Modeling a manipulator with vision is achieved by the following procedure.

Since the camera is located on the fixed link 0, shown in the digraph as node c, where node c is located between the end-effector's node (m&e) and the part node p (Figure 2.33).

Based on the camera input, two matrices and the relation between them are defined next.

- The matrix $\mathbf{R}_c$ with the orientation and position of the camera's frame O_c as viewed in the global frame O in Equation (2.128) is defined by a translation of frame O_c and a rotation between the two frames:
- The origin of O_c is offset from the origin of O by the translation vector $\mathbf{r}^0_{O_c}$.
- The camera's axes (x_c, y_c, z_c) are viewed as rotated relative to the fixed frame (x_0, y_0, z_0) from the rotation matrix $\mathbf{D}_{O_c}$:

$$\mathbf{R}_c = \left[\begin{array}{c|c} \mathbf{D}_{O_c} & \mathbf{r}^0_{O_c} \\ \hline 0 & 1 \end{array}\right] \tag{2.128}$$

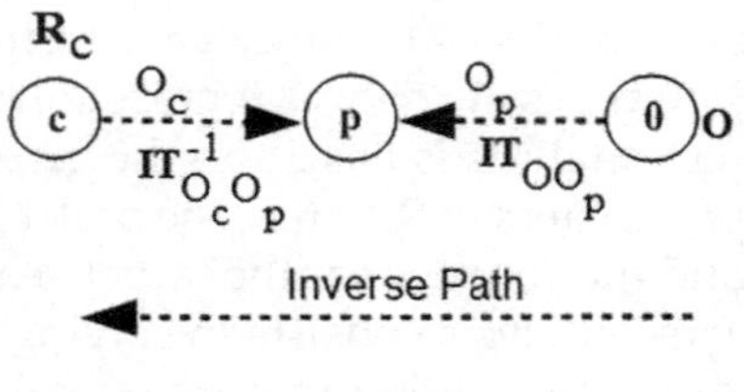

FIGURE 2.33
Part positioned from camera.

This matrix with constant entries is used for *calibration of the camera position.*

- The matrix $\mathbf{IT}_{O_cO_p}$, obtained after taking a picture of the part, shows the orientation and position of part's frame O_p as viewed in the camera's frame O_c, is defined as follows:

$$\mathbf{IT}_{O_cO_p} = \left[\begin{array}{c|c} \mathbf{D}_{O_p} & \mathbf{L}^{c}_{O_cO_p} \\ \hline 0 & 1 \end{array}\right] \tag{2.129}$$

- The origin O_p is offset from the origin O_c, as defined by translation vector $\mathbf{L}_{O_cO_p}$ with the provided components (Figure 2.30).
- The part's axes (O_p, x_p, y_p, z_p) are viewed as rotated relative to the camera's frame (O_c, x_c, y_c, z_c), as shown in the rotation matrix $\mathbf{D}_{O_p}$.

Along an inverse path, starting from right side of point O in the digraph, the Equation (2.91) is written as

$$\mathbf{R}_c = \mathbf{IT}_{OO_p} * \mathbf{IT}^{-1}_{O_cO_p} \tag{2.130}$$

The "–1" superscript indicates that it is the inverse matrix of $\mathbf{IT}_{O_cO_p}$, since the path's direction is from O_p to O_c, whereas the above matrix $\mathbf{IT}_{O_cO_p}$ is assigned to an edge with inverse orientation.

Equation (2.130) is post-multiplied by $\mathbf{IT}_{O_cO_p}$, which yields Equation (2.131):

$$\mathbf{IT}_{OO_p} = \mathbf{R}_c * \mathbf{IT}_{O_cO_p} \tag{2.131}$$

The information from the camera is read in the program as:

The command for coding

$$\text{READ: } \mathbf{IT}_{O_cO_p} \text{ from Equation } \mathbf{IT}_{OO_p} = \mathbf{R}_c * \mathbf{IT}_{O_cO_p} \tag{2.132}$$

2.15.6 Example for Orientation and Positional Analysis of a 4 DOF Robotic Mechanism with Vision

The TRRT mechanism with a camera included in the workstation is shown in Figure 2.34 (Animation 2.2).

The following matrices are defined from the camera input.

- The matrix $\mathbf{R}_c$ for the orientation and position of the camera's frame, O_c, as viewed in the global frame, O, is defined from
  - Translation of frame O_c: $x_{O_c} = 120''; y_{O_c} = 100''; z_{O_c} = -50''$

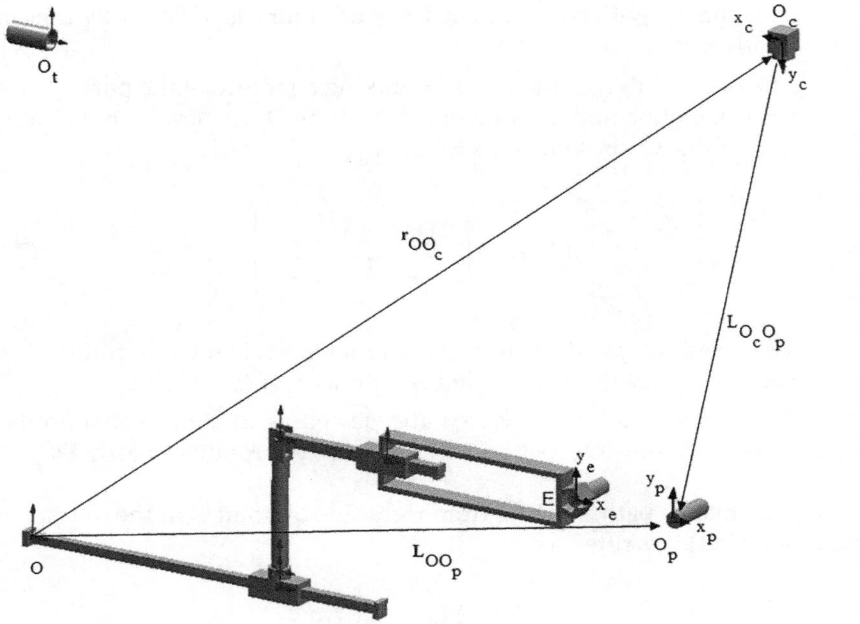

FIGURE 2.34
Part positioned from camera.

- Rotation between the two frames from the rotation matrix $\mathbf{D}_{O_c}$, with

$$x_c \uparrow\downarrow x_0; x_c \perp y_0; x_c \perp z_0, \quad y_c \perp x_0; y_c \uparrow\downarrow y_0;$$

$$y_c \perp z_0, \quad z_c \perp x_0; z_c \perp y_0; z_c \parallel z_0$$

where the notation ↑↓ indicates two antiparallel axes (i.e., 180° angle between them):

$$\mathbf{R}_c = \left[\begin{array}{c|c} \mathbf{D}_{O_c} & \mathbf{r}^0_{O_c} \\ \hline 0 & 1 \end{array}\right] = \begin{array}{c} \\ x_0 \\ y_0 \\ z_0 \\ \\ \end{array} \begin{array}{c} \begin{array}{ccc} i_c & j_c & k_c \end{array} \\ \left[\begin{array}{ccc|c} C180^\circ & C90^\circ & C90^\circ & 120 \\ C90^\circ & C180^\circ & C90^\circ & 100 \\ C90^\circ & C90^\circ & C0^\circ & -50 \\ \hline 0 & 0 & 0 & 1 \end{array}\right] \end{array} \quad (2.133)$$

$$= \left[\begin{array}{ccc|c} -1 & 0 & 0 & 120 \\ 0 & -1 & 0 & 100 \\ 0 & 0 & 1 & -50 \\ \hline 0 & 0 & 0 & 1 \end{array}\right]$$

This constant entries' matrix is used for *calibration of the camera position*.

- The matrix $\mathbf{IT}_{O_cO_p}$ obtained after taking a picture of the part shows the orientation and position of the part's frame, O_p, as viewed in the camera's frame O_c:
- The origin O_p is offset from the origin O_c by the translation vector $\mathbf{L}_{O_cO_p}$ with the components:

$$x^c_{O_cO_p} = -9.5'';y^c_{O_cO_p} = 72'';z^c_{O_cO_p} = 50''$$

- The part's axes (O_p, x_p, y_p, z_p) are viewed as rotated relative to the camera's frame (O_c, x_c, y_c, z_c), as shown below in the rotation matrix $\mathbf{D}_{O_p}$:

$$x_p \uparrow\downarrow x_c;x_p \perp y_c;x_p \perp z_c, \quad y_p \perp x_c;y_p \uparrow\downarrow y_c;y_p \perp z_c,$$

$$z_p \perp x_c;z_p \perp y_c;z_p \parallel z_c$$

$$\mathbf{IT}_{O_cO_p} = \left[\begin{array}{c:c} \mathbf{D}_{O_p} & \mathbf{L}^c_{O_cO_p} \\ \hdashline 0 & 1 \end{array}\right] = \begin{array}{c} \\ x_c \\ y_c \\ z_c \\ \\ \end{array} \begin{array}{c} \begin{array}{cccc} i_p & j_p & k_p & \end{array} \\ \left[\begin{array}{ccc:c} C180^\circ & C90^\circ & C90^\circ & -9.5 \\ C90^\circ & C180^\circ & C90^\circ & 72 \\ C90^\circ & C90^\circ & C0^\circ & 50 \\ \hdashline 0 & 0 & 0 & 1 \end{array}\right] \end{array}$$

$$= \left[\begin{array}{ccc:c} -1 & 0 & 0 & -9.5 \\ 0 & -1 & 0 & 72 \\ 0 & 0 & 1 & 50 \\ \hdashline 0 & 0 & 0 & 1 \end{array}\right] \tag{2.134}$$

Equation (2.131) becomes

$$\mathbf{IT}_{OO_p} = \mathbf{R}_c * \mathbf{IT}_{O_cO_p} = \left[\begin{array}{ccc:c} -1 & 0 & 0 & 120 \\ 0 & -1 & 0 & 100 \\ 0 & 0 & 1 & -50 \\ \hdashline 0 & 0 & 0 & 1 \end{array}\right] * \left[\begin{array}{ccc:c} -1 & 0 & 0 & -9.5 \\ 0 & -1 & 0 & 72 \\ 0 & 0 & 1 & 50 \\ \hdashline 0 & 0 & 0 & 1 \end{array}\right]$$

$$= \left[\begin{array}{ccc:c} 1 & 0 & 0 & 129.5 \\ 0 & 1 & 0 & 28 \\ 0 & 0 & 1 & 0 \\ \hdashline 0 & 0 & 0 & 1 \end{array}\right] \tag{2.135}$$

Note: The homogeneous matrix for the part location and orientation $\mathbf{IT}_{OO_p}$ coincides with that from Equation (2.98), without the camera in the workstation.

2.16 Direct Orientation and Positional Analysis for Planar Open Cycle Mechanisms

Specifically, for the 2D planar mechanisms, the motion of their links is within the same plane. Their relative rotation is about parallel axes which are all perpendicular to the plane. In the case of a 2D mechanism designed with a constant shift between links, along the axes of rotation, all links still move within parallel planes.

2.16.1 Governing Equation for Links' Orientation for Planar Open Cycle Mechanisms

For planar mechanisms, the last row and last column in relative and absolute rotation matrices are dropped. All entries in the last row and the last column do not change during matrix multiplications because they are all rotations about z-axes. The link m rotates about axis z_m; therefore, Equation (2.98) holds:

$$\mathbf{D}_m = \begin{bmatrix} C\theta_m & -S\theta_m & 0 \\ S\theta_m & C\theta_m & 0 \\ 0 & 0 & 1 \end{bmatrix} \Rightarrow \mathbf{D}_m = \begin{bmatrix} C\theta_m & -S\theta_m \\ S\theta_m & C\theta_m \end{bmatrix} \tag{2.136}$$

Equation (2.20) considers the trigonometric formula for the sum of two angles when performing the multiplication of two planar rotation matrices. Therefore,

$$\mathbf{D}_m = \mathbf{D}_{m-1} * \mathbf{D}_Z = \begin{bmatrix} C\theta_{m-1} & -S\theta_{m-1} \\ S\theta_{m-1} & C\theta_{m-1} \end{bmatrix} * \begin{bmatrix} C\theta_Z & -S\theta_Z \\ S\theta_Z & C\theta_Z \end{bmatrix}$$
$$= \begin{bmatrix} C(\theta_{m-1}+\theta_Z) & -S(\theta_{m-1}+\theta_Z) \\ S(\theta_{m-1}+\theta_Z) & C(\theta_{m-1}+\theta_Z) \end{bmatrix} \tag{2.137}$$

From Statement 2.2 and Equation (2.20), the absolute matrix at node m is determined as a post multiplication of all relative matrices of the edges located before node m:

$$\mathbf{D}_m = \mathbf{D}_A|_A * \mathbf{D}_B|_B * \mathbf{D}_C|_C * \cdots * \mathbf{D}_Z|_Z = \mathbf{D}_{m-1} * \mathbf{D}_Z \tag{2.138}$$

Thus, to determine $\mathbf{D}_1$, the term on the left side of the bracket $|_A$ is considered. To determine $\mathbf{D}_2$, all the terms on the left side of bracket $|_B$ are considered.

Considering Equations (2.137) and (2.138), then Equation (2.139) is validated, see Statement 2.10.

Statement 2.10

The absolute angular displacement at node (link) m is the sum of the relative angular displacements at the edges (joints) located before node m (Equation (2.139)):

$$\theta_m = \theta_A|_A + \theta_B|_B + \theta_C|_C + \cdots + \theta_Z|_Z = \theta_{m-1} + \theta_Z \tag{2.139}$$

Equation (2.139) is the *governing equation for a link's orientation* for planar open cycle mechanisms.

Thus, to determine θ_1, the term on the left side of bracket $|_A$ is considered. To determine θ_2, all the terms on the left side of bracket $|_B$ are considered.

2.16.2 Relations between Absolute and Relative Angular Displacements

Equation (2.139) is recurrently written along a path starting at node 0 and ending at node m:

$$\begin{array}{ll} \theta_1 = \theta_A; & \text{Path ending at node m} = 1 \\ \theta_2 = \theta_A + \theta_B; & \text{Path ending at edge A} \\ \theta_3 = \theta_A + \theta_B + \theta_C; & \text{Path ending at edge B} \\ \theta_m = \theta_A + \theta_B + \theta_C + \cdots + \theta_Z; & \text{Path ending at edge Z} \end{array} \tag{2.140}$$

The components of Equation (2.140) are shown in matrix form in Equation (2.141):

$$\overbrace{\left\{\begin{array}{c} \theta_1 \\ \theta_2 \\ \theta_3 \\ \cdots \\ \theta_m \end{array}\right\}}^{\theta_m} = \left[\begin{array}{ccccc} 1 & 0 & 0 & 0 & 0 \\ 1 & 1 & 0 & 0 & 0 \\ 1 & 1 & 1 & 0 & 0 \\ \cdots & \cdots & \cdots & \cdots & \cdots \\ 1 & 1 & 1 & 1 & 1 \end{array}\right] * \overbrace{\left\{\begin{array}{c} \theta_A \\ \theta_B \\ \theta_C \\ \cdots \\ \theta_Z \end{array}\right\}}^{\theta_a} \tag{2.141}$$

The links' absolute angular displacements are in the leftmost column, noted $\boldsymbol{\theta}_m$. The joints' relative angular displacements are in the rightmost column, noted θ_a. The coefficients' matrix is the n × m square matrix that contains only ones and zeros; this can be said because in the digraph, the number of arcs a is the same as the number of nodes m.

The transposed path matrix $\mathbf{Z}^T$ can be implemented for an automatic determination of the coefficients in Equation (2.141), which will be discussed in Section 2.16.3.

2.16.3 The Path Matrix and Its Transposed, $\mathbf{Z}^T$

In Equation (1.17), the *path matrix* Z for a digraph is defined as a square a × m matrix whose entries are the numbers +1, –1, or 0. These numbers are assigned to edges present and not present in a path initiated at node 0 and ending at node m. Thus,

- +1 is for the edge present in the path and in the path direction
- –1 is for the edge present in the path and in the opposite direction
- 0 is for edges not in the path.

For the general digraph in Figure 2.7, Equation (2.142) shows the path matrix and its transposed maxtrix:

Nodes digraph (mobile links)

$$\mathbf{Z} = \begin{array}{c} \begin{array}{ccccc} 1 & 2 & 3 & \cdots & m \end{array} \\ \left[\begin{array}{c|c|c|c|c} 1 & 1 & 1 & \cdots & 1 \\ 0 & 1 & 1 & \cdots & 1 \\ 0 & 0 & 1 & \cdots & 1 \\ 0 & 0 & 0 & \cdots & 1 \\ 0 & 0 & 0 & \cdots & 1 \end{array}\right] \end{array} \begin{array}{l} \\ \text{A} \quad \text{Arcs} \\ \text{B} \quad \text{in.} \\ \text{C} \quad \text{Path}; \\ \cdots \quad \downarrow \\ \text{Z} \end{array} \quad \mathbf{Z}^T = \begin{array}{c} \begin{array}{ccccc} A & B & C & \cdots & Z \end{array} \\ \left[\begin{array}{ccccc} 1 & 0 & 0 & \cdots & 0 \\ \hline 1 & 1 & 0 & \cdots & 0 \\ \hline 1 & 1 & 1 & \cdots & 0 \\ \hline \cdots & \cdots & \cdots & \cdots & 0 \\ \hline 1 & 1 & 1 & \cdots & 0 \end{array}\right] \end{array} \begin{array}{c} \\ 1 \\ 2 \\ 3 \\ \cdots \\ m \end{array} \tag{2.142}$$

One could notice that the coefficients' matrix in Equation (2.141) is the transposed matrix of the path matrix: $\mathbf{Z}^T$

Statement 2.11

For any planar mechanism, the absolute and relative angular displacements are related using Equation (2.143):

$$\theta_m = \mathbf{Z}^T * \theta_a \tag{2.143}$$

Note: The matrix $\mathbf{Z}^T$ is used further in the velocity analyses of spatial and planar mechanisms with single or multiple cycles [4,5].

2.16.4 Example of Simulation for Planar Open Cycle Manipulator with 3 DOF

The mechanism in Figure 2.35 (Animation 2.3) illustrates the planar TRT mechanism, which possesses one revolute joint and two prismatic joints: A (prismatic), B (revolute), and C (prismatic). Link 3 has a point of interest E, the end-effector.

2.16.4.1 Link and Joint Labeling

The links are labeled with numbers, starting with 0 which is assigned to the fixed link. There are three mobile links, m = 3, which have been assigned the numbers 1, 2, and 3. There are three joints, which are labeled with capital letters: A, B, and C. All joints are 1 DOF; therefore, $j_1 = 3$. There are no joints with 2 DOF, so $j_2 = 0$.

The total number of joints is

$$j = j_1 + j_2 = 3 + 0 = 3 \text{ joints} \tag{2.144}$$

2.16.4.2 Home Position of Mechanism

Figure 2.36 shows the mechanism in the initial configuration. The linear displacements are measured from the home position. Point A_0 on link 0 and

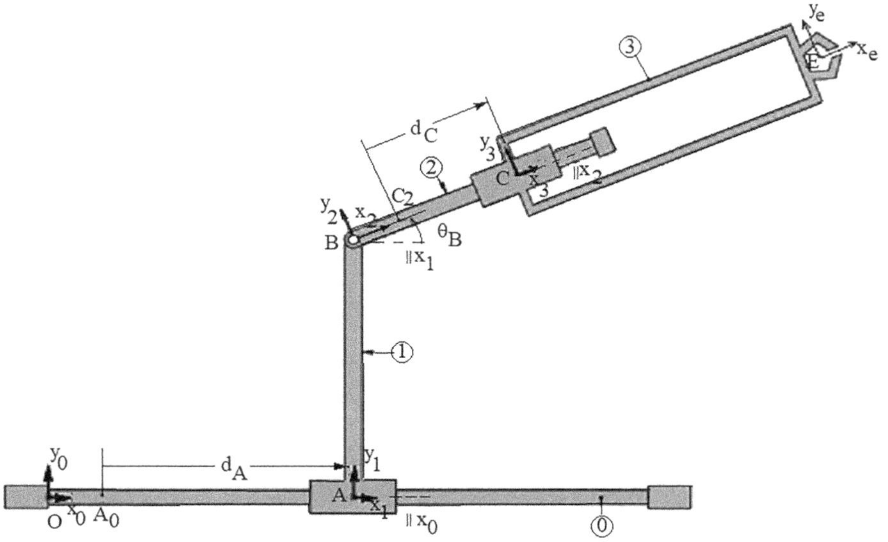

FIGURE 2.35
The planar TRT mechanisms and the IT notation for relative displacements.

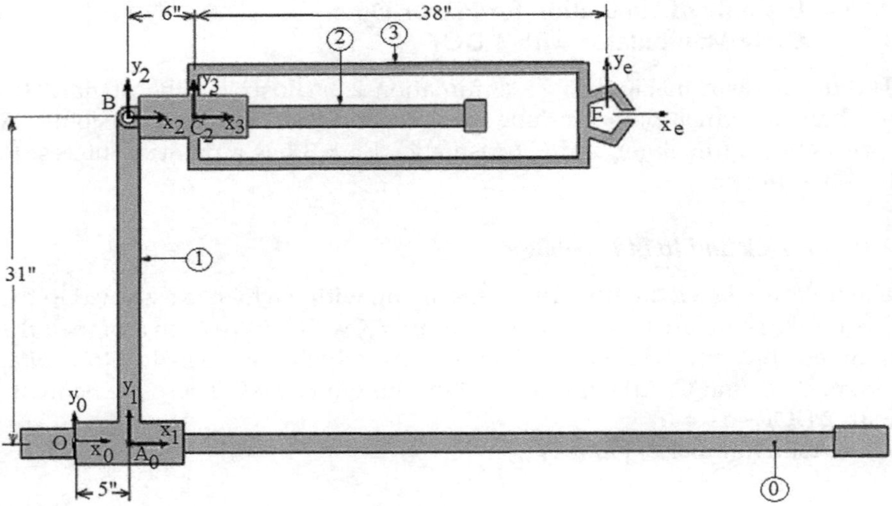

FIGURE 2.36
Home position.

point C_2 on link 2 define the initial positions of sliders 1 and 3. Based on the CAD model, the constant lengths are

- $L_{OA_0} = 5''$ is the location of slider 1 at home position A_0
- $L_{AB} = 31''$ is the constant length of vector L_{AB} between the two points A and B located on link 1
- $L_{BC_2} = 6''$ is the location of slider 3 at home position C_2
- $L_{CE} = 38''$ is the constant length of link 3, between the two points C and E located on link 3

The home position for the orientation of link 2 is at 0° (measured from the fixed x_0-axis).

2.16.4.3 Mechanism's Digraph for Open Cycle is a Spanning Tree

The number of cycles is determined from Equation (2.145):

$$c = j - m = 4 - 4 = 0 \text{ (no cycles).} \tag{2.145}$$

Since there are no cycles, the digraph is a spanning tree. In general, it is characteristic for the digraphs of open cycle mechanisms to be spanning trees. The digraph attached to the mechanism is illustrated in Figure 2.37a,

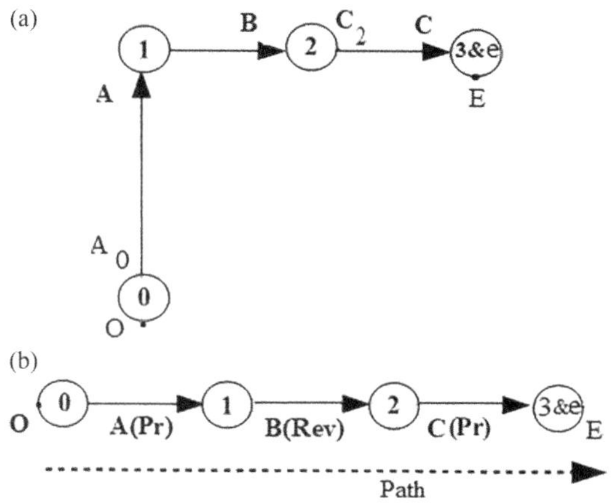

FIGURE 2.37
(a) Spanning tree; (b) path.

where its $n = m + 1 = 4$ nodes correspond to the mechanism's links and $j = 3$ joints correspond to the mechanism's edges.

The digraph's edges are oriented from the lower node number to the higher node number.

For further orientation and positional analyses, the path along the tree (sequence of nodes and edges) is shown in Figure 2.37b.

2.16.4.4 Notation for Frames Based on Digraph's Nodes

Figure 2.36 illustrates the frames and their origins as follows:

O is the origin of global frame (O, x_0, y_0, z_0)
A is the origin of frame (A, x_1, y_1, z_1), attached to link 1
B is the origin of frame (B, x_2, y_2, z_2), attached to link 2
C is the origin of frame (C, x_3, y_3, z_3), attached to link 3

2.16.4.5 Mechanism's Mobility

For planar mechanisms, the mobility M can be predicted with the Gruebler formula (Equation (1.28)):

$$M = 3 \cdot m - 2 \cdot j_1 - j_2 = 3 \cdot 3 - 2 \cdot 3 - 0 = 3 \tag{2.146}$$

Therefore, three motors are required for actuation in time t. Figure 2.38 shows the M = 3 parameters that are variable in time: $d_A(t)$ [in.], $\theta_B(t)$ [deg],

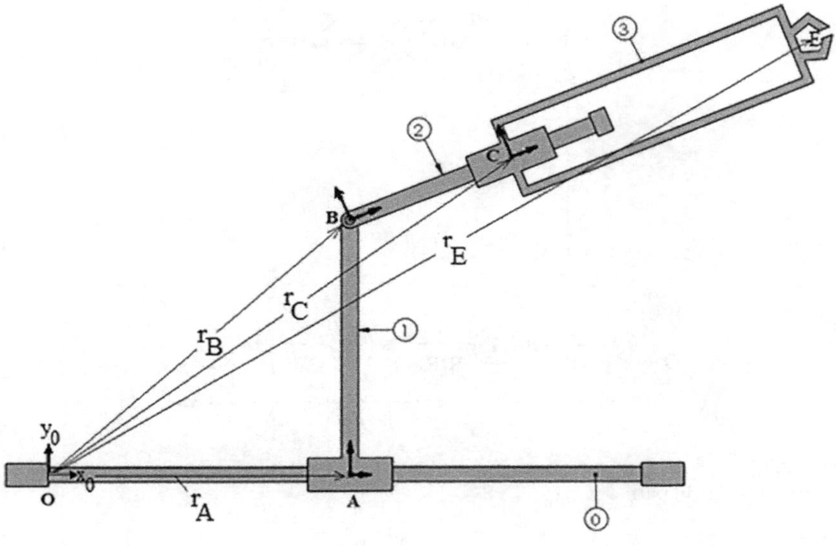

FIGURE 2.38
Position vectors.

and $d_C(t)$ [in.]. The three actuating displacements are chosen as quadratic functions of time within the time interval $0 \le t \le 1$ s:

$$q_M(t) = \begin{Bmatrix} d_A(t) \\ \theta_B(t) \\ d_C(t) \end{Bmatrix} = \begin{Bmatrix} 60 \cdot t^2 \\ 270 \cdot t^2 \\ 20 \cdot t^2 \end{Bmatrix} \tag{2.147}$$

The above actuating displacements labeled with the IT notation are as follows:

- $d_A(t)$ is the linear displacement at A of link 1 displaced relative to the fixed link 0 and measured from home position A_0.
- $\theta_B(t)$ is the angular displacement at B of link 2 rotated relative to link 1. Figure 2.35 illustrates this angle between x_2 and $||x_1$ (parallel to x_1, built at B), positive from $||x_1$ toward x_2.
- $d_C(t)$ is the linear displacement at C of link 3 displaced relative to link 2 and measured from home position C_2.

2.16.4.6 Constraint Equations for Angular and Linear Displacements

Three null displacement equations are expressed by the constraint Equation (2.148):

$$\begin{aligned}
&\theta_A = 0°; && \text{joint A is prismatic, then there is null relative rotation} \\
& && \text{between link 1 and link 0} \\
&d_B = 0''; && \text{joint B is prismatic, then there is null relative linear} \\
& && \text{displacement between link 2 and link 1} \\
&\theta_C = 0°; && \text{joint C is prismatic, then there is null relative rotation} \\
& && \text{between link 3 and link 2}
\end{aligned} \tag{2.148}$$

2.16.4.7 Relation between Absolute and Relative Angular Displacements

Equation (2.143) is recurrently written along a path initiated at node 0 and ending at node 3 for the planar TRT mechanism:

$$\boldsymbol{\theta}_m = \mathbf{Z}^T * \boldsymbol{\theta}_a$$

$$\begin{Bmatrix} \theta_1 \\ \theta_2 \\ \theta_3 \end{Bmatrix} = \begin{bmatrix} 1 & 0 & 0 \\ 1 & 1 & 0 \\ 1 & 1 & 1 \end{bmatrix} * \begin{Bmatrix} \theta_A \\ \theta_B \\ \theta_C \end{Bmatrix} \tag{2.149}$$

where the path matrix $\mathbf{Z}$ and its transposed matrix $\mathbf{Z}^T$ are

$$\mathbf{Z} = \begin{array}{c} \begin{array}{ccc} 1 & 2 & 3 \end{array} \\ \left[\begin{array}{c|c|c} 1 & 1 & 1 \\ 0 & 1 & 1 \\ 0 & 0 & 1 \end{array}\right] \end{array} \begin{array}{c} \\ A \\ B \\ C \end{array}; \quad \mathbf{Z}^T = \begin{array}{c} \begin{array}{ccc} A & B & C \end{array} \\ \left[\begin{array}{ccc} 1 & 0 & 0 \\ \hline 1 & 1 & 0 \\ \hline 1 & 1 & 1 \end{array}\right] \end{array} \begin{array}{c} \\ 1 \\ 2 \\ 3 \end{array} \tag{2.150}$$

The mechanism's relative angular displacement matrix $\boldsymbol{\theta}_a$, considering the joints' constraints from Equation (2.148) is

$$\boldsymbol{\theta}_a = \begin{Bmatrix} \theta_A \\ \theta_B \\ \theta_C \end{Bmatrix} = \begin{Bmatrix} 0 \\ \theta_B(t) \\ 0 \end{Bmatrix} \tag{2.151}$$

where a denotes the joint labels: A, B, and C.

The links' absolute angular displacement $\boldsymbol{\theta}_m$ determined from Equation (2.149) is

$$\boldsymbol{\theta}_m = \begin{Bmatrix} \theta_1 \\ \theta_2 \\ \theta_3 \end{Bmatrix} = \begin{Bmatrix} 0 \\ \theta_B(t) \\ \theta_B(t) \end{Bmatrix} \tag{2.152}$$

2.16.4.8 The Relative and Absolute Rotation Matrices

- *Edge A (prismatic joint)*: There is a translation of the frame (A, x_1, y_1, z_1) with respect to frame (O, x_0, y_0, z_0); therefore, a constraint rotation, $\theta_A = 0$:

$$\mathbf{D}_A = \left[\begin{array}{c|c} 1 & 0 \\ 0 & 1 \end{array}\right] = \mathbf{U}_{3\times 3} \tag{2.153a}$$

- *Node m = 1*: The orientation of link 1's frame (A, x_1, y_1, z_1) with respect to the fixed frame results from its absolute matrix (Equation (2.20)):

$$\mathbf{D}_1 = \mathbf{D}_A = \begin{array}{c} \\ i_0 \\ j_0 \end{array} \begin{array}{c} \begin{array}{cc} i_1 & j_1 \end{array} \\ \left[\begin{array}{c|c} 1 & 0 \\ 0 & 1 \end{array}\right] \end{array} \tag{2.153b}$$

- *Edge B (revolute joint)*: There is a relative rotation of the frame (B, x_2, y_2, z_2) about z_2 with respect to frame (A, x_1, y_1, z_1) (Equation (2.11)):

$$\mathbf{D}_B = \left[\begin{array}{c|c} C\theta_B & -S\theta_B \\ S\theta_B & C\theta_B \end{array}\right] \tag{2.153c}$$

- *Node m = 2*: The orientation of link 2's frame (B, x_2, y_2, z_2) with respect to the fixed frame results from its absolute matrix (Equation (2.138)):

$$\mathbf{D}_2 = \mathbf{D}_A * \mathbf{D}_B = \begin{array}{c} \\ i_0 \\ j_0 \end{array} \begin{array}{c} \begin{array}{cc} i_2 & j_2 \end{array} \\ \left[\begin{array}{c|c} C(0^\circ + \theta_B) & -S(0^\circ + \theta_B) \\ S(0^\circ + \theta_B) & C(0^\circ + \theta_B) \end{array}\right] \end{array} = \left[\begin{array}{c|c} C\theta_B & -S\theta_B \\ S\theta_B & C\theta_B \end{array}\right] \tag{2.153d}$$

- *Edge C (prismatic joint)*: There is a translation of the frame (C, x_3, y_3, z_3) with respect to frame (B, x_2, y_2, z_2); therefore, a constraint null rotation, $\theta_C = 0$:

$$\mathbf{D}_C = \left[\begin{array}{c|c} 1 & 0 \\ 0 & 1 \end{array}\right] \tag{2.153e}$$

- *Node m = 3*: The orientation of link 3's frame (C, x_3, y_3, z_3) with respect to the fixed frame results from its absolute matrix (Equation (2.138)):

$$\mathbf{D}_3 = \mathbf{D}_A * \mathbf{D}_B * \mathbf{D}_C = \begin{matrix} \\ i_0 \\ j_0 \end{matrix} \begin{matrix} i_3 & j_3 \\ \left[\begin{array}{c|c} C(0^\circ + \theta_B + 0^\circ) & -S(0^\circ + \theta_B + 0^\circ) \\ S(0^\circ + \theta_B + 0^\circ) & C(0^\circ + \theta_B + 0^\circ) \end{array}\right] \end{matrix} \tag{2.153f}$$

- *End-effector E*: There is a translation of the frame (E, x_e, y_e, z_e) relative to frame (C, x_3, y_3, z_3); therefore, there is a null rotation matrix, $\theta_E = 0$:

$$\mathbf{D}_E = \begin{matrix} \\ i_3 \\ j_3 \end{matrix} \begin{matrix} i_e & j_e \\ \left[\begin{array}{c|c} 1 & 0 \\ 0 & 1 \end{array}\right] \end{matrix} \tag{2.153g}$$

The orientation of the end-effector's frame (E, x_e, y_e, z_e) with respect to the fixed frame (O, x_0, y_0, z_0) is

$$\mathbf{D}_e = \mathbf{D}_A * \mathbf{D}_B * \mathbf{D}_C * \mathbf{D}_E = \left[\begin{array}{c|c} C\theta_B & -S\theta_B \\ S\theta_B & C\theta_B \end{array}\right] = \begin{matrix} \\ i_0 \\ j_0 \end{matrix} \begin{matrix} i_e & j_e \\ \left[\begin{array}{c|c} \mathrm{Cos}\,\alpha_e^I & \mathrm{Cos}\,\alpha_e^{II} \\ \mathrm{Cos}\,\beta_e^I & \mathrm{Cos}\,\beta_e^{II} \end{array}\right] \end{matrix} \tag{2.153h}$$

The search for independent entries for planar mechanisms is the same as that for spatial mechanisms (Equation (2.24)). For the planar case, the absolute rotation matrix $\mathbf{D}_e$ between frame e and 0 has four entries, and there are three relations among them:

$$\begin{gathered} \left(\mathrm{Cos}\,\alpha_e^I\right)^2 + \left(\mathrm{Cos}\,\beta_e^I\right)^2 = 1, \quad \left(\mathrm{Cos}\,\alpha_e^{II}\right)^2 + \left(\mathrm{Cos}\,\beta_e^{II}\right)^2 = 1, \\ \mathrm{Cos}\,\alpha_e^I \cdot \mathrm{Cos}\,\alpha_e^{II} + \mathrm{Cos}\,\beta_e^I \cdot \mathrm{Cos}\,\beta_e^{II} = 0 \end{gathered} \tag{2.153i}$$

Therefore, only one entry is independent, and it is located on the diagonal, as shown in Equation (2.153j):

$$\mathbf{D}_e = \left[\begin{array}{c|c} C\theta_B & * \\ * & C\theta_B \end{array}\right] = \begin{matrix} \\ i_0 \\ j_0 \end{matrix} \begin{matrix} i_e & j_e \\ \left[\begin{array}{c|c} \mathrm{Cos}\,\alpha_e^I & * \\ \hline * & \mathrm{Cos}\,\beta_e^{II} \end{array}\right] \end{matrix} \tag{2.153j}$$

By identifying the entries on the left and right sides, results that the two angles α_e^I and β_e^{II} are equal:

$$\alpha_e^I = \beta_e^{II} = \theta_B \tag{2.153k}$$

where α_e^I is the angle between x_e and x_0 and β_e^{II} is the angle between y_e and y_0.

2.16.5 Direct Positional Analysis: Position Vectors of Frame Origins and End-Effector

The position vector for each joint and the position for the end-effector are determined from the provided M = 3 actuation parameters: $d_A(t)$, $\theta_B(t)$, and $d_C(t)$.

The output from positional analysis are the joint position vectors, **r**. Based on these vectors, the centers of mass of each position vector, velocities, and accelerations can be determined; their coordinates are with respect to the fixed frame (O, x_0, y_0, z_0) (Figure 2.38).

From the recursion of Equation (2.47), the position vectors of all joints, A, B, and C, are determined as functions of the translation vectors between frame origins:

$$\mathbf{r}_Z = \mathbf{L}_{OA}^0\Big|_A + \mathbf{L}_{AB}^0\Big|_B + \mathbf{L}_{BC}^0\Big|_C \tag{2.154}$$

Equation (2.154) in matrix form is given in Equation (2.155):

$$\mathbf{r}_Z = \mathbf{D}_0 * \left(\mathbf{L}_{OA_0} + \mathbf{d}_A\right)\Big|_A + \mathbf{D}_1 * \left(\mathbf{L}_{AB} + \mathbf{0}\right)\Big|_B + \mathbf{D}_2 * \left(\mathbf{L}_{BC_2} + \mathbf{d}_C\right)\Big|_C \tag{2.155}$$

- *Edge A*: Position vector of the frame's A origin:

$$\mathbf{r}_A = \mathbf{D}_0 * \left(\mathbf{L}_{OA_0} + \mathbf{d}_A\right) = \begin{Bmatrix} 5 + d_A \\ 0 \end{Bmatrix} \tag{2.155a}$$

  where

  $\mathbf{D}_0 = \mathbf{U}_{2\times 2}$; $\mathbf{L}_{OA_0} = \begin{Bmatrix} 5 \\ 0 \end{Bmatrix}$; $\mathbf{d}_A = d_A \cdot \mathbf{u}_A = \begin{Bmatrix} d_A \\ 0 \end{Bmatrix}$ is the linear displacement translation vector from point A_0 to origin A, along the unit vector $\mathbf{u}_A = \begin{Bmatrix} 1 \\ 0 \end{Bmatrix}$

- *Edge B*: Position vector of frame's B origin:

$$\mathbf{r}_B = \mathbf{r}_A + \mathbf{D}_1 * \left(\mathbf{L}_{AB}^1 + \mathbf{0}\right) = \begin{Bmatrix} 5 + d_A \\ 31 \end{Bmatrix} \tag{2.155b}$$

where

$\mathbf{D}_1 = \mathbf{U}_{2\times 2}$; $\mathbf{L}_{AB}^1 = \begin{Bmatrix} 0 \\ 31 \end{Bmatrix}$; $\mathbf{d}_B = \mathbf{0}$ is the constraint equation for revolute joint B (Equation (2.41))

- *Edge* C: Position vector of frame's C origin:

$$\mathbf{r}_C = \mathbf{r}_B + \mathbf{D}_2 * \left(\mathbf{L}_{BC_2}^2 + \mathbf{d}_C^2\right) = \begin{Bmatrix} 5 + d_A + (6 + d_C)\cdot C\theta_B \\ 31 + (6 + d_C)\cdot S\theta_B \end{Bmatrix} \tag{2.155c}$$

where

$\mathbf{D}_2 = \left[\begin{array}{c|c} C\theta_B & -S\theta_B \\ S\theta_B & C\theta_B \end{array}\right]$; $\mathbf{L}_{BC_2}^2 = \begin{Bmatrix} 6 \\ 0 \end{Bmatrix}$; $\mathbf{d}_C^2 = d_C \cdot \mathbf{u}_C^2 = \begin{Bmatrix} d_C \\ 0 \end{Bmatrix}$ is the linear displacement vector from point C_2 to origin C, along the unit vector $\mathbf{u}_C^2 = \begin{Bmatrix} 1 \\ 0 \end{Bmatrix}$

2.16.5.1 Position Vector Matrix

The position of all joints in time is shown in the position vector matrix:

$$\mathbf{r} = \left[\begin{array}{c|c|c} \mathbf{r}_A & \mathbf{r}_B & \mathbf{r}_C \end{array}\right] \tag{2.155d}$$

With the scalar components as

$$\mathbf{r} = \left[\begin{array}{c|c|c} x_A & x_B & x_C \\ y_A & y_B & y_C \end{array}\right] = \left[\begin{array}{c|c|c} 5 + d_A & 5 + d_A & 5 + d_A + (6 + d_C)\cdot C\theta_B \\ 0 & 31 & 31 + (6 + d_C)\cdot S\theta_B \end{array}\right] \tag{2.155e}$$

2.16.5.2 End-Effector Position Vector

- *Edge E*: Position vector of frame's E origin:

$$\mathbf{r}_E = \mathbf{r}_C + \mathbf{D}_3 * \left(\mathbf{L}_{CE}^3 + \mathbf{0}\right) = \begin{Bmatrix} 5 + d_A + (44 + d_C)\cdot C\theta_B \\ 31 + (44 + d_C)\cdot S\theta_B \end{Bmatrix} \tag{2.155f}$$

where

$$\mathbf{D}_3 = \begin{bmatrix} C\theta_B & -S\theta_B \\ S\theta_B & C\theta_B \end{bmatrix}; \quad \mathbf{L}_{CE} = \begin{Bmatrix} 38 \\ 0 \end{Bmatrix}; \quad \mathbf{d}_E = \mathbf{0} \text{ is the null relative}$$

displacement between origins C and E; the distance between them remains constant in time.

2.16.6 Center of Mass Position Vectors

Velocity, acceleration, static, and dynamic analyses require the position vectors for links' centers of mass (COM) G_1, G_2, and G_3 (Figure 2.39). The coordinates for G_1, G_2, and G_3 in the local frames A, B, and C are determined from the CAD model:

$$\mathbf{L}^1_{AG_1} = \begin{Bmatrix} x_{AG_1} \\ y_{AG_1} \end{Bmatrix} = \begin{Bmatrix} 0 \\ 11.117 \end{Bmatrix}; \quad \mathbf{L}^2_{BG_2} = \begin{Bmatrix} x_{BG_2} \\ y_{BG_2} \end{Bmatrix} = \begin{Bmatrix} 17.183 \\ 0 \end{Bmatrix};$$

$$\mathbf{L}^3_{CG_3} = \begin{Bmatrix} x_{CG_3} \\ y_{CG_3} \end{Bmatrix} = \begin{Bmatrix} 15.796 \\ 0 \end{Bmatrix} \tag{2.156}$$

Thus, the G_1, G_2, and G_3 positions in time, expressed in the fixed frame 0, are determined by applying Equation (2.157) for different links:

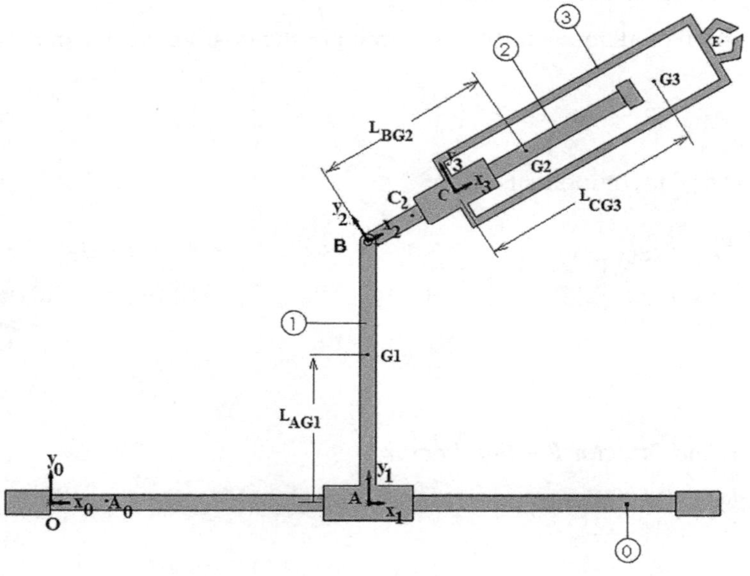

FIGURE 2.39
Centers of mass location.

$$\mathbf{r}_{G_m} = \mathbf{r}_Z + \mathbf{D}_m * \mathbf{L}^m_{ZG_m} \tag{2.157}$$

as follows:

- $m = 1$ for link 1:

$$\mathbf{r}_{G_1} = \mathbf{r}_A + \mathbf{D}_1 * \mathbf{L}^1_{AG_1} = \begin{Bmatrix} 5 + d_A \\ 11.117 \end{Bmatrix} \tag{2.157a}$$

- $m = 2$ for link 2:

$$\mathbf{r}_{G_2} = \mathbf{r}_B + \mathbf{D}_2 * \mathbf{L}^2_{BG_2} = \begin{Bmatrix} 5 + d_A + 17.183 \cdot \mathrm{Cos}\,\theta_B \\ 31 + 17.183 \cdot \mathrm{Sin}\,\theta_B \end{Bmatrix} \tag{2.157b}$$

- $m = 3$ for link 3:

$$\mathbf{r}_{G_3} = \mathbf{r}_C + \mathbf{D}_3 * \mathbf{L}^3_{CG_3} = \begin{Bmatrix} 5 + d_A + (21.796 + d_C) \cdot C\theta_B \\ 31 + (21.796 + d_C) \cdot S\theta_B \end{Bmatrix} \tag{2.157c}$$

2.16.7 The Matroid Method: Equations for Inverse Orientation and Positional Analysis for Planar Open Cycle Mechanisms

The data is for the end-effector's orientation and position parameters (x_E, y_E, α_e^I) from a desired task for which the actuation parameters d_A, θ_B, and d_C are determined.

2.16.7.1 Equations for Inverse Orientation

In the absolute rotation matrix $\mathbf{D}_e$, Equation (2.153h), only the diagonal entries are considered:

$$\mathbf{D}_e = \begin{bmatrix} C\theta_B & * \\ * & C\theta_B \end{bmatrix} = \begin{matrix} \\ i_0 \\ j_0 \end{matrix} \begin{matrix} i_e & j_e \\ \begin{bmatrix} \mathrm{Cos}\,\alpha_e^I & * \\ * & \mathrm{Cos}\,\beta_e^{II} \end{bmatrix} \end{matrix} \tag{2.158a}$$

Two equations for orientation are obtained by isolating the entries from the left and right diagonals:

$$\mathrm{Cos}\,\theta_B = \mathrm{Cos}\,\alpha_e^I = \mathrm{Cos}\,\beta_e^{II} \tag{2.158b}$$

Therefore, there is one independent equation for the orientation:
Equation 1:

$$\theta_B = \alpha_e^I \tag{2.158c}$$

2.16.7.2 Equations for Inverse Positional Analysis

The analysis of the relative position between the frame origins is developed by generating the entries in the Latin matrix. The Latin matrix $\mathbf{\Delta L}$ is a 1×4 matrix with vector entries defined based on the cycle (Equation (2.158d)):

$$\begin{matrix} & 0 & 1 & 2 & 3 \\ \mathbf{\Delta L} = & \Big[\ \mathbf{L}_{EA}^0 & \mathbf{L}_{AB}^0 & \mathbf{L}_{BC}^0 & \mathbf{L}_{CE}^0\ \Big] \end{matrix} = \left[\ \mathbf{r}_A - \mathbf{r}_E \ \vdots\ \mathbf{r}_B - \mathbf{r}_A \ \vdots\ \mathbf{r}_C - \mathbf{r}_B \ \vdots\ \mathbf{r}_E - \mathbf{r}_C\ \right] \tag{2.158d}$$

The entries in the Latin matrix are automatically generated from Equation (2.77) as a difference between position vectors of the frame origins:
Column 0 vector:

$$\mathbf{L}_{EA}^0 = \left\{ \begin{matrix} x_{EA} \\ y_{EA} \end{matrix} \right\} = \mathbf{r}_A - \mathbf{r}_E = \left\{ \begin{matrix} -(44 + d_C) \cdot C\theta_B \\ -31 - (44 + d_C) \cdot S\theta_B \end{matrix} \right\} \tag{2.158e}$$

Column 1 vector:

$$\mathbf{L}_{AB}^0 = \left\{ \begin{matrix} x_{AB} \\ y_{AB} \end{matrix} \right\} = \mathbf{r}_B - \mathbf{r}_A = \left\{ \begin{matrix} 0 \\ 31 \end{matrix} \right\} \tag{2.158f}$$

Column 2 vector:

$$\mathbf{L}_{BC}^0 = \left\{ \begin{matrix} x_{BC} \\ y_{BC} \end{matrix} \right\} = \mathbf{r}_C - \mathbf{r}_B = \left\{ \begin{matrix} (6 + d_C) \cdot C\theta_B \\ (6 + d_C) \cdot S\theta_B \end{matrix} \right\} \tag{2.158g}$$

Column 3 vector:

$$\mathbf{L}_{CE}^0 = \left\{ \begin{matrix} x_{CE} \\ y_{CE} \end{matrix} \right\} = \mathbf{r}_E - \mathbf{r}_C = \left\{ \begin{matrix} 38 \cdot C\theta_B \\ 38 \cdot S\theta_B \end{matrix} \right\} \tag{2.158h}$$

Equation (2.158d) with scalar entries is shown in Equation (2.158i):

$$\begin{array}{l}
\qquad\quad 0 \qquad\quad 1 \qquad\quad 2 \qquad\quad 3 \\
\Delta\mathbf{L} = \left[\begin{array}{c|c|c|c} x_{EA} & x_{AB} & x_{BC} & x_{CE} \\ y_{EA} & y_{AB} & y_{BC} & y_{CE} \end{array}\right] \\
= \left[\begin{array}{c|c|c|c} -(44+d_C)\cdot C\theta_B & 0 & (6+d_C)\cdot C\theta_B & 38\cdot C\theta_B \\ -31-(44+d_C)\cdot S\theta_B & 31 & (6+d_C)\cdot S\theta_B & 38\cdot S\theta_B \end{array}\right] \\
= \left[\begin{array}{c|c|c|c} 5+d_A-x_E & 0 & x_{BC} & x_{CE} \\ -y_E & 31 & y_{BC} & y_{CE} \end{array}\right]
\end{array} \tag{2.158i}$$

2.16.7.2.1 Equations for Latin Matrix Rows

Equation 2: The sum of the x entries is zero

$$5+d_A-x_E+x_{BC}+x_{CE}=0 \tag{2.158j}$$

Equation 3: The sum of the y entries is zero

$$-y_E+31+y_{BC}+y_{CE}=0 \tag{2.158k}$$

2.16.7.2.2 Equations for Latin Matrix Columns

A planar vector from each column could be defined by two parameters (magnitude and direction), two Cartesian components, or one Cartesian component and its magnitude. The elimination of the direction parameter (described by an angular displacement trigonometric function) is desired in equations for inverse analysis as follows:

- The vector's $\mathbf{L}_{EA}^0$ components are expressed as a function of the end-effector coordinates x_E, y_E:

$$\begin{array}{l}
\mathbf{L}_{EA}^0 = \left\{\begin{array}{c} x_{EA} \\ y_{EA} \end{array}\right\} = \left\{\begin{array}{c} -(44+d_C)\cdot C\theta_B \\ -31-(44+d_C)\cdot S\theta_B \end{array}\right\} \\
= \mathbf{r}_A - \mathbf{r}_E = \left\{\begin{array}{c} 5+d_A-x_E \\ -y_E \end{array}\right\}
\end{array} \tag{2.158l}$$

The two equations extraced from vector L_{EA} are as follows:

Equation 4: This equation is for the x_{EA} component, which is expressed as a function of the end-effector's orientation and position. The variable angular displacement in joint B (Equation 1) is substituted in Equation (2.158m) in conjunction with the provided task orientation of the end-effector:

$$-(44+d_C)\cdot C\alpha_e^I = 5+d_A - x_E \tag{2.158m}$$

Equation 5: For a vector with unknown magnitude L_{EA}:

$$(5+d_A - x_E)^2 + (-y_E)^2 = (L_{EA})^2 \tag{2.158n}$$

Equation 6: For the magnitude of vector $\mathbf{L}_{BC}^0$ with the unknown components x_{BC} and y_{BC}:

$$(x_{BC})^2 + (y_{BC})^2 = (6+d_C)^2 \tag{2.158o}$$

Equation 7: For the magnitude of vector $\mathbf{L}_{CE}^0$:

$$(x_{CE})^2 + (y_{CE})^2 = 38^2 \tag{2.158p}$$

Equation 8: For 0° constant angle between the two vectors $\mathbf{L}_{BC}^0$ and $\mathbf{L}_{CE}^0$:

$$x_{BC}\cdot x_{CE} + y_{BC}\cdot y_{CE} = 38\cdot(6+d_C) \tag{2.158q}$$

2.16.7.3 Solution of Nonlinear System of Equations

The nonlinear system with eight equations is solved for eight unknowns. The eight unknowns are

- *Actuating parameters*:

$$\mathbf{q}_M = \left\{ d_A, \quad \theta_B, \quad d_C \right\}^T \tag{2.159}$$

- *Latin matrix entries*:

$$\left(L_{EA}, x_{BC}, y_{BC}, x_{CE}, y_{CE}\right) \tag{2.160}$$

2.16.8 Application for Inverse Analysis: The Required Manipulator's Joint Displacements to Place the End-Effector E in Three Task Orientation Positions

The manipulator is used for a packaging application. The initial location of the end-effector is the home position h, the part to be grasped is located at position pg, and the target release for the part is located at position tr:

$$\left\{\begin{bmatrix} x_E^h \\ y_E^h \\ \alpha_e^I \end{bmatrix}\right\} = \left\{\begin{bmatrix} 49.000 \\ 31.000 \\ 0° \end{bmatrix}\right\}; \quad \left\{\begin{bmatrix} x_E^{pg} \\ y_E^{pg} \\ \alpha_e^{pg} \end{bmatrix}\right\} = \left\{\begin{bmatrix} 52.052 \\ 44.135 \\ 16.9° \end{bmatrix}\right\}; \quad \left\{\begin{bmatrix} x_E^{tr} \\ y_E^{tr} \\ \alpha_e^{tr} \end{bmatrix}\right\} = \left\{\begin{bmatrix} 38.751 \\ 76.270 \\ 67.5° \end{bmatrix}\right\} \tag{2.161}$$

2.16.8.1 EES for Inverse Orientation and Positional Analysis of the TRT 3 DOF Manipulator

- In the Equations Window, the user types the equations as illustrated in Figure 2.40.

  Highlight Equation 1 in Equations Window and right click to *select Comment* { } in order to not consider the equation between the brackets in calculations. This is done because Equation 1 is substituted in Equation 4.

  In Figure 2.40, the numerical values included between instruction $IFNOT Parametric Table and $endif can be changed when solving the equations for the three different sets of data: home position, part grasp, and target release.
- From the EES menu, select Options/Variable Info, shown in Figure 2.41.
- From the Variable Information window, the user inserts the initial (guess) values in addition to the lower and upper limits of the variables. Solving the system with the initial (guess) values and variable contraints increases the convergence of the solution for nonlinear equations.

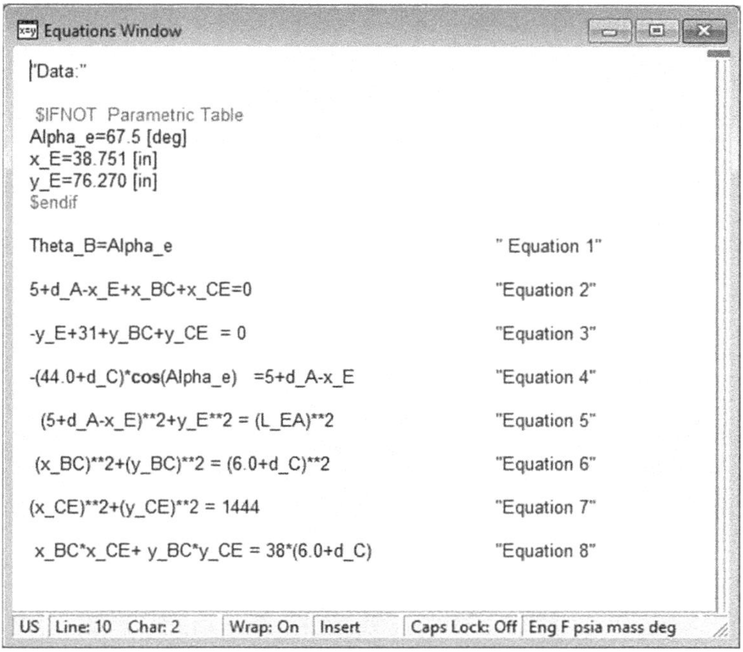

FIGURE 2.40
Equations Window.

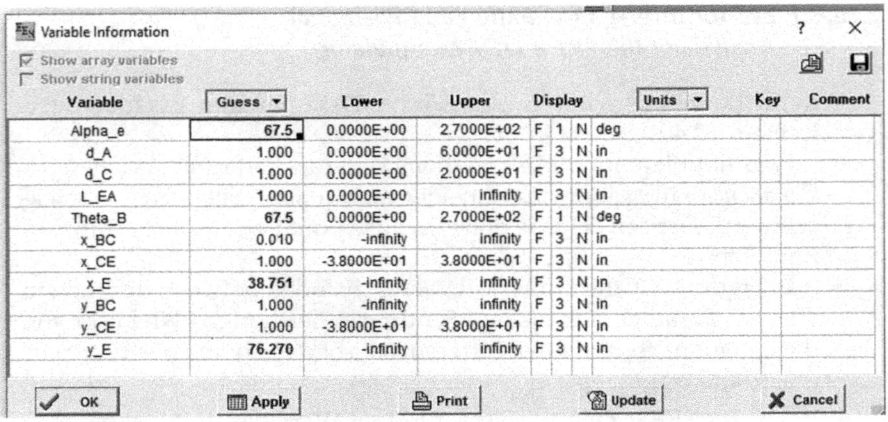

Variable	Guess	Lower	Upper	Display	Units	Key	Comment
Alpha_e	67.5	0.0000E+00	2.7000E+02	F 1 N	deg		
d_A	1.000	0.0000E+00	6.0000E+01	F 3 N	in		
d_C	1.000	0.0000E+00	2.0000E+01	F 3 N	in		
L_EA	1.000	0.0000E+00	infinity	F 3 N	in		
Theta_B	67.5	0.0000E+00	2.7000E+02	F 1 N	deg		
x_BC	0.010	-infinity	infinity	F 3 N	in		
x_CE	1.000	-3.8000E+01	3.8000E+01	F 3 N	in		
x_E	38.751	-infinity	infinity	F 3 N	in		
y_BC	1.000	-infinity	infinity	F 3 N	in		
y_CE	1.000	-3.8000E+01	3.8000E+01	F 3 N	in		
y_E	76.270	-infinity	infinity	F 3 N	in		

FIGURE 2.41
Variable information.

An example of lower and upper limits for linear displacements are $d_A \in [0'';60'']$ and $d_D \in [0'';20'']$. Both limits are from the stroke specifications for the manipulator. The variables x_{CE}, y_{CE} are in the interval [–38, 38] because the components of this vector cannot exceed the vector's magnitude $\mathbf{L}^0_{CE} = 38''$ (Equation 7). Since L_{EA} is the magnitude of vector $\mathbf{L}_{EA}$, its limits are selected to be within the interval [0, infinity).

- *Home-position state*:
  Input: Desired task parameters: $\alpha_e = 0°$, $x_E = 49.000''$, $y_E = 31.000''$
  Solution: The solution from Figure 2.42 is
  - *Actuating joint coordinates*:

$$\mathbf{q}_M = \left\{ \begin{array}{ccc} d_A, & \theta_B, & d_C \end{array} \right\}^T = \left\{ \begin{array}{ccc} 0'', & 0°, & 0'' \end{array} \right\}^T \tag{2.162}$$

  - *Latin matrix entries*: $(L_{EA}, x_{BC}, y_{BC}, x_{CE}, y_{CE})$ = (53.824, 6.000, 0.000, 38.000, 0.000) in.
- *Part-grasped state*:
  Input: Desired task parameters: $\alpha_e = 16.9°$, $x_E = 52.052''$, $y_E = 44.135''$
  Solution: The solution from Figure 2.43 is
  - *Actuating joint coordinates*:

$$\mathbf{q}_M = \left\{ \begin{array}{ccc} d_A, & \theta_B, & d_C \end{array} \right\}^T = \left\{ \begin{array}{ccc} 3.820'', & 16.9°, & 1.184'' \end{array} \right\} \tag{2.163}$$

  - *Latin matrix entries*: $(L_{EA}, x_{BC}, y_{BC}, x_{CE}, y_{CE})$ = (61.781, 6.872, 2.093, 36.361, 11.042) in.

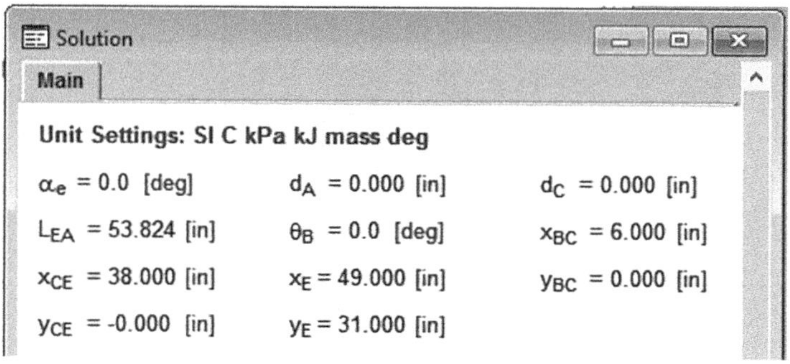

FIGURE 2.42
Home position.

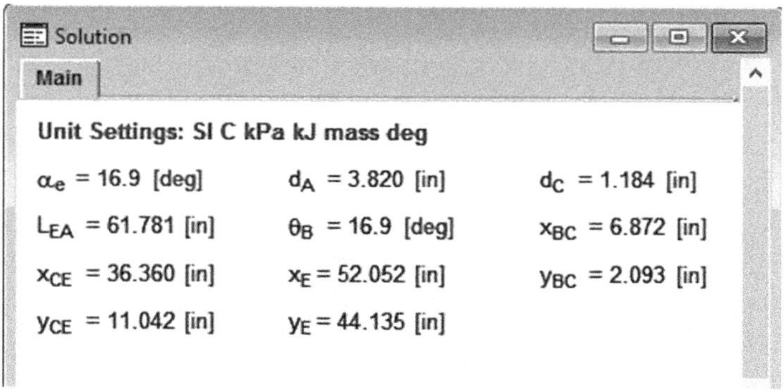

FIGURE 2.43
Part grasp.

- *Target-released state*:
  Input: Desired task parameters: $\alpha_e = 67.5°$, $x_E = 38.751''$, $y_E = 76.270''$
  Solution: The solution from Figure 2.44 is
  - *Actuating joint coordinates*:

$$\mathbf{q}_M = \left\{ \begin{array}{ccc} d_A, & \theta_B, & d_C \end{array} \right\}^T = \left\{ \begin{array}{ccc} 15.000'', & 67.5°, & 5.000'' \end{array} \right\}^T \quad (2.164)$$

  - *Latin matrix entries*: $(L_{EA}, x_{BC}, y_{BC}, x_{CE}, y_{CE})$ = (78.541, 4.200, 10.167, 14.552, 35.103) in.

From the inverse analysis of the TRT manipulator with 3 DOF, the target orientation and position x_E, y_E of the manipulator's end-effector E are provided. Then, the angular displacements of the linear motors in joints A and C and the revolute motor in joint B can be

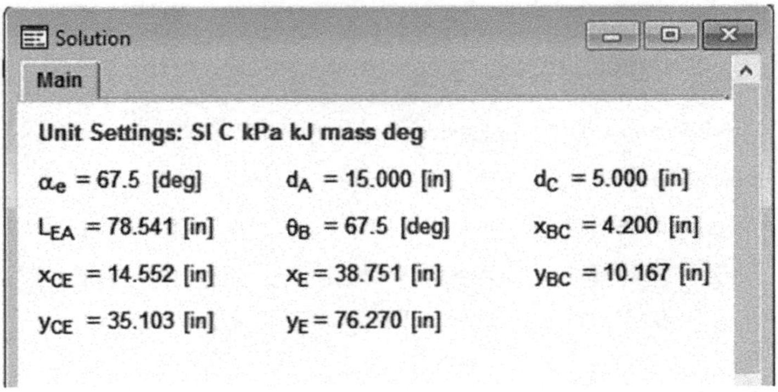

FIGURE 2.44
Target release.

determined. The results for the link's coordinates x_{BC}, y_{BC}, x_{CE}, y_{CE} are used for further velocity and acceleration analyses.

2.16.8.2 EES for Direct Positional Analysis of the TRT 3 DOF Manipulator

The results from the above inverse positional analysis are compared with the results from the direct (forward) positional analysis. For the direct positional analysis, the displacements for actuated joints A, B, and C are shown in Equation (2.147) as quadratic functions of time t. Then, the end-effector's Cartesian coordinates x_E, y_E are determined from Equation (2.155f).

- In the Equations Window in Figure 2.45, the user types the Equation (2.155f).
- In Tables/New Parametric Table, insert 25 for the number of runs and First Value: 0 and Last (linear): 1 for the time interval. Then, from Calculate select Solve Table.

  Appendix 2.5 shows the EES Parametric Table with the displacements of joints A, B, and C in addition to the end-effector's coordinates for the time interval of t = 1 s.
- In Plots/New Plot ⟶ Window/X–Y Plot, select t for x-Axis and x_E for the Y-Axis; repeat for y_E. The two previously generated graphs can be displayed on the same plot using Plots/Overlay Plot, which is illustrated in Figure 2.46.
- In Tables/New Lookup Table, copy and paste the EES columns with the x_E, y_E values from the Parametric Table. In addition, copy and paste the Excel columns with the x_E, y_E values from the SW Motion.

  Appendix 2.6 tabulates the calculated values (using EES) and simulated values (using SW) for the end-effector's coordinates x_E and y_E for the TRT planar mechanism.

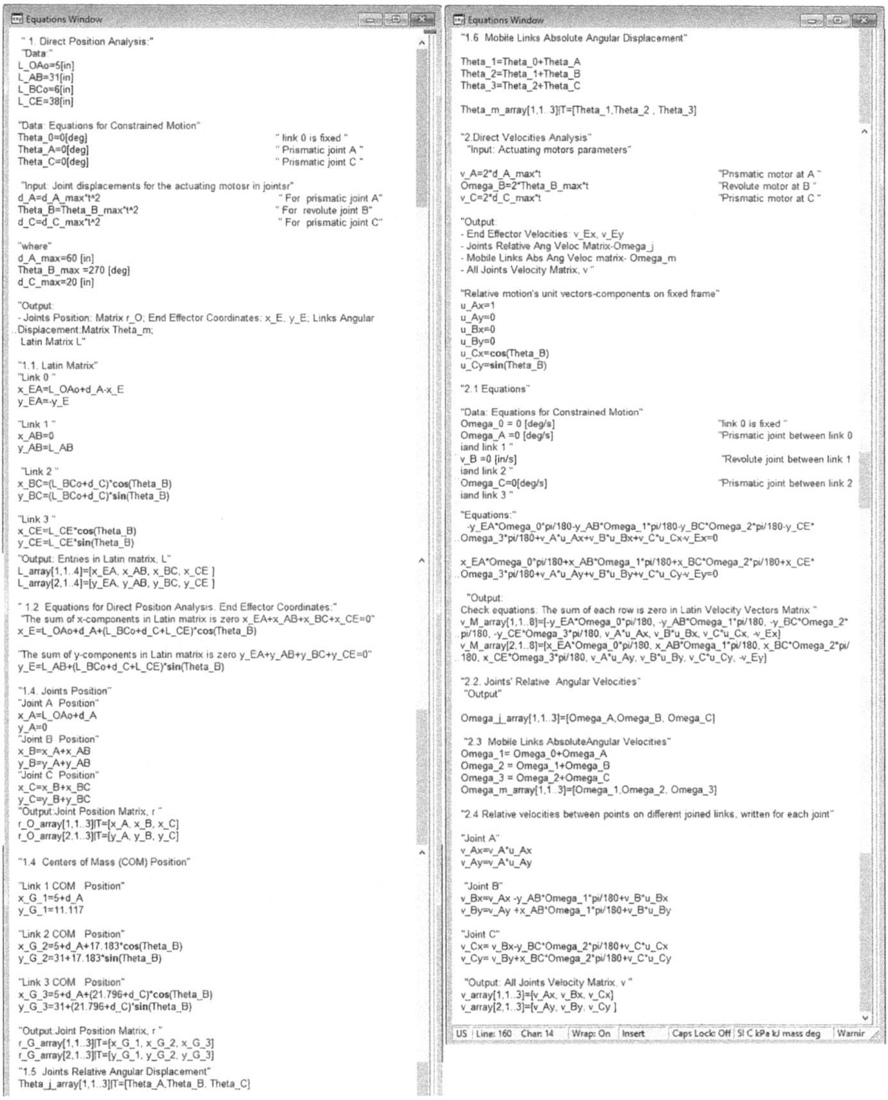

```
" 1. Direct Position Analysis:"
"Data:"
L_OAo=5[in]
L_AB=31[in]
L_BCo=6[in]
L_CE=38[in]

"Data: Equations for Constrained Motion"
Theta_0=0[deg]                                  " link 0 is fixed "
Theta_A=0[deg]                                  " Prismatic joint A "
Theta_C=0[deg]                                  " Prismatic joint C "

"Input: Joint displacements for the actuating motosr in jointsr"
d_A=d_A_max*t^2                                 " For  prismatic joint A"
Theta_B=Theta_B_max*t^2                         " For  revolute joint B"
d_C=d_C_max*t^2                                 " For  prismatic joint C"

"where"
d_A_max=60 [in]
Theta_B_max =270 [deg]
d_C_max=20 [in]

"Output:
- Joints Position: Matrix r_O; End Effector Coordinates: x_E, y_E; Links Angular
..Displacement:Matrix Theta_m;
 Latin Matrix L"

"1.1. Latin Matrix"
"Link 0 "
x_EA=L_OAo+d_A-x_E
y_EA=-y_E

"Link 1 "
x_AB=0
y_AB=L_AB

"Link 2 "
x_BC=(L_BCo+d_C)*cos(Theta_B)
y_BC=(L_BCo+d_C)*sin(Theta_B)

"Link 3 "
x_CE=L_CE*cos(Theta_B)
y_CE=L_CE*sin(Theta_B)
"Output: Entries in Latin matrix, L"
L_array[1,1..4]=[x_EA, x_AB, x_BC, x_CE ]
L_array[2,1..4]=[y_EA, y_AB, y_BC, y_CE ]

" 1.2  Equations for Direct Position Analysis. End Effector Coordinates:"
 "The sum of x-components in Latin matrix is zero x_EA+x_AB+x_BC+x_CE=0"
x_E=L_OAo+d_A+(L_BCo+d_C+L_CE)*cos(Theta_B)

"The sum of y-components in Latin matrix is zero y_EA+y_AB+y_BC+y_CE=0"
y_E=L_AB+(L_BCo+d_C+L_CE)*sin(Theta_B)

"1.4. Joints Position"
"Joint A  Position"
x_A=L_OAo+d_A
y_A=0
"Joint B  Position"
x_B=x_A+x_AB
y_B=y_A+y_AB
"Joint C  Position"
x_C=x_B+x_BC
y_C=y_B+y_BC
"Output:Joint Position Matrix, r "
r_O_array[1,1..3]|T=[x_A, x_B, x_C]
r_O_array[2,1..3]|T=[y_A, y_B, y_C]

"1.4  Centers of Mass (COM) Position"

"Link 1 COM   Position"
x_G_1=5+d_A
y_G_1=11.117

"Link 2 COM   Position"
x_G_2=5+d_A+17.183*cos(Theta_B)
y_G_2=31+17.183*sin(Theta_B)

"Link 3 COM   Position"
x_G_3=5+d_A+(21.796+d_C)*cos(Theta_B)
y_G_3=31+(21.796+d_C)*sin(Theta_B)

"Output:Joint Position Matrix, r "
r_G_array[1,1..3]|T=[x_G_1, x_G_2, x_G_3]
r_G_array[2,1..3]|T=[y_G_1, y_G_2, y_G_3]
"1.5  Joints Relative Angular Displacement"
Theta_j_array[1,1..3]|T=[Theta_A,Theta_B, Theta_C]
```

```
"1.6  Mobile Links Absolute Angular Displacement"

Theta_1=Theta_0+Theta_A
Theta_2=Theta_1+Theta_B
Theta_3=Theta_2+Theta_C

Theta_m_array[1,1..3]|T=[Theta_1,Theta_2 , Theta_3]

"2.Direct Velocities Analysis"
 "Input: Actuating motors parameters"

v_A=2*d_A_max*t                                 "Prismatic motor at A "
Omega_B=2*Theta_B_max*t                         "Revolute motor at B "
v_C=2*d_C_max*t                                 "Prismatic motor at C "

"Output:
- End Effector Velocities: v_Ex, v_Ey
- Joints Relative Ang Veloc Matrix-Omega_j
- Mobile Links Abs Ang Veloc matrix- Omega_m
- All Joints Velocity Matrix, v "

"Relative motion's unit vectors-components on fixed frame"
u_Ax=1
u_Ay=0
u_Bx=0
u_By=0
u_Cx=cos(Theta_B)
u_Cy=sin(Theta_B)

"2.1 Equations"

"Data: Equations for Constrained Motion"
Omega_0 = 0 [deg/s]                             "link 0 is fixed "
Omega_A =0 [deg/s]                              "Prismatic joint between link 0
iand link 1 "
v_B =0 [in/s]                                   "Revolute joint between link 1
iand link 2 "
Omega_C=0[deg/s]                                "Prismatic joint between link 2
iand link 3 "

"Equations:"
 -y_EA*Omega_0*pi/180-y_AB*Omega_1*pi/180-y_BC*Omega_2*pi/180-y_CE*
..Omega_3*pi/180+v_A*u_Ax+v_B*u_Bx+v_C*u_Cx-v_Ex=0

x_EA*Omega_0*pi/180+x_AB*Omega_1*pi/180+x_BC*Omega_2*pi/180+x_CE*
..Omega_3*pi/180+v_A*u_Ay+v_B*u_By+v_C*u_Cy-v_Ey=0

 "Output:
Check equations: The sum of each row is zero in Latin Velocity Vectors Matrix "
v_M_array[1,1..8]=[-y_EA*Omega_0*pi/180, -y_AB*Omega_1*pi/180, -y_BC*Omega_2*
..pi/180, -y_CE*Omega_3*pi/180, v_A*u_Ax, v_B*u_Bx, v_C*u_Cx, -v_Ex]
v_M_array[2,1..8]=[x_EA*Omega_0*pi/180, x_AB*Omega_1*pi/180, x_BC*Omega_2*pi/
..180, x_CE*Omega_3*pi/180, v_A*u_Ay, v_B*u_By, v_C*u_Cy, -v_Ey]

"2.2. Joints' Relative  Angular Velocities"
 "Output"

Omega_j_array[1,1..3]=[Omega_A,Omega_B, Omega_C]

 "2.3  Mobile Links AbsoluteAngular Velocities"
Omega_1= Omega_0+Omega_A
Omega_2 = Omega_1+Omega_B
Omega_3 = Omega_2+Omega_C
Omega_m_array[1,1..3]=[Omega_1,Omega_2, Omega_3]

"2.4 Relative velocities between points on different joined links, written for each joint"

"Joint A"
v_Ax=v_A*u_Ax
v_Ay=v_A*u_Ay

 "Joint B"
v_Bx=v_Ax -y_AB*Omega_1*pi/180+v_B*u_Bx
v_By=v_Ay +x_AB*Omega_1*pi/180+v_B*u_By

"Joint C"
v_Cx= v_Bx-y_BC*Omega_2*pi/180+v_C*u_Cx
v_Cy= v_By+x_BC*Omega_2*pi/180+v_C*u_Cy

 "Output: All Joints Velocity Matrix, v "
v_array[1,1..3]=[v_Ax, v_Bx, v_Cx]
v_array[2,1..3]=[v_Ay, v_By, v_Cy ]
```

US | Line: 160 Char: 14 | Wrap: On | Insert | Caps Lock: Off | SI C kPa kJ mass deg | Warnir

FIGURE 2.45
Equations Window.

Note: Both method produce the same values for the end-effector's coordinates for the TRT planar mechanism.

- Using Plots/New Plot ⟶ Window/X–Y Plot, plot the two sets of data.

  Figure 2.47 illustrates the EES plot for the two sets of data: calculated and simulated values.

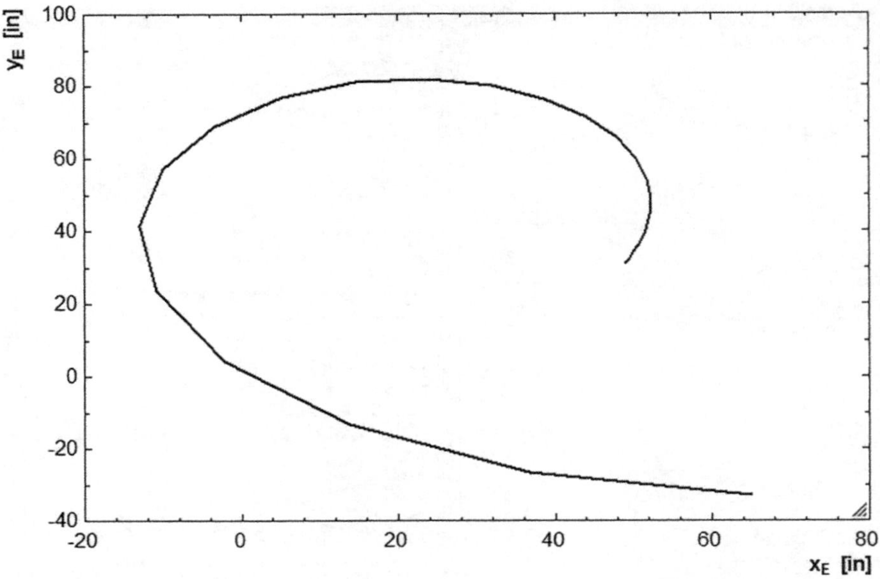

FIGURE 2.46
End-effector's path.

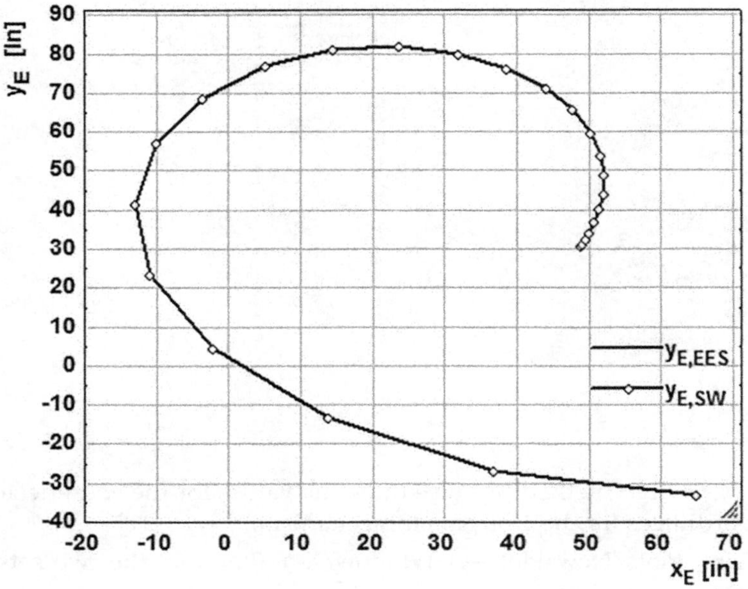

FIGURE 2.47
End-effector's path for calculated data (EES) and simulated values (SW).

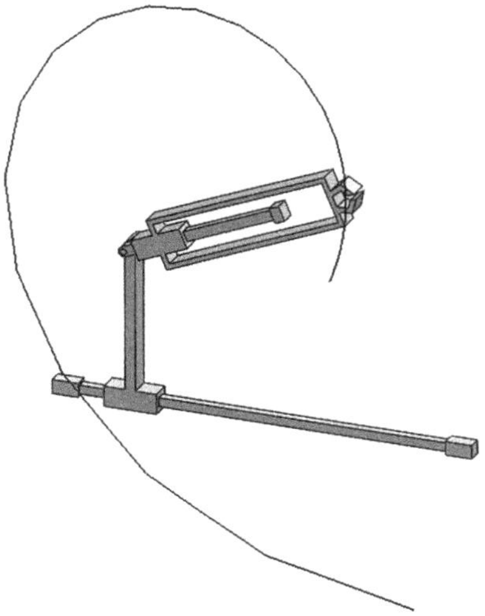

FIGURE 2.48
End-effector's path from SW Motion simulation.

Notice that the path in Figure 2.47 from the EES plot is coincident wit the path in Figure 2.48 from SW Motion/Results/Trace Path.

- In Tables/New Parametric Table, insert 25 for the number of runs and First Value: 0 and Last (linear): 1 for the time interval. Then, use Calculate/Solve Table to solve for the angular displacements of joints A, B, and C, and the absolute angular orientation of links 1, 2, and 3, shown in Appendix 2.7.
- In Tables/New Parametric Table, insert 25 for the number of runs and First Value: 0 and Last (linear): 1 for the time interval. Then, use Calculate/Solve Table to solve for the x and y coordinates of joints A, B, and C, shown in Appendix 2.8.
- In Tables/New Parametric Table, insert 25 for the number of runs and First Value: 0 and Last (linear): 1 for the time interval. Then, use Calculate/Solve Table to solve for the x and y entries in the Latin matrix, shown in Appendix 2.9.

  These tabulated values are the coefficients in the velocity and acceleration equations that are developed in Chapters 2 and 3.
- In Tables/New Parametric Table, insert 25 for the number of runs and First Value: 0 and Last (linear): 1 for the time interval. Then, use Calculate/Solve Table to solve for the x and y coordinates of each link's COM G_1, G_2, and G_3, shown in Appendix 2.10.

  These centers of mass coordinates are used in the dynamic analysis of mechanisms in Chapter 4 to locate the application points for the force of gravity. The second part of the program in Figure 2.45 refers to the velocity analysis of open cycle mechanisms that are explored further in Chapter 2.

2.16.9 The IT Method of Relative Homogeneous Matrices: Absolute Homogeneous Matrix and Relative Homogeneous Matrix for Planar Mechanisms

For planar mechanisms, the equations for orientation and position can be combined into a single set of equations by introducing a 3 × 3 matrix, which is a reduced size from the 4 × 4 homogeneous matrix used for spatial mechanisms. Thus, the nodes are assigned the absolute homogeneous matrices and the edges are assigned the relative homogeneous matrices.

2.16.9.1 Absolute Homogeneous Matrix for Planar Mechanisms, $\mathbf{R}_m$

A 2 × 3 non-square matrix is created from the following two matrices:

- $\mathbf{D}_m$ is the 2 × 2 absolute rotation matrix (Equation (2.136))
- $\mathbf{r}_Z$ is the 2 × 1 joint Z's position vector (Equation (2.155))

Equation (2.165) shows a 3 × 3 square matrix with a third row added to the previously mentioned 2 × 3 non-square matrix. In this third row, the first two entries are zeroes, and the third entry is 1:

$$\mathbf{R}_m = \left[\begin{array}{c|c} \mathbf{D}_m & \mathbf{r}_Z \\ \hline 0 & 1 \end{array}\right] = \left[\begin{array}{cc|c} \text{Cos}\,\alpha_m^{I} & \text{Cos}\,\alpha_m^{II} & x_Z \\ \text{Cos}\,\beta_m^{I} & \text{Cos}\,\alpha_m^{II} & y_Z \\ \hline 0 & 0 & 1 \end{array}\right] \tag{2.165}$$

2.16.9.2 IT Relative Homogeneous Matrix for Planar Mechanisms, $\mathbf{IT}_{YZ}$

A 2 × 3 non-square matrix is created from two matrices:

- $\mathbf{D}_Z$ is the 2 × 2 relative rotation matrix
- $\mathbf{L}_{YZ}^{m-1}$ is the 2 × 1 translation vector between the frame origins Y and Z. This vector is an entry in the Latin matrix expressed in local frame m−1

Equation (2.166) shows a 3 × 3 square matrix with a third row added to the previously mentioned 2 × 3 matrix. Within this additional third row, the first two entries are 0 and the third entry is 1:

$$\mathbf{IT}_{YZ} = \left[\begin{array}{c|c} \mathbf{D}_Z & \mathbf{L}_{YZ}^{m-1} \\ \hline 0 & 1 \end{array}\right] = \left[\begin{array}{c|c} \mathbf{D}_Z & \mathbf{L}_{YZ_{m-1}}^{m-1} + \mathbf{d}_Z^{m-1} \\ \hline 0 & 1 \end{array}\right] = \left[\begin{array}{cc|c} \text{Cos}\,\alpha_Z^{I} & \text{Cos}\,\alpha_Z^{II} & x_{YZ}^{m-1} \\ \text{Cos}\,\beta_Z^{I} & \text{Cos}\,\beta_Z^{II} & y_{YZ}^{m-1} \\ \hline 0 & 0 & 1 \end{array}\right] \tag{2.166}$$

The matrix in Equation (2.166) is partitioned into two sub-matrices using the same notation that is used for spatial mechanisms, shown in Equation (2.167):

$$\mathbf{I} = \left[\begin{array}{c} \mathbf{D}_Z \\ \hline 0 \end{array} \right] \text{and } \mathbf{T} = \left[\begin{array}{c} \mathbf{L}_{YZ}^{m-1} \\ \hline 1 \end{array} \right] \tag{2.167}$$

2.16.9.3 *Planar Mechanisms: Relations between Absolute and Relative Homogeneous Matrices*

Equation (2.83) for link m's absolute matrix $\mathbf{R}_m$ is defined as a multiplication of two homogeneous matrices, shown in Equation (2.168):

$$\mathbf{R}_m = \mathbf{R}_{m-1} * \mathbf{IT}_{YZ} \tag{2.168}$$

Equation (2.84) is reproduced here for planar mechanisms as Equation (2.169). It states that:

For planar mechanisms, the absolute homogeneous matrix at node m is determined from the post multiplication of all relative matrices located before the node m in the path:

$$\mathbf{R}_m = \mathbf{IT}_{OA}\big|_{m=1} * \mathbf{IT}_{AB}\big|_{m=2} * \mathbf{IT}_{BC}\big|_{m=3} * \cdots * \mathbf{IT}_{YZ}\big|_{m} \tag{2.169}$$

The recursion process is developed in detail in the next example for the TRT planar mechanism.

2.16.10 Example for Orientation and Positional Analysis of a 3 DOF Planar Manipulator

The TRT mechanism in Figure 2.35 has m = 3 links, j = 3 joints, and mobility M = 3.

Data: Information for direct positional analysis is provided

- The vectors with constant magnitude (from CAD model) written with respect to local frames are

$$\mathbf{L}_{OA_0} = \left\{ \begin{array}{c} 5 \\ 0 \end{array} \right\}; \ \mathbf{L}_{AB}^{1} = \left\{ \begin{array}{c} 0 \\ 31 \end{array} \right\}; \ \mathbf{L}_{BC_2}^{2} = \left\{ \begin{array}{c} 6 \\ 0 \end{array} \right\}; \ \mathbf{L}_{CE}^{3} = \left\{ \begin{array}{c} 38 \\ 0 \end{array} \right\}$$

- The unit vectors for linear displacements d_A and d_C at prismatic joints A and D are

$$\mathbf{u}_{A_{0,1}}^{0} = \left\{ \begin{array}{c} 1 \\ 0 \end{array} \right\}; \ \mathbf{u}_{C_{2,3}}^{2} = \left\{ \begin{array}{c} 1 \\ 0 \end{array} \right\}$$

- The relative constraint equations for the revolute joints B and the end-effector E are

$$\mathbf{d}_B = \mathbf{d}_E = \mathbf{0}$$

- The M = 3 actuation parameters:

$$d_A(t); \quad \theta_B(t); \quad d_C(t)$$

- The home position is shown in Figure 2.36.
  It is the initial position of mechanism when t = 0 s, where all actuating positional parameters are zero:

$$d_A = 0\,[\text{in.}]; \quad \theta_B = 0\,[°]; \quad d_C = 0$$

Objective: For direct positional analysis, the following are determined:

1. Each link's orientation matrix and frame position vectors at any time
2. Absolute end-effector orientation matrix and frame's E position vector at any time
3. Mechanism's equations for direct orientation and position at any time
4. Each link's orientation and frame position vectors at the home position
5. End-effector orientation and frame position vector at the home position

Solution:

1. Each link's orientation matrix and frame position vectors at any time
   We consider the path in the tree, see Figure 2.49.

   Node 0: Absolute homogeneous matrix for the fixed link 0 is the unit 3 × 3 matrix:

$$\mathbf{R}_0 = \left[\begin{array}{c|c} \mathbf{D}_0 & \mathbf{r}_O \\ \hline 0 & 1 \end{array}\right] = \mathbf{U}_{3\times3} = \left[\begin{array}{cc|c} 1 & 0 & 0 \\ 0 & 1 & 0 \\ \hline 0 & 0 & 1 \end{array}\right] \tag{2.170a}$$

   Edge A (prismatic): Relative homogeneous matrix at joint A is defined as

$$\mathbf{IT}_{OA} = \left[\begin{array}{c|c} \mathbf{D}_A & \mathbf{L}^0_{OA_0} + \mathbf{d}^0_A \\ \hline 0 & 1 \end{array}\right] = \left[\begin{array}{cc|c} 1 & 0 & 5 + d_A \\ 0 & 1 & 0 \\ \hline 0 & 0 & 1 \end{array}\right] \tag{2.170b}$$

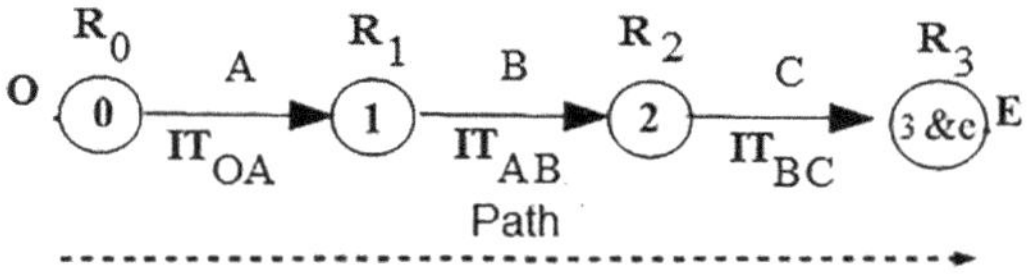

FIGURE 2.49
Absolute and relative homogeneous matrices along the tree for TRT planar mechanism.

Node 1: Absolute homogeneous matrix for link m = 1 is determined from

$$\mathbf{R}_1 = \mathbf{R}_0 * \mathbf{IT}_{OA} = \left[\begin{array}{cc|c} 1 & 0 & 5+d_A \\ 0 & 1 & 0 \\ \hline 0 & 0 & 1 \end{array}\right] \tag{2.170c}$$

Output: Orientation matrix $\mathbf{D}_1$ and position vector $\mathbf{r}_A$

Edge B (revolute): Relative homogeneous matrix at joint B is defined as

$$\mathbf{IT}_{AB} = \left[\begin{array}{c|c} \mathbf{D}_B & \mathbf{L}^1_{AB} + \mathbf{0} \\ \hline 0 & 1 \end{array}\right] = \left[\begin{array}{cc|c} C\theta_B & -S\theta_B & 0 \\ S\theta_B & C\theta_B & 31 \\ \hline 0 & 0 & 1 \end{array}\right] \tag{2.170d}$$

Node 2: Absolute homogeneous matrix for link m = 2 is determined from

$$\mathbf{R}_2 = \mathbf{R}_1 * \mathbf{IT}_{AB} = \left[\begin{array}{cc|c} C\theta_B & -S\theta_B & 5+d_A \\ S\theta_B & C\theta_B & 31 \\ \hline 0 & 0 & 1 \end{array}\right] \tag{2.170e}$$

Output: Orientation matrix $\mathbf{D}_2$ and position vector $\mathbf{r}_B$

Edge C (prismatic): Relative homogeneous matrix at joint C is defined as

$$\mathbf{IT}_{BC} = \left[\begin{array}{c|c} \mathbf{D}_C & \mathbf{L}^2_{BC_2} + \mathbf{0} \\ \hline 0 & 1 \end{array}\right] = \left[\begin{array}{cc|c} 1 & 0 & 6+d_C \\ 0 & 1 & 0 \\ \hline 0 & 0 & 1 \end{array}\right] \tag{2.170f}$$

Node 3: Absolute homogeneous matrix for link m = 3 is determined from

$$\mathbf{R}_3 = \mathbf{R}_2 * \mathbf{IT}_{BC} = \left[\begin{array}{cc|c} C\theta_B & -S\theta_B & 5+d_A+(6+d_C)\cdot C\theta_B \\ S\theta_B & C\theta_B & 31+(6+d_C)\cdot S\theta_B \\ \hline 0 & 0 & 1 \end{array}\right] \tag{2.170g}$$

Output: Orientation matrix $\mathbf{D}_3$ and position vector $\mathbf{r}_C$

2. Absolute end-effector orientation matrix and frame's E position vector at any time.

   Relative homogeneous matrix at origin of the end-effector is defined as

$$\mathbf{IT}_{CE} = \left[\begin{array}{c|c} \mathbf{D}_E & \mathbf{L}_{CE}^3 + \mathbf{0} \\ \hline 0 & 1 \end{array}\right] = \left[\begin{array}{cc|c} 1 & 0 & 38 \\ 0 & 1 & 0 \\ \hline 0 & 0 & 1 \end{array}\right] \tag{2.170h}$$

   where

   - the orientation of frame E relative to frame C is defined by the unit matrix, $\mathbf{D}_E = \mathbf{U}$
   - $\mathbf{d}_E = \mathbf{0}$, given a null displacement vector at E

     End-effector E: Absolute homogeneous matrix for the end-effector is determined from

$$\mathbf{R}_e^{mech} = \mathbf{R}_4 * \mathbf{IT}_{DE} = \left[\begin{array}{cc|c} C\theta_B & -S\theta_B & 5 + d_A + (44 + d_C)\cdot C\theta_B \\ S\theta_B & C\theta_B & 31 + (44 + d_C)\cdot S\theta_B \\ \hline 0 & 0 & 1 \end{array}\right] \tag{2.170i}$$

     Output: Orientation matrix $\mathbf{D}_e$ and position vector $\mathbf{r}_E$:

$$\mathbf{D}_e = \begin{array}{c} \\ i_0 \\ j_0 \end{array} \begin{array}{c} \begin{array}{cc} i_e & j_e \end{array} \\ \begin{bmatrix} C\alpha_e^{I} & C\alpha_e^{II} \\ C\beta_e^{I} & C\beta_e^{II} \end{bmatrix} \end{array} = \begin{bmatrix} C\theta_B & -S\theta_B \\ -S\theta_B & C\theta_B \end{bmatrix} \tag{2.170j}$$

$$\begin{Bmatrix} x_E \\ y_E \end{Bmatrix} = \begin{Bmatrix} 5 + d_A + (44 + d_C)\cdot C\theta_B \\ 31 + (44 + d_C)\cdot S\theta_B \end{Bmatrix} \tag{2.170k}$$

3. Mechanism's equations for direct orientation and position at any time

   - *Input actuating parameters*: M = 3 angular and linear joint displacements within the matrix:

$$q_M(t) = \left\{ \begin{array}{c|c|c} d_A & \theta_B & d_C \end{array} \right\}^T \tag{2.170l}$$

   - *Equations for orientation*: Identify the entries located on the diagonal in Equation (2.158a):

$$\operatorname{Cos}\alpha_e^{I} = \operatorname{Cos}\theta_B; \quad \operatorname{Cos}\beta_e^{II} = \operatorname{Cos}\theta_B \tag{2.170m}$$

Then, the end-effector's frame shows the angles in Equation (2.170n) with respect to the fixed frame's axes:

$$\alpha_e^I = \beta_e^{II} = \text{Arccos}(\text{Cos}\,\theta_B) = \theta_B \tag{2.170n}$$

Since α_e^I and β_e^{II} are direction angles, the inverse trigonometric function "Arccos" has values within the interval 0°–180°.

- *Equations for position*:

$$\begin{aligned} x_E &= 5 + d_A + (44 + d_C)\cdot C\theta_B \\ y_E &= 31 + (44 + d_C)\cdot S\theta_B \end{aligned} \tag{2.170o}$$

4. Each link's orientation and frame position vectors at the home position.

   At the home position in Figure (2.33), the angular and linear displacements are zero:

$$d_A = 0\,[\text{in.}];\quad \theta_B = 0[°];\quad d_C = 0\,[\text{in.}]$$

The absolute $\mathbf{R}_m$ matrices from Equation (2.170c), (2.170e), and (2.170g) are

$$\mathbf{R}_1 = \left[\begin{array}{cc|c} 1 & 0 & 5 \\ 0 & 1 & 0 \\ \hline 0 & 0 & 1 \end{array}\right];\quad \mathbf{R}_2 = \left[\begin{array}{cc|c} 1 & 0 & 5 \\ 0 & 1 & 31 \\ \hline 0 & 0 & 1 \end{array}\right];\quad \mathbf{R}_3 = \left[\begin{array}{cc|c} 1 & 0 & 11 \\ 0 & 1 & 31 \\ \hline 0 & 0 & 1 \end{array}\right] \tag{2.170p}$$

From Equation (2.170p), the orientation matrices are unit matrices $\mathbf{D}_m = \mathbf{U}$, where m = 1, 2, and 3. Therefore, all frames have the axes parallel to the fixed frame axes, see Figure (2.36).

Identify the entries in the last columns of Equation (2.170p), then the position vectors can be deduced (Equation (2.170q)):

$$\begin{aligned} &\mathbf{r}_A = \begin{Bmatrix} x_A \\ y_A \end{Bmatrix} = \begin{Bmatrix} 5 \\ 0 \end{Bmatrix};\quad \mathbf{r}_B = \begin{Bmatrix} x_B \\ y_B \end{Bmatrix} = \begin{Bmatrix} 5 \\ 31 \end{Bmatrix}; \\ &\mathbf{r}_C = \begin{Bmatrix} x_C \\ y_C \end{Bmatrix} = \begin{Bmatrix} 11 \\ 31 \end{Bmatrix} \end{aligned} \tag{2.170q}$$

5. End-effector orientation and frame position vector at the home position

Because all the angular and linear displacements are zero, the matrix in Equation (2.170i) becomes

$$\mathbf{R}_e = \left[\begin{array}{cc|c} 1 & 0 & 49 \\ 0 & 1 & 31 \\ \hline 0 & 0 & 1 \end{array}\right] \tag{2.170r}$$

The orientation and position are determined by entries' identification in Equation (2.170s):

$$\left[\begin{array}{cc|c} \mathrm{Cos}\,\alpha_e^{I} & * & x_E \\ * & \mathrm{Cos}\,\beta_e^{II} & y_E \\ \hline 0 & 0 & 1 \end{array}\right] = \left[\begin{array}{cc|c} 1 & 0 & 49 \\ 0 & 1 & 31 \\ \hline 0 & 0 & 1 \end{array}\right] \tag{2.170s}$$

- *Orientation*: *Identify* the diagonal entries in the first two columns in Equation (2.170s), then the orientations can be found $\alpha_e^{I} = \beta_e^{II} = \mathrm{Arccos}(1) = 0°$. Therefore, at home position, the mechanism's end-effector's frame axes are parallel to the fixed frame axes, see Figure (2.33).
- *Position*: *Identify* the entries within the last column in Equation (2.170s), then

$$\mathbf{r}_E = \begin{Bmatrix} x_E \\ y_E \end{Bmatrix} = \begin{Bmatrix} 49 \\ 31 \end{Bmatrix} \tag{2.170t}$$

The end-effector coordinates correspond with those of point E in Figure (2.36).

2.16.10.1 The Inverse of Homogeneous Matrix

In the digraph of a mechanism, if an edge is opposite to the path, then an inverse homogeneous matrix is assigned to such an edge:

$$\mathbf{R}_m^{-1} = \left[\begin{array}{c|c} \mathbf{D}_m^{T} & -\mathbf{D}_m^{T} * \mathbf{r}_Z \\ \hline 0 & 1 \end{array}\right]; \quad \mathbf{IT}_{YZ}^{-1} = \left[\begin{array}{c|c} \mathbf{D}_Z^{T} & -\mathbf{D}_Z^{T} * \mathbf{L}_{YZ}^{m-1} \\ \hline 0 & 1 \end{array}\right] \tag{2.171}$$

An inverse is determined as a function of the transposed rotation matrix $\mathbf{D}_m^{T}$ because the following property holds:

$$\mathbf{R}_m * \mathbf{R}_m^{-1} = \mathbf{U}_{3\times3}; \quad \mathbf{IT}_{YZ} * \mathbf{IT}_{YZ}^{-1} = \mathbf{U}_{3\times3} \tag{2.172}$$

2.17 Velocity Analysis

The previous sections are referring to displacement vectors: relative if expressed with respect to local frames and absolute if expressed with respect to the fixed frame 0. In this section, we say *relative velocity* (angular or linear) when the vector is written with respect to the local frame and *absolute velocity* (angular or linear) when the vector is written with respect to the fixed frame 0 (global).

2.17.1 Direct Angular Velocity Analysis for Open Cycle Spatial Mechanisms

Based on solid rigid kinematics, the vector difference between two links' absolute velocities is the *relative angular velocity* (Equation (2.173)):

$$\boldsymbol{\omega}_{0,m} - \boldsymbol{\omega}_{0,m-1} = \boldsymbol{\omega}_{Z_{m-1,m}} \tag{2.173a}$$

where

m = 1, 2,…, is the mobile link mechanism

- $\boldsymbol{\omega}_{0,m-1}$, $\boldsymbol{\omega}_{0,m}$ are the absolute angular velocities for link m−1 and m, expressed in the fixed frame 0 and measured in [rad/s] or [°/s]
- $\boldsymbol{\omega}_{Z_{m-1,m}}$ is the relative angular velocity at joint Z, for link m in rotation relative to link m−1

The joint Z's relative angular velocity is expressed in the fixed frame 0 with scalar $\dot{\theta}_Z$. For direct velocity analysis, the scalar $\dot{\theta}_Z$ is the joint's time rate of change of angular displacement. It is determined from the derivative of angular displacement. For inverse velocity analysis, the unknown scalars are determined from the equations written one for each cycle, as shown further in Section 2.17.6:

$$\boldsymbol{\omega}^0_{Z_{m-1,m}} = \mathbf{u}^0_{Z_{m-1,m}} \cdot \dot{\theta}_Z \tag{2.173b}$$

- $\mathbf{u}^m_Z$ is the unit vector for joint Z's axis of rotation. The relative velocity is expressed in its local frame m and then in frame 0 by a left-side multiplication by the absolute rotation matrix:

$$\mathbf{u}^0_Z = \mathbf{D}_{0,m} * \mathbf{u}^m_{Z_{m-1,m}} \tag{2.173c}$$

Notation: Once defined on the digraph, the subscripts can be dropped for in calculations for simplicity. Thus,

- The 0-subscript is dropped in: $\boldsymbol{\omega}_{0,m} = \boldsymbol{\omega}_m$ and $\boldsymbol{\omega}_{0,m-1} = \boldsymbol{\omega}_{m-1}$. Since they are absolute vectors, with respect to the fixed frame 0, the first subscript is omitted.

- For $\boldsymbol{\omega}^0_{Z_{m-1,m}} = \omega_Z$, the relative angular velocity at joint Z where are omitted the subscripts for Z. Since it is written with respect to frame 0, the upper script is omitted.

Using the new notation, the previous three equations can be written as

$$\omega_m - \omega_{m-1} = \omega_Z \tag{2.173d}$$

where

$$\omega^0_Z = \mathbf{u}^0_Z \cdot \dot{\theta}_Z \text{ and } \mathbf{u}^0_Z = \mathbf{D}_m * \mathbf{u}^m_Z \tag{2.173e}$$

Refering to Figure 2.50, an absolute velocity $\boldsymbol{\omega}_m$ is assigned to each node m in the digraph and represents the mobile link m's angular velocity relative to the fixed frame. An observer on the ground sees the link m in rotation at $\boldsymbol{\omega}_m$. A relative velocity $\boldsymbol{\omega}_Z$ is assigned to each edge Z in the digraph. An observer on link m−1 observes the link m in rotation at $\boldsymbol{\omega}_Z$.

The absolute velocity at node m is the result of adding the previous node's m−1 absolute angular velocity to the relative angular velocity at the edge Z between the two adjacent nodes m−1 and m, as shown in the recursion Equation (2.173f):

$$\omega_m = \omega_{m-1} + \omega_Z \tag{2.173f}$$

A recursive procedure: m = 1, 2, 3,..., m from Equation (2.173f) leads to the following conclusion.

Statement 2.12

The absolute velocity at node m is the result of the addition of all relative velocity of the edges located before the node m, as shown in Equation (2.174):

$$\omega_m = \omega_A\big|_A + \omega_B\big|_B + \omega_C\big|_C + \cdots + \omega_Z\big|_Z = \omega_{m-1} + \omega_Z \tag{2.174}$$

Considering Equation (2.173e) then

$$\omega_m = \mathbf{u}^0_A \cdot \dot{\theta}_A\big|_A + \mathbf{u}^0_B \cdot \dot{\theta}_B\big|_B + \mathbf{u}^0_C \cdot \dot{\theta}_C\big|_C + \cdots + \mathbf{u}^0_Z \cdot \dot{\theta}_Z\big|_Z = \omega_{m-1} + \omega_Z \tag{2.175}$$

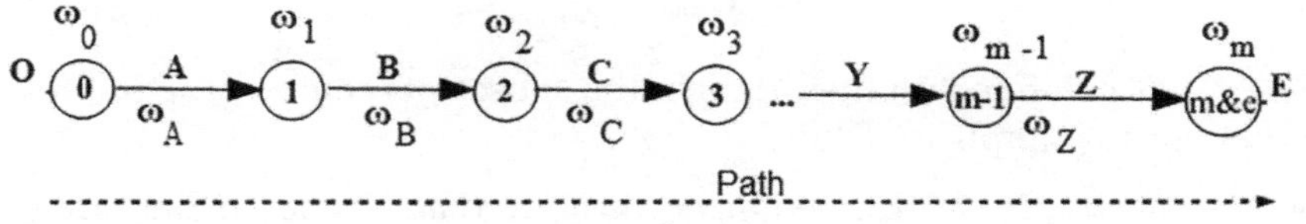

FIGURE 2.50
Relation between absolute and relative angular velocities along the tree.

Thus, to determine $\boldsymbol{\omega}_1$, then the term on the left side of bracket $|_A$ is considered. To determine $\boldsymbol{\omega}_2$, then all the terms on the left side of bracket $|_B$ are considered.

Notation:

- Notice the analogy between Equations (2.175) and (2.20) from the orientation analysis. Equation (2.175) is developed from Equation (2.20) when matrix **D** is replaced by vector $\boldsymbol{\omega}$ and the multiplication sign (*) is replaced by the addition sign (+).

  The recursive procedure along the path in Figure 2.50 is developed in Equation (2.176).
- *Node 0*: The link 0 is fixed, and its absolute angular velocity is zero:

$$\boldsymbol{\omega}_0 = 0 \tag{2.176a}$$

- *Edge A*: Assigned relative angular velocity:

$$\boldsymbol{\omega}_A = \mathbf{u}_A^0 \cdot \dot{\theta}_A; \quad \text{from Equation (2.173e)} \tag{2.176b}$$

- *Node m = 1*: The link 1's absolute angular velocity:

$$\boldsymbol{\omega}_1 = \boldsymbol{\omega}_0 + \mathbf{u}_A^0 \cdot \dot{\theta}_A \tag{2.176c}$$

- *Edge B*: Assigned relative angular velocity:

$$\boldsymbol{\omega}_B = \mathbf{u}_B^0 \cdot \dot{\theta}_B; \quad \text{from Equation (2.173e)} \tag{2.176d}$$

- *Node m* = 2: The link 2's absolute angular velocity:

$$\boldsymbol{\omega}_2 = \boldsymbol{\omega}_1 + \mathbf{u}_B^0 \cdot \dot{\theta}_B \tag{2.176e}$$

- *Edge C*: Assigned relative angular velocity:

$$\boldsymbol{\omega}_C = \mathbf{u}_C^0 \cdot \dot{\theta}_C; \quad \text{from Equation (2.173e)} \tag{2.176f}$$

- *Node m = 3*: The link 3's orientation matrix:

$$\boldsymbol{\omega}_3 = \boldsymbol{\omega}_2 + \mathbf{u}_C^0 \cdot \dot{\theta}_C \tag{2.176g}$$

- *Edge Z*: Assigned relative matrix:

$$\boldsymbol{\omega}_Z; \quad \text{from Equation (2.173e)} \tag{2.176h}$$

- *Node m*: The link m's orientation matrix from Equation (2.174):

$$\boldsymbol{\omega}_m = \boldsymbol{\omega}_{m-1} + \boldsymbol{\omega}_Z = \mathbf{u}_A^0 \cdot \dot{\theta}_A + \mathbf{u}_B^0 \cdot \dot{\theta}_B + \mathbf{u}_C^0 \cdot \dot{\theta}_C + \cdots + \mathbf{u}_Z^0 \cdot \dot{\theta}_Z \tag{2.176i}$$

2.17.2 Automatic Generation of Mobile Links' Angular Velocities from the Path Matrix

This matrix was introduced in Equation (1.17) as $a \times m$ square matrix in which each row is assigned to an arc in tree and each column is assigned to an open path initiated at node 0 toward each node m. Its entries, noted $z_{a,m}$, are

- $z_{a,m} = +1$, if arc a (joint) is on the path originating at node 0 (fixed link) and ending at node (mobile link) m, and it has the same direction as the path
- $z_{a,m} = -1$, if arc a is on the path and opposite to the path
- $z_{a,m} = 0$, if arc a is not on the path
  Note: The matrix Z for the TRRT mechanism is an upper triangular matrix, where all z entries are +1. For multiple-cycle mechanisms, shown in Chapter 3, these entries are –1 (if the edge is oriented inverse to the path), 0, or +1.
  The automatic generation of mobile links' absolute angular velocities from Equation (2.176) is shown in Equation (2.177) [4,5]:

$$\boldsymbol{\omega}_m = \mathbf{Z}^T * \boldsymbol{\omega}_a \tag{2.177}$$

2.17.3 Example of Direct Angular Velocity Analysis for Open Cycle TRRT Spatial Mechanism with 4 DOF

For the direct velocity analysis of spatial TRRT mechanism, the links' absolute angular velocities are determined as a function of joints' relative angular velocities.

We consider the paths in Figure 2.51, where the relative angular velocities are assigned at edges, and the absolute angular velocities are assigned at nodes. Then, the absolute angular velocities are determined as the sum of relative velocities in the joints situated within the path originated from the fixed link 0.

- The path matrix is shown in Equation (2.178):

$$\mathbf{Z} = \begin{matrix} \\ A \\ B \\ C \\ D \end{matrix} \begin{matrix} \begin{matrix} 1 & \quad 2 & \quad 3 & \quad 4 \end{matrix} \\ \begin{bmatrix} z_{A,1} & z_{A,2} & z_{A,3} & z_{A,4} \\ 0 & z_{B,2} & z_{B,3} & z_{B,4} \\ 0 & 0 & z_{C,3} & z_{C,4} \\ 0 & 0 & 0 & z_{D,4} \end{bmatrix} \end{matrix} = \begin{bmatrix} +1 & +1 & +1 & +1 \\ 0 & +1 & +1 & +1 \\ 0 & 0 & +1 & +1 \\ 0 & 0 & 0 & +1 \end{bmatrix} \tag{2.178}$$

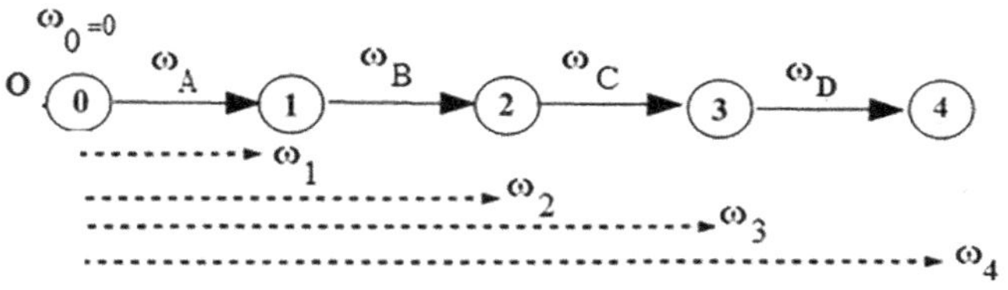

FIGURE 2.51
Relative and absolute velocities along the tree for TRRT mechanism.

- Then, Equation (2.177) holds:

$$\begin{Bmatrix} \omega_1 \\ \omega_2 \\ \omega_3 \\ \omega_4 \end{Bmatrix} = \begin{matrix} 1 \\ 2 \\ 3 \\ 4 \end{matrix} \overset{\begin{matrix} A & \quad B & \quad C & \quad D \end{matrix}}{\begin{bmatrix} z_{A,1} & 0 & 0 & 0 \\ z_{A,2} & z_{B,2} & 0 & 0 \\ z_{A,3} & z_{B,3} & z_{C,3} & 0 \\ z_{A,4} & z_{B,4} & z_{C,4} & z_{D,4} \end{bmatrix}} * \begin{Bmatrix} \omega_A \\ \omega_B \\ \omega_C \\ \omega_D \end{Bmatrix} \tag{2.178a}$$

$$\text{where} \begin{Bmatrix} \omega_A \\ \omega_B \\ \omega_C \\ \omega_D \end{Bmatrix} = \begin{Bmatrix} \mathbf{u}_A^0 \cdot \dot{\theta}_A \\ \mathbf{u}_B^0 \cdot \dot{\theta}_B \\ \mathbf{u}_C^0 \cdot \dot{\theta}_C \\ \mathbf{u}_D^0 \cdot \dot{\theta}_D \end{Bmatrix}$$

For the prismatic joints A and D, the relative angular velocities are $\dot{\theta}_A = \dot{\theta}_D = 0$.

The result of matrix multiplications in Equation (2.178b) is

$$\begin{Bmatrix} \omega_1 \\ \omega_2 \\ \omega_3 \\ \omega_4 \end{Bmatrix} = \begin{Bmatrix} z_{A,1} \cdot \omega_A \\ z_{A,2} \cdot \omega_A + z_{B,2} \cdot \omega_B \\ z_{A,3} \cdot \omega_A + z_{B,3} \cdot \omega_B + z_{C,3} \cdot \omega_C \\ z_{A,4} \cdot \omega_A + z_{B,4} \cdot \omega_B + z_{C,4} \cdot \omega_C + z_{D,4} \cdot \omega_D \end{Bmatrix} \tag{2.178b}$$

Each row in Equation (2.178b) determines the link's absolute angular velocity as a function of joints' relative velocities in the joints situated within the path originated from the fixed link 0.

- *Joints' relative angular velocities (rates)*: For M = 4 actuating joints: A (prism), B (rev), C (rev), and D (prism), the joint rates are assigned from Equation (2.178c):

$$\dot{\theta}_a = \left\{ \dot{\theta}_A \;\vdots\; \dot{\theta}_B \;\vdots\; \dot{\theta}_C \;\vdots\; \dot{\theta}_D \right\}^T = \left\{ 0 \;\vdots\; \mathbf{u}_B^0 \cdot \dot{\theta}_B \;\vdots\; \mathbf{u}_C^0 \cdot \dot{\theta}_C \;\vdots\; 0 \right\}^T \tag{2.178c}$$

Since joints A and D are prismatic joints, they are constrained for rotation: $\dot{\theta}_A = \dot{\theta}_D = 0$.

The angular rates at time t for revolute joints B and C are determined from the angular displacements (Equation (2.57b)):

$$\dot{\theta}_B(t) = \begin{cases} 0 & 0 \le t \le 1 \\ 0 & 1 < t \le 2 \\ 90 & 2 < t \le 3 \\ 0 & 3 < t \le 4 \end{cases} \; [°/s]; \quad \dot{\theta}_C(t) = \begin{cases} 0 & 0 \le t \le 1 \\ 0 & 1 < t \le 2 \\ 90 & 2 < t \le 3 \\ 0 & 3 < t \le 4 \end{cases} \; [°/s] \tag{2.178d}$$

- The components of unit vectors are determined as follows:
  - For joint A (prismatic along x_1), the entries in the first column in $\mathbf{D}_1$ are

$$\mathbf{u}_A^0 = \mathbf{D}_1 * \mathbf{u}_A^1 = \begin{bmatrix} 1 & 0 & 0 \\ 0 & 1 & 0 \\ 0 & 0 & 1 \end{bmatrix} * \begin{Bmatrix} 1 \\ 0 \\ 0 \end{Bmatrix} = \begin{Bmatrix} 1 \\ 0 \\ 0 \end{Bmatrix} \tag{2.178e}$$

  - For joint B (revolute about y_2), the entries in the second column in $\mathbf{D}_2$ are

$$\mathbf{u}_B^0 = \mathbf{D}_2 * \mathbf{u}_B^2 = \begin{bmatrix} C\theta_{B1,2} & 0 & S\theta_{B1,2} \\ 0 & 1 & 0 \\ -S\theta_{B1,2} & 0 & C\theta_{B1,2} \end{bmatrix} * \begin{Bmatrix} 0 \\ 1 \\ 0 \end{Bmatrix} = \begin{Bmatrix} 0 \\ 1 \\ 0 \end{Bmatrix} \tag{2.178f}$$

  - For joint C (revolute about z_3), the entries in the third column in $\mathbf{D}_3$ are

$$\mathbf{u}_C^0 = \mathbf{D}_3 * \mathbf{u}_C^3 = \begin{bmatrix} C\theta_B \cdot C\theta_C & -C\theta_B \cdot S\theta_C & S\theta_B \\ S\theta_C & C\theta_C & 0 \\ -S\theta_B \cdot C\theta_C & S\theta_B \cdot S\theta_C & C\theta_B \end{bmatrix} * \begin{Bmatrix} 0 \\ 0 \\ 1 \end{Bmatrix} = \begin{Bmatrix} S\theta_B \\ 0 \\ C\theta_B \end{Bmatrix} \tag{2.178g}$$

  - For joint D (prismatic along x_4), the entries in the first column in $\mathbf{D}_4$ are

$$\mathbf{u}_D^0 = \mathbf{D}_4 * \mathbf{u}_D^4 = \begin{bmatrix} C\theta_B \cdot C\theta_C & -C\theta_B \cdot S\theta_C & S\theta_B \\ S\theta_C & C\theta_C & 0 \\ -S\theta_B \cdot C\theta_C & S\theta_B \cdot S\theta_C & C\theta_B \end{bmatrix} * \begin{Bmatrix} 0 \\ 0 \\ 1 \end{Bmatrix} = \begin{Bmatrix} C\theta_B \cdot C\theta_C \\ S\theta_C \\ -S\theta_B \cdot C\theta_C \end{Bmatrix} \tag{2.178h}$$

Links' Absolute Angular Velocities

- Link 1's absolute angular velocity from row 1 in Equation (2.178b):

$$\boldsymbol{\omega}_1 = z_{A,1} \cdot \mathbf{0} = \mathbf{0} \tag{2.178i}$$

  - Magnitude link 1's absolute angular velocity: $|\boldsymbol{\omega}_1| = 0$, shown in Figure 2.52a

- Link 2's absolute angular velocity from row 2 in Equation (2.178b):

$$\boldsymbol{\omega}_2 = z_{A,2} \cdot \mathbf{0} + z_{B,2} \cdot \left(\mathbf{u}_B^0 \cdot \dot{\theta}_B\right) = \begin{Bmatrix} 0 \\ \dot{\theta}_B \\ 0 \end{Bmatrix} \tag{2.178j}$$

  - Magnitude link 2's absolute angular velocity: $|\boldsymbol{\omega}_2| = |\dot{\theta}_B|$, shown in Figure 2.52b

- Link 3's absolute angular velocity from row 3 in Equation (2.178b):

$$\boldsymbol{\omega}_3 = z_{A,3} \cdot \mathbf{0} + z_{B,3} \cdot \left(\mathbf{u}_B^0 \cdot \dot{\theta}_B\right) + z_{C,3} \cdot \left(\mathbf{u}_C^0 \cdot \dot{\theta}_C\right) = \begin{Bmatrix} S\theta_B \cdot \dot{\theta}_C \\ \dot{\theta}_B \\ C\theta_B \cdot \dot{\theta}_C \end{Bmatrix} \tag{2.178k}$$

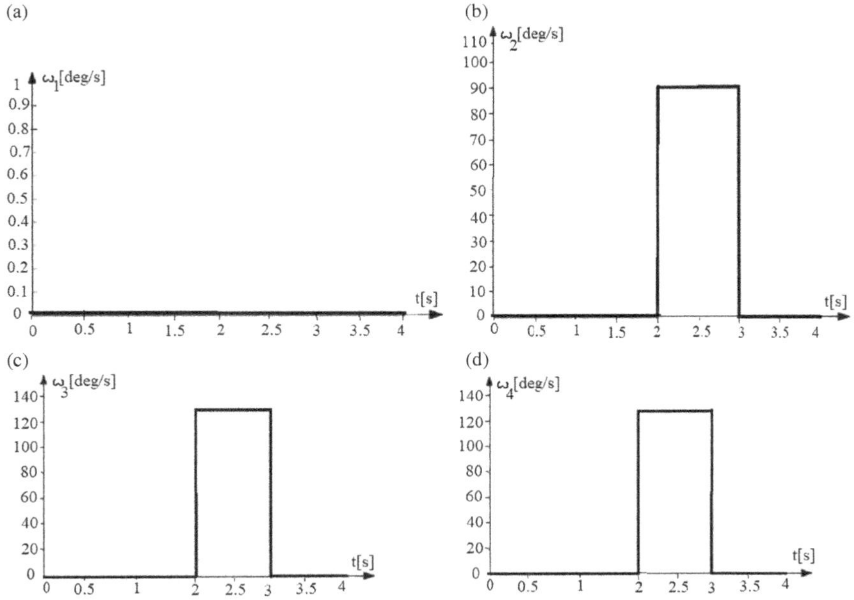

FIGURE 2.52
Absolute angular velocities for TRRT mechanism: (a) Link 1; (b) Link 2; (c) Link 3; (d) Link 4.

- Magnitude link 3's absolute angular velocity:
  $|\omega_3| = \sqrt{\dot{\theta}_B^2 + \dot{\theta}_C^2}$, shown in Figure 2.52c
- Link 4's absolute angular velocity from row 4 in Equation (2.178b):

$$\omega_4 = z_{A,4} \cdot \mathbf{0} + z_{B,4} \cdot \left(\mathbf{u}_B^0 \cdot \dot{\theta}_B\right) + z_{C,4} \cdot \left(\mathbf{u}_C^0 \cdot \dot{\theta}_C\right) + z_{D,4} \cdot \mathbf{0} = \begin{Bmatrix} S\theta_B \cdot \dot{\theta}_C \\ \dot{\theta}_B \\ C\theta_B \cdot \dot{\theta}_C \end{Bmatrix} \tag{2.178l}$$

- Magnitude link 4's absolute angular velocity:
  $|\omega_4| = \sqrt{\dot{\theta}_B^2 + \dot{\theta}_C^2}$, shown in Figure 2.52d

2.17.4 Example of Direct Angular Velocity Analysis for Open Cycle TRRTR Spatial Mechanism with 5 DOF

The TRRTR mechanism with 5 DOF in Figure 2.12a is generated from the TRRT mechanism with addition of a revolute joint E and a link 5 (flange). The extra revolute joint E allows rotation about axis x_5. The tool frame T is located on link 5 (node 5) (Figure 2.53).

- *Joints' relative angular velocities (rates)*: For M = 5 actuating joints: A (prism), B (rev), C (rev), D (prism), and E (rev), the joint rates are assigned from Equation (2.179a):

$$\begin{aligned} \dot{\theta}_a &= \left\{ \dot{\theta}_A \;\vdots\; \dot{\theta}_B \;\vdots\; \dot{\theta}_C \;\vdots\; \dot{\theta}_D \;\vdots\; \dot{\theta}_E \right\}^T \\ &= \left\{ 0 \;\vdots\; \mathbf{u}_B^0 \cdot \dot{\theta}_B \;\vdots\; \mathbf{u}_C^0 \cdot \dot{\theta}_C \;\vdots\; 0 \;\vdots\; \mathbf{u}_E^0 \cdot \dot{\theta}_E \right\}^T \end{aligned} \tag{2.179a}$$

Since joints A and D are prismatic joints, they are constrained for rotation: $\dot{\theta}_A = \dot{\theta}_D = 0$.

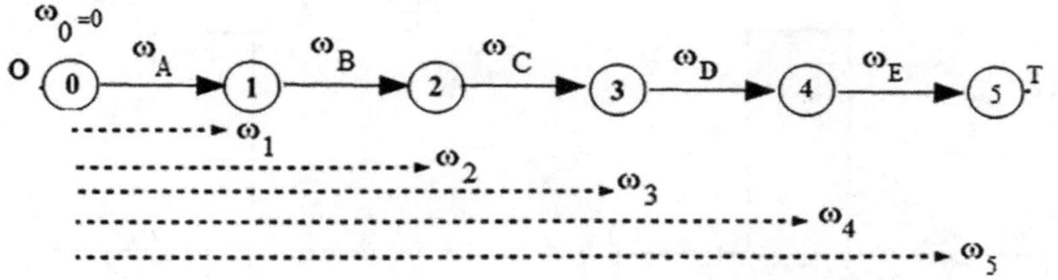

FIGURE 2.53
Relative and absolute velocities along the tree for TRRTR mechanism.

From Equation (2.179b), the rates for revolute joints B, C, and E at time t are

$$\dot{\theta}_B(t) = \begin{cases} 0 & 0 \le t \le 1 \\ 0 & 1 < t \le 2 \\ 90 & 2 < t \le 3 \\ 0 & 3 < t \le 4 \end{cases} [°/s];\dot{\theta}_C(t) = \begin{cases} 0 & 0 \le t \le 1 \\ 0 & 1 < t \le 2 \\ 90 & 2 < t \le 3 \\ 0 & 3 < t \le 4 \end{cases} [°/s];$$

$$\dot{\theta}_E(t) = \begin{cases} 0 & 0 \le t \le 1 \\ 0 & 1 < t \le 2 \\ 180 & 2 < t \le 3 \\ 0 & 3 < t \le 4 \end{cases} [°/s] \tag{2.179b}$$

- Then, Equation (2.177) holds:

$$\begin{Bmatrix} \omega_1 \\ \omega_2 \\ \omega_3 \\ \omega_4 \\ \omega_5 \end{Bmatrix} = \begin{matrix} \\ 1 \\ 2 \\ 3 \\ 4 \\ 5 \end{matrix} \begin{matrix} \begin{matrix} A & B & C & D & E \end{matrix} \\ \begin{bmatrix} z_{A,1} & 0 & 0 & 0 & 0 \\ z_{A,2} & z_{B,2} & 0 & 0 & 0 \\ z_{A,3} & z_{B,3} & z_{C,3} & 0 & 0 \\ z_{A,4} & z_{B,4} & z_{C,4} & z_{D,4} & 0 \\ z_{A,5} & z_{B,5} & z_{C,5} & z_{D,5} & z_{E,5} \end{bmatrix} \end{matrix} * \begin{Bmatrix} \omega_A \\ \omega_B \\ \omega_C \\ \omega_D \\ \omega_E \end{Bmatrix} \tag{2.179c}$$

where

$$\begin{Bmatrix} \omega_A \\ \omega_B \\ \omega_C \\ \omega_D \\ \omega_E \end{Bmatrix} = \begin{Bmatrix} \mathbf{u}_A^0 \cdot \dot{\theta}_A \\ \mathbf{u}_B^0 \cdot \dot{\theta}_B \\ \mathbf{u}_C^0 \cdot \dot{\theta}_C \\ \mathbf{u}_D^0 \cdot \dot{\theta}_D \\ \mathbf{u}_E^0 \cdot \dot{\theta}_E \end{Bmatrix} \tag{2.179d}$$

The result of matrix multiplications in Equation (2.179) is

$$\begin{Bmatrix} \omega_1 \\ \omega_2 \\ \omega_3 \\ \omega_4 \\ \omega_5 \end{Bmatrix} = \begin{Bmatrix} z_{A,1} \cdot \omega_A \\ z_{A,2} \cdot \omega_A + z_{B,2} \cdot \omega_B \\ z_{A,3} \cdot \omega_A + z_{B,3} \cdot \omega_B + z_{C,3} \cdot \omega_C \\ z_{A,4} \cdot \omega_A + z_{B,4} \cdot \omega_B + z_{C,4} \cdot \omega_C + z_{D,4} \cdot \omega_D \\ z_{A,5} \cdot \omega_A + z_{B,5} \cdot \omega_B + z_{C,5} \cdot \omega_C + z_{D,5} \cdot \omega_D + z_{E,5} \cdot \omega_E \end{Bmatrix} \tag{2.179e}$$

With all z-coefficients as +1

- The components of unit vectors are determined as follows:
  - From Equations (2.178e) to (2.178h), the unit vectors for joints A, B, C, and D are

$$\mathbf{u}_A^0 = \begin{Bmatrix} 1 \\ 0 \\ 0 \end{Bmatrix}; \quad \mathbf{u}_B^0 = \begin{Bmatrix} 1 \\ 0 \\ 0 \end{Bmatrix}; \quad \mathbf{u}_C^0 = \begin{Bmatrix} S\theta_B \\ 0 \\ C\theta_B \end{Bmatrix}; \quad \mathbf{u}_D^0 = \begin{Bmatrix} C\theta_B \cdot C\theta_C \\ S\theta_C \\ -S\theta_B \cdot C\theta_C \end{Bmatrix} \tag{2.179f}$$

  - For joint E (revolute about x_5), the entries in the third column in $\mathbf{D}_5$ from Equation (2.36c) are

$$\mathbf{u}_E^0 = \mathbf{D}_5 * \mathbf{u}_E^4$$

$$= \begin{bmatrix} C\theta_B \cdot C\theta_C & -C\theta_B \cdot S\theta_C \cdot C\theta_E + S\theta_B \cdot S\theta_E \cdot & C\theta_B \cdot S\theta_C \cdot S\theta_E + S\theta_B \cdot C\theta_E \\ S\theta_C & C\theta_C \cdot C\theta_E & -C\theta_C \cdot S\theta_E \\ -S\theta_B \cdot C\theta_C & S\theta_B \cdot S\theta_C \cdot C\theta_E + C\theta_B \cdot S\theta_E & -S\theta_B \cdot S\theta_C \cdot S\theta_E + C\theta_B \cdot C\theta_E \end{bmatrix}$$

$$* \begin{Bmatrix} 1 \\ 0 \\ 0 \end{Bmatrix} = \begin{Bmatrix} C\theta_B \cdot C\theta_C \\ S\theta_C \\ -S\theta_B \cdot C\theta_C \end{Bmatrix} \tag{2.179g}$$

- Link 1's, 2's, 3's, and 4's absolute angular velocities were developed in Equations (2.178i)–(2.178l) for the TRRT mechanism:

$$\boldsymbol{\omega}_1 = \mathbf{0}; \quad \boldsymbol{\omega}_2 = \begin{Bmatrix} 0 \\ \dot{\theta}_B \\ 0 \end{Bmatrix}; \quad \boldsymbol{\omega}_3 = \begin{Bmatrix} S\theta_B \cdot \dot{\theta}_C \\ \dot{\theta}_B \\ C\theta_B \cdot \dot{\theta}_C \end{Bmatrix}; \quad \boldsymbol{\omega}_4 = \begin{Bmatrix} S\theta_B \cdot \dot{\theta}_C \\ \dot{\theta}_B \\ C\theta_B \cdot \dot{\theta}_C \end{Bmatrix} \tag{2.179h}$$

  Link 5's absolute angular velocity from row 5 in Equation (2.179e) is

$$\boldsymbol{\omega}_5 = \boldsymbol{\omega}_A + \boldsymbol{\omega}_B + \boldsymbol{\omega}_C + \boldsymbol{\omega}_D + \boldsymbol{\omega}_E = \begin{Bmatrix} S\theta_B \cdot \dot{\theta}_C + C\theta_B \cdot C\theta_C \cdot \dot{\theta}_E \\ \dot{\theta}_B + S\theta_C \cdot \dot{\theta}_E \\ C\theta_B \cdot \dot{\theta}_C - S\theta_B \cdot C\theta_C \cdot \dot{\theta}_E \end{Bmatrix} \tag{2.179i}$$

  - Magnitude link 5's absolute angular velocity:

$$|\boldsymbol{\omega}_5| = \sqrt{\dot{\theta}_B^2 + \dot{\theta}_C^2 + \dot{\theta}_E^2 + 2 \cdot S\theta_C \cdot \dot{\theta}_B \cdot \dot{\theta}_E}$$

2.17.5 Example of Direct Angular Velocity Analysis for Open Cycle TRRTRT Spatial Mechanism with 6 DOF

The TRRTRT mechanism with 6 DOF in Figure 2.13a is generated from the TRRTR mechanism with addition of a prismatic joint F and a link 6. The extra prismatic joint F allows linear motion along axis y_6. The tool frame T is located on link 6 (node 6) (Figure 2.54).

- *Joints' relative angular velocities (rates)*: For M = 6 actuating joints: A (prism), B (rev), C (rev), D (prism), E (rev), and F (prism), the joint rates are assigned from Equation (2.180a):

$$\begin{aligned}\theta_a &= \left\{ \begin{array}{c:c:c:c:c:c} \dot{\theta}_A & \dot{\theta}_B & \dot{\theta}_C & \dot{\theta}_D & \dot{\theta}_E & \dot{\theta}_F \end{array} \right\}^T \\ &= \left\{ \begin{array}{c:c:c:c:c:c} 0 & \mathbf{u}_B^0 \cdot \dot{\theta}_B & \mathbf{u}_C^0 \cdot \dot{\theta}_C & 0 & \mathbf{u}_E^0 \cdot \dot{\theta}_E & 0 \end{array} \right\}^T\end{aligned} \tag{2.180a}$$

Since joints A, D, and F are prismatic joints, they are constrained for rotation: $\dot{\theta}_A = \dot{\theta}_D = \dot{\theta}_F = 0$.

The rates for revolute joints B, C, and E at time t are shown in Equation (2.179b):

$$\dot{\theta}_B(t) = \begin{cases} 0 & 0 \le t \le 1 \\ 0 & 1 < t \le 2 \\ 90 & 2 < t \le 3 \\ 0 & 3 < t \le 4 \end{cases} \left[{}^{\circ}/s\right]; \dot{\theta}_C(t) = \begin{cases} 0 & 0 \le t \le 1 \\ 0 & 1 < t \le 2 \\ 90 & 2 < t \le 3 \\ 0 & 3 < t \le 4 \end{cases} \left[{}^{\circ}/s\right]$$

$$\dot{\theta}_E(t) = \begin{cases} 0 & 0 \le t \le 1 \\ 0 & 1 < t \le 2 \\ 180 & 2 < t \le 3 \\ 0 & 3 < t \le 4 \end{cases} \left[{}^{\circ}/s\right] \tag{2.180b}$$

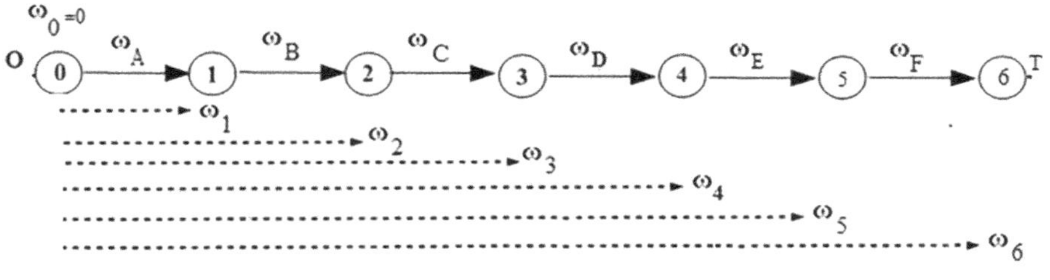

FIGURE 2.54
Relative and absolute velocities along the tree for TRRTRT mecha nism.

- Then, Equation (2.177) holds:

$$\begin{Bmatrix} \omega_1 \\ \omega_2 \\ \omega_3 \\ \omega_4 \\ \omega_5 \\ \omega_6 \end{Bmatrix} = \begin{matrix} 1 \\ 2 \\ 3 \\ 4 \\ 5 \\ 6 \end{matrix} \overset{\begin{matrix} A & B & C & D & E & F \end{matrix}}{\begin{bmatrix} z_{A,1} & 0 & 0 & 0 & 0 & 0 \\ z_{A,2} & z_{B,2} & 0 & 0 & 0 & 0 \\ z_{A,3} & z_{B,3} & z_{C,3} & 0 & 0 & 0 \\ z_{A,4} & z_{B,4} & z_{C,4} & z_{D,4} & 0 & 0 \\ z_{A,5} & z_{B,5} & z_{C,5} & z_{D,5} & z_{E,5} & 0 \\ z_{A,6} & z_{B,6} & z_{C,6} & z_{D,6} & z_{E,6} & z_{F,6} \end{bmatrix}} * \begin{Bmatrix} \omega_A \\ \omega_B \\ \omega_C \\ \omega_D \\ \omega_E \\ \omega_F \end{Bmatrix} \tag{2.180c}$$

where

$$\begin{Bmatrix} \omega_A \\ \omega_B \\ \omega_C \\ \omega_D \\ \omega_E \\ \omega_F \end{Bmatrix} = \begin{Bmatrix} \mathbf{u}_A^0 \cdot \dot{\theta}_A \\ \mathbf{u}_B^0 \cdot \dot{\theta}_B \\ \mathbf{u}_C^0 \cdot \dot{\theta}_C \\ \mathbf{u}_D^0 \cdot \dot{\theta}_D \\ \mathbf{u}_E^0 \cdot \dot{\theta}_E \\ \mathbf{u}_F^0 \cdot \dot{\theta}_F \end{Bmatrix} \tag{2.180d}$$

For the prismatic joints A, D, and F, the relative angular velocities are $\dot{\theta}_A = \dot{\theta}_D = \dot{\theta}_F = 0$.

The result of matrix multiplications in Equation (2.180c) is

$$\begin{Bmatrix} \omega_1 \\ \omega_2 \\ \omega_3 \\ \omega_4 \\ \omega_5 \\ \omega_6 \end{Bmatrix} = \begin{Bmatrix} z_{A,1} \cdot \omega_A \\ z_{A,2} \cdot \omega_A + z_{B,2} \cdot \omega_B \\ z_{A,3} \cdot \omega_A + z_{B,3} \cdot \omega_B + z_{C,3} \cdot \omega_C \\ z_{A,4} \cdot \omega_A + z_{B,4} \cdot \omega_B + z_{C,4} \cdot \omega_C + z_{D,4} \cdot \omega_D \\ z_{A,5} \cdot \omega_A + z_{B,5} \cdot \omega_B + z_{C,5} \cdot \omega_C + z_{D,5} \cdot \omega_D + z_{E,5} \cdot \omega_E \\ z_{A,6} \cdot \omega_A + z_{B,6} \cdot \omega_B + z_{C,6} \cdot \omega_C + z_{D,6} \cdot \omega_D + z_{E,6} \cdot \omega_E + z_{F,6} \cdot \omega_F \end{Bmatrix} \tag{2.180e}$$

with all z-coefficients as +1.

- The unit vectors for joints A, B, C, D, and E were developed in Equations (2.179f) and (2.179g) for the TRRTR mechanism:

$$\mathbf{u}_A^0 = \begin{Bmatrix} 1 \\ 0 \\ 0 \end{Bmatrix}; \quad \mathbf{u}_B^0 = \begin{Bmatrix} 1 \\ 0 \\ 0 \end{Bmatrix}; \quad \mathbf{u}_C^0 = \begin{Bmatrix} S\theta_B \\ 0 \\ C\theta_B \end{Bmatrix};$$

$$\mathbf{u}_D^0 = \begin{Bmatrix} C\theta_B \cdot C\theta_C \\ S\theta_C \\ -S\theta_B \cdot C\theta_C \end{Bmatrix}; \quad \mathbf{u}_E^0 = \begin{Bmatrix} C\theta_B \cdot C\theta_C \\ S\theta_C \\ -S\theta_B \cdot C\theta_C \end{Bmatrix} \tag{2.180f}$$

- For joint F (prismatic about y_6), the entries in the second column in $\mathbf{D}_6$ from Equation (2.37c) are

$$\mathbf{u}_F^0 = \mathbf{D}_6 * \mathbf{u}_F^5$$

$$= \left[\begin{array}{c:c:c} C\theta_B \cdot C\theta_C & -C\theta_B \cdot S\theta_C \cdot C\theta_E + S\theta_B \cdot S\theta_E \cdot & C\theta_B \cdot S\theta_C \cdot S\theta_E + S\theta_B \cdot C\theta_E \\ S\theta_C & C\theta_C \cdot C\theta_E & -C\theta_C \cdot S\theta_E \\ -S\theta_B \cdot C\theta_C & S\theta_B \cdot S\theta_C \cdot C\theta_E + C\theta_B \cdot S\theta_E & -S\theta_B \cdot S\theta_C \cdot S\theta_E + C\theta_B \cdot C\theta_E \end{array}\right]$$

$$* \begin{Bmatrix} 1 \\ 0 \\ 0 \end{Bmatrix} = \begin{Bmatrix} -C\theta_B \cdot S\theta_C \cdot C\theta_E + S\theta_B \cdot S\theta_E \\ C\theta_C \cdot C\theta_E \\ S\theta_B \cdot S\theta_C \cdot C\theta_E + C\theta_B \cdot S\theta_E \end{Bmatrix} \tag{2.180g}$$

- Link 1's, 2's, 3's, 4's, and 5's absolute angular velocities were developed in Equations (2.179h) and (2.179i) for the TRRT mechanism:

$$\boldsymbol{\omega}_1 = \mathbf{0}; \quad \boldsymbol{\omega}_2 = \begin{Bmatrix} 0 \\ \dot{\theta}_B \\ 0 \end{Bmatrix}; \quad \boldsymbol{\omega}_3 = \begin{Bmatrix} S\theta_B \cdot \dot{\theta}_C \\ \dot{\theta}_B \\ C\theta_B \cdot \dot{\theta}_C \end{Bmatrix}; \quad \boldsymbol{\omega}_4 = \begin{Bmatrix} S\theta_B \cdot \dot{\theta}_C \\ \dot{\theta}_B \\ C\theta_B \cdot \dot{\theta}_C \end{Bmatrix};$$

$$\boldsymbol{\omega}_5 = \begin{Bmatrix} S\theta_B \cdot \dot{\theta}_C + C\theta_B \cdot C\theta_C \cdot \dot{\theta}_E \\ \dot{\theta}_B + S\theta_C \cdot \dot{\theta}_E \\ C\theta_B \cdot \dot{\theta}_C - S\theta_B \cdot C\theta_C \cdot \dot{\theta}_E \end{Bmatrix} \tag{2.180h}$$

Link 6's absolute angular velocity from row 6 in Equation (2.180e) is

$$\boldsymbol{\omega}_6 = \boldsymbol{\omega}_A + \boldsymbol{\omega}_B + \boldsymbol{\omega}_C + \boldsymbol{\omega}_D + \boldsymbol{\omega}_E + \boldsymbol{\omega}_F = \begin{Bmatrix} S\theta_B \cdot \dot{\theta}_C + C\theta_B \cdot C\theta_C \cdot \dot{\theta}_E \\ \dot{\theta}_B + S\theta_C \cdot \dot{\theta}_E \\ C\theta_B \cdot \dot{\theta}_C - S\theta_B \cdot C\theta_C \cdot \dot{\theta}_E \end{Bmatrix} \tag{2.180i}$$

- Magnitude link 6's absolute angular velocity:

$$|\omega_6| = \sqrt{\dot{\theta}_B^2 + \dot{\theta}_C^2 + \dot{\theta}_E^2 + 2 \cdot S\theta_C \cdot \dot{\theta}_B \cdot \dot{\theta}_E}$$

2.17.6 The Matroid Method for Inverse Angular Velocity Analysis on a Closed Path

The above direct analysis finds the links' angular velocities based on angular velocities at actuating joints. The inverse analysis finds the actuating angular velocities to move the end-effector in a desired angular velocity task, required in the manipulator's control.

In the closed path shown in Figure 2.55, the cut-edge is oriented by the same rule as all the other edges in the digraph, that is, from the lower node number 0 to the higher node number m. The cut-edge is associated with a *fictitious joint* at E between node 0 and node m. Placed on the fixed link (node) 0, an observer is observing frame E on link m in rotation with the relative angular velocity $\omega_{E0,m}^{0} = \omega_E^{task}$.

- For open cycle mechanisms, the components in the absolute angular velocity of end-effector (desired task) are known, $\omega_E^{task} = \left\{ \begin{matrix} \omega_E^x & \omega_E^y & \omega_E^z \end{matrix} \right\}^T$.

  Since $\omega_E^{task} = \omega_E^{task} - \omega_0 = \omega_E^{task}$, then

$$\begin{Bmatrix} \omega_E^{x,task} \\ \omega_E^{y,task} \\ \omega_E^{z,task} \end{Bmatrix} = \begin{Bmatrix} \omega_E^{x,task} \\ \omega_E^{y,task} \\ \omega_E^{z,task} \end{Bmatrix} \tag{2.181}$$

- For closed-cycle mechanisms, analyzed further in Chapter 3, the cut-edge E is not anymore a fictitious joint; it is rather one of the mechanism's joints, and its unknown ω_E^0 is determined from the equations for inverse velocity analysis.

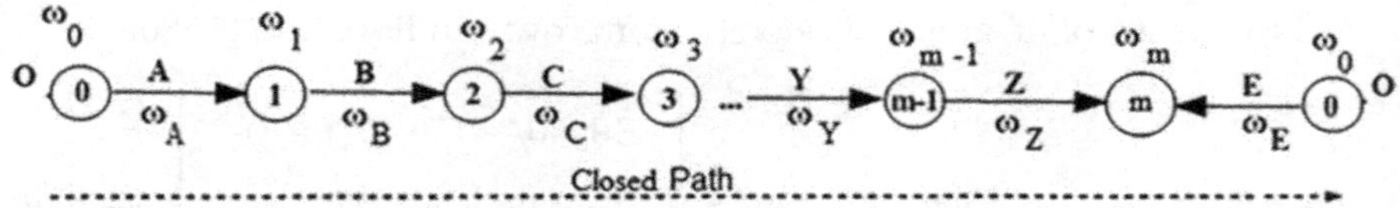

FIGURE 2.55
Relation between absolute and relative angular velocities along the cycle.

2.17.7 Cycle Basis Matrix (Matroid)

A closed path (cycle) is generated when c = 1 cut-edge is added to a given spanning tree. The direction of a path along the cycle's edges is arbitrarily. A superscript "–" sign is assigned to an edge which is contrary to the cycle's direction and "+" if the edge is in the path's direction. The set of edges which form the cycle is $C_E = \{A^+, B^+, C^+, \ldots, Y^+, Z^+, E^-\}$. The subscript E shows the chord closing the cycle.

The cycle matroid is the basis incidence matrix, C, and was defined as a $c \times j$ matrix in which one row is assigned to a single cycle and each column is assigned to an edge:

$$\begin{array}{c} \text{Edges} \\ \begin{array}{cccccccc} & A & B & C & & Y & Z & E \end{array} \\ C = \left[\begin{array}{c:c:c:c:c:c:c} +1 & +1 & +1 & \ldots & +1 & +1 & -1 \end{array}\right] \\ = \left[\begin{array}{c:c:c:c:c:c:c} c_{E,A} & c_{E,B} & c_{E,C} & \ldots & c_{E,Y} & c_{E,Z} & c_{E,E} \end{array}\right] \end{array} \tag{2.182}$$

2.17.8 The Relative Angular Velocity Matrix, ω_j

The relative angular velocities, assigned to joints (edges) on the cycle, have the direction of joints' unit vectors:

$$\begin{array}{c} \begin{array}{ccccc} A & B & C & & E \end{array} \\ \omega_j = \left[\begin{array}{ccccc} +1\cdot\omega_A^0 & +1\cdot\omega_B^0 & +1\cdot\omega_C^0 & \ldots & -1\cdot\omega_E^0 \end{array}\right] \\ = \left[\begin{array}{c:c:c:c:c} c_{E,A}\cdot\left(\mathbf{u}_A^0\cdot\dot{\theta}_A\right) & c_{E,B}\cdot\left(\mathbf{u}_B^0\cdot\dot{\theta}_B\right) & c_{E,C}\cdot\left(\mathbf{u}_C^0\cdot\dot{\theta}_C\right) & \ldots & -\omega_E^0 \end{array}\right] \end{array} \tag{2.183}$$

Statement 2.13

The sum of the entries on each row in the relative angular velocity matrix ω_j is zero:

$$\sum c_{E,j}\cdot\omega_j^0 = 0 \tag{2.184}$$

Proof: If Equation (2.173a) substitutes each entry in the left side of Equation (2.183), then matrix ω_j has the entries shown in Equation (2.185):

$$\omega_j = \left[\begin{array}{c:c:c:c:c:c} \omega_1 - \omega_0 & \omega_2 - \omega_1 & \omega_3 - \omega_2 & \omega_m - \omega_{m-1} & \ldots & -(\omega_m - \omega_0) \end{array}\right] \tag{2.185}$$

Notice that each vector shows twice with opposite sign; therefore, the sum of entries is a zero vector.

- *Analogy with forces in statics*: Let us consider the analogy: relative angular velocities at joints $\boldsymbol{\omega}_Z^0$ are analog to forces $\mathbf{F}_Z^0$ in statics. Then, the relative angular velocity at the chord E is analog to the resultant of forces:

$$c_{E,A}\cdot\boldsymbol{\omega}_A^0 + c_{E,B}\cdot\boldsymbol{\omega}_B^0 + c_{E,C}\cdot\boldsymbol{\omega}_C^0 + \cdots + c_{E,Z}\cdot\boldsymbol{\omega}_Z^0 = \boldsymbol{\omega}_E^0 \tag{2.186}$$

- *Scalar equations*: Equation (2.187) generates three scalar equations, as shown in the next example:

$$\left[\begin{array}{c|c|c|c|c} c_{E,A}\cdot\mathbf{u}_A^0 & c_{E,B}\cdot\mathbf{u}_B^0 & c_{E,C}\cdot\mathbf{u}_C^0 & \dots & c_{E,Z}\cdot\mathbf{u}_Z^0 \end{array}\right] * \begin{Bmatrix} \dot{\theta}_A \\ \dot{\theta}_B \\ \dot{\theta}_C \\ \dots \\ \dot{\theta}_Z \end{Bmatrix} = \boldsymbol{\omega}_E^0 \tag{2.187}$$

- *Joint constraints*: $\dot{\theta}_Z = 0$ for a prismatic joint
- *Geometric Jacobian for angular velocities*: The a-columns with vector entries in Equation (2.187), equal to the number of joints (arcs tree), form the columns in a $3\times a$ scalar matrix named *geometric Jacobian sub-matrix* for angular velocities $\mathbf{J}_a^{av}$. The joint rates are the entries in the $a\times 1$ column matrix $\dot{\mathbf{q}}_a$.

  Therefore, Equation (2.187) holds:

$$\mathbf{J}_a^{av} * \dot{\mathbf{q}}_a = \boldsymbol{\omega}_E^{task} \tag{2.188}$$

Example: For the TRRT spatial mechanism's inverse angular velocity analysis, the task absolute angular velocity is provided.

Data:

- The task absolute angular velocity at end-effector's frame $\boldsymbol{\omega}_4^{task} = \boldsymbol{\omega}_4^{task} - \boldsymbol{\omega}_0 = \boldsymbol{\omega}_4^{task}$, with the scalar components $\boldsymbol{\omega}_E^{task} = \begin{Bmatrix} \omega_E^{x,task} & \omega_E^{y,task} & \omega_E^{z,task} \end{Bmatrix}^T = \begin{Bmatrix} \omega_4^{x,task} & \omega_4^{y,task} & \omega_4^{z,task} \end{Bmatrix}^T$

  – The matrix Equation (2.187):

$$\begin{array}{cccc} A & B & C & D \end{array}$$

$$\left[\begin{array}{c|c|c|c} +1\cdot\mathbf{u}_A^0 & +1\cdot\mathbf{u}_B^0 & +1\cdot\mathbf{u}_C^0 & +1\cdot\mathbf{u}_D^0 \end{array}\right] * \begin{Bmatrix} \dot{\theta}_A \\ \dot{\theta}_B \\ \dot{\theta}_C \\ \dot{\theta}_D \end{Bmatrix} = \boldsymbol{\omega}_E^0 \tag{2.189a}$$

Considering the components of unit vectors from Equations (2.178e) to (2.178h), then

$$\left[\begin{array}{c:c:c:c} 1 & 0 & S\theta_B & C\theta_B \cdot C\theta_C \\ 0 & 1 & 0 & S\theta_C \\ 0 & 0 & C\theta_B & -S\theta_B \cdot C\theta_C \end{array}\right] * \begin{Bmatrix} \dot{\theta}_A \\ \dot{\theta}_B \\ \dot{\theta}_C \\ \dot{\theta}_D \end{Bmatrix} = \begin{Bmatrix} \omega_4^{x,task} \\ \omega_4^{y,task} \\ \omega_4^{z,task} \end{Bmatrix} \tag{2.189b}$$

- *Joint constraints*: Since joints A and D are prismatic joints, their relative angular velocities are zero
- *Scalar equations*: The three scalar equations for inverse angular velocity analysis from Equation (2.189b) are

$$\begin{aligned} &\text{Equation 1}: S\theta_B \cdot \dot{\theta}_C = \omega_4^{x,task} \\ &\text{Equation 2}: \dot{\theta}_B = \omega_4^{y,task} \\ &\text{Equation 3}: C\theta_B \cdot \dot{\theta}_C = \omega_4^{z,task} \end{aligned} \tag{2.189c}$$

The system of three linear equations has two unknowns $\dot{\theta}_B$ and $\dot{\theta}_C$. Since the number of equations is larger than the number of unknowns, in general, there are no actuation rates to move for any desired angular velocities: $\omega_4^{x,task}$, $\omega_4^{y,task}$, $\omega_4^{z,task}$. Solution exists only if it satisfies the condition:

$$S\theta_B \cdot \omega_4^{z,task} = C\theta_B \cdot \omega_4^{x,task} \tag{2.189d}$$

- *Solution*:

$$\begin{aligned} \dot{\theta}_B &= \omega_4^{y,task} \\ \dot{\theta}_C &= \frac{1}{C\theta_B} \omega_4^{z,task} \end{aligned} \tag{2.189e}$$

Singularity positions: For $\theta = 90°$, the mechanism has a singularity position in equation for $\dot{\theta}_C$

– *Geometric Jacobian for angular velocities*: It is the matrix of coefficients in Equation (2.189c) written as functions of joints rates as:

$$\left[\begin{array}{c:c:c:c} 0 & 0 & S\theta_B & 0 \\ 0 & 1 & 0 & 0 \\ 0 & 0 & C\theta_B & 0 \end{array}\right] * \left\{\begin{array}{c} \dot{d}_A \\ \dot{\theta}_B \\ \dot{\theta}_C \\ \dot{d}_D \end{array}\right\} = \left\{\begin{array}{c} \boldsymbol{\omega}_4^{x,task} \\ \boldsymbol{\omega}_4^{y,task} \\ \boldsymbol{\omega}_4^{z,task} \end{array}\right\}$$

where

$$\mathbf{J}_a^{av} = \left[\begin{array}{c:c:c:c} 0 & 0 & S\theta_B & 0 \\ 0 & 1 & 0 & 0 \\ 0 & 0 & C\theta_B & 0 \end{array}\right] \tag{2.189f}$$

2.17.9 Inverse Angular Velocity Analysis Equations for TRRTR Spatial Mechanism

- The matrix Equation (2.187):

A B C D E

$$\left[\begin{array}{c:c:c:c:c} +1\cdot\mathbf{u}_A^0 & +1\cdot\mathbf{u}_B^0 & +1\cdot\mathbf{u}_C^0 & +1\cdot\mathbf{u}_D^0 & -1\cdot\mathbf{u}_E^0 \end{array}\right] * \left\{\begin{array}{c} \dot{\theta}_A \\ \dot{\theta}_B \\ \dot{\theta}_C \\ \dot{\theta}_D \\ \dot{\theta}_E \end{array}\right\} = \boldsymbol{\omega}_5^0 \tag{2.190a}$$

Considering the components of unit vectors from Equations (2.179f) to (2.179g), then

$$\left[\begin{array}{c:c:c:c:c} 1 & 0 & S\theta_B & C\theta_B\cdot C\theta_C & C\theta_B\cdot C\theta_C \\ 0 & 1 & 0 & S\theta_C & C\theta_B\cdot C\theta_C \\ 0 & 0 & C\theta_B & -S\theta_B\cdot C\theta_C & -S\theta_B\cdot C\theta_C \end{array}\right]$$

$$* \left\{\begin{array}{c} \dot{\theta}_A \\ \dot{\theta}_B \\ \dot{\theta}_C \\ \dot{\theta}_D \\ \dot{\theta}_E \end{array}\right\} = \left\{\begin{array}{c} \boldsymbol{\omega}_5^{x,task} \\ \boldsymbol{\omega}_5^{y,task} \\ \boldsymbol{\omega}_5^{z,task} \end{array}\right\} \tag{2.190b}$$

- *Joint constraints*: Since joints A and D are prismatic joints, their relative angular velocities are zero
- *Scalar equations*: The three scalar equations for inverse angular velocity analysis are

$$\begin{aligned}
&\text{Equation 1}: S\theta_B \cdot \dot{\theta}_C + C\theta_B \cdot C\theta_C \cdot \dot{\theta}_E = \omega_5^{x,task}\\
&\text{Equation 2}: \theta_B + S\theta_C \cdot \dot{\theta}_E = \omega_4^{y,task}\\
&\text{Equation 3}: C\theta_B \cdot \dot{\theta}_C - S\theta_B \cdot C\theta_C \cdot \dot{\theta}_E = \omega_4^{z,task}
\end{aligned} \tag{2.190c}$$

$$\begin{bmatrix} 0 & S\theta_B & C\theta_B \cdot C\theta_C \\ 1 & 0 & S\theta_C \\ 0 & C\theta_B & -S\theta_B \cdot C\theta_C \end{bmatrix} * \begin{Bmatrix} \dot{\theta}_B \\ \dot{\theta}_C \\ \dot{\theta}_E \end{Bmatrix} = \begin{Bmatrix} \omega_5^{x,task} \\ \omega_5^{y,task} \\ \omega_5^{z,task} \end{Bmatrix}$$

Solution:

$$\begin{aligned}
\dot{\theta}_B &= \frac{1}{\Delta} \cdot \left(S\theta_B\, S\theta_C \omega_5^{z,task} + C\theta_C \omega_5^{y,task} - C\theta_B\, S\theta_C \omega_5^{x,task} \right)\\
\dot{\theta}_C &= \frac{1}{\Delta} \cdot \left(S\theta_B\, C\theta_C \omega_5^{x,task} + C\theta_B\, C\theta_C \omega_5^{z,task} \right)\\
\dot{\theta}_E &= \frac{1}{\Delta} \cdot \left(C\theta_B \omega_5^{x,task} - S\theta_B \omega_5^{z,task} \right)
\end{aligned} \tag{2.190d}$$

where

$$\Delta = C\theta_C \tag{2.190e}$$

Singularity positions: For $\Delta = 0$, the mechanism has a singularity position for $\dot{\theta}_B$, $\dot{\theta}_C$, and $\dot{\theta}_E$.

- *Geometric Jacobian for angular velocities*: It is the matrix of coefficients in Equation (2.190c) written as functions of joint rates as:

$$\left[\begin{array}{c:c:c:c:c} 0 & 0 & S\theta_B & 0 & C\theta_B \cdot C\theta_C \\ 0 & 1 & 0 & 0 & S\theta_C \\ 0 & 0 & C\theta_B & 0 & -S\theta_B \cdot C\theta_C \end{array}\right] * \begin{Bmatrix} \dot{d}_A \\ \dot{\theta}_B \\ \dot{\theta}_C \\ \dot{d}_D \\ \dot{\theta}_E \end{Bmatrix} = \begin{Bmatrix} \omega_5^{x,task} \\ \omega_5^{y,task} \\ \omega_5^{z,task} \end{Bmatrix} \tag{2.190f}$$

where

$$J^{av} = \left[\begin{array}{c:c:c:c:c} 0 & 0 & S\theta_B & 0 & C\theta_B \cdot C\theta_C \\ \mathbf{0} & \mathbf{1} & 0 & 0 & S\theta_C \\ 0 & 0 & C\theta_B & 0 & -S\theta_B \cdot C\theta_C \end{array}\right]$$

2.17.10 Inverse Angular Velocity Analysis Equations for TRRTRT Spatial Mechanism

- The matrix Equation (2.187):

$$\begin{matrix} \quad A \quad & \quad B \quad & \quad C \quad & \quad D \quad & \quad E \quad & \quad F \quad \end{matrix}$$

$$\left[+1\cdot\mathbf{u}_A^0 \;\vdots\; +1\cdot\mathbf{u}_B^0 \;\vdots\; +1\cdot\mathbf{u}_C^0 \;\vdots\; +1\cdot\mathbf{u}_D^0 \;\vdots\; +1\cdot\mathbf{u}_E^0 \;\vdots\; +1\cdot\mathbf{u}_F^0 \right] * \begin{Bmatrix} \dot{\theta}_A \\ \dot{\theta}_B \\ \dot{\theta}_C \\ \dot{\theta}_D \\ \dot{\theta}_E \\ \dot{\theta}_F \end{Bmatrix} = \boldsymbol{\omega}_6^0 \tag{2.191a}$$

Considering the components of unit vectors from Equations (2.179f) to (2.179g), then

$$\begin{bmatrix} 1 & 0 & S\theta_B & C\theta_B\cdot C\theta_C & C\theta_B\cdot C\theta_C & C\theta_B\cdot C\theta_C \\ 0 & 1 & 0 & S\theta_C & S\theta_C & S\theta_C \\ 0 & 0 & C\theta_B & -S\theta_B\cdot C\theta_C & -S\theta_B\cdot C\theta_C & -S\theta_B\cdot C\theta_C \end{bmatrix}$$

$$* \begin{Bmatrix} \dot{\theta}_A \\ \dot{\theta}_B \\ \dot{\theta}_C \\ \dot{\theta}_D \\ \dot{\theta}_E \\ \dot{\theta}_F \end{Bmatrix} = \begin{Bmatrix} \omega_6^{x,task} \\ \omega_6^{y,task} \\ \omega_6^{z,task} \end{Bmatrix} \tag{2.191b}$$

- *Joint constraints*: Since joints A, D, and F are prismatic joints, their relative angular velocities are zero
- *Scalar equations*: The three scalar equations for inverse angular velocity analysis are

$$\begin{aligned} &\text{Equation 1}: S\theta_B\cdot\dot{\theta}_C + C\theta_B\cdot C\theta_C\cdot\dot{\theta}_E = \omega_6^{x,task} \\ &\text{Equation 2}: \dot{\theta}_B + S\theta_C\cdot\dot{\theta}_E = \omega_6^{y,task} \\ &\text{Equation 3}: C\theta_B\cdot\dot{\theta}_C - S\theta_B\cdot C\theta_C\cdot\dot{\theta}_E = \omega_6^{z,task} \end{aligned} \tag{2.191c}$$

$$\begin{bmatrix} 0 & S\theta_B & C\theta_B \cdot C\theta_C \\ 1 & 0 & S\theta_C \\ 0 & C\theta_B & -S\theta_B \cdot C\theta_C \end{bmatrix} * \begin{Bmatrix} \dot{\theta}_B \\ \dot{\theta}_C \\ \dot{\theta}_E \end{Bmatrix} = \begin{Bmatrix} \omega_6^{x,task} \\ \omega_6^{y,task} \\ \omega_6^{z,task} \end{Bmatrix}$$

Solution:

$$\begin{aligned} \dot{\theta}_B &= \frac{1}{\Delta} \cdot \left(S\theta_B\, S\theta_C \omega_6^{z,task} + C\theta_C \omega_6^{y,task} - C\theta_B\, S\theta_C \omega_6^{x,task} \right) \\ \dot{\theta}_C &= \frac{1}{\Delta} \cdot \left(S\theta_B\, C\theta_C \omega_6^{x,task} + C\theta_B\, C\theta_C \omega_6^{z,task} \right) \\ \dot{\theta}_E &= \frac{1}{\Delta} \cdot \left(C\theta_B \omega_6^{x,task} - S\theta_B \omega_6^{z,task} \right) \end{aligned} \tag{2.191d}$$

where

$$\Delta = C\theta_C \tag{2.191e}$$

Critical positions: From Equation (2.188d), the values for actuating rates $\dot{\theta}_B$, $\dot{\theta}_C$, and $\dot{\theta}_E$ are undefined for $\Delta = 0$. The mechanism has critical positions for angular velocities $\theta_C = 90°;270°$.

- *Geometric Jacobian for angular velocities*: It is the matrix of coefficients in Equation (2.191d) written as functions of joint rates as:

$$\left[\begin{array}{c|c|c|c|c|c} 0 & 0 & S\theta_B & 0 & C\theta_B \cdot C\theta_C & 0 \\ 0 & 1 & 0 & 0 & S\theta_C & 0 \\ 0 & 0 & C\theta_B & 0 & -S\theta_B \cdot C\theta_C & 0 \end{array}\right] * \begin{Bmatrix} \dot{d}_A \\ \dot{\theta}_B \\ \dot{\theta}_C \\ \dot{d}_D \\ \dot{\theta}_E \\ \dot{d}_F \end{Bmatrix} = \begin{Bmatrix} \omega_6^{x,task} \\ \omega_6^{y,task} \\ \omega_6^{z,task} \end{Bmatrix}$$

where

$$J_a^{av} = \left[\begin{array}{c|c|c|c|c|c} 0 & 0 & S\theta_B & 0 & C\theta_B \cdot C\theta_C & 0 \\ 0 & 1 & 0 & 0 & S\theta_C & 0 \\ 0 & 0 & C\theta_B & 0 & -S\theta_B \cdot C\theta_C & 0 \end{array}\right] \tag{2.191f}$$

2.17.11 Direct Linear Velocity Analysis for Open Cycle Spatial Mechanisms

From linear positional analysis, the difference between two frame origins' position vectors is defined as the Latin matrix vector, $\mathbf{L}_{YZ}^0$ (Equation (2.192)):

$$\mathbf{r}_{Z_m} - \mathbf{r}_{Y_{m-1}} = \mathbf{L}_{YZ}^0 \tag{2.192}$$

A *Latin velocity vector,* $\dot{\mathbf{L}}^0_{YZ}$, is defined as the difference between two frame origins' linear velocities, and noted with an upper dot:

$$\mathbf{v}^0_{Z_m} - \mathbf{v}^0_{Y_{m-1}} = \dot{\mathbf{L}}^0_{YZ} \tag{2.193}$$

The right-side vector shows in a digraph edge and has two components according to the two vector equations from solid rigid kinematics:

- The relation between velocities of two points located on link m − 1

$$\mathrm{v}^0_{Z_{m-1}} - \mathrm{v}^0_{Y_{m-1}} = \omega^0_{m-1} \times \mathrm{L}^0_{YZ} \tag{2.194}$$

- The relation between velocities of two points located on different links m−1 and m, joined at Z

$$\mathbf{v}^0_{Z_m} - \mathbf{v}^0_{Z_{m-1}} = \dot{\mathbf{d}}^0_{Z_{m-1,m}} \tag{2.195}$$

Adding the terms on the left and right sides of Equations (2.194) and (2.195), and considering (2.192), then the Latin velocity vector is

$$\dot{\mathbf{L}}^0_{YZ} = \boldsymbol{\omega}^0_{m-1} \times \mathbf{L}^0_{YZ} + \dot{\mathbf{d}}^0_{Z_{m-1,m}} \tag{2.196}$$

where

- m = 1, 2,…, is the mobile link mechanism
- $\mathbf{v}^0_{Y_{m-1}}$, $\mathbf{v}^0_{Z_m}$ are the absolute linear velocities of link m−1's origin Y and link m's origin Z, measured in [ft/s] or [m/s]
- $\boldsymbol{\omega}^0_{m-1}$ is the absolute angular velocity for link m−1 expressed in the fixed frame 0, measured in [rad/s] or [°/s]
- $\dot{\mathbf{d}}^0_{Z_{m-1,m}}$ is the relative linear velocity at sliding joint Z, for link m sliding relative to link m−1
  The joint Z's relative linear velocity with respect to the fixed frame 0, and its scalar is noted $\dot{d}_Z$. For direct velocity analysis, the scalar $\dot{d}_Z$ is the joint's rate of change in time of linear displacement. It is determined from the derivative with respect to time from linear displacement $d_Z(t)$. For inverse velocity analysis, the scalars are unknown, determined from the equations written one for each cycle:

$$\dot{\mathbf{d}}^0_{Z_{m-1,m}} = \mathbf{u}^0_{Z_{m-1,m}} \cdot \dot{d}_Z \tag{2.197}$$

- $\mathbf{u}^m_Z$ is the unit vector for joint Z's axis of rotation. The relative velocity is expressed for easiness in its local frame m and then in frame 0 by left-side multiplication with absolute rotation matrix:

$$\mathbf{u}^0_Z = \mathbf{D}_m * \mathbf{u}^0_{Z_{m-1,m}} \tag{2.198}$$

Notation: Once defined on digraph for their relative meaning, the subscripts do not show in vector notations.

Thus,

- For $\mathbf{v}^0_{Y_{m-1}}$ and $\mathbf{v}^0_{Z_m}$, the subscripts m−1 and m are dropped. Since they are absolute vectors, with respect to the fixed frame 0, the upper script is omitted.
- For $\dot{\mathbf{d}}^0_{Z_{m-1,m}} = \dot{d}_Z$, relative linear velocity at joint Z does not show the two subscripts for the two links. Since is written with respect to frame 0, the upper script is omitted.

Thus, the previous equations are

$$\mathbf{v}_Z - \mathbf{v}_Z = \dot{\mathbf{L}}^0_{YZ}; \quad \text{difference between Y and Z frame origins' linear velocities} \tag{2.199a}$$

$$\dot{\mathbf{L}}^0_{YZ} = \omega_{m-1} \times \mathbf{L}^0_{YZ} + \dot{\mathbf{d}}_Z; \quad \text{Latin velocity vector} \tag{2.199b}$$

$$\dot{\mathbf{L}}^0_{YZ} = -\tilde{\mathbf{L}}^0_{YZ} * \omega_{m-1} + \dot{\mathbf{d}}_Z; \quad \text{matrix form for Latin velocity vector} \tag{2.199c}$$

The symbol "~" denotes a skew-symmetric matrix, and the symbol "*" represents a multiplication between a matrix and a column vector.

$$\dot{\mathbf{d}}^0_Z = \mathbf{u}^0_Z \cdot \dot{d}_Z; \quad \text{relative linear velocity at joint Z with respect to the fixed frame} \tag{2.199d}$$

$$\mathbf{u}^0_Z = \mathbf{D}_m * \mathbf{u}^m_Z; \quad \text{unit vector for relative linear velocity at joint Z} \tag{2.199e}$$

Refering to Figure 2.56, an absolute velocity $\mathbf{v}_Z$ shows at each node m in the digraph and represents the frame origin's Z linear velocity with respect to the fixed frame. An observer on the ground observes this origin moving with the linear velocity $\mathbf{v}_Z$. A relative velocity $\dot{\mathbf{d}}^0_Z$ shows at each edge Z in the digraph. An observer on link m−1 observes the link m in translation at the $\dot{\mathbf{d}}^0_Z$ rate.

The link m frame origin's linear velocity is the result of previous link m−1 frame origin's linear velocity added to joint Z's relative velocity. This shows in the recursion Equation (2.200):

$$\mathbf{v}_Z = \mathbf{v}_Y + \dot{\mathbf{L}}^0_{YZ} \tag{2.200}$$

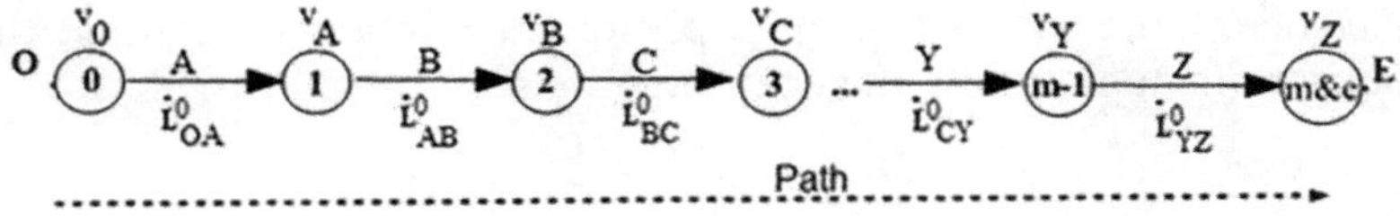

FIGURE 2.56
Relation between absolute and relative angular velocities along the tree.

A recursive procedure: m = 1, 2, 3,..., m from Equation (2.200) leads to the following conclusion:

Statement 2.14

The link m (node m-digraph) frame origin's linear velocity is the result of addition of all Latin velocity vectors of the edges located before the node m, as shown in Equation (2.201):

$$\mathbf{v}_Z = \dot{\mathbf{L}}^0_{OA}\Big|_A + \dot{\mathbf{L}}^0_{OB}\Big|_B + \dot{\mathbf{L}}^0_{OC}\Big|_C + \cdots + \dot{\mathbf{L}}^0_{YZ}\Big|_Z \tag{2.201}$$

Thus, to determine $\mathbf{v}_A$, then the term on the left side of bracket $|_A$ is considered. To determine $\mathbf{v}_B$, all the terms on the left side of bracket $|_B$ are considered.

Considering Equation (2.199e), the vector form of previous equation holds:

$$\begin{aligned}\mathbf{v}_Z = &\left(\boldsymbol{\omega}_0 \times \mathbf{L}^0_{OA} + \dot{\mathbf{d}}_A\right)\Big|_A + \left(\boldsymbol{\omega}_1 \times \mathbf{L}^0_{AB} + \dot{\mathbf{d}}_B\right)\Big|_B \\ &+ \left(\boldsymbol{\omega}_2 \times \mathbf{L}^0_{BC} + \dot{\mathbf{d}}_C\right)\Big|_C + \cdots + \left(\boldsymbol{\omega}_{m-1} \times \mathbf{L}^0_{YZ} + \dot{\mathbf{d}}_Z\right)\Big|_Z\end{aligned} \tag{2.202}$$

The matrix form of Equation (2.202) shows in Equation (2.203):

$$\begin{aligned}\mathbf{v}_Z = &\left(-\tilde{\mathbf{L}}^0_{OA} * \boldsymbol{\omega}_0 + \mathbf{u}^0_A \cdot \dot{d}_A\right)\Big|_A \left(-\tilde{\mathbf{L}}^0_{AB} * \boldsymbol{\omega}_1 + \mathbf{u}^0_B \cdot \dot{d}_B\right)\Big|_B \left(-\tilde{\mathbf{L}}^0_{BC} * \boldsymbol{\omega}_2 + \mathbf{u}^0_C \cdot \dot{d}_C\right)\Big|_C \\ &+ \cdots + \left(-\tilde{\mathbf{L}}^0_{YZ} * \boldsymbol{\omega}_{m-1} + \mathbf{u}^0_Z \cdot \dot{d}_Z\right)\Big|_Z\end{aligned} \tag{2.203}$$

The recursive procedure along the path in Figure 2.56 shows in Equations (2.204):

- *Node 0*: The link 0 is fixed, and its absolute angular and linear velocities are zero:

$$\boldsymbol{\omega}_0 = 0; \quad \mathbf{v}_O = 0; \tag{2.204a}$$

- *Edge A*: Assigned Latin velocity vector:

$$\dot{\mathbf{L}}^0_{OA} = -\tilde{\mathbf{L}}^0_{OA} * \boldsymbol{\omega}_0 + \mathbf{u}^0_A \cdot \dot{d}_A \tag{2.204b}$$

- *Node m = 1*: The link 1's frame origin absolute linear velocity:

$$\mathbf{v}_A = \mathbf{v}_O + \dot{\mathbf{L}}^0_{OA} = \dot{\mathbf{L}}^0_{OA} \tag{2.204c}$$

- *Edge B*: Assigned Latin velocity vector:

$$\dot{\mathbf{L}}^0_{AB} = -\tilde{\mathbf{L}}^0_{AB} * \boldsymbol{\omega}_1 + \mathbf{u}^0_B \cdot \dot{d}_B \tag{2.204d}$$

- *Node m = 2*: The link 2's frame origin absolute linear velocity:

$$\mathbf{v}_B = \mathbf{v}_A + \dot{\mathbf{L}}^0_{AB} = \dot{\mathbf{L}}^0_{OA} + \dot{\mathbf{L}}^0_{AB} \tag{2.204e}$$

- *Edge C*: Assigned relative angular velocity:

$$\dot{\mathbf{L}}^0_{BC} = -\tilde{\mathbf{L}}^0_{BC} * \boldsymbol{\omega}_2 + \mathbf{u}^0_C \cdot \dot{d}_C \tag{2.204f}$$

- *Node m = 3*: The link 3's frame origin absolute linear velocity:

$$\mathbf{v}_C = \mathbf{v}_B + \dot{\mathbf{L}}^0_{BC} = \dot{\mathbf{L}}^0_{OA} + \dot{\mathbf{L}}^0_{AB} + \dot{\mathbf{L}}^0_{BC} \tag{2.204g}$$

- …*Edge Z*: Assigned relative matrix:

$$\dot{\mathbf{L}}^0_{YZ} = -\tilde{\mathbf{L}}^0_{YZ} * \boldsymbol{\omega}_{m-1} + \mathbf{u}^0_Z \cdot \dot{d}_Z \tag{2.204h}$$

- *Node m*: The link m's frame origin absolute linear velocity:

$$\begin{aligned}\mathbf{v}_Z &= \mathbf{v}_Y + \dot{\mathbf{L}}^0_{YZ} = \dot{\mathbf{L}}^0_{OA} + \dot{\mathbf{L}}^0_{AB} + \dot{\mathbf{L}}^0_{BC} + \cdots + \dot{\mathbf{L}}^0_{YZ} \\ &= -\tilde{\mathbf{L}}^0_{OA} * \boldsymbol{\omega}_0 - \tilde{\mathbf{L}}^0_{AB} * \boldsymbol{\omega}_1 - \tilde{\mathbf{L}}^0_{BC} * \boldsymbol{\omega}_2 - \cdots - \tilde{\mathbf{L}}^0_{YZ} * \boldsymbol{\omega}_{m-1} \\ &\quad + \mathbf{u}^0_A \cdot \dot{d}_A + \mathbf{u}^0_B \cdot \dot{d}_B + \mathbf{u}^0_C \cdot \dot{d}_C + \cdots + \mathbf{u}^0_Z \cdot \dot{d}_Z\end{aligned} \tag{2.204i}$$

- *End-effector E*:

$$\begin{aligned}\mathbf{v}_E &= \mathbf{v}_Z + \dot{\mathbf{L}}^0_{ZE} = \dot{\mathbf{L}}^0_{OA} + \dot{\mathbf{L}}^0_{AB} + \dot{\mathbf{L}}^0_{BC} + \cdots + \dot{\mathbf{L}}^0_{YZ} + \cdots + \dot{\mathbf{L}}^0_{ZE} \\ &= -\tilde{\mathbf{L}}^0_{OA} * \boldsymbol{\omega}_0 - \tilde{\mathbf{L}}^0_{AB} * \boldsymbol{\omega}_1 - \tilde{\mathbf{L}}^0_{BC} * \boldsymbol{\omega}_2 - \cdots - \tilde{\mathbf{L}}^0_{YZ} * \boldsymbol{\omega}_{m-1} \\ &\quad - \tilde{\mathbf{L}}^0_{ZE} * \boldsymbol{\omega}_m + \mathbf{u}^0_A \cdot \dot{d}_A + \mathbf{u}^0_B \cdot \dot{d}_B + \mathbf{u}^0_C \cdot \dot{d}_C + \cdots + \mathbf{u}^0_Z \cdot \dot{d}_Z\end{aligned} \tag{2.204j}$$

Note: The vector form of Equation (2.204j) generated from a digraph represents the general equation in kinematics of a system of m mobile rigid links connected through a = A, B, C,…, Z joints. Since $\boldsymbol{\omega}_0 = \mathbf{0}$ for the fixed link 0, then

$$\mathbf{v}_E = \sum_{m=1}^{m} \boldsymbol{\omega}_m \times \mathbf{L}_{YZ}^0 + \sum_{a=A}^{Z} \dot{\mathbf{d}}_a \tag{2.204k}$$

2.17.12 Automatic Generation of All Mobile Links' Linear Velocities from the Path Matrix

This matrix was introduced in Equation (1.17) as a × m square matrix in which each row is assigned to an arc in tree and each column is assigned to an open path initiated at node 0 toward each node m. Its entries, noted $z_{a,m}$, are

- $z_{a,m} = +1$, if arc a (joint) is on the path originating at node 0 (fixed link) and ending at node (mobile link) m, and it has the same direction as the path
- $z_{a,m} = -1$, if arc a is on the path and opposite to the path
- $z_{a,m} = 0$, if arc a is not on the path

Note: The matrix Z for the TRRT mechanism in Figure 2.1 is an upper triangular matrix, where all z entries are +1.

For multiple-cycle mechanisms, shown in Chapter 3, these entries are –1, 0, or +1.

The automatic generation of mobile links' linear absolute velocities from Equation (2.197) shows in Equation (2.205):

$$\mathbf{v}_m = \mathbf{Z}^T * \dot{\mathbf{L}}_a^0 \tag{2.205}$$

where

$$\mathbf{v}_m = \begin{Bmatrix} \mathbf{v}_A \\ \mathbf{v}_B \\ \mathbf{v}_C \\ \cdots \\ \mathbf{v}_Z \end{Bmatrix} \text{and } \dot{\mathbf{L}}_a^0 = \begin{Bmatrix} \dot{\mathbf{L}}_{OA} \\ \dot{\mathbf{L}}_{AB} \\ \dot{\mathbf{L}}_{BC} \\ \cdots \\ \dot{\mathbf{L}}Z \end{Bmatrix} \tag{2.206}$$

2.17.13 Example of Direct Linear Velocity Analysis for Open Cycle TRRT Spatial Mechanism with 4 DOF

For the direct velocity analysis of spatial TRRT mechanism, the links' absolute linear velocities are determined as a function of joints' linear velocities.

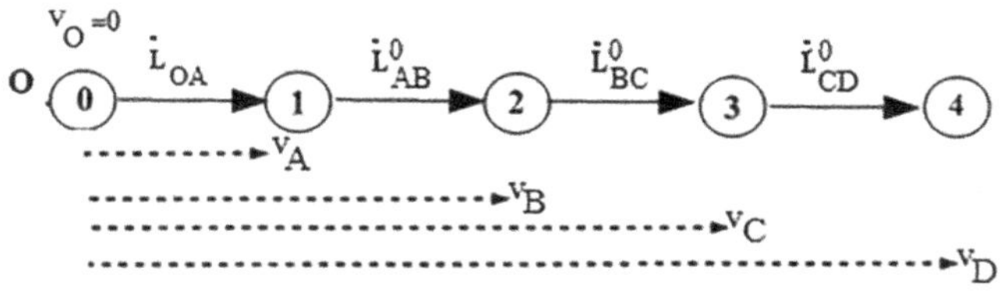

FIGURE 2.57
Relative and absolute velocities along the tree for TRRT mechanism.

We consider the paths in Figure 2.57, where the relative Latin velocities are shown at edges and the absolute linear velocities are shown at nodes. Then, the absolute velocities are determined as the sum of relative velocities in the joints situated within the path originated from the fixed link 0.

- The path matrix is shown in Equation (2.207a):

$$
\mathbf{Z} = \begin{matrix} & \begin{matrix} 1 & 2 & 3 & 4 \end{matrix} \\ \begin{matrix} A \\ B \\ C \\ D \end{matrix} & \begin{bmatrix} +1 & +1 & +1 & +1 \\ 0 & +1 & +1 & +1 \\ 0 & 0 & +1 & +1 \\ 0 & 0 & 0 & +1 \end{bmatrix} \end{matrix} = \begin{bmatrix} z_{A,1} & z_{A,2} & z_{A,3} & z_{A,4} \\ 0 & z_{B,2} & z_{B,3} & z_{B,4} \\ 0 & 0 & z_{C,3} & z_{C,4} \\ 0 & 0 & 0 & z_{D,4} \end{bmatrix} \tag{2.207a}
$$

Equation (2.205) holds:

$$
\begin{Bmatrix} \mathbf{v}_A \\ \mathbf{v}_B \\ \mathbf{v}_C \\ \mathbf{v}_D \end{Bmatrix} = \begin{matrix} & \begin{matrix} A & B & C & D \end{matrix} \\ \begin{matrix} 1 \\ 2 \\ 3 \\ 4 \end{matrix} & \begin{bmatrix} z_{A,1} & 0 & 0 & 0 \\ z_{A,2} & z_{B,2} & 0 & 0 \\ z_{A,3} & z_{B,3} & z_{C,3} & 0 \\ z_{A,4} & z_{B,4} & z_{C,4} & z_{D,4} \end{bmatrix} \end{matrix} * \begin{Bmatrix} \dot{\mathbf{L}}_{OA} \\ \dot{\mathbf{L}}_{AB} \\ \dot{\mathbf{L}}_{BC} \\ \dot{\mathbf{L}}_{CD} \end{Bmatrix} \tag{2.207b}
$$

The result of matrix multiplications in Equation (2.207b) is

$$
\begin{Bmatrix} \mathbf{v}_A \\ \mathbf{v}_B \\ \mathbf{v}_C \\ \mathbf{v}_D \end{Bmatrix} = \begin{Bmatrix} \dot{\mathbf{L}}_{OA} \\ \dot{\mathbf{L}}_{OA} + \dot{\mathbf{L}}_{AB} \\ \dot{\mathbf{L}}_{OA} + \dot{\mathbf{L}}_{AB} + \dot{\mathbf{L}}_{BC} \\ \dot{\mathbf{L}}_{OA} + \dot{\mathbf{L}}_{AB} + \dot{\mathbf{L}}_{BC} + \dot{\mathbf{L}}_{CD} \end{Bmatrix} \tag{2.207c}
$$

Each row in Equation (2.207c) determines the link's absolute angular velocity as a function of joints' relative velocities in the joints situated within the path originated from the fixed link 0.

where

$$\begin{Bmatrix} \dot{\mathbf{L}}_{OA} \\ \dot{\mathbf{L}}_{AB} \\ \dot{\mathbf{L}}_{BC} \\ \dot{\mathbf{L}}_{CD} \end{Bmatrix} = \begin{Bmatrix} \boldsymbol{\omega}_0^0 \times \mathbf{L}_{OA}^0 + \dot{\mathbf{d}}_A^0 \\ \boldsymbol{\omega}_1^0 \times \mathbf{L}_{AB}^0 + \dot{\mathbf{d}}_B^0 \\ \boldsymbol{\omega}_2^0 \times \mathbf{L}_{BC}^0 + \dot{\mathbf{d}}_C^0 \\ \boldsymbol{\omega}_3^0 \times \mathbf{L}_{CD}^0 + \dot{\mathbf{d}}_D^0 \end{Bmatrix} = \begin{Bmatrix} -\tilde{\mathbf{L}}_{OA}^0 * \boldsymbol{\omega}_0^0 + \mathbf{u}_A^0 \cdot \dot{\mathbf{d}}_A \\ -\tilde{\mathbf{L}}_{AB}^0 * \boldsymbol{\omega}_1^0 + \mathbf{u}_B^0 \cdot \dot{\mathbf{d}}_B \\ -\tilde{\mathbf{L}}_{BC}^0 * \boldsymbol{\omega}_2^0 + \mathbf{u}_C^0 \cdot \dot{\mathbf{d}}_C \\ -\tilde{\mathbf{L}}_{CD}^0 * \boldsymbol{\omega}_3^0 + \mathbf{u}_D^0 \cdot \dot{\mathbf{d}}_D \end{Bmatrix} \quad (2.207d)$$

- *Joints' relative linear velocities (rates)*: For M = 4 actuating joints: A (prism), B (rev), C (rev), and D (prism), the linear relative displacements for all joints (arcs a-digraph) shown in Equation (2.181e):

$$\begin{aligned} \dot{\mathrm{d}}_a &= \left\{ \dot{\mathrm{d}}_A \;\vdots\; \dot{\mathrm{d}}_B \;\vdots\; \dot{\mathrm{d}}_C \;\vdots\; \dot{\mathrm{d}}_D \right\}^T \\ &= \left\{ \mathbf{u}_A^0 \cdot \dot{\mathrm{d}}_A \;\vdots\; \mathbf{u}_B^0 \cdot \dot{\mathrm{d}}_B \;\vdots\; \mathbf{u}_C^0 \cdot \dot{\mathrm{d}}_C \;\vdots\; \mathbf{u}_D^0 \cdot \dot{\mathrm{d}}_D \right\}^T \end{aligned} \quad (2.207e)$$

Since joints B and C are revolute joints, they are constrained for linear displacement: $\dot{d}_B = \dot{d}_C = 0$. Therefore,

$$\dot{\mathrm{d}}_a = \left\{ \dot{\mathrm{d}}_A \;\vdots\; \dot{\mathrm{d}}_B \;\vdots\; \dot{\mathrm{d}}_C \;\vdots\; \dot{\mathrm{d}}_D \right\}^T = \left\{ \mathbf{u}_A^0 \cdot \dot{\mathrm{d}}_A \;\vdots\; \mathbf{0} \;\vdots\; \mathbf{0} \;\vdots\; \mathbf{u}_D^0 \cdot \dot{\mathrm{d}}_D \right\}^T \quad (2.207f)$$

For the prismatic joints A and D from Equation (2.57b), the rates shown in Equation (2.207g) are

$$\dot{\mathrm{d}}_A(t) = \begin{cases} 60 & 0 \le t \le 1 \\ 0 & 1 < t \le 2 \\ -40 & 2 < t \le 3 \\ -20 & 3 < t \le 4 \end{cases} \ [\text{in./s}];$$

$$\dot{\mathrm{d}}_D(t) = \begin{cases} 0 & 0 \le t \le 1 \\ 20 & 1 < t \le 2 \\ 0 & 2 < t \le 3 \\ 0 & 3 < t \le 4 \end{cases} \ [\text{in./s}] \quad (2.207g)$$

- The unit vectors for actuated revolute joints B (about y_2) and C (about z_3) are

$$\mathbf{u}_A^0 = \mathbf{D}_1 * \mathbf{u}_A^1 = \left[\begin{array}{c:c:c} 1 & 0 & 0 \\ 0 & 1 & 0 \\ 0 & 0 & 1 \end{array}\right] * \begin{Bmatrix} 1 \\ 0 \\ 0 \end{Bmatrix} = \begin{Bmatrix} 1 \\ 0 \\ 0 \end{Bmatrix}$$

$$\mathbf{u}_B^0 = \mathbf{D}_2 * \mathbf{u}_B^2 = \left[\begin{array}{c:c:c} C\theta_{B_{1,2}} & 0 & S\theta_{B_{1,2}} \\ 0 & 1 & 0 \\ -S\theta_{B_{1,2}} & 0 & C\theta_{B_{1,2}} \end{array}\right] * \begin{Bmatrix} 0 \\ 1 \\ 0 \end{Bmatrix} = \begin{Bmatrix} 0 \\ 1 \\ 0 \end{Bmatrix}$$

$$\mathbf{u}_C^0 = \mathbf{D}_3 * \mathbf{u}_C^3 = \left[\begin{array}{c:c:c} C\theta_B \cdot C\theta_C & -C\theta_B \cdot S\theta_C & S\theta_B \\ S\theta & C\theta_C & 0 \\ -S\theta_B \cdot C\theta_C & S\theta_B \cdot S\theta_C & C\theta_B \end{array}\right] * \begin{Bmatrix} 0 \\ 0 \\ 1 \end{Bmatrix} = \begin{Bmatrix} S\theta_B \\ 0 \\ C\theta_B \end{Bmatrix}$$

$$\mathbf{u}_D^0 = \mathbf{D}_4 * \mathbf{u}_D^4 = \left[\begin{array}{c:c:c} C\theta_B \cdot C\theta_C & -C\theta_B \cdot S\theta_C & S\theta_B \\ S\theta_C & C\theta_C & 0 \\ -S\theta_B \cdot C\theta_C & S\theta_B \cdot S\theta_C & C\theta_B \end{array}\right] * \begin{Bmatrix} 1 \\ 0 \\ 0 \end{Bmatrix} = \begin{Bmatrix} C\theta_B \cdot C\theta_C \\ S\theta_C \\ -S\theta_B \cdot C\theta_C \end{Bmatrix}$$

(2.207h)

- *Latin vector-velocities*:

$$\begin{Bmatrix} \dot{\mathbf{L}}_{OA} \\ \dot{\mathbf{L}}_{AB} \\ \dot{\mathbf{L}}_{BC} \\ \dot{\mathbf{L}}_{CD} \end{Bmatrix} = \begin{Bmatrix} -\tilde{\mathbf{L}}_{OA}^0 * \boldsymbol{\omega}_0^0 + \mathbf{u}_A^0 \cdot \dot{d}_A \\ -\tilde{\mathbf{L}}_{AB}^0 * \boldsymbol{\omega}_1^0 + \mathbf{u}_B^0 \cdot \dot{d}_B \\ -\tilde{\mathbf{L}}_{BC}^0 * \boldsymbol{\omega}_2^0 + \mathbf{u}_C^0 \cdot \dot{d}_C \\ -\tilde{\mathbf{L}}_{CD}^0 * \boldsymbol{\omega}_3^0 + \mathbf{u}_D^0 \cdot \dot{d}_D \end{Bmatrix} = \begin{Bmatrix} \mathbf{u}_A^0 \cdot \dot{d}_A \\ \mathbf{0} \\ -\tilde{\mathbf{L}}_{BC}^0 * \boldsymbol{\omega}_2^0 \\ -\tilde{\mathbf{L}}_{CD}^0 * \boldsymbol{\omega}_3^0 + \mathbf{u}_D^0 \cdot \dot{d}_D \end{Bmatrix}$$

(2.207i)

  - Considering the absolute angular velocities from Equations (2.178i) to (2.178l), shown in Equation (2.207j):

$$\boldsymbol{\omega}_0 = 0; \quad \boldsymbol{\omega}_1 = 0; \quad \boldsymbol{\omega}_2 = \begin{Bmatrix} 0 \\ \dot{\theta}_B \\ 0 \end{Bmatrix};$$

$$\boldsymbol{\omega}_3 = \begin{Bmatrix} S\theta_B \cdot \dot{\theta}_C \\ \dot{\theta}_B \\ C\theta_B \cdot \dot{\theta}_C \end{Bmatrix}; \quad \boldsymbol{\omega}_4 = \begin{Bmatrix} S\theta_B \cdot \dot{\theta}_C \\ \dot{\theta}_B \\ C\theta_B \cdot \dot{\theta}_C \end{Bmatrix} \tag{2.207j}$$

  - Considering the frame origins' position vectors from Equation (2.56), then the Latin vectors are

$$\mathbf{L}_{OA} = \mathbf{r}_A = \begin{Bmatrix} 5+d_A \\ 0 \\ 0 \end{Bmatrix}; \quad \mathbf{L}_{AB} = \mathbf{r}_B - \mathbf{r}_A = \begin{Bmatrix} 0 \\ 3 \\ 0 \end{Bmatrix}; \quad \mathbf{L}_{BC} = \mathbf{r}_C - \mathbf{r}_B = \begin{Bmatrix} 0 \\ 25 \\ 0 \end{Bmatrix}$$

$$\mathbf{L}_{CD} = \mathbf{r}_D - \mathbf{r}_C = \begin{Bmatrix} (6.5+d_D)\cdot C\theta_B \cdot C\theta_C \\ (6.5+d_D)\cdot S\theta_C \\ -(6.5+d_D)\cdot S\theta_B \cdot C\theta_C \end{Bmatrix} = \begin{Bmatrix} x_{CD} \\ y_{CD} \\ z_{CD} \end{Bmatrix};$$

$$\mathbf{L}^0_{DE} = \mathbf{r}_E - \mathbf{r}_D = \begin{Bmatrix} 38\cdot C\theta_B \cdot C\theta_C \\ 38\cdot S\theta_C \\ -38\cdot S\theta_B \cdot C\theta_C \end{Bmatrix} = \begin{Bmatrix} x_{DE} \\ y_{DE} \\ z_{DE} \end{Bmatrix} \tag{2.207k}$$

– Their skew-symmetric matrices are

$$\tilde{\mathbf{L}}^0_{OA} = \begin{bmatrix} 0 & 0 & 0 \\ 0 & 0 & -(5+d_A) \\ 0 & (5+d_A) & 0 \end{bmatrix}; \quad \tilde{\mathbf{L}}^0_{AB} = \begin{bmatrix} 0 & 0 & 3 \\ 0 & 0 & 0 \\ -3 & 0 & 0 \end{bmatrix};$$

$$\tilde{\mathbf{L}}^0_{BC} = \begin{bmatrix} 0 & 0 & 25 \\ 0 & 0 & 0 \\ -25 & 0 & 0 \end{bmatrix}$$

$$\tilde{\mathbf{L}}^0_{CD} = \begin{bmatrix} 0 & -z_{CD} & y_{CD} \\ z_{CD} & 0 & -x_{CD} \\ -y_{CD} & x_{CD} & 0 \end{bmatrix}; \quad \tilde{\mathbf{L}}^0_{DE} = \begin{bmatrix} 0 & -z_{DE} & y_{DE} \\ z_{DE} & 0 & -x_{DE} \\ -y_{DE} & x_{DE} & 0 \end{bmatrix} \tag{2.207l}$$

Links' Absolute Linear Velocities

- *Node 0*: The link 0 is fixed, and its absolute angular and linear velocities are zero:

$$\mathbf{v}_O = 0 \tag{2.207m}$$

- *Node 1*: The link 1's frame origin absolute linear velocity from row 1 in Equation (2.207c):

$$\mathbf{v}_A = \dot{\mathbf{L}}_{OA}^0 = \begin{Bmatrix} \dot{d}_A \\ 0 \\ 0 \end{Bmatrix} \tag{2.207n}$$

- *Node 2*: The link 2's frame origin absolute linear velocity from row 2 in Equation (2.207c):

$$\mathbf{v}_B = \mathbf{v}_A + \dot{\mathbf{L}}_{AB}^0 = \begin{Bmatrix} \dot{d}_A \\ 0 \\ 0 \end{Bmatrix} + \begin{Bmatrix} 0 \\ 0 \\ 0 \end{Bmatrix} = \begin{Bmatrix} \dot{d}_A \\ 0 \\ 0 \end{Bmatrix} \tag{2.207o}$$

- *Node 3*: Link 3's absolute angular velocity from row 3 in Equation (2.207c):

$$\mathbf{v}_C = \mathbf{v}_B + \dot{\mathbf{L}}_{BC}^0 = \mathbf{v}_B - \tilde{\mathbf{L}}_{BC}^0 * \boldsymbol{\omega}_2^0 = \begin{Bmatrix} \dot{d}_A \\ 0 \\ 0 \end{Bmatrix}$$

$$- \begin{bmatrix} 0 & 0 & 25 \\ 0 & 0 & 0 \\ -25 & 0 & 0 \end{bmatrix} * \begin{Bmatrix} 0 \\ \dot{\theta}_B \\ 0 \end{Bmatrix} = \begin{Bmatrix} \dot{d}_A \\ 0 \\ 0 \end{Bmatrix} \tag{2.207p}$$

- *Node 4*: Link 4's absolute angular velocity from row 4 in Equation (2.207c):

$$\mathbf{v}_D = \mathbf{v}_C + \dot{\mathbf{L}}_{CD} = \mathbf{v}_C - \tilde{\mathbf{L}}_{CD}^0 * \boldsymbol{\omega}_3^0 + \mathbf{u}_D^0 \cdot \dot{d}_D$$

$$= \begin{Bmatrix} \dot{d}_A \\ 0 \\ 0 \end{Bmatrix} - \begin{bmatrix} 0 & -z_{CD} & y_{CD} \\ z_{CD} & 0 & -x_{CD} \\ -y_{CD} & x_{CD} & 0 \end{bmatrix} * \begin{Bmatrix} S\theta_B \cdot \dot{\theta}_C \\ \dot{\theta}_B \\ C\theta_B \cdot \dot{\theta}_C \end{Bmatrix} + \begin{Bmatrix} C\theta_B \cdot C\theta_C \\ S\theta_C \\ -S\theta_B \cdot C\theta_C \end{Bmatrix} \cdot \dot{d}_D$$

$$= \begin{Bmatrix} \dot{d}_A - (6.5 + d_D) \cdot S\theta_B \cdot C\theta_C \cdot \dot{\theta}_B - (6.5 + d_D) \cdot S\theta_C \cdot C\theta_B \cdot \dot{\theta}_C + C\theta_B \cdot C\theta_C \cdot \dot{d}_D \\ (6.5 + d_D) \cdot C\theta_C \cdot \dot{\theta}_C + S\theta_C \cdot \dot{d}_D \\ -(6.5 + d_D) \cdot C\theta_B \cdot C\theta_C \cdot \dot{\theta}_B + (6.5 + d_D) \cdot S\theta_C \cdot S\theta_B \cdot \dot{\theta}_C - S\theta_B \cdot C\theta_C \cdot \dot{d}_D \end{Bmatrix} \tag{2.207q}$$

- End-effector's E linear velocity is determined from Equation (2.194), for the two points D and E located on the same link 4:

$$\mathbf{v}_E^0 - \mathbf{v}_D^0 = \boldsymbol{\omega}_4^0 \times \mathbf{L}_{DE}^0 \tag{2.207r}$$

Therefore,

$$\mathbf{v}_E = \mathbf{v}_D + \dot{\mathbf{L}}_{DE} = \mathbf{v}_D - \tilde{\mathbf{L}}_{DE}^0 * \boldsymbol{\omega}_4^0$$

$$= \begin{Bmatrix} \dot{d}_A - (44.5 + d_D) \cdot S\theta_B \cdot C\theta_C \cdot \dot{\theta}_B - (44.5 + d_D) \cdot S\theta_C \cdot C\theta_B \cdot \dot{\theta}_C + C\theta_B \cdot C\theta_C \cdot \dot{d}_D \\ (44.5 + d_D) \cdot C\theta_C \cdot \dot{\theta}_C + S\theta_C \cdot \dot{d}_D \\ -(44.5 + d_D) \cdot C\theta_B \cdot C\theta_C \cdot \dot{\theta}_B + (44.5 + d_D) \cdot S\theta_C \cdot S\theta_B \cdot \dot{\theta}_C - S\theta_B \cdot C\theta_C \cdot \dot{d}_D \end{Bmatrix} = \begin{Bmatrix} v_E^x \\ v_E^y \\ v_E^z \end{Bmatrix} \tag{2.207s}$$

For the actuation shown in Figure 2.17, Figure 2.58a–c illustrates the frame's E origin velocity components obtained from SW Motion/Study/Results imported in an Excel spreadsheet.

2.17.14 Inverse Linear Velocity Analysis of Open Cycle Mechanisms with Equation Functions of Absolute Angular Velocities

The open path starts at origin O and ends at E. If we imagine the path to end at the initial origin O, then it generates a closed path, named "cycle." This closed path is a fictitious cycle, created by addition to the open path of a fictitious edge, named chord E. The chord is associated with a fictitious joint E between node 0 and node m.

Equation (2.195) extends from point E to the starting point O:

$$\mathbf{v}_O = \dot{\mathbf{L}}_{OA}^0 + \dot{\mathbf{L}}_{AB}^0 + \dot{\mathbf{L}}_{BC}^0 + \cdots + \dot{\mathbf{L}}_{YZ}^0 + \dot{\mathbf{L}}_{ZE}^0 - \dot{\mathbf{L}}_{OE}^0 = \mathbf{0} \tag{2.208}$$

In Figure 2.59, the edge E is oriented from node 0 to node m (low to higher number), therefore inverse to the direction of the path. Thus, the vector $\dot{\mathbf{L}}_{OE}^0$ shows with minus sign.

Since $\dot{\mathbf{L}}_{OA}^0 - \dot{\mathbf{L}}_{OE}^0 = \dot{\mathbf{L}}_{EA}^0$

Then, Equation (2.208) becomes

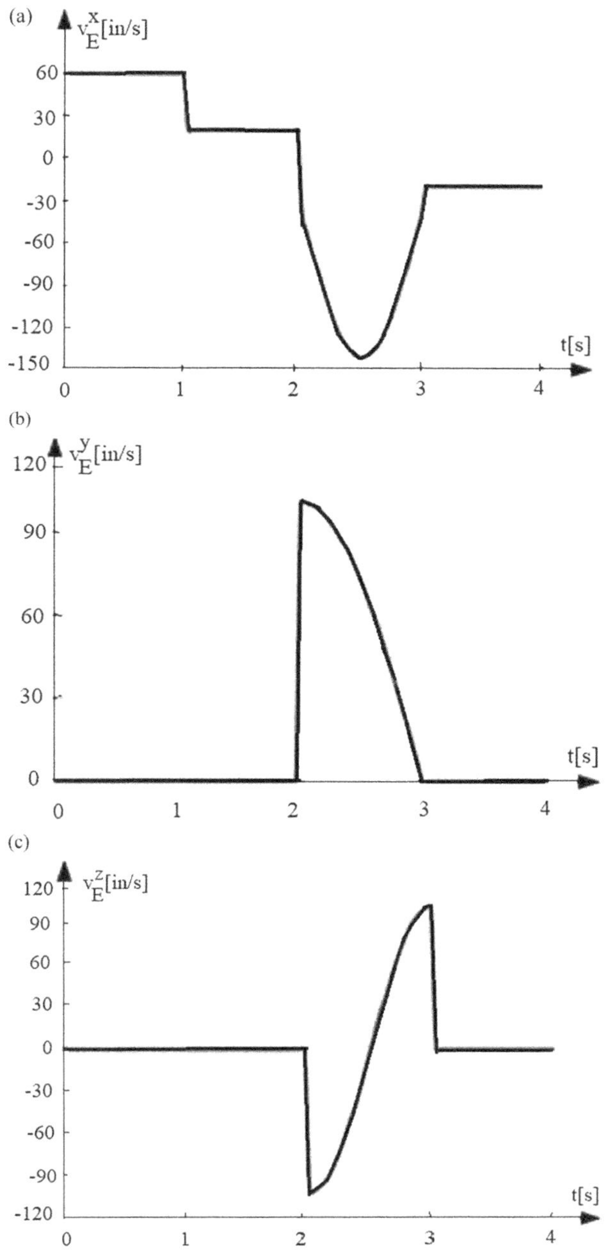

FIGURE 2.58
Components of end-effector velocity for TRRT mechanism: (a) x-component; (b) y-component; (c) z-component.

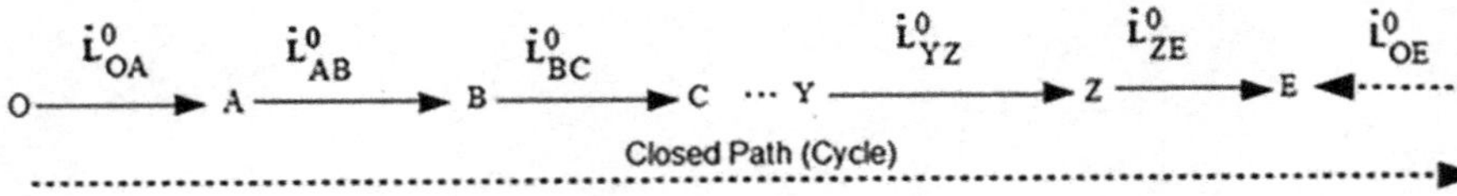

FIGURE 2.59
Latin velocity vectors between frame origins along the closed path.

$$\dot{\mathbf{L}}_{EA} + \dot{\mathbf{L}}_{AB} + \dot{\mathbf{L}}_{BC} + \dot{\mathbf{L}}_{CD} + \cdots + \dot{\mathbf{L}}_{ZE} = \mathbf{0} \tag{2.209}$$

- The velocity Latin matrix

  The Latin matrix, Equation (2.70), was defined as a matrix whose columns are the translation vectors between frame origins along a closed path. The number of columns in this matrix is m+1, that is, the number of nodes in the path. Moreover, the sum of entries in the Latin matrix is zero for a closed cycle in Equation (2.69).

  The velocity Latin matrix, $\dot{\mathbf{L}}$, has as columns the velocity vectors from Equation (2.209). These velocity vectors are generated from the entry vectors in Latin matrix. For example, an entry in the Latin matrix: $\mathbf{L}^0_{EA} = \mathbf{r}_A - \mathbf{r}_E$ generates a velocity Latin vector: $\dot{\mathbf{L}}^0_{EA} = \mathbf{v}_A - \mathbf{v}_E$:

$$\begin{aligned} & \qquad\quad\; 0 \qquad\;\; 1 \qquad\;\; 2 \quad\; \cdots \quad\; m \\ \dot{\mathbf{L}} &= \left[\; \dot{\mathbf{L}}^0_{EA} \;\vdots\; \dot{\mathbf{L}}^0_{AB} \;\vdots\; \dot{\mathbf{L}}^0_{BC} \;\vdots\; \dots \;\vdots\; \dot{\mathbf{L}}^0_{ZE} \;\right] \\ &= \left[\; \mathbf{v}_A - \mathbf{v}_E \;\vdots\; \mathbf{v}_B - \mathbf{v}_A \;\vdots\; \mathbf{v}_C - \mathbf{v}_B \;\vdots\; \dots \;\vdots\; \mathbf{v}_E - \mathbf{v}_Z \;\right] \end{aligned} \tag{2.210}$$

Statement 2.15

The sum of vector entries in Latin velocity matrix $\dot{\mathbf{L}}$ is zero:

$$\dot{\mathbf{L}}_{EA} + \dot{\mathbf{L}}_{AB} + \dot{\mathbf{L}}_{BC} + \dot{\mathbf{L}}_{CD} + \cdots + \dot{\mathbf{L}}_{ZE} = \mathbf{0} \tag{2.211}$$

Proof: In the right-side matrix, each vector shows twice and opposite sign; therefore, the sum of entries is zero.

- Equations for inverse velocity analysis for open-cycle spatial mechanisms

  Considering in Equation (2.190) that velocity vector of frame 0 is a zero vector, $\mathbf{v}_O = \mathbf{0}$, then

$$\mathbf{v}_{\mathrm{E}} - \mathbf{v}_{\mathrm{O}} = \dot{\mathbf{L}}_{\mathrm{OE}}^{0} \Rightarrow \mathbf{v}_{\mathrm{E}} = \dot{\mathbf{L}}_{\mathrm{OE}}^{0} \tag{2.212}$$

From Equation (2.208), then Equation (2.213) holds:

$$\dot{\mathbf{L}}_{\mathrm{OA}}^{0} + \dot{\mathbf{L}}_{\mathrm{AB}}^{0} + \dot{\mathbf{L}}_{\mathrm{BC}}^{0} + \cdots + \dot{\mathbf{L}}_{\mathrm{ZE}}^{0}\Big|_{\mathrm{E}} = \dot{\mathbf{L}}_{\mathrm{OE}}^{0} \tag{2.213}$$

- *The task velocity vector* is the end-effector linear velocity vector with the provided components for a specific task:

$$\dot{\mathbf{L}}_{\mathrm{OE}}^{0} = \mathbf{v}_{\mathrm{E}}^{\mathrm{task}} \tag{2.214}$$

It shows in Figure 2.59 as the edge's E velocity vector when a path is initiated at 0 from the right to left direction (inverse path).

Equation (2.215) represents the inverse velocity equation for open cycle mechanisms:

$$\dot{\mathbf{L}}_{\mathrm{OA}}^{0} + \dot{\mathbf{L}}_{\mathrm{AB}}^{0} + \dot{\mathbf{L}}_{\mathrm{BC}}^{0} + \cdots + \dot{\mathbf{L}}_{\mathrm{ZE}}^{0}\Big|_{\mathrm{E}} = \mathbf{v}_{\mathrm{E}}^{\mathrm{task}} \tag{2.215}$$

The sum on the left side is $\mathbf{v}_{\mathrm{E}}^{\mathrm{mech}}$, velocity of point E, determined as a function of actuation velocities in Equation (2.207s). Then, Equation (2.215) for the inverse velocity analysis is

$$\mathbf{v}_{\mathrm{E}}^{\mathrm{mech}} = \mathbf{v}_{\mathrm{E}}^{\mathrm{task}} \tag{2.216}$$

The linear actuation velocities are determined by equating the entries within the left (function of actuating linear displacements) and right (task-provided) matrix. In general, this leads to a system of linear equations.

Considering Equation (2.199c), Equation (2.215) holds:

$$\begin{aligned}&\left(-\tilde{\mathbf{L}}_{\mathrm{OA}}^{0} * \omega_0 + \dot{\mathbf{d}}_{\mathrm{A}}^{0}\right) + \left(-\tilde{\mathbf{L}}_{\mathrm{AB}}^{0} * \omega_1 + \dot{\mathbf{d}}_{\mathrm{B}}^{0}\right) + \left(-\tilde{\mathbf{L}}_{\mathrm{BC}}^{0} * \omega_2 + \dot{\mathbf{d}}_{\mathrm{C}}^{0}\right) + \cdots \\ &+ \left(-\tilde{\mathbf{L}}_{\mathrm{YZ}}^{0} * \omega_{m-1} + \dot{\mathbf{d}}_{\mathrm{Z}}^{0}\right) - \omega_m * \tilde{\mathbf{L}}_{\mathrm{ZE}}^{0} = \mathbf{v}_{\mathrm{E}}^{\mathrm{task}}\end{aligned} \tag{2.217}$$

Considering the angular velocity vector of frame 0 is a zero vector, $\boldsymbol{\omega}_0 = \mathbf{0}$, then

$$\begin{aligned}&\left(-\tilde{\mathbf{L}}_{\mathrm{AB}}^{0} * \boldsymbol{\omega}_1 - \tilde{\mathbf{L}}_{\mathrm{BC}}^{0} * \boldsymbol{\omega}_2 - \tilde{\mathbf{L}}_{\mathrm{YZ}}^{0} * \boldsymbol{\omega}_{m-1} - \boldsymbol{\omega}_m * \tilde{\mathbf{L}}_{\mathrm{ZE}}^{0}\right) \\ &+ \left(\dot{\mathbf{d}}_{\mathrm{A}}^{0} + \dot{\mathbf{d}}_{\mathrm{B}}^{0} + \dot{\mathbf{d}}_{\mathrm{C}}^{0} + \cdots + \dot{\mathbf{d}}_{\mathrm{Z}}^{0}\right) = \mathbf{v}_{\mathrm{E}}^{\mathrm{task}}\end{aligned} \tag{2.218}$$

Therefore, Equation (2.218) is symbollicaly written as

$$\left[\begin{array}{c:c:c:c:c} -\tilde{\mathbf{L}}_{AB}^{0} & -\tilde{\mathbf{L}}_{BC}^{0} & \ldots & -\tilde{\mathbf{L}}_{YZ}^{0} & -\tilde{\mathbf{L}}_{ZE}^{0} \end{array}\right] * \left\{\begin{array}{c} \omega_1 \\ \omega_2 \\ \ldots \\ \omega_{m-1} \\ \omega_m \end{array}\right\}$$

$$+\left[\begin{array}{c:c:c:c:c} \mathbf{u}_{A}^{0} & \mathbf{u}_{B}^{0} & \mathbf{u}_{C}^{0} & \ldots & \mathbf{u}_{Z}^{0} \end{array}\right] * \left\{\begin{array}{c} \dot{d}_A \\ \dot{d}_B \\ \dot{d}_C \\ \ldots \\ \dot{d}_Z \end{array}\right\} = \mathbf{v}_E^{task} \tag{2.219}$$

- *Link and joint constraints*:
  $\boldsymbol{\omega}_{m-1} = \boldsymbol{\omega}_m$, if Z is prismatic joint, links m−1 and m have the same angular velocity vectors.
  $\mathbf{d}_Z = 0$, if Z is revolute joint.
  Equation (2.213) represents the inverse velocity equation as a function of absolute angular velocities for open cycle mechanisms.

2.17.15 Example of Inverse Velocity Analysis for Open Cycle TRRT Spatial Mechanism with Equation Functions of Absolute Angular Velocities

For the TRRT mechanism, Equation (2.219) holds:

$$\left[\begin{array}{c:c:c:c} -\tilde{\mathbf{L}}_{AB}^{0} & -\tilde{\mathbf{L}}_{BC}^{0} & -\tilde{\mathbf{L}}_{CD}^{0} & -\tilde{\mathbf{L}}_{DE}^{0} \end{array}\right] * \left\{\begin{array}{c} \omega_1 \\ \omega_2 \\ \omega_3 \\ \omega_4 \end{array}\right\}$$

$$+\left[\begin{array}{c:c:c:c} \mathbf{u}_{A}^{0} & \mathbf{u}_{B}^{0} & \mathbf{u}_{C}^{0} & \mathbf{u}_{D}^{0} \end{array}\right] * \left\{\begin{array}{c} \dot{d}_A \\ \dot{d}_B \\ \dot{d}_C \\ \dot{d}_D \end{array}\right\} = \mathbf{v}_E^{task} \tag{2.220a}$$

- *Link and joint constraints*:

$$\begin{aligned} \omega_1 = \boldsymbol{\omega}_0 = 0; &\quad \text{joint A prismatic} \\ \omega_3 = \omega_4; &\quad \text{joint D prismatic} \\ \dot{\mathbf{d}}_B^0 = \dot{\mathbf{d}}_C^0 = \mathbf{0}; &\quad \text{joints C and D are revolute joints} \end{aligned} \tag{2.220b}$$

- *Scalar equations*: Therefore, after the constraints implementation, Equation (2.220a) holds:

$$\left[\begin{array}{c:c:c:c} -\tilde{\mathbf{L}}_{AB}^0 & -\tilde{\mathbf{L}}_{BC}^0 & -\tilde{\mathbf{L}}_{CD}^0 & -\tilde{\mathbf{L}}_{DE}^0 \end{array} \right] * \left\{ \begin{array}{c} \omega_1(=0) \\ \omega_2 \\ \omega_3 \\ \omega_4(=\omega_3) \end{array} \right\}$$

$$+\left[\begin{array}{c:c:c:c} \mathbf{u}_A^0 & \mathbf{u}_B^0 & \mathbf{u}_C^0 & \mathbf{u}_D^0 \end{array} \right] * \left\{ \begin{array}{c} \dot{d}_A \\ \dot{d}_B(=0) \\ \dot{d}_C(=0) \\ \dot{d}_D \end{array} \right\} = \mathbf{v}_E^{task}$$

Therefore, after matrix multiplications, Equation (2.220c) holds:

$$\left(\mathbf{0} - \tilde{\mathbf{L}}_{BC}^0 * \boldsymbol{\omega}_2 - \tilde{\mathbf{L}}_{CD}^0 * \omega_3 - \tilde{\mathbf{L}}_{DE}^0 * \omega_3\right) + \left(\dot{\mathbf{d}}_A^0 + \mathbf{0} + \mathbf{0} + \dot{\mathbf{d}}_D^0\right) = \mathbf{v}_E^{task} \tag{2.220c}$$

where

- From Equation (2.179h), the absolute angular velocities as a function of angular joint rates are

$$\omega_1 = \mathbf{0}; \quad \omega_2 = \left\{ \begin{array}{c} 0 \\ \dot{\theta}_B \\ 0 \end{array} \right\}; \quad \omega_3 = \omega_4 = \left\{ \begin{array}{c} S\theta_B \cdot \dot{\theta}_C \\ \dot{\theta}_B \\ C\theta_B \cdot \dot{\theta}_C \end{array} \right\} \tag{2.220d}$$

- From Equation (2.179f), the unit vectors for relative motion, linear or angular, at joints are

$$\mathbf{u}_A^0 = \left\{ \begin{array}{c} 1 \\ 0 \\ 0 \end{array} \right\}; \quad \mathbf{u}_B^0 = \left\{ \begin{array}{c} 0 \\ 1 \\ 0 \end{array} \right\}; \quad \mathbf{u}_C^0 = \left\{ \begin{array}{c} S\theta_B \\ 0 \\ C\theta_B \end{array} \right\}; \quad \mathbf{u}_D^0 = \left\{ \begin{array}{c} C\theta_B \cdot C\theta_C \\ S\theta_C \\ -S\theta_B \cdot C\theta_C \end{array} \right\} \tag{2.220e}$$

Equation (2.220c) generates three scalar equations:

$$\begin{Bmatrix} \dot{d}_A - (44.5 + d_D)\cdot S\theta_B \cdot C\theta_C \cdot \dot{\theta}_B - (44.5 + d_D)\cdot S\theta_C \cdot C\theta_B \cdot \dot{\theta}_C + C\theta_B \cdot C\theta_C \cdot \dot{d}_D \\ (44.5 + d_D)\cdot C\theta_C \cdot \dot{\theta}_C + S\theta_C \cdot \dot{d}_D \\ -(44.5 + d_D)\cdot C\theta_B \cdot C\theta_C \cdot \dot{\theta}_B + (44.5 + d_D)\cdot S\theta_C \cdot S\theta_B \cdot \dot{\theta}_C - S\theta_B \cdot C\theta_C \cdot \dot{d}_D \end{Bmatrix}$$

$$= \begin{Bmatrix} v_E^{x,task} \\ v_E^{y,task} \\ v_E^{z,task} \end{Bmatrix} \tag{2.220f}$$

- *Combined equations*: From inverse angular velocity analysis, Equation (2.187c) becomes

$$\begin{Bmatrix} S\theta_B \cdot \dot{\theta}_C \\ \dot{\theta}_B \\ C\theta_B \cdot \dot{\theta}_C \end{Bmatrix} = \begin{Bmatrix} \omega_4^{x,task} \\ \omega_4^{y,task} \\ \omega_4^{z,task} \end{Bmatrix} \tag{2.220g}$$

Equations (2.220f) and (2.220g) form a system of six linear equations with four unknowns, the actuation rates: $\dot{\theta}_B$, $\dot{\theta}_C$, $\dot{d}_A$, and $\dot{d}_D$. Since the number of equations is larger than the number of unknowns, in general, there are no actuation rates to control the mechanism for any desired velocities: $\omega_4^{x,task}$, $\omega_4^{y,task}$, $\omega_4^{z,task}$, $v_E^{x,task}$, $v_E^{y,task}$, and $v_4^{z,task}$. Although, particular positions exist only if it satisfies the following two equations:

- From the first and third entries in Equation (2.220g):

$$S\theta_B \cdot \omega_4^{z,task} = C\theta_B \cdot \omega_4^{x,task} \tag{2.220h}$$

- From the second and third entries in Equation (2.220f):

$$\left(v_E^{y,task} - \frac{(44.5 + d_D)\cdot C\theta_C}{C\theta_B} \cdot \omega_4^{z,task} \right) \cdot \frac{S\theta_B \cdot C\theta_C}{S\theta_C} = -v_E^{z,task} \\ -(44.5 + d_D)\cdot C\theta_B \cdot C\theta_C \cdot \omega_4^{y,task} + (44.5 + d_D)\cdot S\theta_C \cdot \omega_4^{x,task} \tag{2.220i}$$

2.17.16 The Inverse Linear Velocity Analysis: Governing Equation Functions of Relative Angular Velocities

Considering Equation (2.177), we express absolute as functions of relative angular velocities:

$$\begin{Bmatrix} \omega_1 \\ \omega_2 \\ \dots \\ \omega_{m-1} \\ \omega_m \end{Bmatrix} = \mathbf{Z}^T * \begin{Bmatrix} \omega_A \\ \omega_B \\ \dots \\ \omega_Y \\ \omega_Z \end{Bmatrix}; \quad \text{where} \begin{Bmatrix} \omega_A \\ \omega_B \\ \dots \\ \omega_Y \\ \omega_Z \end{Bmatrix} = \mathbf{Z}^T * \begin{Bmatrix} \mathbf{u}_A^0 \cdot \dot{\theta}_A \\ \mathbf{u}_B^0 \cdot \dot{\theta}_B \\ \dots \\ \mathbf{u}_Y^0 \cdot \dot{\theta}_Y \\ \mathbf{u}_Z^0 \cdot \dot{\theta}_Z \end{Bmatrix} \tag{2.221}$$

Since

$$\begin{bmatrix} 1 & 2 & 3 & \dots & m-1 & m \\ -\tilde{\mathbf{L}}_{AB}^0 & -\tilde{\mathbf{L}}_{BC}^0 & -\tilde{\mathbf{L}}_{CD}^0 & \dots & -\tilde{\mathbf{L}}_{YZ}^0 & -\tilde{\mathbf{L}}_{ZE}^0 \end{bmatrix}$$

$$* \begin{array}{c} \\ 1 \\ 2 \\ 3 \\ \dots \\ m-1 \\ m \end{array} \begin{array}{c} \begin{array}{cccccc} A & B & C & \dots & Y & Z \end{array} \\ \begin{bmatrix} 1 & 0 & 0 & \dots & 0 & 0 \\ 1 & 1 & 0 & \dots & 0 & 0 \\ 1 & 1 & 1 & \dots & 0 & 0 \\ \dots & \dots & \dots & \dots & \dots & \dots \\ 1 & 1 & 1 & \dots & 1 & 0 \\ 1 & 1 & 1 & \dots & 1 & 1 \end{bmatrix} \end{array} \tag{2.222}$$

$$\begin{array}{cccccc} A & B & C & \dots & Y & Z \end{array}$$

$$= \left[\begin{array}{c:c:c:c:c:c} \tilde{\mathbf{L}}_{EA}^0 & \tilde{\mathbf{L}}_{EB}^0 & \tilde{\mathbf{L}}_{EC}^0 & \dots & \tilde{\mathbf{L}}_{EY}^0 & \tilde{\mathbf{L}}_{EZ}^0 \end{array} \right]$$

Then, Equation (2.219) becomes

$$\left[\begin{array}{c:c:c:c:c:c} \tilde{\mathbf{L}}_{EA}^0 & \tilde{\mathbf{L}}_{EB}^0 & \tilde{\mathbf{L}}_{EC}^0 & \dots & \tilde{\mathbf{L}}_{EY}^0 & \tilde{\mathbf{L}}_{EZ}^0 \end{array} \right] * \begin{Bmatrix} \omega_A \\ \omega_B \\ \dots \\ \omega_Y \\ \omega_Z \end{Bmatrix}$$

$$+ \left[\begin{array}{c:c:c:c:c} \mathbf{u}_A^0 & \mathbf{u}_B^0 & \mathbf{u}_{CA}^0 & \dots & \mathbf{u}_Z^0 \end{array} \right] * \begin{Bmatrix} \dot{d}_A \\ \dot{d}_B \\ \dot{d}_C \\ \dots \\ \dot{d}_Z \end{Bmatrix} = \mathbf{v}_E^{task} \tag{2.223}$$

- *Scalar equations*: Considering Equation (2.221), then Equation (2.223) becomes Equation (2.224), from which the three scalar equations are determined:

$$\left[\begin{array}{c:c:c:c} \tilde{\mathbf{L}}_{EA}^{0} * \mathbf{u}_{A}^{0} & \tilde{\mathbf{L}}_{EB}^{0} * \mathbf{u}_{B}^{0} & \tilde{\mathbf{L}}_{EC}^{0} * \mathbf{u}_{C}^{0} & \tilde{\mathbf{L}}_{ED}^{0} * \mathbf{u}_{D}^{0} \end{array}\right] * \overbrace{\left\{\begin{bmatrix} \dot{\theta}_A \\ \dot{\theta}_B \\ \dot{\theta}_C \\ \dot{\theta}_D \end{bmatrix}\right\}}^{\dot{\theta}_a} + \left[\begin{array}{c:c:c:c} \mathbf{u}_{A}^{0} & \mathbf{u}_{B}^{0} & \mathbf{u}_{C}^{0} & \mathbf{u}_{D}^{0} \end{array}\right] * \overbrace{\left\{\begin{bmatrix} \dot{d}_A \\ \dot{d}_B \\ \dot{d}_C \\ \dot{d}_D \end{bmatrix}\right\}}^{\dot{d}_a} = \mathbf{v}_{E}^{task} \tag{2.224}$$

- *Joint constraints*: Two types of joint constraints apply to the entries in the scalar matrices $\dot{\theta}_a$ and $\dot{d}_a$:
  - $\dot{\theta}_Z = 0$, for a prismatic joint
  - $\dot{d}_Z = 0$, for a revolute joint

  Equation (2.224) has the form, which is shown in Equation (2.225):

$$\begin{aligned} &+1\cdot\left(\tilde{\mathbf{L}}_{EA}^{0} * \mathbf{u}_{A}^{0}\cdot\dot{\theta}_A + \mathbf{u}_{A}^{0}\cdot\dot{d}_A\right) + 1\cdot\left(\tilde{\mathbf{L}}_{EB}^{0} * \mathbf{u}_{B}^{0}\cdot\dot{\theta}_B + \mathbf{u}_{B}^{0}\cdot\dot{d}_B\right) \\ &+\cdots+1\cdot\left(\tilde{\mathbf{L}}_{EZ}^{0} * \mathbf{u}_{Z}^{0}\cdot\dot{\theta}_Z + \mathbf{u}_{Z}^{0}\cdot\dot{d}_Z\right) = \mathbf{v}_{E}^{task} \end{aligned} \tag{2.225}$$

  Notice the coefficients in front of brackets, noted $c_{E,A}$, one for each oriented arc a in the tree from Equation (2.225) are +1:

$$c_{E,A} = c_{E,B} = \cdots = c_{E,Z} = 1 \tag{2.226}$$

There are mechanisms where there exist arcs oriented inverse as the path, and the coefficients are –1. The coefficients: +1, –1, or 0 are determined next as entries in the spaning tree matrix T.

2.17.16.1 The Spanning Tree Matrix, T

The open path starts at A and ends at chord E. For the tree in Figure 2.56, the following edge set belonging to the three paths: $T_E = \{A^+, B^+, C^+, \ldots, Y^+, Z^+\}$ will be generated.

In Section 1.2.12, the c × a matrix T has c = 1 row for chord E which represents the cut-edge, and the columns represent the arcs a that belong to the tree.

The entries $c_{E,a}$ are defined as follows:

$$c_{E,a} = \left\{ \begin{array}{rl} -1; & \text{if arc a is on the path and opposite to the path} \\ +1; & \text{if arc a is onto the path and same direction as the path} \\ 0; & \text{if arc a is not on the path} \end{array} \right\}$$

Example: The TRRT mechanism in Figure 2.1, the tree shows in Figure 2.3. The edge set from the tree's path is $T_E = \{A^+, B^+, C^+, D^+\}$ and the spanning tree matrix is

$$\mathbf{T} = \begin{array}{c} \quad A \quad\; B \quad\; C \quad\; D \\ \left[\begin{array}{c:c:c:c} c_{E,A} & c_{E,B} & c_{E,C} & c_{E,D} \end{array}\right] \end{array} = \left[\begin{array}{c:c:c:c} +1 & +1 & +1 & +1 \end{array}\right] \tag{2.227}$$

Equation (1.20) shows automatic calculation of entries in matrix T from the multiplication of two digraph matrices:

$$\mathbf{T} = \mathbf{G}_c^T * \mathbf{Z}^T \tag{2.228}$$

The two factors are as follows: $\mathbf{G}_c^T$ from Equation (1.11) and $\mathbf{Z}^T$ from Equation (1.17). The superscript T designates the transpose operation.

2.17.16.2 The Analogy to Moment of a Force and Couple from Statics

Equation (2.225) for the linear velocity at the chord E shows as

$$\sum_{Z=A}^{Z} c_{E,A} \cdot \left(\tilde{\mathbf{L}}_{EZ}^0 * \omega_Z + \dot{\mathbf{d}}_Z^0\right) = \mathbf{v}_E \tag{2.229}$$

Let us notice the analogy between relative angular velocities ω_Z at joints as "static forces" acting at joints (considering the mechanism as a static frame in equilibrium). In addition, it is an analogy between relative linear velocities $\dot{\mathbf{d}}_Z^0$ at joints and "static couples." Then, from Equation (2.229), the linear velocity $\mathbf{v}_E$ is the resultant of "moments of forces" with respect to the chord E, $\tilde{\mathbf{L}}_{EZ}^0 * \omega_Z$, and "static couples" $\dot{\mathbf{d}}_Z^0$.

- *Geometric Jacobian for linear velocities*: There are a brackets in Equation (2.225), equal to the number of joints (arcs tree), which form the columns in a 3 × a matrix named *geometric Jacobian sub-matrix* for linear velocities $\mathbf{J}_a^{lv}$. The joint rates are the entries in the a × 1 column matrix $\dot{\mathbf{q}}_a$.

Therefore, Equation (2.230) holds:

$$\mathbf{J}_a^{lv} * \dot{\mathbf{q}}_a = \mathbf{v}_E^{task} \tag{2.230}$$

2.17.16.3 The Jacobean Matrix from the Combined Equations for Inverse Angular and Inverse Linear Velocities

The two Jacobean sub-matrices $\mathbf{J}^{av}$ from Equation (2.188) and $\mathbf{J}^{lv}$ from Equation (2.230) when assembled generate a 6 × a matrix, named geometric Jacobean, **J**:

$$\mathbf{J}_{6\times a} = \begin{bmatrix} \mathbf{J}^{av} \\ \text{------} \\ \mathbf{J}^{lv} \end{bmatrix} \tag{2.231}$$

The combined equations for inverse angular and inverse linear velocities were shown in Equation (2.232):

$$\mathbf{J}_{6\times a} * \dot{\mathbf{q}}_{a\times 1} = \begin{bmatrix} \boldsymbol{\omega}_E^{task} \\ \text{--------} \\ \mathbf{v}_E^{task} \end{bmatrix}_{6\times 1} \tag{2.232}$$

This matrix provides the relation input-actuating joint rates and output-desired angular and linear velocities of end-effector. In the case of a square and nonsingular matrix, it is invertible, therefore solving from Equation (2.232) for actuation is required in the control. The name geometric Jacobean shows that it provides qualitative information only on the potential output for a given input, and the forces are not included in analysis.

2.17.16.4 Example of Inverse Velocity Analysis for Open Cycle TRRT Spatial Mechanism with Equation Functions of Relative Angular Velocities

Example: For the TRRT mechanism in Figure 2.1, the Equation (2.223) holds:

$$\begin{bmatrix} \tilde{\mathbf{L}}_{EA}^0 * \mathbf{u}_A^0 & \vdots & \tilde{\mathbf{L}}_{EB}^0 * \mathbf{u}_B^0 & \vdots & \tilde{\mathbf{L}}_{EC}^0 * \mathbf{u}_C^0 & \vdots & \tilde{\mathbf{L}}_{ED}^0 * \mathbf{u}_D^0 \end{bmatrix} * \begin{Bmatrix} \dot{\theta}_A(=0) \\ \dot{\theta}_B \\ \dot{\theta}_C \\ \dot{\theta}_D(=0) \end{Bmatrix}$$

$$+ \begin{bmatrix} \mathbf{u}_A^0 & \vdots & \mathbf{u}_B^0 & \vdots & \mathbf{u}_C^0 & \vdots & \mathbf{u}_D^0 \end{bmatrix} * \begin{Bmatrix} \dot{d}_A \\ \dot{d}_B(=0) \\ \dot{d}_C(=0) \\ \dot{d}_D \end{Bmatrix} = \mathbf{v}_E^{task} \tag{2.233a}$$

- *Joint constraints*:
  - Constraints for the prismatic joints A and D: $\dot{\theta}_A = \dot{\theta}_D = 0$.
  - Constraints for revolute joints C and D: $\dot{d}_B = \dot{d}_C = 0$, apply to entries in column matrices.

$$\begin{Bmatrix} \dot{\theta}_A \\ \dot{\theta}_B \\ \dot{\theta}_C \\ \dot{\theta}_D \end{Bmatrix} = \begin{Bmatrix} 0 \\ \dot{\theta}_B \\ \dot{\theta}_C \\ 0 \end{Bmatrix}; \text{ and } \begin{Bmatrix} \dot{d}_A \\ \dot{d}_B \\ \dot{d}_C \\ \dot{d}_D \end{Bmatrix} = \begin{Bmatrix} \dot{d}_A \\ 0 \\ 0 \\ \dot{d}_D \end{Bmatrix}$$

- *Scalar equations*:

$$\mathbf{u}_A^0 \cdot \dot{d}_A + \left(\tilde{\mathbf{L}}_{EB}^0 * \mathbf{u}_B^0\right) \cdot \dot{\theta}_B + \left(\tilde{\mathbf{L}}_{EC}^0 * \mathbf{u}_C^0\right) \cdot \dot{\theta}_C + \mathbf{u}_D^0 \cdot \dot{d}_D = \mathbf{v}_E^{task} \tag{2.233b}$$

where

- The unit vectors from Equation (2.179f) are reproduced here:

$$\mathbf{u}_A^0 = \begin{Bmatrix} 1 \\ 0 \\ 0 \end{Bmatrix}; \quad \mathbf{u}_B^0 = \begin{Bmatrix} 0 \\ 1 \\ 0 \end{Bmatrix}; \quad \mathbf{u}_C^0 = \begin{Bmatrix} S\theta_B \\ 0 \\ C\theta_B \end{Bmatrix}; \quad \mathbf{u}_D^0 = \begin{Bmatrix} C\theta_B \cdot C\theta_C \\ S\theta_C \\ -S\theta_B \cdot C\theta_C \end{Bmatrix} \tag{2.233c}$$

- The position vectors with respect to the chord E and their skew-symmetric matrices are

$$\mathbf{L}_{EB} = \mathbf{r}_B - \mathbf{r}_E = \begin{Bmatrix} -(44.5 + d_D) \cdot C\theta_B \cdot C\theta_C \\ -25 - (44.5 + d_D) \cdot S\theta_C \\ (44.5 + d_D) \cdot S\theta_B \cdot C\theta_C \end{Bmatrix} = \begin{Bmatrix} x_{EB} \\ y_{EB} \\ z_{EB} \end{Bmatrix};$$

$$\mathbf{L}_{EC} = \mathbf{r}_C - \mathbf{r}_E = \begin{Bmatrix} -(44.5 + d_D) \cdot C\theta_B \cdot C\theta_C \\ -25 - (44.5 + d_D) \cdot S\theta_C \\ (44.5 + d_D) \cdot S\theta_B \cdot C\theta_C \end{Bmatrix} = \begin{Bmatrix} x_{EC} \\ y_{EC} \\ z_{EC} \end{Bmatrix}$$

$$\tilde{\mathbf{L}}_{EB}^0 * \mathbf{u}_B^0 = \left[\begin{array}{c:c:c} 0 & -z_{EB} & y_{EB} \\ z_{EB} & 0 & -x_{EB} \\ -y_{EB} & x_{EB} & 0 \end{array}\right] * \begin{Bmatrix} 0 \\ 1 \\ 0 \end{Bmatrix} = \begin{Bmatrix} -z_{EB} \\ 0 \\ x_{EB} \end{Bmatrix};$$

$$\tilde{\mathbf{L}}_{EC}^0 * \mathbf{u}_C^0 = \left[\begin{array}{c:c:c} 0 & -z_{EC} & y_{EC} \\ z_{EC} & 0 & -x_{EC} \\ -y_{EC} & x_{EC} & 0 \end{array}\right] * \begin{Bmatrix} S\theta_B \\ 0 \\ C\theta_B \end{Bmatrix} = \begin{Bmatrix} y_{EC} \cdot C\theta_B \\ z_{EC} \cdot S\theta_B - x_{EC} \cdot C\theta_B \\ -x_{EC} \cdot S\theta_B \end{Bmatrix} \tag{2.233d}$$

- Therefore, the system of three scalar equations is generated, identical to that from Equation (2.207s), which was generated based on absolute angular velocities:

$$\left\{\begin{array}{c} \dot{d}_A - (44.5 + d_D)\cdot S\theta_B \cdot C\theta_C \cdot \dot{\theta}_B - (44.5 + d_D)\cdot S\theta_C \cdot C\theta_B \cdot \dot{\theta}_C + C\theta_B \cdot C\theta_C \cdot \dot{d}_D \\ (44.5 + d_D)\cdot C\theta_C \cdot \dot{\theta}_C + S\theta_C \cdot \dot{d}_D \\ -(44.5 + d_D)\cdot C\theta_B \cdot C\theta_C \cdot \dot{\theta}_B + (44.5 + d_D)\cdot S\theta_C \cdot S\theta_B \cdot \dot{\theta}_C - S\theta_B \cdot C\theta_C \cdot \dot{d}_D \end{array}\right\}$$

$$= \left\{\begin{array}{c} v_E^x \\ v_E^y \\ v_E^z \end{array}\right\} \tag{2.233e}$$

- *Geometric Jacobean for linear velocities*: Equation (2.229) in matrix form is

$$\underbrace{\left[\begin{array}{cccc} A & B & C & D \\ \mathbf{u}_A^0 \vdots & \tilde{\mathbf{L}}_{EB}^0 * \mathbf{u}_B^0 \vdots & \tilde{\mathbf{L}}_{EC}^0 * \mathbf{u}_C^0 \vdots & \mathbf{u}_D^0 \end{array}\right]}_{\mathbf{J}_a^{lv}} * \underbrace{\left\{\begin{array}{c} \dot{d}_A \\ \dot{\theta}_B \\ \dot{\theta}_C \\ \dot{d}_D \end{array}\right\}}_{\dot{q}_a} = v_E^{task} \tag{2.233f}$$

Therefore, the Jacobean for linear velocities is

$$\mathbf{J}_a^{lv} = \begin{array}{c} \begin{array}{cccc} A & B & C & D \end{array} \\ \left[\begin{array}{c:c:c:c} 1 & -z_{EB} & y_{EC}\cdot C\theta_B & C\theta_B \cdot C\theta_C \\ 0 & 0 & z_{EC}\cdot S\theta_B - x_{EC}\cdot C\theta_B & S\theta_C \\ 0 & x_{EB} & -x_{EC}\cdot S\theta_B & -S\theta_B \cdot C\theta_C \end{array}\right] \end{array} \tag{2.233g}$$

- *Combined equations for inverse angular and inverse linear velocities*: The two Jacobean sub-matrices $\mathbf{J}^{av}$ from Equation (2.189f) and $\mathbf{J}^{lv}$ from Equation (2.233g) were assembled in a 6 × a matrix, which was shown in Equation (2.233h):

$$\underbrace{\begin{array}{c} \begin{array}{cccc} A & B & C & D \end{array} \\ \left[\begin{array}{c:c:c:c} 0 & 0 & S\theta_B & 0 \\ 0 & 1 & 0 & 0 \\ 0 & 0 & C\theta_B & 0 \\ \hdashline 1 & -z_{EB} & y_{EC}\cdot C\theta_B & C\theta_B \cdot C\theta_C \\ 0 & 0 & z_{EC}\cdot S\theta_B - x_{EC}\cdot C\theta_B & S\theta_C \\ 0 & x_{EB} & -x_{EC}\cdot S\theta_B & -S\theta_B \cdot C\theta_C \end{array}\right] \end{array}}_{\mathbf{J}} * \underbrace{\left\{\begin{array}{c} \dot{d}_A \\ \dot{\theta}_B \\ \dot{\theta}_C \\ \dot{d}_D \end{array}\right\}}_{\dot{q}} = \left\{\begin{array}{c} \omega_E^x \\ \omega_E^y \\ \omega_E^z \\ \hdashline v_E^x \\ v_E^y \\ v_E^z \end{array}\right\} \tag{2.233h}$$

2.17.16.4.1 Redundant and Nonredundant System

The set of equations for inverse positional analysis is solved for relative joint rates, angular and linear. If the number of linear equations is the same as the number of unknown rates in $\dot{\mathbf{q}}_a$, where a is the number of actuated joints, then the system is considered *nonredundant*, and the system of equations could be symbolically or numerically solved. In case there are more unknown rates, then the system is *redundant*. The degree of redundancy is the difference between the number of parameters and the number of equations.

2.17.17 Combined Equations for Inverse Velocity Analysis Based on Twists

Equations (2.186) and (2.229) are combined and shown in Equation (2.234):

$$\sum_{Z=A}^{c} c_{E,Z} \cdot \left\{ \begin{array}{c} \boldsymbol{\omega}_Z \\ \hline \tilde{\mathbf{L}}_{EZ}^{0} * \boldsymbol{\omega}_Z + \dot{\mathbf{d}}_Z \end{array} \right\} = \left[\begin{array}{c} \boldsymbol{\omega}_E^{task} \\ \hline \mathbf{v}_E^{task} \end{array} \right] \tag{2.234}$$

For any open mechanism with m links and c joints, then Equation (2.234) becomes

$$c_{E,A} \cdot \hat{\mathbf{t}}_{E,A} + c_{E,B} \cdot \hat{\mathbf{t}}_{E,A} + c_{E,C} \cdot \hat{\mathbf{t}}_{E,A} + \cdots + c_{E,Z} \cdot \hat{\mathbf{t}}_{E,Z} = \left[\begin{array}{c} \boldsymbol{\omega}_E^{task} \\ \hline \mathbf{v}_E^{task} \end{array} \right] \tag{2.234a}$$

A twist, $\hat{\mathbf{t}}_{EZ}$, is a 6×1 column matrix attached to each joint, that is, the bracket {*} in Equation (2.234). The upper vector in the bracket is the joint's relative angular velocity, whereas the lower vector represents a linear velocity:

$$\begin{aligned} \hat{\mathbf{t}}_{E,Z} &= \left\{ \begin{array}{c} \boldsymbol{\omega}_Z \\ \hline \tilde{\mathbf{L}}_{EZ}^{0} * \boldsymbol{\omega}_Z + \dot{\mathbf{d}}_Z \end{array} \right\} = \left\{ \begin{array}{c} \mathbf{u}_Z^0 \cdot \dot{\theta}_Z \\ \hline \tilde{\mathbf{L}}_{EZ}^{0} * \mathbf{u}_Z^0 \cdot \dot{\theta}_Z + \mathbf{u}_Z^0 \cdot \dot{d}_Z \end{array} \right\} \\ &= \left\{ \begin{array}{c} \mathbf{u}_Z^0 \\ \hline \tilde{\mathbf{L}}_{EZ}^{0} * \mathbf{u}_Z^0 \end{array} \right\} \cdot \dot{\theta}_Z + \left\{ \begin{array}{c} 0 \\ \hline \mathbf{u}_Z^0 \end{array} \right\} \cdot \dot{d}_Z \end{aligned} \tag{2.235}$$

2.17.17.1 Twists for Joints with Single and Multiple DOF

- *Twist for revolute joint*: The constraint $\dot{d}_Z = 0$ for Z a revolute joint applies in Equation (2.235) and generates the twist:

$$\hat{\mathbf{t}}_{E,Z} = \left\{ \begin{array}{c} \mathbf{u}_Z^0 \\ \hline \tilde{\mathbf{L}}_{EZ}^{0} * \mathbf{u}_Z^0 \end{array} \right\} \cdot \dot{\theta}_Z = \hat{\mathbf{u}}_{E,Z} \cdot \dot{\theta}_Z \tag{2.236}$$

where

- $\hat{\mathbf{u}}_{E,Z} = \left\{ \begin{array}{c} \mathbf{u}_Z^0 \\ \hline \tilde{\mathbf{L}}_{EZ}^0 * \mathbf{u}_Z^0 \end{array} \right\}$ is a 6 × 1 matrix named *unit screw* of a revolute joint (2.237)

  The unit screw, $\hat{\mathbf{u}}_{E,Z}$, is a dual vector which has as entries the six *Plucker coordinates* of its line support. These entries, in a 6 × 1 column matrix, are shown in Equation (2.238):

$$\hat{\mathbf{u}}_{E,Z} = \left\{ \begin{array}{c} l_Z \\ m_Z \\ n_Z \\ \hline P_{EZ} \\ Q_{EZ} \\ R_{EZ} \end{array} \right\} = \left\{ \begin{array}{c} l_Z \\ m_Z \\ n_Z \\ \hline -z_{EZ} \cdot m_Z + y_{EZ} \cdot n_Z \\ z_{EZ} \cdot l_Z - x_{EZ} \cdot n_Z \\ -y_{EZ} \cdot l_Z + x_{EZ} \cdot m_Z \end{array} \right\} \tag{2.238}$$

  There are two relations between the six Plucker coordinates, shown in Equations (2.239) and (2.241); therefore, only four of them are independent. The fifth parameter associate to the screw is the *pitch.*

  – The upper vector $\mathbf{u}_Z^0$ is a unit vector, which defines the direction of screw; thus, the first relation between Plucker coordinates holds:

$$\mathbf{u}_Z^0 = \left\{ \begin{array}{c} l_Z \\ m_Z \\ n_Z \end{array} \right\} \Rightarrow l_Z^2 + m_Z^2 + n_Z^2 = 1 \tag{2.239}$$

  The absolute matrix $\mathbf{D}_m$, Equation (2.18), generates the components for upper vector $\mathbf{u}_Z^0$ from equation:

$$\mathbf{u}_Z^0 = \mathbf{D}_m * \mathbf{u}_Z^m.$$

  Thus,

- For revolute joint with local x-axis of rotation, the components are from column (I) in matrix $\mathbf{D}_m$:

$$\mathbf{u}_Z^I = \left[\begin{array}{c|c|c} \mathrm{Cos}\,\alpha_m^I & \mathrm{Cos}\,\alpha_m^{II} & \mathrm{Cos}\,\alpha_m^{III} \\ \mathrm{Cos}\,\beta_m^I & \mathrm{Cos}\,\beta_m^{II} & \mathrm{Cos}\,\beta_m^{III} \\ \mathrm{Cos}\,\gamma_m^I & \mathrm{Cos}\,\gamma_m^{II} & \mathrm{Cos}\,\gamma_m^{III} \end{array} \right] * \left\{ \begin{array}{c} 1 \\ 0 \\ 0 \end{array} \right\} = \left\{ \begin{array}{c} \mathrm{Cos}\,\alpha_m^I \\ \mathrm{Cos}\,\beta_m^I \\ \mathrm{Cos}\,\gamma_m^I \end{array} \right\} = \left\{ \begin{array}{c} l_m^I \\ m_m^I \\ n_m^I \end{array} \right\} \tag{2.239a}$$

- For revolute joint with local y-axis of rotation, the components are from column (II) in matrix $\mathbf{D}_m$:

$$\mathbf{u}_Z^{II} = \begin{bmatrix} \cos\alpha_m^I & \cos\alpha_m^{II} & \cos\alpha_m^{III} \\ \cos\beta_m^I & \cos\beta_m^{II} & \cos\beta_m^{III} \\ \cos\gamma_m^I & \cos\gamma_m^{II} & \cos\gamma_m^{III} \end{bmatrix} * \begin{Bmatrix} 0 \\ 1 \\ 0 \end{Bmatrix} = \begin{Bmatrix} \cos\alpha_m^{II} \\ \cos\beta_m^{II} \\ \cos\gamma_m^{II} \end{Bmatrix} = \begin{Bmatrix} l_m^{II} \\ m_m^{II} \\ n_m^{II} \end{Bmatrix} \tag{2.239b}$$

- For revolute joint with local z-axis of rotation, the components are from column (III) in matrix $\mathbf{D}_m$:

$$\mathbf{u}_Z^{III} = \begin{bmatrix} \cos\alpha_m^I & \cos\alpha_m^{II} & \cos\alpha_m^{III} \\ \cos\beta_m^I & \cos\beta_m^{II} & \cos\beta_m^{III} \\ \cos\gamma_m^I & \cos\gamma_m^{II} & \cos\gamma_m^{III} \end{bmatrix} * \begin{Bmatrix} 0 \\ 0 \\ 1 \end{Bmatrix} = \begin{Bmatrix} \cos\alpha_m^{III} \\ \cos\beta_m^{III} \\ \cos\gamma_m^{III} \end{Bmatrix} = \begin{Bmatrix} l_m^{III} \\ m_m^{III} \\ n_m^{III} \end{Bmatrix} \tag{2.239c}$$

- The lower vector defines the position of the line support, provided by the "moment" $\mathbf{L}_{EZ} \times \mathbf{u}_Z$ of its unit vector with respect to the chord E:

$$\tilde{\mathbf{L}}_{EZ}^0 * \mathbf{u}_Z^0 = \begin{Bmatrix} P_{EZ} \\ Q_{EZ} \\ R_{EZ} \end{Bmatrix} = \begin{Bmatrix} -z_{EZ} \cdot m_Z + y_{EZ} \cdot n_Z \\ z_{EZ} \cdot l_Z - x_{EZ} \cdot n_Z \\ -y_{EZ} \cdot l_Z + x_{EZ} \cdot m_Z \end{Bmatrix} \tag{2.240}$$

with $\mathbf{L}_{EZ}$ is joint Z's position vector originating from reference point E.

From the mixed product zero, then the second relation between Plucker coordinates holds:

$$\mathbf{u}_Z^0 \cdot \left(\mathbf{L}_{EZ} \times \mathbf{u}_Z^0\right) = 0 \Rightarrow l_Z \cdot P_{EZ} + m_Z \cdot Q_{EZ} + n_Z \cdot R_{EZ} = 0 \tag{2.241}$$

$$\dot{\theta}_Z \quad \text{named twist intensity} \tag{2.242}$$

It is a scalar to measure the rate of angular displacement of link m relative to link $m-1$

The upper vector in the twist is the joint's relative angular velocity, whereas the lower vector represents a linear velocity.

The ratio between their sizes is the pitch. The two vectors being orthogonal, Equation (2.241), and the projection of second vector on the unit screw direction is zero. Therefore, for a revolute joint, there is no sliding along the axis of rotation. The revolute joint is therefore a *zero-pitch screw.*

- *Twist for prismatic joint*: The constraint $\dot{\theta}_Z = 0$ for Z a prismatic joint applies in Equation (2.235) and generates the twist:

$$\hat{\mathbf{t}}_{E,Z} = \left\{ \begin{array}{c} 0 \\ \hdashline \dot{d}_Z \end{array} \right\} = \left\{ \begin{array}{c} 0 \\ \hdashline \mathbf{u}_Z^0 \end{array} \right\} \cdot \dot{d}_Z = \hat{\mathbf{u}}_{E,Z} \cdot \dot{d}_Z \tag{2.243}$$

where

$$\hat{\mathbf{u}}_{E,Z} = \left\{ \begin{array}{c} 0 \\ \hdashline \mathbf{u}_Z^0 \end{array} \right\} \quad \text{is a } 6 \times 1 \text{ matrix named } \textit{unit screw} \text{ of a prismatic joint} \tag{2.244}$$

The unit screw, $\hat{\mathbf{u}}_{E,Z}$, is a dual vector with entries the *Plucker coordinates* of its line support, shown in Equation (2.245). The prismatic joint at A has assigned a unit screw $\hat{\mathbf{u}}_{E,Z}$, which has its upper vector a zero vector, and the lower vector is the unit vector $\mathbf{u}_Z^0$:

$$\hat{\mathbf{u}}_{E,Z} = \left\{ \begin{array}{c} 0 \\ 0 \\ 0 \\ \hdashline l_Z \\ m_Z \\ n_Z \end{array} \right\} \tag{2.245}$$

$$\dot{d}_Z; \quad \text{scalar, named } \textit{twist intensity} \tag{2.246}$$

For a prismatic joint with local x-, y-, or z-axis of translation, the components are from column (I), (II), or (III) in matrix $\mathbf{D}_m$, respectively.

This scalar measures the rate of linear displacement of link m relative to link $m-1$.

The upper vector in the twist is the null joint's relative angular velocity, whereas the lower vector represents a linear velocity. The ratio between their sizes results an *infinity pitch*. Therefore, the prismatic joint has assigned a screw with its support as a *line at infinity*.

- *Twists for spherical joint, Figure 1.3f*: The spherical joint with 3 DOF has assigned three twists, around the three-unit screws $\hat{\mathbf{u}}_{E,Z}^{I}$, $\hat{\mathbf{u}}_{E,Z}^{II}$, and $\hat{\mathbf{u}}_{E,Z}^{II}$ corresponding to the three rotations about local frame's axes. The three unit screws are the three columns in the 6 × 3 matrix:

$$\left[\begin{array}{c} \mathbf{D}_m \\ \hdashline \tilde{\mathbf{L}}_{EZ}^0 * \mathbf{D}_m \end{array} \right] = \left[\begin{array}{c:c:c} \hat{\mathbf{u}}_{E,Z}^{I} & \hat{\mathbf{u}}_{E,Z}^{II} & \hat{\mathbf{u}}_{E,Z}^{III} \end{array} \right] \tag{2.247}$$

The spherical joint Z shows in the sum of twists, Equation (2.234a), as

$$\hat{\mathbf{t}}_{E,Z} = \hat{\mathbf{u}}^{I}_{E,Z} \cdot \dot{\theta}^{x}_{Z} + \hat{\mathbf{u}}^{II}_{E,Z} \cdot \dot{\theta}^{y}_{Z} + \hat{\mathbf{u}}^{III}_{E,Z} \cdot \dot{\theta}^{z}_{Z} \tag{2.248}$$

- *Twists for cylindrical joint*: The cylindrical joint is modeled as a set of two twists, around the two-unit screws, which correspond to a rotation and translation about one of local frame's axes. The screws are the two columns one from each of the two terms in the sum (Equation (2.249)):

$$\hat{\mathbf{t}}_{E,Z} = \begin{Bmatrix} \mathbf{u}^{0}_{Z} \\ \text{-----------} \\ \tilde{\mathbf{L}}^{0}_{EZ} * \mathbf{u}^{0}_{Z} \end{Bmatrix} \cdot \dot{\theta}_{Z} + \begin{Bmatrix} 0 \\ \text{------} \\ \mathbf{u}^{0}_{Z} \end{Bmatrix} \cdot \dot{d}_{Z} \tag{2.249}$$

- *Twists for a screw joint, Figure 1.3c*: The screw joint A allows two relative motions. However, since one complete revolution of the screw causes a linear advance p, named the screw's lead, only one motion is independent.

  Therefore, the screw joint has $f = 1$ DOF, its twist is from Equation (2.249), where the equation of dependence between the two relative motions is considered:

$$\dot{d}_{Z} = p \cdot \dot{\theta}_{Z} \tag{2.250}$$

Notes:

- The theory of screws applies to mechanisms for the screw-line configurations. Such line configurations predict the mechanism singularities from a geometric perspective. The reader interested in the theory of screws can consult additional topic in [6–10].
- This text introduces the screws and twists about screws from a digraph approach, writing the lower vector in screw with respect to a cut-joint (chord). This brings a simplification in generating equations for the inverse of multiple-cycle mechanisms, presented in Chapter 3 [5].
- For open cycle mechanisms, the geometric Jacobean entries result from the entries in spanning tree matrix, T, and twists at joints, shown next.

2.17.17.2 Geometric Jacobean Based on Twists along the Path in Tree

Notice in Equation (2.234a) that the entries in spanning tree matrix T from Equation (2.228) multiply the twists at joints. Therefore, the spanning tree matrix becomes function of twists, noted as $\mathbf{T}(\hat{\mathbf{t}})$. The twist intensities are entries in joint rates' column matrix $\dot{\mathbf{q}}_a$. Equation (2.234a) becomes

$$\mathbf{T}(\hat{\mathbf{t}}) * \dot{\mathbf{q}}_{a\times 1} = \begin{Bmatrix} \boldsymbol{\omega}^{task}_{E} \\ \text{---------} \\ \mathbf{v}^{task}_{E} \end{Bmatrix} \tag{2.251}$$

Comparing Equations (2.234a) and (2.251), then

$$\mathbf{T}\left(\hat{\mathbf{t}}\right) = \mathbf{J}_{6\times c} \tag{2.252}$$

2.17.17.3 Example of Inverse Velocity Analysis for Open Cycle TRRT Spatial Mechanism with Equation Based on Twists

Equation (2.234a) as a function of twists in the joints A (pr), B (rev), C (rev), and D (pr) of TRRT mechanism holds:

$$\hat{\mathbf{t}}_{E,A} + \hat{\mathbf{t}}_{E,B} + \hat{\mathbf{t}}_{E,C} + \hat{\mathbf{t}}_{E,D} = \left\{ \begin{array}{c} \boldsymbol{\omega}_E^{task} \\ \hline \mathbf{v}_E^{task} \end{array} \right\} \tag{2.253}$$

with the twists from Equations (2.236) and (2.244), then

$$\left\{ \begin{array}{c} 0 \\ \hline \mathbf{u}_A^0 \end{array} \right\} \cdot \dot{d}_A + \left\{ \begin{array}{c} \mathbf{u}_B^0 \\ \hline \tilde{\mathbf{L}}_{EB}^0 * \mathbf{u}_B^0 \end{array} \right\} \cdot \dot{\theta}_B + \left\{ \begin{array}{c} \mathbf{u}_C^0 \\ \hline \tilde{\mathbf{L}}_{EC}^0 * \mathbf{u}_C^0 \end{array} \right\} \cdot \dot{\theta}_C + \left\{ \begin{array}{c} 0 \\ \hline \mathbf{u}_D^0 \end{array} \right\} \cdot \dot{d}_D = \left\{ \begin{array}{c} \boldsymbol{\omega}_E^{task} \\ \hline \mathbf{v}_E^{task} \end{array} \right\} \tag{2.254}$$

Equation (2.251) holds:

$$\underbrace{\left\{ \begin{array}{c|c|c|c} \overset{A}{0} & \overset{B}{\mathbf{u}_B^0} & \overset{C}{\mathbf{u}_C^0} & \overset{D}{0} \\ \hline \mathbf{u}_A^0 & \tilde{\mathbf{L}}_{EB}^0 * \mathbf{u}_B^0 & \tilde{\mathbf{L}}_{EC}^0 * \mathbf{u}_C^0 & \mathbf{u}_D^0 \end{array} \right\}}_{\mathbf{T}\left(\hat{\mathbf{t}}\right)} * \underbrace{\left[\begin{array}{c} \dot{d}_A \\ \dot{\theta}_B \\ \dot{\theta}_C \\ \dot{d}_D \end{array} \right]}_{\dot{\mathbf{q}}} = \left\{ \begin{array}{c} \boldsymbol{\omega}_E^{task} \\ \hline \mathbf{v}_E^{task} \end{array} \right\} \tag{2.255}$$

Considering Equation (2.233c) for the components of unit vectors: $\mathbf{u}_A^0$, $\mathbf{u}_B^0$, $\mathbf{u}_C^0$, and $\mathbf{u}_D^0$, and Equation (2.233d) for the components of skew-symmetric products $\tilde{\mathbf{L}}_{EB}^0 * \mathbf{u}_B^0$ and $\tilde{\mathbf{L}}_{EC}^0 * \mathbf{u}_C^0$, then

$$\underbrace{\left[\begin{array}{c|c|c|c} \overset{A}{0} & \overset{B}{0} & \overset{C}{S\theta_B} & \overset{D}{0} \\ 0 & 1 & 0 & 0 \\ 0 & 0 & C\theta_B & 0 \\ \hline 1 & -z_{EB} & y_{EC} \cdot C\theta_B & C\theta_B \cdot C\theta_C \\ 0 & 0 & z_{EC} \cdot S\theta_B - x_{EC} \cdot C\theta_B & S\theta_C \\ 0 & x_{EB} & -x_{EC} \cdot S\theta_B & -S\theta_B \cdot C\theta_C \end{array} \right]}_{\mathbf{J}} * \underbrace{\left[\begin{array}{c} \dot{d}_A \\ \dot{\theta}_B \\ \dot{\theta}_C \\ \dot{d}_D \end{array} \right]}_{\dot{\mathbf{q}}} = \left\{ \begin{array}{c} \omega_E^x \\ \omega_E^y \\ \omega_E^z \\ \hline \mathbf{v}_E^x \\ \mathbf{v}_E^y \\ \mathbf{v}_E^z \end{array} \right\} \tag{2.256}$$

Therefore, the combined equations for inverse analysis of angular and linear velocities are

$$\left\{\begin{array}{c} S\theta_B\cdot\dot{\theta}_C=\omega_E^x \\ \dot{\theta}_B=\omega_E^y \\ C\theta_B\cdot\dot{\theta}_C=\omega_E^z \\ \hdashline \dot{d}_A-(44.5+d_D)\cdot S\theta_B\cdot C\theta_C\cdot\dot{\theta}_B-(44.5+d_D)\cdot S\theta_C\cdot C\theta_B\cdot\dot{\theta}_C+C\theta_B\cdot C\theta_C\cdot\dot{d}_D=v_E^x \\ (44.5+d_D)\cdot C\theta_C\cdot\dot{\theta}_C+S\theta_C\cdot\dot{d}_D=v_E^y \\ -(44.5+d_D)\cdot C\theta_B\cdot C\theta_C\cdot\dot{\theta}_B+(44.5+d_D)\cdot S\theta_C\cdot S\theta_B\cdot\dot{\theta}_C-S\theta_B\cdot C\theta_C\cdot\dot{d}_D=v_E^z \end{array}\right. \tag{2.257}$$

Notice that Equation (2.257) is identical to those previously generated based on the relative angular velocity approach.

2.17.18 Example of Velocity Analysis for Planar Open Cycle Mechanisms with Equation Functions of Absolute Velocities

From Statement 2.15, Equation (2.211), the sum of vector entries in Latin velocity matrix $\dot{\mathbf{L}}$ is zero:

$$\dot{\mathbf{L}}_{EA}+\dot{\mathbf{L}}_{AB}+\dot{\mathbf{L}}_{BC}+\dot{\mathbf{L}}_{CD}+\cdots+\dot{\mathbf{L}}_{ZE}=0 \tag{2.258}$$

Example: For the TRT planar manipulator in Figure 2.35, the Equation (2.258) holds:

$$\begin{gathered} -\tilde{\mathbf{L}}_{EA}^0*\boldsymbol{\omega}_0-\tilde{\mathbf{L}}_{AB}^0*\boldsymbol{\omega}_1-\tilde{\mathbf{L}}_{BC}^0*\boldsymbol{\omega}_2-\tilde{\mathbf{L}}_{CE}^0*\boldsymbol{\omega}_3 \\ +\mathbf{u}_A^0\cdot\dot{d}_A+\mathbf{u}_B^0\cdot\dot{d}_B+\mathbf{u}_C^0\cdot\dot{d}_C-\mathbf{v}_E=\mathbf{0} \end{gathered} \tag{2.259}$$

For a (x_0, y_0) planar mechanism, the direction of all links' angular velocities is the z_0-axis:

$$\boldsymbol{\omega}_0=\begin{Bmatrix}0\\0\\0\end{Bmatrix};\quad \boldsymbol{\omega}_1=\begin{Bmatrix}0\\0\\\omega_1\end{Bmatrix};\quad \boldsymbol{\omega}_2=\begin{Bmatrix}0\\0\\\omega_2\end{Bmatrix};\quad \boldsymbol{\omega}_3=\begin{Bmatrix}0\\0\\\omega_3\end{Bmatrix}$$

Therefore, the cross vector multiplications have a simplified form:

$$-\tilde{\mathbf{L}}_{AB}^0*\boldsymbol{\omega}_1=-\begin{bmatrix}0&-z_{AB}&y_{AB}\\z_{AB}&0&-x_{AB}\\-y_{AB}&x_{AB}&0\end{bmatrix}*\begin{Bmatrix}0\\0\\\omega_1\end{Bmatrix}=\begin{Bmatrix}-y_{AB}\\x_{AB}\\0\end{Bmatrix}\cdot\omega_1 \tag{2.260}$$

The linear motion for prismatic joints on the plane (x_0, y_0), that is, the linear velocity vectors have all the components along z_0 as zero:

$$\mathbf{u}_A^0 \cdot \dot{d}_A = \begin{Bmatrix} \mathbf{u}_A^x \\ \mathbf{u}_A^x \\ 0 \end{Bmatrix} \cdot \dot{d}_A \tag{2.261}$$

Considering Equations (2.260) and (2.261), there are only two scalar equations generated from Equation (2.259):

$$\begin{bmatrix} -y_{EA} & -y_{AB} & -y_{BC} & -y_{CE} \\ x_{EA} & x_{AB} & x_{BC} & x_{CE} \end{bmatrix} * \begin{Bmatrix} \omega_0(=0) \\ \omega_1(=\omega_0) \\ \omega_2 \\ \omega_3(=\omega_2) \end{Bmatrix} + \begin{bmatrix} u_A^x & u_B^x & u_C^x & -\mathbf{v}_E^x \\ u_A^y & u_B^y & u_C^y & -\mathbf{v}_E^y \end{bmatrix} * \begin{Bmatrix} \dot{d}_A \\ \dot{d}_B(=0) \\ \dot{d}_C \\ 1 \end{Bmatrix} = \begin{Bmatrix} 0 \\ 0 \end{Bmatrix} \tag{2.262}$$

where the scalar equations are

$$\begin{cases} -y_{EA} \cdot \omega_0 - y_{AB} \cdot \omega_1 - y_{BC} \cdot \omega_2 - y_{CE} \cdot \omega_3 + u_A^x \cdot \dot{d}_A + u_B^x \cdot \dot{d}_B + u_C^x \cdot \dot{d}_C - v_E^x = 0 \\ x_{EA} \cdot \omega_0 + x_{AB} \cdot \omega_1 + x_{BC} \cdot \omega_2 + x_{CE} \cdot \omega_3 + u_A^y \cdot \dot{d}_A + u_B^y \cdot \dot{d}_B + u_C^y \cdot \dot{d}_C - v_E^y = 0 \end{cases} \tag{2.263}$$

- *Joint and link constraints for motion*: The constraints apply to entries in column matrices and shown between round brackets in Equation (2.262):

$$\begin{aligned} &\text{Link 0 is fixed} && : \omega_0 = 0 \\ &\text{Constraints for the prismatic joints A and C} && : \omega_1 = \omega_0 \text{ and } \omega_3 = \omega_2 = \dot{\theta}_B \\ &\text{Constraints for revolute joint B} && : \dot{d}_B = 0 \end{aligned} \tag{2.264}$$

- *Positional constraints* apply to entries in column matrices

  The matrix multiplying the angular velocity vector in Equation (2.262), noted $\tilde{\mathbf{L}}^0$, suggests that its entries are generated from the Latin matrix vectors, interchanging the x-components with the y-components and then changing the sign for the y-components:

$$\tilde{\mathbf{L}}^0 = \left[\begin{array}{c:c:c:c} -y_{EA} & -y_{AB} & -y_{BC} & -y_{CE} \\ x_{EA} & x_{AB} & x_{BC} & x_{CE} \end{array}\right] = \left[\begin{array}{c:c:c:c} -y_{EA} & -y_{AB} & 0 & 0 \\ x_{EA} & 0 & x_{BC} & x_{CE} \end{array}\right] \tag{2.265}$$

- *Positional constraints* apply to entries in matrix $\tilde{\mathbf{L}}^0$
  - $x_{AB} = 0$—Joints A and B are aligned on y_1-axis
  - $x_{BC} + x_{CE} = x_{BE}$ and $y_{BC} = y_{CE} = 0$—Joints B, C, and E are collinear along x_3-axis

The unit vectors in prismatic joints, Equations (2.153f), are

$$\mathbf{u}_A^0 = \begin{Bmatrix} \mathbf{u}_A^x \\ \mathbf{u}_A^y \end{Bmatrix} = \begin{Bmatrix} 1 \\ 0 \end{Bmatrix}; \quad \mathbf{u}_C^0 = \mathbf{D}_3 * \mathbf{u}_C^3 = \left[\begin{array}{c:c} C\theta_B & -S\theta_B \\ S\theta_B & C\theta_B \end{array}\right]\begin{Bmatrix} 1 \\ 0 \end{Bmatrix} = \begin{Bmatrix} C\theta_B \\ S\theta_B \end{Bmatrix} \tag{2.266}$$

2.17.18.1 Inverse Velocity Analysis of the TRT 3 DOF Manipulator

Equation (2.263) generates a system of linear equations. If the constraints from Equation (2.264) are substituted in equations, then the system has the form:

$$\begin{cases} \dot{d}_A + C\theta_B \cdot \dot{d}_C = v_E^x \\ S\theta_B \cdot \dot{d}_C = v_E^y + x_{EB} \cdot \omega_2 \end{cases} \tag{2.267}$$

- Input data for the inverse velocity analysis is the task angular velocity $\omega_3 = \omega_2 = \dot{\theta}_B$ and components v_E^x, v_E^y of end-effector's linear velocity.

  The combined equations, Equations (2.264) and (2.267), are written in the form:

$$\begin{cases} \dot{\theta}_B = \omega_3 \\ \dot{d}_A + C\theta_B \cdot \dot{d}_C = v_E^x \\ S\theta_B \cdot \dot{d}_C = v_E^y + x_{EB} \cdot \omega_2 \end{cases} \tag{2.268}$$

The equations are written in matrix form as a function of joint rates matrix $\dot{\mathbf{q}}$:

$$\begin{bmatrix} 0 & 1 & 0 \\ 1 & 0 & C\theta_B \\ 0 & 0 & S\theta_B \end{bmatrix} * \begin{Bmatrix} \dot{d}_A \\ \dot{\theta}_B \\ \dot{d}_C \end{Bmatrix} = \begin{Bmatrix} \omega_3 \\ v_E^x \\ v_E^y + x_{EB} \cdot \omega_2 \end{Bmatrix} \tag{2.269}$$

The determinant for the matrix in the left side is $\Delta = S\,\theta_B$.
If $\Delta \neq 0$, then the solution is

$$\dot{\theta}_B = \omega_3; \dot{d}_A = v_E^x - C\theta_B \cdot \dot{d}_C; \quad \dot{d}_C = \frac{v_E^y + x_{EB} \cdot \omega_3}{S\theta_B} \tag{2.270}$$

2.17.18.2 Singularities for Inverse Velocity Analysis

For $\theta_B = 0°$ or $\theta_B = 180°$, then $\Delta = S\,\theta_B = 0$ and the manipulator in these positions has a *critical position* for velocities. In EES and SW simulations, a guess value might be required; for example, for an EES simulation, consider in the Variable Info window a guess value $\theta_B = 0.01°$, in case the manipulator starts from initial (home) position where $\theta_B = 0.0°$.

Algebraic interpretation: The matrix of coefficients is singular for a critical position, and the system cannot be solved for the unknown rates. Since the last row has all entries zeroes, the matrix with rank 3 becomes a matrix with its rank at most 2.

2.17.19 Example of Velocity Analyses for Planar Open Cycle Mechanisms with Equation Functions of Twists

Equation (2.234a) as a function of twists in the joints A (pr), B (rev), and C (pr) of the TRT planar manipulator holds:

$$+1 \cdot \hat{\mathbf{t}}_{E,A} + 1 \cdot \hat{\mathbf{t}}_{E,B} + 1 \cdot \hat{\mathbf{t}}_{E,C} = \begin{Bmatrix} \omega_E \\ \hline \mathbf{v}_E \end{Bmatrix} \tag{2.271}$$

with the twists from Equations (2.236) and (2.244), then

$$\begin{Bmatrix} \mathbf{0} \\ \hline \mathbf{u}_A^0 \end{Bmatrix} \cdot \dot{d}_A + \begin{Bmatrix} \mathbf{u}_B^0 \\ \hline \tilde{\mathbf{L}}_{EB}^0 * \mathbf{u}_B^0 \end{Bmatrix} \cdot \dot{\theta}_B + \begin{Bmatrix} \mathbf{0} \\ \hline \mathbf{u}_C^0 \end{Bmatrix} \cdot \dot{d}_C = \begin{Bmatrix} \omega_E \\ \hline \mathbf{v}_E \end{Bmatrix} \tag{2.272}$$

For inverse analysis, the 6 × 1 task vector in the right side is provided.
Equation (2.272) holds:

$$\underbrace{\begin{Bmatrix} \overset{A}{\mathbf{0}} & \overset{B}{\mathbf{u}_B^0} & \overset{C}{\mathbf{0}} \\ \hline \mathbf{u}_A^0 & \tilde{\mathbf{L}}_{EB}^0 * \mathbf{u}_B^0 & \mathbf{u}_C^0 \end{Bmatrix}}_{\mathbf{T}(\hat{\mathbf{t}})} * \underbrace{\begin{Bmatrix} \dot{d}_A \\ \dot{\theta}_B \\ \dot{d}_C \end{Bmatrix}}_{\dot{\mathbf{q}}} = \begin{Bmatrix} \omega_E \\ \hline \mathbf{v}_E \end{Bmatrix} \tag{2.273}$$

Considering the 3 × 1 unit vectors

$$\mathbf{u}_A^0 = \begin{Bmatrix} 1 \\ 0 \\ 0 \end{Bmatrix}; \quad \mathbf{u}_B^0 = \begin{Bmatrix} 0 \\ 0 \\ 1 \end{Bmatrix}; \quad \mathbf{u}_C^0 = \begin{Bmatrix} C\theta_B \\ S\theta_B \\ 0 \end{Bmatrix}$$

With the skew-symmetric product shown in Equation (2.274):

$$\tilde{\mathbf{L}}_{EB}^{0} * \mathbf{u}_{B}^{0} = \begin{bmatrix} 0 & -z_{EB} & y_{EB} \\ z_{EB} & 0 & -x_{EB} \\ -y_{EB} & x_{EB} & 0 \end{bmatrix} * \begin{Bmatrix} 0 \\ 0 \\ 1 \end{Bmatrix} = \begin{Bmatrix} 0 \\ -x_{EB} \\ 0 \end{Bmatrix} \quad (2.274)$$

Then, Equation (2.273) becomes

$$\underbrace{\begin{bmatrix} \overset{A}{0} & \overset{B}{0} & \overset{C}{0} \\ 0 & 0 & 0 \\ 0 & 1 & 0 \\ \hline 1 & 0 & C\theta_B \\ 0 & -x_{EB} & S\theta_B \\ 0 & 0 & 0 \end{bmatrix}}_{\mathbf{J}} * \underbrace{\begin{Bmatrix} \dot{d}_A \\ \dot{\theta}_B \\ \dot{d}_C \end{Bmatrix}}_{\dot{\mathbf{q}}} = \begin{Bmatrix} \omega_E^x \\ \omega_E^y \\ \omega_E^z \\ \hline v_E^x \\ v_E^y \\ v_E^z \end{Bmatrix} \quad (2.275)$$

2.17.19.1 Capability of Motion for the TRT Manipulator

$$\begin{cases} 0 = \omega_E^0 \\ 0 = \omega_E^y \\ \dot{\theta}_B = \omega_E^z \\ \hline \dot{d}_A + C\theta_B \cdot \dot{d}_C = v_E^x \\ -x_{EB} \cdot \dot{\theta}_B + S\theta_B \cdot \dot{d}_C = v_E^y \\ 0 = v_E^z \end{cases} \quad (2.276)$$

Since ω_E^x, ω_E^y, and v_E^z have zero values, then the actuators at A, B, and C (equal to the mechanism's DOF) cannot provide rotation about x_0, y_0, or linear motion along z_0.

Note:

- Equation (2.268) generated from absolute angular velocities is identical to those previously generated based on twists (Equation (2.276)).

Geometric interpretation of equations based on twists: Since the first two and the last rows in matrix **J** have all entries zeroes, its rank is at most 3.

Eliminating the zeroes' rows, the determinant of the 3 × 3 reduced matrix $\mathbf{J}^{red}$ is

$$\det J^{red} = -S\theta_B \quad (2.277)$$

The entries in the matrix are the Plucker coordinates for the support lines for the unit screws $\hat{u}_{EA}$, $\hat{u}_{EB}$, and $\hat{u}_{EC}$. The geometric configuration of their support lines is a set of two fascicles of lines: one fascicle of parallel lines, normal to the plane of motion (parallel to z_0)—for revolute joints, and the second fascicle of coplanar and intersecting lines in the plane of motion (x_0, y_0)—for prismatic joints. Therefore, the rank for this configuration of lines is at most 3 for θ_B other than $\theta_B = 0°$ or $\theta_B = 180°$ (Equation (2.277)).

For $\theta_B = 0°$ or $\theta_B = 180°$, the configuration of the support lines changes into a set of two fascicles: one fascicle of parallel lines, normal to the plane of motion, and the second fascicle of coplanar and parallel lines is in the plane of motion (x_0, y_0). The rank for this configuration of lines is at most 2. In general, if the line supports of unit screws change its rank, and then the critical positions are detected.

There is a relation between the configuration of lines and the mechanism's DOF. The next chapter is presented the formula for DOF based on the rank of matrix $\mathbf{J} = \mathbf{T}(\hat{t})$—for open cycles, and $\mathbf{J} = \mathbf{C}(\hat{t})$—for multiple closed-cycle mechanisms. The screws' coefficients: –1, 0, +1, or –1 are entries in the tree matrix T and cycle matroid matrix C.

2.17.19.2 EES for Direct Velocity Analysis of the TRT 3 DOF Manipulator

- The rates for displacements in actuated joints A, B, and C are determined from Equations (2.147) as linear functions of time t:

$$\dot{q}_M(t) = \begin{Bmatrix} \dot{d}_A(t) \\ \dot{\theta}_B(t) \\ \dot{d}_C(t) \end{Bmatrix} = \begin{Bmatrix} 120 \cdot t\,[\text{in./s}] \\ 540 \cdot t\,[°/\text{s}] \\ 40 \cdot t\,[\text{in./s}] \end{Bmatrix} = \begin{Bmatrix} 120 \cdot t\,[\text{in./s}] \\ 540 \cdot t \cdot (\pi/180)\,[\text{rad/s}] \\ 40 \cdot t\,[\text{in./s}] \end{Bmatrix} \quad (2.278)$$

- The end-effector's E linear velocity components are determined from Equation (2.276).
- The links' angular velocity scalars are determined from Equation (2.178a), as functions of joint rates:

$$\omega_m = \mathbf{Z}^T * \dot{\theta}_a$$

$$\begin{Bmatrix} \omega_1 \\ \omega_2 \\ \omega_3 \end{Bmatrix} = \begin{bmatrix} 1 & 0 & 0 \\ 1 & 1 & 0 \\ 1 & 1 & 1 \end{bmatrix} * \begin{Bmatrix} \dot{\theta}_A(=0) \\ \dot{\theta}_B \\ \dot{\theta}_C(=0) \end{Bmatrix} = \begin{Bmatrix} 0 \\ \dot{\theta}_B \\ \dot{\theta}_B \end{Bmatrix} \quad (2.279)$$

- Latin vector velocities $\dot{\mathbf{L}}$ are determined from Equation (2.207i):

$$\begin{Bmatrix} \dot{\mathbf{L}}_{OA} \\ \dot{\mathbf{L}}_{AB} \\ \dot{\mathbf{L}}_{BC} \end{Bmatrix} = \begin{Bmatrix} -\tilde{\mathbf{L}}_{OA}^{0} * \boldsymbol{\omega}_0^0 + \mathbf{u}_A^0 \cdot \dot{d}_A \\ -\tilde{\mathbf{L}}_{AB}^{0} * \boldsymbol{\omega}_1^0 + \mathbf{u}_B^0 \cdot \dot{d}_B \\ -\tilde{\mathbf{L}}_{BC}^{0} * \boldsymbol{\omega}_2^0 + \mathbf{u}_C^0 \cdot \dot{d}_C \end{Bmatrix} = \begin{Bmatrix} \mathbf{u}_A^0 \cdot \dot{d}_A \\ 0 \\ -\tilde{\mathbf{L}}_{BC}^{0} * \boldsymbol{\omega}_2^0 + \mathbf{u}_C^0 \cdot \dot{d}_C \end{Bmatrix} \quad (2.280)$$

- All joints' absolute linear velocities are determined from
  - *Node m = 1*: The link 1's frame origin absolute linear velocity:

$$\mathbf{v}_A = \dot{\mathbf{L}}_{OA}^0 \Rightarrow \begin{Bmatrix} v_A^x \\ v_A^y \end{Bmatrix} = \begin{Bmatrix} 1 \\ 0 \end{Bmatrix} \cdot \dot{d}_A = \begin{Bmatrix} \dot{d}_A \\ 0 \end{Bmatrix} \quad (2.281)$$

  - *Edge B*: The Latin velocity vector (Equation (2.280)):

$$\dot{\mathbf{L}}_{AB}^0 = \mathbf{0} \quad (2.282)$$

  - *Node m = 2*: The link 2's frame origin absolute linear velocity:

$$\mathbf{v}_B = \mathbf{v}_A + \dot{\mathbf{L}}_{AB}^0 \Rightarrow \begin{Bmatrix} v_B^x \\ v_B^y \end{Bmatrix} = \begin{Bmatrix} v_A^x \\ v_A^y \end{Bmatrix} = \begin{Bmatrix} \dot{d}_A \\ 0 \end{Bmatrix} \quad (2.283)$$

  - *Edge C*: The Latin velocity vector (Equation (2.280)):

$$\dot{\mathbf{L}}_{BC}^0 = -\tilde{\mathbf{L}}_{BC}^0 * \boldsymbol{\omega}_2 + \mathbf{u}_C^0 \cdot \dot{d}_C \quad (2.284)$$

  - *Node m = 3*: The link 3's frame origin absolute linear velocity:

$$\mathbf{v}_C = \mathbf{v}_B + \dot{\mathbf{L}}_{BC}^0 \Rightarrow \begin{Bmatrix} v_C^x \\ v_C^y \end{Bmatrix} = \begin{Bmatrix} v_B^x \\ v_B^y \end{Bmatrix} + \begin{Bmatrix} -y_{BC} \cdot \omega_2 \\ x_{BC} \cdot \omega_2 \end{Bmatrix} + \begin{Bmatrix} C\theta_B \cdot \dot{d}_C \\ S\theta_B \cdot \dot{d}_C \end{Bmatrix} \quad (2.285)$$

- The components for the end-effector linear velocity are determined from Equation (2.263).
  - In Equations Window, Appendix 2.11, the user types the Equation (2.263); constraints from Equation (2.264); links' angular velocities from Equation (2.279); and joints from Equations (2.281), (2.283), and (2.285).
  - In Equations Window, type the text for an Array to display the links' angular velocities $\boldsymbol{\omega}_m$ as single row matrix.
  - In Equations Window, type the text for an Array to display the all joints' absolute linear velocities v as two rows matrix.

Figure 2.60 shows the EES Formatted Equations.

Formatted Equations

2.Direct Velocities Analysis

Input: Actuating motors parameters

$\dot{d}_A = 2 \cdot d_{A,max} \cdot t$ Prismatic motor at A

$\dot{\theta}_B = 2 \cdot \theta_{B,max} \cdot t$ Revolute motor at B

$\dot{d}_C = 2 \cdot d_{C,max} \cdot t$ Prismatic motor at C

Output:

- End Effector Velocities: v_{Ex}, v_{Ey}; Joints Relative Ang Veloc Matrix-Theta$_j$,Mobile Links Abs Ang Veloc matrix- Omega$_m$;
- All Joints Absolute Velocity Matrix, v ,Relative motion's unit vectors-components on fixed frame

$u_{Ax} = 1$

$u_{Ay} = 0$

$u_{Bx} = 0$

$u_{By} = 0$

$u_{Cx} = \cos(\theta_B)$

$u_{Cy} = \sin(\theta_B)$

2.1 Equations

Data: Equations for Constrained Motion

$\omega_0 = 0$ [deg/s] link 0 is fixed

$\dot{\theta}_A = 0$ [deg/s] Prismatic joint between link 0 iand link 1

$\dot{d}_B = 0$ [in/s] Revolute joint between link 1 iand link 2

$\dot{\theta}_C = 0$ [deg/s] Prismatic joint between link 2 iand link 3

Equations:

$$-y_{EA} \cdot \omega_0 \cdot \frac{\pi}{180} - y_{AB} \cdot \omega_1 \cdot \frac{\pi}{180} - y_{BC} \cdot \omega_2 \cdot \frac{\pi}{180} - y_{CE} \cdot \omega_3 \cdot \frac{\pi}{180} + \dot{d}_A \cdot u_{Ax} + \dot{d}_B \cdot u_{Bx} + \dot{d}_C \cdot u_{Cx} - v_{Ex} = 0$$

$$x_{EA} \cdot \omega_0 \cdot \frac{\pi}{180} + x_{AB} \cdot \omega_1 \cdot \frac{\pi}{180} + x_{BC} \cdot \omega_2 \cdot \frac{\pi}{180} + x_{CE} \cdot \omega_3 \cdot \frac{\pi}{180} + \dot{d}_A \cdot u_{Ay} + \dot{d}_B \cdot u_{By} + \dot{d}_C \cdot u_{Cy} - v_{Ey} = 0$$

2.3 Mobile Links AbsoluteAngular Velocities

$\omega_1 = \omega_0 + \dot{\theta}_A$

$\omega_2 = \omega_1 + \dot{\theta}_B$

$\omega_3 = \omega_2 + \dot{\theta}_C$

$\omega_{m,array,1,1..3}$ = [Omega_1, Omega_2, Omega_3]

2.4 Relative velocities between points on different joined links, written for each joint

Joint A

$v_{Ax} = \dot{d}_A \cdot u_{Ax}$

$v_{Ay} = \dot{d}_A \cdot u_{Ay}$

Joint B

$$v_{Bx} = v_{Ax} - y_{AB} \cdot \omega_1 \cdot \frac{\pi}{180} + \dot{d}_B \cdot u_{Bx}$$

$$v_{By} = v_{Ay} + x_{AB} \cdot \omega_1 \cdot \frac{\pi}{180} + \dot{d}_B \cdot u_{By}$$

Joint C

$$v_{Cx} = v_{Bx} - y_{BC} \cdot \omega_2 \cdot \frac{\pi}{180} + \dot{d}_C \cdot u_{Cx}$$

$$v_{Cy} = v_{By} + x_{BC} \cdot \omega_2 \cdot \frac{\pi}{180} + \dot{d}_C \cdot u_{Cy}$$

Output: All Joints Velocity Matrix, v

$v_{array,1,1..3}$ = [v_Ax, v_Bx, v_Cx]

$v_{array,2,1..3}$ = [v_Ay, v_By, v_Cy]

FIGURE 2.60
Program for direct velocity analysis of TRT manipulator-EES Formatted Equations.

- In Tables/New Parametric Table, insert 25 for the number of runs and the time interval First Value: 0 and Last (linear): 1. Then, from Calculate select Solve Table.

  Appendix 2.12 shows the EES Parametric Table with the joints A, B, and C rates, and the end-effector's angular and linear velocity components for the time interval of 1 s
- In Plots/New Plot Window/X–Y Plot, select t for x-Axis and x_E for Y-Axis. Repeat for y_E. The two previously generated graphs show on the same plot using Plots/Overlay Plot, illustrated in Figure 2.61.
- In Plots/New Plot Window/X–Y Plot, select t for x-Axis and ω_3 for Y-Axis. The end-effector angular velocity, Figure 2.62, shows a linear variation during the simulation time of 1 s.
- In Tables/New Lookup Table, copy and paste the EES columns from Parametric Table for v_E^x, v_E^y, and ω_3 values. In addition, copy and paste the Excel columns for v_E^x, v_E^y, and ω_3 values from the SW Motion Appendix 2.13 tabulates the calculated values (using EES) and simulated values (using SW) of the end-effector linear velocity components and its angular velocity.

  Note: All values are coincidental.

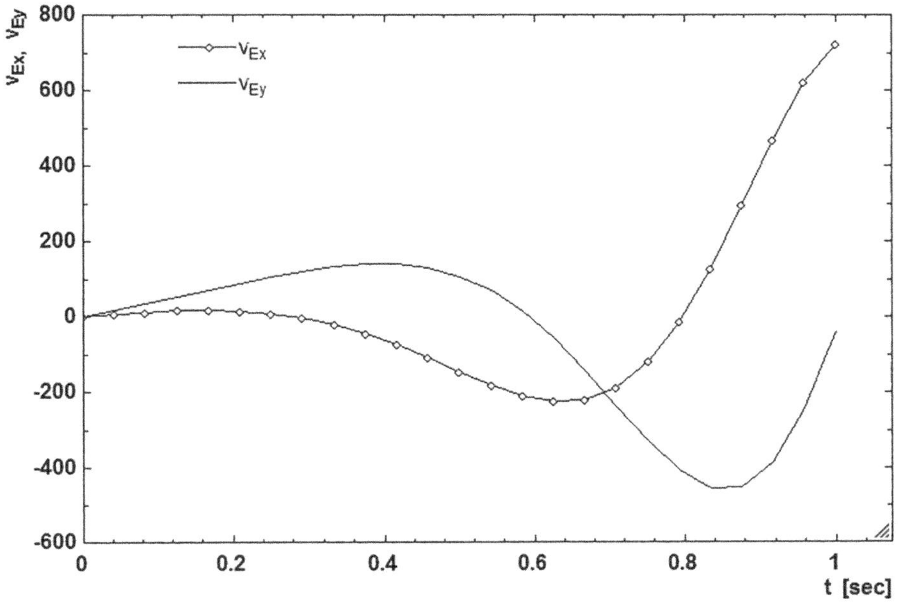

FIGURE 2.61
End-effector's linear velocity components.

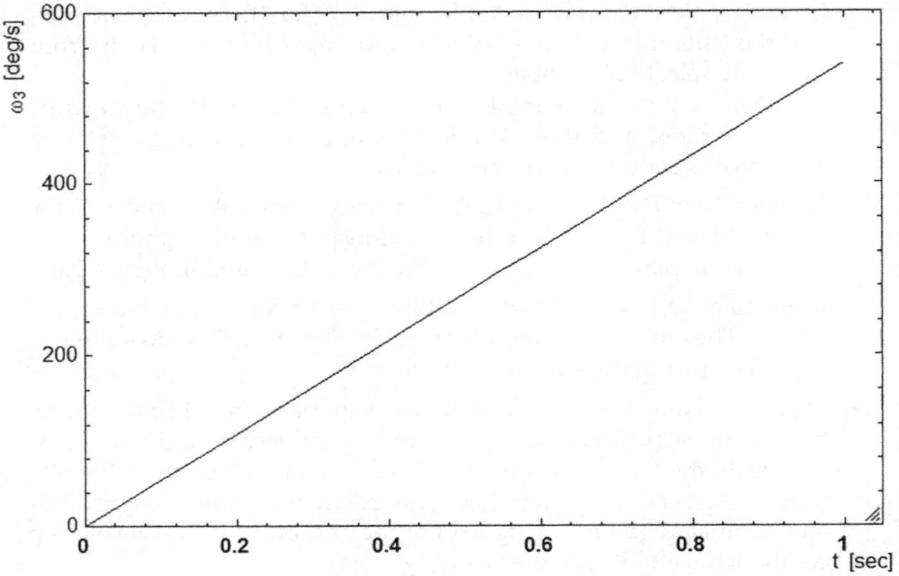

FIGURE 2.62
End-effector's angular velocity.

2.18 Velocity Analysis of Planar Open Cycle Mechanisms with All Revolute Joints

For mechanisms with revolute joints only, all relative linear rates $\dot{\mathbf{d}}$ are zero, and Equation (2.262) has the form shown in Equation (2.286):

$$\tilde{\mathbf{L}}^0 * \boldsymbol{\omega} = \mathbf{v}_E \tag{2.286}$$

Problems

P2.1: The TRRT robotic mechanism in Figure 2.1 (Animation P2.1). For the actuated joint displacements shown in Figure P2.1a,b, determine

a. The analytic functions for joint displacements, if the time interval is 3 s
b. The relative rotation matrices
c. The absolute rotation matrix for the end-effector, $\mathbf{D}_e$

d. The equations for the position vectors of joints A, B, C, D, and end-effector E

e. Create a spreadsheet with 25 rows for end-effector coordinates x_E, y_E, and z_E. Graph the coordinates versus time.

P2.2–P2.5: In Figures P2.2–P2.5, the TRRT mechanism shows at different positions. The constant lengths are from Figure 2.1 and the actuated joint displacements from Figure P2.1.

For each position, at time t = 0, 1, 2, and 3s, by using the homogeneous matrices, determine

a. The absolute and relative homogeneous matrices
b. The coordinates for joints A, B, C, and D
c. The end-effector E orientation and position: $\alpha^{I}_{0,e}$, $\beta^{II}_{0,e}$, $\gamma^{III}_{0,e}$, x_E, y_E, and z_E

(a)

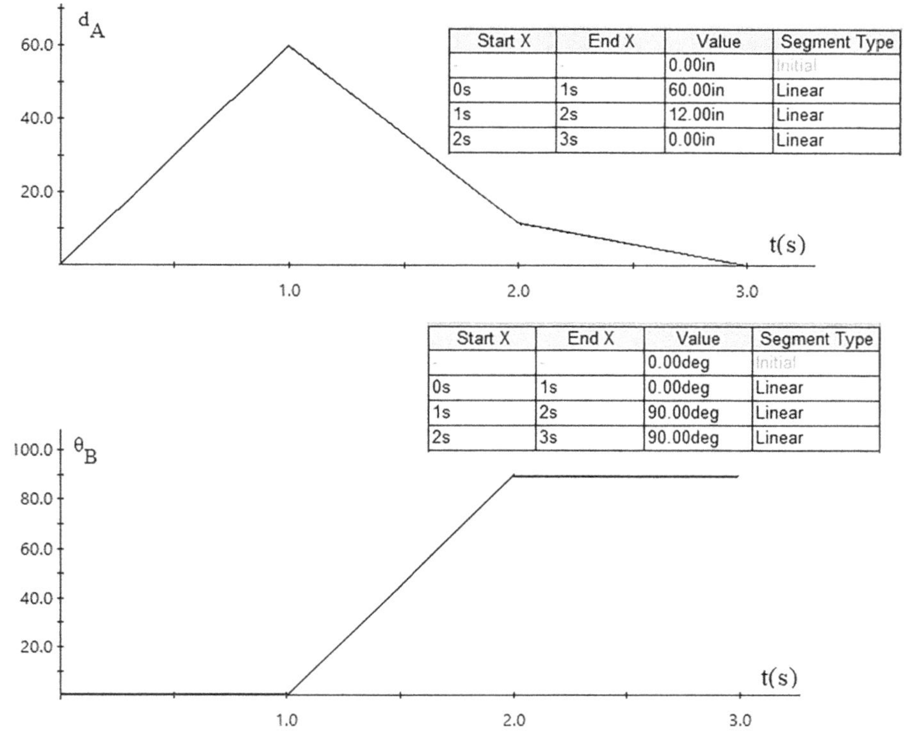

Start X	End X	Value	Segment Type
-	-	0.00in	Initial
0s	1s	60.00in	Linear
1s	2s	12.00in	Linear
2s	3s	0.00in	Linear

Start X	End X	Value	Segment Type
-	-	0.00deg	Initial
0s	1s	0.00deg	Linear
1s	2s	90.00deg	Linear
2s	3s	90.00deg	Linear

FIGURE P2.1
(a) Joint A—linear displacement and joint B—angular displacement. (b) Joint C—angular displacement and joint D—linear displacement.

(Continued)

(b)

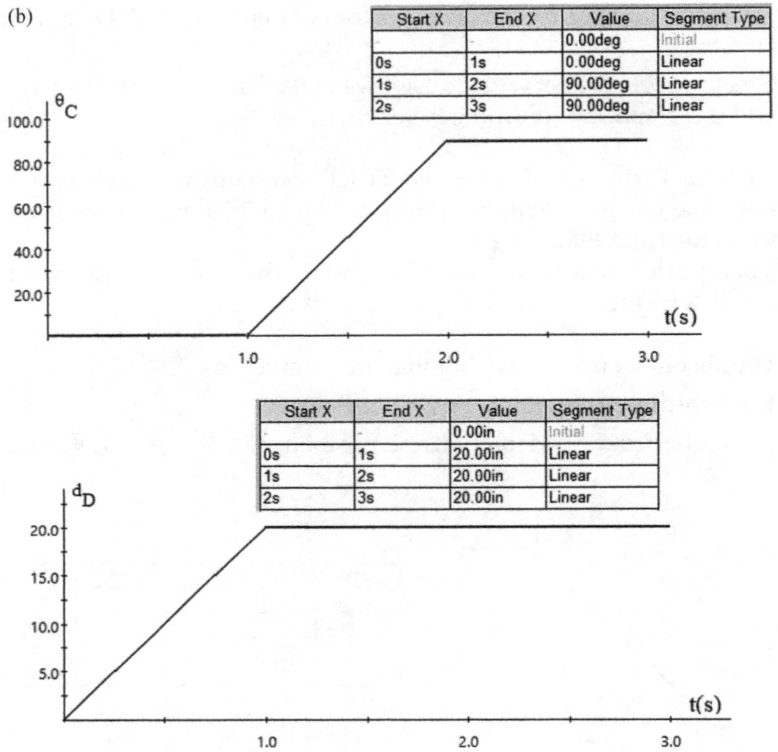

Start X	End X	Value	Segment Type
-	-	0.00deg	Initial
0s	1s	0.00deg	Linear
1s	2s	90.00deg	Linear
2s	3s	90.00deg	Linear

Start X	End X	Value	Segment Type
-	-	0.00in	Initial
0s	1s	20.00in	Linear
1s	2s	20.00in	Linear
2s	3s	20.00in	Linear

FIGURE P2.1 (CONTINUED)
(a) Joint A—linear displacement and joint B—angular displacement. (b) Joint C—angular displacement and joint D—linear displacement.

P2.2 At $t = 0\,s$, Figure P2.2

P2.3 At $t = 1\,s$, Figure P2.3

P2.4 At $t = 2\,s$, Figure P2.4

P2.5 At $t = 3\,s$, Figure P2.5

P2.6–P2.9: In Figures P2.2–P2.5, the TRRT mechanism shows at different positions. Consider the constant lengths from the mechanism in Figure 2.1.
For inverse positional analysis, determine

a. The system of nonlinear equations
b. Using an equation solver, find the actuated joint displacements: d_A, θ_B, θ_C, and d_D to reach the following end-effector's orientation and position parameters, provided from P2.2–P2.5.

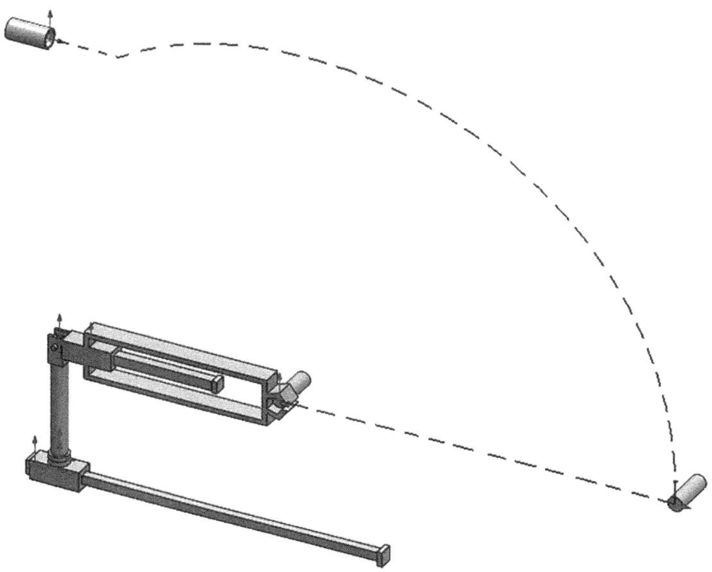

FIGURE P2.2
Home position.

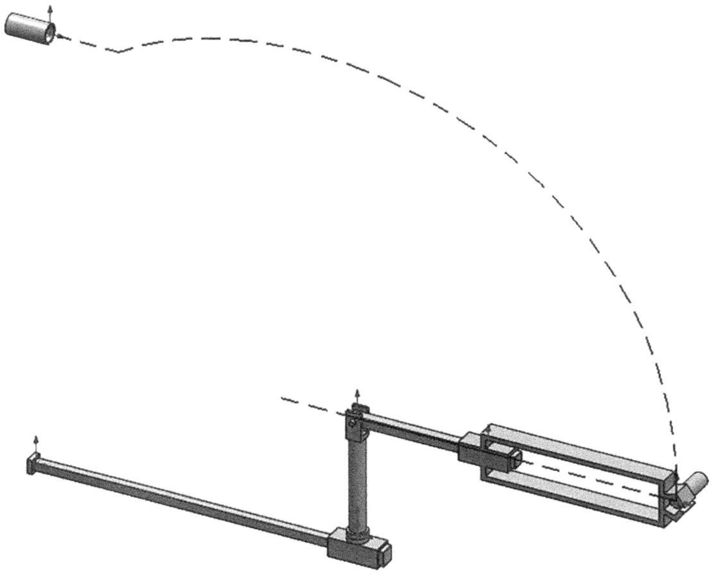

FIGURE P2.3
Part-grasped position.

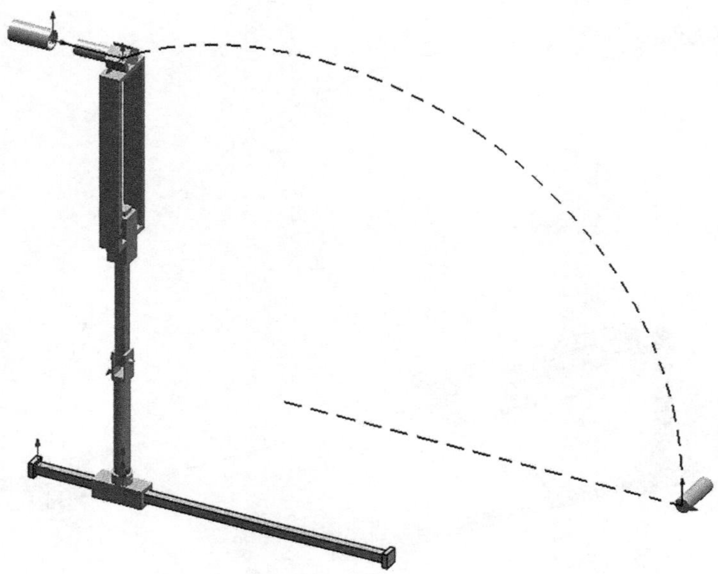

FIGURE P2.4
Target-approach position.

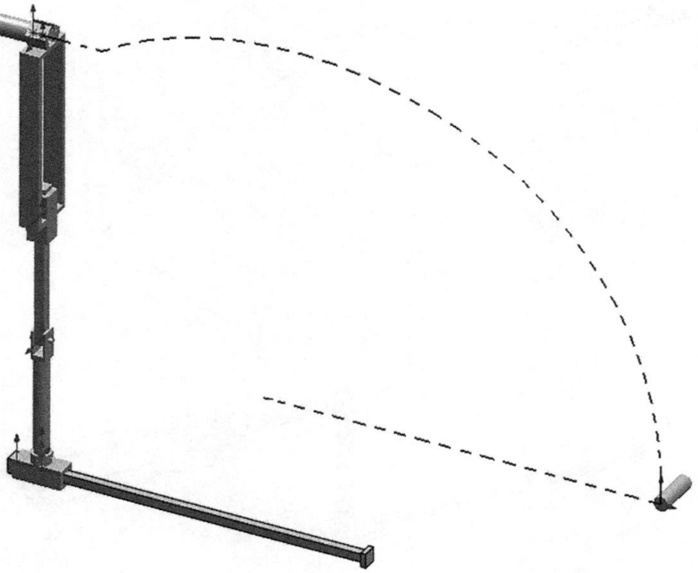

FIGURE P2.5
Target-reached position.

P2.6 The end-effector's orientation and position ($\alpha^{I}_{0,e}$, $\beta^{II}_{0,e}$, $\gamma^{III}_{0,e}$, x_E, y_E, z_E) from P2.2

P2.7 The end-effector's orientation and position ($\alpha^{I}_{0,e}$, $\beta^{II}_{0,e}$, $\gamma^{III}_{0,e}$, x_E, y_E, z_E) from P2.3

P2.8 The end-effector's orientation and position ($\alpha^{I}_{0,e}$, $\beta^{II}_{0,e}$, $\gamma^{III}_{0,e}$, x_E, y_E, z_E) from P2.4

P2.9 The end-effector's orientation and position ($\alpha^{I}_{0,e}$, $\beta^{II}_{0,e}$, $\gamma^{III}_{0,e}$, x_E, y_E, z_E) from P2.5

P2.10: For the TRRT robotic mechanism in Figure 2.1, with the actuated joint displacements shown in Figure P2.1, and a time interval is 3 s, determine

a. The joint rates in time $\dot{d}_A$, $\dot{\theta}_B$, $\dot{\theta}_C$, and $\dot{d}_D$
b. The links' angular velocity magnitudes
c. The joints' linear velocity magnitudes
d. The end-effector's E linear velocity components

P2.11–P2.14: In Figures P2.2–P2.5, the TRRT mechanism shows the end-effector at different positions. The constant lengths are from Figure 2.1 and the actuated joint displacements from Figure P2.1.

For each position, at time t = 0, 1, 2, and 3 s, determine

a. The joint rates in time $\dot{d}_A$, $\dot{\theta}_B$, $\dot{\theta}_C$, and $\dot{d}_D$
b. The links' angular velocity magnitudes
c. The joints' linear velocity magnitudes
d. The end-effector's E linear velocity components

P2.11 At t = 0 s, Figure P2.2

P2.12 At t = 1 s, Figure P2.3

P2.13 At t = 2 s, Figure P2.4

P2.14 At t = 3 s, Figure P2.5

P2.15: The cylindrical robotic mechanism RTTR (Animation P2.2) has the actuated joints A (revolute about y_1), B (prismatic along y_2), C (prismatic along z_3), and D (revolute about x_4). Figure P2.6a,b shows the joint displacements for an interval of 1 s. Figure P2.6c shows the robot in the initial (home position), and Figure P2.6d at t = 0.5 s.

Determine

a. Draw the digraph (tree) and label the tree's nodes and arcs
b. Calculate the mechanism's DOF

c. Create a spreadsheet with 25 rows for end-effector coordinates x_E, y_E, and z_E. Graph the variation of coordinates for a period of 1 s
d. The end-effector's orientation and position ($\alpha^{I}_{0,e}$, $\beta^{II}_{0,e}$, $\gamma^{III}_{0,e}$, x_E, y_E, z_E) at t = 0.5 s

P2.16: Figure P2.6d shows the orientation and position of a cylindrical robot at t = 0.5 s.

For inverse positional analysis, determine

a. The system of nonlinear equations
b. Using an equation solver, find the actuated joint displacements: θ_A, d_B, d_C, and θ_D to reach the end-effector E provided orientation and position parameters.

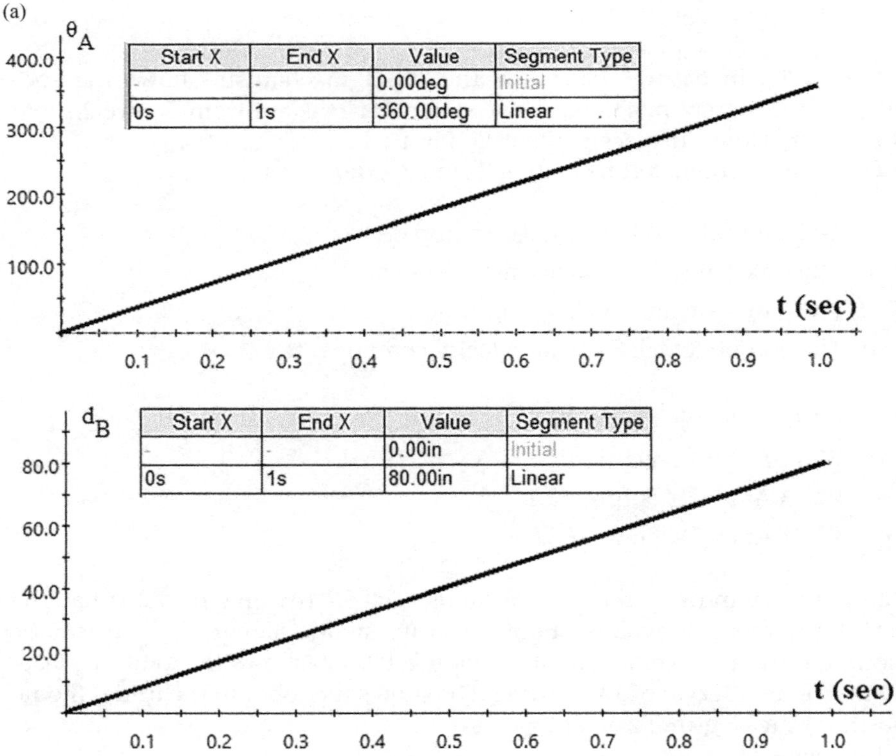

FIGURE P2.6
The cylindrical robotic mechanism RTTR: (a) joint A and B displacements; (b) joint C and D displacements; (c) home position; (d) position at t = 0.5 s.

(Continued)

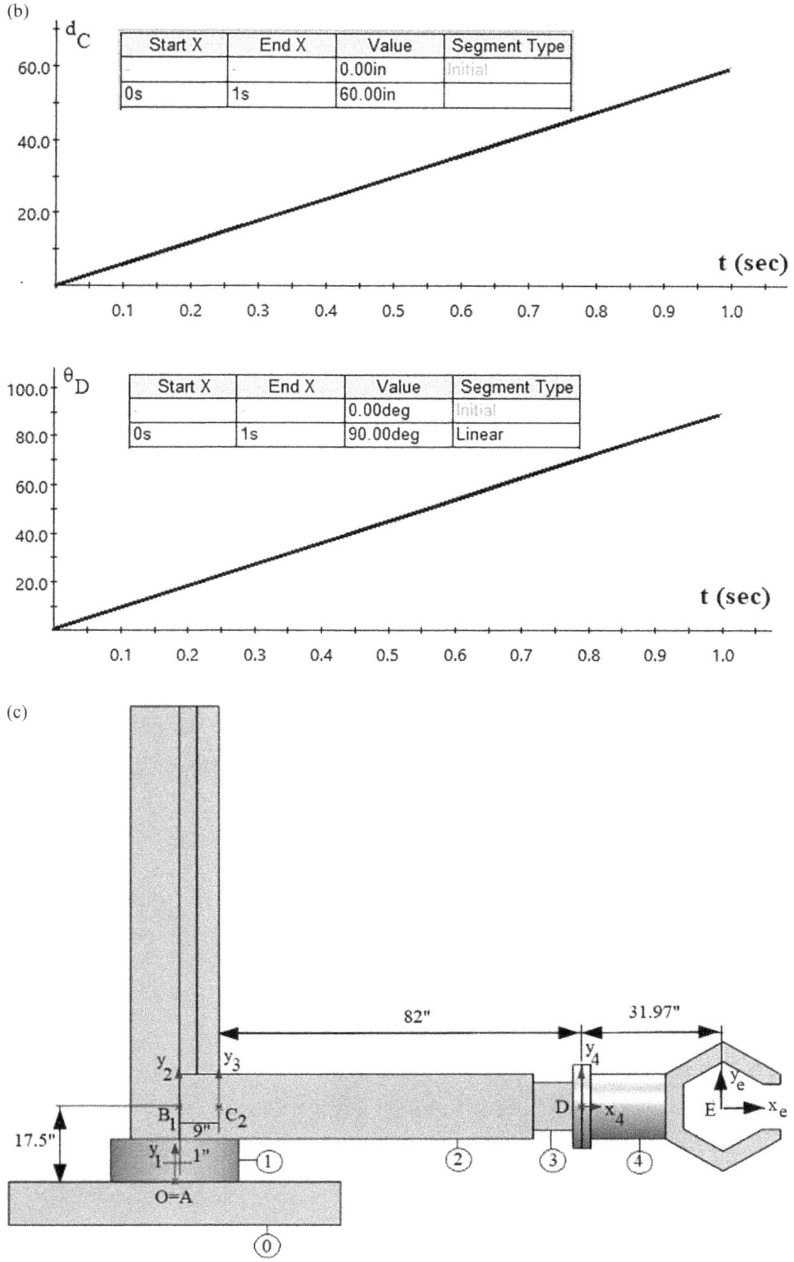

FIGURE P2.6 (CONTINUED)
The cylindrical robotic mechanism RTTR: (a) joint A and B displacements; (b) joint C and D displacements; (c) home position; (d) position at t = 0.5 s.

(*Continued*)

(d)

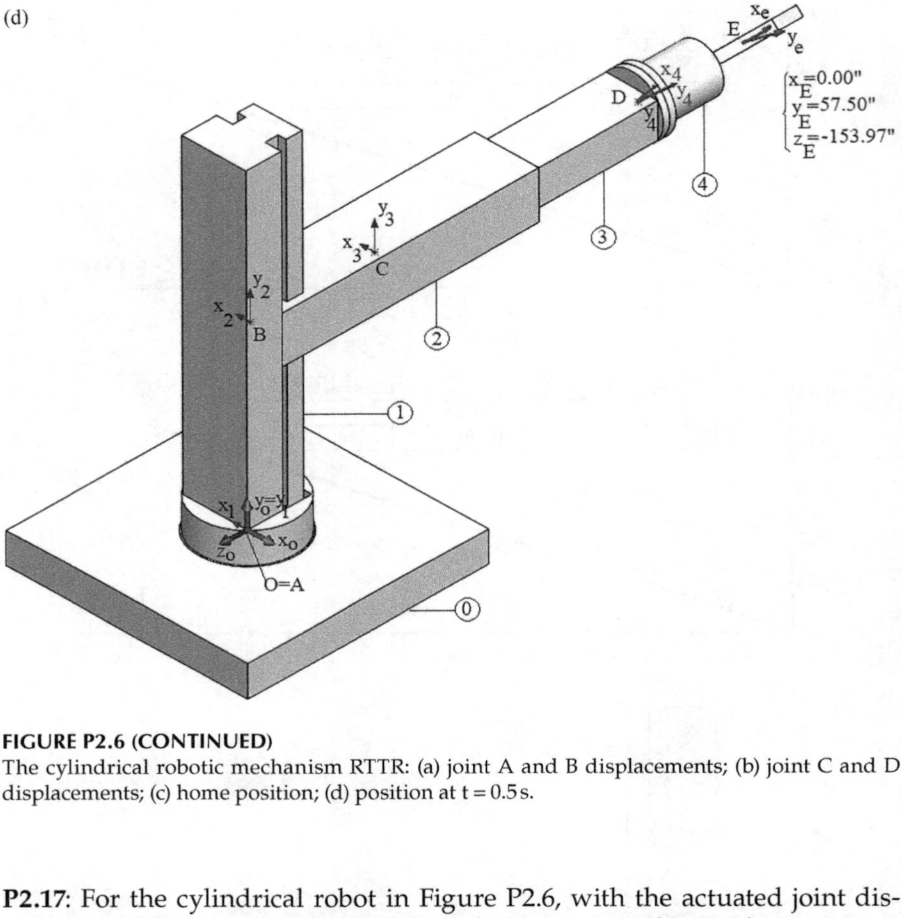

FIGURE P2.6 (CONTINUED)
The cylindrical robotic mechanism RTTR: (a) joint A and B displacements; (b) joint C and D displacements; (c) home position; (d) position at t = 0.5 s.

P2.17: For the cylindrical robot in Figure P2.6, with the actuated joint displacements shown in Figure P2.6a,b, and a time interval is 1 s, determine

a. The joint rates in time $\dot{\theta}_A$, $\dot{d}_B$, $\dot{d}_C$, and $\dot{\theta}_D$
b. The links' angular velocity magnitudes
c. The joints' linear velocity magnitudes
d. The end-effector's E linear velocity components

P2.18: The cylindrical robot at t = 0.5 s is in position shown in Figure P2.6d. *Determine*

a. The joint rates $\dot{\theta}_A$, $\dot{d}_B$, $\dot{d}_C$, and $\dot{\theta}_D$
b. The links' angular velocity magnitudes
c. The joints' linear velocity magnitudes
d. The end-effector's E linear velocity components and its magnitude

P2.19: The spherical robot RRT (Animation P2.3) is shown in Figure 2.7a in the initial (home position) and in Figure 2.7b at t = 0.1 s. The actuated joint displacements at A (revolute about y_1), B (revolute about z_2), and C (prismatic along x_3) shown in Figure 2.7c.

Determine

a. Draw the digraph (tree) and label the tree's nodes and arcs
b. Calculate the mechanism's DOF
c. Create a spreadsheet with 25 rows for end-effector's coordinates x_E, y_E, and z_E. Graph the variation of coordinates for a period of 1 s
d. The end-effector's orientation and position ($\alpha^{I}_{0,e}$, $\beta^{II}_{0,e}$, $\gamma^{III}_{0,e}$, x_E, y_E, z_E) at t = 0.5 s

P2.20: Figure P2.7b shows the orientation and position of a spherical robot at t = 0.1 s.

For inverse positional analysis, determine

a. The system of nonlinear equations
b. Using an equation solver, find the actuated joint displacements: θ_A, d_B, d_C, and θ_D to reach the end-effector E provided orientation and position parameters.

(a)

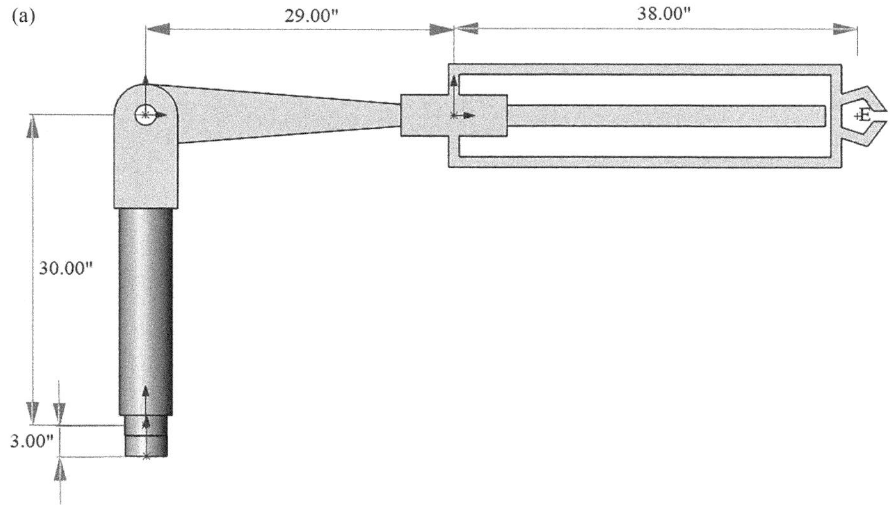

FIGURE P2.7
(a) The spherical robot RRT at home position; (b) position of spherical robot at t = 0.1 s; (c) joint displacements.

(Continued)

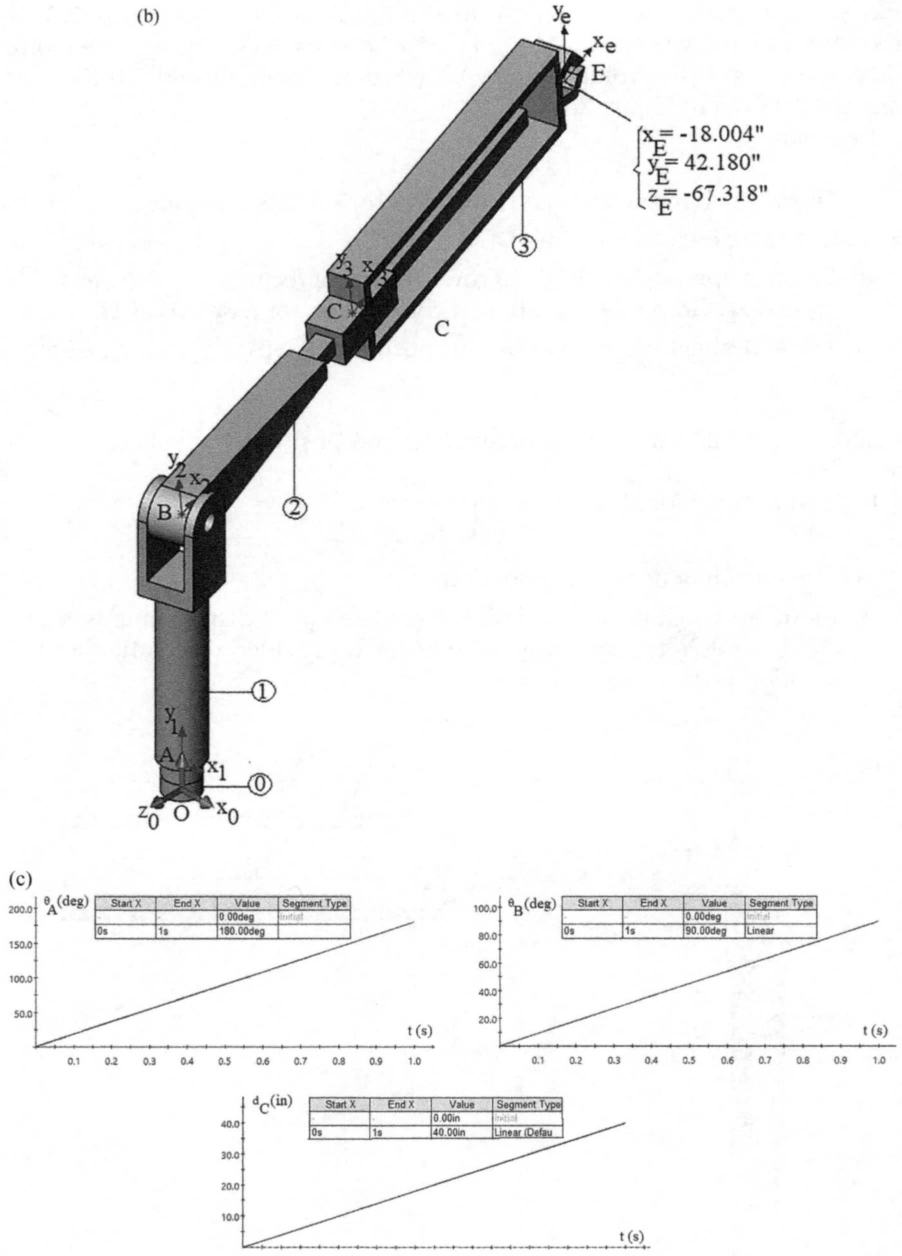

FIGURE P2.7 (CONTINUED)
(a) The spherical robot RRT at home position; (b) position of spherical robot at t = 0.1 s; (c) joint displacements.

P2.21: For the spherical robot in Figure P2.7, with the actuated joint displacements shown in Figure P2.7c, and a time interval is 1 s, determine

a. The joint rates in time $\dot{\theta}_A$, $\dot{d}_B$, $\dot{d}_C$, and $\dot{\theta}_D$
b. The links' angular velocity magnitudes
c. The joints' linear velocity magnitudes
d. The end-effector's E linear velocity components

P2.22: The spherical robot at t = 0.1 s is in position shown in Figure P2.7. Determine

a. The joint rates $\dot{\theta}_A$, $\dot{d}_B$, $\dot{d}_C$, and $\dot{\theta}_D$
b. The links' angular velocity magnitudes
c. The joints' linear velocity magnitudes
d. The end-effector's E linear velocity components and its magnitude

P2.23: The vertical articulated robotic mechanism RRRR, Figure P2.8 and (Animation P2.4), has the actuated A (revolute about y_1), B (revolute about z_2), C (revolute about z_3), and D (revolute about z_4). Figure P2.8a shows the robot in the initial (home position), and at t = 0.5 s. Figure P2.8b,c shows the joint displacements for an interval of 1 s. Determine

a. Draw the digraph (tree) and label the tree's nodes and arcs
b. Calculate the mechanism's DOF
c. Create a spreadsheet with 25 rows for end-effector's coordinates x_E, y_E, and z_E. Graph the variation of coordinates for a period of 1 s.
d. The end-effector's orientation and position ($\alpha^{I}_{0,e}$, $\beta^{II}_{0,e}$, $\gamma^{III}_{0,e}$, x_E, y_E, z_E) at t = 0.5 s.

P2.24: Figure P2.8a shows the orientation and position of the vertical articulated robotic mechanism at t = 0.5 s. For inverse positional analysis, determine

a. The system of nonlinear equations
b. Using an equation solver, find the actuated joint displacements: θ_A, θ_B, θ_C, and θ_D to reach the end-effector E provided orientation and position parameters.

P2.25: For the vertical articulated robotic mechanism in Figure P2.8, with the actuated joint displacements shown in Figure P2.8b,c, and a time interval is 1 s, determine

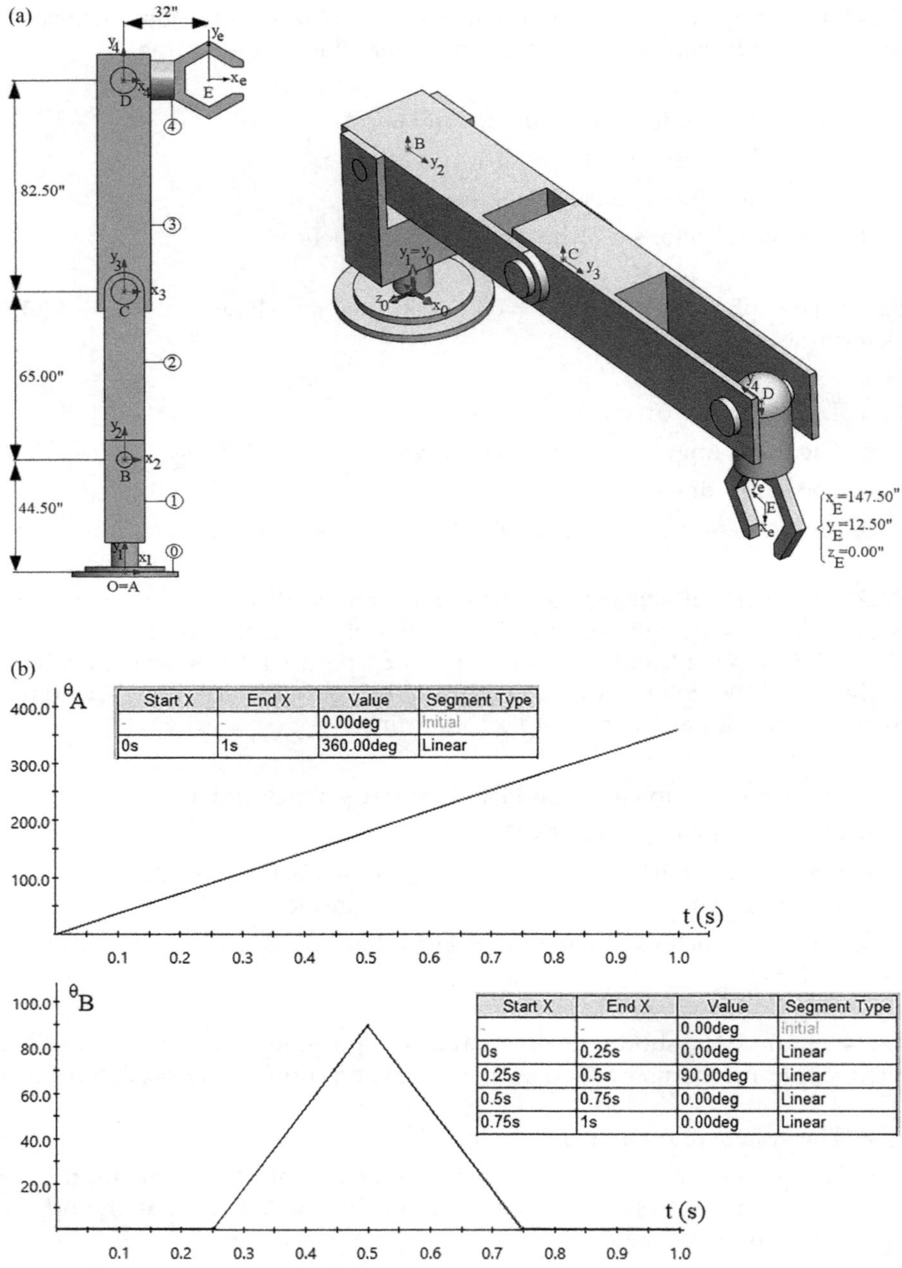

Start X	End X	Value	Segment Type
-	-	0.00deg	Initial
0s	1s	360.00deg	Linear

Start X	End X	Value	Segment Type
-	-	0.00deg	Initial
0s	0.25s	0.00deg	Linear
0.25s	0.5s	90.00deg	Linear
0.5s	0.75s	0.00deg	Linear
0.75s	1s	0.00deg	Linear

FIGURE P2.8
The vertical articulated spatial robotic mechanism RRRR: (a) home position and position at 0.4 s; (b) joint A and B displacements; (c) joint C and D displacements.

(Continued)

(c)

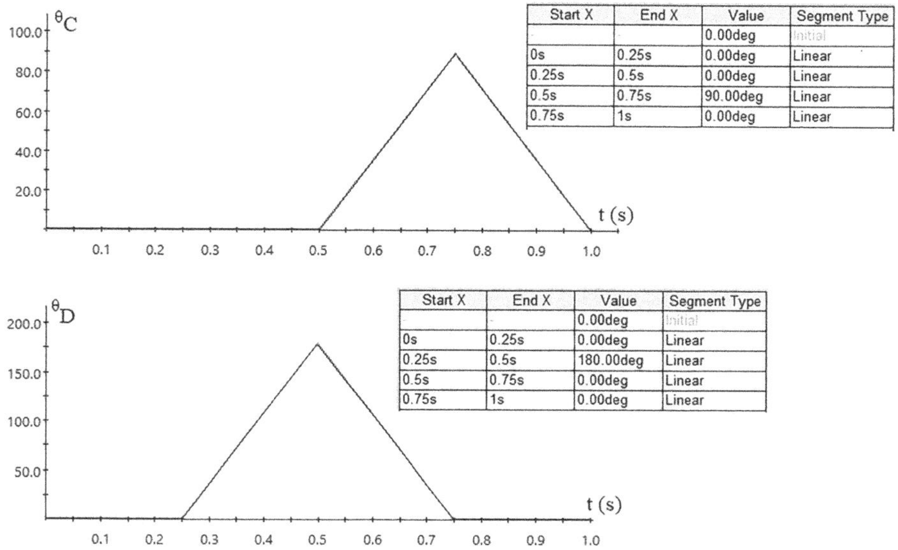

Start X	End X	Value	Segment Type
-	-	0.00deg	Initial
0s	0.25s	0.00deg	Linear
0.25s	0.5s	0.00deg	Linear
0.5s	0.75s	90.00deg	Linear
0.75s	1s	0.00deg	Linear

Start X	End X	Value	Segment Type
-	-	0.00deg	Initial
0s	0.25s	0.00deg	Linear
0.25s	0.5s	180.00deg	Linear
0.5s	0.75s	0.00deg	Linear
0.75s	1s	0.00deg	Linear

FIGURE P2.8 (CONTINUED)
The vertical articulated spatial robotic mechanism RRRR: (a) home position and position at 0.4 s; (b) joint A and B displacements; (c) joint C and D displacements.

a. The joint rates in time $\dot{\theta}_A$, $\dot{\theta}_B$, $\dot{\theta}_C$, and $\dot{\theta}_D$
b. The links' angular velocity magnitudes
c. The joints' linear velocity magnitudes
d. The end-effector's E linear velocity components

P2.26: The vertical articulated robotic mechanism at t = 0.5 s is in position shown in Figure P2.8a.
Determine

a. The joint rates $\dot{\theta}_A$, $\dot{\theta}_B$, $\dot{\theta}_C$, and $\dot{\theta}_D$
b. The links' angular velocity magnitudes
c. The joints' linear velocity magnitudes
d. The end-effector's E linear velocity components and its magnitude

P2.27: The horizontal articulated Scara robot RRT, Figure P2.9 and (Animation P2.5), has the actuated A (revolute about y_1), B (revolute about y_2), and C (prismatic along y_3). Figure P2.9a shows the robot in the initial (home position), and Figure P2.9b at t = 0.25 s. Figure P2.9c shows the joint displacements for an interval of 1 s. Determine

(a)

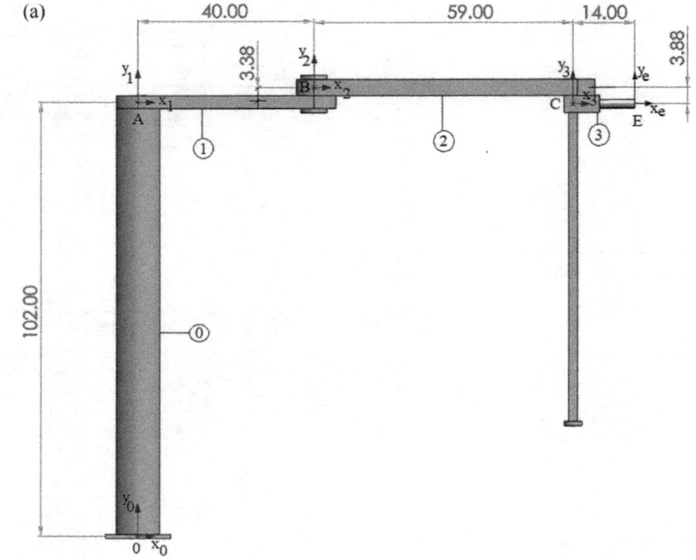

(b)

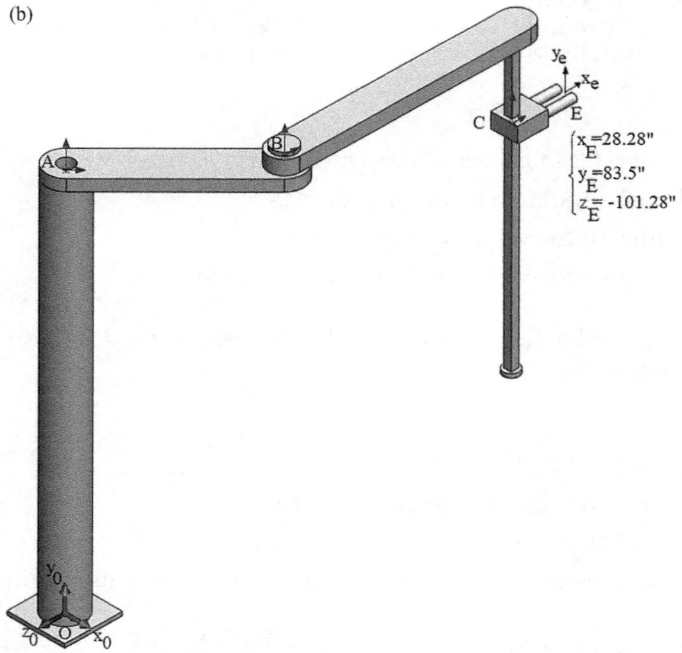

FIGURE P2.9
The horizontal articulated robot: (a) robot in the home position; (b) position at 0.25 s; (c) joint displacements.

(Continued)

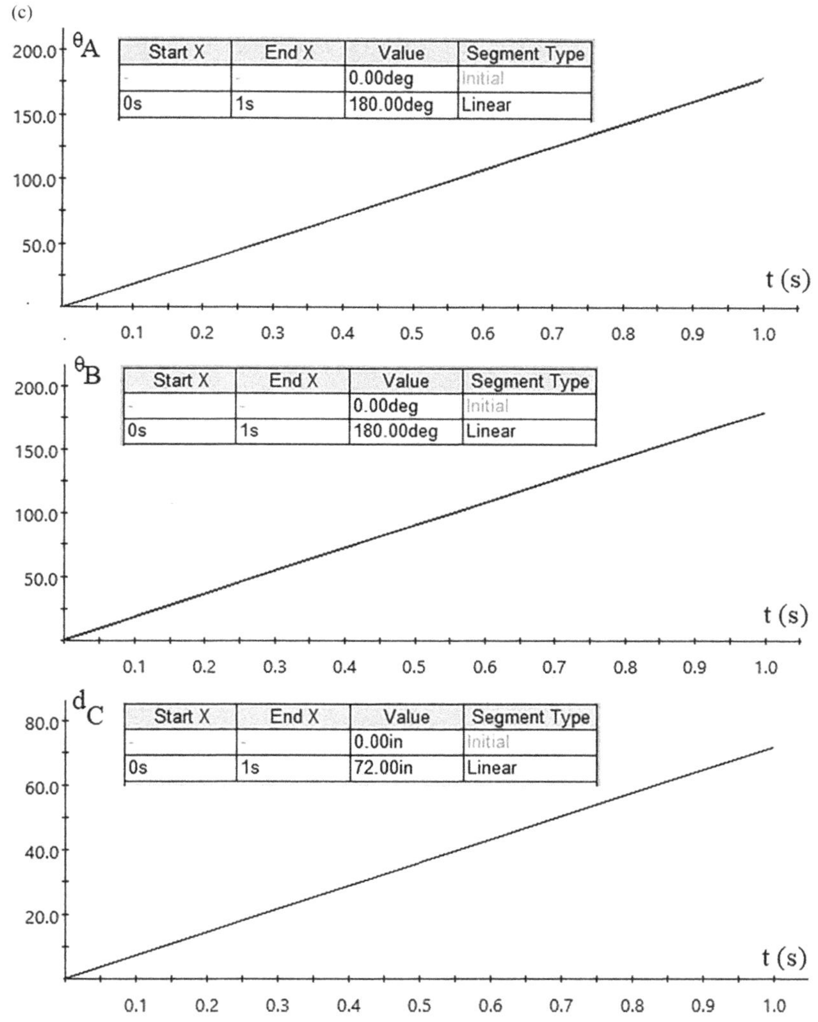

FIGURE P2.9 (CONTINUED)
The horizontal articulated robot: (a) robot in the home position; (b) position at 0.5 s; (c) joint displacements.

a. Draw the digraph (tree) and label the tree's nodes and arcs
b. Calculate the mechanism's DOF
c. Create a spreadsheet with 25 rows for end-effector's coordinates x_E, y_E, and z_E. Graph the variation of coordinates for a period of 1 s
d. The end-effector's orientation and position ($\alpha^{I}_{0,e}$, $\beta^{II}_{0,e}$, $\gamma^{III}_{0,e}$, x_E, y_E, z_E) at $t = 0.25$ s

P2.28: Figure P2.9b shows the orientation and position of the horizontal articulated robotic mechanism at t = 0.25 s. For inverse positional analysis, determine

a. The system of nonlinear equations
b. Using an equation solver, find the actuated joint displacements: θ_A, θ_B, and d_C to reach the end-effector E provided orientation and position parameters.

P2.29: For the horizontal articulated robotic mechanism in Figure P2.9, with the actuated joint displacements shown in Figure P2.9c, and a time interval is 1 s, determine

a. The joint rates in time $\dot{\theta}_A$, $\dot{\theta}_B$, and $\dot{d}_C$.
b. The links' angular velocity magnitudes
c. The joints' linear velocity magnitudes
d. The end-effector's E linear velocity components

P2.30: The horizontal articulated robotic mechanism at t = 0.25 s is in position shown in Figure P2.9b. Determine

a. The joint rates $\dot{\theta}_A$, $\dot{\theta}_B$, and $\dot{d}_C$
b. The links' angular velocity magnitudes
c. The joints' linear velocity magnitudes
d. The end-effector's E linear velocity components and its magnitude

P2.31: The TRC spatial mechanism, Figure P2.10 (Animation P2.6) used in manipulation, has the actuated joints A (T-prismatic along the direction which makes 45° with y_1), B (R-revolute about z_2), and C (C-2 DOF cylindrical joint: revolute about x_3 and sliding along x_3). Figure P2.10a shows the mechanism in the initial position and Figure P2.10b at t = 0.25 s. Figure P2.10c shows the joint displacements for an interval of 1 s. A point of interest noted E is located on the last link 3, where it is also considered as a frame e translated from frame 3. Determine

a. Draw the digraph (tree) and label the tree's nodes and arcs
b. Calculate the mechanism's DOF
c. Create a spreadsheet with 25 rows for the point of interest E coordinates x_E, y_E, and z_E. Graph the variation of coordinates for a period of 1 s
d. The end-effector orientation and position ($\alpha^{I}_{0,e}$, $\beta^{II}_{0,e}$, $\gamma^{III}_{0,e}$, x_E, y_E, z_E) at t = 0.25 s

P2.32: Figure P2.10b shows the orientation and position of the TRC spatial mechanism at t = 0.25 s. For inverse positional analysis, determine

a. The system of nonlinear equations
b. Using an equation solver, find the actuated joint displacements: d_A, θ_B, θ_C and d_C to reach the point of interest E's orientation and positional parameters.

(a)

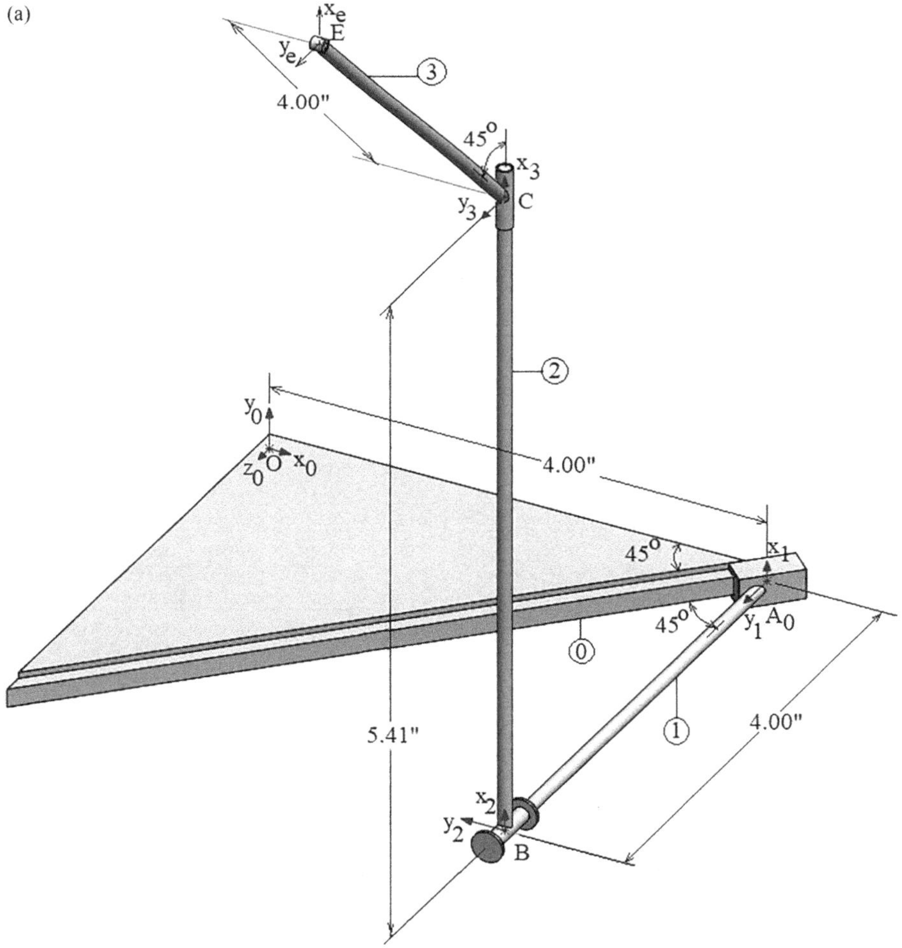

FIGURE P2.10
Open cycle TRC spatial mechanism with revolute, prismatic, and cylindrical joints: (a) initial position; (b) position at 0.25 s; (c) joint displacements.

(Continued)

(b)

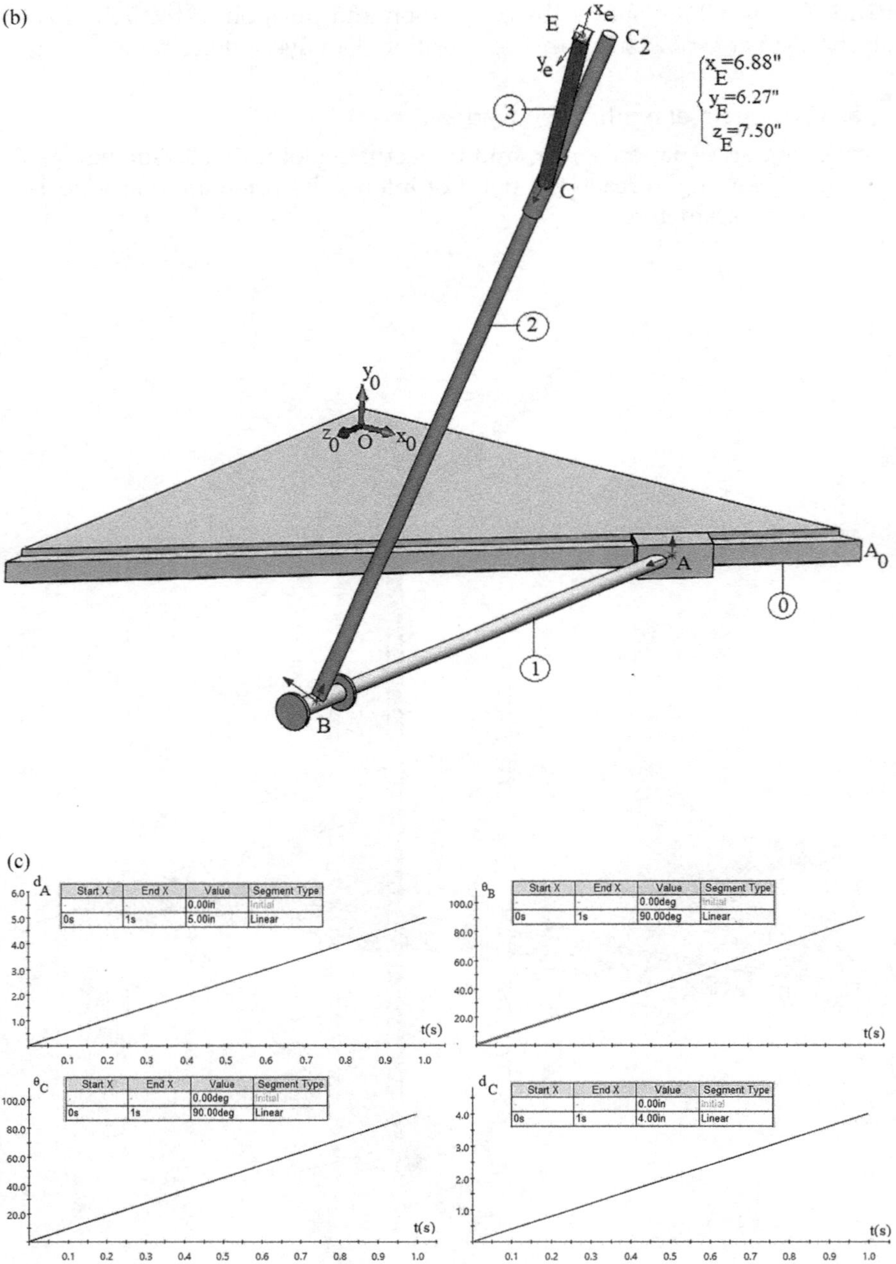

FIGURE P2.10 (CONTINUED)
Open cycle TRC spatial mechanism with revolute, prismatic, and cylindrical joints: (a) initial position; (b) position at 0.25 s; (c) joint displacements.

P2.33: For the TRC spatial mechanism in Figure P2.10, with the actuated joint displacements shown in Figure P2.10c, and a time interval is 1 s, determine

a. The joint rates in time $\dot{d}_A$, $\dot{\theta}_B$, $\dot{\theta}_C$, and $\dot{d}_D$
b. The joint twists
c. The links' angular velocity magnitudes
d. The joints' linear velocity magnitudes
e. The point of interest E linear velocity components

P2.34: The the TRC spatial mechanism at t = 0.25 s is in position shown in Figure P2.10b. Determine

a. The joint rates $\dot{d}_A$, $\dot{\theta}_B$, $\dot{\theta}_C$, and $\dot{d}_D$
b. The links' angular velocity magnitudes
c. The joints' linear velocity magnitudes
d. The point of interest E linear velocity components and its magnitude

P2.35: Figure P2.11a (Animation P2.7) shows the planar exoskeleton mechanism for applications in biomechanics. Figure P2.11b shows the seven links, seven joints, and the frames attached to links. Figure P2.11c shows the mechanism in the initial position. The actuated joints are A (prismatic along x_1), and six revolute joints B, C, D, E, F, and G, all about z-axes. The horizontal link 1 is in contact with the ground for an interval of 0.5 s, and then translated horizontally. For a forward motion, Figure P2.11d–f shows the direction of linear and angular displacements: d_A (in the positive direction of axis x_1), θ_B (counterclockwise), θ_C (clockwise), θ_D (clockwise), θ_E (counterclockwise), θ_F (clockwise), and θ_G (clockwise). Determine

a. Draw the digraph (tree) and label the tree's nodes and arcs
b. The mechanism's DOF
c. The equations for orientation $\alpha^I_{0,h}$ of frame's h, and the coordinates x_H, y_H for the point of interest H, with respect to the fixed frame
d. The orientation $\alpha^I_{0,h}$ of frame's h, and the coordinates x_H, y_H, at t = 1 s.

P2.36: At t = 0.5 s, for the planar exoskeleton mechanism, determine

a. The joint rates
b. The links' angular velocity magnitudes
c. The point's H components on x and y for the linear velocity, and its magnitude

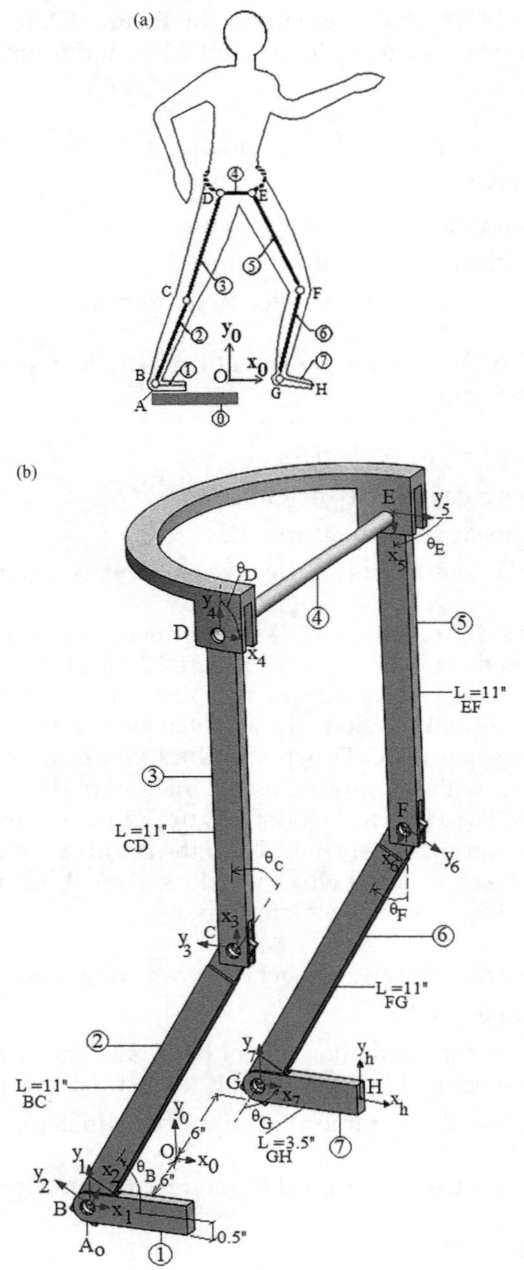

FIGURE P2.11
(a) Planar exoskeleton mechanism; (b) frames; (c) initial position; (d) joint A and B displacements; (e) joint C and D displacements; (f) joint E, F, and G displacements.

(*Continued*)

(c)

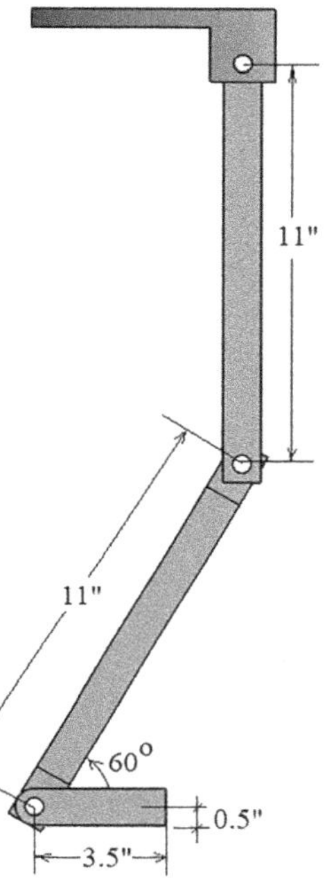

(d)

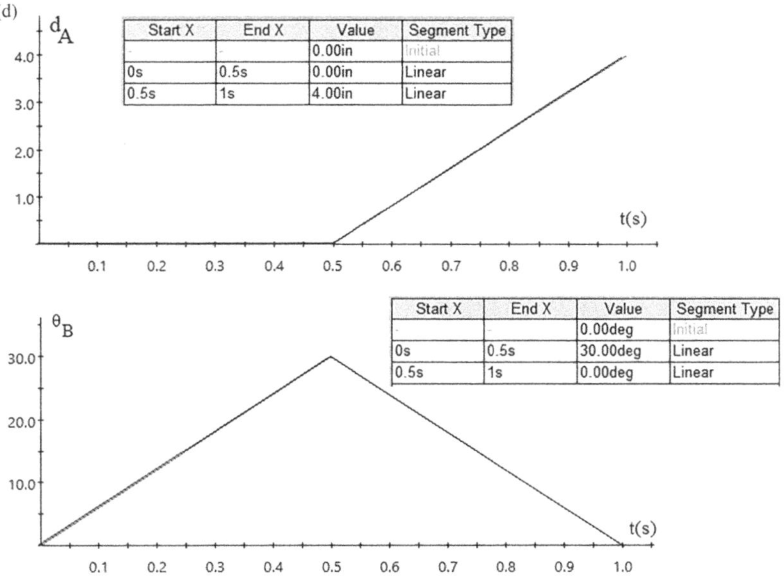

FIGURE P2.11 (CONTINUED)
(a) Planar exoskeleton mechanism; (b) frames; (c) initial position; (d) joint A and B displacements; (e) joint C and D displacements; (f) joint E, F, and G displacements.

(Continued)

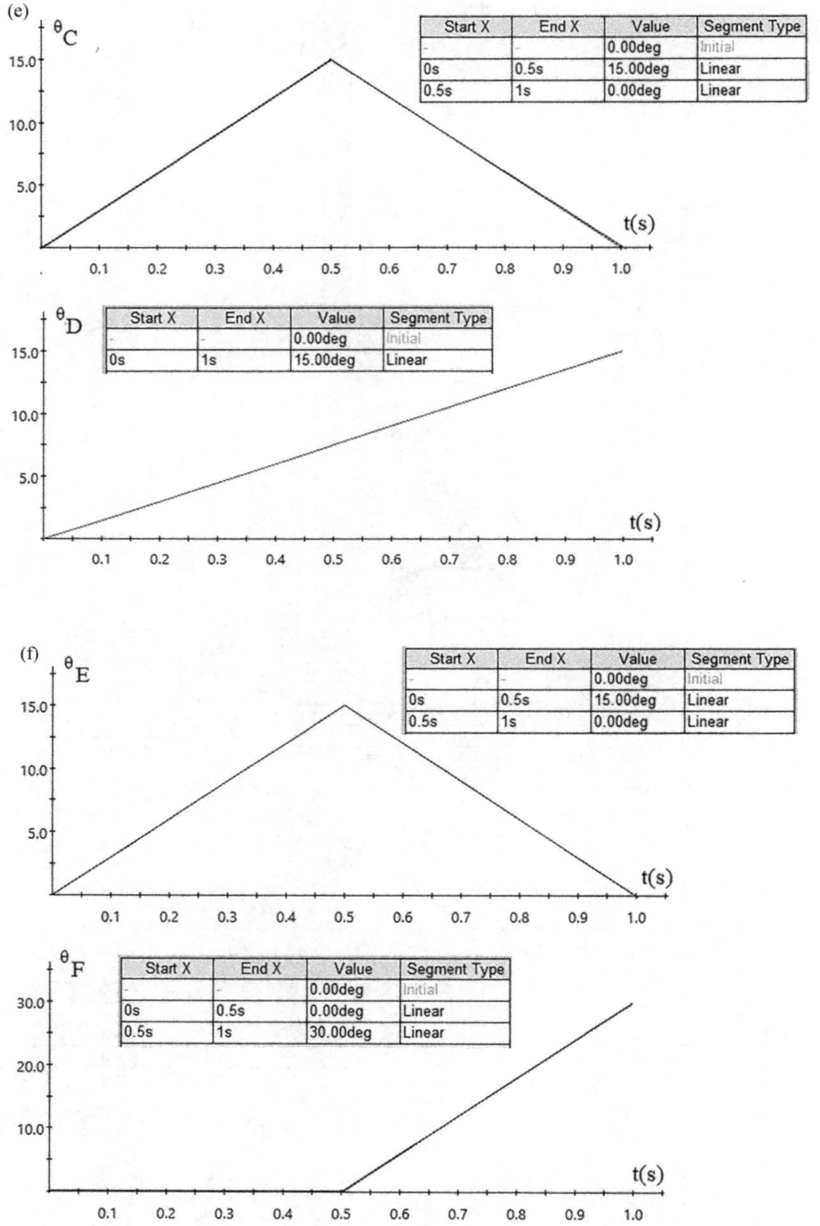

FIGURE P2.11 (CONTINUED)
(a) Planar exoskeleton mechanism; (b) frames; (c) initial position; (d) joint A and B displacements; (e) joint C and D displacements; (f) joint E, F, and G displacements.

(Continued)

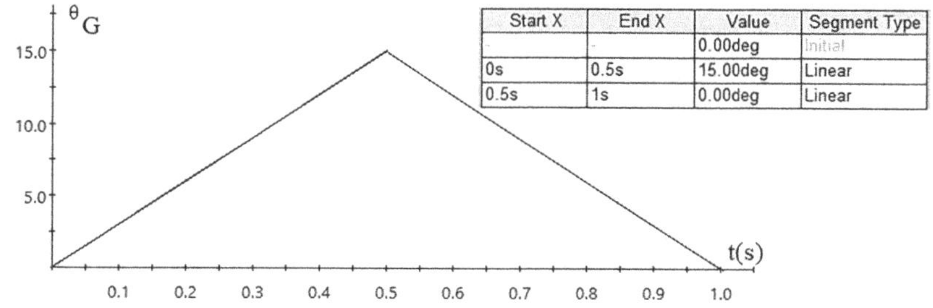

FIGURE P2.11 (CONTINUED)
(a) Planar exoskeleton mechanism; (b) frames; (c) initial position; (d) joint A and B displacements; (e) joint C and D displacements; (f) joint E, F, and G displacements.

Appendix 2.1: IT Notation for Unit Vectors of x_m, y_m, and z_m Axes of Rotation

Rotation About x_m-Axis

Considering the unit vectors expressed in the m-frame from Figure 2.4:

$$\mathbf{j}_{m-1} = \begin{Bmatrix} 0 \\ C\theta_Z \\ -S\theta_Z \end{Bmatrix}; \quad \mathbf{j}_m = \begin{Bmatrix} 0 \\ 1 \\ 0 \end{Bmatrix}$$

Then, the cross product of these two vectors is calculated with a symbolic determinant:

$$\mathbf{j}_{m-1} \times \mathbf{j}_m = \begin{bmatrix} \mathbf{i}_m & \mathbf{j}_m & \mathbf{k}_m \\ 0 & C\theta_Z & -S\theta_Z \\ 0 & 1 & 0 \end{bmatrix} = S\theta_Z \cdot \mathbf{i}_m$$

Therefore, the unit vector $\mathbf{j}_{m-1}$ (on the reference axis $||y_{m-1}$) is rotated ccw with θ_Z to define the unit vector $\mathbf{j}_m$ (on axis y_m), according to the right-hand rule:

Rotation About y_m-Axis

Considering the unit vectors expressed in the m-frame from Figure 2.5:

$$\mathbf{i}_{m-1} = \begin{Bmatrix} C\theta_Z \\ 0 \\ S\theta_Z \end{Bmatrix}; \quad \mathbf{i}_m = \begin{Bmatrix} 1 \\ 0 \\ 0 \end{Bmatrix}$$

Then, the cross product of these two vectors is calculated with a symbolic determinant:

$$\mathbf{i}_{m-1} \times \mathbf{i}_m = \begin{bmatrix} \mathbf{i}_m & \mathbf{j}_m & \mathbf{k}_m \\ C\theta_Z & 0 & S\theta_Z \\ 1 & 0 & 0 \end{bmatrix} = S\theta_Z \cdot \mathbf{j}_m$$

Therefore, the unit vector $\mathbf{i}_{m-1}$ (on the reference axis $||x_{m-1}$) is rotated ccw θ_Z to define the unit vector $\mathbf{i}_m$ (on axis x_m), according to the right-hand rule.

Rotation About z_m-Axis

Considering the unit vectors expressed in the m-frame from Figure 2.6:

$$\mathbf{i}_{m-1} = \begin{Bmatrix} C\theta_Z \\ -S\theta_Z \\ 0 \end{Bmatrix}; \quad \mathbf{i}_m = \begin{Bmatrix} 1 \\ 0 \\ 0 \end{Bmatrix}$$

Then, the cross product of these two vectors is calculated with a symbolic determinant

$$\mathbf{i}_{m-1} \times \mathbf{i}_m = \begin{bmatrix} \mathbf{i}_m & \mathbf{j}_m & \mathbf{k}_m \\ C\theta_Z & -S\theta_Z & 0 \\ 1 & 0 & 0 \end{bmatrix} = S\theta_Z \cdot \mathbf{k}_m$$

Therefore, the unit vector $\mathbf{i}_{m-1}$ (on the reference axis $||x_{m-1}$) is rotated ccw θ_Z to define the unit vector $\mathbf{i}_m$ (on axis x_m), according to the right-hand rule.

Appendix 2.2: Properties of a Rotation Matrix

Let us consider a rotation between the frames x_m, y_m, z_m and x_{m-1}, y_{m-1}, z_{m-1}.

The matrix of rotation $\mathbf{D}_Z$ has nine entries. Since a spatial rotation has only 3 DOF, six relations of dependence hold between the nine entries.

1. The matrix $\mathbf{D}$ is an *orthonormal* matrix based on the following six equations
   - The sum of squared entries on each column is 1, since on each column are the components of a unit vector with magnitude 1:

$$|\mathbf{i}_m|^2 = 1; \quad |\mathbf{j}_m|^2 = 1; \quad |\mathbf{k}_m|^2 = 1 \tag{a}$$

   - The unit vectors are perpendicular (orthogonal) to each other:

$$\mathbf{i}_m \cdot \mathbf{j}_m = 0; \quad \mathbf{j}_m \cdot \mathbf{k}_m = 0; \quad \mathbf{k}_m \cdot \mathbf{i}_m = 0 \tag{b}$$

2. The inverse, $\mathbf{D}_Z^{-1}$, which corresponds to a negative rotation $(-\theta_Z)$, is its transposed matrix:

$$\mathbf{D}_Z^{-1} = \mathbf{D}_Z^T \tag{c}$$

Proof:

$$\mathbf{D}_Z^{-1} = \begin{bmatrix} C(-\theta_Z) & -S(-\theta_Z) & 0 \\ S(-\theta_Z) & C(-\theta_Z) & 0 \\ 0 & 0 & 1 \end{bmatrix} = \begin{bmatrix} C\theta_Z & S\theta_Z & 0 \\ -S\theta_Z & C\theta_Z & 0 \\ 0 & 0 & 1 \end{bmatrix} = \mathbf{D}_Z^T$$

3. The product of orthogonal matrix and its transpose is a 3×3 unit matrix:

$$\mathbf{D}_Z * \mathbf{D}_Z^T = \mathbf{U}_{3\times3} \tag{d}$$

   This property is a result of Equation (c), since the product of a matrix and its inverse is the unit matrix. The advantage of using orthogonal matrices in calculations is that there is no need to find its inverse; instead, the inverse is replaced by its transposed matrix (obtained by changing its rows into columns).

4. The orthogonal matrix with interchanged subscripts is the transposed:

$$\mathbf{D}_{Z_{m,m-1}} = \mathbf{D}_{Z_{m-1,m}}^T \tag{e}$$

Since a negative rotation is equivalent to a positive rotation with subscripts interchanged, $\theta_{Z_{m,m-1}} = -\theta_{Z_{m-1,m}}$, and considering Equation (c), then Equation (e) holds.

5. The product between an orthogonal matrix and that with subscripts interchanged is the unit matrix:

$$\mathbf{D}_{Z_{m-1,m}} * \mathbf{D}_{Z_{m,m-1}} = \mathbf{U}_{3\times 3} \tag{f}$$

This property is a consequence of Equations (d) and (e).

6. det $\mathbf{D}_Z = 1$

Proof: If we consider a rotation about z-axis, then

$$\det\begin{bmatrix} C\theta_Z & -S\theta_Z & 0 \\ S\theta_Z & C\theta_Z & 0 \\ 0 & 0 & 1 \end{bmatrix} = \begin{vmatrix} C\theta_z & -S\theta_z \\ S\theta_z & C\theta_z \end{vmatrix} = 1$$

7. The null rotation is a translation. Indeed, if two frames are in *translation* relative to each other, then $\theta_Z = 0$ and the orthonormal matrix becomes the 3×3 unit matrix (Equation (g)):

$$\mathbf{D}_Z = \begin{matrix} & i_m & j_m & k_m \\ x_{m-1} & 1 & 0 & 0 \\ y_{m-1} & 0 & 1 & 0 \\ z_{m-1} & 0 & 0 & 1 \end{matrix} \tag{g}$$

Appendix 2.3: End-Effector's Coordinates for TRRT Spatial Mechanism

t (sec)	x_E (in)	y_E (in)	z_E (in)	t (sec)	x_E (in)	y_E (in)	z_E (in)
0.00	49.500	28.000	0.000	2.04	127.557	32.219	-4.209
0.04	52.000	28.000	0.000	2.08	125.068	36.419	-8.347
0.08	54.500	28.000	0.000	2.13	122.045	40.583	-12.342
0.13	57.000	28.000	0.000	2.17	118.513	44.694	-16.125
0.17	59.500	28.000	0.000	2.21	114.502	48.733	-19.633
0.21	62.000	28.000	0.000	2.25	110.054	52.683	-22.804
0.25	64.500	28.000	0.000	2.29	105.216	56.528	-25.586
0.29	67.000	28.000	0.000	2.33	100.042	60.250	-27.929
0.33	69.500	28.000	0.000	2.38	94.592	63.834	-29.795
0.38	72.000	28.000	0.000	2.42	88.930	67.265	-31.151
0.42	74.500	28.000	0.000	2.46	83.126	70.528	-31.974
0.46	77.000	28.000	0.000	2.50	77.250	73.608	-32.250
0.50	79.500	28.000	0.000	2.54	71.374	76.494	-31.974
0.54	82.000	28.000	0.000	2.58	65.570	79.171	-31.151
0.58	84.500	28.000	0.000	2.63	59.908	81.630	-29.795
0.63	87.000	28.000	0.000	2.67	54.458	83.859	-27.929
0.67	89.500	28.000	0.000	2.71	49.284	85.848	-25.586
0.71	92.000	28.000	0.000	2.75	44.446	87.590	-22.804
0.75	94.500	28.000	0.000	2.79	39.998	89.077	-19.633
0.79	97.000	28.000	0.000	2.83	35.987	90.302	-16.125
0.83	99.500	28.000	0.000	2.88	32.455	91.261	-12.342
0.88	102.000	28.000	0.000	2.92	29.432	91.948	-8.347
0.92	104.500	28.000	0.000	2.96	26.943	92.362	-4.209
0.96	107.000	28.000	0.000	3.00	25.000	92.500	0.000
1.00	109.500	28.000	0.000	3.04	24.167	92.500	0.000
1.04	110.333	28.000	0.000	3.08	23.333	92.500	0.000
1.08	111.167	28.000	0.000	3.13	22.500	92.500	0.000
1.13	112.000	28.000	0.000	3.17	21.667	92.500	0.000
1.17	112.833	28.000	0.000	3.21	20.833	92.500	0.000
1.21	113.667	28.000	0.000	3.25	20.000	92.500	0.000
1.25	114.500	28.000	0.000	3.29	19.167	92.500	0.000
1.29	115.333	28.000	0.000	3.33	18.333	92.500	0.000
1.33	116.167	28.000	0.000	3.38	17.500	92.500	0.000
1.38	117.000	28.000	0.000	3.42	16.667	92.500	0.000
1.42	117.833	28.000	0.000	3.46	15.833	92.500	0.000
1.46	118.667	28.000	0.000	3.50	15.000	92.500	0.000
1.50	119.500	28.000	0.000	3.54	14.167	92.500	0.000
1.54	120.333	28.000	0.000	3.58	13.333	92.500	0.000
1.58	121.167	28.000	0.000	3.63	12.500	92.500	0.000
1.63	122.000	28.000	0.000	3.67	11.667	92.500	0.000
1.67	122.833	28.000	0.000	3.71	10.833	92.500	0.000
1.71	123.667	28.000	0.000	3.75	10.000	92.500	0.000
1.75	124.500	28.000	0.000	3.79	9.167	92.500	0.000
1.79	125.333	28.000	0.000	3.83	8.333	92.500	0.000
1.83	126.167	28.000	0.000	3.87	7.500	92.500	0.000
1.88	127.000	28.000	0.000	3.92	6.667	92.500	0.000
1.92	127.833	28.000	0.000	3.96	5.833	92.500	0.000
1.96	128.667	28.000	0.000	4.00	5.000	92.500	0.000
2.00	129.500	28.000	0.000				

Appendix 2.4: TRRT Direct Positional Analysis—EES Parametric Table

Parametric Table

Table 1-End Effector Coordinates for TRRT Robot

1..99	t [sec]	x_E [in]	y_E [in]	z_E [in]
Run 1	0.000	49.500	28.000	0.000
Run 2	0.041	51.949	28.000	0.000
Run 3	0.082	54.398	28.000	0.000
Run 4	0.122	56.847	28.000	0.000
Run 5	0.163	59.296	28.000	0.000
Run 6	0.204	61.745	28.000	0.000
Run 7	0.245	64.194	28.000	0.000
Run 8	0.286	66.643	28.000	0.000
Run 9	0.327	69.092	28.000	0.000
Run 10	0.367	71.541	28.000	0.000
Run 11	0.408	73.990	28.000	0.000
Run 12	0.449	76.439	28.000	0.000
Run 13	0.490	78.888	28.000	0.000
Run 14	0.531	81.337	28.000	0.000
Run 15	0.571	83.786	28.000	0.000
Run 16	0.612	86.235	28.000	0.000
Run 17	0.653	88.684	28.000	0.000
Run 18	0.694	91.133	28.000	0.000
Run 19	0.735	93.582	28.000	0.000
Run 20	0.776	96.031	28.000	0.000
Run 21	0.816	98.480	28.000	0.000
Run 22	0.857	100.929	28.000	0.000
Run 23	0.898	103.378	28.000	0.000
Run 24	0.939	105.827	28.000	0.000
Run 25	0.980	108.276	28.000	0.000
Run 26	1.020	109.908	28.000	0.000
Run 27	1.061	110.724	28.000	0.000
Run 28	1.102	111.541	28.000	0.000
Run 29	1.143	112.357	28.000	0.000
Run 30	1.184	113.173	28.000	0.000
Run 31	1.224	113.990	28.000	0.000
Run 32	1.265	114.806	28.000	0.000
Run 33	1.306	115.622	28.000	0.000
Run 34	1.347	116.439	28.000	0.000
Run 35	1.388	117.255	28.000	0.000
Run 36	1.429	118.071	28.000	0.000
Run 37	1.469	118.888	28.000	0.000
Run 38	1.510	119.704	28.000	0.000
Run 39	1.551	120.520	28.000	0.000
Run 40	1.592	121.337	28.000	0.000
Run 41	1.633	122.153	28.000	0.000
Run 42	1.673	122.969	28.000	0.000
Run 43	1.714	123.786	28.000	0.000
Run 44	1.755	124.602	28.000	0.000
Run 45	1.796	125.418	28.000	0.000
Run 46	1.837	126.235	28.000	0.000
Run 47	1.878	127.051	28.000	0.000
Run 48	1.918	127.867	28.000	0.000
Run 49	1.959	128.684	28.000	0.000
Run 50	2.000	129.500	28.000	0.000
Run 51	2.041	127.603	32.133	-4.124
Run 52	2.082	125.180	36.248	-8.180
Run 53	2.122	122.245	40.330	-12.102
Run 54	2.163	118.819	44.361	-15.826
Run 55	2.204	114.932	48.324	-19.289
Run 56	2.245	110.621	52.205	-22.436
Run 57	2.286	105.929	55.986	-25.214
Run 58	2.327	100.907	59.651	-27.578
Run 59	2.367	95.610	63.187	-29.490
Run 60	2.408	90.099	66.578	-30.917
Run 61	2.449	84.438	69.811	-31.837
Run 62	2.490	78.692	72.872	-32.233
Run 63	2.531	72.929	75.748	-32.101
Run 64	2.571	67.217	78.428	-31.441
Run 65	2.612	61.622	80.901	-30.266
Run 66	2.653	56.211	83.157	-28.593
Run 67	2.694	51.044	85.186	-26.451
Run 68	2.735	46.181	86.980	-23.874
Run 69	2.776	41.673	88.531	-20.905
Run 70	2.816	37.569	89.834	-17.593
Run 71	2.857	33.908	90.883	-13.993
Run 72	2.898	30.725	91.673	-10.162
Run 73	2.939	28.044	92.202	-6.165
Run 74	2.980	25.883	92.467	-2.066
Run 75	3.020	24.592	92.500	0.000
Run 76	3.061	23.776	92.500	0.000
Run 77	3.102	22.959	92.500	0.000
Run 78	3.143	22.143	92.500	0.000
Run 79	3.184	21.327	92.500	0.000
Run 80	3.224	20.510	92.500	0.000
Run 81	3.265	19.694	92.500	0.000
Run 82	3.306	18.878	92.500	0.000
Run 83	3.347	18.061	92.500	0.000
Run 84	3.388	17.245	92.500	0.000
Run 85	3.429	16.429	92.500	0.000
Run 86	3.469	15.612	92.500	0.000
Run 87	3.510	14.796	92.500	0.000
Run 88	3.551	13.980	92.500	0.000
Run 89	3.592	13.163	92.500	0.000
Run 90	3.633	12.347	92.500	0.000
Run 91	3.673	11.531	92.500	0.000
Run 92	3.714	10.714	92.500	0.000
Run 93	3.755	9.898	92.500	0.000
Run 94	3.796	9.082	92.500	0.000
Run 95	3.837	8.265	92.500	0.000
Run 96	3.878	7.449	92.500	0.000
Run 97	3.918	6.633	92.500	0.000
Run 98	3.959	5.816	92.500	0.000
Run 99	4.000	5.000	92.500	0.000

Appendix 2.5: Joint Displacements and End-Effector Coordinates Displacements for TRT Planar Manipulator—EES Parametric Table

Parametric Table

Joint displacements and end effector coordinates | Joint angular displac

1..25	t [sec]	d_A [in]	θ_B [deg]	d_C [in]	x_E [in]	y_E [in]
Run 6	0.21	2.604	11.7	0.868	51.537	40.113
Run 7	0.25	3.750	16.9	1.250	52.052	44.135
Run 8	0.29	5.104	23.0	1.701	52.182	48.834
Run 9	0.33	6.667	30.0	2.222	51.696	54.111
Run 10	0.38	8.438	38.0	2.813	50.342	59.801
Run 11	0.42	10.417	46.9	3.472	47.868	65.648
Run 12	0.46	12.604	56.7	4.201	44.055	71.296
Run 13	0.50	15.000	67.5	5.000	38.751	76.270
Run 14	0.54	17.604	79.2	5.868	31.932	79.988
Run 15	0.58	20.417	91.9	6.806	23.754	81.778
Run 16	0.63	23.438	105.5	7.813	14.618	80.936
Run 17	0.67	26.667	120.0	8.889	5.222	76.803
Run 18	0.71	30.104	135.5	10.035	-3.415	68.894
Run 19	0.75	33.750	151.9	11.250	-9.976	57.045
Run 20	0.79	37.604	169.2	12.535	-12.933	41.575
Run 21	0.83	41.667	187.5	13.889	-10.727	23.444
Run 22	0.88	45.938	206.7	15.313	-2.042	4.332
Run 23	0.92	50.417	226.9	16.806	13.850	-13.380
Run 24	0.96	55.104	248.0	18.368	36.709	-26.814
Run 25	1.00	60.000	270.0	20.000	65.000	-33.000

Appendix 2.6: End-Effector Coordinates Calculated (EES) and Simulated (SW) for TRT Planar Mechanism

Lookup Table

Direct -End Eff Veloc-Calc and Sim | End effector coordinates from EES and SW

Paste Special	t [sec]	$x_{E,EES}$	$x_{E,SW}$ [in]	$y_{E,EES}$ [in]	$y_{E,SW}$
Row 1	0.00	49.000	49.000	31.000	31.000
Row 2	0.04	49.137	49.137	31.360	31.360
Row 3	0.08	49.532	49.532	32.444	32.444
Row 4	0.13	50.130	50.130	34.260	34.260
Row 5	0.17	50.841	50.841	36.816	36.816
Row 6	0.21	51.537	51.537	40.113	40.113
Row 7	0.25	52.052	52.052	44.135	44.135
Row 8	0.29	52.182	52.182	48.834	48.834
Row 9	0.33	51.696	51.696	54.111	54.111
Row 10	0.38	50.342	50.342	59.801	59.801
Row 11	0.42	47.868	47.868	65.648	65.648
Row 12	0.46	44.055	44.055	71.296	71.296
Row 13	0.50	38.751	38.751	76.270	76.270
Row 14	0.54	31.932	31.932	79.988	79.988
Row 15	0.58	23.754	23.754	81.778	81.778
Row 16	0.63	14.618	14.618	80.936	80.936
Row 17	0.67	5.222	5.222	76.803	76.803
Row 18	0.71	-3.415	-3.415	68.894	68.894
Row 19	0.75	-9.976	-9.976	57.045	57.045
Row 20	0.79	-12.933	-12.933	41.575	41.575
Row 21	0.83	-10.727	-10.727	23.444	23.444
Row 22	0.88	-2.042	-2.042	4.332	4.332
Row 23	0.92	13.850	13.850	-13.380	-13.380
Row 24	0.96	36.709	36.709	-26.814	-26.814
Row 25	1.00	65.000	65.000	-33.000	-33.000

Appendix 2.7: Joint Angular Displacemens and Links' Angular Absolute Angular Orientation for TRT Manipulator—EES Parametric Table

Parametric Table

Joint angular displacements and links orientation

1..25	t [sec]	θ_A [deg]	θ_B [deg]	θ_C [deg]	θ_1 [deg]	θ_2 [deg]	θ_3 [deg]
Run 1	0.00	0.0	0.0	0.0	0.0	0.0	0.0
Run 2	0.04	0.0	0.5	0.0	0.0	0.5	0.5
Run 3	0.08	0.0	1.9	0.0	0.0	1.9	1.9
Run 4	0.13	0.0	4.2	0.0	0.0	4.2	4.2
Run 5	0.17	0.0	7.5	0.0	0.0	7.5	7.5
Run 6	0.21	0.0	11.7	0.0	0.0	11.7	11.7
Run 7	0.25	0.0	16.9	0.0	0.0	16.9	16.9
Run 8	0.29	0.0	23.0	0.0	0.0	23.0	23.0
Run 9	0.33	0.0	30.0	0.0	0.0	30.0	30.0
Run 10	0.38	0.0	38.0	0.0	0.0	38.0	38.0
Run 11	0.42	0.0	46.9	0.0	0.0	46.9	46.9
Run 12	0.46	0.0	56.7	0.0	0.0	56.7	56.7
Run 13	0.50	0.0	67.5	0.0	0.0	67.5	67.5
Run 14	0.54	0.0	79.2	0.0	0.0	79.2	79.2
Run 15	0.58	0.0	91.9	0.0	0.0	91.9	91.9
Run 16	0.63	0.0	105.5	0.0	0.0	105.5	105.5
Run 17	0.67	0.0	120.0	0.0	0.0	120.0	120.0
Run 18	0.71	0.0	135.5	0.0	0.0	135.5	135.5
Run 19	0.75	0.0	151.9	0.0	0.0	151.9	151.9
Run 20	0.79	0.0	169.2	0.0	0.0	169.2	169.2
Run 21	0.83	0.0	187.5	0.0	0.0	187.5	187.5
Run 22	0.88	0.0	206.7	0.0	0.0	206.7	206.7
Run 23	0.92	0.0	226.9	0.0	0.0	226.9	226.9
Run 24	0.96	0.0	248.0	0.0	0.0	248.0	248.0
Run 25	1.00	0.0	270.0	0.0	0.0	270.0	270.0

Appendix 2.8: Coordinates x and y for the TRT Manipulator's Joints—EES Parametric Table

Parametric Table

Coordinates x and y for the TRT manipulator joints

1..25	t [sec]	x_A [in]	y_A [in]	x_B [in]	y_B [in]	x_C [in]	y_C [in]
Run 1	0.00	5.000	0.000	5.000	31.000	11.000	31.000
Run 2	0.04	5.104	0.000	5.104	31.000	11.139	31.049
Run 3	0.08	5.417	0.000	5.417	31.000	11.552	31.201
Run 4	0.13	5.938	0.000	5.938	31.000	12.233	31.464
Run 5	0.17	6.667	0.000	6.667	31.000	13.166	31.856
Run 6	0.21	7.604	0.000	7.604	31.000	14.329	32.395
Run 7	0.25	8.750	0.000	8.750	31.000	15.688	33.105
Run 8	0.29	10.104	0.000	10.104	31.000	17.195	34.005
Run 9	0.33	11.667	0.000	11.667	31.000	18.787	35.111
Run 10	0.38	13.438	0.000	13.438	31.000	20.385	36.422
Run 11	0.42	15.417	0.000	15.417	31.000	21.892	37.913
Run 12	0.46	17.604	0.000	17.604	31.000	23.202	39.528
Run 13	0.50	20.000	0.000	20.000	31.000	24.210	41.163
Run 14	0.54	22.604	0.000	22.604	31.000	24.824	42.659
Run 15	0.58	25.417	0.000	25.417	31.000	24.998	43.799
Run 16	0.63	28.438	0.000	28.438	31.000	24.754	44.312
Run 17	0.67	31.667	0.000	31.667	31.000	24.222	43.894
Run 18	0.71	35.104	0.000	35.104	31.000	23.674	42.245
Run 19	0.75	38.750	0.000	38.750	31.000	23.537	39.132
Run 20	0.79	42.604	0.000	42.604	31.000	24.397	34.467
Run 21	0.83	46.667	0.000	46.667	31.000	26.948	28.404
Run 22	0.88	50.938	0.000	50.938	31.000	31.901	21.418
Run 23	0.92	55.417	0.000	55.417	31.000	39.827	14.355
Run 24	0.96	60.104	0.000	60.104	31.000	50.963	8.411
Run 25	1.00	65.000	0.000	65.000	31.000	65.000	5.000

Appendix 2.9: Latin Matrix Entries for TRT Manipulator—EES Parametric Table

Parametric Table

Coordinates x and y for the TRT manipulator joints | **Latin matrix entries**

1..25	1 t [sec]	2 x_{EA} [in]	3 x_{AB} [in]	4 x_{BC} [in]	5 x_{CE} [in]	6 y_{EA} [in]	7 y_{AB} [in]	8 y_{BC} [in]	9 y_{CE} [in]
Run 1	0.00	-44.000	0.000	6.000	38.000	-31.000	31.000	0.000	0.000
Run 2	0.04	-44.033	0.000	6.035	37.999	-31.360	31.000	0.049	0.311
Run 3	0.08	-44.115	0.000	6.136	37.980	-32.444	31.000	0.201	1.243
Run 4	0.13	-44.192	0.000	6.295	37.897	-34.260	31.000	0.464	2.795
Run 5	0.17	-44.174	0.000	6.499	37.675	-36.816	31.000	0.856	4.960
Run 6	0.21	-43.933	0.000	6.725	37.208	-40.113	31.000	1.395	7.718
Run 7	0.25	-43.302	0.000	6.938	36.364	-44.135	31.000	2.105	11.031
Run 8	0.29	-42.078	0.000	7.091	34.987	-48.834	31.000	3.005	14.829
Run 9	0.33	-40.030	0.000	7.121	32.909	-54.111	31.000	4.111	19.000
Run 10	0.38	-36.904	0.000	6.947	29.957	-59.801	31.000	5.422	23.379
Run 11	0.42	-32.452	0.000	6.475	25.977	-65.648	31.000	6.913	27.735
Run 12	0.46	-26.450	0.000	5.598	20.852	-71.296	31.000	8.528	31.768
Run 13	0.50	-18.751	0.000	4.210	14.542	-76.270	31.000	10.163	35.107
Run 14	0.54	-9.328	0.000	2.220	7.108	-79.988	31.000	11.659	37.329
Run 15	0.58	1.662	0.000	-0.419	-1.243	-81.778	31.000	12.799	37.980
Run 16	0.63	13.819	0.000	-3.684	-10.135	-80.936	31.000	13.312	36.623
Run 17	0.67	26.444	0.000	-7.444	-19.000	-76.803	31.000	12.894	32.909
Run 18	0.71	38.520	0.000	-11.431	-27.089	-68.894	31.000	11.245	26.649
Run 19	0.75	48.726	0.000	-15.213	-33.513	-57.045	31.000	8.132	17.913
Run 20	0.79	55.537	0.000	-18.208	-37.329	-41.575	31.000	3.467	7.108
Run 21	0.83	57.394	0.000	-19.719	-37.675	-23.444	31.000	-2.596	-4.960
Run 22	0.88	52.979	0.000	-19.037	-33.943	-4.332	31.000	-9.582	-17.085
Run 23	0.92	41.566	0.000	-15.590	-25.977	13.380	31.000	-16.645	-27.735
Run 24	0.96	23.395	0.000	-9.141	-14.254	26.814	31.000	-22.589	-35.225
Run 25	1.00	-0.000	0.000	0.000	0.000	33.000	31.000	-26.000	-38.000

Appendix 2.10: Links' COM Coordinates for TRT Manipulator—EES Parametric Table

Parametric Table

Coordinates x and y for the TRT manipulator joints | Latin matrix entries | Links cent

1..25	1 t [sec]	2 $x_{G,1}$ [in]	3 $y_{G,1}$ [in]	4 $x_{G,2}$ [in]	5 $y_{G,2}$ [in]	6 $x_{G,3}$ [in]	7 $y_{G,3}$ [in]
Run 1	0.00	5.000	11.117	22.183	31.000	26.796	31.000
Run 2	0.04	5.104	11.117	22.287	31.141	26.934	31.179
Run 3	0.08	5.417	11.117	22.590	31.562	27.340	31.718
Run 4	0.13	5.938	11.117	23.074	32.264	27.986	32.626
Run 5	0.17	6.667	11.117	23.703	33.243	28.827	33.917
Run 6	0.21	7.604	11.117	24.429	34.490	29.796	35.603
Run 7	0.25	8.750	11.117	25.193	35.988	30.804	37.690
Run 8	0.29	10.104	11.117	25.925	37.705	31.739	40.169
Run 9	0.33	11.667	11.117	26.548	39.592	32.467	43.009
Run 10	0.38	13.438	11.117	26.984	41.572	32.838	46.140
Run 11	0.42	15.417	11.117	27.163	43.541	32.690	49.442
Run 12	0.46	17.604	11.117	27.033	45.365	31.870	52.733
Run 13	0.50	20.000	11.117	26.576	46.875	30.254	55.756
Run 14	0.54	22.604	11.117	25.818	47.880	27.779	58.176
Run 15	0.58	25.417	11.117	24.854	48.174	24.481	59.586
Run 16	0.63	28.438	11.117	23.855	47.561	20.541	59.536
Run 17	0.67	31.667	11.117	23.075	45.881	16.324	57.574
Run 18	0.71	35.104	11.117	22.855	43.050	12.413	53.323
Run 19	0.75	38.750	11.117	23.596	39.100	9.606	46.578
Run 20	0.79	42.604	11.117	25.724	34.214	8.879	37.422
Run 21	0.83	46.667	11.117	29.631	28.757	11.287	26.342
Run 22	0.88	50.938	11.117	35.589	23.274	17.791	14.316
Run 23	0.92	55.417	11.117	43.671	18.459	29.029	2.826
Run 24	0.96	60.104	11.117	53.659	15.072	45.038	-6.231
Run 25	1.00	65.000	11.117	65.000	13.817	65.000	-10.796

Appendix 2.11: EES Equation Window for Direct Velocity Analysis of the TRT Manipulator

Equations Window

```
"2.Direct Velocities Analysis"
 "Input: Actuating motors parameters"

d_dot_A=2*d_A_max*t                          "Prismatic motor at A "
Theta_dot_B=2*Theta_B_max*t                  "Revolute motor at B "
d_dot_C=2*d_C_max*t                          "Prismatic motor at C "

"Output:"

"- End Effector Velocities: v_Ex, v_Ey ; Joints Relative Ang Veloc Matrix-Theta_dot_j ,Mobile Links Abs Ang Veloc matrix- Omega_m;"

"- All Joints Absolute Velocity Matrix, v ,Relative motion's unit vectors-components on fixed frame"

u_Ax=1
u_Ay=0
u_Bx=0
u_By=0
u_Cx=cos(Theta_B)
u_Cy=sin(Theta_B)

"2.1 Equations"

"Data: Equations for Constrained Motion"
Omega_0 = 0 [deg/s]                          "link 0 is fixed "
Theta_dot_A =0 [deg/s]                       "Prismatic joint between link 0 iand link 1 "
d_dot_B =0 [in/s]                            "Revolute joint between link 1 iand link 2 "
Theta_dot_C=0[deg/s]                         "Prismatic joint between link 2 iand link 3 "

"Equations:"
 -y_EA*Omega_0*pi/180-y_AB*Omega_1*pi/180-y_BC*Omega_2*pi/180-y_CE*Omega_3*pi/180+d_dot_A*u_Ax+d_dot_B*u_Bx+d_dot_C*u_Cx-v_Ex=0

x_EA*Omega_0*pi/180+x_AB*Omega_1*pi/180+x_BC*Omega_2*pi/180+x_CE*Omega_3*pi/180+d_dot_A*u_Ay+d_dot_B*u_By+d_dot_C*u_Cy-v_Ey=0

"2.2. Joints' Relative  Angular Velocities"
 "Output"

Theta_dot_j_array[1,1..3]=[Theta_dot_A,Theta_dot_B, Theta_dot_C]

 "2.3  Mobile Links AbsoluteAngular Velocities"
Omega_1= Omega_0+Theta_dot_A
Omega_2 = Omega_1+Theta_dot_B
Omega_3 = Omega_2+Theta_dot_C
Omega_m_array[1,1..3]=[Omega_1,Omega_2, Omega_3]

"2.4 Relative velocities between points on different joined links, written for each joint"

"Joint A"
v_Ax=d_dot_A*u_Ax
v_Ay=d_dot_A*u_Ay

 "Joint B"
v_Bx=v_Ax -y_AB*Omega_1*pi/180+d_dot_B*u_Bx
v_By=v_Ay +x_AB*Omega_1*pi/180+d_dot_B*u_By

"Joint C"
v_Cx= v_Bx-y_BC*Omega_2*pi/180+d_dot_C*u_Cx
v_Cy= v_By+x_BC*Omega_2*pi/180+d_dot_C*u_Cy

 "Output: All Joints Velocity Matrix, v "
v_array[1,1..3]=[v_Ax, v_Bx, v_Cx]
v_array[2,1..3]=[v_Ay, v_By, v_Cy ]
```

US | Line: 129 Char: 2 | Wrap: On | Insert | Caps Lock: Off | SI C kPa kJ mass deg | Warnings: On | Unit Chk: On | Complex: Off | Syntax Highlight:On

Appendix 2.12: Input Joint Rates and Output End-Effector Linear and Angular Velocities for the TRT Manipulator—EES Parametric Table

Parametric Table

Joint rates vs end effector linear and angular velocity

1..25	t [sec]	$\dot{d}_A$	$\dot{\theta}_B$	$\dot{d}_C$	ω_3 [deg/s]	v_{Ex} [in/s]	v_{Ey} [in/s]
Run 1	0.00	0.000	0.0	0.000	0.000	0.000	0.000
Run 2	0.04	5.000	22.5	1.667	22.500	6.525	17.305
Run 3	0.08	10.000	45.0	3.333	45.000	12.197	34.757
Run 4	0.13	15.000	67.5	5.000	67.500	16.146	52.431
Run 5	0.17	20.000	90.0	6.667	90.000	17.474	70.259
Run 6	0.21	25.000	112.5	8.333	112.500	15.266	87.955
Run 7	0.25	30.000	135.0	10.000	135.000	8.620	104.930
Run 8	0.29	35.000	157.5	11.667	157.500	-3.282	120.221
Run 9	0.33	40.000	180.0	13.333	180.000	-21.059	132.423
Run 10	0.38	45.000	202.5	15.000	202.500	-44.964	139.660
Run 11	0.42	50.000	225.0	16.667	225.000	-74.670	139.602
Run 12	0.46	55.000	247.5	18.333	247.500	-109.005	129.584
Run 13	0.50	60.000	270.0	20.000	270.000	-145.677	106.842
Run 14	0.54	65.000	292.5	21.667	292.500	-181.034	68.906
Run 15	0.58	70.000	315.0	23.333	315.000	-209.932	14.182
Run 16	0.63	75.000	337.5	25.000	337.500	-225.813	-57.307
Run 17	0.67	80.000	360.0	26.667	360.000	-221.123	-143.061
Run 18	0.71	85.000	382.5	28.333	382.500	-188.177	-237.282
Run 19	0.75	90.000	405.0	30.000	405.000	-120.557	-330.283
Run 20	0.79	95.000	427.5	31.667	427.500	-15.014	-408.452
Run 21	0.83	100.000	450.0	33.333	450.000	126.297	-455.119
Run 22	0.88	105.000	472.5	35.000	472.500	293.656	-452.640
Run 23	0.92	110.000	495.0	36.667	495.000	468.349	-385.868
Run 24	0.96	115.000	517.5	38.333	517.500	622.800	-246.840
Run 25	1.00	120.000	540.0	40.000	540.000	723.186	-40.000

Appendix 2.13: End-Effector's Linear Velocity Components and the Angular Velocity-Calculated (EES) and Simulated (SW) for TRT Planar Mechanism

Lookup Table

End effector coordinates from EES and SW | **Linear and angular velocity from EES and SW**

Paste Special	t [sec]	$\omega_{3,EES}$ [deg/s]	$\omega_{3,SW}$ [de/s]	$V_{E,x,EES}$ [in/s]	$V_{E,x,SW}$ [in/s]	$V_{E,y,EES}$ [in/s]	$V_{E,y,SW}$ [in/s]
Row 1	0.00	0.000	0.000	0.000	0.000	0.000	0.000
Row 2	0.04	22.500	22.500	6.525	6.525	17.305	17.305
Row 3	0.08	45.000	45.000	12.197	12.197	34.757	34.757
Row 4	0.13	67.500	67.500	16.146	16.146	52.431	52.431
Row 5	0.17	90.000	90.000	17.474	17.474	70.259	70.259
Row 6	0.21	112.500	112.500	15.266	15.266	87.955	87.955
Row 7	0.25	135.000	135.000	8.620	8.620	104.930	104.930
Row 8	0.29	157.500	157.500	-3.282	-3.282	120.221	120.221
Row 9	0.33	180.000	180.000	-21.059	-21.059	132.423	132.423
Row 10	0.38	202.500	202.500	-44.964	-44.964	139.660	139.660
Row 11	0.42	225.000	225.000	-74.670	-74.670	139.602	139.602
Row 12	0.46	247.500	247.500	-109.005	-109.005	129.584	129.584
Row 13	0.50	270.000	270.000	-145.677	-145.677	106.842	106.842
Row 14	0.54	292.500	292.500	-181.034	-181.034	68.906	68.906
Row 15	0.58	315.000	315.000	-209.932	-209.932	14.182	14.182
Row 16	0.63	337.500	337.500	-225.813	-225.813	-57.307	-57.307
Row 17	0.67	360.000	360.000	-221.123	-221.123	-143.061	-143.061
Row 18	0.71	382.500	382.500	-188.177	-188.177	-237.282	-237.282
Row 19	0.75	405.000	405.000	-120.557	-120.557	-330.283	-330.283
Row 20	0.79	427.500	427.500	-15.014	-15.014	-408.452	-408.452
Row 21	0.83	450.000	450.000	126.297	126.297	-455.119	-455.119
Row 22	0.88	472.500	472.500	293.656	293.656	-452.640	-452.640
Row 23	0.92	495.000	495.000	468.349	468.349	-385.868	-385.868
Row 24	0.96	517.500	517.500	622.800	622.800	-246.840	-246.840
Row 25	1.00	540.000	540.000	723.186	723.186	-40.000	-40.000

References

1. Paul, R.P., *The Robot Manipulators: Mathematics, Programming and Control*, The MIT Press, Cambridge, MA, 1981.
2. N-Nagy, F., and Siegler, A., *Engineering Foundations of Robotics*, Prentice Hall: Upper Saddle River, NJ, 1987.
3. Angeles, J., *Fundamentals of Robotic Mechanical Systems*, Springer: New York, 1997.
4. Talpasanu, I., Optimisation in kinematic and kinetostatic analysis of rigid systems with applications in machine design, *PhD Dissertation*, University Politehnica, Bucharest, 1991.
5. Talpasanu, I., A general method for kinematic analysis of robotic wrist mechanisms, *ASME Journal of Mechanisms and Robotics*, 7(3), 2015.
6. Ball, R.S, *A Treatise on the Theory of Screws*, Cambridge University Press, Cambridge, UK, 1998.
7. Dimentberg, F.M., and Kislitsyn, S.G., Applications of screw calculus to the analysis of three-dimensional mechanisms, *Proceedings of the Second All Union Conference on Basic Problems in the Theory of Machines and Mechanisms*, Moscow, 1960, pp. 55–56.
8. Voinea, R, and Atanasiu, M., *New Analytical Methods in Theory of Mechanisms*, Edit. Tehnica: Bucharest, 1964.
9. Yuan, M.S.C., Freudenstein, F., and Woo, L.S., Kinematic analysis of spatial mechanisms by means of screw coordinates, part I: Screw coordinates, *ASME Journal of Engineering for Industry*, 93, 1971.
10. Pandrea, N., *Elements of the Mechanics of Solid Rigid in Plückerian Coordinates*, The Publishing House of the Romanian Academy: Bucharest, 2000.

3

Kinematics of Single and Multiple Closed Cycle Mechanisms

3.1 Coordinate Systems for Planar Mechanism

Two types of coordinate systems are required for kinematic analysis:

- A fixed coordinate system $x_0y_0z_0$, attached to the ground 0, also called inertial or world frame.
- Mobile coordinate systems $x_my_mz_m$, rigidly attached to each mobile link m. The mobile frame origins are located at the joints.

Example: Figure 3.1 (Animation 3.1) illustrates the mobile frames for an $x - y$ planar four-bar mechanism.

- Fixed $Ox_0y_0z_0$ frame, attached to link 0, with the origin chosen at O, located at the revolute joint A
- Mobile $Ax_1y_1z_1$ frame, attached to link 1, with the origin in the revolute joint A
- Mobile $Bx_2y_2z_2$ frame, attached to link 2, with the origin in the revolute joint B
- Mobile $Dx_3y_3z_3$ frame, attached to link 3, with the origin in the revolute joint D

Example: Figure 3.2 (Animation 3.2) illustrates the mobile frames for a crank slider x–y planar mechanism.

- Fixed $Ox_0y_0z_0$ frame, attached to link 0, with the origin chosen at O, coincident with the revolute joint A
- Mobile $Ax_1y_1z_1$ frame, attached to link 1, with the origin in the revolute joint A

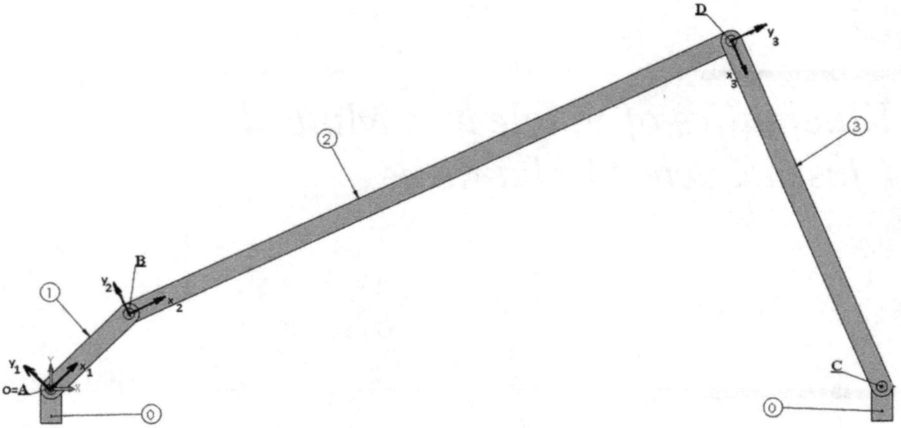

FIGURE 3.1
Frames for the four-bar mechanism.

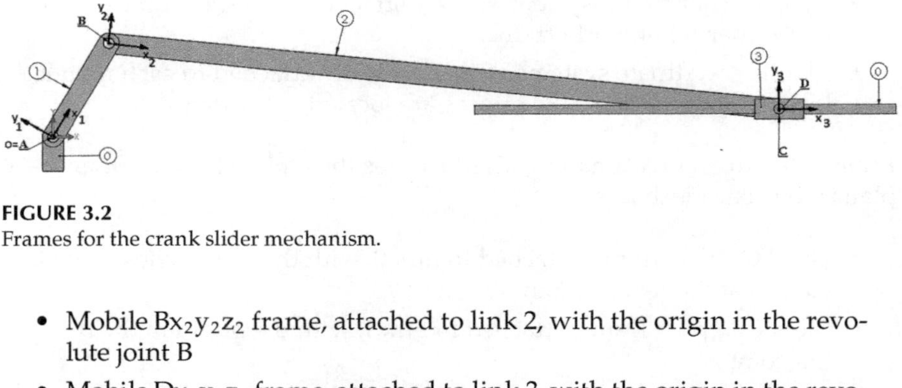

FIGURE 3.2
Frames for the crank slider mechanism.

- Mobile $Bx_2y_2z_2$ frame, attached to link 2, with the origin in the revolute joint B
- Mobile $Dx_3y_3z_3$ frame, attached to link 3, with the origin in the revolute joint D

Example: Figure 3.3 (Animation 3.3) illustrates the frames for a x–y planar crank slider attached to a four-bar mechanism with two cycles.

3.2 Enumeration of Planar Mechanisms Based on the Number of Cycles

The system of two equations, Equation (1.7): $c = j - m$ and Equation (1.28): $M = 3 \cdot m - 2 \cdot j_1 - j_2$, is used to solve the two variables m and j_1, which is given in Equation (3.1):

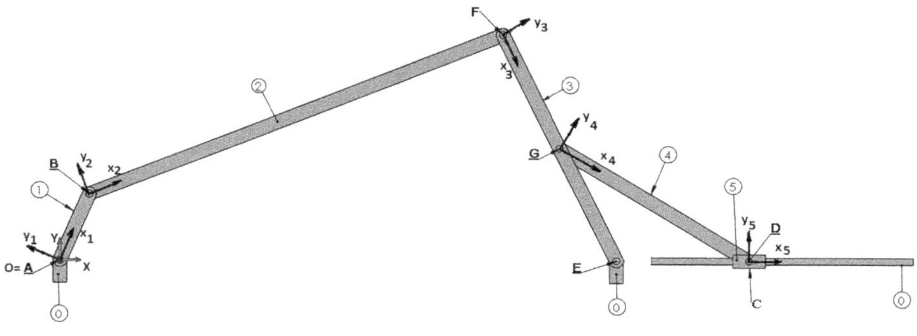

FIGURE 3.3
Frames for a planar mechanism with two cycles.

$$\begin{cases} m = 2 \cdot c + M - j_2 \\ j_1 = 3 \cdot c + M - 2 \cdot j_2 \end{cases} \tag{3.1}$$

Since $m > 0$ and $j_1 \geq 0$, the search for mechanisms starts with assigned values for M mobility and c cycles, and continues with possible j_2 values from Equation (3.2):

$$\begin{cases} j_2 < 2 \cdot c + M \\ 2 \cdot j_2 \leq 3 \cdot c + M \end{cases} \tag{3.2}$$

The search ends with the calculated values for m and j_1 from Equation (3.1).

Mechanisms with M = 1, 2, and 3 mobilities with c = 1 up to five cycles are further tabulated.

The search for mechanisms with mobilities larger than three or larger than five cycles can be conducted also using Equations (3.1) and (3.2).

Mechanisms with one mobility, M = 1

Mechanisms in this category require a single actuation: ac or dc electric motor, pneumatic, or hydraulic. Actuators are usually connected to the fixed link.

The search for single mobility mechanisms is based on Equation (3.3):

$$\begin{cases} j_2 < 2 \cdot c + 1 \\ 2 \cdot j_2 \leq 3 \cdot c + 1 \end{cases}$$

$$m = 2 \cdot c + 1 - j_2 \tag{3.3}$$

$$j_1 = 3 \cdot c + 1 - 2 \cdot j_2$$

Appendix 3.1 illustrates the search results for single mobility mechanisms with one to five cycles.

Mechanisms with two mobilities, M = 2

Mechanisms in this category require a double actuation. To design such mechanisms, one can include brakes to block one or several mobile links, for example, parallel axes gear trains, where the mobility can be reduced to M = 1.

The search for mechanisms with two mobilities is based on Equation (3.4):

$$\begin{cases} j_2 < 2 \cdot c + 2 \\ 2 \cdot j_2 \le 3 \cdot c + 2 \end{cases}$$

$$m = 2 \cdot c + 2 - j_2 \tag{3.4}$$

$$j_1 = 3 \cdot c + 2 - 2 \cdot j_2$$

Appendix 3.2 illustrates the search results for mechanisms with two mobilities, and one to five cycles.

Mechanisms with three mobilities, M = 3

Mechanisms in this category require three actuators.

The search for mechanisms with three mobilities is based on Equation (3.5):

$$\begin{cases} j_2 < 2 \cdot c + 3 \\ 2 \cdot j_2 \le 3 \cdot c + 3 \end{cases}$$

$$m = 2 \cdot c + 3 - j_2 \tag{3.5}$$

$$j_1 = 3 \cdot c + 3 - 2 \cdot j_2$$

Appendix 3.3 illustrates the search results for mechanisms with three mobilities, and one to five cycles.

3.2.1 Parallel Axes Gear Trains with Gear and Revolute Joints

For mechanisms with gears, parallel axes, and bevel gears, a good design requires that the following condition should be met:

Statement 3.1

The number j_2 of meshing joints is the same as the number c of cycles:

$$j_2 = c \tag{3.6}$$

Proof: By equating $m = j_1$ in Equation (1.7), the result $j_2 = c$ is derived. The result is independent of M, therefore valid for any type of geared mechanism.

Note: Since $m > 0$, $j_1 \geq 0$, and $m = j_1$, Equation (3.2) is equivalent with $0 < c + M$, which is valid for any c and M values.

Equation (3.1) has the simplified form:

$$\begin{cases} m = c + M \\ j_1 = c + M \end{cases} \tag{3.7}$$

The search for mechanisms with any number of c cycles and any number of M mobilities is conducted using Equations (3.6) and (3.7).

The geared mechanisms with M = 1, 2, and 3 mobilities and c = 1 up to c = 10, cycles are further tabulated.

Mechanisms with one mobility, M = 1

Mechanisms in this category require a single actuation and are tabulated in Appendix 3.4.

Equations (3.6) and (3.7) are equivalent to Equation (3.8), which is used to tabulate the mechanisms with M = 1:

$$\begin{cases} j_2 = c \\ m = c + 1 \\ j_1 = c + 1 \end{cases} \tag{3.8}$$

Mechanisms with two mobilities, M = 2

Mechanisms in this category require a double actuation. Equations (3.6) and (3.7) are equivalent to Equation (3.9), which is used to tabulate the mechanisms with M = 2 in Appendix 3.5:

$$\begin{cases} j_2 = c \\ m = c + 2 \\ j_1 = c + 2 \end{cases} \tag{3.9}$$

Mechanisms with three mobilities, M = 3

Mechanisms in this category require a triple actuation.

Equations (3.6) and (3.7) are equivalent to Equation (3.10), which is used to tabulate the mechanisms with M = 3 (Appendix 3.6):

$$\begin{cases} j_2 = c \\ m = c + 3 \\ j_1 = c + 3 \end{cases} \tag{3.10}$$

Note: The carriers in parallel axes planetary mechanisms are designed with single or multiple identical planet gears. The multiple planet gear solution has the advantage of better balancing, noise reduction, and lower transmitted force per tooth.

Statement 3.2

The mobility of geared mechanisms can be calculated using Equation (3.11):

$$M = m - c = j_1 - c \tag{3.11}$$

In the case of planetary mechanisms by adding extra identical planet gears to the carrier, the mobility does not change, and the number of cycles increases with the number of extra added planet gears.

Proof: Solving for M, Equation (3.7) results in Equation (3.11).

Indeed, addition to the mechanism of m_p planet gears will increase the following:

- The number of mobile links to $m + m_p$
- The number of cycles to $c + m_p$, since m_p is the same as the number of gear joints
- The number of revolute joints to $j_1 + m_p$

The mobility after the addition of extra m_p planets is calculated using Equation (3.11): $M_p = (m + m_p) - (c + m_p) = m - c = M$, which is the same as the mobility before the addition of the m_p planets. Therefore, the kinematics does not change; moreover, it has the advantage of distributing the tangential force per multiple planets and increasing the life of the gear teeth.

Example: The planetary mechanism in Figure P3.10 has $c = 1$, $j_2 = 1$, $m = 3$, and $j_1 = 3$.

The mobility from Equation (3.11) is $M = 2$ and corresponds to the mechanism in row 1 from Appendix 3.5.

Instead of one planet, this mechanism is redesigned to have three planet gears on the planet carrier 2; this will add two mobile links to the mechanism. The new mechanism is illustrated in Figure P3.11 and has $c = 3$, $j_2 = 3$, $m = 5$, $j_1 = 5$, and same mobility $M = 2$, and corresponds to the mechanism in row 3 from Appendix 3.5.

3.3 Position Analysis for Single-Cycle Planar Mechanisms with Revolute Joints

Example: The four-bar mechanism, Figure 3.4a (Animation 3.1) with revolute joints only, where

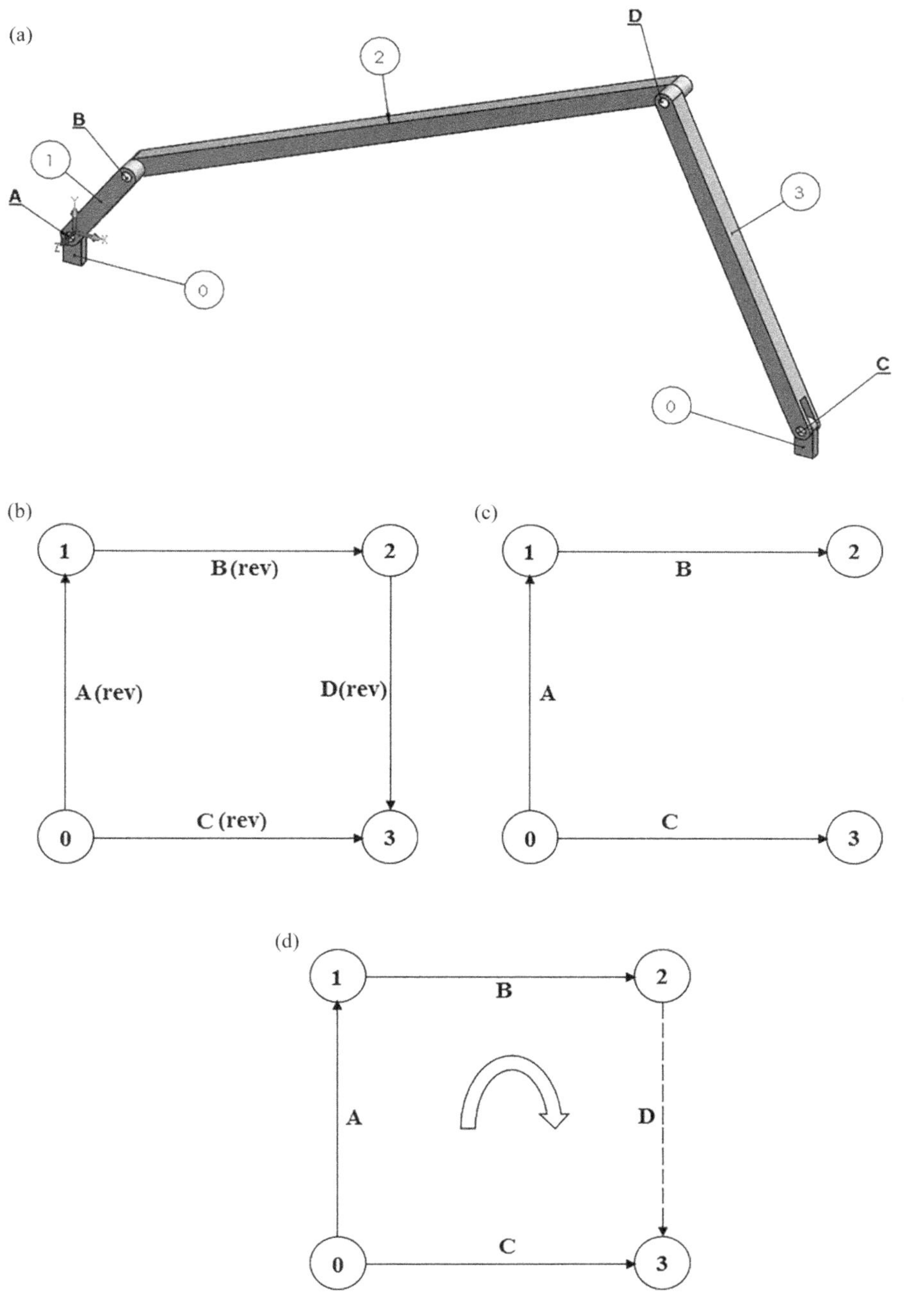

FIGURE 3.4
(a) Planar four-bar mechanism with revolute joints and one cycle; (b) digraph; (c) spanning tree; (d) cycle.

- The number of links and joints: The links are labeled with numbers, starting with 0 which is assigned to the fixed link. There are three mobile links, m = 3, which have been assigned the numbers 1, 2, and 3. There are four joints labeled with capital letters: A, B, C, and D. All mechanisms' joints are 1 degrees of freedom (DOF) each: revolute (description "rev" at A, B, C, and D), therefore $j_1 = 4$. There are no joints with 2 DOF, therefore $j_2 = 0$.
- The total number of joints from Equation (1.27) is

$$j = j_1 + j_2 = 4 + 0 = 4 \text{ joints.}$$

- Mechanism's digraph: The digraph attached to the mechanism is illustrated in Figure 3.4b, with its n = m + 1 = 4 nodes corresponding to the mechanism's links and the edges to the mechanism's j = 4 joints.
  The digraph's edges are oriented from the lower number node to the higher number node.
- Spanning tree: Figure 3.4c shows the spanning tree cutting the chord D. The chord D is labeled with the last letter.
- Digraph's number of independent cycles: The number of cycles is determined from Equation (1.7):

$$c = j - m = 4 - 3 = 1 \text{ cycle.}$$

  Figure 3.4d shows the cycle oriented in the direction of the chord D (clockwise).
- Mechanism's mobility: The mobility (DOF) of the planar four-bar mechanism is determined from Equation (1.28):

$$M = 3 \cdot m - 2 \cdot j_1 - j_2 = 3 \cdot 3 - 2 \cdot 4 - 0 = 1$$

3.3.1 The Incidence Nodes–Edges Matrix, G

Considering its definition (Equation 1.15), the matrix $\underline{G}$ has each row assigned to a link and each column assigned to a joint, as shown in Equation (3.12).

For each column, there has assigned a pair of two numbers: –1 assigned for the lower number link and +1 assigned for the higher number link:

$$\begin{array}{c} \text{Joints} \\ \begin{array}{cccc} A & B & C & D \end{array} \end{array}$$

$$[\underline{G}] = \begin{bmatrix} -1 & 0 & -1 & 0 \\ 1 & -1 & 0 & 0 \\ 0 & 1 & 0 & -1 \\ 0 & 0 & 1 & 1 \end{bmatrix} \begin{matrix} 0 \\ 1 \\ 2 \\ 3 \end{matrix} \quad \text{Links} \tag{3.12}$$

3.3.2 The Cycle Basis Matroid Matrix, C

Considering its definition (Equation 1.25), the matrix **C** has one row assigned to one cycle and each column assigned to a joint, as shown in Equation (3.13).

The *cycle's direction* clockwise or counterclockwise is assigned for each cycle. A –1 is assigned to a joint which is contrary to the cycle's direction and 1 if the joint is in the cycle's direction.

$$\begin{matrix} & \text{A} & \text{B} & \text{C} & \text{D} \end{matrix}$$

$$[\text{C}] = \begin{bmatrix} 1 & 1 & -1 & 1 \end{bmatrix} = \begin{bmatrix} c_A & c_B & c_C & c_D \end{bmatrix} \tag{3.13}$$

3.3.3 Joint Position Vectors Matrix, $\mathbf{r}_j$:

Direct position analysis is conducted in the determination of the $\mathbf{r}_j$ joint position vectors. The $\mathbf{r}_j$, where j = A, B, C, D, position vectors have the components written on the fixed frame (*global frame*), assigned to the fixed link 0.

Joint position matrix, **r**, is defined as a one-row matrix with each column assigned to a joint, as shown in Equation (3.14):

$$\begin{matrix} & \text{A} & \text{B} & \text{C} & \text{D} \end{matrix}$$

$$\left[\mathbf{r}_j\right] = \begin{bmatrix} \mathbf{r}_A^0 & \mathbf{r}_B^0 & \mathbf{r}_C^0 & \mathbf{r}_D^0 \end{bmatrix} \tag{3.14}$$

Vectors **r** in Equation (3.14) for mechanisms with four joints (the number of columns in the matrix) are written as functions of their x and y components, as shown in Equation (3.15). For any planar mechanism with more than four joints, the number of columns in the matrix will be the number of joints:

$$\begin{matrix} & \text{A} & \text{B} & \text{C} & \text{D} \end{matrix}$$

$$[\mathbf{r}] = \begin{bmatrix} x_A & x_B & x_C & x_D \\ y_A & y_B & y_C & y_D \end{bmatrix} \tag{3.15}$$

The matrix **r** is useful to visualize the mechanism's position on the fixed frame.

The entries in the joint position matrix **r** starting from a joint A on the base and continuing to all of the cycle's joints are calculated from the entries in the Latin matrix (explained in Section 3.3.5) (Equation (3.16)).

Joint A (data): Since A is the origin of the coordinate system, then $\mathbf{r}_A = 0$:

$$\mathbf{r}_A = \begin{Bmatrix} x_A \\ y_A \end{Bmatrix} = \begin{Bmatrix} 0 \\ 0 \end{Bmatrix}$$

Joint B:

$$\mathbf{r}_B = \mathbf{r}_A + \mathbf{L}_{AB} = \begin{Bmatrix} 0 \\ 0 \end{Bmatrix} + \begin{Bmatrix} x_{AB} \\ y_{AB} \end{Bmatrix} = \begin{Bmatrix} x_{AB} \\ y_{AB} \end{Bmatrix} \tag{3.16}$$

Joint D:

$$\mathbf{r}_D = \mathbf{r}_B + \mathbf{L}_{BD} = \begin{Bmatrix} x_{AB} + x_{BD} \\ y_{AB} + y_{BD} \end{Bmatrix}$$

Joint C:

$$\mathbf{r}_C = \mathbf{r}_D + \mathbf{L}_{DC} = \begin{Bmatrix} x_{AB} + x_{BD} + x_{DC} \\ y_{AB} + y_{BD} + y_{DC} \end{Bmatrix}$$

Joint C (data): The obtained coordinates in the previous equation should match the ones provided (checkup equation):

$$\mathbf{r}_C = \begin{Bmatrix} x_C \\ y_C \end{Bmatrix}$$

3.3.4 Digraph Joint Position Matrix, $\mathbf{r}_{c,j}$

The *digraph joint position matrix*, $\mathbf{r}_{c,j}$, with vector entries, is defined like the matrix **C** from the digraph:

$$\begin{array}{c} \text{Edges} \\ \left[\mathbf{r}_{c,j}\right] = \begin{bmatrix} * & \cdots & * \\ \vdots & \mathbf{r}_{c,j} & \vdots \\ * & \cdots & * \end{bmatrix} \text{Cycles} \end{array} \tag{3.17}$$

where

$$\mathbf{r}_{c,j} = \left\{ \begin{array}{ll} -\mathbf{r}_j & \text{if joint j is on the cycle and opposite to the cycle} \\ +\mathbf{r}_j & \text{if joint j is on the cycle and same direction as the cycle} \\ 0 & \text{if the joint is not on the cycle} \end{array} \right\} \tag{3.18}$$

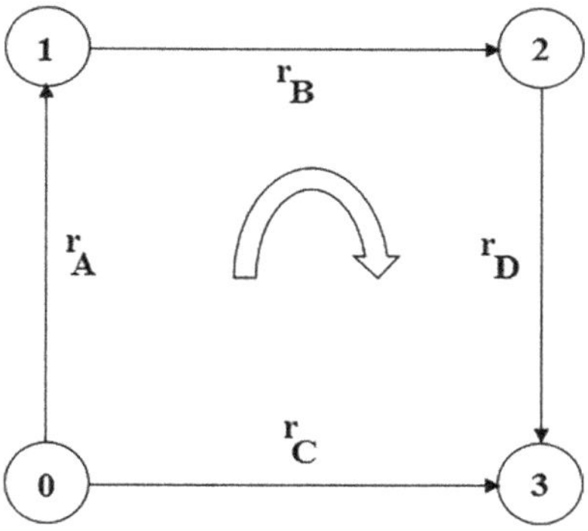

FIGURE 3.5
Digraph joint position vectors, around the cycle, form the matrix $\mathbf{r}_{c,j}$.

The joint position matrix is shown in Equation (3.19) and Figure 3.5:

$$\begin{array}{c} \\ [\mathbf{r}_{c,j}] = \end{array} \begin{array}{cccc} A & B & C & D \\ \left[c_A \cdot \mathbf{r}_A \right. & c_B \cdot \mathbf{r}_B & c_C \cdot \mathbf{r}_C & \left. c_D \cdot \mathbf{r}_D \right] \end{array} \begin{array}{c} \\ = \left[\mathbf{r}_A \quad \mathbf{r}_B \quad -\mathbf{r}_C \quad \mathbf{r}_D \right] \end{array} \tag{3.19}$$

and expressed as a function of their x and y components, as shown in Equation (3.20):

$$[\mathbf{r}_{cj}] = \begin{array}{c} \begin{array}{cccc} A & B & C & D \end{array} \\ \begin{bmatrix} x_A & x_B & -x_C & x_D \\ y_A & y_B & -y_C & y_D \end{bmatrix} \end{array} \tag{3.20}$$

3.3.5 Latin Matrix Method for Positional Analysis

The *Latin matrix*, **L**, Equation (3.21), is a (c × n) matrix with vector entries defined around the links in the digraph and written in the direction of each cycle (i.e., clockwise in Figure 3.6), [1,2]:

$$\begin{array}{c} \text{Links} \\ \begin{array}{cccc} 0 & 1 & 2 & 3 \end{array} \\ [\mathbf{L}] = \left[\mathbf{L}_{CA} \quad \mathbf{L}_{AB} \quad \mathbf{L}_{BD} \quad \mathbf{L}_{DC} \right] \text{Cycle} \end{array} \tag{3.21}$$

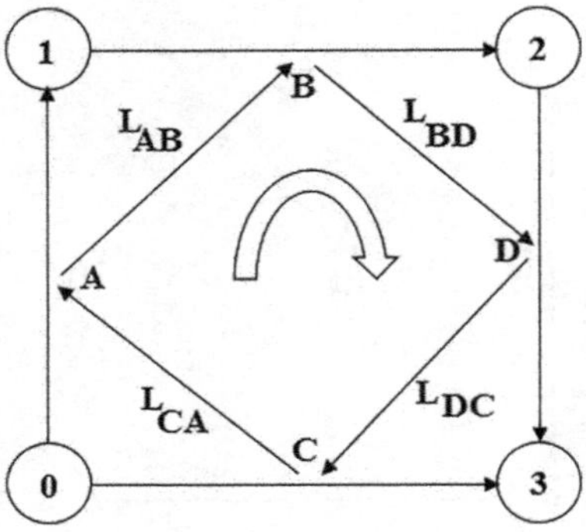

FIGURE 3.6
The Latin matrix vectors, all with a constant magnitude, connecting revolute joints, around the cycle.

The scalar components in the Latin matrix, Equation (3.22), contain a number of 4c = 4 unknowns for the positional analysis, as shown further:

$$[\mathbf{L}] = \begin{array}{c} \begin{matrix} 0 & 1 & 2 & 3 \end{matrix} \\ \begin{bmatrix} x_{CA} & x_{AB} & x_{BD} & x_{DC} \\ y_{CA} & y_{AB} & y_{BD} & y_{DC} \end{bmatrix} \end{array} \tag{3.22}$$

where

$$\mathbf{L}_{CA} = \begin{Bmatrix} x_{CA} \\ y_{CA} \end{Bmatrix} = \begin{Bmatrix} x_A - x_C \\ y_A - y_C \end{Bmatrix}; \mathbf{L}_{AB} = \begin{Bmatrix} x_{AB} \\ y_{AB} \end{Bmatrix} = \begin{Bmatrix} x_B - x_A \\ y_B - y_A \end{Bmatrix};$$
$$\mathbf{L}_{BD} = \begin{Bmatrix} x_{BD} \\ y_{BD} \end{Bmatrix} = \begin{Bmatrix} x_D - x_B \\ y_D - y_B \end{Bmatrix}; \mathbf{L}_{DC} = \begin{Bmatrix} x_{DC} \\ y_{DC} \end{Bmatrix} = \begin{Bmatrix} x_C - x_D \\ y_C - y_D \end{Bmatrix} \tag{3.23}$$

Statement 3.3

The entries in the Latin matrix are automatically generated from the digraph matrix $\underline{\mathbf{G}}$:

$$-\left[\mathbf{r}_{c,j}\right] \cdot \left[\underline{G}\right]^T = \mathbf{L} \tag{3.24}$$

Example: If we consider the digraph position vector matrix (Equation (3.19)):

$$\left[\mathbf{r}_{c,j}\right] = \begin{array}{c} \begin{matrix} A & B & C & D \end{matrix} \\ \begin{bmatrix} \mathbf{r}_A & \mathbf{r}_B & -\mathbf{r}_C & \mathbf{r}_D \end{bmatrix} \end{array}$$

and the transposed of $\underline{\mathbf{G}}$ incidence links–joints matrix from Equation (3.12):

$$
\begin{array}{c}
\text{Links} \\
\begin{array}{cccc} 0 & 1 & 2 & 3 \end{array}
\end{array}
$$

$$
[\underline{G}]^T = \begin{bmatrix} -1 & 1 & 0 & 0 \\ 0 & -1 & 1 & 0 \\ -1 & 0 & 0 & 1 \\ 0 & 0 & -1 & 1 \end{bmatrix} \begin{matrix} A \\ B \\ C \\ D \end{matrix} \quad \text{Joints}
$$

Then, Equation (3.23) generates the entries in the Latin matrix [1,2]:

$$
-\left[\mathbf{r}_{c,j}\right]\cdot\left[\underline{G}\right]^T = \left[\begin{array}{cccc} \{\mathbf{r}_A - \mathbf{r}_C\} & \{\mathbf{r}_B - \mathbf{r}_A\} & \{\mathbf{r}_D - \mathbf{r}_B\} & \{\mathbf{r}_C - \mathbf{r}_D\} \end{array}\right]
$$

$$
= \left[\begin{array}{cccc} \mathbf{L}_{CA} & \mathbf{L}_{AB} & \mathbf{L}_{BD} & \mathbf{L}_{DC} \end{array}\right] \tag{3.25}
$$

Statement 3.4

The sum of entries on each row in the Latin matrix is zero:

$$
\mathbf{L}_{CA} + \mathbf{L}_{AB} + \mathbf{L}_{BD} + \mathbf{L}_{DC} = 0 \tag{3.26}
$$

Proof: In Equation (3.25), each position vector appears twice and with the opposite sign, then causing the sum to be zero.

Statement 3.5

For positional analysis of planar mechanisms, a nonlinear system of 4c independent equations is generated with 4c independent unknowns.

NONLINEAR SYSTEM OF EQUATIONS FOR POSITIONAL ANALYSIS

Type I equations: The sum of vector components is zero from each row in the Latin matrix

A set of c vector equations or 2c scalar equations (Equation 3.27) with 4c unknowns are generated based on the previous statement.

Example: four-bar mechanism:

$$
\begin{cases} x_{CA} + x_{AB} + x_{BD} + x_{DC} = 0 \\ x_{CA} + x_{AB} + x_{BD} + x_{DC} = 0 \end{cases} \tag{3.27}
$$

Type II equations: The magnitude of vectors from columns in the Latin matrix

The second set of 2c equations with 4c unknowns are written based on the given magnitude between joints.

$$\begin{cases} x_{BD}^2 + y_{BD}^2 = L_{BD}^2 \\ x_{DC}^2 + y_{DC}^2 = L_{DC}^2 \end{cases} \tag{3.28}$$

Type III equations (isometric equations): Constraint equations that preserve the angle between vectors

In the case of complex links (see Figure 1.2), a third joint P in addition to A and B will exist.

During the motion, the constant length vectors L_{AP}, L_{AB} keep a constant angle between them.

The *isometric* Equation 3.29 reflects the conservation of length and angle [1,2]:

$$\begin{Bmatrix} x_{AP} \\ y_{AP} \end{Bmatrix} = \frac{L_{AP}}{L_{AB}} \cdot \begin{bmatrix} \mathrm{Cos}(\Phi) & -\mathrm{Sin}(\Phi) \\ \mathrm{Sin}(\Phi) & \mathrm{Cos}(\Phi) \end{bmatrix} * \begin{Bmatrix} x_{AB} \\ y_{AB} \end{Bmatrix} \tag{3.29}$$

The point P (point of interest) is considered to be any third point on the link, like the center of mass, labeled G.

Vector $\mathbf{L}_{AP}$ coordinates are not independent, since they are determined as a function of vector $\mathbf{L}_{AB}$ coordinates. If Φ is the angle between the positive directions of the two vectors, then a positive angle is considered for the case: $\mathbf{L}_{AB}$ rotates counterclockwise toward $\mathbf{L}_{AP}$, and at a negative angle for a clockwise rotation.

Examples:

a. Two vectors that are collinear and oriented in the same direction, Equation (3.29) for $\Phi = 0°$ becomes

$$\begin{Bmatrix} x_{AP} \\ y_{AP} \end{Bmatrix} = \frac{L_{AP}}{L_{AB}} \cdot \begin{bmatrix} \mathrm{Cos}(0°) & -\mathrm{Sin}(0°) \\ \mathrm{Sin}(0°) & \mathrm{Cos}(0°) \end{bmatrix} * \begin{Bmatrix} x_{AB} \\ y_{AB} \end{Bmatrix} = \frac{L_{AP}}{L_{AB}} \cdot \begin{Bmatrix} x_{AB} \\ y_{AB} \end{Bmatrix} \tag{3.29a}$$

b. Two vectors that are collinear and oriented in the opposite direction, Equation (3.29) for $\Phi = 180°$ becomes

$$\begin{Bmatrix} x_{AP} \\ y_{AP} \end{Bmatrix} = \frac{L_{AP}}{L_{AB}} \cdot \begin{bmatrix} \mathrm{Cos}(180°) & -\mathrm{Sin}(180°) \\ \mathrm{Sin}(180°) & \mathrm{Cos}(180°) \end{bmatrix} * \begin{Bmatrix} x_{AB} \\ y_{AB} \end{Bmatrix} = \frac{L_{AP}}{L_{AB}} \cdot \begin{Bmatrix} -x_{AB} \\ -y_{AB} \end{Bmatrix} \tag{3.29b}$$

c. Two vectors that are in directions perpendicular to each other, Equation (3.29) for $\Phi = 90°$ becomes

$$\begin{Bmatrix} x_{AP} \\ y_{AP} \end{Bmatrix} = \frac{L_{AP}}{L_{AB}} \cdot \begin{bmatrix} \text{Cos}(90°) & -\text{Sin}(90°) \\ \text{Sin}(90°) & \text{Cos}(90°) \end{bmatrix} * \begin{Bmatrix} x_{AB} \\ y_{AB} \end{Bmatrix} = \frac{L_{AP}}{L_{AB}} \cdot \begin{Bmatrix} -y_{AB} \\ x_{AB} \end{Bmatrix} \tag{3.29c}$$

Note: Equation (3.29) represents a *planar rotation* of $\mathbf{u}_{AB} = \mathbf{L}_{AB}/L_{AB}$, the unit vector with magnitude 1, toward the unit vector $\mathbf{u}_{AP} = \mathbf{L}_{AP}/L_{AP}$.

Indeed, Equation (3.29) could be written in the form:

$$\begin{Bmatrix} u_{AP}^{x} \\ u_{AP}^{y} \end{Bmatrix} = \begin{bmatrix} \text{Cos}(\Phi) & -\text{Sin}(\Phi) \\ \text{Sin}(\Phi) & \text{Cos}(\Phi) \end{bmatrix} * \begin{Bmatrix} u_{AB}^{x} \\ u_{AB}^{y} \end{Bmatrix} \tag{3.30}$$

The unit vector $\mathbf{u}_{AB}$ after rotation coincides with the unit vector $\mathbf{u}_{AP}$ and conserves its magnitude (*isometric* operation).

For *multiple-cycle* mechanisms, more than one vector, that is, $\mathbf{L}_{AB}$, $\mathbf{L}_{AC}$, $\mathbf{L}_{BC}$, will show on the same column in the Latin matrix. Then, the extra vectors $\mathbf{L}_{AC}$, $\mathbf{L}_{BC}$ are dependent, and two isometric equations (3.29) are written. Indeed, the three vectors are located on the same mobile link (same column in the Latin matrix), and then two constant angle equations will be written.

Input data: Vectors with constant magnitude in the Latin matrix

- The components for $\mathbf{L}_{CA}$ (fixed link 0):

$$\mathbf{L}_{CA} = \begin{Bmatrix} x_{CA} \\ y_{CA} \end{Bmatrix} = \begin{Bmatrix} x_A - x_C \\ y_A - y_C \end{Bmatrix} \tag{3.31}$$

- The components for $\mathbf{L}_{AB}$ (crank, link 1):

$$\mathbf{L}_{AB} = \begin{Bmatrix} x_{AB} \\ y_{AB} \end{Bmatrix} = \begin{Bmatrix} L_{AB} \cdot \text{Cos}(\theta_1) \\ L_{AB} \cdot \text{Sin}(\theta_1) \end{Bmatrix} \tag{3.32}$$

Output data: Latin matrix entries for variable crank angle θ_1

The nonlinear system of 4c = 4 equations (*Type I*: Equation 3.27 and *Type II*: Equation 3.28) is solved for the 4c = 4 independent variables ($x_{BD}, y_{BD}, x_{DC}, y_{DC}$):

- The components for $\mathbf{L}_{BD}$ (link 2)
- The components for $\mathbf{L}_{DC}$ (link 3)

Therefore, all entries in the Latin matrix are found as functions of the variable crank angle θ_1

3.3.6 Centers of Mass Position Vector Matrix, $\mathbf{r}_{G_m}$

The centers' of mass G_1, G_2, and G_3 coordinates, measured in the fixed frame, are stored in the r_{G_m} matrix, where m = 1, 2, and 3 are the link labels for the mobile links. This matrix is useful for further velocity, acceleration, and dynamic analyses:

$$
\begin{matrix} & 1 & 2 & 3 \end{matrix}
$$

$$
\left[\mathbf{r}_{G_m}\right]=\begin{bmatrix} x_{G_1} & x_{G_2} & x_{G_3} \\ y_{G_1} & y_{G_2} & y_{G_3} \end{bmatrix} \tag{3.33}
$$

If the location for centers of mass on each link is provided by L_{AG_1}, L_{BG_2}, and L_{DG_3}, the *isometric* type III Equation (3.29) is also given, which defines the rotations ($\mathbf{L}_{AB}$ toward $\mathbf{L}_{AG_1}$), ($\mathbf{L}_{BD}$ toward $\mathbf{L}_{BG_2}$), and ($\mathbf{L}_{DC}$ toward $\mathbf{L}_{DG_3}$), then Equation (3.34) holds:

$$
\mathbf{r}_{G_1}=\mathbf{r}_A+\mathbf{L}_{AG_1}=\begin{Bmatrix} 0 \\ 0 \end{Bmatrix}+\frac{L_{AG_1}}{L_{AB}}\cdot\begin{bmatrix} \mathrm{Cos}(0^\circ) & -\mathrm{Sin}(0^\circ) \\ \mathrm{Sin}(0^\circ) & \mathrm{Cos}(0^\circ) \end{bmatrix}\cdot\begin{Bmatrix} x_{AB} \\ y_{AB} \end{Bmatrix}
$$

$$
=\begin{Bmatrix} \dfrac{L_{AG_1}}{L_{AB}}\cdot x_{AB} \\ \dfrac{L_{AG_1}}{L_{AB}}\cdot y_{AB} \end{Bmatrix}
$$

$$
\mathbf{r}_{G_2}=\mathbf{r}_B+\mathbf{L}_{BG_2}=\begin{Bmatrix} x_{AB} \\ y_{AB} \end{Bmatrix}+\frac{L_{BG_2}}{L_{BD}}\cdot\begin{bmatrix} \mathrm{Cos}(0^\circ) & -\mathrm{Sin}(0^\circ) \\ \mathrm{Sin}(0^\circ) & \mathrm{Cos}(0^\circ) \end{bmatrix}\cdot\begin{Bmatrix} x_{BD} \\ y_{BD} \end{Bmatrix}
$$

$$
=\begin{Bmatrix} x_{AB}+\dfrac{L_{BG_2}}{L_{BD}}\cdot x_{BD} \\ y_{AB}+\dfrac{L_{BG_2}}{L_{BD}}\cdot y_{BD} \end{Bmatrix}
$$

$$\mathbf{r}_{G_3} = \mathbf{r}_D + \mathbf{L}_{DG_3} = \begin{Bmatrix} x_{AB} + x_{BD} \\ y_{AB} + y_{BD} \end{Bmatrix} + \frac{L_{DG_3}}{L_{DC}} \cdot \begin{bmatrix} \text{Cos}(0°) & -\text{Sin}(0°) \\ \text{Sin}(0°) & \text{Cos}(0°) \end{bmatrix} \cdot \begin{Bmatrix} x_{DC} \\ y_{DC} \end{Bmatrix}$$

$$= \begin{Bmatrix} x_{AB} + x_{BD} + \dfrac{L_{BG_3}}{L_{DC}} \cdot x_{DC} \\ y_{AB} + y_{BD} + \dfrac{L_{BG_3}}{L_{DC}} \cdot y_{DC} \end{Bmatrix} \tag{3.34}$$

3.3.7 Center of Mass to Joint Position Matrix, $\mathbf{L}_{G_j}$

Equation (3.35) is the matrix stored for further velocity, acceleration, and dynamic analyses:

$$\left[\mathbf{L}_{G_j}\right] = \begin{bmatrix} x_{G_1A} & x_{G_1B} & x_{G_2B} & x_{G_2D} & x_{G_3D} & x_{G_3C} \\ y_{G_1A} & y_{G_1B} & y_{G_2B} & y_{G_2D} & y_{G_3D} & y_{G_3C} \end{bmatrix} \tag{3.35}$$

where its entries are determined using Equation (3.36):

$$\begin{aligned}
\mathbf{L}_{G_1A} &= \mathbf{r}_A - \mathbf{r}_{G_1} = \begin{Bmatrix} x_A - x_{G_1} \\ y_A - y_{G_1} \end{Bmatrix} \\
\mathbf{L}_{G_1B} &= \mathbf{r}_B - \mathbf{r}_{G_1} = \begin{Bmatrix} x_B - x_{G_1} \\ y_B - y_{G_1} \end{Bmatrix} \\
\mathbf{L}_{G_2B} &= \mathbf{r}_B - \mathbf{r}_{G_2} = \begin{Bmatrix} x_B - x_{G_2} \\ y_B - y_{G_2} \end{Bmatrix} \\
\mathbf{L}_{G_2D} &= \mathbf{r}_D - \mathbf{r}_{G_2} = \begin{Bmatrix} x_D - x_{G_2} \\ y_D - y_{G_2} \end{Bmatrix} \\
\mathbf{L}_{G_3D} &= \mathbf{r}_D - \mathbf{r}_{G_3} = \begin{Bmatrix} x_D - x_{G_3} \\ y_D - y_{G_3} \end{Bmatrix} \\
\mathbf{L}_{G_3C} &= \mathbf{r}_C - \mathbf{r}_{G_3} = \begin{Bmatrix} x_C - x_{G_3} \\ y_C - y_{G_3} \end{Bmatrix}
\end{aligned} \tag{3.36}$$

3.3.8 Absolute Links' Orientation, θ_n

Absolute angular orientation of each link is provided by the angles θ_n, (n = 0, 1, 2, 3), measured from the fixed x_0-axis, which is given in Equation (3.37):

$$\begin{matrix} & 0 & 1 & 2 & 3 \\ [\theta_n] = [& \theta_0 & \theta_1 & \theta_2 & \theta_3 &] \end{matrix} \tag{3.37}$$

Link 0:

$$\theta_0 = 0 \text{ (fixed link)}$$

Link 1:

$$\theta_1(t) \text{ is variable with time}$$

Link 2: Determined from the Latin matrix with trigonometric function "tan." When writing Engineering Equation Solver (EES) code, a test for x and y sign displays the angle in the right quadrant:

$$\theta_2 = \arctan\left(\frac{y_{BD}}{x_{BD}}\right) \tag{3.37a}$$

Link 3: Determined from the Latin matrix with trigonometric function "tan." A test for x and y sign displays the angle in the right quadrant:

$$\theta_3 = \arctan\left(\frac{y_{DC}}{x_{DC}}\right) \tag{3.37b}$$

3.3.9 Relative Links' Orientation, θ_j

The *relative links' orientation matrix* is defined as one row matrix with each column assigned to a joint, as shown in Equation (3.38):

$$\begin{matrix} & A & B & C & D \\ [\theta_j] = [& \theta_{A_{01}} & \theta_{B_{12}} & \theta_{C_{03}} & \theta_{D_{23}} &] = [\; \theta_A \;\; \theta_B \;\; \theta_C \;\; \theta_D \;] \end{matrix} \tag{3.38}$$

Notation: The subscripts for relative parameters can be dropped, that is, $\theta_{A_{01}} = \theta_A$

The relative angles between two joined links, shown in Figure 3.7a, are the entries in matrix θ_j and are related to the absolute angles θ_n, as listed in Equation (3.39):

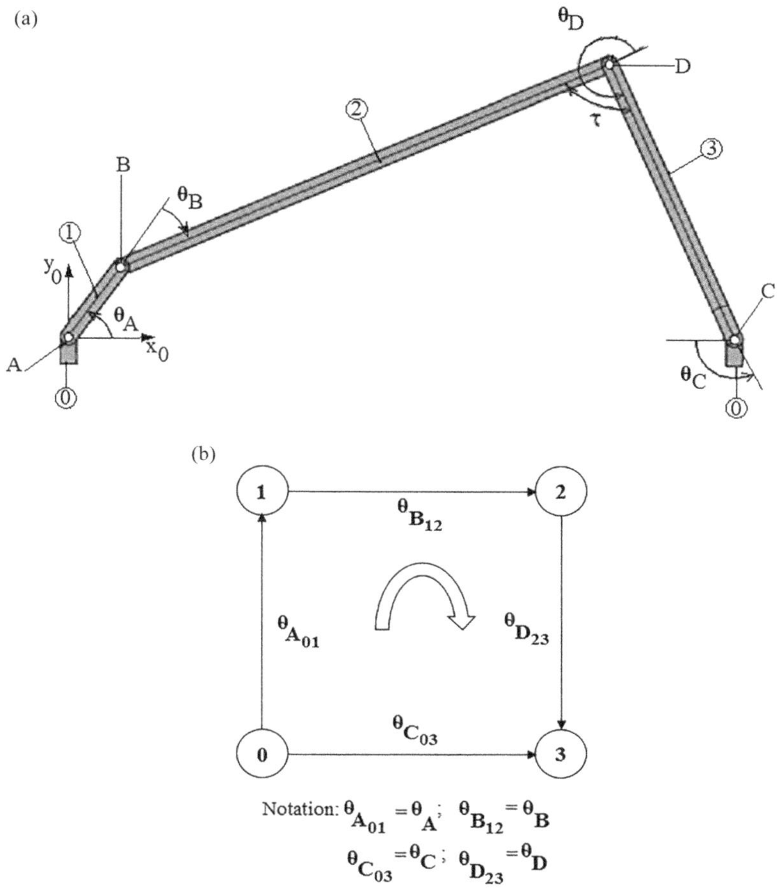

FIGURE 3.7
(a) Relative angles between two joined links; (b) relative angles assigned to edges in digraph.

$$
\begin{aligned}
\theta_0 &= 0 \text{ (fixed link)} \\
\theta_A &= \theta_1 - \theta_0 \\
\theta_B &= \theta_2 - \theta_1 \\
\theta_C &= \theta_3 - \theta_0 \\
\theta_D &= \theta_3 - \theta_2
\end{aligned}
\tag{3.39}
$$

One could visualize the relative angles by considering as "frozen" the first of the two links (the first link in the subscript). Thus, θ_A, the angle of link 1 if 0 is frozen; θ_B, the angle of link 2 if 1 is frozen; θ_C, the angle of link 3 if 0 is frozen; and θ_D, the angle of link 3 if 2 is frozen.

3.3.10 Relative Links' Orientation from Digraph, $\Theta_{c,j}$

The *relative links' orientation* is a (c × j) matrix, from Equation (3.40), where each row is assigned to a cycle and each column is assigned to a joint, and is generated along the edges in the cycle, as shown in Figure 3.7b. A '–1' is assigned to a joint which is contrary to the cycle's direction and a '+1' is assigned if the joint is in the same direction as the cycle:

$$\begin{matrix} & A & B & C & D \\ \left[\Theta_{c,j}\right]=\Big[& c_A\cdot\theta_A & c_B\cdot\theta_B & c_C\cdot\theta_C & c_D\cdot\theta_D \end{matrix}\Big]=\left[\begin{matrix}\theta_A & \theta_B & -\theta_C & \theta_D\end{matrix}\right] \tag{3.40}$$

Equation (3.40) is automatically generated from the digraph's matrix $\left[\underline{G}\right]$:

$$\left[\Theta_{c,j}\right]=\left[\theta_n\right]*\left[\underline{G}\right] \tag{3.41}$$

Example:

$$\left[\Theta_{c,j}\right]=\left[\begin{matrix}\theta_0 & \theta_1 & \theta_2 & \theta_3\end{matrix}\right]*\left[\begin{matrix}-1 & 0 & -1 & 0\\ 1 & -1 & 0 & 0\\ 0 & 1 & 0 & -1\\ 0 & 0 & 1 & 1\end{matrix}\right]$$

$$=\left[\begin{matrix}\theta_A & \theta_B & -\theta_C & \theta_D\end{matrix}\right]$$

Statement 3.6

The sum of digraph relative angular displacements is zero

$$\theta_A+\theta_B-\theta_C+\theta_D=0 \tag{3.42}$$

Proof: If addition is performed in Equation (2.40), then clearly the sum of entries on the left side is zero.

3.3.11 Transmission Angle

For a four-bar mechanism, the τ transmission angle is defined by Alt as the angle between the centerlines of links 2 and 3 (Figure 3.7a):

$$\tau=\text{abs}\left(\theta_D-180°\right) \tag{3.43}$$

The transmission angle, also, can be determined using Equation (3.44), as the angle between the Latin matrix vectors $\mathbf{L}_{BD}$ and $\mathbf{L}_{DC}$:

$$\tau = 180^\circ - \arccos\left(\frac{x_{BD} \cdot x_{DC} + y_{BD} \cdot y_{DC}}{L_{BD} \cdot L_{DC}}\right) \tag{3.44}$$

This variable angle plays an important role in force and torque transmission from the driving link 2 to the output link 3. The optimum value is 90°, although the tolerance ±50° is acceptable in the design of mechanisms [3].

3.3.12 Input to Output Relation

For a given input $\theta_1(t)$, crank angle, the output $\theta_3(t)$ is desired. Such input–output relation is important for both tasks: control of output motion or synthesis of mechanisms. In a further section, synthesis consists in finding the links' lengths for which the mechanism's output follows a desired function.

3.3.13 Dead Centers

The dead center position is defined as the mechanism's positions when the crank and the coupler are in extension. Equation (3.45) represents the condition of collinearity between vectors $\mathbf{L}_{AB}$ and $\mathbf{L}_{BD}$:

$$x_{AB} \cdot y_{BD} - y_{AB} \cdot x_{BD} = 0 \tag{3.45}$$

Mechanisms might have no dead center points, one, or up to a maximum of two solutions for the crank angle θ_1 from the condition of collinearity between vectors $\mathbf{L}_{AB}$ (link 1) and $\mathbf{L}_{BD}$ (link 2).

At dead centers, the relative angle between the two links 1 and 2 is 0° or 180°:

$$\begin{aligned} &\theta_B = 0^\circ \text{ when } \mathbf{L}_{AB} \text{ is oriented in the same direction as } \mathbf{L}_{BD} \\ &\theta_B = 180^\circ \text{ when } \mathbf{L}_{AB} \text{ is oriented in the opposite direction as } \mathbf{L}_{BD} \end{aligned} \tag{3.46}$$

One could notice that the determinant with the columns 2 and 3 in the Latin matrix vanishes at dead-point positions. The entries in the two rows are proportional; therefore, the determinant is zero:

$$\begin{array}{cc} 1 & 2 \end{array}$$

$$\begin{vmatrix} x_{AB} & x_{BD} \\ y_{AB} & y_{BD} \end{vmatrix} = 0$$

The previous condition was determined using Equation (3.45).

The system of five equations, Equations (3.27), (3.28), and (3.45), provides the solution for five unknowns: θ_1, x_{BD}, y_{BD}, x_{DC}, and y_{DC} (Figure 3.8).

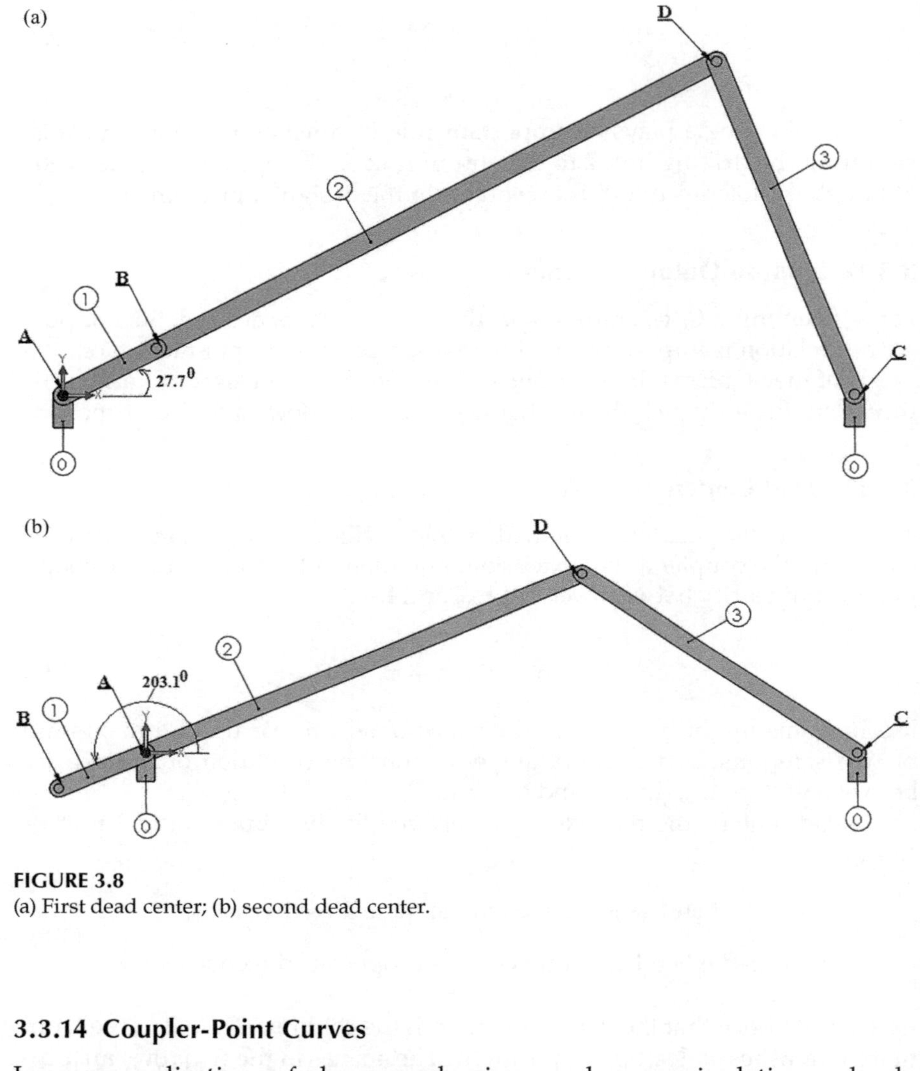

FIGURE 3.8
(a) First dead center; (b) second dead center.

3.3.14 Coupler-Point Curves

In many applications of planar mechanisms such as manipulation and palletizing, an extension is designed and attached to link 2 (coupler) in order to reach the coordinates of a specific point P of interest:

$$\mathbf{r}_P = \begin{Bmatrix} x_P \\ y_P \end{Bmatrix}$$

In Figure 3.9 (Animation 3.4), a point of interest P attached to link 2 is defined by the two variables L_{BP} and Φ. The angle Φ is defined as the angle between the positive directions of vectors $\mathbf{L}_{BD}$, and $\mathbf{L}_{BP}$, attached to link 2.

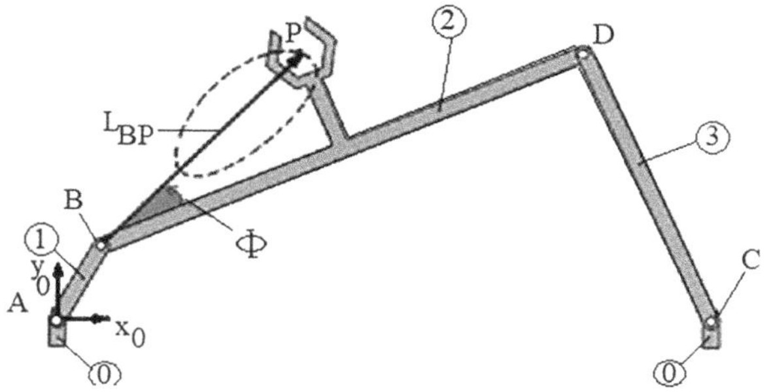

FIGURE 3.9
Coupler-point P and its path (coupler-point curve).

The *isometric* type III Equation (3.47) is used to define the rotation $\mathbf{L}_{BD}$ toward $\mathbf{L}_{BP}$, and therefore the coupler-point curves are given by

$$\mathbf{r}_P = \mathbf{r}_B + \mathbf{L}_{BP} = \begin{Bmatrix} x_B \\ y_B \end{Bmatrix} + \frac{L_{BP}}{L_{BD}} \cdot \begin{bmatrix} \text{Cos}\,\Phi & -\text{Sin}\,\Phi \\ \text{Sin}\,\Phi & \text{Cos}\,\Phi \end{bmatrix} * \begin{Bmatrix} x_{BD} \\ y_{BD} \end{Bmatrix}$$

$$= \begin{Bmatrix} x_B + \dfrac{L_{BP}}{L_{BD}} \cdot \text{Cos}\,\Phi \cdot x_{BD} - \dfrac{L_{BP}}{L_{BD}} \cdot \text{Sin}\,\Phi \cdot y_{BD} \\ y_B + \dfrac{L_{BP}}{L_{BD}} \cdot \text{Sin}\,\Phi \cdot x_{BD} + \dfrac{L_{BP}}{L_{BD}} \cdot \text{Cos}\,\Phi \cdot y_{BD} \end{Bmatrix} \tag{3.47}$$

3.3.15 Mechanism Branches

The nonlinear system of four equations, Equations (3.27) and (3.28), converges for two sets of solutions for unknowns: x_{BD}, y_{BD}, x_{DC}, and y_{DC} for each crank angle θ_1. The two configurations are named the mechanism *branches*. The mechanism does not change from one branch to another one during the motion, unless disassembled from the first and assembled back in the second branch (Figure 3.10).

3.3.16 Grashof's Criterion for the Four-Bar Mechanisms and Mechanism Inversions

The link lengths and their connection to the ground could generate different types of four-bar mechanisms. Grashof's criterion defines the type of mechanism and when one of its links is allowed to undergo a full rotation. A *crank* is considered as the actuated link, in full revolution. The link with the desired output motion is labeled as *follower*. The *coupler* link connects the crank and the follower. The connection between the fixed revolute joints on the ground is the *base* link.

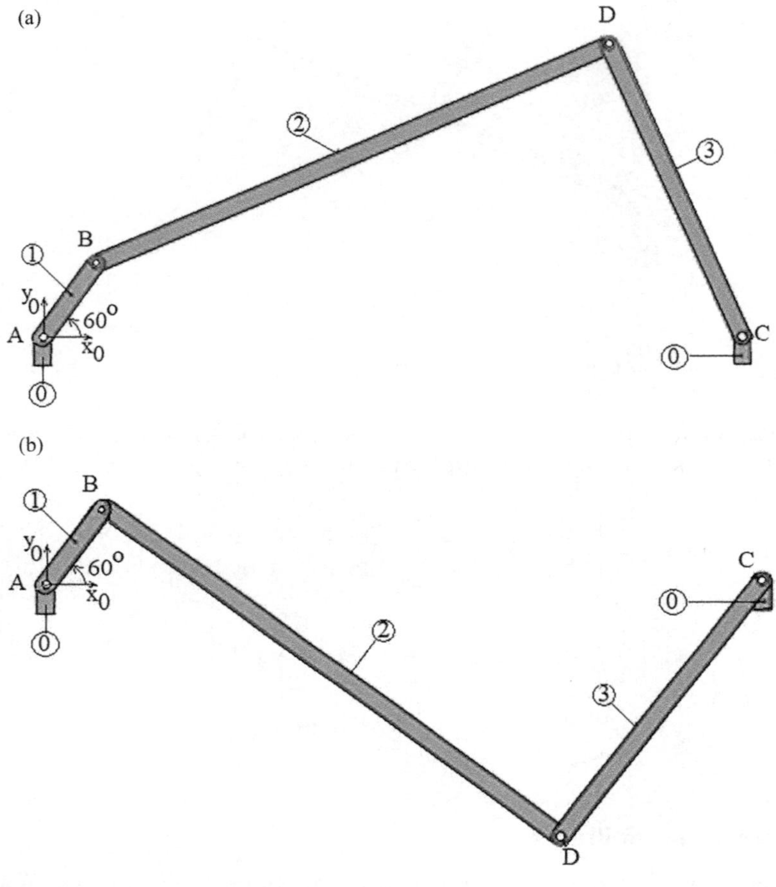

FIGURE 3.10
(a) The first branch; (b) the second branch.

Given a set of four link lengths L_{AB}, L_{BC}, L_{CD}, and L_{DA}, the following labeling is performed before using the Grashof's criterion (Equation (3.48)) [4]:

S = the shortest link length
L = the largest link length
P, Q = the intermediate link lengths

Grashof's criterion: The four-bar mechanism has at least one fully rotating link, if

$$S + L \leq P + Q \tag{3.48}$$

There are several types of mechanisms:

If $S+L<P+Q$ and S is located as a side, then it is a *crank-rocker* mechanism.

If $S+L<P+Q$ and S is located as a base, then it is a *double-crank* mechanism.

If $S+L<P+Q$ and S is located as a coupler link, then it is a *double-rocker* mechanism.

If $S+L=P+Q$ and S in any location, then it is a *change-point* mechanism.

In the case of $S+L>P+Q$, then it is a *triple-rocker* mechanism, there is no fully rotating link, and all three mobile links will rock. The condition for the mechanism to be assembled is

$$L \leq S+P+Q \tag{3.49}$$

which holds for each of the following type of mechanisms:

Example: Crank-rocker four-bar mechanism, Figure 3.11 (Animation 3.1).

The mechanism with given $L_{AB}=4''$, $L_{BD}=24''$, $L_{DC}=14''$, and $L_{CA}=30''$ has $S=L_{AB}=4''$ and $L=L_{CA}=30''$. Since $S+L<P+Q$ and S is located as a side, then we conclude that it is a *crank-rocker* mechanism. The shortest link 1 is the crank undergoing full rotation, whereas link 3 is the rocker (oscillating link).

Example: Double-crank four-bar mechanism, Figure 3.12 (Animation 3.5).

The mechanism with given $L_{AB}=30''$, $L_{BD}=24''$, $L_{DC}=14''$, and $L_{CA}=4''$ has $S=L_{CA}=4''$ and $L=L_{AB}=30''$. Since $S+L<P+Q$ and S is located as a base, then it is a *double-crank (named also drag link)* mechanism. Both links, 1 and 3, undergo full revolution.

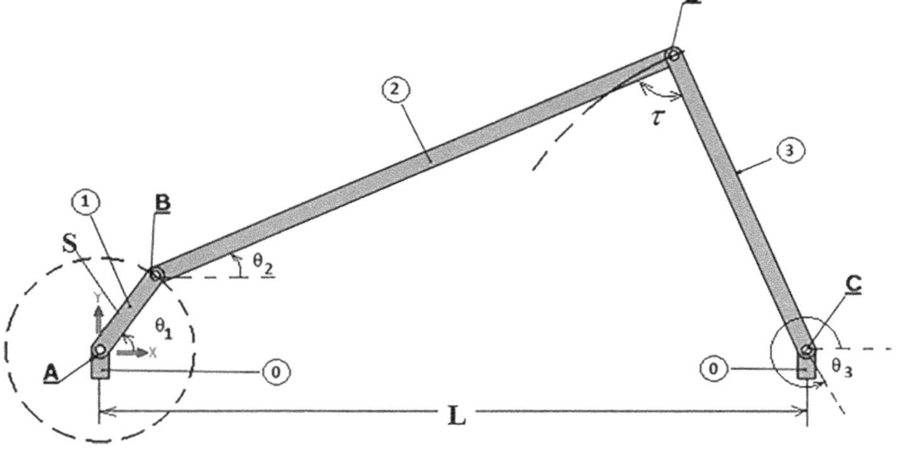

FIGURE 3.11
Crank-rocker four-bar mechanism.

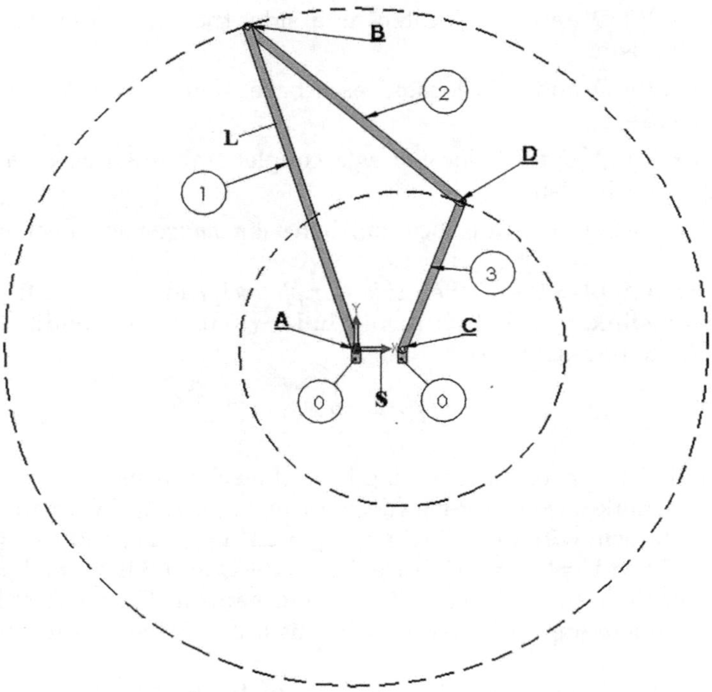

FIGURE 3.12
Double-crank four-bar mechanism.

Example: Double-rocker four-bar mechanism, Figure 3.13 (Animation 3.6).

The mechanism with given $L_{AB} = 14''$, $L_{BD} = 4''$, $L_{DC} = 30''$, and $L_{CA} = 24''$ has $S = L_{BD} = 4''$ and $L = L_{DC} = 30''$. Since $S + L < P + Q$ and S is located as a coupler link, then it is a *double-rocker* mechanism. Both links 1 and 3 are rocking (oscillating links).

Example: Change-point four-bar mechanism, Figure 3.14 (Animation 3.7).

The mechanism with given $L_{AB} = 4''$, $L_{BD} = 24''$, $L_{DC} = 10''$, and $L_{CA} = 30''$ has $S = L_{AB} = 4''$ and $L = L_{CA} = 30''$. Since $S + L < P + Q$ and S is in any location (here chosen as side), then it is a *change-point* mechanism. It is a crank-rocker mechanism but has "change points", twice per each crank rotation. At change-point configuration, the mechanism has all links collinear.

Note: The change-point position (named *singularity*) should be avoided as the start of a simulation.

Analytically, when the links are collinear, the two rows in the Latin matrix are proportional, and therefore its rank is at most $2c - 1$ (c = number of cycles).

To avoid it, for EES simulations, the start can be set with $\theta_1 = 0.001°$, instead of $\theta_1 = 0°$.

In dynamic analysis, when the forces and moments are considered, the flywheel has the role of taking away the mechanism from the singular positions.

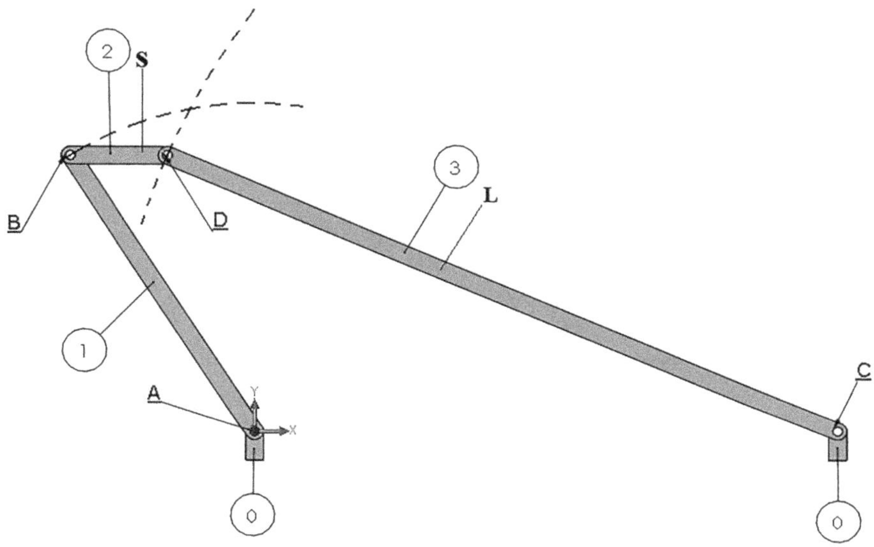

FIGURE 3.13
Double-rocker four-bar mechanism.

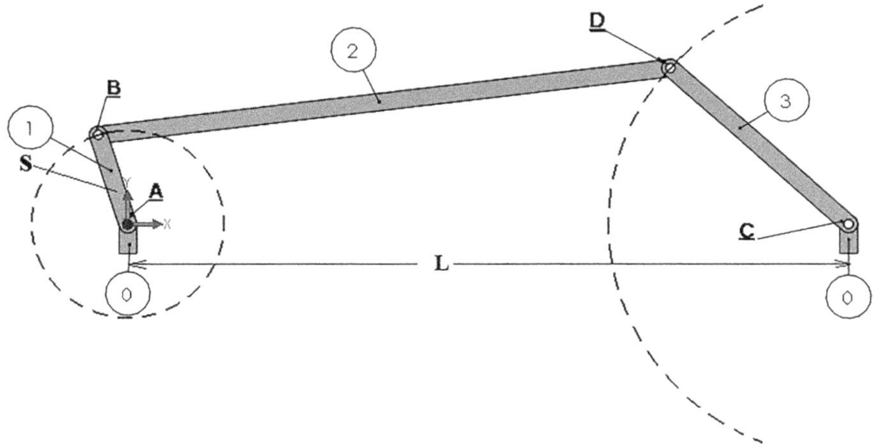

FIGURE 3.14
Change-point four-bar mechanism.

In SolidWorks (SW) simulations, the same setting can be applied in order to obtain the results.

Inversion is the process of obtaining different mechanisms (with different input–output relations) by making different links fixed. The four-bar mechanisms in Figures 3.11–3.14 were derived by inversion, considering different links as the fixed links.

- EES for planar mechanisms

  In this text, finding the solution of nonlinear systems was done with EES software. The calculated values from equations using EES are compared with those from SW motion simulations, which has the capability to simulate a mechanism (in assembly mode) for positions, velocities, accelerations, and forces. The SW simulated values have been labeled with "sim."

  As presented in Chapter 2, within the EES command Windows/Equations, the equations are typed using the keyboard, then proper formatting can be checked within Windows/Formatted equations.

  The EES command Options/Variable Info allows one to set the units for each variable. Setting the lower and upper limits for variables, illustrated in Equation (3.50) is not required, default settings can be used.

  The system of equations for positional analysis is nonlinear, and multiple solutions exist. Although not required, setting the lower/upper bounds for variables, Equation (3.50), EES will solve the multiple solutions for the mechanism's configurations. We are interested in finding the solutions corresponding to all mechanisms' link lengths L within the interval [0, infinity]:

$$\begin{aligned} -L_{BD} \le x_{BD} \le L_{BD};\ -L_{BD} \le y_{BD} \le L_{BD} \\ -L_{DC} \le x_{DC} \le L_{DC};\ -L_{DC} \le y_{DC} \le L_{DC} \end{aligned} \tag{3.50}$$

A guess value is required when solving nonlinear systems. In case operations like square root of a negative number encounter during simulation, then another guess value is chosen in Variable Info window.

Example: Positional analysis for crank-rocker mechanism (Figure 3.11)

Input data: Coordinates of joints on the fixed link 0

$$\begin{Bmatrix} x_A \\ y_A \end{Bmatrix} = \begin{Bmatrix} 0 \\ 0 \end{Bmatrix}; \begin{Bmatrix} x_C \\ y_C \end{Bmatrix} = \begin{Bmatrix} 30 \\ 0 \end{Bmatrix}$$

Vectors with constant magnitude in the Latin matrix:

$$L_{AB} = 4''; L_{BD} = 24''; L_{DC} = 14''; L_{CA} = 30''$$

According to Grashof's criterion, this mechanism is a *crank rocker*, and link 1 is the crank

- Nonlinear system of equations for positional analysis
  - Type I equations from the rows in the Latin matrix:

$$x_{CA} + x_{AB} + x_{BD} + x_{DC} = 0 \tag{3.51}$$

$$y_{CA} + y_{AB} + y_{BD} + y_{DC} = 0 \tag{3.52}$$

  - Type II equations from the columns in the Latin matrix:
    - Column 0 (link 0):

$$\mathbf{L}_{CA} = \begin{Bmatrix} x_{CA} \\ y_{CA} \end{Bmatrix} = \begin{Bmatrix} x_A - x_C \\ y_A - y_C \end{Bmatrix} = \begin{Bmatrix} -30 \\ 0 \end{Bmatrix}; \text{given coordinates}$$

    - Column 1 (link 1):

$$\mathbf{L}_{AB} = \begin{Bmatrix} x_{AB} \\ y_{AB} \end{Bmatrix} = \begin{Bmatrix} 4.0 \cdot \mathrm{Cos}(\theta_1) \\ 4.0 \cdot \mathrm{Sin}(\theta_1) \end{Bmatrix}; \text{the vector is along link 1 (the crank)}$$

    - Column 2 (link 2):

$$\mathbf{L}_{BD} = \begin{Bmatrix} x_{BD} \\ y_{BD} \end{Bmatrix} \tag{3.53}$$

$x_{BD}^2 + y_{BD}^2 = 24^2$; constant magnitude vector, type II equation

    - Column 3 (link 3):

$$\mathbf{L}_{DC} = \begin{Bmatrix} x_{DC} \\ y_{DC} \end{Bmatrix} \tag{3.54}$$

$x_{DC}^2 + y_{DC}^2 = 14^2$; constant magnitude vector, type II equation

  - The mechanism's number of equations and the number of unknowns

The nonlinear system of 4c = 4 independent equations (3.51–3.54) shown in Equation (3.55) is solved for the four independent variables: x_{BD}, y_{BD}, x_{DC}, and y_{DC}:

$$\begin{cases} -30 + 4 \cdot \mathrm{Cos}(\theta_1) + x_{BD} + x_{DC} = 0 \\ 4 \cdot \mathrm{Sin}(\theta_1) + y_{BD} + y_{DC} = 0 \\ x_{BD}^2 + y_{BD}^2 = 24^2 \\ x_{DC}^2 + y_{DC}^2 = 14^2 \end{cases} \tag{3.55}$$

Output data: Latin matrix entries for any crank angle θ_1:

$$[\mathbf{L}] = \begin{bmatrix} x_{CA} & x_{AB} & x_{BD} & x_{DC} \\ y_{CA} & y_{AB} & y_{BD} & y_{DC} \end{bmatrix} = \begin{bmatrix} -30 & 4\cdot\mathrm{Cos}(\theta_1) & x_{BD} & x_{DC} \\ 0 & 4\cdot\mathrm{Sin}(\theta_1) & y_{BD} & y_{DC} \end{bmatrix} \quad (3.56)$$

(columns 0, 1, 2, 3)

In Appendix 3.7, the values are tabulated from the command Tables/ New Parametric Table.

The mechanism has 1 DOF, for which the parameter is chosen as the crank angle. The range of values is $0° \le \theta_1 \le 360°$, with an incremental value of 15°.

- Checking the results for positional analysis:

Statement 3.7

The sum of x and y entries on each row in the Latin matrix is zero Indeed, for each row in Appendix 3.7, the Statement 3.7 holds.

- Joint position matrix, **r**

$$[\mathbf{r}] = \begin{bmatrix} x_A & x_B & x_C & x_D \\ y_A & y_B & y_C & y_D \end{bmatrix} \quad (3.57)$$

(columns A, B, C, D)

The entries in joint position matrix **r** are calculated using Equation (3.16), from the previous entries in the Latin matrix, and all joints' coordinates are tabulated in Appendix 3.8 for an incremental crank angle of 15°.

- Absolute links' orientation

Absolute angular orientation of each link is provided by the angles θ_n, (n = 0, 1, 2, 3), measured from the fixed x_0-axis, determined before in Equations (3.37a) and (3.37b):

$$[\theta_n] = \begin{bmatrix} \theta_0 & \theta_1 & \theta_2 & \theta_3 \end{bmatrix} \quad (3.58)$$

(columns 0, 1, 2, 3)

Although the *arctan* function provides an output angle $-90° < \theta_2 < 90°$, the required quadrant pointed to by each of the vectors $\mathbf{L}_{BD}$ and $\mathbf{L}_{DC}$ is found by checking the sign of the components (x_{BD}, y_{BD}) and (x_{DC}, y_{DC}) with an *IF* statement for the output angle within $0° < \theta_2 < 360°$.

In Appendix 3.9, two sets of data are tabulated: one set with the calculated values (from equations, using EES) and the second set with the simulated values (from SW simulations, labeled "sim"), for an incremental crank angle of 15°. The values for the two sets coincide.

In Figure 3.15, the EES plots are shown on the same graph with calculated and simulated values for absolute links' orientations versus time, for an incremental crank angle of 15°.

- Transmission angle: In Figure 3.16, the calculated and simulated values are shown on the same EES plot for the transmission angle tabulated in Appendix 3.9, for an incremental crank angle of 15°.
- Input–output: In Figure 3.17, the calculated and simulated values are shown on the same EES plot from Appendix 3.9 for the link 3's output angle as a function of the link 1's input angle (incremental of 15°).
- Mechanism branches: For the crank angle $\theta_1 = 60°$, Figure 3.10a,b illustrate the two branches.

From the EES command Calculate/Solve, the nonlinear system (Equation 3.55) converges for two sets of solutions, in this case, the mechanism *branches*. The two sets of numerical entries for positional analysis are labeled with "+" for the first branch and "–" for the second branch.

First branch: From the EES command Calculate/Solve, the first branch is obtained by setting the initial guess values for y_{DC} to be negative, Figure 3.10a.

Latin matrix:

$$\left[\mathbf{L}^-\right] = \begin{bmatrix} -30 & 2 & 22.145 & 5.855 \\ 0 & 3.464 & 9.253 & -12.717 \end{bmatrix} \tag{3.59}$$

Transmission angle: $\tau^- = 87.95°$

Second branch: From the EES command Calculate/Solve, the second solution, Figure 3.10b, is obtained by setting the initial guess values for y_{DC} to be positive.

Latin matrix:

$$\left[\mathbf{L}^+\right] = \begin{bmatrix} -30 & 2 & 19.222 & 8.778 \\ 0 & 3.464 & -14.370 & 10.906 \end{bmatrix} \tag{3.60}$$

Transmission angle: $\tau^+ = 92°$

- Dead centers

First dead center: From Options/Variable Info, set a positive guess value for $0 < y_{AB} < 4$

$$\text{Output: } \theta_1 = 27.7°$$

Second dead center: From Options/Variable Info, set a negative guess value for $-4 < y_{AB} < 0$

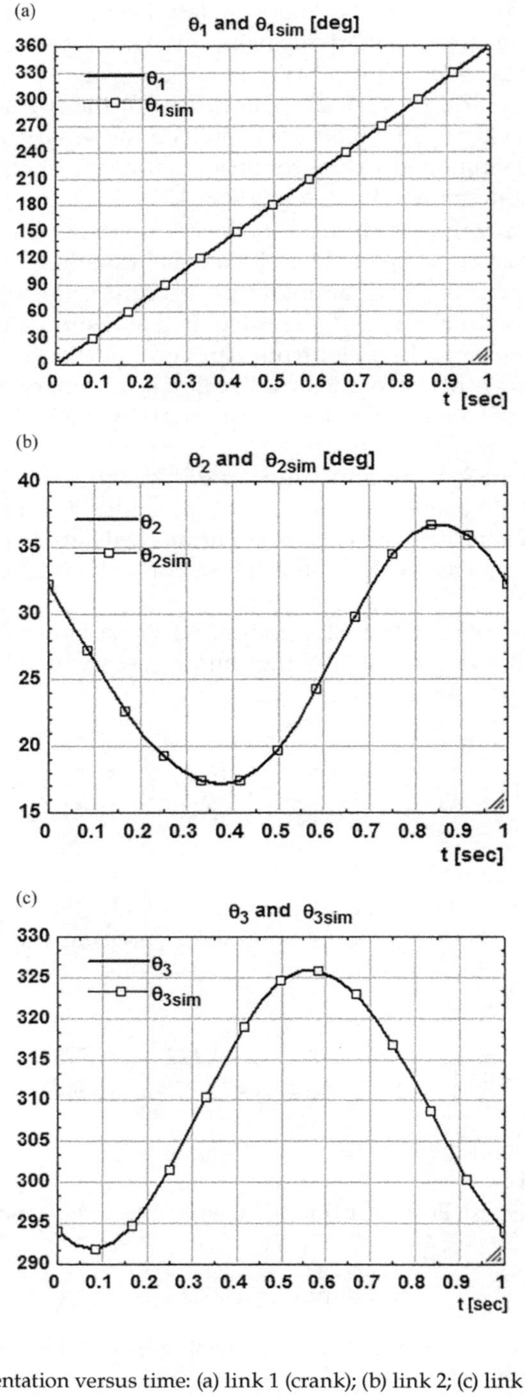

FIGURE 3.15
Absolute links' orientation versus time: (a) link 1 (crank); (b) link 2; (c) link 3.

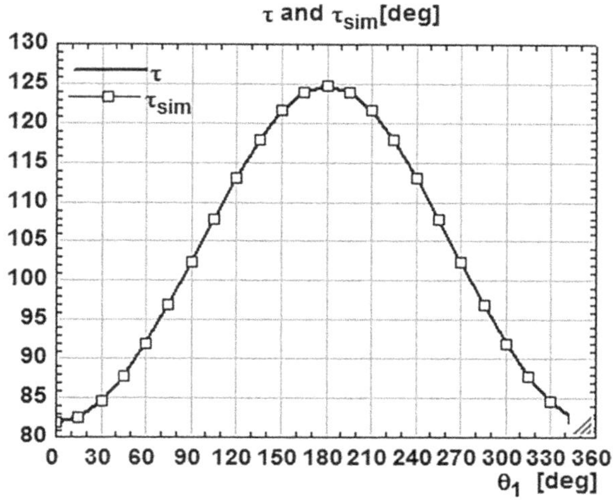

FIGURE 3.16
Transmission angle at different positions of the crank.

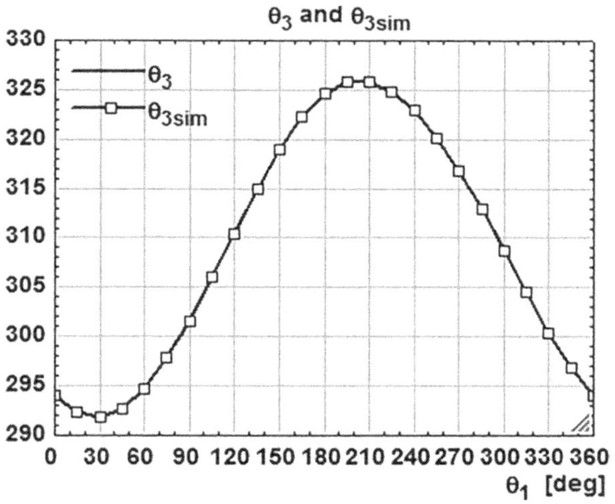

FIGURE 3.17
Input–output relation.

$$\text{Output: } \theta_1 = 203.1^\circ$$

- Coupler-point curves

  In Figure 3.18, two coupler-point curves are shown for

$$\left(L_{BP} = 13''; \Phi = 22.62^\circ\right) \text{ and } \left(L_{BP} = 25''; \Phi = 30^\circ\right)$$

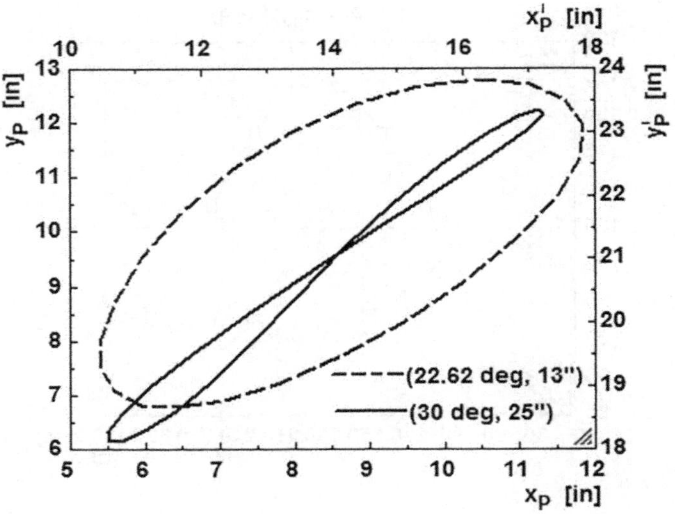

FIGURE 3.18
Coupler-point curves.

3.4 Single-Cycle Planar Mechanisms with Revolute and Prismatic Joints' Position Analysis

We first present a single-cycle mechanism with revolute and prismatic joints followed by further examples of multiple-cycle mechanisms.

3.4.1 The Planar Crank Slider Mechanism

In Figure 3.19a (Animation 3.8), the planar crank slider mechanism with three revolute joints and one prismatic joint is illustrated, where

- The number of links and joints:

  The links are labeled with numbers, starting with 0 which is assigned to the fixed link. There are three mobile links, m = 3, which have been assigned the numbers 1, 2, and 3. There are four joints labeled with capital letters: A, B, C, and D. All mechanisms' joints are 1 DOF each: revolute (description "rev" at A, B, D, and "pr" at C), therefore $j_1 = 4$. There are no joints with 2 DOF, therefore $j_2 = 0$.

  The total number of joints from Equation (1.27) in Chapter 1 is

$$j = j_1 + j_2 = 4 + 0 = 4 \text{ joints}$$

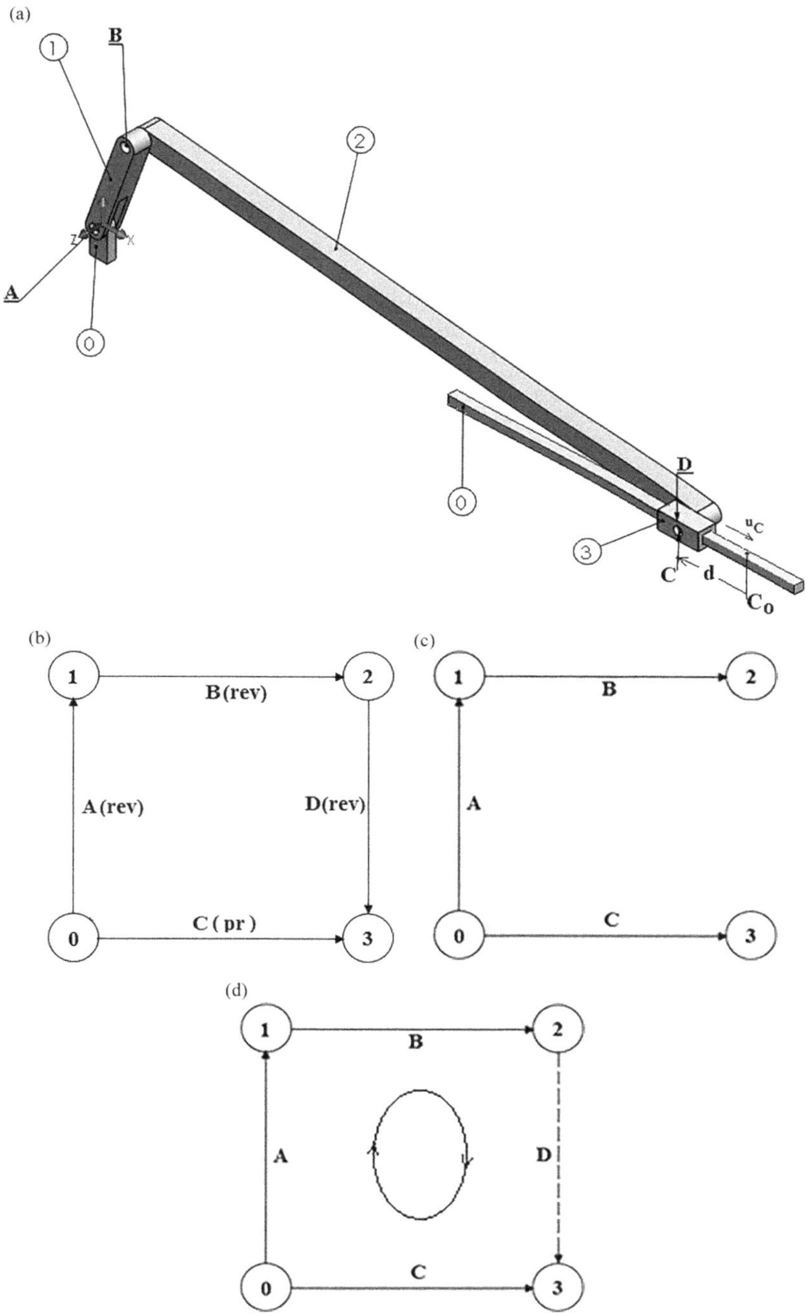

FIGURE 3.19
(a) Planar crank slider mechanism with revolute joints and one cycle; (b) digraph; (c) spanning tree; (d) cycle.

- Mechanism's digraph:
  The digraph attached to the mechanism is illustrated in Figure 3.19b, with its $n = m + 1 = 4$ nodes corresponding to the mechanism's links and the edges to the mechanism's $j = 4$ joints.
  The digraph's edges are oriented from the lower number node to the higher number node.
  One could notice that the digraph is similar to the digraph of the four-bar mechanism.
- Spanning tree:
  Figure 3.19c shows the spanning tree cutting the chord D. The chord D is labeled with the last letter.
  The digraph is similar to the four-bar mechanism's digraph.
- Digraph's number of independent cycles:
  The number of cycles is determined from Equation (1.7) in Chapter 1:

$$c = j - m = 4 - 3 = 1 \text{ cycle} \tag{3.61}$$

  Figure 3.19d shows the cycle oriented in the direction of the chord D (clockwise).
- Mechanism's mobility:
  The mobility of the planar four-bar mechanism is determined from Equation (1.28) in Chapter 1:

$$M = 3 \cdot m - 2 \cdot j_1 - j_2 = 3 \cdot 3 - 2 \cdot 4 - 0 = 1 \tag{3.62}$$

- The incidence links–joints matrix, $\underline{G}$:
  Considering its definition Equation (1.15) in Chapter 1, the 4×4 matrix $\underline{G}$ has each row assigned to a link and each column assigned to a joint. Each column has assigned a pair [–1, +1] of two numbers: –1 assigned for the lower number link and +1 assigned for the higher number link:

$$\begin{array}{c} \text{Joints} \\ \begin{array}{cccc} A & B & C & D \end{array} \end{array}$$

$$[\underline{G}] = \begin{bmatrix} -1 & 0 & -1 & 0 \\ 1 & -1 & 0 & 0 \\ 0 & 1 & 0 & -1 \\ 0 & 0 & 1 & 1 \end{bmatrix} \begin{matrix} 0 \\ 1 \\ 2 \\ 3 \end{matrix} \quad \text{Links} \tag{3.63}$$

- The cycle basis matroid matrix, **C**:
  Considering its definition Equation (1.25) in Chapter 1, the (1×4) cycle matroid matrix **C** has one row assigned to one cycle and each

column assigned to a joint. A − 1 is assigned to a joint which is contrary to the cycle's direction and 1 if the joint is in the cycle's direction.

$$\begin{matrix} & A & B & C & D \\ [C]= & [\,1 & 1 & -1 & 1\,] \end{matrix} \tag{3.64}$$

The vector *components for the Latin matrix* are shown in Equation (3.65) (Figure 3.20):

$$\begin{matrix} & 0 & 1 & 2 & 3 \\ [\mathbf{L}]= & [\,\mathbf{L}_{CA} & \mathbf{L}_{AB} & \mathbf{L}_{BD} & \mathbf{L}_{DC}\,] \end{matrix} \tag{3.65}$$

and their scalar components are shown in Equation (3.66) (Figure 3.20):

$$[L]=\begin{bmatrix} x_{CA} & x_{AB} & x_{BD} & x_{DC} \\ y_{CA} & y_{AB} & y_{BD} & y_{DC} \end{bmatrix} \tag{3.66}$$

Input data:

- Coordinates of joints on the fixed link 0:

$$\begin{Bmatrix} x_A \\ y_A \end{Bmatrix}=\begin{Bmatrix} 0 \\ 0 \end{Bmatrix} \tag{3.67}$$

- Offset distance: $y_{C_0} = 1''$
- Vectors with constant magnitude:

$$L_{AB} = 4''; L_{BD} = 24''; L_{DC} = 0''$$

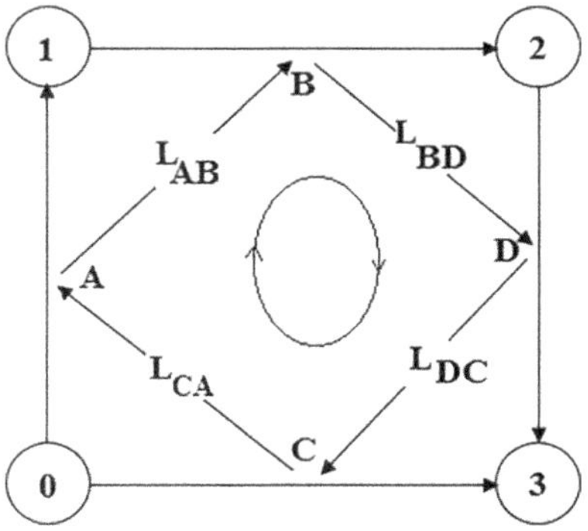

FIGURE 3.20
The Latin matrix vectors around the cycle.

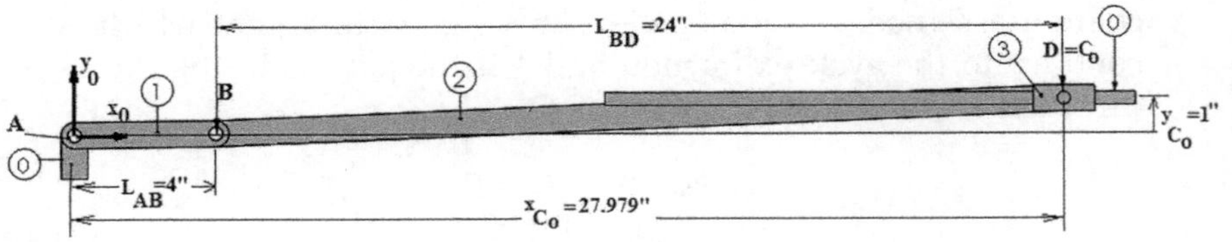

FIGURE 3.21
Determination of the home position.

- Home position, C_0

  The crank at position $\theta_1 = 0°$ brings the slider at "home position," C_0. The x-coordinate of point C_0 is determined geometrically (Figure 3.21):

$$x_{C_0} = L_{AB} + \sqrt{L_{BD}^2 - y_{C_0}^2} = 27.979''$$

$$\begin{Bmatrix} x_{C_0} \\ y_{C_0} \end{Bmatrix} = \begin{Bmatrix} 27.979 \\ 1 \end{Bmatrix} \tag{3.68}$$

  Note: In the case of curvilinear sliding, the home position C_0 is chosen as the center of curvature.

- Nonlinear system of equations for positional analysis
  - Type I equations from the rows in the Latin matrix:

$$x_{CA} + x_{AB} + x_{BD} + x_{DC} = 0$$

$$y_{CA} + y_{AB} + y_{BD} + y_{DC} = 0$$

  Column matrix 0 (node digraph 0): Located on the first column is the vector $\mathbf{L}_{CA}$ with a variable magnitude

$$\mathbf{L}_{CA} = \begin{Bmatrix} x_{CA} \\ y_{CA} \end{Bmatrix}$$

  - *Type IV equation* (the slider is constraint to be displaced along the slide)

    Around node 0 in the digraph, Figure 3.22, vector $\mathbf{L}_{CA}$ is resolved in two vectors:

$$\mathbf{L}_{CA} = \mathbf{L}_{CC_0} + \mathbf{L}_{C_0A} \tag{3.69}$$

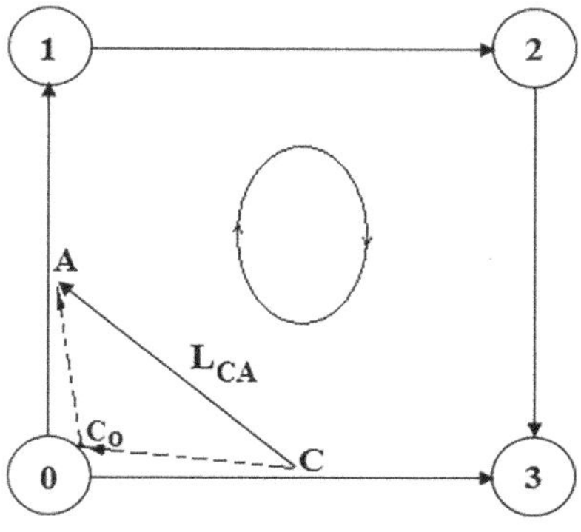

FIGURE 3.22
Illustration of a constraint type IV equation.

– One vector along the sliding direction:

$$\mathbf{L}_{CC_0} = d_C \cdot \mathbf{u}_C$$

where

$\mathbf{u}_C$ = vector's $\mathbf{L}_{CC_0}$ unit vector, with the fixed direction $\alpha_C = 0°$

$$u_C = \begin{Bmatrix} u_C^x \\ u_C^y \end{Bmatrix} = \begin{Bmatrix} \text{Cos}(\alpha_C) \\ \text{Sin}(\alpha_C) \end{Bmatrix} = \begin{Bmatrix} 1 \\ 0 \end{Bmatrix}$$

and

d_C = displacement of point C on link 3 relative to point C_0 on link 0 (home position)

If: $d_C = 0$, the slider is at home position; $d_C > 0$, the slider is moving in the direction of the unit vector $\mathbf{u}_C$; $d_C < 0$, the slider is moving in the opposite direction of the unit vector $\mathbf{u}_C$.

– One vector locating the slider's home position:

$$\mathbf{L}_{C_0A} = \begin{Bmatrix} x_{C_0A} \\ y_{C_0A} \end{Bmatrix} = \begin{Bmatrix} x_A - x_{C_0} \\ y_A - y_{C_0} \end{Bmatrix} = \begin{Bmatrix} -27.979 \\ -1 \end{Bmatrix}$$

Therefore, the vector's $\mathbf{L}_{CA}$ scalar components are

$$\begin{Bmatrix} x_{CA} \\ y_{CA} \end{Bmatrix} = \begin{Bmatrix} \text{Cos}(\alpha_C) \\ \text{Sin}(\alpha_C) \end{Bmatrix} \cdot d_C \cdot + \begin{Bmatrix} x_A - x_{C_0} \\ y_A - y_{C_0} \end{Bmatrix} \tag{3.70}$$

- Number of equations and number of unknowns for positional analysis:

  The nonlinear system of 4c = 4 equations: *Type I* (Equation 3.71), *Type II* (Equation 3.72), which have been inserted the constraint *Type IV* (Equation 3.73), is solved for the 4c = 4 variables (x_{BD}, y_{BD}, d_C, and L_{CA}).

  Type I: 2c = 2 equations

$$\begin{cases} x_{CA} + x_{AB} + x_{BD} + x_{DC} = 0 \\ y_{CA} + y_{AB} + y_{BD} + y_{DC} = 0 \end{cases} \tag{3.71}$$

Type II: 2c = 2 equations

$$\begin{aligned} x_{BD}^2 + y_{BD}^2 &= L_{BD}^2 \\ x_{CA}^2 + y_{CA}^2 &= L_{CA}^2 \end{aligned} \tag{3.72}$$

Type IV equation

$$\begin{Bmatrix} x_{CA} \\ y_{CA} \end{Bmatrix} = \begin{Bmatrix} d_C - 27.979 \\ -1 \end{Bmatrix} \tag{3.73}$$

The nonlinear system, Equation (3.74), provides the solution for mechanism's positional analysis:

$$\begin{cases} (d_C - 27.979) + 4 \cdot \mathrm{Cos}(\theta_1) + x_{BD} = 0 \\ -1 + 4 \cdot \mathrm{Sin}(\theta_1) + y_{BD} = 0 \\ x_{BD}^2 + y_{BD}^2 = 24^2 \\ (d_C - 27.979)^2 + (-1)^2 = L_{CA}^2 \end{cases} \tag{3.74}$$

For any mechanism, the entries in matrix **L**, Equation (3.75), are stored for further velocity, acceleration, and dynamic analyses.

Output data: Latin matrix entries as a function of the variable crank angle θ_1

$$[\mathbf{L}] = \begin{bmatrix} \overset{0}{x_{CA}} & \overset{1}{x_{AB}} & \overset{2}{x_{BD}} & \overset{3}{x_{DC}} \\ y_{CA} & y_{AB} & y_{BD} & y_{DC} \end{bmatrix} = \begin{bmatrix} (d_C - 27.979) & 4 \cdot \mathrm{Cos}(\theta_1) & x_{BD} & 0 \\ -1 & 4 \cdot \mathrm{Sin}(\theta_1) & y_{BD} & 0 \end{bmatrix} \tag{3.75}$$

In Appendix 3.10, the EES calculated values are tabulated for the Latin matrix components for an incremental crank angle of 15°.

- Checking the results for positional analysis:
  Indeed, for each row in Appendix 3.10, the Statement 3.7 holds, that is, the sum of x and y entries on each row in the Latin matrix is zero.
- Joint position matrix, **r**:

$$[\mathbf{r}] = \begin{array}{c} \\ \left[\begin{array}{cccc} A & B & C & D \\ x_A & x_B & x_C & x_D \\ y_A & y_B & y_C & y_D \end{array} \right] \end{array} \tag{3.76}$$

The entries in the joint position matrix **r** are calculated using Equation (2.16) from the previous entries in the Latin matrix, as illustrated in Appendix 3.11.

- *Joint A*: (data)
  Since A is the origin of the coordinate system, then $\mathbf{r}_A = 0$:

$$\mathbf{r}_A = \begin{Bmatrix} x_A \\ y_A \end{Bmatrix} = \begin{Bmatrix} 0 \\ 0 \end{Bmatrix}$$

- *Joint B*:

$$\mathbf{r}_B = \mathbf{r}_A + \mathbf{L}_{AB} = \begin{Bmatrix} 0 \\ 0 \end{Bmatrix} + \begin{Bmatrix} x_{AB} \\ y_{AB} \end{Bmatrix} = \begin{Bmatrix} x_{AB} \\ y_{AB} \end{Bmatrix} \tag{3.77}$$

- *Joint D*:

$$\mathbf{r}_D = \mathbf{r}_B + \mathbf{L}_{BD} = \begin{Bmatrix} x_{AB} + x_{BD} \\ y_{AB} + y_{BD} \end{Bmatrix}$$

- *Joint C*:

$$\mathbf{r}_C = \mathbf{r}_D + 0 = \begin{Bmatrix} x_{AB} + x_{BD} \\ y_{AB} + y_{BD} \end{Bmatrix}; \text{ since C and D are geometrically coincident}$$

The calculated (EES) and simulated (SW) values for the x-axis displacement of prismatic joint C are shown in Figure 3.23.

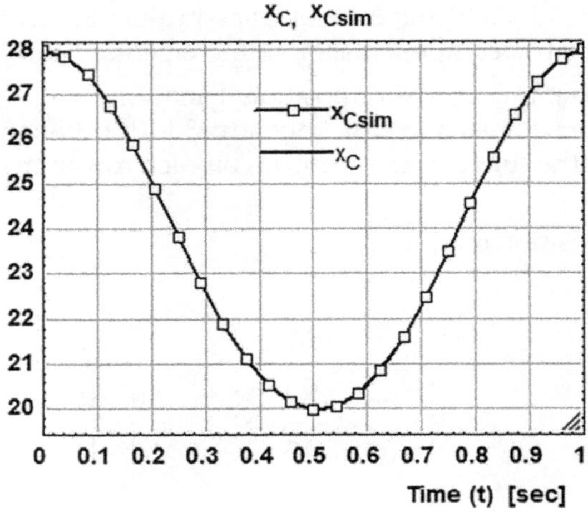

FIGURE 3.23
The x-axis displacement of prismatic joint C.

- Absolute links' orientation:
  Absolute angular orientation of each link is provided by the angles θ_n, (n = 0, 1, 2, 3), measured from the fixed x_0-axis, determined before in Equation (2.37):

$$\begin{matrix} & 0 & 1 & 2 & 3 \\ [\theta_n] = [& \theta_0 & \theta_1 & \theta_2 & \theta_3] \end{matrix} \tag{3.78}$$

– *Link 0*:

$$\theta_0 = 0 \text{ (fixed link)} \tag{3.79}$$

– *Link 1*:

$\theta_1(t)$; the crank angle variable in time is the input for the mechanism motion (3.80)

– *Link 2*: Determined from the Latin matrix with trigonometric function "tan." A test for x and y sign displays the angle in the right quadrant:

$$\theta_2 = \arctan\left(\frac{y_{BD}}{x_{BD}}\right) \tag{3.81}$$

Although the *arctan* function provides an output angle $-90° < \theta_2 < 90°$, the required quadrant pointed by the vector

L_{BD} is found by checking the sign of the components (x_{BD}, y_{BD}) using an *IF* statement.

- *Link* 3: Determined from the Latin matrix with the trigonometric function "tan." A test for x and y sign displays the angle in the right quadrant:

$$\theta_3 = 0 \text{ (constraint equation for link 3 in linear motion, it does not rotate)} \tag{3.82}$$

In Appendix 3.12, two sets of data are tabulated: one set with the calculated (from equations, using EES) and the second set simulated (from SW simulations, labeled "sim"), for an incremental crank angle of 15°. The values for the two sets coincide. Figure 3.24a–c shows two superimposed data sets, calculated EES and SW simulated data for mobile links 1, 2, and 3.

- Digraph relative links orientation, $\theta_{c,j}$

The relative links' orientation matrix is defined in Equation (3.83) as one row matrix, with each column assigned to a joint:

$$\begin{matrix} & & \text{A} & \text{B} & \text{C} & \text{D} \\ \left[\theta_{c,j}\right] & = & [\ \theta_A & \theta_B & -\theta_C & \theta_D\ ] \end{matrix} \tag{3.83}$$

The relative angles between two joined links in matrix θ_j are related to the absolute angles θ_n, as shown in Equation (3.84).

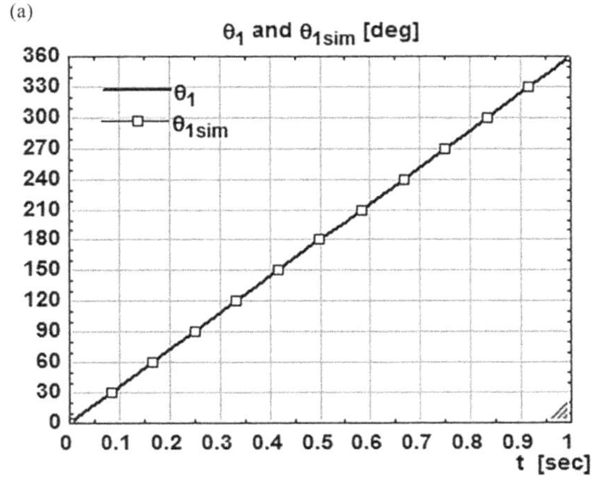

FIGURE 3.24
Absolute links' orientation versus time for the planar crank slider mechanism: (a) link 1; (b) link 2; (c) link 3.

(Continued)

(b)

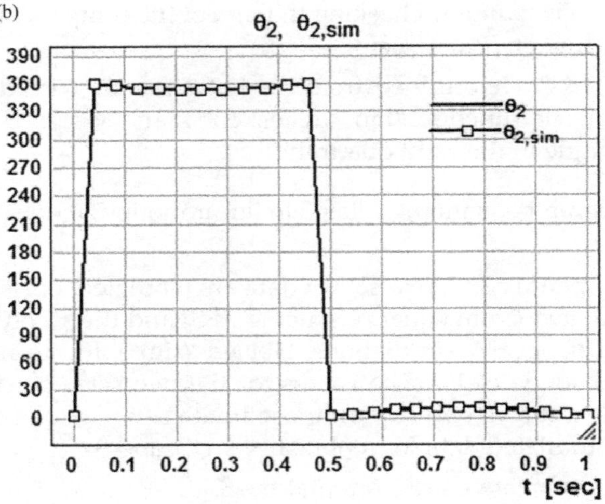

(c)

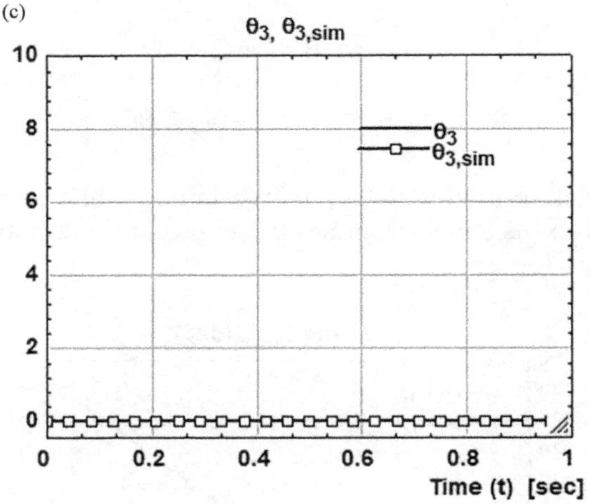

FIGURE 3.24 (CONTINUED)
Absolute links' orientation versus time for the planar crank slider mechanism: (a) link 1; (b) link 2; (c) link 3.

$$\begin{aligned}
&\theta_0 = 0 \text{ (fixed link)} \\
&\theta_{A_{01}} = \theta_1 - \theta_0 = \theta_1 \\
&\theta_{B_{12}} = \theta_2 - \theta_1 \\
&\theta_{C_{03}} = 0 \text{ (there is not relative rotation between slide 3 and slide 0} \\
&\theta_{D_{23}} = \theta_3 - \theta_2 = -\theta_2
\end{aligned} \tag{3.84}$$

- Transmission angle:

  For a crank slider mechanism, the τ transmission angle was defined by Alt as the angle between the centerlines of links 2 and 3, which is given in equation (3.85):

$$\tau = \text{abs}(\theta_D - 90°) \tag{3.85}$$

Since the linear displacement is horizontal, the transmission angle, also, is defined as the angle between the vertical plane and link 2, and can be determined based on the Latin matrix vector $\mathbf{L}_{DC}$ (Figure 3.25):

$$\tau = \left(\text{abs}\,90° - \text{arcos}\left(x_{BD}/L_{BD}\right)\right) \tag{3.86}$$

The transmission angle is a variable angle, which describes how efficient the force and torque are transmitted from the driving link 2 to output link 3 (pumps, compressors), or from the driving link 3 to output link 1 (internal combustion engines). The optimum value is 90°.

In Appendix 3.13, the two sets of data are tabulated: one set with the calculated transmission angles (from equations, using EES) and the second set simulated (from SW simulations, labeled "sim"), for an incremental crank angle of 15°. The values for the two sets coincide.

In Figure 3.26, the calculated and simulated values are shown on the same EES plot for the transmission angle, for an incremental crank angle of 15°, based on the values listed in Appendix 3.13.

- Input–output:

  For a given input $\theta_1(t)$, crank angle, the output d_C, displacement of link 3 (piston), is desired.

  Inversely, if given the input d_C, displacement, then the output $\theta_1(t)$, crank angle, is desired.

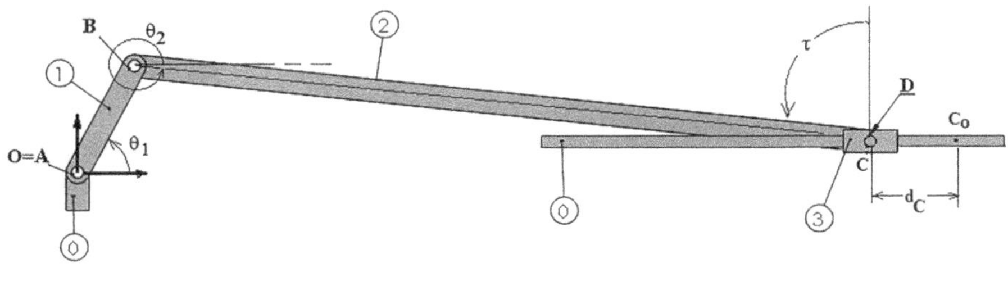

FIGURE 3.25
Transmission angle.

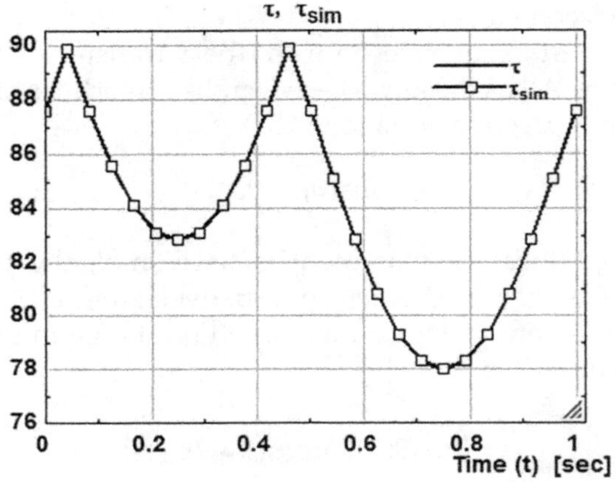

FIGURE 3.26
Transmission angle at different positions of the crank.

Such input–output relation is important for the control of output motion or synthesis of mechanisms.

In Figure 3.27, the calculated and simulated values are shown on the same EES plot for the link 3's output angle as a function of the link 1's input angle (incremental angle of 15°) (Appendix 3.13).

In the case of one-cylinder internal combustion engines, the basic mechanism is a crank slider.

The terminology in this case is as follows: link 0, engine block's cylinder; link 1, crankshaft; link 2, connecting rod; and link 3, piston. The piston slides inside the cylinder from a minimum displacement called "top dead center" (TDC) to a maximum displacement called "bottom dead center" (BDC). The difference between the two values is called *stroke*, which is given in Equation (3.87):

$$s = d_C^{max} - d_C^{min} \tag{3.87}$$

With d_C being the displacement of piston 3 relative to engine block's cylinder 0, in contact at prismatic joint C. For this mechanism, the stroke is 8″.

- Dead centers:

The dead center position is defined when the mechanism's positions have the crank 1 and connecting rod 2 in extension. Equation (3.88) represents the condition of collinearity between vectors $\mathbf{L}_{AB}$ and $\mathbf{L}_{BD}$:

$$x_{AB} \cdot y_{BD} - y_{AB} \cdot x_{BD} = 0 \tag{3.88}$$

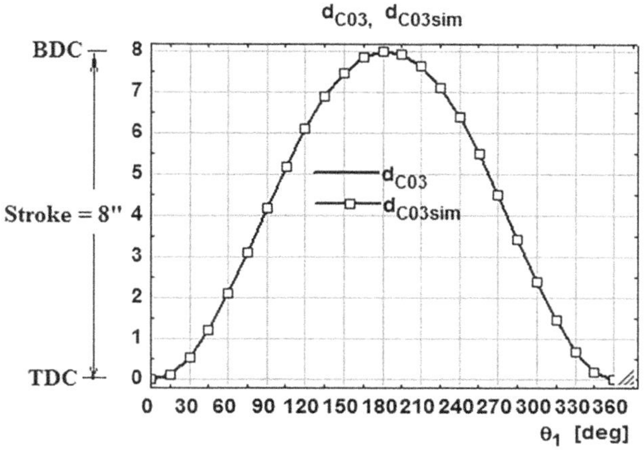

FIGURE 3.27
Input–output relation.

Mechanisms might have no dead center points, one, or up to maximum two solutions for the crank angle θ_1 from the condition of collinearity between vectors $\mathbf{L}_{AB}$ (link 1) and $\mathbf{L}_{BD}$ (link 2).

At dead centers, the relative angle between the two links 1 and 2 is 0° or 180°:

$$\begin{aligned}
&\theta_B = 0^\circ \text{ (the piston located at TDC,} \\
&\qquad \text{where } L_{AB} \text{ is oriented in the same direction as } L_{BD}) \\
&\theta_B = 180^\circ \text{ (the piston located at BDC,} \\
&\qquad \text{where } L_{AB} \text{ is oriented in the opposite direction as } L_{BD})
\end{aligned} \tag{3.89}$$

As it was noticed before, the determinant with the columns 2 and 3 in the Latin matrix vanishes at dead point positions, which is given in Equation (3.90):

$$\begin{array}{cc} 1 & 2 \end{array}$$

$$\begin{vmatrix} x_{AB} & x_{BD} \\ y_{AB} & y_{BD} \end{vmatrix} = 0 \tag{3.90}$$

The system of five equations, Equations (3.74) and (3.90), provides the solution for five unknowns: θ_1, x_{BD}, y_{BD}, d_C, and L_{CA}.

- TDC position:
  In EES menu Options/Variable Info, set a positive guess value from interval: $0 < x_{AB} < 4$ and $0 < y_{AB} < 4$:

$$\text{Output: } \theta_1 = 2.9° \tag{3.91}$$

- BDC position:
  In EES menu Options/Variable Info, set a negative guess value from interval $-4 < x_{AB} < 0$ and $-4 < y_{AB} < 0$:

$$\text{Output: } \theta_1 = 182.9° \tag{3.92}$$

3.4.2 Example: The Planar RRTR Mechanism

In Figure 3.28a (Animation 3.9), the planar RRTR mechanism with three R-revolute joints and one T-prismatic joint is illustrated. The applications for this mechanism include the following: sprinkler mechanism, spray-painting, etc. The following steps are considered for positional analysis:

(a)

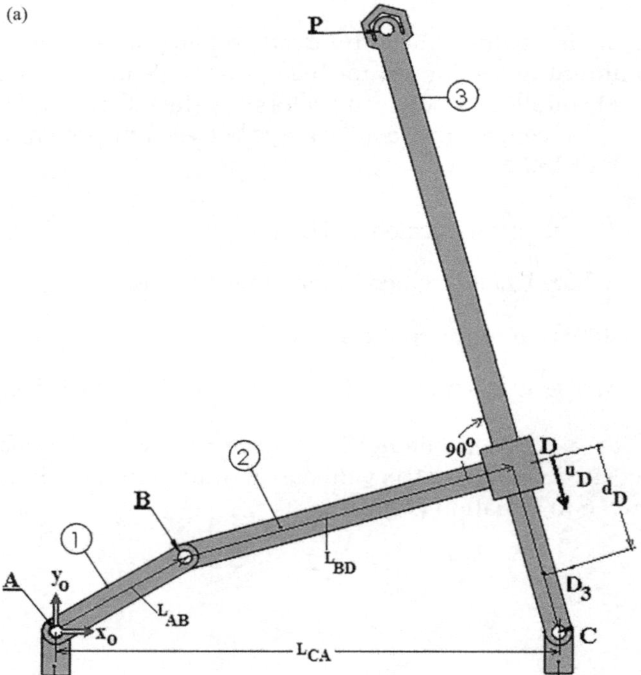

FIGURE 3.28
(a) Planar RRTR planar mechanism with revolute joints and one cycle; (b) digraph; (c) spanning tree; (d) cycle.

(Continued)

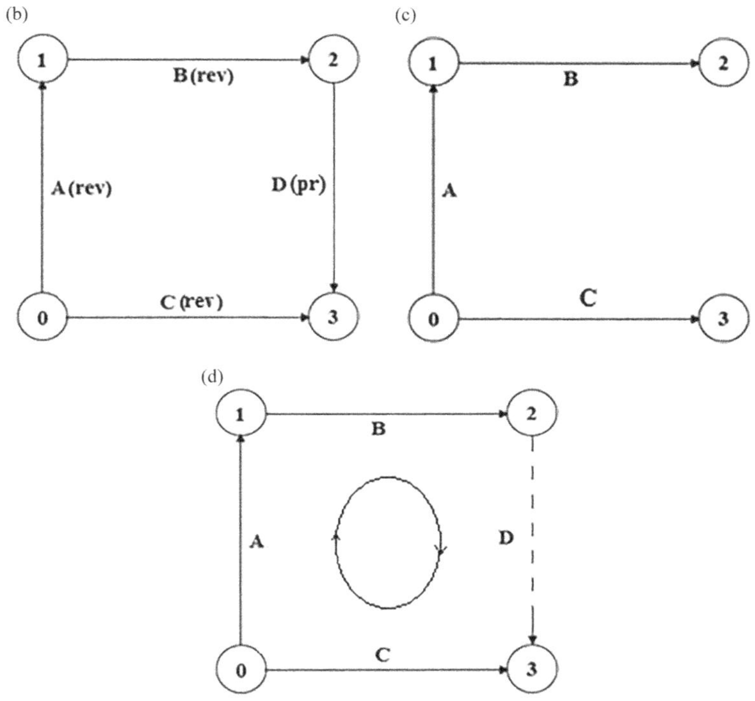

FIGURE 3.28 (CONTINUED)
(a) Planar RRTR planar mechanism with revolute joints and one cycle; (b) digraph; (c) spanning tree; (d) cycle.

- The number of links and joints:

  The links are labeled with numbers, starting with 0 which is assigned to the fixed link. There are three mobile links, m = 3, which have been assigned the numbers 1, 2, and 3. There are four joints labeled with capital letters: A, B, C, and D.

  All the mechanism's joints are 1 DOF each: revolute (description "rev" at A, B, C, and "pr" at D), therefore $j_1 = 4$. There are no joints with 2 DOF, therefore $j_2 = 0$.

  Point D_3, located on link 3, is chosen to locate the slider starting position "home position."

  The crank is at 0° when the slider is at "home." In the digraph, the reference point D_3 shows on the node 3.

- The total number of joints from Equation (1.27) is

$$j = j_1 + j_2 = 4 + 0 = 4 \text{ joints}$$

- Mechanism's digraph:

  The digraph attached to the mechanism is illustrated in Figure 3.28b, with its $n = m + 1 = 4$ nodes corresponding to the mechanism's links and the edges to the mechanism's $j = 4$ joints.

  The digraph's edges are oriented from the lower number node to the higher number node.

  One could notice that the digraph is similar to the digraph of the crank slider mechanism or four-bar mechanisms.
- Spanning tree:

  Figure 3.28c shows the spanning tree cutting the chord D. The chord D is labeled with the last letter. The digraph is similar to the four-bar and crank slider digraphs.
- Digraph's number of independent cycles:

  The number of cycles is determined using Equation (1.7) from Chapter 1:

$$c = j - m = 4 - 3 = 1 \text{ cycle}$$

  Figure 3.28d shows the cycle oriented in the direction of the chord D (clockwise).
- Mechanism's mobility:

  The mobility of the planar four-bar mechanism is determined from Equation (1.28) from Chapter 1:

$$M = 3 \cdot m - 2 \cdot j_1 - j_2 = 3 \cdot 3 - 2 \cdot 4 - 0 = 1$$

  With 1 DOF, the mechanism requires one actuated link, what is chosen here is crank 1.
- The incidence links–joints matrix, $\underline{G}$:

  Considering its definition (Equation 1.15) from Chapter 1, the 4×4 matrix $\underline{G}$ has each row assigned to a link and each column assigned to a joint. Each column has assigned a pair [–1, +1] of two numbers: –1 assigned for the lower number link and +1 assigned for the higher number link:

$$\begin{array}{c} \text{Joints} \\ \begin{array}{cccc} A & B & C & D \end{array} \end{array}$$

$$[\underline{G}] = \begin{bmatrix} -1 & 0 & -1 & 0 \\ 1 & -1 & 0 & 0 \\ 0 & 1 & 0 & -1 \\ 0 & 0 & 1 & 1 \end{bmatrix} \begin{array}{c} 0 \\ 1 \\ 2 \\ 3 \end{array} \quad \text{Links} \tag{3.93}$$

- The cycle basis matroid matrix, **C**:

  Considering its definition in Chapter 1, Equation 1.22, the 1 × 4 matrix **C** has one row assigned to one cycle and each column is assigned to a joint.

  The *direction of the cycle* is assigned in this example as the chord's D direction, clockwise.

  A –1 is assigned to a joint which is contrary to the cycle's direction and 1 if the joint is in the cycle's direction:

$$\begin{matrix} & A & B & C & D \\ [C]= & [\;1 & 1 & -1 & 1\;] \end{matrix} \tag{3.94}$$

- The Latin matrix, **L**:

  The *Latin matrix* **L**, Equation (3.95), is a 1 × 4 matrix with vector entries defined around the links in the digraph and written in the direction of the cycle. The vector $\mathbf{L}_{DC}$ is a variable magnitude vector:

$$\begin{matrix} & 0 & 1 & 2 & 3 \\ [\mathbf{L}]= & [\;\mathbf{L}_{CA} & \mathbf{L}_{AB} & \mathbf{L}_{BD} & \mathbf{L}_{DC}\;] \end{matrix} \tag{3.95}$$

  and in scalar form (Figure 3.29):

$$[L]=\begin{bmatrix} x_{CA} & x_{AB} & x_{BD} & x_{DC} \\ y_{CA} & y_{AB} & y_{BD} & y_{DC} \end{bmatrix} \tag{3.96}$$

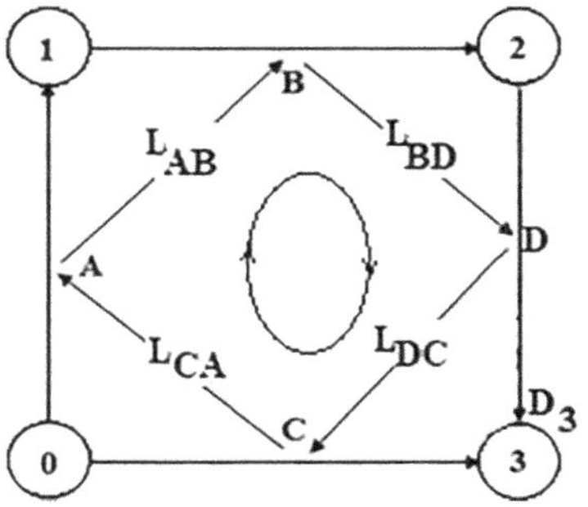

FIGURE 3.29
Latin matrix vectors for the planar RRTR mechanism.

Input data:

- Coordinates of joints A and C on the fixed link 0:

$$\begin{Bmatrix} x_A \\ y_A \end{Bmatrix} = \begin{Bmatrix} 0 \\ 0 \end{Bmatrix}; \begin{Bmatrix} x_C \\ y_C \end{Bmatrix} = \begin{Bmatrix} 14.250 \\ 0 \end{Bmatrix} \quad (3.97)$$

- Vectors with constant magnitude:

$$\begin{aligned} &L_{AB} = 4.250''; L_{BD} = 9.625''; L_{CA} = 14.250''; L_{CP} = 18.250'' \\ &L_{AG_1} = 2.421''; L_{BG_2} = 6.574''; L_{CG_3} = 9.816'' \end{aligned} \quad (3.98)$$

- Home position: The crank at position $\theta_1 = 0°$ brings the slider to the "home position," D_3. The offset of point D_3 from the fixed joint C is determined geometrically from triangle BC D_3 (Figure 3.30):

$$L_{D_3C} = \sqrt{(L_{CA} - L_{AB})^2 - L_{BD_3}^2} = 2.713'' \quad (3.99)$$

- Nonlinear system of equations for positional analysis
  - *Column 0* (node digraph 0): Given the vector's coordinates,

$$\mathbf{L}_{CA} = \begin{Bmatrix} x_{CA} \\ y_{CA} \end{Bmatrix} = \begin{Bmatrix} x_A - x_C \\ y_A - y_C \end{Bmatrix} = \begin{Bmatrix} -14.250 \\ 0 \end{Bmatrix} \quad (3.100a)$$

  - *Column 1* (node digraph 1): the vector is along link 1 (the crank):

$$\mathbf{L}_{AB} = \begin{Bmatrix} x_{AB} \\ y_{AB} \end{Bmatrix} = \begin{Bmatrix} 4.250 \cdot \mathrm{Cos}(\theta_1) \\ 4.250 \cdot \mathrm{Sin}(\theta_1) \end{Bmatrix} \quad (3.100b)$$

  - *Column 2* (node digraph 2):

$$\mathbf{L}_{BD} = \begin{Bmatrix} x_{BD} \\ y_{BD} \end{Bmatrix}$$

$x_{BD}^2 + y_{BD}^2 = 9.625^2$; constant magnitude vector, type II equation (3.100c)

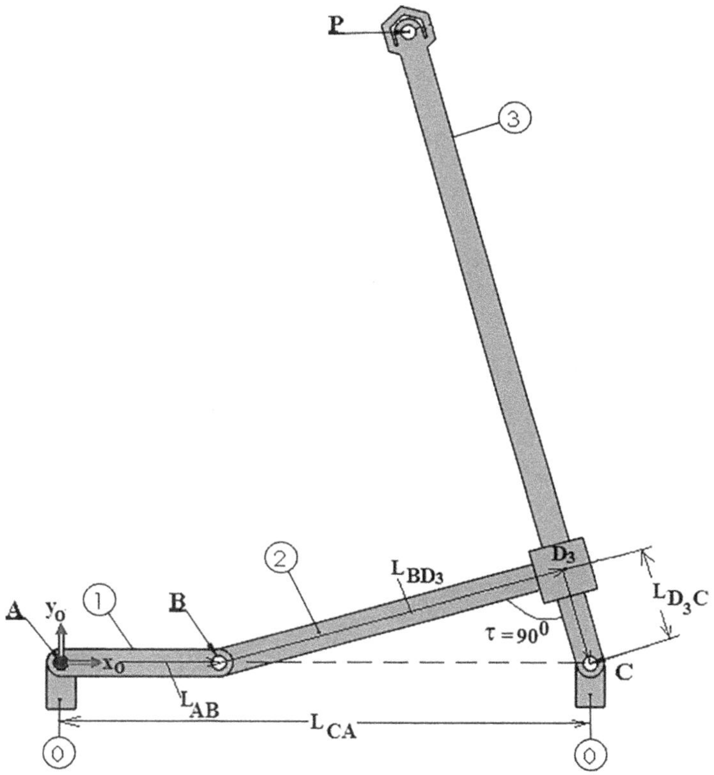

FIGURE 3.30
Determination of the home position, L_{D_3C}.

- *Column 3* (node digraph 3):

$$\mathbf{L}_{DC} = \begin{Bmatrix} x_{DC} \\ y_{DC} \end{Bmatrix}$$

- Constraint type III equation to preserve an angle $\Phi = 90^\circ$ between vectors $\mathbf{L}_{DC}$ and $\mathbf{L}_{BD}$ ($\mathbf{L}_{BD}$ rotates clockwise 90° toward $\mathbf{L}_{DC}$).

$$\begin{Bmatrix} x_{DC} \\ y_{DC} \end{Bmatrix} = \frac{L_{DC}}{L_{BD}} \cdot \begin{bmatrix} \mathrm{Cos}(-90^\circ) & -\mathrm{Sin}(-90^\circ) \\ \mathrm{Sin}(-90^\circ) & \mathrm{Cos}(-90^\circ) \end{bmatrix} * \begin{Bmatrix} x_{BD} \\ y_{BD} \end{Bmatrix} \tag{3.100d}$$

$$= \frac{L_{DC}}{L_{BD}} \cdot \begin{Bmatrix} y_{BD} \\ -x_{BD} \end{Bmatrix}$$

- Constraint type IV equation for the slider's vector displacement $\mathbf{L}_{DD_3} = d_D \cdot \mathbf{u}_D$ to be collinear with $\mathbf{u}_D$, where
  $\mathbf{u}_D$ = the unit vector of vector $\mathbf{L}_{DC}$ which has variable direction, θ_3

$$\mathbf{u}_D = \begin{Bmatrix} u_D^x \\ u_D^y \end{Bmatrix} = \begin{Bmatrix} \text{Cos}(\theta_3) \\ \text{Sin}(\theta_3) \end{Bmatrix} = \frac{1}{L_{DC}} \begin{Bmatrix} x_{DC} \\ y_{DC} \end{Bmatrix} \tag{3.100e}$$

  d_D = displacement (displacement of point D on link 3 relative to point D on link 2), measured from home position D_3

The slider is at its home position if $d_D = 0$; if $d_D > 0$, the slider is moving in the direction of the unit vector $\mathbf{u}_D$; if $d_D < 0$, the slider is moving in the opposite direction of the unit vector $\mathbf{u}_D$.

Around node 3 in the digraph, Figure 3.31, vector $\mathbf{L}_{DC}$ is the resultant of two components:

$$\begin{aligned} &\text{Vector } \mathbf{L}_{DD_3} = d_D \cdot \mathbf{u}_D, \text{ along the sliding direction} \\ &\text{Vector } \mathbf{L}_{D_3C} \text{ locating the slider's home position} \end{aligned} \tag{3.100f}$$

Thus,

$$\mathbf{L}_{DC} = \mathbf{L}_{DD_3} + \mathbf{L}_{D_3C} \tag{3.100g}$$

Considering the unit vector's $\mathbf{u}_D$ orientation, the vector $\mathbf{L}_{DD_3}$ is in the same direction as $\mathbf{u}_D$, whereas the vector's $\mathbf{L}_{D_3C}$ is in the opposite direction. Therefore, the previous equation becomes

$$L_{DC} \cdot \mathbf{u}_D = d_D \cdot \mathbf{u}_D + L_{D_3C} \cdot \mathbf{u}_D$$

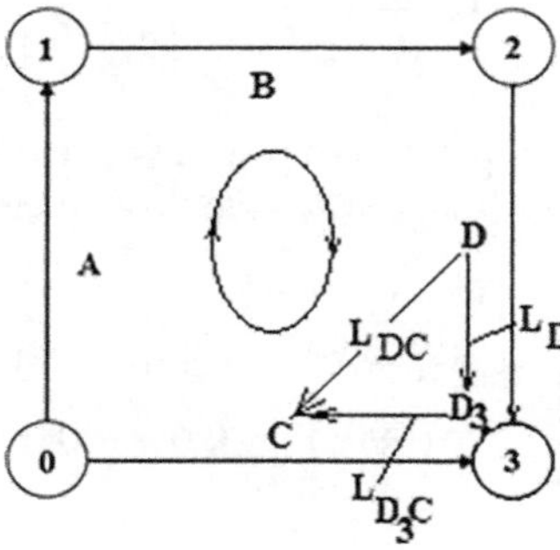

FIGURE 3.31
The constraint equation for slider displacement, d_D.

Since the unit vector has magnitude 1, $\mathbf{u}_D^2 = 1$, then the previous vector equation becomes the *type IV* constraint *equation*:

$$(d_D - L_{DC})^2 = L_{D_3C}^2 \tag{3.100h}$$

The Latin matrix with scalar components is a $(2c \times n) = (2 \times 4)$ matrix:

$$[\mathbf{L}] = \left[\begin{array}{c:c:c:c} -14.250 & 4.250 \cdot \text{Cos}(\theta_1) & x_{BD} & 0.104 \cdot L_{DC} \cdot y_{BD} \\ 0 & 4.250 \cdot \text{Sin}(\theta_1) & y_{BD} & -0.104 \cdot L_{DC} \cdot x_{BD} \end{array}\right] \tag{3.100i}$$

The system of $4c = 4$ equations:
$2c = 2$—Type I equations—the sum of entries in the Latin matrix is zero
$2c = 2$—Type II and IV equations—constraint equations (3.100c) and (3.100e)

$$\begin{cases} -14.250 + 4.250 \cdot \text{Cos}(\theta_1) + x_{BD} + 0.104 \cdot L_{DC} \cdot y_{BD} = 0 \\ 4.250 \cdot \text{Sin}(\theta_1) + y_{BD} - 0.104 \cdot L_{DC} \cdot x_{BD} = 0 \\ x_{BD}^2 + y_{BD}^2 = 9.625^2 \\ (d_D - L_{DC})^2 = L_{D_3C}^2 \end{cases} \tag{3.100j}$$

The system of four equations is solved for $4c = 4$ unknowns: d_D, L_{DC}, x_{BD}, and y_{BD}.

The mechanism has the mobility M = 1; therefore, one independent parameter is provided, the crank angle: $0° \le \theta_1(t) \le 360°$. In this example, θ_1 is chosen to increase linearly with the time t within the interval $0 \le t \le 1$ s.

In Appendix 3.14, the EES calculated values are tabulated for the Latin matrix components for an incremental crank angle of 15°.

- Joint position matrix, **r**:

$$\begin{array}{cccc} A & B & C & D \end{array}$$

$$[\mathbf{r}] = \begin{bmatrix} x_A & x_B & x_C & x_D \\ y_A & y_B & y_C & y_D \end{bmatrix} \tag{3.100k}$$

The entries in the joint position matrix **r** are calculated from the previous entries in the Latin matrix. In Appendix 3.15, the calculated data is tabulated, for an incremental crack angle of 15°.

- Absolute links' orientation:

  Absolute angular orientation of each link is provided by the angles θ_n, (n = 0, 1, 2, 3), measured from the fixed x-axis:

$$\begin{matrix} & 0 & 1 & 2 & 3 \\ [\theta_n]= & [\ \theta_0 & \theta_1 & \theta_2 & \theta_3\ ] \end{matrix} \tag{3.100l}$$

  - *Link 0*:

$$\theta_0 = 0 \text{ (fixed link)}$$

  - *Link 1*:

$$\theta_1(t)\text{; crank angle, variable in time}$$

  - *Link 2 and 3*: Determined from the Latin matrix with the trigonometric function "tan." A test for x and y sign displays the angle in the right quadrant:

$$\theta_2 = \arctan\left(\frac{y_{BD}}{x_{BD}}\right)$$
$$\theta_3 = \arctan\left(\frac{y_{DC}}{x_{DC}}\right) \tag{3.100m}$$

Figure 3.32a, 3.32b, and 3.32c shows the EES plots for the calculated and simulated values for absolute links' orientations versus time, for an incremental crank angle of 15°, as shown in Appendix (3.16).

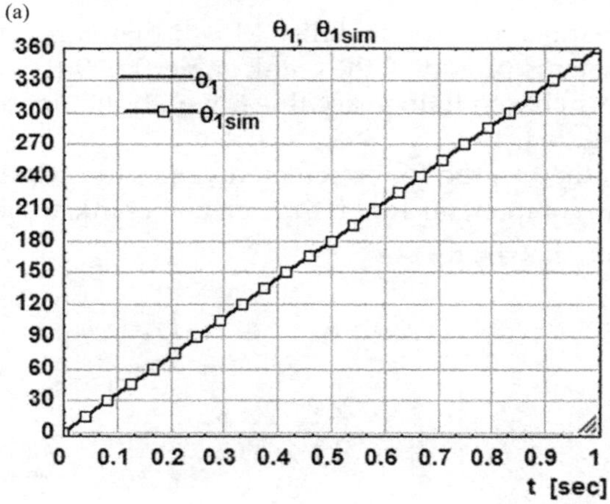

FIGURE 3.32
Absolute links' orientation versus time: (a) link 1; (b) link 2; (c) link 3.

(Continued)

(b)

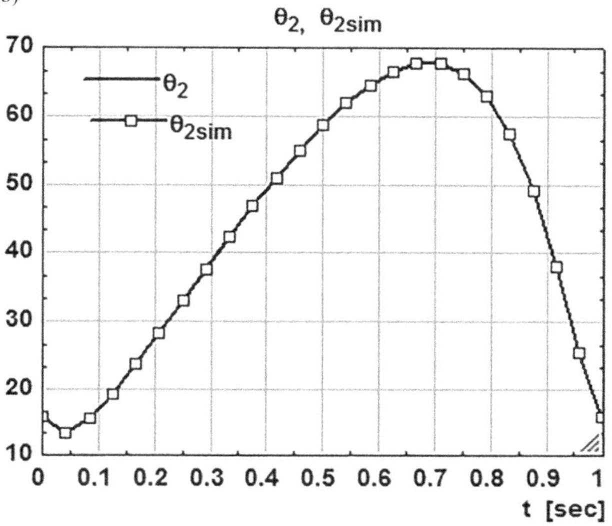

(c)

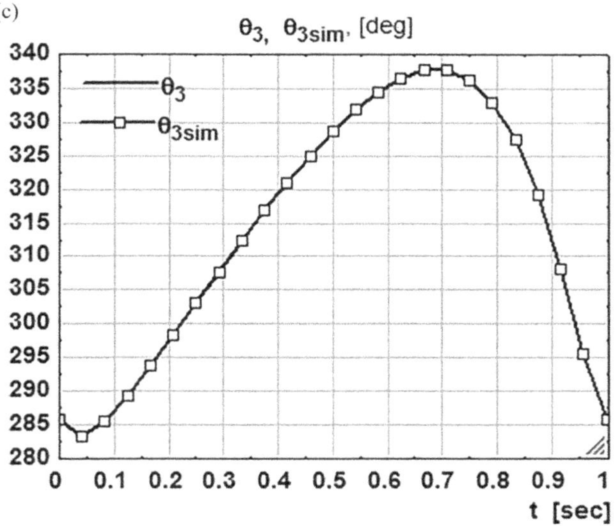

FIGURE 3.32 (CONTINUED)
Absolute links' orientation versus time: (a) link 1; (b) link 2; (c) link 3.

- Transmission angle:

  From the geometry of this mechanism with links 2 and 3 perpendicular to each other, one could notice a constant value of $\tau = 90°$ during the motion (see Appendix 3.16):

$$\tau = 180^\circ - \arccos\left(\frac{x_{BD} \cdot x_{DC} + y_{BD} \cdot y_{DC}}{L_{BD} \cdot L_{DC}}\right) = 90^\circ \tag{3.100n}$$

- Input–output relation:
  For a given input $\theta_1(t)$ crank angle, two different outputs (desired parameters) are to be considered:
  $\theta_3(t)$ = link 3 angle measured from x-axis
  d_D = displacement of point D on link 3 relative to point D on link 2, measured from home position D_3
  The above outputs are important for the control of output motion or for the synthesis of mechanisms.
  In Figure 3.33(a), the calculated and simulated values are shown on the same EES plot. These values are for absolute link 3 orientations versus input crank angle. Figure 3.33(b) shows the calculated and simulated values for slider's displacement, for an incremental crank angle of 15°.
- Dead centers
  The dead center position is defined as the mechanism's positions where the crank and the coupler are in extension.
  Equation (3.100o) represents the condition of collinearity between vectors $\mathbf{L}_{AB}$ and $\mathbf{L}_{BD}$:

$$x_{AB} \cdot y_{BD} - y_{AB} \cdot x_{BD} = 0 \tag{3.100o}$$

Mechanisms might have no dead center points, one, or up to maximum two solutions for the crank angle θ_1 from the condition of collinearity between vectors $\mathbf{L}_{AB}$ (link 1) and $\mathbf{L}_{BD}$ (link 2).

Notice that the determinant within the columns 2 and 3 in the Latin matrix vanishes at dead point positions. The entries in the two rows are proportional; therefore, the determinant is zero:

$$\begin{matrix} 1 & 2 \end{matrix}$$

$$\begin{vmatrix} x_{AB} & x_{BD} \\ y_{AB} & y_{BD} \end{vmatrix} = 0 \tag{3.100p}$$

Figure 3.34 illustrates the mechanism at the two dead center positions:

$$\begin{aligned} &\theta_1 = \theta_2 = 13.2^\circ;\ \text{the first dead center position} \\ &\theta_1 = 247.8^\circ \text{ and } \theta_2 = 67.8^\circ;\ \text{the second dead center position} \end{aligned} \tag{3.100q}$$

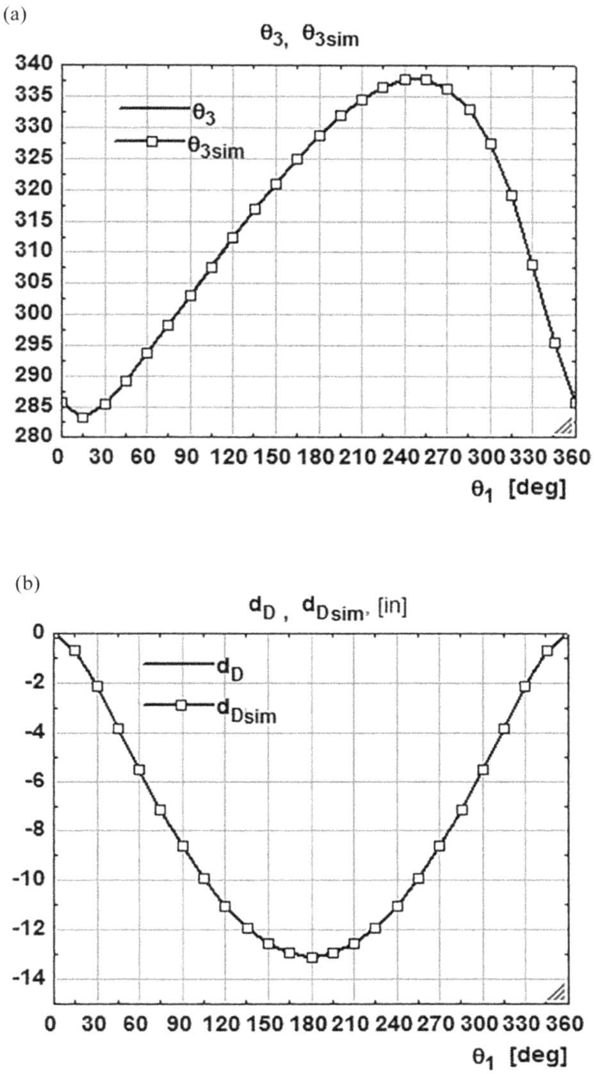

FIGURE 3.33
(a) Output $\theta_3(t)$ versus input $\theta_1(t)$; (b) output d_D versus input $\theta_1(t)$.

- Mechanism branches

  For each value of the crank angle θ_1, the nonlinear system of equations converges for two solutions (mechanism branches). From the EES command Calculate/Solve, the two sets of numerical entries for positional analysis are labeled with "–" for the first branch and "+" for the second branch.

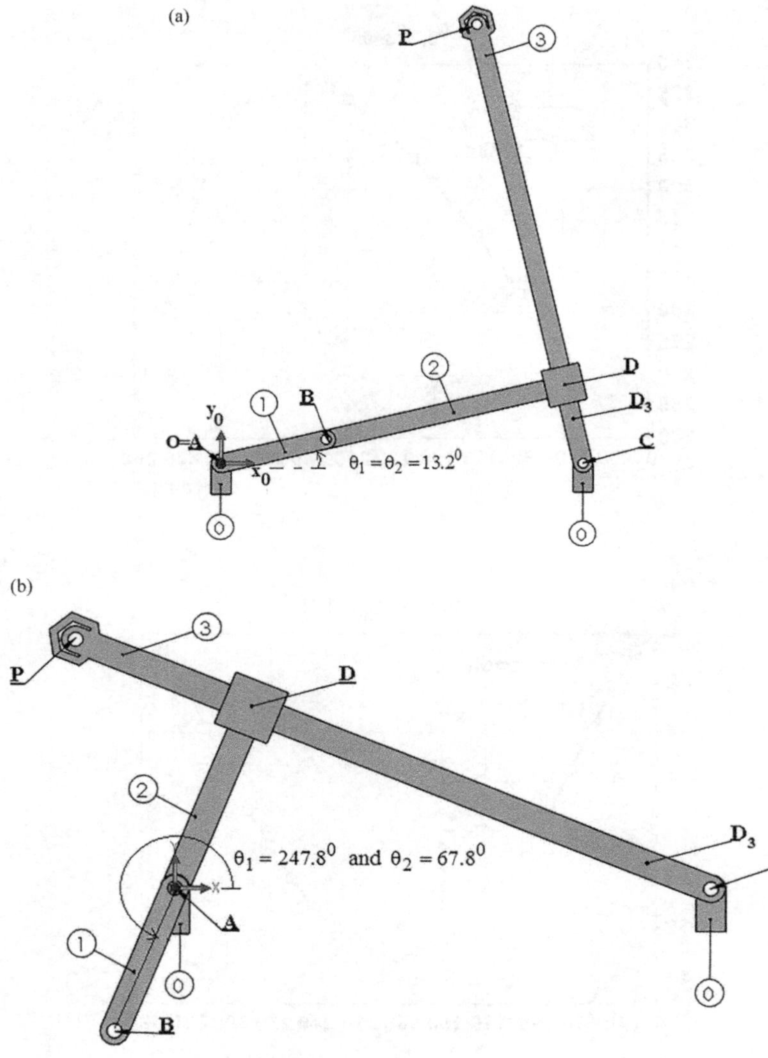

FIGURE 3.34
(a) First dead center position; (b) second dead center position.

Example: For $\theta_1 = 30°$, the two branches are

First branch: From the EES command Calculate/Solve, the first solution is obtained by setting the initial guess values for y_{DC} to be negative.

- Joint position matrix:

$$[\mathbf{r}^-] = \begin{bmatrix} x_A & x_B & x_C & x_D \\ y_A & y_B & y_C & y_D \end{bmatrix} = \begin{bmatrix} 0 & 3.681 & 14.250 & 12.960 \\ 0 & 2.125 & 0 & 4.682 \end{bmatrix} \quad (3.100r)$$

- Absolute links' orientation:

$$[\theta_n^-]=\begin{bmatrix} \theta_0 & \theta_1 & \theta_2 & \theta_3 \end{bmatrix}=\begin{bmatrix} 0° & 30° & 15.4° & 285.4° \end{bmatrix} \quad (3.100s)$$

Second branch: From the EES command Calculate/Solve, the second solution is obtained by setting the initial guess values for y_{DC} to be positive.

- Joint position matrix:

$$[\mathbf{r}^+]=\begin{bmatrix} x_A & x_B & x_C & x_D \\ y_A & y_B & y_C & y_D \end{bmatrix}=\begin{bmatrix} 0 & 3.681 & 14.250 & 11.250 \\ 0 & 2.125 & 0 & -3.820 \end{bmatrix} \quad (3.100t)$$

- Absolute links' orientation:

$$[\theta_n^+]=\begin{bmatrix} \theta_0 & \theta_1 & \theta_2 & \theta_3 \end{bmatrix}=\begin{bmatrix} 0° & 30° & 321.9° & 51.9° \end{bmatrix} \quad (3.100u)$$

- Point of interest P:

  In applications of the RRTR mechanism, the point P is attached to the nozzle of a water sprinkler or a jet painting device. In such applications, the path of interest is that of point P, as shown in Figure 3.35a. Therefore, the coordinates for the point of interest P, at each of the crank angle's values, are to be determined:

$$\mathbf{r}_P=\begin{Bmatrix} x_P \\ y_P \end{Bmatrix} \quad (3.100v)$$

The *isometric* type III Equation (3.29b) is used to determine the coordinates of P:

$$\mathbf{r}_P=\mathbf{r}_C+\mathbf{L}_{CP}=\begin{Bmatrix} x_C \\ y_C \end{Bmatrix}-\frac{L_{CP}}{L_{DC}}\cdot\begin{Bmatrix} x_{DC} \\ y_{DC} \end{Bmatrix} \quad (3.100w)$$

The path from SW is illustrated in Figure 3.35a, and the EES plot in Figure 3.35b shows the calculated and simulated sets of data.

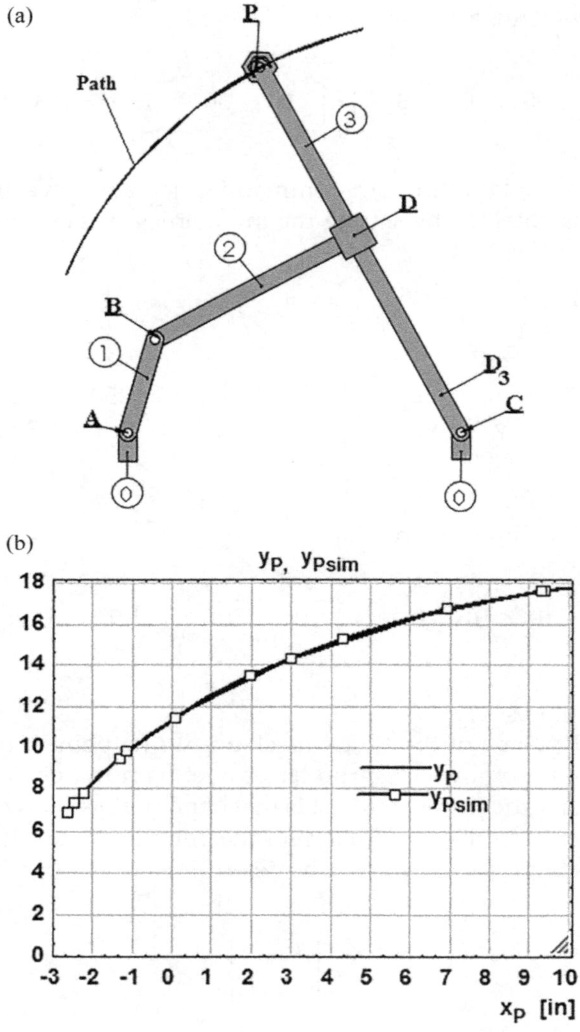

FIGURE 3.35
(a) The path of the point of interest; (b) calculated and simulated coordinates.

3.5 Multiple-Cycle Planar Mechanisms with Revolute and Prismatic Joints' Position Analysis

Example: Two-cycle planar mechanism

Figure 3.36a (Animation 3.10) illustrates the two-cycle planar mechanism with revolute and prismatic joints. The following steps are considered for the positional analysis:

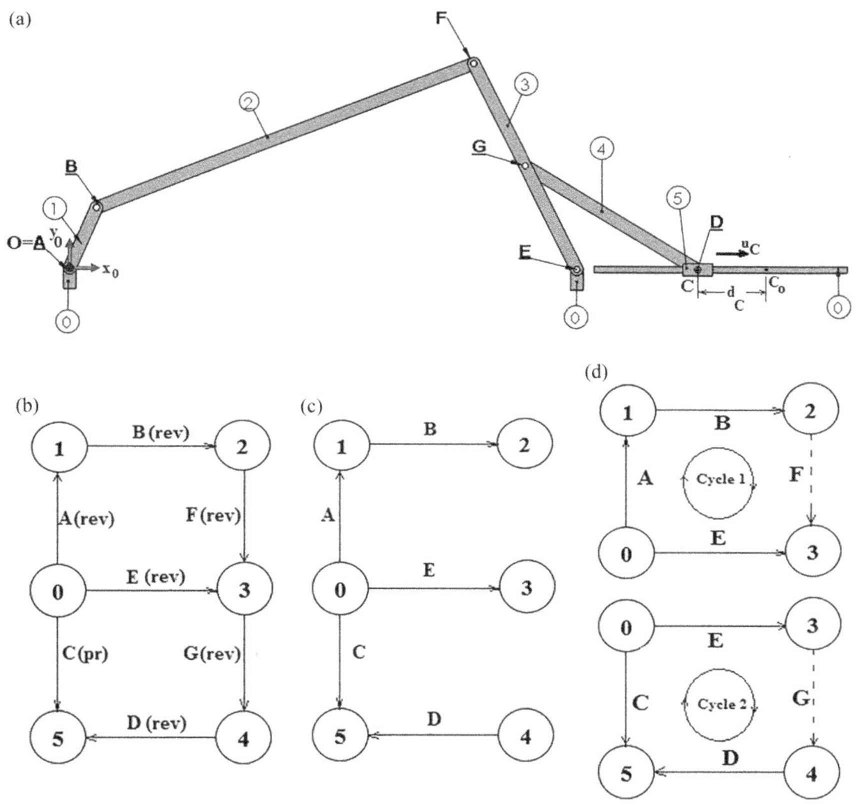

FIGURE 3.36
(a) Planar mechanism with two cycles, and revolute and prismatic joints; (b) graph; (c) tree; (d) cycles.

- The number of links and joints:

  The links are labeled with numbers, starting with 0 which is assigned to the fixed link. There are five mobile links, m = 5, which have been assigned the numbers 1, 2, 3, 4, and 5. There are seven joints labeled with capital letters: A, B, C, D, E, F, and G. All mechanisms' joints are 1 DOF each: revolute at A, B, D, E, F, and G, and one prismatic at C, therefore $j_1 = 7$. There are no joints with 2 DOF, therefore $j_2 = 0$. The total number of joints is

$$j = j_1 + j_2 = 7 + 0 = 7 \text{ joints}$$

- The number of cycles:

  The number of cycles is determined from

$$c = j - m = 7 - 5 = 2 \text{ cycles}$$

- The mechanism's mobility:
  The mobility is determined from

$$M = 3 \cdot m - 2 \cdot j_1 - j_2 = 3 \cdot 5 - 2 \cdot 7 - 0 = 1$$

- The digraph:
  The digraph attached to the mechanism is illustrated in Figure 3.36b, with its nodes corresponding to the mechanism's links and the edges to the mechanism's joints.
- The tree:
  The tree is illustrated in Figure 3.36c, obtained from digraph by cutting the two chords F and G (equal to the number of cycles, $c = 2$).
- The cycles:
  Adding one at a time the chords, the two cycles are generated. In this example, the *direction of each cycle* is coincidental to the chord's direction, that is, both cycle 1 and cycle 2 are clockwise oriented (Figure 3.36d).
- Equations for positional and links' orientation analysis
  - The incidence links–joints matrix, **G̲**
    Considering its definition in Chapter 1, Equation 1.15, the matrix **G̲** has each row assigned to a link and each column assigned to a joint. Each column has assigned a pair [–1, +1] of two numbers, –1 assigned for the lower number link and +1 assigned for the higher number link:

Joints

$$[\underline{G}] = \begin{array}{c} \begin{array}{ccccccc} A & B & C & D & E & F & G \end{array} \\ \left[\begin{array}{ccccccc} -1 & 0 & -1 & 0 & -1 & 0 & 0 \\ 1 & -1 & 0 & 0 & 0 & 0 & 0 \\ 0 & 1 & 0 & -1 & 0 & -1 & 0 \\ 0 & 0 & 0 & 1 & 1 & 1 & -1 \\ 0 & 0 & 0 & -1 & 0 & 0 & 1 \\ 0 & 0 & 1 & 1 & 0 & 0 & 0 \end{array}\right] \end{array} \begin{array}{c} \\ 0 \\ 1 \\ 2 \\ 3 \\ 4 \\ 5 \end{array} \quad \text{Links} \tag{3.101}$$

  - The cycle basis matroid matrix, **C**
    Considering its definition in Chapter 1, Equation 1.22, the matrix **C** has two rows assigned to two cycles, and each column is assigned to a joint.
    A –1 is assigned to a joint which is contrary to the cycle's direction and 1 if the joint is in the cycle's direction:

$$[C]=\begin{array}{c} \\ \\ \end{array}\overset{\begin{matrix} A & B & C & D & E & F & G \end{matrix}}{\left[\begin{array}{ccccc|cc} 1 & 1 & 0 & 0 & -1 & 1 & 0 \\ 0 & 0 & -1 & 1 & 1 & 0 & 1 \end{array}\right]}\begin{matrix} \text{Cycle 1} \\ \text{Cycle 2} \end{matrix} \tag{3.102}$$

Notice that the matrix is partitioned into two matrices:
- The c × a = 2 × 5 tree matrix **T**, as defined in Chapter 1
- The c × c = 2 × 2 diagonal unit matrix, **U**.

Note: If cycle's direction is inverse to chord direction, then the matrix U has a –1 entry

where

c is the number of cycles, equal to the number of joints which were cut: F and G

a is the number of arcs, equal to the number of joint mechanism from the tree: A, B, C, D, and E (Figure 3.37)

- The Latin matrix, **L**

The *Latin matrix* **L**, Equation (2.103), is a 2 × 6 matrix with vector entries defined around the links in the digraph and written in the direction of each cycle:

Vector form:

$$[\mathbf{L}]=\overset{\text{Links}}{\overset{\begin{matrix} 0 & 1 & 2 & 3 & 4 & 5 \end{matrix}}{\begin{bmatrix} \mathbf{L}_{EA} & \mathbf{L}_{AB} & \mathbf{L}_{BF} & \mathbf{L}_{FE} & 0 & 0 \\ \mathbf{L}_{CE} & 0 & 0 & \mathbf{L}_{EG} & \mathbf{L}_{GD} & \mathbf{L}_{DC} \end{bmatrix}}}\begin{matrix} \text{Cycle 1} \\ \text{Cycle 2} \end{matrix} \tag{3.103}$$

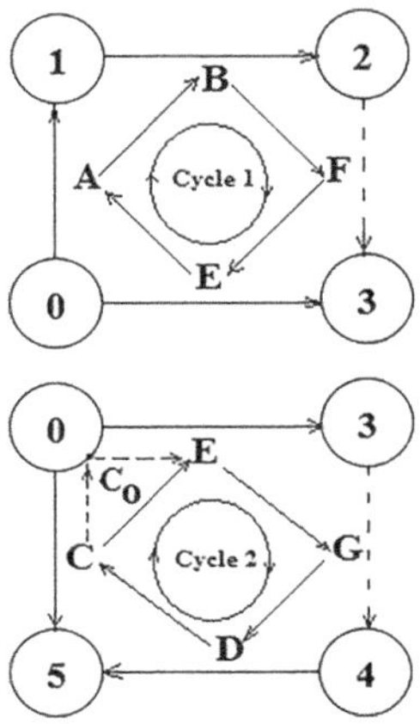

FIGURE 3.37
Latin matrix vectors for the planar mechanism with two cycles.

which, in scalar form, is shown in Equation (3.104):

$$[L]=\begin{array}{c} \\ \end{array}\begin{bmatrix} x_{EA} & x_{AB} & x_{BF} & x_{FE} & 0 & 0 \\ y_{EA} & y_{AB} & y_{BF} & y_{FE} & 0 & 0 \\ x_{CE} & 0 & 0 & x_{EG} & x_{GD} & x_{DC} \\ y_{CE} & 0 & 0 & y_{EG} & y_{GD} & y_{DC} \end{bmatrix} \tag{3.104}$$

(columns numbered 0, 1, 2, 3, 4, 5)

Input data:

- Coordinates of joints A and E on the fixed link 0:

$$\begin{Bmatrix} x_A \\ y_A \end{Bmatrix}=\begin{Bmatrix} 0 \\ 0 \end{Bmatrix};\begin{Bmatrix} x_E \\ y_E \end{Bmatrix}=\begin{Bmatrix} 30 \\ 0 \end{Bmatrix}$$

- The slide is horizontal on link 0. The angle its unit vector $\mathbf{u}_C$ makes with the axis is $\alpha_C = 0°$:

$$\begin{Bmatrix} u_C^x \\ u_C^y \end{Bmatrix}=\begin{Bmatrix} u_C^x \\ u_C^y \end{Bmatrix}=\begin{Bmatrix} \text{Cos}(0°) \\ \text{Sin}(0°) \end{Bmatrix}=\begin{Bmatrix} 1 \\ 0 \end{Bmatrix} \tag{3.105}$$

- Vectors with constant magnitude:

$$L_{AB} = 4''; L_{BF} = 24''; L_{FE} = 14''; L_{EG} = 7''; L_{GD} = 12''; L_{DC} = 0''$$

- Home position:

  The crank at position $\theta_1 = 0°$ brings the slider at "home position," C_0. The coordinate x_{C_0} from the fixed joint A is determined geometric (Figure 3.38):

The cosine law in triangle BFE:

$$\theta_3^\circ = 180° - \arccos\left(\frac{14^2 + 26^2 - 24^2}{2 \cdot 14 \cdot 26}\right) = 114° \tag{3.106}$$

The cos law in triangle C_0GE:

$$12^2 = L_{C_0E}^2 + 7^2 - 2 \cdot 7 \cdot L_{C_0E} \cdot \cos(114°) \Rightarrow L_{C_0E} = 7.308''$$

Therefore, the home position is $x_{C_0} = 37.308''$, and the home position coordinates are

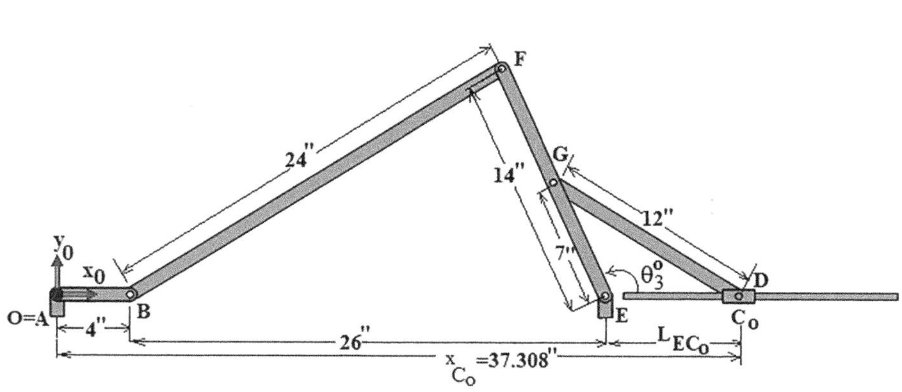

FIGURE 3.38
Determination of the home position, L_{AC_0}.

$$\begin{Bmatrix} x_{C_0} \\ y_{C_0} \end{Bmatrix} = \begin{Bmatrix} 37.308 \\ 0 \end{Bmatrix}$$

- Nonlinear system of equations for positional analysis
  - *Column 0* (link 0):

$$\mathbf{L}_{EA} = \begin{Bmatrix} x_{EA} \\ y_{EA} \end{Bmatrix} = \begin{Bmatrix} x_A - x_E \\ y_A - y_E \end{Bmatrix}$$

$$= \begin{Bmatrix} -30 \\ 0 \end{Bmatrix}; \text{given the coordinates between fixed joints} \tag{3.107a}$$

$$\mathbf{L}_{CE} = \mathbf{L}_{CC_0} + \mathbf{L}_{C_0E} \tag{3.107b}$$

$$\mathbf{L}_{CC_0} = d_C \cdot \mathbf{u}_C = \begin{Bmatrix} d_C \\ 0 \end{Bmatrix}$$

$$\mathbf{L}_{C_0E} = \begin{Bmatrix} x_{C_0E} \\ y_{C_0E} \end{Bmatrix} = \begin{Bmatrix} x_E - x_{C_0} \\ y_E - y_{C_0} \end{Bmatrix} = \begin{Bmatrix} -7.308 \\ 0 \end{Bmatrix} \tag{3.107c}$$

$$\begin{Bmatrix} x_{CE} \\ y_{CE} \end{Bmatrix} = \begin{Bmatrix} d_C - 7.308 \\ 0 \end{Bmatrix}$$

$$(d_C - 7.308)^2 + 0^2 = (L_{CE})^2; \text{constraint type II equation}$$

– *Column 1* (link 1):

$$\mathbf{L}_{AB} = \begin{Bmatrix} x_{AB} \\ y_{AB} \end{Bmatrix} = \begin{Bmatrix} 4 \cdot Cos(\theta_1) \\ 4 \cdot Sin(\theta_1) \end{Bmatrix}; \text{ vector is along the link 1 (crank)} \tag{3.107d}$$

– *Column 2* (link 2):

$$\mathbf{L}_{BF} = \begin{Bmatrix} x_{BF} \\ y_{BF} \end{Bmatrix}$$

Constraint magnitude, *type II* equation:

$$x_{BF}^2 + y_{BF}^2 = 24^2 \tag{3.107e}$$

– *Column 3* (link 3):

$$\mathbf{L}_{FE} = \begin{Bmatrix} x_{FE} \\ y_{FE} \end{Bmatrix}$$

$$x_{FE}^2 + y_{FE}^2 = 14^2; \text{ constraint magnitude, type II equation} \tag{3.107f}$$

$$\mathbf{L}_{EG} = \begin{Bmatrix} x_{EG} \\ y_{EG} \end{Bmatrix}$$

Constraint (isometric) type III equation that preserves the $\Phi = 180°$ angle between vectors $\mathbf{L}_{EG}$ and $\mathbf{L}_{FE}$. Vector $\mathbf{L}_{EG}$ is dependent, since it is expressed as a function of vector $\mathbf{L}_{FE}$:

$$\begin{Bmatrix} x_{EG} \\ y_{EG} \end{Bmatrix} = \frac{L_{EG}}{L_{FE}} \cdot \begin{bmatrix} Cos(180°) & -Sin(180°) \\ Sin(180°) & Cos(180°) \end{bmatrix} * \begin{Bmatrix} x_{FE} \\ y_{FE} \end{Bmatrix} = -0.5 \cdot \begin{Bmatrix} x_{FE} \\ y_{FE} \end{Bmatrix} \tag{3.107g}$$

– *Column 4* (link 4):

$$\mathbf{L}_{GD} = \begin{Bmatrix} x_{GD} \\ y_{GD} \end{Bmatrix}$$

Constraint magnitude, *type II* equation:

$$x_{GD}^2 + y_{GD}^2 = 12^2 \tag{3.107h}$$

– *Column 5* (link 5): Constraint magnitude, $L_{DC} = 0''$, D and C are coincident:

$$\mathbf{L}_{DC} = \begin{Bmatrix} x_{DC} \\ y_{DC} \end{Bmatrix} = \begin{Bmatrix} 0 \\ 0 \end{Bmatrix} \tag{3.107i}$$

The Latin matrix with scalar components is a $2c \times n = 4 \times 6$ matrix:

$$[L] = \begin{bmatrix} -30 & 4\cdot\mathrm{Cos}(\theta_1) & x_{BF} & x_{FE} & 0 & 0 \\ 0 & 4\cdot\mathrm{Sin}(\theta_1) & y_{BF} & y_{FE} & 0 & 0 \\ d_C - 7.308 & 0 & 0 & -0.5x_{FE} & x_{GD} & 0 \\ 0 & 0 & 0 & -0.5y_{FE} & y_{GD} & 0 \end{bmatrix} \tag{3.107j}$$

(columns 0, 1, 2, 3, 4, 5)

Therefore, the nonlinear system of $4c = 8$ nonlinear equations is

$2c = 4$—Type I equations (the sum of entries in the Latin matrix is zero)

$2c = 4$—Type II equations (constraint equations (3.107c), (3.107e), (3.107f), and (3.107h))

$$\begin{cases} -30 + 4\cdot\mathrm{Cos}(\theta_1) + x_{BF} + x_{FE} = 0 \\ 4\cdot\mathrm{Sin}(\theta_1) + y_{BF} + y_{FE} = 0 \\ d_C - 7.308 - 0.5x_{FE} + x_{GD} = 0 \\ -0.5y_{FE} + y_{GD} = 0 \\ (d_C - 7.308)^2 = L_{CE}^2 \\ x_{BF}^2 + y_{BF}^2 = 24^2 \\ x_{FE}^2 + y_{FE}^2 = 14^2 \\ x_{GD}^2 + y_{GD}^2 = 12^2 \end{cases} \tag{3.107k}$$

Solved for $4c = 8$ unknowns: $d_C, L_{CE}, x_{BF}, y_{BF}, x_{FE}, y_{FE}, x_{GD}$, and y_{GD}.

For any $M = 1$ mechanism, one independent parameter is provided, for example, θ_1 (for actuated link 1): $0° \le \theta_1(t) \le 360°$. In this example, θ_1 is chosen to increase linearly with the time t, chosen within the interval $0 \le t \le 1$ s.

- Absolute links' orientation:

  Appendix 3.17 tabulates the calculated and simulated values for links orientation and displacement d_C for the slider, considering an incremental crank angle of 15°. The first set with the

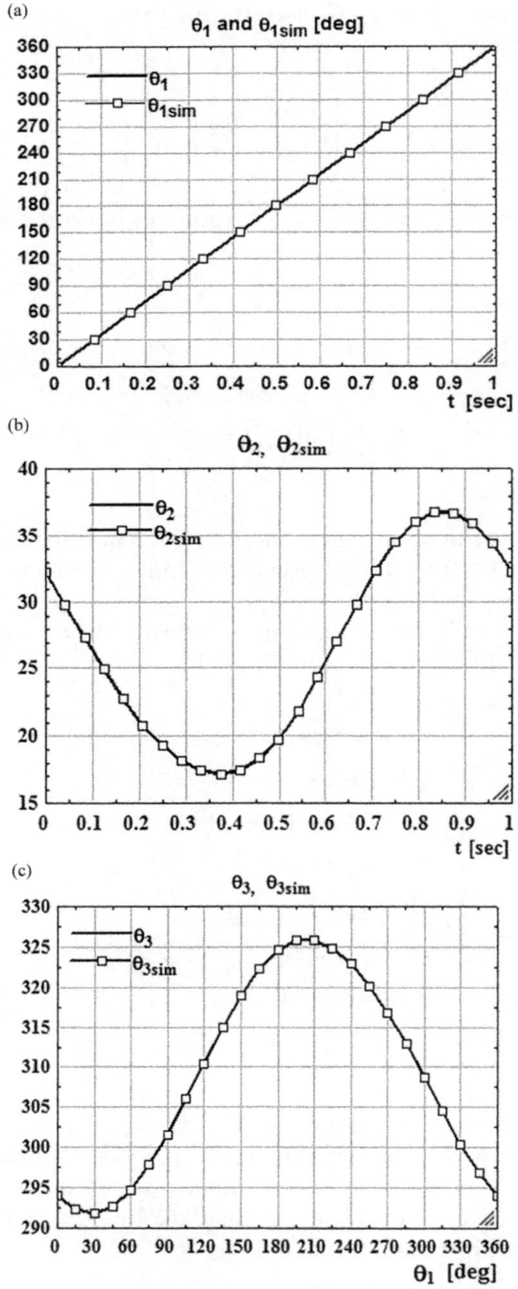

FIGURE 3.39
Absolute links' orientation versus time: (a) link 1; (b) link 2; (c) link 3; (d) link 4.

(Continued)

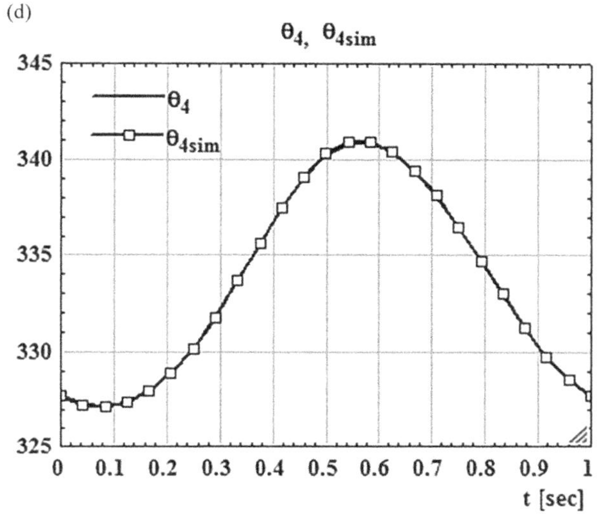

FIGURE 3.39 (CONTINUED)
Absolute links' orientation versus time: (a) link 1; (b) link 2; (c) link 3; (d) link 4.

calculated (output from equations, using EES) data is compared with the second set of simulated data (output from SW simulations, labeled "sim"). The values for the two sets coincide.

Figure 3.39a–d show the EES plots for the calculated and simulated values for absolute links' orientations versus time, for an incremental crank angle of 15°.

- Input–output:

  Figure 3.40a shows the calculated and simulated values $\theta_3(t)$ on the EES plots for absolute link 3 orientations (output) versus input crank angle, $\theta_1(t)$, incremental of 15°.

  Figure 3.40b shows the calculated and simulated values for slider's displacement, d_C, for an incremental crank angle of 15°.

  The above outputs are important for the control of output motion.

3.6 Planar Mechanisms with Cams

3.6.1 Background

Mechanisms that include a *cam* have important applications in engineering. They are used in valve actuation in internal combustion engines. The valve is named a *follower*, which opens and closes periodically, correctly timing the

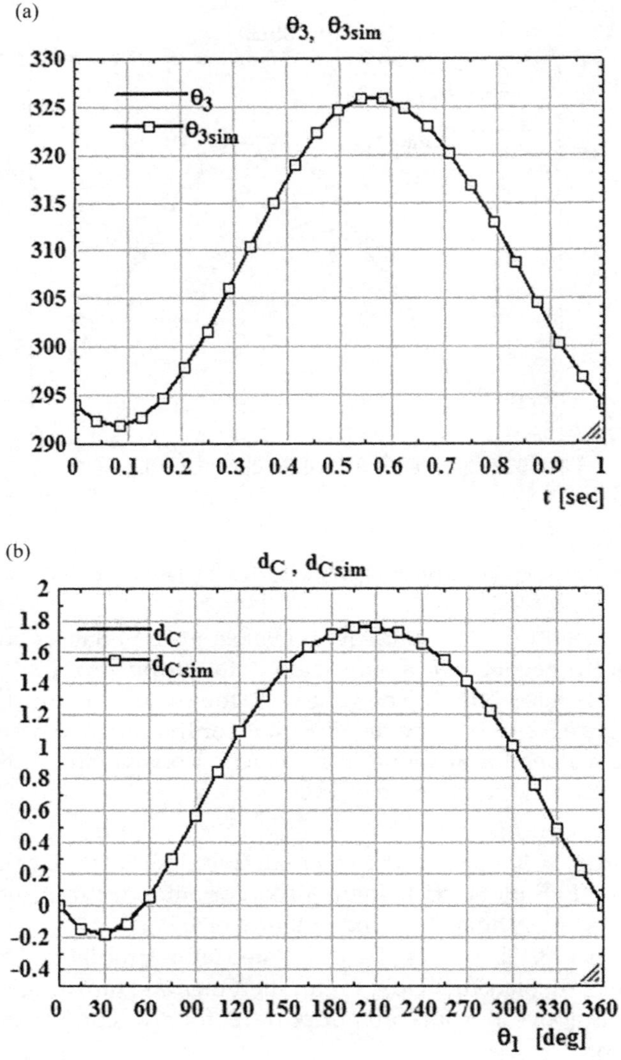

FIGURE 3.40
(a) Output $\theta_3(t)$ versus input $\theta_1(t)$; (b) output d_C versus input $\theta_1(t)$.

intake of an air–fuel mixture inside the combustion chamber, or to exhaust the products of combustion. The cam is machined on the camshaft, which is actuated by the engine. Mechanisms with cams are used also in manufacturing automation, where the motion of a cam engages a link, and the follower periodically performs the same task.

A cam–follower mechanism is considered equivalent to a crank slider, with a variable crank length.

Figure 3.41 illustrates a disk cam. If the cam rotates counterclockwise with θ_1 about its axis at A, then the variable crank in the equivalent crank slider rotates clockwise with $(-\theta_1)$.

- The *home position* is the reference position with the cam at $\theta_1 = 0°$ and the follower with displacement zero: $d_C = 0$. In Figure 3.41, B_1 is the contact point between the cam and the follower at home position.
- The *base circle* is a circle with the *base radius* ρ, centered on the cam's axis and tangent to the cam surface.

  During cam's rotation, the contact between the cam and the follower is successively changed from B_1 to B_2, B_3, B_4, and B_5, and therefore the length L_{AB_i} is variable, where i = 2, 3, 4, and 5.
- The *rise* interval of time, T_r, is the time elapsed during the cam rotation, $\Delta\theta_r$, from B_1 to B_3.

  During this interval, the crank length increases from L_{AB_1} to L_{AB_3}, allowing the follower to rise.
- The *fall* interval of time, T_f, is the time elapsed during the cam rotation, $\Delta\theta_f$, from B_3 to B_5.

  During this interval, the crank length decreases from L_{AB_3} to L_{AB_5}, allowing the follower to fall.
- The *dwell* interval of time, T_d, is the time elapsed during the cam rotation, $\Delta\theta_d$, from B_5 to B_1.

  During this interval, the crank length is constant, equal to the constant base radius: $\rho = L_{AB_5} = L_{AB_1}$, the follower neither falls nor rises.

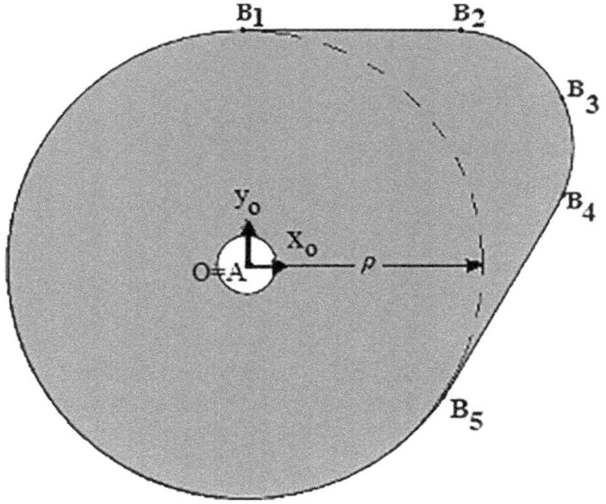

FIGURE 3.41
Planar cam.

- The time for a *full cycle*, T [s]:

$$T = T_r + T_f + T_d \quad (3.108)$$

- Angular velocity of the cam:

$$\omega_1 = \frac{2\pi}{T}[\text{rad/s}] = \frac{360°}{T}[\text{deg/s}] \quad (3.109)$$

- Intervals for cam rotation:

$$\begin{aligned} &\text{Rise interval: } \Delta\theta_r = \omega_1 \cdot T_r \\ &\text{Fall interval: } \Delta\theta_f = \omega_1 \cdot T_f \\ &\text{Dwell interval: } \Delta\theta_d = \omega_1 \cdot T_d \end{aligned} \quad (3.110)$$

Numerical example:

$$\begin{aligned} &\rho = 1''; T_r = 0.33 \text{ s}; T_f = 0.33 \text{ s}; T_d = 1.34 \text{ s}; T = T_f + T_d + T_r = 2 \text{ s}; \\ &\omega_1 = \frac{360°}{T} = 180 \text{ deg/s}; \Delta\theta_r = \omega_1 \cdot T_r = 60°; \Delta\theta_f = \omega_1 \cdot T_f = 60°; \\ &\Delta\theta_d = \omega_1 \cdot T_d = 240° \end{aligned} \quad (3.111)$$

3.6.2 Input–Output Relation

Figure 3.42(a) (Animation 3.11) illustrates a cam–follower mechanism. To find the input–output relation, the following steps are presented:

- The number of links and joints
- Calculate the number of cycles
- Calculate the mechanism's mobility
- Draw the digraph for the mechanism
- Draw the tree and the cycle
- Positional analysis

Solution: The links are labeled with numbers, starting with 0 which is assigned to the fixed link. There are m = 2 mobile links, which have been assigned the numbers 1 and 2. There are three joints labeled with capital letters: A, B, and C.

The mechanism has one revolute joint with 1 DOF (description "rev" at A) and one prismatic joint (description "pr" at C), therefore $j_1 = 2$. There is one cam joint with 2 DOF (description "cam" at B), therefore $j_2 = 1$.

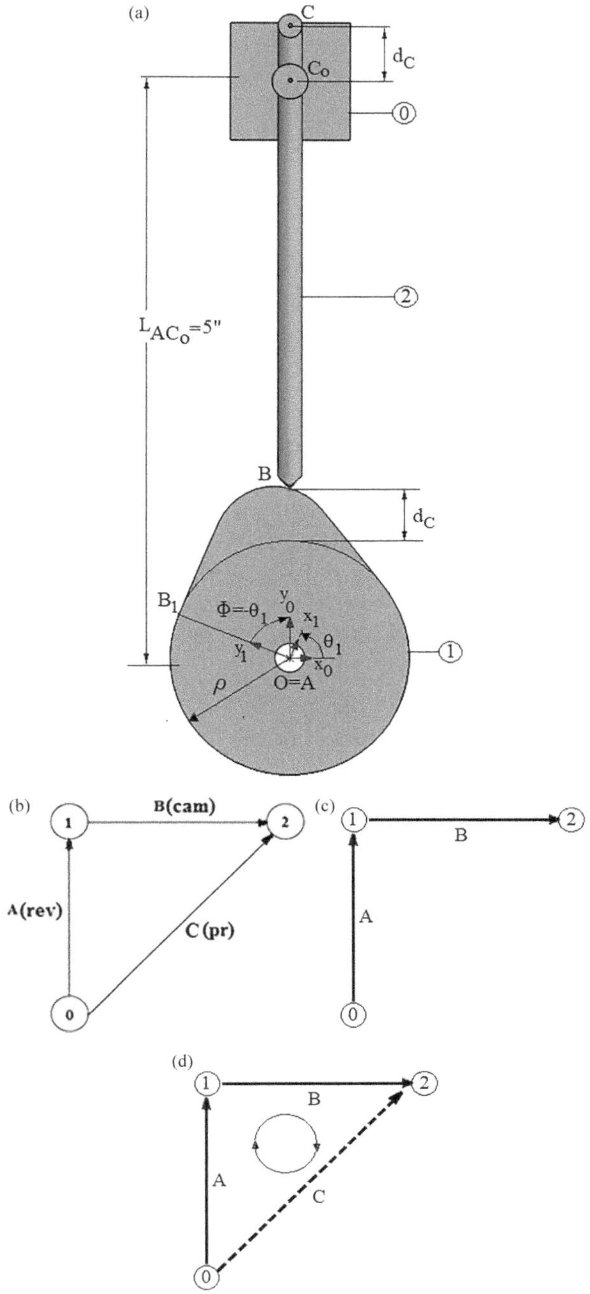

FIGURE 3.42
(a) Planar cam–follower mechanism with one cycle, one revolute, one prismatic, and one cam joint; (b) digraph; (c) spanning tree; (d) cycle.

- The total number of joints from Chapter 1, Equation 1.27:

$$j = j_1 + j_2 = 2 + 1 = 3 \text{ joints}$$

- The number of cycles is determined using Equation 1.7:

$$c = j - m = 3 - 2 = 1 \text{ cycle}$$

- The mobility of the planar four-bar mechanism is determined using Equation 1.28:

$$M = 3 \cdot m - 2 \cdot j_1 - j_2 = 3 \cdot 2 - 2 \cdot 2 - 1 = 1$$

- The mechanism has 1 DOF, and one motor is required for actuation, chosen the revolute joint A (between the cam 1 and the link 0).
- The digraph attached to the mechanism is illustrated in Figure 3.42b, with its nodes corresponding to the mechanism's links and the edges to the mechanism's joints. Cutting the edge C, the tree with two arcs A and B is illustrated in Figure 3.42c. The cut-edge (chord) C is completing the cycle, illustrated in Figure 3.42d. A cycle's direction is selected, in this example, the same as the chord's direction, that is, counterclockwise.
  - The incidence links–joints matrix, $\underline{G}$:
    Considering its definition (Equation 1.11), the 3 × 3 matrix $\underline{G}$ has each row assigned to a link and each column assigned to a joint. Each column is assigned a pair [–1, +1] of two numbers: –1 assigned for the lower number link and +1 assigned for the higher number link:

$$\begin{array}{c} \text{Joints} \\ \begin{array}{ccc} A & B & C \end{array} \end{array}$$

$$[\underline{G}] = \begin{bmatrix} -1 & 0 & -1 \\ 1 & -1 & 0 \\ 0 & 1 & 1 \end{bmatrix} \begin{matrix} 0 \\ 1 \\ 2 \end{matrix} \quad \text{Links} \qquad (3.112)$$

  - The cycle basis matroid matrix, **C**:
    Considering its definition in Chapter 1, Equation (1.22), the (1 × 3) matrix **C** has one row assigned to one cycle and each column assigned to a joint.
    The *direction of the cycle* is assigned inverse to the chord's C direction, clockwise.
    A –1 is assigned to a joint which is contrary to the cycle's direction and 1 if the joint is in the cycle's direction:

$$\begin{array}{c} \quad\; A \quad B \quad C \\ [C] = \begin{bmatrix} 1 & 1 & -1 \end{bmatrix} \end{array} \tag{3.113}$$

- The Latin matrix, **L**:

  The *Latin matrix* **L**, Equation (3.114), is a (1 × 3) matrix with vector entries defined around the links in the digraph and written in the direction of the cycle.

  The vectors with variable magnitudes $\mathbf{L}_{AB}$, $\mathbf{L}_{BC}$, $\mathbf{L}_{CA}$, shown in the digraph from Figure 3.43, are the entries in the matrix:

$$\begin{array}{c} \quad\; 0 \quad\quad 1 \quad\quad 2 \\ [\mathbf{L}] = \begin{bmatrix} \mathbf{L}_{CA} & \mathbf{L}_{AB} & \mathbf{L}_{BC} \end{bmatrix} \end{array} \tag{3.114}$$

Scalar form:

$$[L] = \begin{bmatrix} x_{CA} & x_{AB} & x_{BC} \\ y_{CA} & y_{AB} & y_{BC} \end{bmatrix}$$

Input data:

- Coordinates of joint A and point C_0 on the fixed link 0:

$$\begin{Bmatrix} x_A \\ y_A \end{Bmatrix} = \begin{Bmatrix} 0 \\ 0 \end{Bmatrix}; \begin{Bmatrix} x_{C_0} \\ y_{C_0} \end{Bmatrix} = \begin{Bmatrix} 0 \\ 5 \end{Bmatrix}$$

- Cam's base radius:

$$L_{AB_1} = \rho = 1'' = \text{cam's base radius} \tag{3.115}$$

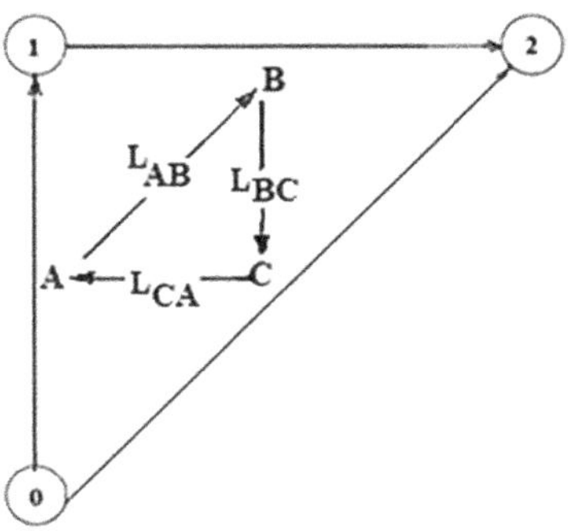

FIGURE 3.43
Latin matrix vectors for the cam-follower mechanism.

– The unit vector for the sliding direction at C:

$$\mathbf{u}_C = \left\{ \begin{array}{c} u_C^x \\ u_C^y \end{array} \right\} = \left\{ \begin{array}{c} \mathrm{Cos}(90°) \\ \mathrm{Sin}(90°) \end{array} \right\} = \left\{ \begin{array}{c} 0 \\ 1 \end{array} \right\}$$

– Home position: Figure 3.44a illustrates the cam 1 at home position B_1, $\theta_1 = 0°$, and the follower 2 at home position C_0, $d_C = 0$.

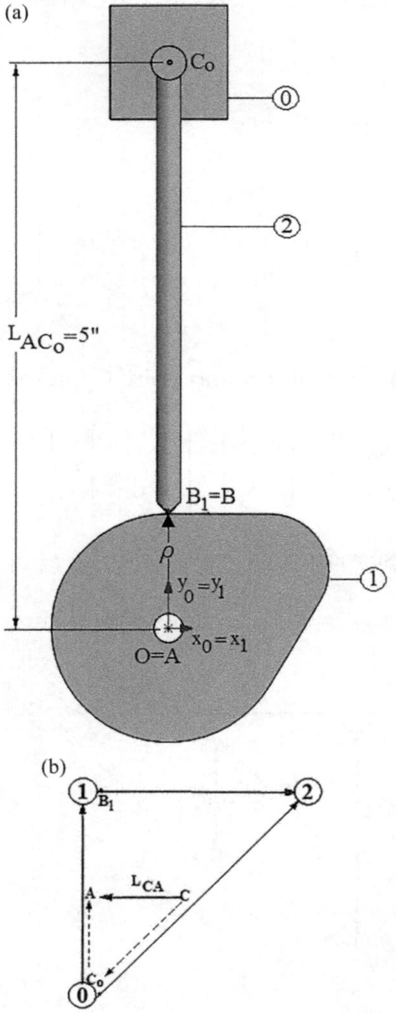

FIGURE 3.44
(a) Cam and follower of the home position; (b) Latin matrix vector $\mathbf{L}_{CA}$.

3.6.3 Equations for Cam Contour

- *Column 0* (node 0):

$$\mathbf{L}_{CA} = \begin{Bmatrix} x_{CA} \\ y_{CA} \end{Bmatrix} \quad (3.116a)$$

Latin matrix vector $\mathbf{L}_{CA}$ is expressed as a function of its home direction $\mathbf{L}_{C_0A}$, shown in Figure 3.44b:

$$\mathbf{L}_{CA} = \mathbf{L}_{CC_0} + \mathbf{L}_{C_0A}\text{; the three vectors are collinear} \quad (3.116b)$$

where

$\mathbf{L}_{C_0C} = d_C \cdot \mathbf{u}_C = \begin{Bmatrix} 0 \\ d_C \end{Bmatrix} \Rightarrow \mathbf{L}_{CC_0} = \begin{Bmatrix} 0 \\ -d_C \end{Bmatrix}$, displacement of point C on the follower measured from home position C_0. For $d_C = 0$, the follower 2 is at the home position. If d_C increases, then the follower rises (moving in the direction of the unit vector $\mathbf{u}_C$). If d_C decreases, then the follower falls (moving in the opposite direction of the unit vector $\mathbf{u}_C$):

$\mathbf{L}_{C_0A} = \begin{Bmatrix} x_A - x_{C_0} \\ y_A - y_{C_0} \end{Bmatrix} = \begin{Bmatrix} 0 \\ -y_{C_0} \end{Bmatrix}$, constant magnitude vector, locates the follower's home position.

$$\Rightarrow \begin{Bmatrix} x_{CA} \\ y_{CA} \end{Bmatrix} = \begin{Bmatrix} 0 \\ -y_{C_0} - d_C \end{Bmatrix} \quad (3.116c)$$

- *Column 1* (node 1):

$$\mathbf{L}_{AB} = \begin{Bmatrix} x_{AB} \\ y_{AB} \end{Bmatrix}\text{; variable magnitude vector that generates the cam's profile} \quad (3.116d)$$

When the cam rotates with an angle θ_1 about its fixed point A, vector $\mathbf{L}_{AB}$ (attached to the cam) is rotating with the same angle; therefore, the following constraint equation holds.

Constraint type III equation: vector $\mathbf{L}_{AB_1}$ is rotated with $\Phi = -\theta_1$ (clockwise, about the fixed point A) to coincide vector $\mathbf{L}_{AB}$:

$$\mathbf{L}_{AB} = \frac{L_{AB}}{L_{AB_1}} \cdot \begin{bmatrix} \mathrm{Cos}(\Phi) & -\mathrm{Sin}(\Phi) \\ \mathrm{Sin}(\Phi) & \mathrm{Cos}(\Phi) \end{bmatrix} * \mathbf{L}_{AB_1} \quad (3.116e)$$

The vector $\mathbf{L}_{AB_1}$ has a constant magnitude ρ, and its components at home position are

$$\mathbf{L}_{AB_1} = \begin{Bmatrix} x_{AB_1} \\ y_{AB_1} \end{Bmatrix} = \begin{Bmatrix} 0 \\ \rho \end{Bmatrix} \tag{3.116f}$$

The vector $\mathbf{L}_{AB}$ has a variable magnitude, $L_{AB} = \rho + d_C$, and the components from (3.116e)

$$\Rightarrow \begin{Bmatrix} x_{AB} \\ y_{AB} \end{Bmatrix} = \frac{\rho + d_C}{\rho} \begin{bmatrix} \mathrm{Cos}(-\theta_1) & -\mathrm{Sin}(-\theta_1) \\ \mathrm{Sin}(-\theta_1) & \mathrm{Cos}(-\theta_1) \end{bmatrix} * \begin{Bmatrix} 0 \\ \rho \end{Bmatrix}$$
$$= \begin{Bmatrix} (\rho + d_C)\cdot \mathrm{Sin}(\theta_1) \\ (\rho + d_C)\cdot \mathrm{Cos}(\theta_1) \end{Bmatrix} \tag{3.116g}$$

- *Column 2* (node 2):

$$\mathbf{L}_{BC} = \begin{Bmatrix} x_{BC} \\ y_{BC} \end{Bmatrix} = \begin{Bmatrix} x_C - x_B \\ y_C - y_B \end{Bmatrix} = \begin{Bmatrix} -x_B \\ y_{C_0} + d_C - y_B \end{Bmatrix} \tag{3.116h}$$

The Latin matrix entries after the substitution of Equations (3.116c), (3.116g), and (3.116h) are

$$[L] = \begin{bmatrix} x_{CA} & x_{AB} & x_{BC} \\ y_{CA} & y_{AB} & y_{BC} \end{bmatrix}$$
$$= \left[\begin{array}{c|c|c} 0 & (\rho + d_C)\cdot \mathrm{Sin}(\theta_1) & -x_B \\ -y_{C_0} - d_C & (\rho + d_C)\cdot \mathrm{Cos}(\theta_1) & y_{C_0} + d_C - y_B \end{array}\right] \tag{3.116i}$$

The system of $2c = 2$ *type I* equations (the sum of entries in the Latin matrix is zero)

$$\begin{cases} x_{AB} + x_{BC} + x_{CA} = 0 \\ y_{AB} + y_{BC} + y_{CA} = 0 \end{cases} \tag{3.116j}$$

is solved for $2c = 2$ unknowns x_B and y_B:

$$x_B = (\rho + d_C) \cdot \mathrm{Sin}(\theta_1)$$
$$y_B = (\rho + d_C) \cdot \mathrm{Cos}(\theta_1) \tag{3.116k}$$

The previous coordinates are used to machine the cam's contour, Figure 3.45, for a desired follower d_C displacement.

Figure 3.46 illustrates the SW motion window for the input revolute motion on the cam as a constant angular velocity of 180 °/s. Figure 3.47a–c shows the output for the follower's displacement, velocity, and acceleration.

Conclusions

Several conclusions can be drawn for the above cam:

- The maximum follower rise, Figure 3.47(a), is $D = d_C^{max} = 0.5''$.
- The very high value for acceleration, for example, at $\theta_1 = 90°$ in Figure 3.47c, induces large values for the cam–follower forces, and therefore vibrations.

To diminish the size of the accelerations, the designer has additional options for cam profiles. Several of such possibilities are presented next.

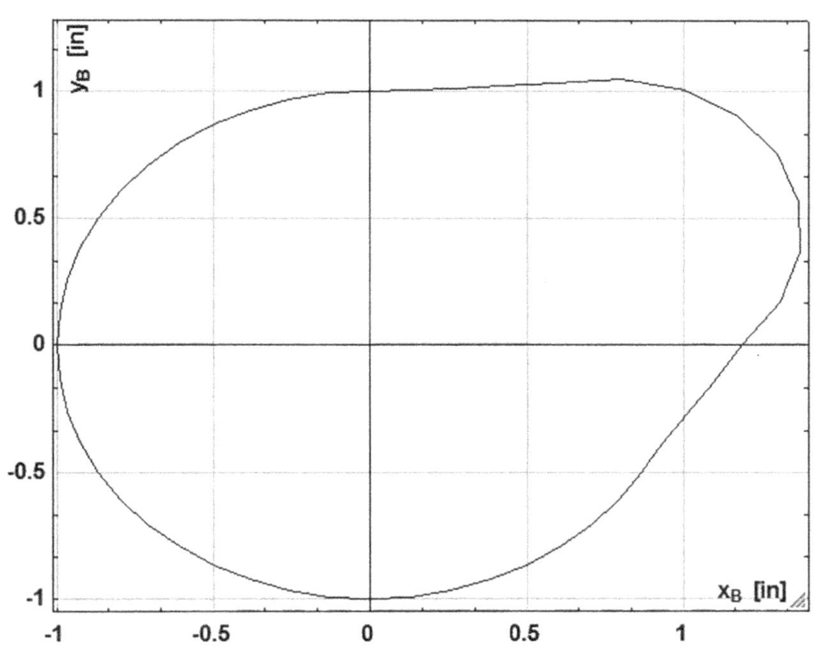

FIGURE 3.45
Cam profile.

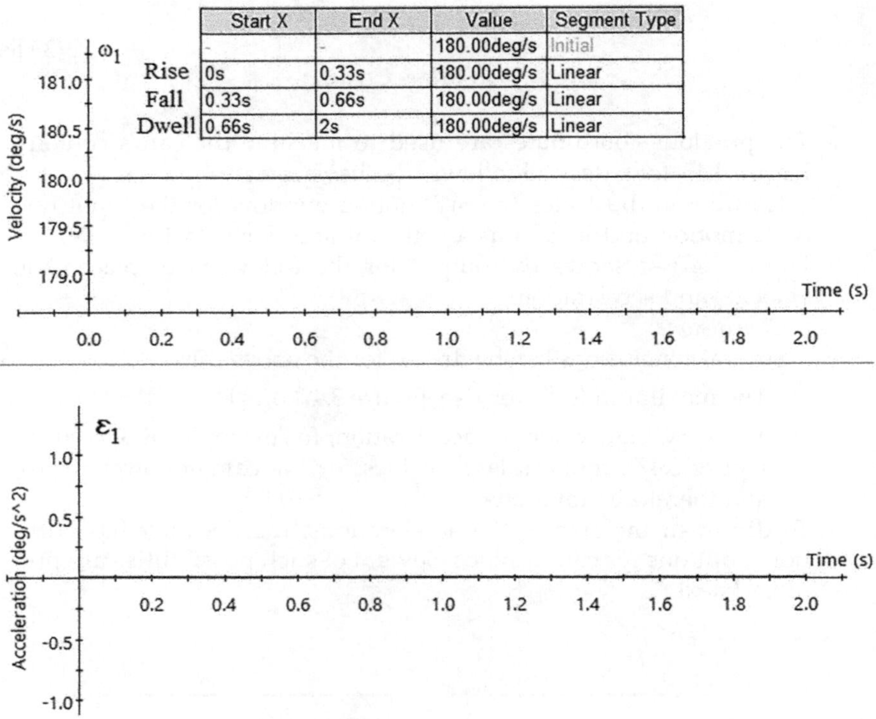

FIGURE 3.46
Input cam's constant angular velocity.

3.6.4 Cam with Constant Velocity Rise or Constant Velocity Fall

This cam provides uniform *linear motion* during the follower's rise or fall (constant velocity motion).

Equations for the constant velocity rise

- The displacement equation for the follower's rise from zero to D is

$$d_C = D \cdot \frac{t_r}{T_r} = D \cdot \frac{\theta_r}{\Delta\theta_r} \tag{3.117a}$$

where

$\theta_r = \theta_1$ is the cam rotation angle during the rise interval $0° \le \theta_r \le \Delta\theta_r$ when the cam rotates with constant angular speed within the interval $0° \le \theta_1 \le \Delta\theta_r$

$D = d_C^{max}$ = follower rise

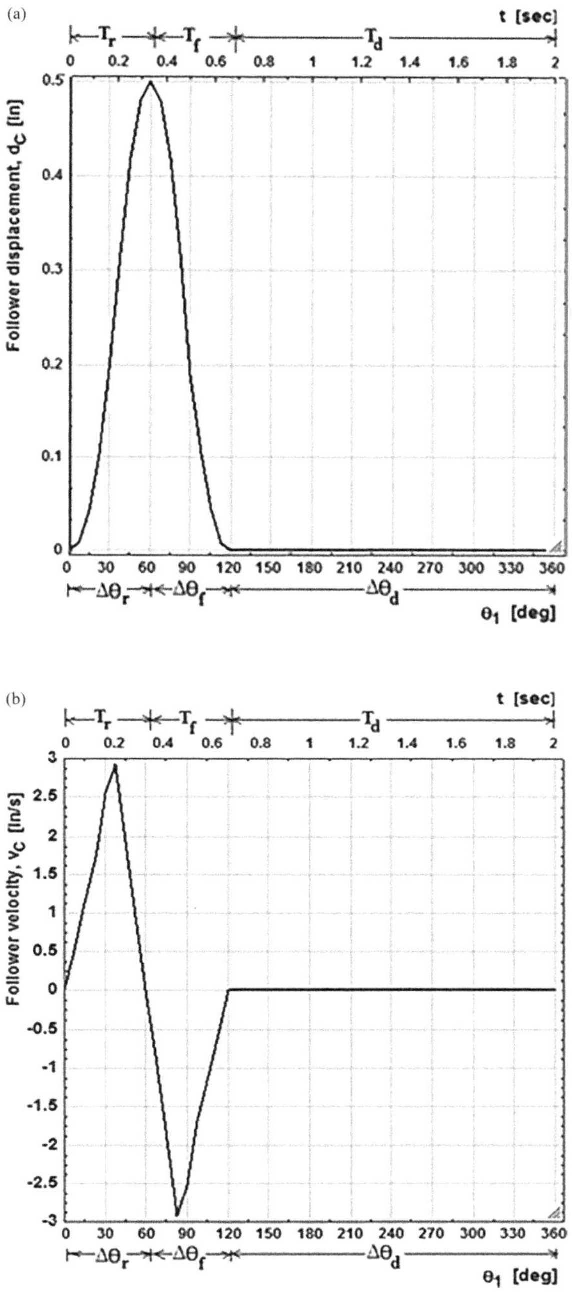

FIGURE 3.47
Cam follower output: (a) follower displacement; (b) follower velocity; (c) follower acceleration.
(Continued)

(c)

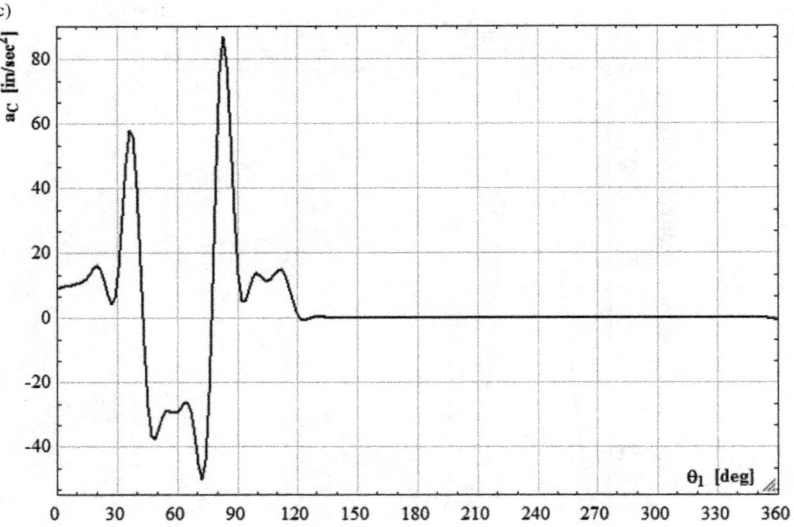

FIGURE 3.47 (CONTINUED)
Cam follower output: (a) follower displacement; (b) follower velocity; (c) follower acceleration.

- The velocity equation for the follower's rise is obtained taking the derivative with time of displacement:

$$v_C = D \cdot \frac{1}{T_r} = D \cdot \frac{\omega_1}{\Delta\theta_r} \tag{3.117b}$$

- The acceleration equation for the follower's rise

$$a_C = 0 \tag{3.117c}$$

Equations for the constant velocity fall

- The displacement equation for the follower fall from D to zero is

$$d_C = D \cdot \left(1 - \frac{t_f}{T_f}\right) = D \cdot \left(1 - \frac{\theta_f}{\Delta\theta_f}\right) \tag{3.118a}$$

where

$\theta_f = \theta_1 - \Delta\theta_r$ is the cam rotation angle within the interval $0° \leq \theta_f \leq \Delta\theta_f$ when the cam rotates with constant angular speed within the interval $\Delta\theta_r < \theta_1 \leq \Delta\theta_r + \Delta\theta_f$.

- The velocity equation for the follower's fall:

$$v_C = D \cdot \left(-\frac{1}{T_f}\right) = D \cdot \left(-\frac{\omega_1}{\Delta\theta_f}\right) \tag{3.118b}$$

- The acceleration equation for the follower's fall is obtained by taking the second derivative of displacement:

$$a_C = 0 \tag{3.118c}$$

Example: A cam with 1″ base radius is to be designed for a prescribed motion in an automated loader as follows:

- Raise the follower 0.5″ in 0.33 s with *constant velocity*
- Lower the follower 0.5″ in 0.33 s with *constant velocity*
- Dwell for 1.34 s
- Repeat the cycle
  Time for one cycle
  From the prescribed timing: $T_r = 0.33$ s; $T_f = 0.33$ s; $T_d = 1.34$ s, the total time is

$$T = T_r + T_f + T_d = 0.33 + 0.33 + 1.34 = 2 \text{ s}$$

Required angular velocity of the cam:

$$\omega_1 = \frac{360^\circ}{T} = 180\ ^\circ/\text{s} \tag{3.119}$$

The cam rotation for each follower interval:

$$\Delta\theta_r = \omega_1 \cdot T_r = 60^\circ;\ \Delta\theta_f = \omega_1 \cdot T_f = 60^\circ;\ \Delta\theta_d = \omega_1 \cdot T_d = 240^\circ$$

The displacement equation for the follower rise from 0″ to $D = d_C^{max} = 0.5''$ is

$$d_C = 0.5 \cdot \frac{\theta_r}{60} \tag{3.119a}$$

where $\theta_r = \theta_1$ is the cam rotation angle during the rise interval $0^\circ \leq \theta_r \leq 60^\circ$ when the cam rotates with constant angular speed within the interval $0^\circ \leq \theta_1 \leq 60^\circ$

The displacement equation for the follower fall from $D = d_C^{max} = 0.5''$ to 0″ is

$$d_C = 0.5 \cdot \left(1 - \frac{\theta_f}{\Delta\theta_f}\right) \tag{3.119b}$$

where $\theta_f = \theta_1 - 60^\circ$ is the cam rotation angle within the interval $0^\circ \leq \theta_f \leq 60^\circ$ when the cam rotates with constant angular speed within the interval $60^\circ < \theta_1 \leq 120^\circ$

The displacement equation for the cam's dwell:

$$d_C = 0''$$

where the cam dwell rotation is within the interval $0° \le \theta_d \le 240°$ when the cam rotates with constant angular speed within the interval:

$$120° < \theta_1 \le 360°$$

The cam profile is based on the coordinates of contact cam–follower point B:

$$x_B = (\rho + d_C) \cdot \text{Sin}(\theta_1)$$
$$y_B = (\rho + d_C) \cdot \text{Cos}(\theta_1) \tag{3.119c}$$

Equations (3.119a)–(3.119c) are used to create a spreadsheet, as shown in Appendix 3.18.

Create a SW motion simulation for the follower, sliding vertically along its guide. A linear motor is selected. In Figure 3.48a, the SW window is illustrated for a linear motor that raises and falls the follower with constant velocity v_C (therefore, a linear displacement d_C).

From motion analysis results for a time period of T = 2 s, although the user can choose a different value, the spreadsheets with results for the follower's displacement, velocity, and acceleration are saved. From the data-spreadsheets are created in Excel their plots. For a plot of the cam contour, then are typed in three different cells the three equations: θ_1, x_B, and y_B.

Next, is presented the procedure by using EES in the design of the cam contour for the follower's prescribed motion.

From EES, in a Window Equations are typed the cam profile coordinates x_B, and y_B, and $\theta_1 = \omega_1 \cdot t$. The data: T_r, T_f, T_d, T, ρ, ω_1 is included in an If Not Parametric statement, which allows generation of different cam alternatives.

From EES create a Parametric Table with a chosen number of runs. In this example 49 runs the same as the number of rows in the SW spreadsheets of data. The time t increases incrementally from 0 to 2 s, and shows in the first column in the Parametric Table. The second column is for cam's angle θ_1. The third column is assigned for follower's displacement d_C. In this column are copied and pasted the Excel data from SW. Additional columns are for the coordinates x_B, and y_B. The Solve Parametric Table command will determine the coordinates x_B, and y_B, used to plot the cam contour, as illustrated in Figure 3.48e.

Note: As an alternative, data for follower's displacement d_C are typed in the EES Equations Window. An IF statement is needed, since d_C has different functions on rise and fall intervals, as shown in Equations (3.119a) and (3.119b)

From EES is created a Lookup Table and inserted the columns. The additional columns for parameters which do not show in Equations Window can be inserted in the Lookup Table. The follower's displacement, velocity, and

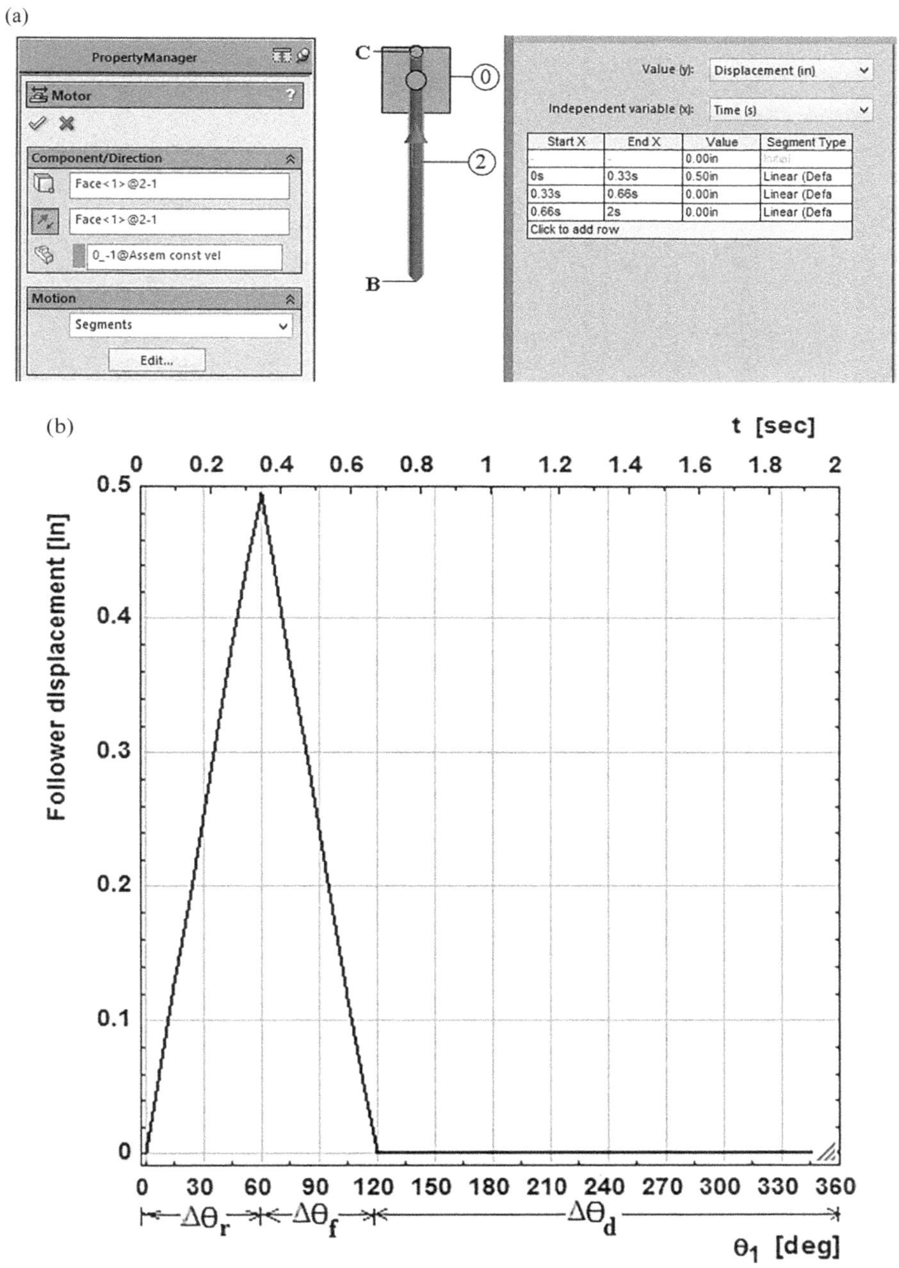

FIGURE 3.48
(a) Follower displacement; (b) follower displacement; (c) follower velocity; (d) follower acceleration; (e) cam contour.

(Continued)

(c)

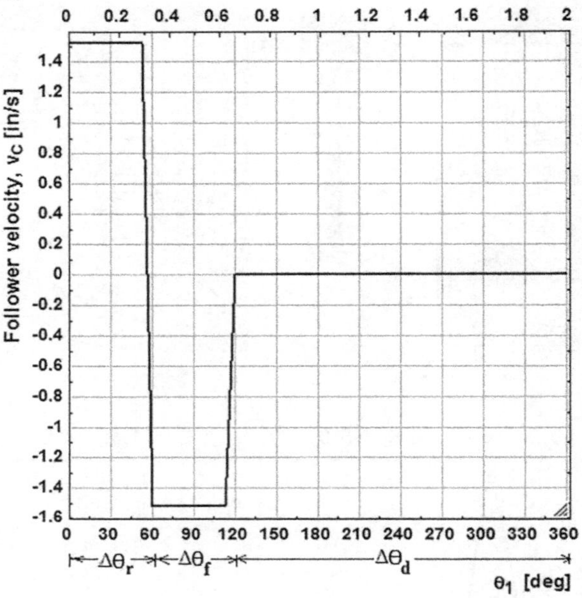

(d)

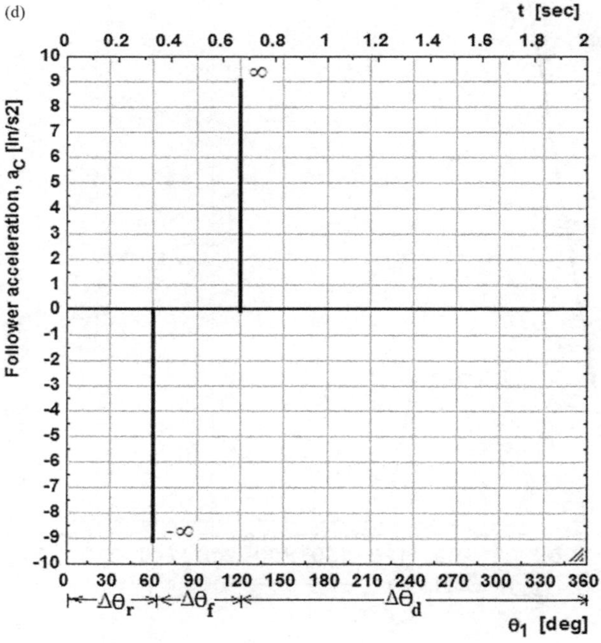

FIGURE 3.48 (CONTINUED)
(a) Follower displacement; (b) follower displacement; (c) follower velocity; (d) follower acceleration; (e) cam contour.

(Continued)

(e)

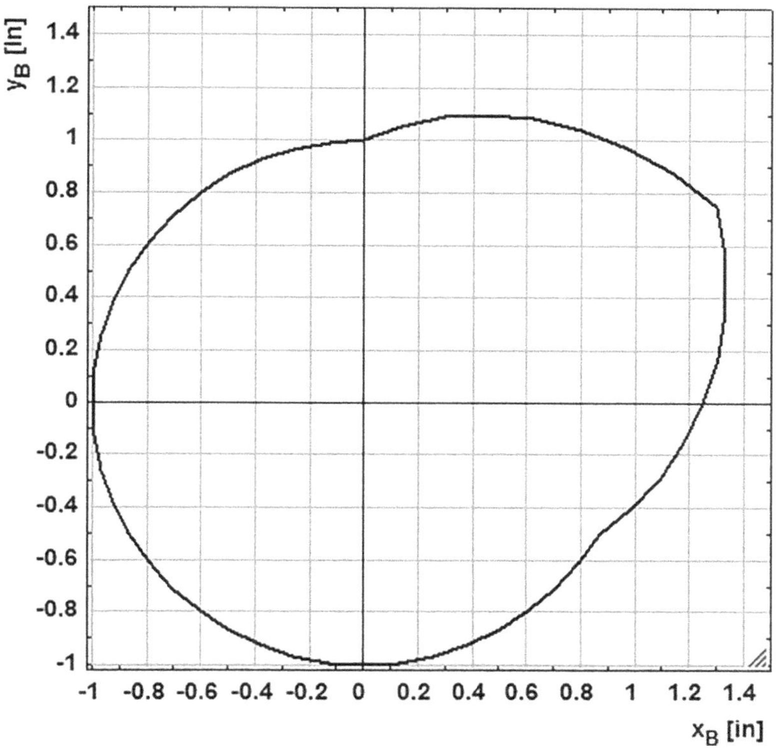

FIGURE 3.48 (CONTINUED)
(a) Follower displacement; (b) follower displacement; (c) follower velocity; (d) follower acceleration; (e) cam contour.

acceleration, from SW-spreadsheets data are copied and pasted in columns inside Lookup Table. The EES plots are illustrated in Figures 3.48b–d. In the plots have been added the intervals for time or cam angle during the follower's rise, fall, and dwell.

3.6.5 Cam with Constant Acceleration Rise or Constant Acceleration Fall

This cam provides a *constant acceleration motion* during the follower's rise or fall (therefore linear velocity profile and parabolic displacement).

Create a SW motion simulation for the follower, sliding vertically along its guide. A linear motor is selected for the follower. In an SW motion window, since the follower raises and falls with constant acceleration a_C, then the displacement d_C is selected as parabolic. The Segment Type is chosen parabolic and the Start and End values from Equations (3.120a) and (3.120b).

- Displacement function for the follower's constant acceleration rise:

$$d_C = \begin{cases} 2 \cdot D \cdot \left(\dfrac{\theta_r}{\Delta\theta_r}\right)^2; & 0^\circ \le \theta_r \le \dfrac{\Delta\theta_r}{2} \\ D \cdot \left[-1 + 4 \cdot \left(\dfrac{\theta_r}{\Delta\theta_r}\right) - 2 \cdot \left(\dfrac{\theta_r}{\Delta\theta_r}\right)^2\right]; & \dfrac{\Delta\theta_r}{2} \le \theta_r \le \Delta\theta_r \end{cases} \qquad (3.120a)$$

Velocity and acceleration equations are obtained by taking the first and second derivatives of the displacement functions.

- Displacement function for the follower's constant acceleration fall:

$$d_C = \begin{cases} D \cdot \left[1 - 2 \cdot \left(\dfrac{\theta_f}{\Delta\theta_f}\right)^2\right]; & 0^\circ \le \theta_f \le \dfrac{\Delta\theta_f}{2} \\ D \cdot \left[2 - 4 \cdot \left(\dfrac{\theta_f}{\Delta\theta_f}\right) + 2 \cdot \left(\dfrac{\theta_f}{\Delta\theta_f}\right)^2\right]; & \dfrac{\Delta\theta_f}{2} \le \theta_f \le \Delta\theta_f \end{cases} \qquad (3.120b)$$

3.6.6 Cam with Harmonic Motion Rise or Fall

This type of cam provides *harmonic motion* during the follower's rise or fall.

Create an SW motion simulation for the follower, sliding vertically along its guide. A linear motor is selected for the follower. In the SW window, Figure 3.48a, the displacement d_C is harmonic function. The Motion Type is chosen oscillating and the Start and End values from Equations (3.121a) and (3.121b).

- Displacement function for the follower's harmonic motion in rise:

$$d_C = \frac{D}{2} \cdot \left(1 - \mathrm{Cos}\frac{\pi \cdot \theta_r}{\Delta\theta_r}\right); 0^\circ \le \theta_r \le \Delta\theta_r \qquad (3.121a)$$

- Displacement function for the follower's harmonic motion in fall:

$$d_C = \frac{D}{2} \cdot \left(1 + \mathrm{Cos}\frac{\pi \cdot \theta_f}{\Delta\theta_f}\right); 0^\circ \le \theta_f \le \Delta\theta_f \qquad (3.121b)$$

3.6.7 Cam with Cycloidal Motion Rise or Fall

This cam provides cycloidal *motion* during the follower's rise or fall.

From EES, a Lookup Table is created and inserted a column for the follower's displacement. Select Alter value/Enter equation. Type Equations (3.122a) and (3.122b).

- Displacement function for the follower's cycloidal motion in rise:

$$d_C = D\cdot\left(\frac{\theta_r}{\Delta\theta_r} - \frac{1}{2\pi}\mathrm{Sin}\frac{2\pi\cdot\theta_r}{\Delta\theta_r}\right);\ 0^\circ \le \theta_r \le \Delta\theta_r \tag{3.122a}$$

- Displacement function for the follower's cycloidal motion in fall:

$$d_C = D\cdot\left(1 - \frac{\theta_f}{\Delta\theta_f} + \frac{1}{2\pi}\mathrm{Sin}\frac{2\pi\cdot\theta_f}{\Delta\theta_f}\right);\ 0^\circ \le \theta_f \le \Delta\theta_f \tag{3.122b}$$

Tables with functions for different types of cams are available in [3,4].

3.7 Velocity Analysis of Single-Cycle Planar Mechanisms

The vector equation for velocities from open cycle can be written for a single closed cycle as

$$\boldsymbol{\omega}_1 \times \mathbf{L}_{AB}^0 + \boldsymbol{\omega}_2 \times \mathbf{L}_{BC}^0 + \boldsymbol{\omega}_3 \times \mathbf{L}_{CD}^0 + \cdots + \boldsymbol{\omega}_m \times \mathbf{L}_{ZE}^0 \cdots + \mathbf{v}_{A_{0,1}} + \mathbf{v}_{B_{1,2}} + \cdots + \mathbf{v}_{E_{m,0}} = 0 \tag{3.123}$$

For planar mechanisms, the direction of links absolute angular velocities is known, that is, all parallel to z_0-axis with links moving in planes parallel to (x_0, y_0):

$$\boldsymbol{\omega}_0 = \begin{Bmatrix} 0 \\ 0 \\ 0 \end{Bmatrix};\ \boldsymbol{\omega}_1 = \begin{Bmatrix} 0 \\ 0 \\ \omega_1 \end{Bmatrix};\ \boldsymbol{\omega}_2 = \begin{Bmatrix} 0 \\ 0 \\ \omega_2 \end{Bmatrix};\ \ldots\ \boldsymbol{\omega}_m = \begin{Bmatrix} 0 \\ 0 \\ \omega_m \end{Bmatrix} \tag{3.124}$$

The cross multiplications of vectors $\boldsymbol{\omega}_1 \times \mathbf{L}_{AB}^0$ have a simplified form when replaced by matrix multiplications. For example,

$$-\tilde{\mathbf{L}}^0_{AB} * \omega_1 = -\begin{bmatrix} 0 & -z_{AB} & y_{AB} \\ z_{AB} & 0 & -x_{AB} \\ -y_{AB} & x_{AB} & 0 \end{bmatrix} * \begin{Bmatrix} 0 \\ 0 \\ \omega_1 \end{Bmatrix} = \begin{Bmatrix} -y_{AB} \cdot \omega_1 \\ x_{AB} \cdot \omega_1 \\ 0 \end{Bmatrix}$$

$$= \begin{Bmatrix} -(y_B - y_A) \cdot \omega_1 \\ (x_B - x_A) \cdot \omega_1 \\ 0 \end{Bmatrix} \tag{3.125}$$

The relative velocity vectors in prismatic joints have components on x_0 and y_0 axes only. For example, a prismatic joint A:

$$\mathbf{v}_{A0,1} = \begin{Bmatrix} v^x_{A0,1} \\ v^y_{A0,1} \\ 0 \end{Bmatrix} \tag{3.126}$$

Equation (3.123) with scalar entries has the form:

$$\begin{cases} -y_{AB} \cdot \omega_1 - \cdots - y_{ZE} \cdot \omega_m + v^x_{A0,1} + v^x_{B1,2} + \cdots + v^x_{Em,0} = 0 \\ +x_{AB} \cdot \omega_1 + \cdots + x_{ZE} \cdot \omega_m + v^y_{A0,1} + v^y_{B1,2} + \cdots + v^y_{Em,0} = 0 \end{cases} \tag{3.127}$$

where the coefficients are expressed as functions of joint coordinates, for example:

$$x_{AB} = x_B - x_A;\ y_{AB} = y_B - y_A, \text{etc.}$$

Note: The kinematic Equation (3.127) was reported in [5,7,8]. They were generated from the vector equations for single and multiple cycles on a non-oriented edges graph. The program VAIT (initials from Voinea, Atanasiu, Iordache, Talpasanu), written in FORTRAN, generated the equations for kinematics of mechanisms, considering the non-oriented graph model [6]. Automatic generation of equations was achieved by using an oriented edge graph (digraph). Introducing matrices from digraphs, the equations for kinematics and dynamics of mechanisms are reported in [1,2,9,11].

Relation Between Latin Matrix L and Velocity Matroidal Matrix $\dot{\mathbf{L}}$

For the positional analysis, the Latin matrix was considered, which is given in Equation (3.128):

$$\begin{matrix} & 0 & 1 & \ldots & m \end{matrix}$$

$$[\mathbf{L}] = \left[\ \mathbf{L}_{EA} \ \vdots\ \mathbf{L}_{AB} \ \vdots\ \ldots \ \vdots\ \mathbf{L}_{ZE}\ \right] = \left[\ \mathbf{r}_A - \mathbf{r}_E \ \vdots\ \mathbf{r}_B - \mathbf{r}_A \ \vdots\ \ldots \ \vdots\ \mathbf{r}_E - \mathbf{r}_Z\ \right] \tag{3.128}$$

The entries in the left side of Equation (3.128) have been determined from a nonlinear algebraic system. The nonlinear system of equations included the equations for the sum of entries on each row from the Latin matrix as being zero, Equation (3.129), in addition to the constraint equations:

$$\mathbf{L}_{EA} + \mathbf{L}_{AB} + \mathbf{L}_{BC} + \mathbf{L}_{CD} + \cdots + \mathbf{L}_{ZE} = 0 \tag{3.129}$$

- **Velocity matroidal matrix, $\dot{\mathbf{L}}$:**
  The *velocity matroidal matrix* $\dot{\mathbf{L}}$ has the same number of rows and columns as the Latin matrix, and developed by replacing in Equation (3.128) the vectors **L** and **r** in the Latin matrix with $\dot{\mathbf{L}}$ and **v**, as shown in Equation (3.130):

$$\begin{matrix} & 0 & 1 & \dots & m \end{matrix}$$

$$[\dot{\mathbf{L}}] = \left[\begin{array}{c:c:c:c} \dot{\mathbf{L}}_{EA} & \dot{\mathbf{L}}_{AB} & \dots & \dot{\mathbf{L}}_{ZE} \end{array}\right] = \left[\begin{array}{c:c:c:c} \mathbf{v}_A - \mathbf{v}_E & \mathbf{v}_B - \mathbf{v}_A & \dots & \mathbf{v}_E - \mathbf{v}_Z \end{array}\right] \tag{3.130}$$

Statement 3.8

The sum of vector entries in the velocity matroidal matrix $\dot{\mathbf{L}}$ is zero:

$$\dot{\mathbf{L}}_{EA} + \dot{\mathbf{L}}_{AB} + \dot{\mathbf{L}}_{BC} + \dot{\mathbf{L}}_{CD} + \cdots + \dot{\mathbf{L}}_{ZE} = 0 \tag{3.131}$$

Proof: The addition of entries in the right side, Equation (3.130) is zero.

The *velocity matroidal vectors*, noted $\dot{\mathbf{L}}$, are the terms within the sum in the previous equation. Each vector is defined as follows:

The vector on column 0 (fixed link, with the origin O):

$$\dot{\mathbf{L}}_{EA} = \mathbf{v}_A - \mathbf{v}_E = \dot{\mathbf{L}}_{OA} - \dot{\mathbf{L}}_{OE} \tag{3.132a}$$

where A is the first joint in the cycle and E the last joint.

- If both are fixed joints, then

$$\dot{\mathbf{L}}_{EA} = 0 \tag{3.132b}$$

- If any of them is a prismatic joint, for example, A is prismatic, then

$$\mathbf{v}_A = \dot{\mathbf{L}}_{OA} = \mathbf{c}_A \cdot \mathbf{u}_A^0 \cdot \dot{d}_A \tag{3.132c}$$

  where
  $\mathbf{u}_A^0$—The unit vector for axis of relative motion in joint A, which joins link 1 to the fixed link 0.
  $\mathbf{c}_A$—The oriented edge's A sign with value 0, 1, or −1 if edge A is not in the cycle, is in the cycle and in the same direction, or inverse to the cycle's direction, respectively.
  $\dot{d}_A$—The scalar for linear relative velocity at joint A.

- A vector $\dot{\mathbf{L}}$ on columns 1, 2, …, m, corresponding to mobile links, is defined as follows:

  In case $\mathbf{L}^0_{AB}$ is a constant magnitude vector between two revolute joints A and B, then

$$\dot{\mathbf{L}}_{AB} = \mathbf{v}_B - \mathbf{v}_A = \boldsymbol{\omega}_1 \times \mathbf{L}^0_{AB} = -\tilde{\mathbf{L}}^0_{AB} * \boldsymbol{\omega}_1 \quad (3.132d)$$

  It represents Euler formula for the difference in velocities of two points A and B located on link 1.

  In case $\mathbf{L}^0_{AG_1}$ is a constant magnitude vector between joint A and center of mass G_1, then

$$\dot{\mathbf{L}}^0_{AG_1} = \mathbf{v}^0_{AG_1} - \mathbf{v}_A = \boldsymbol{\omega}_1 \times \mathbf{L}^0_{AG_1} = -\tilde{\mathbf{L}}^0_{AG_1} * \boldsymbol{\omega}_1 \quad (3.132e)$$

- In case $\mathbf{L}^0_{AB}$ is a variable magnitude vector between revolute joint A and prismatic joint B, then

$$\dot{\mathbf{L}}_{AB} = \mathbf{v}_B - \mathbf{v}_A = \boldsymbol{\omega}_1 \times \mathbf{L}^0_{AB} + \mathbf{c}_B \cdot \mathbf{u}^0_B \cdot \dot{\mathrm{d}}_B = -\tilde{\mathbf{L}}^0_{AB} * \boldsymbol{\omega}_1 + \mathbf{c}_B \cdot \mathbf{u}^0_B \cdot \dot{\mathrm{d}}_B \quad (3.132f)$$

  It represents the formula for the difference in velocities between points: A on link 1 and B on link 2.

  where

  $\mathbf{u}^0_B$—The unit vector for axis of linear motion in prismatic joint B, which joins link 2 and link 1.

  $\mathbf{c}_B$—The oriented edge's B sign with value 0, 1, or –1 if edge B is not in the cycle, is in the cycle and in the same direction, or inverse to the cycle's direction, respectively.

  $\dot{\mathrm{d}}_B$—The scalar for linear relative velocity at joint B, introduced earlier as IT notation.

THE MATROIDAL METHOD FOR VELOCITY EQUATIONS OF MECHANISMS WITH SINGLE AND MULTIPLE CYCLES

During the past two decades, the author's search was focused on which kinematic equations are independent [2,9,11]. Since the kinematic equations for multiple cycles are linear in velocities and accelerations, the search is conducted on the concept of *cycle base matroid*, which is introduced by Whitney [10]. It was adapted to mechanisms by a multiplication of the joint's relative displacement, velocity, or acceleration by the c entries from the cycle matroid matrix **C**. The three matroidal matrices **L** (Latin matrix), $\dot{\mathbf{L}}$ (velocity matroidal), and $\ddot{\mathbf{L}}$ (acceleration matroidal) presented further in Chapter 4 have the sum of entries zero. These are the independent equations, the same as the number of cycles in the cycle basis, which is coincidental with the rank of cycle matroid matrix C. In Chapter 1, the concept of matroid was introduced, and existence of multiple bases, with the same rank. Any base can be chosen to generate the kinematic equations, and the results are identical.

The name *cycle* is adopted in the text instead of similar terms used in the literature such as *loop, circuit,* and *fundamental circuit.*

Equation (3.131) is written in the equivalent form of Equation (3.133) for which the constraint equations apply:

$$\underbrace{\tilde{\mathbf{L}}^0 * \boldsymbol{\omega}_n}_{\text{Nodes}} + \underbrace{\mathbf{C}(u^0) * \dot{\mathbf{d}}_j}_{\text{Edges}} = 0 \tag{3.133}$$

Equation (3.133) generates a linear system of 2c equations for planar mechanisms, which is solved for 2c velocities.

where

n is the number of links (includes the fixed link) and j is the number of joints

$\tilde{\mathbf{L}}^0$ is a matrix with 2c rows and n columns. Its scalar entries are generated from the Latin matrix, interchanging the x-components with the y-components and then changing the sign for the y-components. Positional constraints apply to entries in matrix $\tilde{\mathbf{L}}^0$

$\boldsymbol{\omega}_n$ is a column vector with scalar entries, all the links' angular velocities which are assigned to nodes in digraph. The link constraints for motion are applied to these entries. The scalars are positive for a link in a counterclockwise rotation about z_0 and negative for a clockwise rotation.

$\mathbf{C}(u^0)$ is a matrix with 2c rows and j columns. Its scalar entries are the components for joints' unit vector components multiplied each by the corresponding entry in matrix C.

$\dot{\mathbf{d}}_j$ is a column vector with scalar entries, all the linear relative velocities in joints, which are assigned to edges in digraph. The joint constraints for motion are applied to these entries.

where $\mathbf{c}_A$, $\mathbf{c}_B$, … ,$\mathbf{c}_E$ are the entries in the cycle basis matroid **C**:

$$\begin{matrix} & \text{A} & \text{B} & \dots & \text{E} \\ C_{1,j} = & [\ c_A & | \ c_B & | \ \dots & | \ c_E\ ] \end{matrix} \tag{3.134}$$

For mechanisms with c-cycles, the matrix C is a $(c \times j)$ matrix, as shown in the following examples:

- Joints' linear velocities: Starting from joint A, the linear absolute velocities for B, C, …, D are determined by equating the column entries in the left and right sides of Equation (3.130):

$$\mathbf{v}_E = \mathbf{v}_A - \dot{\mathbf{L}}_{EA}; \ \mathbf{v}_B = \mathbf{v}_A + \dot{\mathbf{L}}_{AB}; \ \mathbf{v}_Z = \mathbf{v}_E - \dot{\mathbf{L}}_{ZE} \tag{3.135}$$

3.7.1 Velocity Analysis for Single-Cycle Planar Mechanisms with Revolute Joints: Example: The Four Bar Mechanism

For the mechanisms with all revolute joints, the constraint equations $\dot{\mathbf{d}}_j = 0_j$ are considered for null linear relative velocity in revolute joints, and $\boldsymbol{\omega}_0 = 0$ for the fixed link. Therefore, Equation (3.133) holds a simplified form since it is written as functions of nodes' angular velocities, only

$$\underbrace{\tilde{\mathbf{L}}^0 * \boldsymbol{\omega}_m}_{\text{Nodes}} = 0 \tag{3.136}$$

where

$\tilde{\mathbf{L}}$ is the 2c × m matrix with coefficients from the Latin matrix, Equation (3.133)

$\boldsymbol{\omega}_m$ is the m × 1 matrix with mobile links' angular velocities

0 is the 2c × 1 zero entries column matrix

The equations for velocities are developed from the following matrix multiplication:

$$\begin{array}{cccc} 1 & 2 & & m \end{array}$$

$$\left[\begin{array}{c|c|c|c} -y_{AB} & -y_{BC} & \cdots & -y_{ZE} \\ x_{AB} & x_{BC} & \cdots & x_{ZE} \end{array}\right] * \left\{\begin{array}{c} \omega_1 \\ \omega_2 \\ \cdots \\ \omega_m \end{array}\right\} = \left\{\begin{array}{c} 0 \\ 0 \end{array}\right\} \tag{3.137}$$

The unknowns are absolute angular velocities assigned to nodes in digraph.

Statement 3.9

For planar mechanisms with c cycles and m mobile links, the system of 2c linear equations has (m–M) unknowns

Proof: If the mechanism has m mobile links from which M of them are actuated, then m–M links have unknown angular velocities. Let us consider the mobility equation: $M = 3m - 2j$, with m the number of mobile links and j the number of revolute joints. Then, $m - M = 2j - 2m = 2c$, which is the number of equations.

Example: The four-bar mechanism in Figure 3.49a has the link lengths: $L_{AB} = 4''$, $L_{BD} = 24''$, $L_{DC} = 14''$, and $L_{CA} = 30''$. The crank 1 rotates counterclockwise with a constant angular velocity $\omega_1 = 360\ °/s = 2\pi$ rad/s.

For the interval of t = 1 s, the crank angle is $\theta_1 = 360 \cdot t\ [°/s] = 2\pi \cdot t\ [rad/s]$

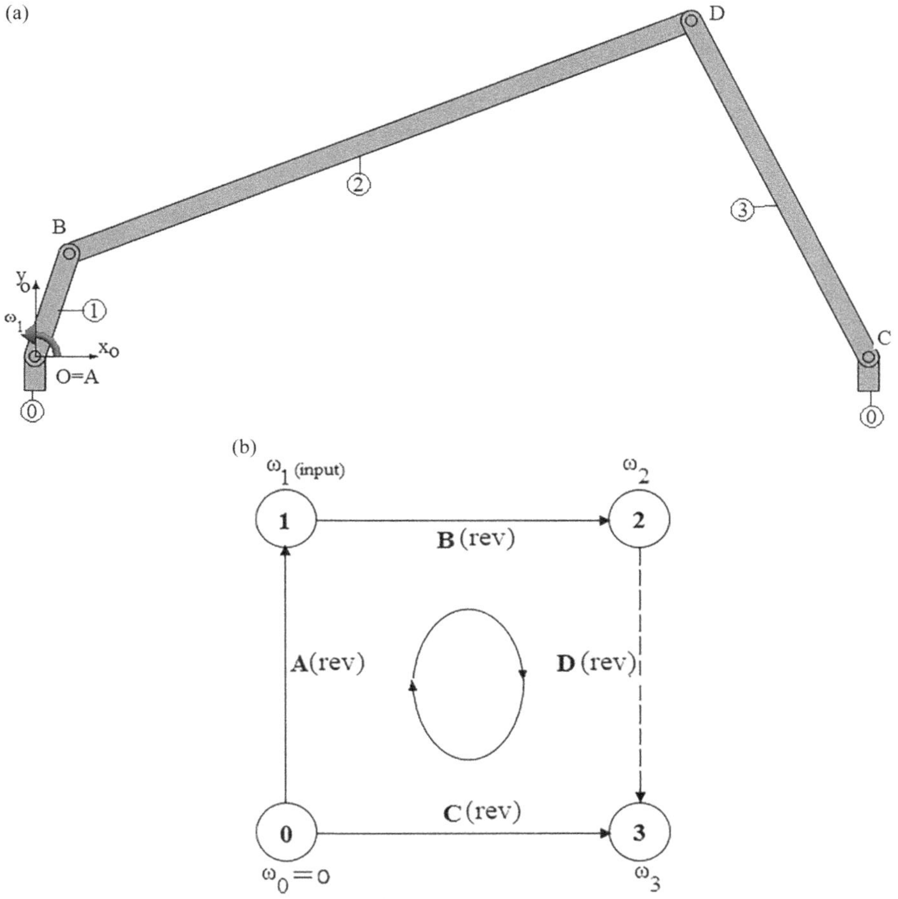

FIGURE 3.49
(a) Four-bar mechanism; (b) the variables in velocity equations are absolute angular velocities assigned to nodes in digraph.

Determine

a. The system of equations for velocity analysis
b. Angular velocities for links 2 and 3
c. Linear velocities of joints A, B, C, and D
d. Linear velocities of centers of mass G_1, G_2, and G_3

Solution

a. The system of equations for velocity analysis

From Equation (3.56), the Latin matrix with vector entries is

$$[L] = \begin{bmatrix} \overset{0}{\mathbf{L}_{CA}} & \vdots & \overset{1}{\mathbf{L}_{AB}} & \vdots & \overset{2}{\mathbf{L}_{BD}} & \vdots & \overset{3}{\mathbf{L}_{DC}} \end{bmatrix}; [L] = \begin{bmatrix} -30 & \vdots & x_{AB} & \vdots & x_{BD} & \vdots & x_{DC} \\ 0 & \vdots & y_{AB} & \vdots & y_{BD} & \vdots & y_{DC} \end{bmatrix} \tag{3.138a}$$

where $x_{AB} = L_{AB} \cdot \mathrm{Cos}(\theta_1)$; $y_{AB} = L_{AB} \cdot \mathrm{Sin}(\theta_1)$

The velocity matroidal matrix is

$$[\dot{L}] = \begin{bmatrix} \overset{0}{\dot{\mathbf{L}}_{CA}} & \vdots & \overset{1}{\dot{\mathbf{L}}_{AB}} & \vdots & \overset{2}{\dot{\mathbf{L}}_{BD}} & \vdots & \overset{3}{\dot{\mathbf{L}}_{DC}} \end{bmatrix}; \tag{3.138b}$$

$$[\dot{L}] = \begin{bmatrix} 0 & \vdots & -y_{AB} \cdot \omega_1 & \vdots & -y_{BD} \cdot \omega_2 & \vdots & -y_{DC} \cdot \omega_3 \\ 0 & \vdots & x_{AB} \cdot \omega_1 & \vdots & x_{BD} \cdot \omega_2 & \vdots & x_{DC} \cdot \omega_3 \end{bmatrix}$$

where the vectors $\dot{\mathbf{L}}$ on columns in the velocity matroidal matrix are expressed as functions of 4c = 4 coordinates from positional equations: x_{BD}, y_{BD}, x_{DC}, y_{DC}, and 2c = 2 unknown angular velocities: ω_2 and ω_3

- *Column 0*: From Equation (3.132b):

$$\dot{\mathbf{L}}_{CA} = \dot{\mathbf{L}}_{OA} - \dot{\mathbf{L}}_{OC} = \begin{Bmatrix} 0 \\ 0 \end{Bmatrix} \tag{3.138c}$$

- *Column 1*: From Equation (3.132d):

$$\dot{\mathbf{L}}_{AB} = -\tilde{\mathbf{L}}_{AB}^{0} * \omega_1 = \begin{Bmatrix} -y_{AB} \cdot \omega_1 \\ x_{AB} \cdot \omega_1 \end{Bmatrix} = \begin{Bmatrix} -4 \cdot \omega_1 \cdot \sin(\theta_1) \\ 4 \cdot \omega_1 \cdot \cos(\theta_1) \end{Bmatrix} \tag{3.138d}$$

- *Column 2*: From Equation (3.132d):

$$\dot{\mathbf{L}}_{BD} = -\tilde{\mathbf{L}}_{BD}^{0} * \omega_2 = \begin{Bmatrix} -y_{BD} \cdot \omega_2 \\ x_{BD} \cdot \omega_2 \end{Bmatrix} \tag{3.138e}$$

- *Column 3*: From Equation (3.132d):

$$\dot{\mathbf{L}}_{DC} = -\tilde{\mathbf{L}}_{DC}^{0} * \omega_3 = \begin{Bmatrix} -y_{DC} \cdot \omega_3 \\ x_{DC} \cdot \omega_3 \end{Bmatrix} \tag{3.138f}$$

The sum of entries on each row in Equation (3.138b) is zero:

$$\begin{Bmatrix} \sum \dot{L}_x = 0 \\ \sum \dot{L}_y = 0 \end{Bmatrix} \Rightarrow \overset{\begin{matrix} 1 & & 2 & & 3 \end{matrix}}{\begin{bmatrix} -y_{AB} & \vdots & -y_{BD} & \vdots & -y_{DC} \\ x_{AB} & \vdots & x_{BD} & \vdots & x_{DC} \end{bmatrix}} * \begin{Bmatrix} \omega_1 \\ \omega_2 \\ \omega_3 \end{Bmatrix} = \begin{Bmatrix} 0 \\ 0 \end{Bmatrix} \quad (3.138g)$$

The unknowns are absolute angular velocities assigned to nodes in digraph (Figure 3.49b).

Therefore, a system of $2c = 2$ linear equations is solved for $m - M = 3 - 1 = 2$ unknowns (ω_2 and ω_3).

b. Angular velocities for links 2 and 3

With the notation:

$$\Delta = x_{BD} \cdot y_{DC} - x_{DC} \cdot y_{BD} \quad (3.138h)$$

The solution is

$$\begin{cases} \omega_2 = \dfrac{4 \cdot \omega_1}{\Delta}\left[x_{DC} \cdot \mathrm{Sin}(\theta_1) - y_{DC} \cdot \mathrm{Cos}(\theta_1)\right] \\ \omega_3 = \dfrac{4 \cdot \omega_1}{\Delta}\left[y_{BD} \cdot \mathrm{Cos}(\theta_1) - x_{BD} \cdot \mathrm{Sin}(\theta_1)\right] \end{cases} \quad (3.138i)$$

Figure 3.50 illustrates the links 2 and 3 angular velocities for a complete revolution of link 1.

Note: For the four-bar mechanisms where $\Delta = 0$ holds, there is a singularity for velocities. This condition is valid if vectors L_{BD} and L_{DC} are collinear.

The mechanism in this example has $L_{AB} = 4''$, $L_{BD} = 24''$, $L_{DC} = 14''$, and $L_{CA} = 30''$.

Since $S + L < P + Q$ and the shortest link $S = L_{AB}$ is located as a side, then it is a *crank-rocker* mechanism, the shortest link 1 is the crank undergoing full rotation, whereas link 3 is the rocker (oscillating link). For this mechanism, the links 2 and 3, with lengths L_{BD} and L_{DC}, do not align.

A geometrically proof: Let us assume L_{BD} and L_{DC} become collinear, then the cosines law in triangle ABC holds:

$$(L_{BD} + L_{DC})^2 = L_{AB}^2 + L_{CA}^2 - 2 \cdot L_{AB} \cdot L_{CA} \cdot \mathrm{Cos}(\theta_1) \quad (3.138j)$$

Equation (3.138j) becomes $\cos(\theta_1) = -2.2$, and since there is no crank angle θ_1 to satisfy this equation, then links L_{BD} and L_{DC} do not align.

The mechanism in Figure 3.13 with given $L_{AB} = 14''$, $L_{BD} = 4''$, $L_{DC} = 30''$, and $L_{CA} = 24''$ has $S = L_{BD} = 4''$, and $L = L_{DC} = 30''$. From Grashof's

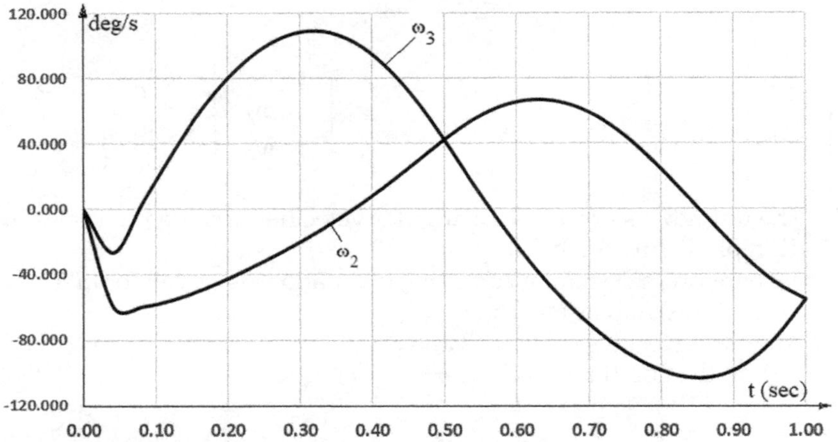

FIGURE 3.50
Links' angular velocities.

criterion: $S+L<P+Q$, and the shortest link $S = L_{BD}$ is located as a coupler link, then it is a *double-rocker* mechanism, both links 1 and 3 are rocking (oscillating links).

Equation (3.138j) becomes $\text{Cos}(\theta_1) = -0.571$, then for crank angle $\theta_1 = 124.82°$, links L_{BD} and L_{DC} are in extension. This position defines the range limit for rocker's L_{DC} rotation which cannot rotate further. In this position, there is a singularity for velocities, due to division by $\Delta = 0$ in Equation (3.138i). In Chapter 4, the reactions at joints become very large in such position is shown. The mechanism designer should be aware about such position. For a motor placed at A, the mechanism will stay locked when L_{BD} and L_{DC} are in extension. As an alternative, the actuation is placed at joint B.

The determinant with columns 1 and 2 entries in matrix $\tilde{\mathbf{L}}^0$ determines the dead positions, where the crank and the coupler are aligned (Equation (3.45)). The determinant Δ has the columns' 2 and 3 entries in matrix $\tilde{\mathbf{L}}^0$, and if becomes zero, locates the singularities for velocities. If both determinants are zero, then the rank 2c of matrix $\tilde{\mathbf{L}}^0$ becomes at most $2c - 1$. Geometrically, the change in rank is viewed as change in configuration of lines for relative motion. Thus, the fascicle of lines for relative revolute motion allowed at joints A, B, C, and D changes the configuration; from a fascicle of all parallel to z_0 lines to all parallel and coplanar (their points of interception with x_0y_0 are collinear). Therefore, positions where the rank of Latin matrix changes locate singular positions.

c. Linear velocities of joints A, B, C, and D

The matroidal velocity with vector entries is

$$
\begin{array}{c} \\ [\dot{\mathbf{L}}] \\ \\ \end{array}
\begin{array}{l} \quad\;\; 0 \qquad\quad 1 \qquad\quad 2 \qquad\quad 3 \\ =\left[\begin{array}{c|c|c|c} \dot{\mathbf{L}}_{CA} & \dot{\mathbf{L}}_{AB} & \dot{\mathbf{L}}_{BD} & \dot{\mathbf{L}}_{DC} \end{array}\right] \\ =\left[\begin{array}{c|c|c|c} \mathbf{v}_A-\mathbf{v}_C & \mathbf{v}_B-\mathbf{v}_A & \mathbf{v}_D-\mathbf{v}_B & \mathbf{v}_C-\mathbf{v}_D \end{array}\right] \end{array}
\tag{3.138k}
$$

And with scalar entries as

$$
\begin{array}{cccc} 0 & 1 & 2 & 3 \end{array}
$$

$$
\left[\begin{array}{c|c|c|c} 0 & -y_{AB}\cdot\omega_1 & -y_{BD}\cdot\omega_2 & -y_{DC}\cdot\omega_3 \\ 0 & x_{AB}\cdot\omega_1 & x_{BD}\cdot\omega_2 & x_{DC}\cdot\omega_3 \end{array}\right]
=\left[\begin{array}{c|c|c|c} v_A^x-v_C^x & v_B^x-v_A^x & v_D^x-v_B^x & v_C^x-v_D^x \\ v_A^y-v_C^y & v_B^y-v_A^y & v_D^y-v_B^y & v_C^y-v_D^y \end{array}\right]
\tag{3.138l}
$$

Starting from origin O coincidental with the fixed joint A ($\mathbf{v}_A = 0$), the linear absolute velocities for B C, and D are determined by equating the column entries in the left and right sides of Equation (3.138l).

Joint A: A is fixed joint:

$$
\mathbf{v}_A = 0;\ \text{vector form}
$$

$$
\left\{\begin{array}{c} v_A^x \\ v_A^y \end{array}\right\} = \left\{\begin{array}{c} 0 \\ 0 \end{array}\right\};\ \text{scalar components} \tag{3.138m}
$$

$$
v_A = \sqrt{\left(v_A^x\right)^2 + \left(v_A^y\right)^2} = 0;\ \text{magnitude}
$$

Joint B:

$$
\mathbf{v}_B = \mathbf{v}_A + \dot{\mathbf{L}}_{AB};\ \text{vector form}
$$

$$
\begin{aligned}
\left\{\begin{array}{c} v_B^x \\ v_B^y \end{array}\right\} &= \left\{\begin{array}{c} v_A^x \\ v_B^y \end{array}\right\} + \left\{\begin{array}{c} -y_{AB}\cdot\omega_1 \\ x_{AB}\cdot\omega_1 \end{array}\right\} \\
&= \left\{\begin{array}{c} -4\cdot\omega_1\cdot \mathrm{Sin}(\theta_1) \\ 4\cdot\omega_1\cdot \mathrm{Cos}(\theta_1) \end{array}\right\};\ \text{scalar components}
\end{aligned}
\tag{3.138n}
$$

$$
v_B = \sqrt{\left(v_B^x\right)^2 + \left(v_B^y\right)^2} = 4\cdot\omega_1;\ \text{magnitude}
$$

Joint D:

$$\mathbf{v}_D = \mathbf{v}_B + \dot{\mathbf{L}}_{BD}\text{; vector form}$$

$$\begin{Bmatrix} v_D^x \\ v_D^y \end{Bmatrix} = \begin{Bmatrix} v_B^x \\ v_B^y \end{Bmatrix} + \begin{Bmatrix} -y_{BD} \cdot \omega_2 \\ x_{BD} \cdot \omega_2 \end{Bmatrix}$$

$$= \begin{Bmatrix} -4 \cdot \omega_1 \cdot \mathrm{Sin}(\theta_1) - y_{BD} \cdot \omega_2 \\ 4 \cdot \omega_1 \cdot \mathrm{Cos}(\theta_1) + x_{BD} \cdot \omega_2 \end{Bmatrix}\text{; scalar components} \tag{3.138o}$$

$$v_D = \sqrt{\left(v_D^x\right)^2 + \left(v_D^y\right)^2}\text{; magnitude}$$

Joint C:

$$\mathbf{v}_C = \mathbf{v}_D + \dot{\mathbf{L}}_{DC}\text{; vector form}$$

$$\begin{Bmatrix} v_C^x \\ v_C^y \end{Bmatrix} = \begin{Bmatrix} v_D^x \\ v_D^y \end{Bmatrix} + \begin{Bmatrix} -y_{DC} \cdot \omega_3 \\ x_{DC} \cdot \omega_3 \end{Bmatrix} \tag{3.138p}$$

$$= \begin{Bmatrix} -4 \cdot \omega_1 \cdot \mathrm{Sin}(\theta_1) - y_{BD} \cdot \omega_2 - y_{DC} \cdot \omega_3 \\ 4 \cdot \omega_1 \cdot \mathrm{Cos}(\theta_1) + x_{BD} \cdot \omega_2 + x_{DC} \cdot \omega_3 \end{Bmatrix} = \begin{Bmatrix} 0 \\ 0 \end{Bmatrix}$$

The scalar components are zero, according to Equation (3.138g):

$$v_C = \sqrt{\left(v_C^x\right)^2 + \left(v_C^y\right)^2} = 0\text{; (fixed joint)}$$

d. Linear velocities of centers of mass G_1, G_2, and G_3

Center of Mass Link 1: Equation (3.132e) between velocities of A and G_1, both located on link 1:

$$\mathbf{v}_{G_1}^0 = \mathbf{v}_A + \dot{\mathbf{L}}_{AG_1}^0 = \mathbf{v}_A - \tilde{\mathbf{L}}_{AG_1}^0 * \omega_1\text{; vector form}$$

$$\begin{Bmatrix} v_{G_1}^x \\ v_{G_1}^y \end{Bmatrix} = \begin{Bmatrix} v_A^x \\ v_A^y \end{Bmatrix} + \begin{Bmatrix} -y_{AG_1} \cdot \omega_1 \\ x_{AG_1} \cdot \omega_1 \end{Bmatrix}$$

$$= \begin{Bmatrix} -y_{AG_1} \cdot \omega_1 \\ x_{AG_1} \cdot \omega_1 \end{Bmatrix}\text{; scalar components} \tag{3.138q}$$

$$v_{G_1} = \sqrt{\left(v_{G_1}^x\right)^2 + \left(v_{G_1}^y\right)^2}\text{; magnitude}$$

Center of Mass Link 2: Equation (3.132e) between velocities of B and G_2, both located on link 2:

$$\mathbf{v}_{G_2}^0 = \mathbf{v}_B - \tilde{\mathbf{L}}_{BG_2}^0 * \boldsymbol{\omega}_2 \text{; vector form}$$

$$\begin{Bmatrix} v_{G_2}^x \\ v_{G_2}^y \end{Bmatrix} = \begin{Bmatrix} v_B^x \\ v_B^y \end{Bmatrix} + \begin{Bmatrix} -y_{BG_2}^0 \cdot \omega_2 \\ x_{BG_2}^0 \cdot \omega_2 \end{Bmatrix}$$

$$= \begin{Bmatrix} -4 \cdot \omega_1 \cdot \mathrm{Sin}(\theta_1) - y_{BG_2}^0 \cdot \omega_2 \\ 4 \cdot \omega_1 \cdot \mathrm{Cos}(\theta_1) + x_{BG_2}^0 \cdot \omega_2 \end{Bmatrix}\text{; scalar components} \tag{3.138r}$$

$$v_{G_2} = \sqrt{\left(v_{G_2}^x\right)^2 + \left(v_{G_2}^y\right)^2}\text{; magnitude}$$

Center of Mass Link 3: Equation (3.132e) between velocities of D and G_3, both located on link 3:

$$\mathbf{v}_{G_3}^0 = \mathbf{v}_D - \tilde{\mathbf{L}}_{DG_3}^0 * \boldsymbol{\omega}_3 \text{; vector form}$$

$$\begin{Bmatrix} v_{G_3}^x \\ v_{G_3}^y \end{Bmatrix} = \begin{Bmatrix} v_D^x \\ v_D^y \end{Bmatrix} + \begin{Bmatrix} -y_{DG_3}^0 \cdot \omega_3 \\ x_{DG_3}^0 \cdot \omega_3 \end{Bmatrix}$$

$$= \begin{Bmatrix} -4 \cdot \omega_1 \cdot \mathrm{Sin}(\theta_1) - y_{BD}^0 \cdot \omega_2 - y_{DG_3}^0 \cdot \omega_3 \\ 4 \cdot \omega_1 \cdot \mathrm{Cos}(\theta_1) + x_{BD}^0 \cdot \omega_2 + x_{DG_3}^0 \cdot \omega_3 \end{Bmatrix}\text{; scalar components}$$

$$v_{G_3} = \sqrt{\left(v_{G_3}^x\right)^2 + \left(v_{G_3}^y\right)^2}\text{; magnitude} \tag{3.138s}$$

3.7.2 Velocity Analysis for Single-Cycle Planar Mechanisms with Revolute and Prismatic Joints: Example: The Crank Slider Mechanism

For planar mechanisms with revolute and prismatic joints, Equation (3.133) applies. The first term is written around digraph's nodes, with the nodes' angular velocities $\boldsymbol{\omega}_n$ and the matrix of coefficients $\tilde{\mathbf{L}}^0$. The second term is written along digraph's edges along the cycle, with the assigned relative linear velocities in prismatic joints $\dot{\mathbf{d}}_j$. The matrix of coefficients $\mathbf{C}\left(\mathbf{u}^0\right)$ are the scalar entries of joint unit vectors:

$$\underbrace{\tilde{\mathbf{L}}^0 * \omega_n}_{\text{Nodes}} + \underbrace{\mathbf{C}(\mathbf{u}^0) * \dot{\mathbf{d}}_j}_{\text{Edges}} = 0 \tag{3.139}$$

The constraint equations apply for the entries in the links' absolute angular velocity matrix and in the prismatic joints' linear velocity matrix $\dot{\mathbf{d}}_j$. According to Statement 3.9, if the planar mechanism has c cycles and n = m + 1 links, then the system of 2c linear equations is solved for 2c unknowns.

Example: The crank slider mechanism, Figure 3.51a, has the crank 1 in counterclockwise rotation with an angular velocity $\omega_1 = 360\ °/s = 2\pi$ rad/s. The link lengths are as follows: $L_{AB} = 4''$, $L_{BD} = 24''$, and $L_{DC} = 0''$.

Determine

a. The system of equations for velocity analysis
b. The angular velocity for link 2 and the relative linear velocity of link 3
c. Linear velocities of joints A, B, C, and D
d. Linear velocities of centers of mass G_1, G_2, and G_3

Solution

a. The system of equations for velocity analysis

A system of 2c = 2 linear equations is solved for m − M = 3 − 1 = 2 unknowns. The variables are absolute angular velocities assigned to nodes in digraph and linear velocities assigned to edges (Figure 3.51b).

The Latin matrix entries are from Equation (3.75) as functions of the variable crank angle θ_1. The crank is considered link 1 with: $x_{AB} = L_{AB} \cdot \text{Cos}(\theta_1)$; $y_{AB} = L_{AB} \cdot \text{Sin}(\theta_1)$. From the positional analysis, the variables x_{BD}, y_{BD} were tabulated as functions of crank angle θ_1:

$$\begin{matrix} & 0 & 1 & 2 & 3 \end{matrix}$$

$$[L] = \left[\begin{array}{c|c|c|c} x_{CA} & x_{AB} & x_{BD} & x_{DC} \\ y_{CA} & y_{AB} & y_{BD} & y_{DC} \end{array}\right] \tag{3.140a}$$

$$= \left[\begin{array}{c|c|c|c} (d_C - 27.979) & x_{AB} & x_{BD} & 0 \\ -1 & y_{AB} & y_{BD} & 0 \end{array}\right]$$

The coefficients' matrix for angular velocities is

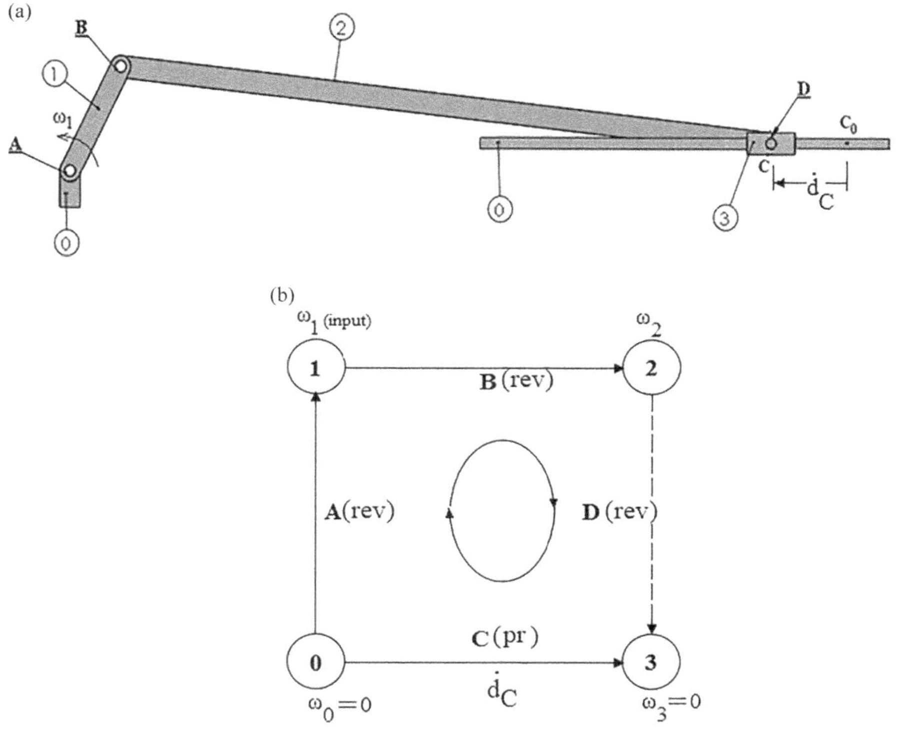

FIGURE 3.51
(a) Crank slider mechanism; (b) the variables in velocity equations are absolute angular velocities assigned to nodes and relative linear velocities assigned to the edge (prismatic joint) in the digraph.

$$[\tilde{L}] = \left[\begin{array}{c|c|c|c} -y_{CA} & -y_{AB} & -y_{BD} & -y_{DC} \\ x_{CA} & x_{AB} & x_{BD} & x_{DC} \end{array}\right]$$

$$= \begin{bmatrix} 1 & -4\cdot \mathrm{Sin}(\theta_1) & -y_{BD} & 0 \\ (d_C - 27.979) & 4\cdot \mathrm{Cos}(\theta_1) & x_{BD} & 0 \end{bmatrix} \tag{3.140b}$$

The cycle basis matroid matrix, **C**, from Equation (3.64) is

$$\begin{matrix} & A & B & C & D \\ [C] = [& 1 & 1 & -1 & 1 &] \end{matrix} \tag{3.140c}$$

The matrix $\mathbf{C}(u^0)$ the components for joints' unit vector components multiplied each by the corresponding entry in matrix **C**:

$$\left[C(u^0)\right]=\begin{bmatrix} u_A^x & u_B^x & -u_C^x & u_D^x \\ u_A^y & u_B^y & -u_C^y & u_D^y \end{bmatrix}=\begin{bmatrix} 0 & 0 & -\text{Cos}\,0° & 0 \\ 0 & 0 & -\text{Sin}\,0° & 0 \end{bmatrix} \quad (3.140d)$$

Equation (3.139) becomes

$$\begin{bmatrix} 1 & -4\cdot\text{Sin}(\theta_1) & -y_{BD} & 0 \\ (d_C-27.979) & 4\cdot\text{Cos}(\theta_1) & x_{BD} & 0 \end{bmatrix} * \begin{Bmatrix} \omega_0(=0) \\ \omega_1(=\text{input}) \\ \omega_2 \\ \omega_3(=0) \end{Bmatrix}$$

$$+\begin{bmatrix} 0 & 0 & -1 & 0 \\ 0 & 0 & 0 & 0 \end{bmatrix} * \begin{Bmatrix} \dot{d}_A(=0) \\ \dot{d}_B(=0) \\ \dot{d}_C \\ \dot{d}_D(=0) \end{Bmatrix} = \begin{Bmatrix} 0 \\ 0 \end{Bmatrix} \quad 3.140e)$$

The terms within brackets illustrate the constraint velocities. Each zero entry in column matrices indicates the deleted column of coefficients. Thus, the first and fourth columns in $\tilde{\mathbf{L}}^0$, and first, second, and fourth columns in **C** are all deleted:

$$\begin{bmatrix} -4\cdot\text{Sin}(\theta_1) & -y_{BD} & -1 \\ 4\cdot\text{Cos}(\theta_1) & x_{BD} & 0 \end{bmatrix} * \begin{Bmatrix} \omega_1(=\text{input}) \\ \omega_2 \\ \dot{d}_C \end{Bmatrix} = \begin{Bmatrix} 0 \\ 0 \end{Bmatrix} \quad (3.140f)$$

The linear system of $2c = 2$ scalar equations with $n - 1 - M = 4 - 1 - 1 = 2$ unknowns (ω_2 and $\dot{d}_C$) is

$$\begin{cases} -y_{BD}\cdot\omega_2 - \dot{d}_C = 4\cdot\omega_1\cdot\text{Sin}(\theta_1) \\ x_{BD}\cdot\omega_2 = -4\cdot\omega_1\cdot\text{Cos}(\theta_1) \end{cases} \quad (3.140g)$$

b. The solution for angular velocity of link 2 and the relative linear velocity of link 3:

$$\begin{cases} \omega_2 = \dfrac{-4\cdot\omega_1\cdot\cos(\theta_1)}{x_{BD}} \\ \dot{d}_C = \dfrac{4\cdot\omega_1\cdot y_{BD}\cdot\cos(\theta_1)}{x_{BD}} - 4\cdot\omega_1\cdot\sin(\theta_1) \end{cases} \quad (3.140h)$$

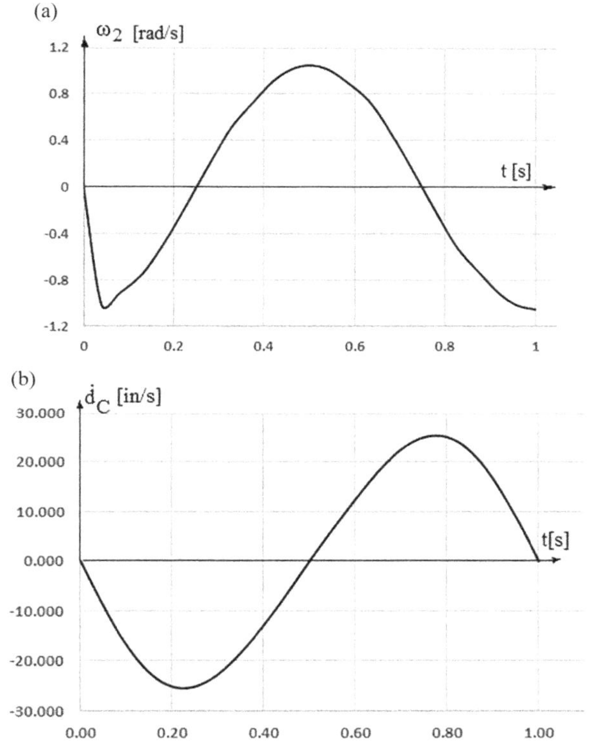

FIGURE 3.52
(a) Angular velocity for link 2; (b) linear velocity for link 3.

Figure 3.52 illustrates the links 2 and 3 angular velocities for a complete revolution of link 1.

c. Linear velocities of joints A, B, C, and D

The velocity matroidal matrix is

$$[\dot{\mathbf{L}}] = \begin{array}{c} \begin{array}{cccc} 0 & 1 & 2 & 3 \end{array} \\ \left[\begin{array}{c:c:c:c} \dot{\mathbf{L}}_{CA} & \dot{\mathbf{L}}_{AB} & \dot{\mathbf{L}}_{BD} & \dot{\mathbf{L}}_{DC} \end{array} \right] \end{array}$$

$$[\dot{\mathrm{L}}] = \left[\begin{array}{c:c:c:c} -\dot{d}_C & -4\cdot\omega_1\cdot\mathrm{Sin}(\theta_1) & -y_{BD}\cdot\omega_2 & 0 \\ 0 & 4\cdot\omega_1\cdot\mathrm{Cos}(\theta_1) & x_{BD}\cdot\omega_2 & 0 \end{array} \right] \quad (3.140i)$$

The vectors' $\dot{\mathbf{L}}$ coordinates, columns in velocity matroidal, are expressed as functions of x_{BD}, y_{BD} (from positional equations) and $2c = 2$ velocities determined from previously determined velocities ω_2 and $\dot{d}_C$.

- *Column 0*: From Equation (3.132a), with revolute joint A (fixed) and prismatic joint C:

$$\dot{\mathbf{L}}_{CA} = \dot{\mathbf{L}}_{OA} - \dot{\mathbf{L}}_{OC} = 0 - \dot{\mathbf{L}}_{OC} = -\mathbf{u}_C \cdot \dot{d}_C = -\begin{Bmatrix} \text{Cos}\,0° \\ \text{Sin}\,0° \end{Bmatrix} \cdot \dot{d}_C = \begin{Bmatrix} -\dot{d}_C \\ 0 \end{Bmatrix} \quad (3.140j)$$

- *Column 1*: Since $\dot{d}_B = 0$ for revolute joint B, then

$$\dot{\mathbf{L}}_{AB} = -\tilde{\mathbf{L}}^0_{AB} * \omega_1 + \mathbf{u}_B \cdot \dot{d}_B = \begin{Bmatrix} -y_{AB} \cdot \omega_1 \\ x_{AB} \cdot \omega_1 \end{Bmatrix} = \begin{Bmatrix} -4 \cdot \omega_1 \cdot \text{Sin}(\theta_1) \\ 4 \cdot \omega_1 \cdot \text{Cos}(\theta_1) \end{Bmatrix} \quad (3.140k)$$

- *Column 2*: Since $\dot{d}_D = 0$ for revolute joint D, then

$$\dot{\mathbf{L}}_{BD} = -\tilde{\mathbf{L}}^0_{BD} * \omega_2 + \mathbf{u}_D \cdot \dot{d}_D = \begin{Bmatrix} -y_{BD} \cdot \omega_2 \\ x_{BD} \cdot \omega_2 \end{Bmatrix} \quad (3.140l)$$

- *Column 3*: The points D and C are coincidental on link 3. The link 3 (slider) is in translation, $\omega_3 = 0$:

$$\dot{\mathbf{L}}_{DC} = -\tilde{\mathbf{L}}^0_{DC} * \omega_3 = \begin{Bmatrix} -y_{DC} \cdot \omega_3 \\ x_{DC} \cdot \omega_3 \end{Bmatrix} = \begin{Bmatrix} 0 \\ 0 \end{Bmatrix} \quad (3.140m)$$

The entries in velocity matroidal are written as functions between linear velocities as

$$\begin{aligned} & \qquad\quad\; 0 \qquad\;\; 1 \qquad\;\; 2 \qquad\;\; 3 \\ [\dot{\mathbf{L}}] &= \left[\begin{array}{c:c:c:c} \dot{\mathbf{L}}_{CA} & \dot{\mathbf{L}}_{AB} & \dot{\mathbf{L}}_{BD} & \dot{\mathbf{L}}_{DC} \end{array}\right] \\ &= \left[\begin{array}{c:c:c:c} \mathbf{v}_A - \mathbf{v}_C & \mathbf{v}_B - \mathbf{v}_A & \mathbf{v}_D - \mathbf{v}_B & \mathbf{v}_C - \mathbf{v}_D \end{array}\right]. \end{aligned} \quad (3.140n)$$

Considering Equation (3.140i) and (3.140n), then

$$\begin{array}{cccc} 0 & 1 & 2 & 3 \end{array}$$

$$\left[\begin{array}{c:c:c:c} -\dot{d}_C & -4 \cdot \omega_1 \cdot \text{Sin}(\theta_1) & -y_{BD} \cdot \omega_2 & 0 \\ 0 & 4 \cdot \omega_1 \cdot \text{Cos}(\theta_1) & x_{BD} \cdot \omega_2 & 0 \end{array}\right]$$

$$= \left[\begin{array}{c:c:c:c} v^x_A - v^x_C & v^x_B - v^x_A & v^x_D - v^x_B & v^x_C - v^x_D \\ v^y_A - v^y_C & v^y_B - v^y_A & v^y_D - v^y_B & v^y_C - v^y_D \end{array}\right] \quad (3.140o)$$

Starting from origin O coincidental with the fixed joint A ($\mathbf{v}_A = 0$), the linear absolute velocities for B, C, and D are determined by equating the column entries in the left and right sides of Equation (3.140o).

Joint A: A is fixed joint:

$$\mathbf{v}_A = 0; \text{ vector form}$$

$$\begin{Bmatrix} v_A^x \\ v_A^y \end{Bmatrix} = \begin{Bmatrix} 0 \\ 0 \end{Bmatrix}; \text{ scalar components} \tag{3.140p}$$

$$v_A = \sqrt{\left(v_A^x\right)^2 + \left(v_A^y\right)^2} = 0; \text{ magnitude}$$

Joint C:

$$\mathbf{v}_C = \mathbf{v}_A - \dot{\mathbf{L}}_{CA}; \text{ vector form}$$

$$\begin{Bmatrix} v_C^x \\ v_C^y \end{Bmatrix} = \begin{Bmatrix} v_A^x \\ v_A^y \end{Bmatrix} - \begin{Bmatrix} -\dot{d}_C \\ 0 \end{Bmatrix} = \begin{Bmatrix} \dot{d}_C \\ 0 \end{Bmatrix}; \text{ scalar components} \tag{3.140q}$$

$$v_C = \sqrt{\left(v_C^x\right)^2 + \left(v_C^y\right)^2} = \left|\dot{d}_C\right|; \text{ magnitude}$$

Joint B:

$$\mathbf{v}_B = \mathbf{v}_A + \dot{\mathbf{L}}_{AB}; \text{ vector form}$$

$$\begin{Bmatrix} v_B^x \\ v_B^y \end{Bmatrix} = \begin{Bmatrix} v_A^x \\ v_A^y \end{Bmatrix} + \begin{Bmatrix} -y_{AB} \cdot \omega_1 \\ x_{AB} \cdot \omega_1 \end{Bmatrix} = \begin{Bmatrix} -4 \cdot \omega_1 \cdot \mathrm{Sin}(\theta_1) \\ 4 \cdot \omega_1 \cdot \mathrm{Cos}(\theta_1) \end{Bmatrix}; \text{ scalar components}$$

$$v_B = \sqrt{\left(v_B^x\right)^2 + \left(v_B^y\right)^2} = 4 \cdot \omega_1; \text{ magnitude}$$

(3.140r)

Joint D:

$$\mathbf{v}_D = \mathbf{v}_B + \dot{\mathbf{L}}_{BD}; \text{ vector form}$$

$$\begin{Bmatrix} v_D^x \\ v_D^y \end{Bmatrix} = \begin{Bmatrix} v_B^x \\ v_B^y \end{Bmatrix} + \begin{Bmatrix} -y_{BD} \cdot \omega_2 \\ x_{BD} \cdot \omega_2 \end{Bmatrix}$$

$$= \begin{Bmatrix} -4 \cdot \omega_1 \cdot \mathrm{Sin}(\theta_1) - y_{BD} \cdot \omega_2 \\ 4 \cdot \omega_1 \cdot \mathrm{Cos}(\theta_1) + x_{BD} \cdot \omega_2 \end{Bmatrix}; \text{ scalar components} \tag{3.140s}$$

$$v_D = \sqrt{\left(v_D^x\right)^2 + \left(v_D^y\right)^2}; \text{ magnitude}$$

Note: A checkup equation results from equating the last columns in Equation (2.140o):

$$\mathbf{v}_C = \mathbf{v}_D + \dot{\mathbf{L}}_{DC}\text{; vector form}$$

$$\left\{\begin{array}{c} v_C^x \\ v_C^y \end{array}\right\} = \left\{\begin{array}{c} v_D^x \\ v_D^y \end{array}\right\} + \left\{\begin{array}{c} 0 \\ 0 \end{array}\right\} = \left\{\begin{array}{c} -4\cdot\omega_1\cdot \mathrm{Sin}(\theta_1) - y_{BD}\cdot\omega_2 \\ 4\cdot\omega_1\cdot \mathrm{Cos}(\theta_1) + x_{BD}\cdot\omega_2 \end{array}\right\} \tag{3.140t}$$

$$= \left\{\begin{array}{c} \dot{d}_C \\ 0 \end{array}\right\}\text{; check-up equation}$$

d. Linear velocities of centers of mass G_1, G_2, and G_3

- *Center of Mass Link 1*: Equation (3.132e) between velocities of A and G_1, both located on link 1:

$$\mathbf{v}_{G_1}^0 = \mathbf{v}_A + \dot{\mathbf{L}}_{AG_1}^0 = \mathbf{v}_A - \tilde{\mathbf{L}}_{AG_1}^0 * \boldsymbol{\omega}_1\text{; vector form}$$

$$\left\{\begin{array}{c} v_{G_1}^x \\ v_{G_1}^y \end{array}\right\} = \left\{\begin{array}{c} v_A^x \\ v_A^y \end{array}\right\} + \left\{\begin{array}{c} -y_{AG_1}\cdot\omega_1 \\ x_{AG_1}\cdot\omega_1 \end{array}\right\} = \left\{\begin{array}{c} -y_{AG_1}\cdot\omega_1 \\ x_{AG_1}\cdot\omega_1 \end{array}\right\}\text{; scalar components}$$

$$v_{G_1} = \sqrt{\left(v_{G_1}^x\right)^2 + \left(v_{G_1}^y\right)^2}\text{; magnitude} \tag{3.140u}$$

- *Center of Mass Link 2*: Equation (3.132e) between velocities of B and G_2, both located on link 2:

$$\mathbf{v}_{G_2}^0 = \mathbf{v}_B - \tilde{\mathbf{L}}_{BG_2}^0 * \boldsymbol{\omega}_2\text{; vector form}$$

$$\left\{\begin{array}{c} v_{G_2}^x \\ v_{G_2}^y \end{array}\right\} = \left\{\begin{array}{c} v_B^x \\ v_B^y \end{array}\right\} + \left\{\begin{array}{c} -y_{BG_2}^0\cdot\omega_2 \\ x_{BG_2}^0\cdot\omega_2 \end{array}\right\}$$

$$= \left\{\begin{array}{c} -4\cdot\omega_1\cdot \mathrm{Sin}(\theta_1) - y_{BG_2}^0\cdot\omega_2 \\ 4\cdot\omega_1\cdot \mathrm{Cos}(\theta_1) + x_{BG_2}^0\cdot\omega_2 \end{array}\right\}\text{; scalar components} \tag{3.140v}$$

$$v_{G_2} = \sqrt{\left(v_{G_2}^x\right)^2 + \left(v_{G_2}^y\right)^2}\text{; magnitude}$$

- *Center of Mass Link 3*: Equation (3.132e) between velocities of D and G_3, both located on slider (link 3) in translation with $\boldsymbol{\omega}_3 = 0$. Then

$$\mathbf{v}^0_{G_3} = \mathbf{v}_D - \tilde{\mathbf{L}}^0_{DG_3} * \omega_3 \text{; vector form}$$

$$\begin{Bmatrix} v^x_{G_3} \\ v^y_{G_3} \end{Bmatrix} = \begin{Bmatrix} v^x_D \\ v^y_D \end{Bmatrix} = \begin{Bmatrix} \dot{d}_C \\ 0 \end{Bmatrix} \text{; scalar component} \quad (3.140w)$$

$$v_{G_3} = \sqrt{\left(v^x_{G_3}\right)^2 + \left(v^y_{G_3}\right)^2} = \left|\dot{d}_C\right| \text{; magnitude}$$

3.8 Velocity Analysis for Multiple-Cycle Planar Mechanisms with Revolute and Prismatic Joints

Example: The two-cycle planar mechanism, Figure 3.53a, has given the link lengths: $L_{EA} = 30''$, $L_{AB} = 4''$, $L_{BF} = 24''$, $L_{FE} = 14''$, $L_{EG} = 7''$, $L_{GD} = 12''$, and $L_{DC} = 0''$.

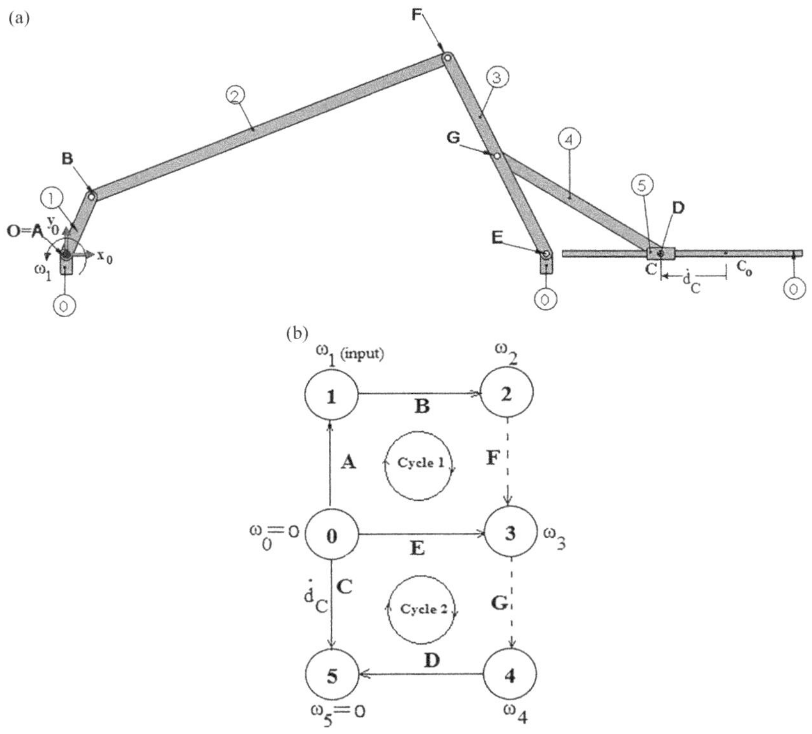

FIGURE 3.53
(a) Two-cycle mechanism; (b) digraph with the assigned absolute angular velocities at nodes and relative linear velocities at edges.

The crank 1 rotates counterclockwise with an angular velocity $\omega_1 = 360\ ^\circ/\text{s} = 2\pi\ \text{rad/s}$.

Determine

a. The system of equations for velocity analysis
b. The angular velocities for links 2, 3, and 4, and the relative linear velocity for link 5
c. Linear velocities of joints B, C, D, E, F, and G
d. Linear velocities of centers of mass G_1, G_2, G_3, G_4, and G_5

Solution

a. The system of equations for velocity analysis

The *Latin matrix* **L**, from Equation (3.107j), is a $2c \times n = 4 \times 6$ matrix with scalar entries defined around the nodes in the digraph and written in the direction of each cycle, as functions of the variable crank angle θ_1:

$$[L] = \begin{array}{c} \begin{array}{cccccc} 0 & 1 & 2 & 3 & 4 & 5 \end{array} \\ \left[\begin{array}{c:c:c:c:c:c} x_{EA} & x_{AB} & x_{BF} & x_{FE} & 0 & 0 \\ y_{EA} & y_{AB} & y_{BF} & y_{FE} & 0 & 0 \\ \hdashline x_{CE} & 0 & 0 & x_{EG} & x_{GD} & x_{DC} \\ y_{CE} & 0 & 0 & y_{EG} & y_{GD} & y_{DC} \end{array}\right] \end{array} \qquad (3.141a)$$

$$= \left[\begin{array}{c:c:c:c:c:c} -30 & 4\cdot \text{Cos}(\theta_1) & x_{BF} & x_{FE} & 0 & 0 \\ 0 & 4\cdot \text{Sin}(\theta_1) & y_{BF} & y_{FE} & 0 & 0 \\ \hdashline d_C - 7.308 & 0 & 0 & -0.5x_{FE} & x_{GD} & 0 \\ 0 & 0 & 0 & -0.5y_{FE} & y_{GD} & 0 \end{array}\right]$$

Therefore, the matrix of coefficients in the velocity equations is

$$[\tilde{L}] = \begin{array}{c} \begin{array}{cccccc} 0 & 1 & 2 & 3 & 4 & 5 \end{array} \\ \left[\begin{array}{c:c:c:c:c:c} 0 & -4\cdot \text{Sin}(\theta_1) & -y_{BF} & -y_{FE} & 0 & 0 \\ -30 & 4\cdot \text{Cos}(\theta_1) & x_{BF} & x_{FE} & 0 & 0 \\ \hdashline 0 & 0 & 0 & 0.5y_{FE} & -y_{GD} & 0 \\ d_C - 7.308 & 0 & 0 & -0.5x_{FE} & x_{GD} & 0 \end{array}\right] \end{array} \qquad (3.141b)$$

- The cycle basis matroid matrix, **C**:

Considering its definition (Equation 3.102), the matrix **C** has two rows assigned to two cycles and each column is assigned to a joint:

$$
\begin{array}{c} \\ [C]= \end{array}
\begin{array}{c}
\begin{array}{ccccccc} A & B & C & D & E & F & G \end{array} \\
\left[\begin{array}{ccccc|cc} 1 & 1 & 0 & 0 & -1 & 1 & 0 \\ 0 & 0 & -1 & 1 & 1 & 0 & 1 \end{array}\right]
\end{array}
\begin{array}{l} \\ \text{Cycle 1} \\ \text{Cycle 2} \end{array}
\quad (3.141c)
$$

Notice that the matrix is partitioned into two matrices:

- The $c \times a = 2 \times 5$ tree matrix **T**, as defined in Chapter 1
- The $c \times c = 2 \times 2$ diagonal matrix, **U**, its entries are 1

where

$c = 2$ is the number of cycles, equal to the number of joints which were cut: F and G

$a = 5$ is the number of arcs, equal to the number of joints' mechanism from the tree: A, B, C, D, and E

The matrix $\mathbf{C}(u^0)$ the components for joints' unit vector components multiplied each by the corresponding entry in matrix **C**:

$$
\begin{array}{c} \\ \left[\mathbf{C}\left(\mathbf{u}^0\right)\right]= \end{array}
\begin{array}{c}
\begin{array}{ccccccc} A & B & C & D & E & F & G \end{array} \\
\left[\begin{array}{ccccc|cc} 1\cdot\mathbf{u}_A & 1\cdot\mathbf{u}_B & 0 & 0 & -1\cdot\mathbf{u}_E & 1\cdot\mathbf{u}_F & 0 \\ 0 & 0 & -1\cdot\mathbf{u}_C & 1\cdot\mathbf{u}_D & 1\cdot\mathbf{u}_E & 0 & 1\cdot\mathbf{u}_G \end{array}\right]
\end{array}
\begin{array}{l} \\ \text{Cycle 1} \\ \text{Cycle 2} \end{array}
\quad (3.141d)
$$

The scalar x, y components are zero for the unit vectors $\mathbf{u}_A$, $\mathbf{u}_B$, $\mathbf{u}_D$, and $\mathbf{u}_E$. The slide for link 5 is the horizontal. The angle its unit vector makes with the axis is $\alpha_C = 0°$:

$$
\mathbf{u}_C = \begin{Bmatrix} u_C^x \\ u_C^y \end{Bmatrix} = \begin{Bmatrix} u_C^x \\ u_C^y \end{Bmatrix} = \begin{Bmatrix} \mathrm{Cos}(0°) \\ \mathrm{Sin}(0°) \end{Bmatrix} = \begin{Bmatrix} 1 \\ 0 \end{Bmatrix}
$$

The scalar components are zero for all the unit vectors of revolute joints $\mathbf{u}_A$, $\mathbf{u}_B$, $\mathbf{u}_D$, $\mathbf{u}_E$, $\mathbf{u}_F$, and $\mathbf{u}_G$:

$$
\begin{array}{c} \\ \left[C\left(u^0\right)\right]= \end{array}
\begin{array}{c}
\begin{array}{ccccccc} A & B & C & D & E & F & G \end{array} \\
\left[\begin{array}{c|c|c|c|c|c|c} 0 & 0 & 0 & 0 & 0 & 0 & 0 \\ 0 & 0 & 0 & 0 & 0 & 0 & 0 \\ \hline 0 & 0 & -1 & 0 & 0 & 0 & 0 \\ 0 & 0 & 0 & 0 & 0 & 0 & 0 \end{array}\right]
\end{array}
\quad (3.141e)
$$

Equation (3.139) becomes

$$[\tilde{L}]*\begin{Bmatrix} \omega_0(=0) \\ \omega_1(=\text{input}) \\ \omega_2 \\ \omega_3 \\ \omega_4 \\ \omega_5(=0) \end{Bmatrix} + [C(u^0)]*\begin{Bmatrix} \dot{d}_A(=0) \\ \dot{d}_B(=0) \\ \dot{d}_C \\ \dot{d}_D(=0) \\ \dot{d}_E(=0) \\ \dot{d}_F(=0) \\ \dot{d}_G(=0) \end{Bmatrix} = \begin{Bmatrix} 0 \\ 0 \\ 0 \\ 0 \end{Bmatrix} \tag{3.141f}$$

The terms within brackets illustrate the constraint velocities. Each zero entry in column matrices indicates the deleted column of coefficients. The entries in matrices $\tilde{\mathbf{L}}$ and $\mathbf{C}(\mathbf{u}^0)$ were determined as functions of the variable crank angle θ_1 at the previous positional analysis. The linear system Equation (3.137f) has 2c = 4 scalar equations with $n - 1 - M = 6 - 1 - 1 = 4$ unknowns: ω_2, ω_3, ω_4, and $\dot{d}_C$:

$$\left[\begin{array}{c:c:c:c:c} -4\cdot\text{Sin}(\theta_1) & -y_{BF} & -y_{FE} & 0 & 0 \\ 4\cdot\text{Cos}(\theta_1) & x_{BF} & x_{FE} & 0 & 0 \\ \hdashline 0 & 0 & 0.5y_{FE} & -y_{GD} & -1 \\ 0 & 0 & -0.5x_{FE} & x_{GD} & 0 \end{array}\right] * \begin{Bmatrix} \omega_1(=\text{input}) \\ \omega_2 \\ \omega_3 \\ \omega_4 \\ \dot{d}_C \end{Bmatrix} = \begin{Bmatrix} 0 \\ 0 \\ 0 \\ 0 \end{Bmatrix} \tag{3.141g}$$

b. The angular velocities for links 2, 3, and 4, and the relative linear velocity for link 5.

The symbolic solution for the system of Equation (3.141g) is

$$\begin{cases} \omega_2 = -\dfrac{4\cdot\omega_1}{\Delta}\left(x_{FE}\cdot\text{Sin}(\theta_1) - y_{FE}\cdot\text{Cos}(\theta_1)\right) \\ \omega_3 = \dfrac{4\cdot\omega_1}{\Delta}\left(x_{BF}\cdot\text{Sin}(\theta_1) - y_{BF}\cdot\text{Cos}(\theta_1)\right) \\ \omega_4 = \dfrac{2\cdot\omega_1\cdot x_{FE}}{\Delta\cdot x_{GD}}\left(x_{BF}\cdot\text{Sin}(\theta_1) - y_{BF}\cdot\text{Cos}(\theta_1)\right) \\ \dot{d}_C = 0.5\cdot x_{FE}\cdot\omega_3 - y_{GD}\cdot\omega_4 \end{cases} \tag{3.141h}$$

where $\Delta = x_{FE}\cdot y_{BF} - y_{FE}\cdot x_{BF}$ is used to detect the potential singularities for velocities.

Figure 3.54 illustrates the solution for a complete revolution of link 1.

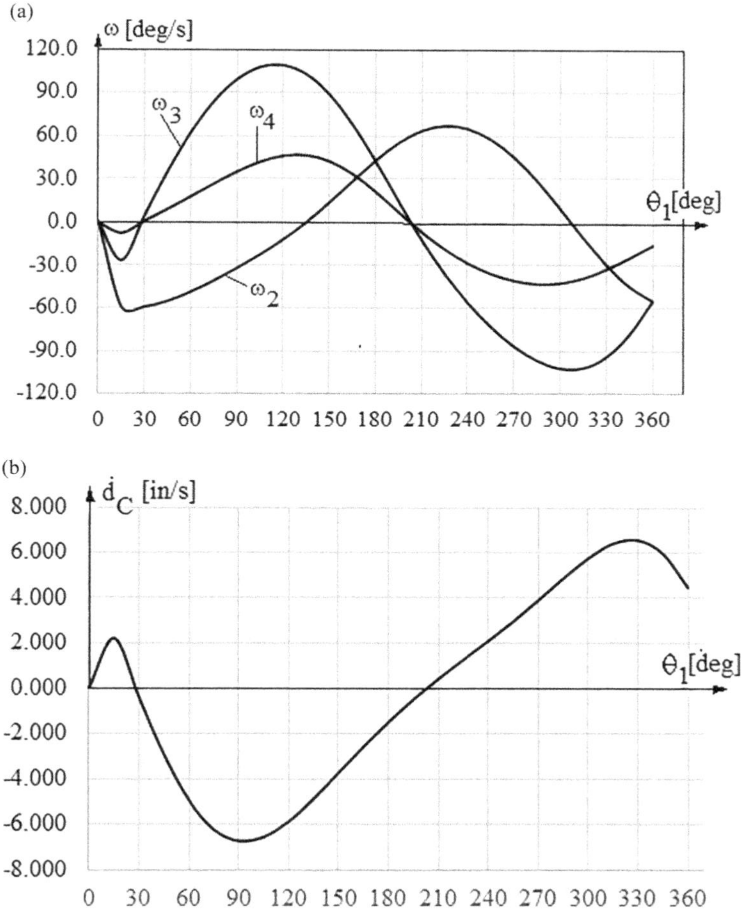

FIGURE 3.54
(a) Links' angular velocities; (b) relative linear velocity at joint C.

c. Linear velocities of joints A, B, C, D, E, F, and G

The velocity matroidal matrix is

$$[\dot{\mathbf{L}}] = \overset{\begin{matrix}0 & 1 & 2 & 3 & 4 & 5\end{matrix}}{\left[\begin{array}{c|c|c|c|c|c} \dot{\mathbf{L}}_{EA} & \dot{\mathbf{L}}_{AB} & \dot{\mathbf{L}}_{BF} & \dot{\mathbf{L}}_{FE} & 0 & 0 \\ \dot{\mathbf{L}}_{CE} & 0 & 0 & \dot{\mathbf{L}}_{EG} & \dot{\mathbf{L}}_{GD} & \dot{\mathbf{L}}_{DC} \end{array}\right]}$$

$$= \left[\begin{array}{c|c|c|c|c|c} \mathbf{v}_A - \mathbf{v}_E & \mathbf{v}_B - \mathbf{v}_A & \mathbf{v}_F - \mathbf{v}_B & \mathbf{v}_E - \mathbf{v}_F & 0 & 0 \\ \mathbf{v}_E - \mathbf{v}_C & 0 & 0 & \mathbf{v}_G - \mathbf{v}_E & \mathbf{v}_D - \mathbf{v}_G & \mathbf{v}_C - \mathbf{v}_D \end{array}\right] \quad (3.141i)$$

The vectors' $\dot{\mathbf{L}}$ coordinates, columns in velocity matroidal, are expressed as functions of x_{BF}, y_{BF}, x_{FE}, y_{FE}, x_{GD}, and y_{GD} (from positional equations), and 2c = 4 velocities determined from previously determined velocities ω_2, ω_3, ω_4, and $\dot{d}_C$.

Column 0: From Equation (3.132a), with the fixed revolute joints A and E:

$$\dot{\mathbf{L}}_{EA} = 0;\ \dot{\mathbf{L}}_{CE} = \dot{\mathbf{L}}_{OE} - \dot{\mathbf{L}}_{OC} = 0 - \dot{\mathbf{L}}_{OC} = -\mathbf{u}_C \cdot \dot{d}_C = -\begin{Bmatrix} \mathrm{Cos}\,0° \\ \mathrm{Sin}\,0° \end{Bmatrix} \cdot \dot{d}_C = \begin{Bmatrix} -\dot{d}_C \\ 0 \end{Bmatrix} \tag{3.141j}$$

Column 1:

$$\dot{\mathbf{L}}_{AB} = -\tilde{\mathbf{L}}^0_{AB} * \boldsymbol{\omega}_1 = \begin{Bmatrix} -y_{AB} \cdot \omega_1 \\ x_{AB} \cdot \omega_1 \end{Bmatrix} = \begin{Bmatrix} -4 \cdot \omega_1 \cdot \mathrm{Sin}(\theta_1) \\ 4 \cdot \omega_1 \cdot \mathrm{Cos}(\theta_1) \end{Bmatrix} \tag{3.141k}$$

Column 2:

$$\dot{\mathbf{L}}_{BF} = -\tilde{\mathbf{L}}^0_{BF} * \omega_2 = \begin{Bmatrix} -y_{BF} \cdot \omega_2 \\ x_{BF} \cdot \omega_2 \end{Bmatrix} \tag{3.141l}$$

Column 3:

$$\dot{\mathbf{L}}_{FE} = -\tilde{\mathbf{L}}^0_{FE} * \omega_3 = \begin{Bmatrix} -y_{FE} \cdot \omega_3 \\ x_{FE} \cdot \omega_3 \end{Bmatrix};$$

$$\dot{\mathbf{L}}_{EG} = -\tilde{\mathbf{L}}^0_{EG} * \omega_3 = \begin{Bmatrix} -y_{EG} \cdot \omega_3 \\ x_{EG} \cdot \omega_3 \end{Bmatrix} = \begin{Bmatrix} 0.5y_{FE} \cdot \omega_3 \\ -0.5x_{FE} \cdot \omega_3 \end{Bmatrix} \tag{3.141m}$$

Column 4:

$$\dot{\mathbf{L}}_{GD} = -\tilde{\mathbf{L}}^0_{GD} * \omega_4 = \begin{Bmatrix} -y_{GD} \cdot \omega_4 \\ x_{GD} \cdot \omega_4 \end{Bmatrix} \tag{3.141n}$$

Column 5:

$$\dot{\mathbf{L}}_{DC} = -\tilde{\mathbf{L}}^0_{DC} * \omega_5 = \begin{Bmatrix} 0 \\ 0 \end{Bmatrix};\ \text{since } \omega_5 = 0 \tag{3.141o}$$

Starting with the fixed joint A ($\mathbf{v}_A = 0$), the linear absolute velocities for B, C, and D, E, F, and G are determined by equating the column entries in the left and right sides of Equation (3.141i):

$$
\begin{array}{c}
\quad\quad 0 \quad\quad 1 \quad\quad 2 \quad\quad 3 \quad\quad 4 \quad\quad 5 \\
[\dot{\mathbf{L}}] = \left[\begin{array}{c|c|c|c|c|c}
\dot{\mathbf{L}}_{EA} & \dot{\mathbf{L}}_{AB} & \dot{\mathbf{L}}_{BF} & \dot{\mathbf{L}}_{FE} & 0 & 0 \\
\dot{\mathbf{L}}_{CE} & 0 & 0 & \dot{\mathbf{L}}_{EG} & \dot{\mathbf{L}}_{GD} & \dot{\mathbf{L}}_{DC}
\end{array}\right] \\
= \left[\begin{array}{c|c|c|c|c|c}
\mathbf{v}_A - \mathbf{v}_E & \mathbf{v}_B - \mathbf{v}_A & \mathbf{v}_F - \mathbf{v}_B & \mathbf{v}_E - \mathbf{v}_F & 0 & 0 \\
\mathbf{v}_E - \mathbf{v}_C & 0 & 0 & \mathbf{v}_G - \mathbf{v}_E & \mathbf{v}_D - \mathbf{v}_G & \mathbf{v}_C - \mathbf{v}_D
\end{array}\right]
\end{array}
\tag{3.141p}
$$

$$\text{Joint A}: \mathbf{v}_A = 0 = \begin{Bmatrix} 0 \\ 0 \end{Bmatrix}; \text{A is a fixed joint}$$

$$\text{Joint E}: \mathbf{v}_E = \mathbf{v}_A - \dot{\mathbf{L}}_{EA} = \begin{Bmatrix} 0 \\ 0 \end{Bmatrix}; \text{E is a fixed joint}$$

$$\text{Joint C}: \mathbf{v}_C = \mathbf{v}_E - \dot{\mathbf{L}}_{CE} = \begin{Bmatrix} \dot{d}_C \\ 0 \end{Bmatrix}$$

$$\text{Joint B}: \mathbf{v}_B = \mathbf{v}_A + \dot{\mathbf{L}}_{AB} = \begin{Bmatrix} -4 \cdot \omega_1 \cdot \mathrm{Sin}(\theta_1) \\ 4 \cdot \omega_1 \cdot \mathrm{Cos}(\theta_1) \end{Bmatrix}$$

$$\text{Joint F}: \mathbf{v}_F = \mathbf{v}_B + \dot{\mathbf{L}}_{BF} = \begin{Bmatrix} -4 \cdot \omega_1 \cdot \mathrm{Sin}(\theta_1) - y_{BF} \cdot \omega_2 \\ 4 \cdot \omega_1 \cdot \mathrm{Cos}(\theta_1) + x_{BF} \cdot \omega_2 \end{Bmatrix}$$

$$\text{Joint E}: \mathbf{v}_E = \mathbf{v}_F + \dot{\mathbf{L}}_{FE} = \begin{Bmatrix} -4 \cdot \omega_1 \cdot \mathrm{Sin}(\theta_1) - y_{BF} \cdot \omega_2 - y_{FE} \cdot \omega_3 \\ 4 \cdot \omega_1 \cdot \mathrm{Cos}(\theta_1) + x_{BF} \cdot \omega_2 + x_{FE} \cdot \omega_3 \end{Bmatrix} \tag{3.141q}$$

$$= \begin{Bmatrix} 0 \\ 0 \end{Bmatrix}; \text{E is a fixed joint}$$

$$\text{Joint G}: \mathbf{v}_G = \mathbf{v}_E + \dot{\mathbf{L}}_{EG} = \begin{Bmatrix} 0.5y_{FE} \cdot \omega_3 \\ -0.5x_{FE} \cdot \omega_3 \end{Bmatrix}$$

$$\text{Joint D}: \mathbf{v}_D = \mathbf{v}_G + \dot{\mathbf{L}}_{GD} = \begin{Bmatrix} 0.5y_{FE} \cdot \omega_3 - y_{GD} \cdot \omega_4 \\ -0.5x_{FE} \cdot \omega_3 + x_{GD} \cdot \omega_4 \end{Bmatrix}$$

Note: A checkup equation results from equating the last columns in Equation (3.141i)

$$\mathbf{v}_C = \mathbf{v}_D + \dot{\mathbf{L}}_{DC}\text{; vector form}$$

$$\begin{Bmatrix} v_C^x \\ v_C^y \end{Bmatrix} = \begin{Bmatrix} v_D^x \\ v_D^y \end{Bmatrix} + \begin{Bmatrix} 0 \\ 0 \end{Bmatrix} = \begin{Bmatrix} \dot{d}_C \\ 0 \end{Bmatrix}\text{; checkup equation}$$

d. Linear velocities of centers of mass G_1, G_2, G_3, G_4, and G_5

Center of Mass Link 1: Equation (3.132e) between velocities of A and G_1, both located on link 1:

$$\mathbf{v}_{G_1}^0 = \mathbf{v}_A + \dot{\mathbf{L}}_{AG_1}^0 = \mathbf{v}_A - \tilde{\mathbf{L}}_{AG_1}^0 * \boldsymbol{\omega}_1 = \begin{Bmatrix} -y_{AG_1} \cdot \omega_1 \\ x_{AG_1} \cdot \omega_1 \end{Bmatrix} \tag{3.141r}$$

Center of Mass Link 2: Equation (3.132e) between velocities of B and G_2, both located on link 2:

$$\mathbf{v}_{G_2}^0 = \mathbf{v}_B + \dot{\mathbf{L}}_{BG_2}^0 = \mathbf{v}_B - \tilde{\mathbf{L}}_{BG_2}^0 * \boldsymbol{\omega}_2 = \begin{Bmatrix} -4 \cdot \omega_1 \cdot \mathrm{Sin}(\theta_1) - y_{BG_2}^0 \cdot \omega_2 \\ 4 \cdot \omega_1 \cdot \mathrm{Cos}(\theta_1) + x_{BG_2}^0 \cdot \omega_2 \end{Bmatrix} \tag{3.141s}$$

Center of Mass Link 3: Equation (3.132e) between velocities of F and G_3, both located on link 3:

$$\mathbf{v}_{G_3}^0 = \mathbf{v}_F + \dot{\mathbf{L}}_{FG_3}^0 = \mathbf{v}_F - \tilde{\mathbf{L}}_{FG_3}^0 * \omega_3 = \begin{Bmatrix} -4 \cdot \omega_1 \cdot \mathrm{Sin}(\theta_1) - y_{BF} \cdot \omega_2 - y_{FG_3} \cdot \omega_3 \\ 4 \cdot \omega_1 \cdot \mathrm{Cos}(\theta_1) + x_{BF} \cdot \omega_2 + x_{FG_3} \cdot \omega_3 \end{Bmatrix} \tag{3.141t}$$

Center of Mass Link 4: Equation (3.132e) between velocities of G and G_4, both located on link 3:

$$\mathbf{v}_{G_4}^0 = \mathbf{v}_G + \dot{\mathbf{L}}_{GG_4}^0 = \mathbf{v}_G - \tilde{\mathbf{L}}_{GG_4}^0 * \boldsymbol{\omega}_4 = \begin{Bmatrix} 0.5y_{FE} \cdot \omega_3 - y_{GG_4} \cdot \omega_4 \\ 0.5x_{FE} \cdot \omega_3 + x_{GG_4} \cdot \omega_4 \end{Bmatrix} \tag{3.141u}$$

Center of Mass Link 5: Equation (3.132e) between velocities of D and G_5, both located on slider (link 5) in translation with $\omega_5 = 0$:

$$\mathbf{v}_{G_5}^0 = \mathbf{v}_D - \tilde{\mathbf{L}}_{DG_5}^0 * \boldsymbol{\omega}_5 = \begin{Bmatrix} \dot{d}_C \\ 0 \end{Bmatrix} \tag{3.141v}$$

3.9 Gears

3.9.1 Parallel Axes Epicyclic Gear Trains

The most commonly used method for kinematic analysis of parallel axes epicyclic gear trains (EGTs) is the Willis method of inversion motion. Analytic methods of generating the kinematic equations of EGTs based on the relative motions within the mechanism are also available as well as graphical methods. The latter methods are difficult to implement in computer algorithms, being however useful to check the results obtained using other analytical methods.

This section illustrates the kinematic equations for gear trains as functions of absolute and relative angular velocities, using the concept of velocity matroidal matrix. The analysis uses the relationships between the number of mobile links, number of joints, and number of cycles in the cycle basis, together with the Latin matrix (whose entries are function of the partial gear ratios of the transmission). Relationships between the output and input angular velocities are then determined for gear trains with large number of gears and multiple DOF.

- **Link and joint numbering**
  For the case of an epicyclic transmission with m mobile links and j joints, link numbering begins with 0 assigned to the fixed link (casing) continuing up to m for the other mobile links. Additionally, the revolute and meshing joints are assigned letters.
  Figure 3.55a,b, Animation 3.12, shows the numbering in a typical EGT mechanism with m = 4 mobile links: 1 (sun gear), 2 (ring gear), 3 (carrier), and 4 (planet gear). There are j = 6 joints (A, B, C, D, E, and F), with j − c = 4 revolute joints (A, B, C, and D) and c = 2 meshing joints (E and F).
- **EGT digraph**
  Figure 3.55c shows its associated digraph. This digraph has n = 5 vertices and j = 6 edges. The spanning tree has m = 5 vertices: {0, 1, 2, 3, 4} and j − c = 4 revolute joints (A, B, C, and D). The edges in digraph are oriented from the lower node number to the higher node number.
- **Spanning tree for gear trains**
  For gear train analysis, the spanning tree obtained from digraph is considered as a result of *cutting the meshing joints*, [11]. The set of cut edges shown in dashed line {E, F} contains c = 2 edges corresponding

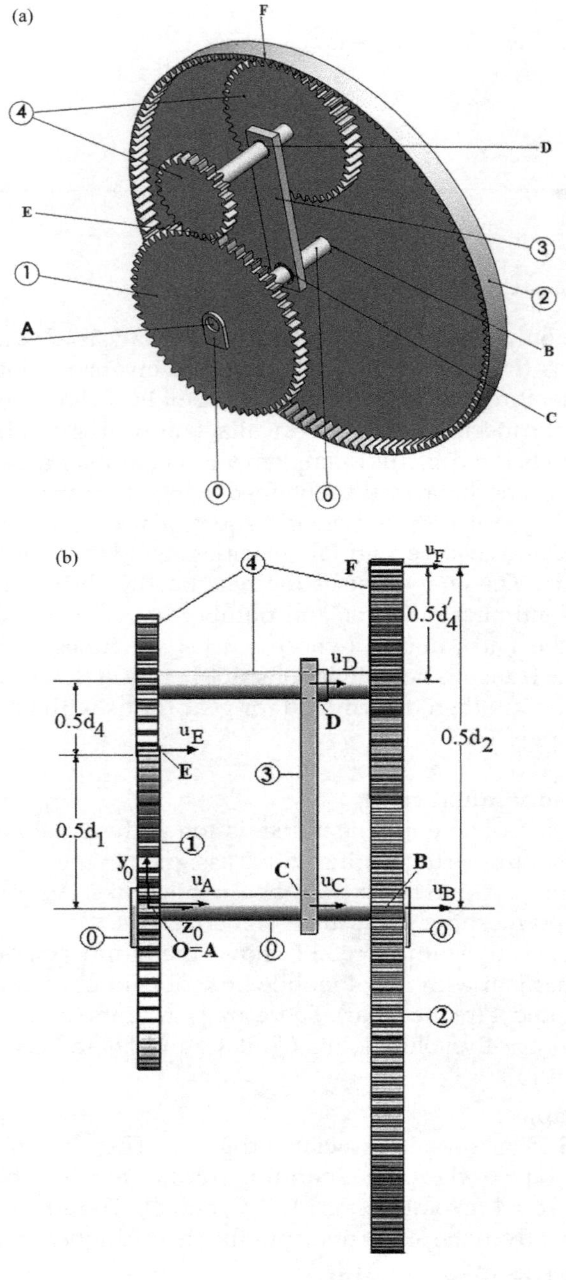

FIGURE 3.55
(a) EGT mechanism; (b) side view for mechanism; (c) digraph; (d) Latin matrix and velocity matroidal entries.

(Continued)

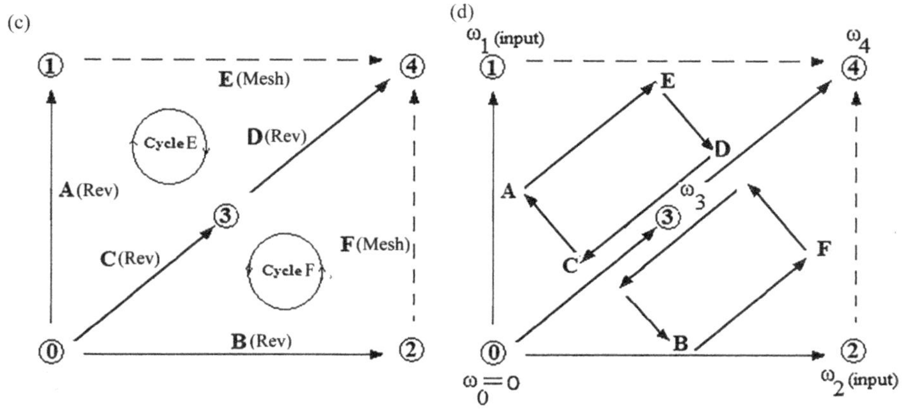

FIGURE 3.55 (CONTINUED)
(a) EGT mechanism; (b) side view for mechanism; (c) digraph; (d) Latin matrix and velocity matroidal entries.

to the meshing joints of the mechanism. The set c = j − m = 2 *independent cycles* is $\{C_E, C_F\}$ is obtained adding one by one the edges from the cut set to the spanning tree, as follows: adding edges labeled E and F, the cycles C_E = {A, E, D, C} and C_F = {B, F, D, C} are obtained. This set $\{C_E, C_F\}$ is a cycle basis.

For the number of cycles in cycle basis, Euler's formula holds:

$$c = j - m = 6 - 4 = 2 \text{ cycles.} \tag{3.142}$$

3.9.1.1 Mobility Formula for Gear Trains Based on the Number of Links and Cycles

Statement 3.10

The mobility for gear trains is determined by subtracting the number of cycles c from the number of mobile links, m, gears, and carriers. This equation applies for all types of geared train mechanisms: parallel planetary axes, fixed axes, or bevel gear trains (BGTs) [11,12]:

$$M = m - c \tag{3.143}$$

Proof: The Gruebler criterion for the total number of DOF of the mechanism is

$$M = 3m - 2j_1 - j_2 \tag{3.144}$$

where j_1 is the number of revolute joints and j_2 is the number of meshing joints. In mechanisms' literature, j_2 is named as half joints. Since for the case of gear trains the number of mobile links equals to the number of revolute joints $j_1 = m$, the total number of joints will therefore be $j = j_1 + j_2 = m + j_2$. By equating j from Euler's equation for the digraph: $j = m + c$, one can conclude that $j_2 = c$ (i.e., the number of meshing joints equals to the number of cycles in cycle basis).

Example: For the EGT with $m = 4$ mobile links and $c = 2$ cycles, then $M = 4 - 2 = 2$ DOF.

The gear trains are different from each other from the configuration of the gears and carrier axes of rotation. Next, the characteristics are presented for the EGT, and then the equations are derived for angular velocities.

- **EGT absolute angular velocity matrix**

  This matrix, noted $\boldsymbol{\omega}_m$, is a column matrix ($m \times 1$) whose entries are the gears' angular velocities, expressed in the fixed frame. The frame (casing) is fixed; therefore, its velocity is zero:

$$\omega_m = \begin{Bmatrix} \omega_1 \\ \omega_2 \\ \omega_3 \\ \omega_4 \end{Bmatrix} \tag{3.145}$$

  In Figure 3.55d, the absolute velocities attached to nodes in digraph are illustrated.

- **EGT relative angular velocity matrix**

  This matrix, noted $\dot{\boldsymbol{\theta}}_j$, is a column matrix ($j \times 1$) whose entries are relative velocities assigned to edges j in digraph to measure the relative rotations between gears: gear (node) m_{head} relative to gear (node) m_{tail}.

  For the EGT mechanism, the relative velocity matrix is

$$\left[\dot{\theta}_j\right] = \left\{ \dot{\theta}_A \quad \dot{\theta}_B \quad \dot{\theta}_C \quad \dot{\theta}_D \quad \dot{\theta}_E \quad \dot{\theta}_F \right\}^T \tag{3.146}$$

  The matrix in Equation (3.146) has the sub-matrix $\dot{\theta}_{rev}$, with the entries corresponding to revolute joints and $\dot{\theta}_{mesh}$ for meshing joints:

$$\left\{\dot{\theta}_j\right\} = \begin{Bmatrix} \dot{\theta}_{rev} \\ \dot{\theta}_{mesh} \end{Bmatrix}; \dot{\theta}_{rev} = \begin{Bmatrix} \dot{\theta}_A \\ \dot{\theta}_B \\ \dot{\theta}_C \\ \dot{\theta}_D \end{Bmatrix}; \dot{\theta}_{mesh} = \begin{Bmatrix} \dot{\theta}_E \\ \dot{\theta}_F \end{Bmatrix} \tag{3.147}$$

  The relative velocities measure the relative motion of rotation between links as shown in Equation (3.148), where the IT notation in digraph is considered. The relative angular velocity vectors are assigned to edges in digraph, and oriented in the direction of unit vectors $\mathbf{u}^0$ (Figure 3.55b):

$$\dot{\theta}_A = \mathbf{u}_A^0 \cdot \dot{\theta}_{0,1};\ \dot{\theta}_B = \mathbf{u}_B^0 \cdot \dot{\theta}_{0,2};\ \dot{\theta}_C = \mathbf{u}_C^0 \cdot \dot{\theta}_{0,3};\ \dot{\theta}_D = \mathbf{u}_D^0 \cdot \dot{\theta}_{3,4};$$
$$\dot{\theta}_E = \mathbf{u}_E^0 \cdot \dot{\theta}_{1,4};\ \dot{\theta}_F = \mathbf{u}_F^0 \cdot \dot{\theta}_{2,4} \tag{3.148}$$

- **The relative rotation matrices**

  In Chapter 2, the relative rotation matrices are introduced. For the EGT mechanism, the matrices are rotations about z-axes.

$$\mathbf{D}_A = \begin{bmatrix} C\theta_A & -S\theta_A & 0 \\ S\theta_A & C\theta_A & 0 \\ 0 & 0 & 1 \end{bmatrix};\ \mathbf{D}_B = \begin{bmatrix} C\theta_B & -S\theta_B & 0 \\ S\theta_B & C\theta_B & 0 \\ 0 & 0 & 1 \end{bmatrix};$$
$$\mathbf{D}_C = \begin{bmatrix} C\theta_C & -S\theta_C & 0 \\ S\theta_C & C\theta_C & 0 \\ 0 & 0 & 1 \end{bmatrix};\ \mathbf{D}_D = \begin{bmatrix} C\theta_D & -S\theta_D & 0 \\ S\theta_D & C\theta_D & 0 \\ 0 & 0 & 1 \end{bmatrix}; \tag{3.149}$$
$$\mathbf{D}_E = \begin{bmatrix} C\theta_E & -S\theta_E & 0 \\ S\theta_E & C\theta_E & 0 \\ 0 & 0 & 1 \end{bmatrix};\ \mathbf{D}_F = \begin{bmatrix} C\theta_F & -S\theta_F & 0 \\ S\theta_F & C\theta_F & 0 \\ 0 & 0 & 1 \end{bmatrix}$$

- **The gears and carrier absolute rotation matrices**

  Also, in Chapter 2, the absolute rotation matrices are determined. For the EGT mechanism, the matrices are

$$\mathbf{D}_{0,1} = \mathbf{D}_A = \begin{bmatrix} C\theta_A & -S\theta_A & 0 \\ S\theta_A & C\theta_A & 0 \\ 0 & 0 & 1 \end{bmatrix};\ \mathbf{D}_{0,2} = \mathbf{D}_B = \begin{bmatrix} C\theta_B & -S\theta_B & 0 \\ S\theta_B & C\theta_B & 0 \\ 0 & 0 & 1 \end{bmatrix};$$
$$\mathbf{D}_{0,3} = \mathbf{D}_C = \begin{bmatrix} C\theta_C & -S\theta_C & 0 \\ S\theta_C & C\theta_C & 0 \\ 0 & 0 & 1 \end{bmatrix};\ \mathbf{D}_{0,4} = \mathbf{D}_{0,3} * \mathbf{D}_{3,4} = \mathbf{D}_C * \mathbf{D}_D$$
$$= \begin{bmatrix} C(\theta_C + \theta_D) & -S(\theta_C + \theta_D) & 0 \\ S(\theta_C + \theta_D) & C(\theta_C + \theta_D) & 0 \\ 0 & 0 & 1 \end{bmatrix} \tag{3.150}$$

- **The unit vectors for axes of rotation with respect to the gears frame**
  If the local frames attached to the gears and if all frames and carriers have their axes parallel to z, then the unit vectors have the form shown in Equation (3.151):

$$\mathbf{u}_A^1 = \mathbf{u}_B^2 = \mathbf{u}_C^3 = \mathbf{u}_D^4 = \mathbf{u}_E^4 = \mathbf{u}_F^4 = \begin{Bmatrix} 0 \\ 0 \\ 1 \end{Bmatrix} \tag{3.151}$$

- **The unit vectors for axes of rotation with respect to the fixed frame (casing)**
  The unit vectors $\mathbf{u}_Z^0$ with respect to the fixed frame are expressed as functions of $\mathbf{u}_Z^m$ from Equation (3.151), and absolute rotation matrices from Equation (3.150):

$$\mathbf{u}_Z^0 = \mathbf{D}_{0,m} * \mathbf{u}_Z^m;\ \text{where } Z = A, B, C, D, E, F, \text{ and } m = 1,2,3,4 \tag{3.152}$$

Example: For the EGT in Figure 3.55b,

$$\mathbf{u}_A^0 = \mathbf{u}_B^0 = \mathbf{u}_C^0 = \mathbf{u}_D^0 = \mathbf{u}_E^0 = \mathbf{u}_F^0 = \begin{Bmatrix} 0 \\ 0 \\ 1 \end{Bmatrix}$$

3.9.1.2 Equations Based on Absolute Angular Velocities: The Matroidal Method for Gear Trains

- **The Latin matrix**
  In Figure 3.55d, the vector entries in the Latin matrix written around the nodes are illustrated. As stated before in the text, it has the property that the sum of entries on each cycle (row) row in matrix **L** is zero. In matrix form,

$$\begin{matrix} & 0 & 1 & 2 & 3 & 4 & \\ [\mathbf{L}] = & \Big[\ \mathbf{L}_{CA} & \mathbf{L}_{AE} & 0 & \mathbf{L}_{DC} & \mathbf{L}_{ED}\ \Big] & \text{Cycle } C_E \\ & \Big[\ \mathbf{L}_{CB} & 0 & \mathbf{L}_{BF} & \mathbf{L}_{DC} & \mathbf{L}_{FD}\ \Big] & \text{Cycle } C_F \end{matrix} \tag{3.153}$$

A zero entry denotes that the node is not on the cycle. Each vector has components on y_0 and z_0:

$$\mathbf{L}_{AE} = \begin{Bmatrix} 0 \\ y_{AE} \\ z_{AE} \end{Bmatrix} = \begin{Bmatrix} 0 \\ y_E - y_A \\ z_E - z_A \end{Bmatrix} \tag{3.154}$$

- **The velocity matroidal matrix**

  For the EGT, the velocity matroidal matrix is

$$
\begin{array}{c} \\ [\dot{\mathbf{L}}] = \end{array}
\begin{array}{c}
\begin{array}{ccccc} \quad 0 & \quad 1 & \quad 2 & \quad 3 & \quad 4 \end{array} \\
\left[\begin{array}{ccccc}
\dot{\mathbf{L}}_{CA} & \dot{\mathbf{L}}_{AE} & 0 & \dot{\mathbf{L}}_{DC} & \dot{\mathbf{L}}_{ED} \\
\dot{\mathbf{L}}_{CB} & 0 & \dot{\mathbf{L}}_{BF} & \dot{\mathbf{L}}_{DC} & \dot{\mathbf{L}}_{FD}
\end{array} \right]
\end{array}
\begin{array}{l} \\ \text{Cycle } C_E \\ \text{Cycle } C_F \end{array}
\tag{3.155}
$$

From Statement 3.8: The sum of vector entries in velocity matroidal matrix $\dot{\mathbf{L}}$ is zero, then

$$
\left\{ \begin{array}{l}
\dot{\mathbf{L}}_{CA} + \dot{\mathbf{L}}_{AE} + \dot{\mathbf{L}}_{DC} + \dot{\mathbf{L}}_{ED} = 0 \\
\dot{\mathbf{L}}_{CB} + \dot{\mathbf{L}}_{BF} + \dot{\mathbf{L}}_{DC} + \dot{\mathbf{L}}_{FD} = 0
\end{array} \right.
\tag{3.156}
$$

Since $\dot{\mathbf{d}}_E = \dot{\mathbf{d}}_F = 0$, that is, neglecting the linear displacement at meshing joints, and then the general form of velocity matroidal vector is

$$
\dot{\mathbf{L}}_{AE} = \boldsymbol{\omega}_1 \times \mathbf{L}_{AE} + \dot{\mathbf{d}}_E = \boldsymbol{\omega}_1 \times \mathbf{L}_{AE}
\tag{3.157}
$$

The cross product of vectors in the system of equations:

$$
\left\{ \begin{array}{l}
\boldsymbol{\omega}_0 \times \mathbf{L}_{CA} + \boldsymbol{\omega}_1 \times \mathbf{L}_{AE} + \boldsymbol{\omega}_3 \times \mathbf{L}_{DC} + \boldsymbol{\omega}_4 \times \mathbf{L}_{ED} = 0 \\
\boldsymbol{\omega}_0 \times \mathbf{L}_{CB} + \boldsymbol{\omega}_2 \times \mathbf{L}_{BF} + \boldsymbol{\omega}_3 \times \mathbf{L}_{DC} + \boldsymbol{\omega}_4 \times \mathbf{L}_{FD} = 0
\end{array} \right.
\tag{3.158}
$$

is written as functions of matrix multiplications as

$$
\left\{ \begin{array}{l}
-\tilde{\mathbf{L}}_{CA} * \boldsymbol{\omega}_0 - \tilde{\mathbf{L}}_{AE} * \boldsymbol{\omega}_1 - \tilde{\mathbf{L}}_{DC} * \boldsymbol{\omega}_3 - \tilde{\mathbf{L}}_{ED} * \boldsymbol{\omega}_4 = 0 \\
-\tilde{\mathbf{L}}_{CB} * \boldsymbol{\omega}_0 - \tilde{\mathbf{L}}_{BF} * \boldsymbol{\omega}_2 - \tilde{\mathbf{L}}_{DC} * \boldsymbol{\omega}_3 - \tilde{\mathbf{L}}_{FD} * \boldsymbol{\omega}_4 = 0
\end{array} \right.
\tag{3.159}
$$

The factors in the previous equations for EGT have the scalar components developed as in the following example:

$$
\begin{array}{l}
-\tilde{\mathbf{L}}_{AE} * \boldsymbol{\omega}_1 = -\tilde{\mathbf{L}}_{AE} * \mathbf{u}_1^0 \cdot \omega_1 = - \underbrace{\left[\begin{array}{ccc}
0 & -z_{AE} & y_{AE} \\
z_{AE} & 0 & -x_{AE} \\
-y_{AE} & x_{AE} & 0
\end{array} \right]}_{\tilde{\mathbf{L}}_{AE}} * \\
\underbrace{\left\{ \begin{array}{c} 0 \\ 0 \\ 1 \end{array} \right\} \cdot \omega_1}_{\boldsymbol{\omega}_1} = \left\{ \begin{array}{c} -y_{AE \cdot \omega_1} \\ 0 \\ 0 \end{array} \right\}
\end{array}
\tag{3.160}
$$

- **Equations based on absolute angular velocities**

  The c = 2 equations for EGT in scalar form and functions of absolute angular velocities are

$$\begin{cases} -y_{CA} \cdot \omega_0 - y_{AE} \cdot \omega_1 - y_{DC} \cdot \omega_3 - y_{ED} \cdot \omega_4 = 0 \\ -y_{CB} \cdot \omega_0 - y_{BF} \cdot \omega_2 - y_{DC} \cdot \omega_3 - y_{FD} \cdot \omega_4 = 0 \end{cases} \tag{3.161}$$

After multiplication by –1 in both equations, and since $\omega_0 = 0$, then Equation (3.161) is written as

$$\begin{cases} (y_E - y_A) \cdot \omega_1 + (y_C - y_D) \cdot \omega_3 + (y_D - y_E) \cdot \omega_4 = 0 \\ (y_F - y_B) \cdot \omega_2 + (y_C - y_D) \cdot \omega_3 + (y_D - y_F) \cdot \omega_4 = 0 \end{cases} \tag{3.161a}$$

and written in matrix form as

$$[L_m(y)] * \{\omega_m\} = 0 \tag{3.161b}$$

$$\begin{matrix} 1 & 2 & 3 & 4 \end{matrix}$$

$$\begin{bmatrix} y_{AE} & 0 & y_{DC} & y_{ED} \\ 0 & y_{BF} & y_{DC} & y_{FD} \end{bmatrix} * \begin{Bmatrix} \omega_1 \\ \omega_2 \\ \omega_3 \\ \omega_4 \end{Bmatrix} = \begin{Bmatrix} 0 \\ 0 \end{Bmatrix} \tag{3.161c}$$

where

$[L_m(y)]$ is the (c × m) matrix with y-components from Latin matrix (Equation (3.153))

$\{\omega_m\}$ is the (m × 1) matrix with scalar components of gears and carrier angular velocities (Equation (3.145))

Equation (3.161) for EGT can be written directly along the cycles. This type of equations derived from vector equations in mechanics along a non-oriented graph was reported in early 1960 by Voinea and Atanasiu [7]. For an oriented graph, Talpasanu [1,2,11] adopted in kinematics the digraph matrices, and based on matrix manipulations, developed the matroidal method as presented in this text.

- **Equations based on gear ratios**
  - The joints' y-coordinates

    Considering the joints' y-coordinates in the following table

Joint	A	B	C	D	E	F
y	0	0	0	A	$0.5 \cdot d_1$	$0.5 \cdot d_2$

where A is the length of carrier and the distance between parallel shafts

$$A = 0.5 \cdot (d_1 + d_4) = 0.5 \cdot (d_2 - d_4')$$

– The gear ratios between two nodes in digraph, corresponding to the two meshing joints, are defined as

$$\begin{aligned} i_E = i_{E1,4} &= \frac{d_4(in)}{d_1(in)} = \frac{n_1(rpm)}{n_4(rpm)} = \frac{\omega_1(rad/s)}{\omega_4(rad/s)} = \frac{N_4(\#\,teeth)}{N_1(\#\,teeth)} \\ i_F = i_{F2,4} &= \frac{d_4'(in)}{d_2(in)} = \frac{n_2(rpm)}{n_4(rpm)} = \frac{\omega_2(rad/s)}{\omega_4(rad/s)} = \frac{N_4'(\#\,teeth)}{N_2(\#\,teeth)} \end{aligned} \tag{3.162}$$

Then, Equation (3.161) becomes

$$\begin{cases} -\omega_1 + (1 + i_E) \cdot \omega_3 - i_E \cdot \omega_4 = 0 \\ -\omega_2 + (1 - i_F) \cdot \omega_3 + i_F \cdot \omega_4 = 0 \end{cases} \tag{3.163}$$

$$\begin{bmatrix} -1 & 0 & \vdots & 1 + i_E & -i_E \\ 0 & -1 & \vdots & 1 - i_F & i_F \end{bmatrix} * \begin{Bmatrix} \omega_1 \\ \omega_2 \\ \text{------} \\ \omega_3 \\ \omega_4 \end{Bmatrix} = \begin{Bmatrix} 0 \\ 0 \end{Bmatrix} \tag{3.163a}$$

- **Number of equations and number of unknown angular velocities**
  The number of equations is c, equal to the rows in the Latin matrix for EGT mechanism.

Statement 3.11

The number of unknown angular velocities for EGT is the number c of cycles

Proof: The EGT with M DOF has M actuated gears and carriers; therefore, there are m–M unknowns.

Consider Equation (3.143) written as $m - M = c$, therefore the number of equations equals the number of unknowns, which is the number of cycles c

Solution: The linear system of $c = 2$ equations, solved for $c = 2$ unknowns, holds the solution:

$$\omega_3 = \frac{i_F}{i_E + i_F}\omega_1 + \frac{i_E}{i_E + i_F}\omega_2; \; \omega_4 = \frac{1 - i_F}{i_E + i_F}\omega_1 + \frac{1 + i_E}{i_E + i_F}\omega_2 \tag{3.164}$$

Numerical example: The EGT has metric gears with

$$d_1 = 120\ \text{mm}; d_2 = 260\ \text{mm}; d_4 = 50\ \text{mm}; d_4' = 90\ \text{mm}; \omega_1 = \omega_2 = 360\ °/\text{s};$$
$$i_E = \frac{d_4}{d_1} = 0.417; i_F = \frac{d_4'}{d_2} = 0.346 \quad (3.165)$$

Solution:

$$\omega_3 = 360\ °/\text{s}; \omega_4 = 360\ °/\text{s} \quad (3.166)$$

The SW simulation results: $\omega_3^{SW} = 360\ °/\text{s}; \omega_4^{SW} = 360\ °/\text{s}$ are coincidental with the above solution.

3.9.1.3 Gears' Number of Teeth

From a preferred list of modules m, the modules $m_E = m_F = 2$ mm are selected. Then, the gears' number of teeth is

$$N_1 = \frac{d_1}{m_E} = 60\ \text{teeth}; N_2 = \frac{d_2}{m_F} = 130\ \text{teeth}; N_4 = \frac{d_4}{m_E} = 25\ \text{teeth};$$
$$N_4' = \frac{d_4'}{m_F} = 45\ \text{teeth} \quad (3.167)$$

- Analogy to Willis equations: If Equation (3.163) is solved for gear rations, with the solution

$$i_E = -\frac{\omega_1 - \omega_3}{\omega_4 - \omega_3}; i_F = +\frac{\omega_2 - \omega_3}{\omega_4 - \omega_3} \quad (3.168)$$

Note:

- The two equations (generated from the sum of entries in the matroidal velocity matrix) become the Willis equations (generated from a set of two gears and a carrier for which is assigned an inverse motion, $-\omega_3$) for the carrier.
- The negative sign in front of ratio i_E indicates a set of external meshing at joint E, whereas the positive sign in front of ratio i_F indicates a set of internal meshing at joint F.

3.9.2 Gear Trains with the Fixed Parallel Axes (GT)

In Figure 3.56a,b, Animation 3.13, a typical GT is illustrated.

The analytic approach, presented previously for EGT applies for GT as follows.

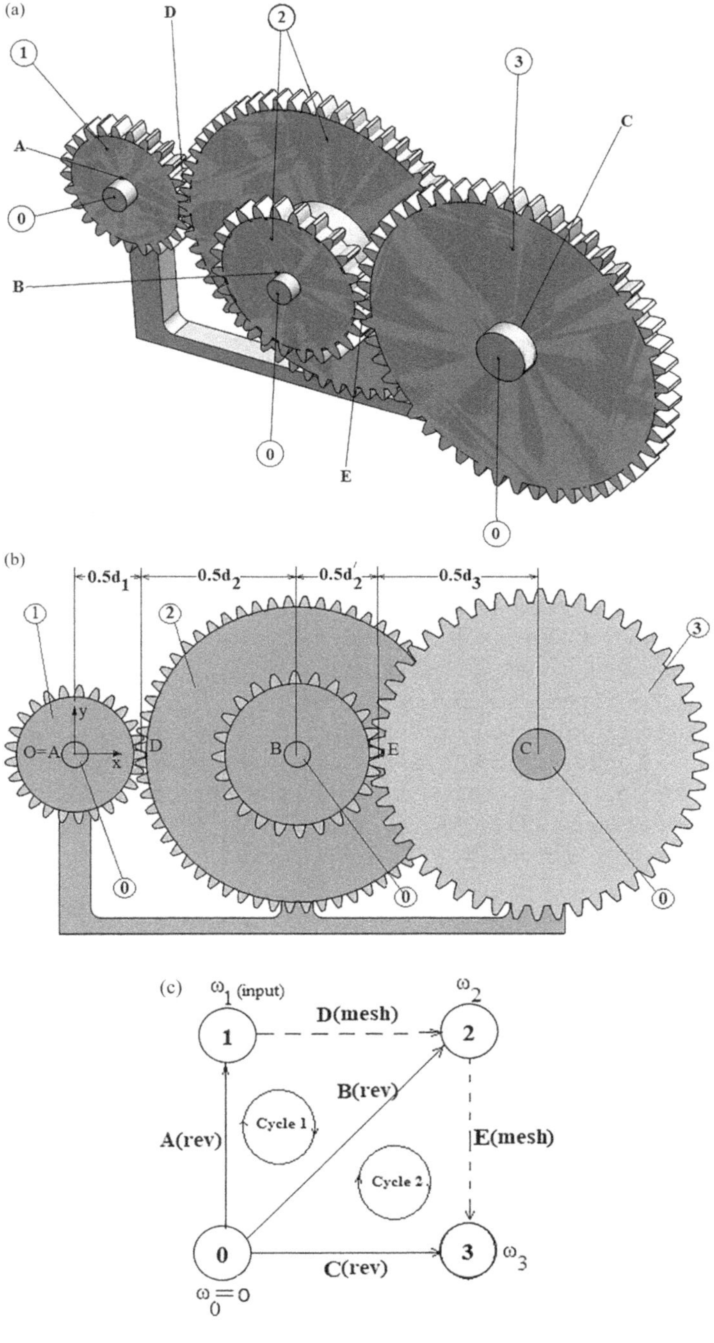

FIGURE 3.56
(a) Gear train with parallel fixed axes; (b) front view; (c) digraph.

- **Link and joint numbering**
  The links are assigned numbers, starting with 0 which is assigned to the fixed link. There are three mobile links, m = 3, which have been assigned the numbers 1, 2, and 3. Link 2 is a compound of two gears, and they rotate as a whole. There are five joints labeled with capital letters: A, B, C, D, and E. The mechanism has three revolute joints with 1 DOF each (description "rev" at A, B, and C), therefore $j_1 = 3$. There are two meshing joints with 2 DOF each (description "mesh" at D and E), therefore $j_2 = c = 2$. These two meshing joints are labeled with the last letters D and E. The total number of joints is

$$j = j_1 + j_2 = 3 + 2 = 5 \text{ joints}. \tag{3.169}$$

- **GT digraph**
  Figure 3.56c shows its associated graph. This graph has n = 4 nodes and j = 5 edges. The edges in digraph are oriented from the lower node number to the higher node number.
- **Spanning tree and the number of cycles for gear train**
  For gear train analysis, the spanning tree obtained from digraph is considered as a result of *cutting the meshing joints*. The spanning tree has n = 4 nodes: {0, 1, 2, 3}, and j – c = 3 revolute joints (A, B, C). The set of cut edges shown in dashed line {D, E} contains c = 2 edges corresponding to the meshing joints of the mechanism. The set c = j – m = 2 *independent cycles* is $\{C_D, C_F\}$ and is obtained adding one by one the edges from the cut set to the spanning tree, as follows: adding edges labeled D and E, the cycles $C_D = \{A, D, B\}$ and $C_E = \{B, E, C\}$ are obtained. This set $\{C_D, C_E\}$ is a cycle basis.

- **Mobility**
  There are m = 3 gears, and from digraph c = 2 cycles. The mobility for gear trains is determined by subtracting the number of cycles c from the number of mobile links (Equation (3.143)):

$$M = m - c = 3 - 2 = 1 \text{ DOF}. \tag{3.170}$$

- **The unit vectors for axes of rotation with respect to the fixed frame (casing)**
  The unit vectors $\mathbf{u}^0$ with respect to the fixed frame are all parallel with z_0-axis:

$$\mathbf{u}_A^0 = \mathbf{u}_B^0 = \mathbf{u}_C^0 = \mathbf{u}_D^0 = \mathbf{u}_E^0 = \begin{Bmatrix} 0 \\ 0 \\ 1 \end{Bmatrix} \tag{3.171}$$

3.9.2.1 *Equations Based on Absolute Angular Velocities: The Matroidal Method for GT*

- **The Latin matrix**

  The vector entries in the Latin matrix are written around the nodes in digraph. As stated before in the text, the sum of entries on each cycle, row in matrix **L**, is zero:

$$
\begin{array}{c} \\ [\mathbf{L}] = \end{array}
\begin{array}{c} \begin{array}{cccc} 0 & 1 & 2 & 3 \end{array} \\ \begin{bmatrix} \mathbf{L}_{BA} & \mathbf{L}_{AD} & \mathbf{L}_{DB} & 0 \\ \mathbf{L}_{CB} & 0 & \mathbf{L}_{BE} & \mathbf{L}_{EC} \end{bmatrix} \end{array}
\begin{array}{l} \\ \text{Cycle } C_D \\ \text{Cycle } C_E \end{array} \qquad (3.172)
$$

A zero entry denotes that the node is not on the cycle. Each vector has components on x_0:

$$
\mathbf{L}_{AD} = \begin{Bmatrix} x_{AD} \\ 0 \\ 0 \end{Bmatrix} = \begin{Bmatrix} x_D - x_A \\ 0 \\ 0 \end{Bmatrix} \qquad (3.173)
$$

- **Velocity matroidal matrix**

  For the GT, the velocity matroidal matrix is

$$
\begin{array}{c} \\ [\dot{\mathbf{L}}] = \end{array}
\begin{array}{c} \begin{array}{cccc} 0 & 1 & 2 & 3 \end{array} \\ \begin{bmatrix} \dot{\mathbf{L}}_{BA} & \dot{\mathbf{L}}_{AD} & \dot{\mathbf{L}}_{DB} & 0 \\ \dot{\mathbf{L}}_{CB} & 0 & \dot{\mathbf{L}}_{BE} & \dot{\mathbf{L}}_{EC} \end{bmatrix} \end{array}
\begin{array}{l} \\ \text{Cycle } C_D \\ \text{Cycle } C_E \end{array} \qquad (3.174)
$$

From Statement 3.7: The sum of vector entries in velocity matroidal matrix $\dot{\mathbf{L}}$ is zero, then

$$
\begin{cases} \dot{\mathbf{L}}_{BA} + \dot{\mathbf{L}}_{AD} + \dot{\mathbf{L}}_{DB} = 0 \\ \dot{\mathbf{L}}_{CB} + \dot{\mathbf{L}}_{BE} + \dot{\mathbf{L}}_{EC} = 0 \end{cases} \qquad (3.175)
$$

Since is neglected the linear displacement at meshing joints, then the general form of velocity matroidal vector is $\dot{\mathbf{L}}_{AD} = \boldsymbol{\omega}_1 \times \mathbf{L}_{AD}$

The cross product of vectors in the system of equations:

$$
\begin{cases} \boldsymbol{\omega}_0 \times \mathbf{L}_{BA} + \boldsymbol{\omega}_1 \times \mathbf{L}_{AD} + \boldsymbol{\omega}_2 \times \mathbf{L}_{DB} = 0 \\ \boldsymbol{\omega}_0 \times \mathbf{L}_{CB} + \boldsymbol{\omega}_2 \times \mathbf{L}_{BE} + \boldsymbol{\omega}_3 \times \mathbf{L}_{EC} = 0 \end{cases} \qquad (3.176)
$$

is written as functions of matrix multiplications:

$$\begin{cases} -\tilde{\mathbf{L}}_{BA} * \boldsymbol{\omega}_0 - \tilde{\mathbf{L}}_{AD} * \boldsymbol{\omega}_1 - \tilde{\mathbf{L}}_{DB} * \boldsymbol{\omega}_2 = 0 \\ -\tilde{\mathbf{L}}_{CB} * \boldsymbol{\omega}_0 - \tilde{\mathbf{L}}_{BE} * \boldsymbol{\omega}_2 - \tilde{\mathbf{L}}_{EC} * \boldsymbol{\omega}_3 = 0 \end{cases} \tag{3.177}$$

The factors in previous equations for GT have the scalar components developed considering Equation (3.173), where vector $\mathbf{L}_{AD}$ has a non-zero component x_{AD}:

$$-\tilde{\mathbf{L}}_{AD} * \boldsymbol{\omega}_1 = -\tilde{\mathbf{L}}_{AD} * \mathbf{u}_1^0 \cdot \omega_1 = -\underbrace{\begin{bmatrix} 0 & 0 & 0 \\ 0 & 0 & -x_{AD} \\ 0 & x_{AD} & 0 \end{bmatrix}}_{\tilde{\mathbf{L}}_{AD}} * \underbrace{\begin{Bmatrix} 0 \\ 0 \\ 1 \end{Bmatrix} \cdot \omega_1}_{\boldsymbol{\omega}_1} \tag{3.178}$$

$$= \begin{Bmatrix} x_{AD} \cdot \omega_1 \\ 0 \\ 0 \end{Bmatrix}$$

- **Equations based on absolute angular velocities**

  The c = 2 equations for GT in scalar form and functions of absolute angular velocities are

$$\begin{cases} x_{BA} \cdot \omega_0 + x_{AD} \cdot \omega_1 + x_{DB} \cdot \omega_2 = 0 \\ x_{CB} \cdot \omega_0 + x_{BE} \cdot \omega_2 + x_{EC} \cdot \omega_3 = 0 \end{cases} \tag{3.179}$$

and considering $\omega_0 = 0$, they are written in matrix form as

$$[L_m(x)] * \{\omega_m\} = 0 \tag{3.180a}$$

$$\begin{matrix} & 1 & 2 & 3 \end{matrix}$$

$$\begin{bmatrix} x_{AD} & x_{DB} & 0 \\ 0 & x_{BE} & x_{EC} \end{bmatrix} * \begin{Bmatrix} \omega_1 \\ \omega_2 \\ \omega_3 \end{Bmatrix} = \begin{Bmatrix} 0 \\ 0 \end{Bmatrix} \tag{3.180b}$$

where

$[L_m(x)]$ is the $(c \times m)$ matrix with x-components from Latin matrix (Equation (3.172))

$\{\omega_m\}$ is the $(m \times 1)$ matrix with scalar components of gears

- **Equations based on gear ratios**
  - **The joints' x-coordinates**

    Considering the joints' y-coordinates in the following table:

Joint	A	B	C	D	E
x	0	$0.5\cdot(d_1+d_2)$	$0.5\cdot(d_1+d_2+d_2'+d_3)$	$0.5\cdot d_1$	$0.5\cdot(d_1+d_2+d_2')$

Where A_1 and A_2 are the distances between parallel shafts

$$A_1 = 0.5\cdot(d_1+d_2) \text{ and } A_2 = 0.5\cdot(d_2'+d_3)$$

– The gear ratios between two nodes in digraph, corresponding to the two meshing joints, are defined as

$$\begin{aligned} i_D = i_{D1,2} &= \frac{d_2(mm)}{d_1(mm)} = \frac{n_1(rpm)}{n_2(rpm)} = \frac{\omega_1(rad/s)}{\omega_2(rad/s)} = \frac{N_2(\#\,teeth)}{N_1(\#\,teeth)} \\ i_E = i_{E2,3} &= \frac{d_3(mm)}{d_2'(mm)} = \frac{n_2(rpm)}{n_3(rpm)} = \frac{\omega_2(rad/s)}{\omega_3(rad/s)} = \frac{N_3(\#\,teeth)}{N_2'(\#\,teeth)} \end{aligned} \tag{3.181}$$

Then, Equation (3.180b) becomes

$$\begin{cases} 0.5\cdot d_1\cdot\omega_1 + 0.5\cdot d_2\cdot\omega_2 = 0 \\ 0.5\cdot d_2'\cdot\omega_2 + 0.5\cdot d_3\cdot\omega_3 = 0 \end{cases} \tag{3.182}$$

$$\begin{bmatrix} 1 & i_D & 0 \\ 0 & 1 & i_E \end{bmatrix} * \begin{Bmatrix} \omega_1 \\ \omega_2 \\ \omega_3 \end{Bmatrix} = \begin{Bmatrix} 0 \\ 0 \end{Bmatrix} \tag{3.182a}$$

- **Number of equations and number of unknown angular velocities**
  - The number of equations is $c = 2$, equal to the rows in the Latin matrix.
  - From Statement 3.11, the number of unknown angular velocities for GT is the number c of cycles, $c = 2$ unknowns ω_2 and ω_3

- *Solution*

  The linear system of $c = 2$ equations, solved for $c = 2$ unknowns, holds the solution:

$$\omega_2 = -\frac{1}{i_D}\omega_1;\ \omega_3 = \frac{1}{i_D\cdot i_E}\omega_1 \tag{3.183}$$

3.9.2.2 GT Velocity Ratio

The GT *velocity ratio* $e_{1,3}$ is defined as the ratio of output gear's angular velocity to the input gear's angular velocity. Considering the results from the matroidal method solution, Equation (3.183), then

$$e_{1,3} = \frac{n_3(\text{output gear rpm})}{n_1(\text{input gear rpm})} = \frac{\omega_3}{\omega_1} = \frac{n_3}{n_1} = \frac{1}{i_D \cdot i_E} = \frac{N_1 \cdot N_2'(\text{driving gears \# teeth})}{N_2 \cdot N_3(\text{driven gears \# teeth})} \tag{3.184}$$

- **Analogy to tabular method**

  Note:

  The above equation (generated from the sum of entries in matroidal velocity matrix) becomes the formula used in the tabular method for GT, [13].

  Numerical example: The GT has metric gears with

$$d_1 = 50\,\text{mm};\ d_2 = 120\ \text{mm};\ d_3 = 125\ \text{mm};\ d_2' = 60\ \text{mm};$$

$$\omega_1 = 360\ °/\text{s} \Rightarrow n_1 = 60\ \text{rpm}$$

$$i_D = \frac{d_2(\text{mm})}{d_1(\text{mm})} = 2.4;\ i_E = \frac{d_3(\text{mm})}{d_2'(\text{mm})} = 2.083 \tag{3.185}$$

Solution: $\omega_2 = -150\ °/\text{s};\ \omega_3 = 72.01\ °/\text{s}$

The SW simulation: $\omega_2^{SW} = -150\ °/\text{s};\ \omega_3^{SW} = 72.01\ °/\text{s}$ are coincidental with the above solution.

3.9.2.3 Gears' Number of Teeth

From a preferred list of modules m, the modules $m_D = 2$ mm and $m_E = 2.5$ mm are selected. Then, the gears' number of teeth is

$$N_1 = \frac{d_1}{m_D} = 25\ \text{teeth};\ N_2 = \frac{d_2}{m_D} = 60\ \text{teeth};\ N_3 = \frac{d_3}{m_E} = 50\ \text{teeth};$$

$$N_2' = \frac{d_2'}{m_E} = 24\ \text{teeth} \tag{3.186}$$

- **The GT velocity ratio**

$$e_{1,3} = \frac{\omega_3}{\omega_1} = \frac{n_3}{n_1} = \frac{25 \cdot 24}{60 \cdot 50} = 0.2$$

- Since the first gear is in rotation with $n_1 = 60$ rpm, the last gear in the TG is in rotation with $n_3 = e_{1,3} \cdot n_1 = 12$ rpm, therefore the GT is a reducer.

3.9.3 Bevel Gear Trains

- **Link and joint numbering**
  For the case of a bevel gear train (BGT) with m mobile links and j joints, link numbering begins with 0 assigned to the fixed link (casing) continuing up to m for the other mobile links. Additionally, the revolute and meshing joints are assigned letters.
  Figure 2.57a,b, (Animation 3.14), shows the numbering in a typical BGT mechanism with m = 4 mobile links labeled as: 1 (carrier), 2 (planet bevel gear), and 3 and 4 (bevel gears).There are j = 6 joints (A, B, C, D, E, and F), with c = 2 meshing joints (E and F) and j – c = 4 revolute joints (A, B, C, and D).

3.9.3.1 Equations Based on Twists: The Matroidal Method for BGT

- **BGT digraph**
  Figure 3.57c shows its associated digraph. This graph has n = m + 1 = 5 vertices and j = 6 edges. The spanning tree has n = 5 vertices: {0, 1, 2, 3, 4} and j – c = 4 revolute joints (A, B, C, and D). The edges in digraph are oriented from the lower node number to the higher node number.
- **Cycle basis and spanning tree for BGTs**
  For gear train analysis, the spanning tree obtained from digraph is considered as a result of *cutting the meshing joints*. The same procedure was used previously for parallel axes gear trains. The set of cut edges shown in dashed line {E, F} contains c = 2 edges corresponding to the meshing joints of the mechanism. The set c = j – m = 2 *independent cycles* is $\{C_E, C_F\}$ and is obtained adding one by one the edges from the cut set to the spanning tree, as follows: adding edges labeled E and F, the cycles $C_E = \{A, B, E, C\}$ and $C_F = \{A, B, F, D\}$ are obtained. This set $\{C_E, C_F\}$ is a cycle basis.
  For the number of cycles in cycle basis, Euler's formula holds:

$$c = j - m = 6 - 4 = 2 \text{ cycle.} \tag{3.187}$$

- **Matrix description of the digraph**
  For an automatic generation of coefficients in velocity equations, the following matrices on the digraph are considered [9]:
  - **Incidence nodes–edges matrix, G**
    This matrix, introduced in Chapter 1, illustrates the incidence nodes–edges in digraph: each column j will have a –1 and a +1

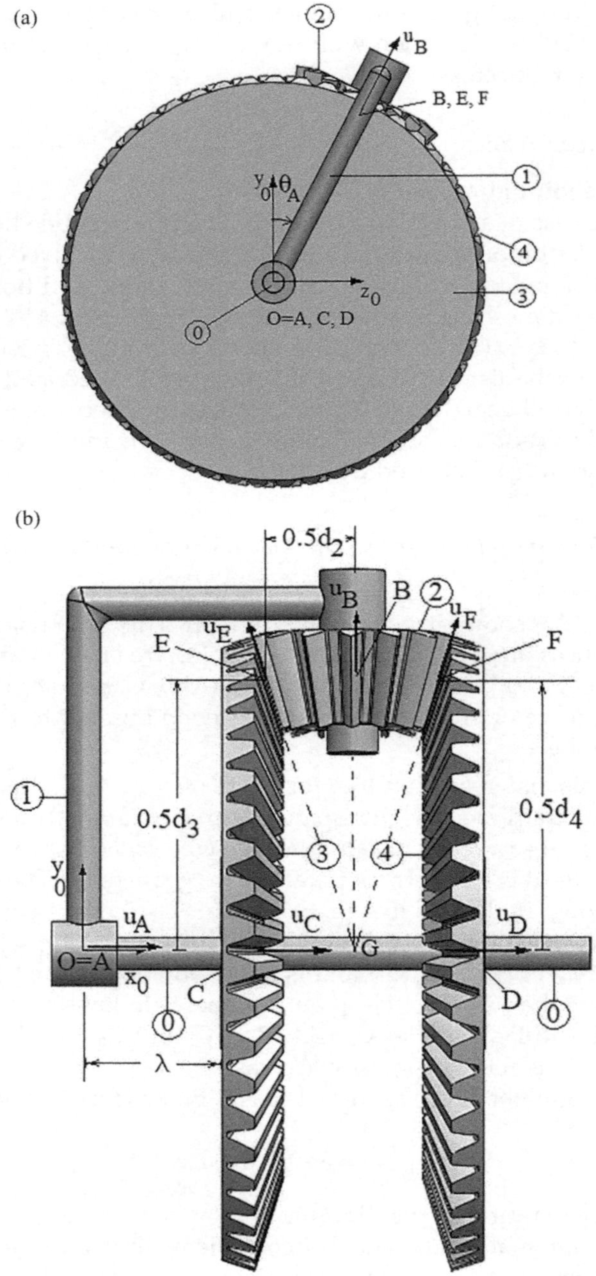

FIGURE 3.57
(a) BGT; (b) home (initial) position; (c) digraph and cycles; (d) spanning tree.

(Continued)

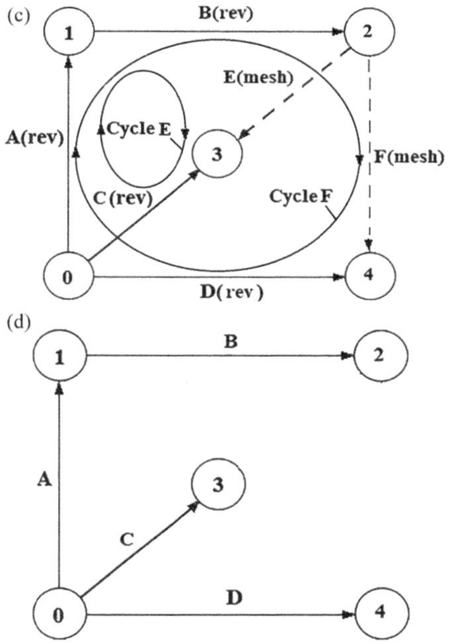

FIGURE 3.57 (CONTINUED)
(a) BGT; (b) home (initial) position; (c) digraph and cycles; (d) spanning tree.

entry corresponding to the two nodes connected by the respective edge. For the digraph in Figure 3.57c, the incidence matrix will therefore be

$$
\underline{\mathbf{G}} = \begin{array}{c} \begin{array}{cccccc} A & B & C & D & E & F \end{array} \\ \left[\begin{array}{cccc:cc} -1 & 0 & -1 & -1 & 0 & 0 \\ \hdashline 1 & -1 & 0 & 0 & 0 & 0 \\ 0 & 1 & 0 & 0 & -1 & -1 \\ 0 & 0 & 1 & 0 & 1 & 0 \\ 0 & 0 & 0 & 1 & 0 & 1 \end{array}\right] \begin{array}{c} 0 \\ 1 \\ 2 \\ 3 \\ 4 \end{array} \end{array} \qquad (3.188)
$$

- **Reduced incidence nodes–edges matrix and its partitions**
  This $(m \times j)$ matrix $\mathbf{G}$ is obtained by deleting from the incidence nodes–edges matrix $\underline{\mathbf{G}}$ the first row corresponding to the fixed link. The reduced incidence matrix is partitioned into two submatrices: a $(m \times m)$ square matrix $\mathbf{G}_a$, with the number of columns equal to the number of revolute joints (arcs of the spanning tree) and a $(m \times c)$ matrix $\mathbf{G}_c$, with the number of columns equal to the number of meshing joints. The reduced incidence matrix is

$$\mathbf{G} = \left[\ \mathbf{G}\ \vdots\ \mathbf{G}_c\ \right]; \mathbf{G} = \begin{array}{c} \begin{matrix} A & B & C & D \end{matrix} \\ \left[\begin{matrix} 1 & -1 & 0 & 0 \\ 0 & 1 & 0 & 0 \\ 0 & 0 & 1 & 0 \\ 0 & 0 & 0 & 1 \end{matrix}\right] \end{array} \begin{matrix} 1 \\ 2 \\ 3 \\ 4 \end{matrix}; \quad \mathbf{G}_c = \begin{array}{c} \begin{matrix} E & F \end{matrix} \\ \left[\begin{matrix} 0 & 0 \\ -1 & -1 \\ 1 & 0 \\ 0 & 1 \end{matrix}\right] \end{array} \begin{matrix} 1 \\ 2 \\ 3 \\ 4 \end{matrix} \tag{3.189}$$

- **The path matrix, Z**
  This (m × m) matrix **Z**, constructed from the spanning tree, has its entries $z_{m,j}$ defined as follows:
  $z_{m,j} = -1$ if edge j belongs to the path from vertex 0 (ground) to vertex m and is oriented as the path
  $z_{m,j} = +1$ if edge j belongs to the path from the vertex 0 to vertex n and is oriented inverse
  $z_{m,j} = 0$ if edge j does not belong to the path
  The path matrix for the tree in Figure 3.56d is

$$\mathbf{Z} = \begin{matrix} A \\ B \\ C \\ D \end{matrix} \begin{array}{c} \begin{matrix} 1 & 2 & 3 & 4 \end{matrix} \\ \left[\begin{matrix} 1 & 1 & 0 & 0 \\ 0 & 1 & 0 & 0 \\ 0 & 0 & 1 & 0 \\ 0 & 0 & 0 & 1 \end{matrix}\right] \end{array} \tag{3.190}$$

- **The spanning tree matrix, T**
  This (c × m) matrix **T** is defined based on the incidence matrices **Z** and $\mathbf{G}_c$ according to the following relation:

$$\mathbf{T} = \mathbf{G}_c^T * \mathbf{Z}^T \tag{3.191}$$

Therefore,

$$\mathbf{T} = \begin{array}{c} \begin{matrix} A & B & C & D \end{matrix} \\ \left[\begin{matrix} 1 & 1 & -1 & 0 \\ 1 & 1 & 0 & -1 \end{matrix}\right] \end{array}$$

- **The cycle basis matroid matrix, C**
  This (c × j) = 2 × 6 matrix **C** has c = 2 rows equal to the number of cycles and j columns corresponding to the edges of the digraph. Since j = m + c, the matrix **C** is partitioned into two

sub-matrices: the (c × m) matrix **T**, with the number of columns equal to the number of revolute joints (edges of spanning tree) and a (c × c) unit matrix **U** with the number of columns equal to the number of meshing joints:

$$\mathbf{C} = \left[\begin{array}{c|c} \mathbf{T}_{c,m} & \mathbf{U}_{c,c} \end{array}\right] \tag{3.192}$$

An $c_{c,j}$ entry could be either equal to –1, +1, or 0 as follows. A –1 entry denotes the column label of the edge belonging to the cycle and oriented in the opposite direction to the closed path (clockwise or counterclockwise). A +1 entry denotes the column label of the edge belonging to the cycle and having the same direction as the path. A 0 entry indicates that the respective edge does not belong to the cycle. The digraph in Figure 3.57c has a set of two independent cycles (the two rows in matrix **C**). Therefore, its cycle basis matrix C is

$$\begin{array}{c} \\ \mathbf{C} = \end{array} \begin{array}{c} \begin{array}{cccccc} A & B & C & D & E & F \end{array} \\ \left[\begin{array}{cccc|cc} 1 & 1 & -1 & 0 & 1 & 0 \\ 1 & 1 & 0 & -1 & 0 & 1 \end{array}\right] \end{array} \begin{array}{c} \\ C_E \\ C_F \end{array} \tag{3.193}$$

Note:

- The entries in cycle basis matroid **C** can be written directly along the cycles in digraph and are the coefficients in kinematic equations.
- The entries in cycle basis matroid **C** are automatically generated from Equation (3.191), as needed in programming.

- **BGT absolute angular velocity matrix**

  This matrix, noted $\boldsymbol{\omega}_m = \boldsymbol{\omega}_{0,m}$, is a column matrix (m × 1) whose entries are the gears' angular velocity vectors, expressed with respect to the fixed frame. The frame (casing) is fixed; therefore, its velocity is zero:

$$\{\boldsymbol{\omega}_m\} = \left\{\begin{array}{c} \boldsymbol{\omega}_1 \\ \boldsymbol{\omega}_2 \\ \boldsymbol{\omega}_3 \\ \boldsymbol{\omega}_4 \end{array}\right\} \tag{3.194}$$

- **BGT relative (joint rates) angular velocity matrix**

  This matrix, noted $\dot{\boldsymbol{\theta}}_j$, is a column matrix (j × 1) whose entries are relative velocity vectors assigned to edges j in digraph to measure the relative rotations between gears and between gears and carriers: the gear (node) m_{head} relative to gear (node) m_{tail}. For the BGT mechanism, the relative velocity matrix is

$$\begin{aligned}\{\dot{\theta}_j\} &= \left\{ \begin{array}{cccc:cc} \dot{\theta}_A & \dot{\theta}_B & \dot{\theta}_C & \dot{\theta}_D & \dot{\theta}_E & \dot{\theta}_F \end{array} \right\}^T \\ &= \left\{ \begin{array}{cccc:cc} \dot{\theta}_{0,1} & \dot{\theta}_{0,2} & \dot{\theta}_{0,3} & \dot{\theta}_{0,4} & \dot{\theta}_{2,3} & \dot{\theta}_{2,4} \end{array} \right\}^T \end{aligned} \tag{3.195}$$

The IT notation in digraph is considered for the relative velocities. They measure the relative motion of rotation between links as shown in Equation (3.195). Thus, $\dot{\theta}_A = \dot{\theta}_{0,1}$ is the angular velocity of carrier 1 in rotation relative to the fixed frame; $\dot{\theta}_B = \dot{\theta}_{1,2}$ is the angular velocity of planet 2 in rotation relative to carrier 1, etc.

The relative angular velocity vectors are assigned to edges in digraph and oriented in the direction of unit vectors $\mathbf{u}^0$ (Figure 3.57b):

$$\begin{aligned}&\dot{\theta}_A = \mathbf{u}_A^0 \cdot \dot{\theta}_{0,1};\ \dot{\theta}_B = \mathbf{u}_B^0 \cdot \dot{\theta}_{1,2};\ \dot{\theta}_C = \mathbf{u}_C^0 \cdot \dot{\theta}_{0,3};\ \dot{\theta}_D = \mathbf{u}_D^0 \cdot \dot{\theta}_{0,4};\\ &\dot{\theta}_E = \mathbf{u}_E^0 \cdot \dot{\theta}_{2,3};\ \dot{\theta}_F = \mathbf{u}_F^0 \cdot \dot{\theta}_{2,4}\end{aligned} \tag{3.196}$$

The matrix in Equation (3.195) has the sub-matrix $\dot{\theta}_{rev}$, with the entries corresponding to revolute joints on the spanning tree, and $\dot{\theta}_{mesh}$ for the meshing joints:

$$\{\dot{\theta}_j\} = \begin{Bmatrix} \dot{\theta}_{rev} \\ \dot{\theta}_{mesh} \end{Bmatrix};\ \dot{\theta}_{rev} = \begin{Bmatrix} \dot{\theta}_A \\ \dot{\theta}_B \\ \dot{\theta}_C \\ \dot{\theta}_D \end{Bmatrix};\ \dot{\theta}_{mesh} = \begin{Bmatrix} \dot{\theta}_E \\ \dot{\theta}_F \end{Bmatrix} \tag{3.197}$$

- **The relative rotation matrices**

  In Chapter 2, the relative rotation matrices required in the further calculation of absolute rotation matrices are introduced. For the revolute joints at A, C, and D, rotation matrices about local x-axes are considered, whereas at B a rotation about y-axis is considered.

$$\mathbf{D}_{A0,1} = \begin{bmatrix} 1 & 0 & 0 \\ 0 & C\theta_A & -S\theta_A \\ 0 & S\theta_A & C\theta_A \end{bmatrix}; \mathbf{D}_{B1,2} = \begin{bmatrix} C\theta_B & 0 & S\theta_B \\ 0 & 1 & 0 \\ -S\theta_B & 0 & C\theta_B \end{bmatrix};$$
$$\mathbf{D}_{C0,3} = \begin{bmatrix} 1 & 0 & 0 \\ 0 & C\theta_C & -S\theta_C \\ 0 & S\theta_C & C\theta_C \end{bmatrix}; \mathbf{D}_{D0,4} = \begin{bmatrix} 1 & 0 & 0 \\ 0 & C\theta_D & -S\theta_D \\ 0 & S\theta_D & C\theta_D \end{bmatrix} \tag{3.198}$$

- **The gears and carrier absolute rotation matrices**
  The absolute rotation matrices for the mobile links, expressed as functions of the above relative rotation matrices, are

$$\mathbf{D}_1 = \mathbf{D}_{A0,1} = \begin{bmatrix} 1 & 0 & 0 \\ 0 & C\theta_A & -S\theta_A \\ 0 & S\theta_A & C\theta_A \end{bmatrix};$$
$$\mathbf{D}_2 = \mathbf{D}_{A0,1} * \mathbf{D}_{B1,2}$$
$$= \begin{bmatrix} 1 & 0 & 0 \\ 0 & C\theta_A & -S\theta_A \\ 0 & S\theta_A & C\theta_A \end{bmatrix} * \begin{bmatrix} C\theta_B & 0 & S\theta_B \\ 0 & 1 & 0 \\ -S\theta_B & 0 & C\theta_B \end{bmatrix} \tag{3.199}$$
$$= \begin{bmatrix} C\theta_B & 0 & S\theta_B \\ S\theta_A S\theta_B & C\theta_A & -S\theta_A C\theta_B \\ -C\theta_A S\theta_B & S\theta_A & C\theta_A C\theta_B \end{bmatrix};$$
$$\mathbf{D}_3 = \mathbf{D}_{C0,3} = \begin{bmatrix} 1 & 0 & 0 \\ 0 & C\theta_C & -S\theta_C \\ 0 & S\theta_A & C\theta_C \end{bmatrix}; \mathbf{D}_4 = \mathbf{D}_{C0,4} = \begin{bmatrix} 1 & 0 & 0 \\ 0 & C\theta_D & -S\theta_D \\ 0 & S\theta_D & C\theta_D \end{bmatrix}$$

- **The unit vectors for axes of rotation with respect to the gears frame**
  For the local frames attached to the gears and carriers, then the unit vectors have the form shown in Equation (3.151):

$$\mathbf{u}_A^1 = \begin{Bmatrix} 1 \\ 0 \\ 0 \end{Bmatrix}; \mathbf{u}_B^2 = \begin{Bmatrix} 0 \\ 1 \\ 0 \end{Bmatrix}; \mathbf{u}_C^3 \begin{Bmatrix} 1 \\ 0 \\ 0 \end{Bmatrix}; \mathbf{u}_D^4 = \begin{Bmatrix} 1 \\ 0 \\ 0 \end{Bmatrix} \tag{3.200}$$

- **The screws at joints**
  A *twist*, $\hat{\mathbf{t}}_{EZ}$, is a 6 × 1 column matrix assigned to each joint. The upper vector is the joint's relative angular velocity, whereas the lower vector represents a linear velocity:

$$\hat{\mathbf{t}}_{E,Z} = \begin{Bmatrix} \dot{\boldsymbol{\theta}}_Z \\ \text{-----------} \\ \tilde{\mathbf{L}}^0_{EZ} * \dot{\boldsymbol{\theta}}_Z + \dot{\mathbf{d}}_Z \end{Bmatrix} = \begin{Bmatrix} \mathbf{u}^0_Z \cdot \dot{\theta}_Z \\ \text{-----------} \\ \tilde{\mathbf{L}}^0_{EZ} * \mathbf{u}^0_Z \cdot \dot{\theta}_Z \end{Bmatrix} = \begin{Bmatrix} \mathbf{u}^0_Z \\ \text{-----------} \\ \tilde{\mathbf{L}}^0_{EZ} * \mathbf{u}^0_Z \end{Bmatrix} \cdot \dot{\theta}_Z = \hat{\mathbf{u}}_{E,Z} \cdot \dot{\theta}_Z \tag{3.201}$$

The constraint in Equation (3.202) applies for gear trains where the slippage in revolute and meshing joints is neglected:

$$\dot{\mathbf{d}}_Z = 0 \tag{3.202}$$

Note: The lower letters E and F in the twist notation show that the lower vector in the twist is referenced with respect to E and F (meshing joints). The 6×1 column matrix unit screw, $\hat{\mathbf{u}}_{E,Z}$, is a dual vector with its entries the *Plucker coordinates* of its line support, shown in Equation (3.203):

$$\hat{\mathbf{u}}_{E,Z} = \begin{Bmatrix} \mathbf{u}^0_Z \\ \text{-----------} \\ \tilde{\mathbf{L}}^0_{EZ} * \mathbf{u}^0_Z \end{Bmatrix} = \begin{Bmatrix} l_Z \\ m_Z \\ n_Z \\ \text{------} \\ P_{EZ} \\ Q_{EZ} \\ R_{EZ} \end{Bmatrix} = \begin{Bmatrix} l_Z \\ m_Z \\ n_Z \\ \text{------} \\ -z_{EZ} \cdot m_Z + y_{EZ} \cdot n_Z \\ z_{EZ} \cdot l_Z - x_{EZ} \cdot n_Z \\ -y_{EZ} \cdot l_Z + x_{EZ} \cdot m_Z \end{Bmatrix} = \begin{Bmatrix} 0 \\ 0 \\ 1 \\ \text{------} \\ y_{EZ} \\ 0 \\ 0 \end{Bmatrix} \tag{3.203}$$

- **The unit screw's upper vector with respect to the fixed frame**
  The unit vectors $\mathbf{u}^0_Z$ are the upper vectors in the twists, $\hat{\mathbf{u}}$, and written with respect to the fixed frame, Equation (3.204):

$$\mathbf{u}^0_Z = \mathbf{D}_m * \mathbf{u}^m_Z; \text{ where } Z = A, B, C, \text{ and } D, \text{ and } m = 1, 2, 3, 4 \tag{3.204}$$

For the BGT in Figure 3.57a, considering Equations (3.199) and (3.200):

$$\mathbf{u}^0_A = \mathbf{D}_1 * \mathbf{u}^1_A = \begin{Bmatrix} 1 \\ 0 \\ 0 \end{Bmatrix}; \mathbf{u}^0_B = \mathbf{D}_2 * \mathbf{u}^2_B = \begin{Bmatrix} 0 \\ C\theta_A \\ S\theta_A \end{Bmatrix}; \mathbf{u}^0_C = \mathbf{D}_3 * \mathbf{u}^3_C = \begin{Bmatrix} 1 \\ 0 \\ 0 \end{Bmatrix};$$

$$\mathbf{u}^0_D = \mathbf{D}_4 * \mathbf{u}^4_D = \begin{Bmatrix} 1 \\ 0 \\ 0 \end{Bmatrix}; \mathbf{u}^0_E = \begin{Bmatrix} l_E \\ m_E \\ n_E \end{Bmatrix}; \mathbf{u}^0_F = \begin{Bmatrix} l_F \\ m_F \\ n_F \end{Bmatrix} \tag{3.204a}$$

- **The unit screw's lower vector with respect to meshing joint**
  The lower vectors (moments of unit vectors) are determined using Equation (3.205):

$$\tilde{\mathbf{L}}_{EZ}^{0} * \mathbf{u}_{Z}^{0} = \begin{Bmatrix} P_{EZ} \\ Q_{EZ} \\ R_{EZ} \end{Bmatrix} = \begin{Bmatrix} -z_{EZ} \cdot m_Z + y_{EZ} \cdot n_Z \\ z_{EZ} \cdot l_Z - x_{EZ} \cdot n_Z \\ -y_{EZ} \cdot l_Z + x_{EZ} \cdot m_Z \end{Bmatrix};$$

$$\text{where } \mathbf{u}_{Z}^{0} = \begin{Bmatrix} l_Z \\ m_Z \\ n_Z \end{Bmatrix}; \mathbf{L}_{EZ}^{0} = \mathbf{r}_{Z}^{0} - \mathbf{r}_{E}^{0} = \begin{Bmatrix} x_{EZ} \\ y_{EZ} \\ z_{EZ} \end{Bmatrix} \tag{3.205}$$

– *Cycle C_E position vectors with respect to O and E:*

$$\mathbf{r}_{A}^{0} = \begin{Bmatrix} 0 \\ 0 \\ 0 \end{Bmatrix}; \mathbf{r}_{B}^{0} = \begin{Bmatrix} \lambda + 0.5 \cdot d_2 \\ 0.5 \cdot d_3 \cdot C\theta_A \\ 0.5 \cdot d_3 \cdot S\theta_A \end{Bmatrix}; \mathbf{r}_{C}^{0} = \begin{Bmatrix} \lambda \\ 0 \\ 0 \end{Bmatrix}; \mathbf{r}_{E}^{0} = \begin{Bmatrix} \lambda \\ 0.5 \cdot d_3 \cdot C\theta_A \\ 0.5 \cdot d_3 \cdot S\theta_A \end{Bmatrix};$$

$$\mathbf{L}_{EA}^{0} = \mathbf{r}_{A}^{0} - \mathbf{r}_{E}^{0} = \begin{Bmatrix} -\lambda \\ -0.5 \cdot d_3 \cdot C\theta_A \\ -0.5 \cdot d_3 \cdot S\theta_A \end{Bmatrix}; \mathbf{L}_{EB}^{0} = \mathbf{r}_{B}^{0} - \mathbf{r}_{E}^{0} = \begin{Bmatrix} 0.5 \cdot d_2 \\ 0 \\ 0 \end{Bmatrix};$$

$$\mathbf{L}_{EC}^{0} = \mathbf{r}_{C}^{0} - \mathbf{r}_{E}^{0} = \begin{Bmatrix} 0 \\ -0.5 \cdot d_3 \cdot C\theta_A \\ -0.5 \cdot d_3 \cdot S\theta_A \end{Bmatrix}; \mathbf{L}_{EE}^{0} = \mathbf{r}_{E}^{0} - \mathbf{r}_{E}^{0} = \begin{Bmatrix} 0 \\ 0 \\ 0 \end{Bmatrix}$$

– *Cycle C_E moments of unit vectors with respect to E:*

$$\tilde{\mathbf{L}}_{EA}^{0} * \mathbf{u}_{A}^{0} = \begin{Bmatrix} P_{EA} \\ Q_{EA} \\ R_{EA} \end{Bmatrix} = \begin{Bmatrix} 0 \\ -0.5 \cdot d_3 \cdot S\theta_A \\ 0.5 \cdot d_3 \cdot C\theta_A \end{Bmatrix}; \tilde{\mathbf{L}}_{EB}^{0} * \mathbf{u}_{B}^{0} = \begin{Bmatrix} P_{EB} \\ Q_{EB} \\ R_{EB} \end{Bmatrix}$$

$$= \begin{Bmatrix} 0 \\ -0.5 \cdot d_2 \cdot S\theta_A \\ 0.5 \cdot d_2 \cdot C\theta_A \end{Bmatrix};$$

$$\tilde{\mathbf{L}}_{EC}^{0} * \mathbf{u}_{C}^{0} = \begin{Bmatrix} P_{EC} \\ Q_{EC} \\ R_{EC} \end{Bmatrix} = \begin{Bmatrix} 0 \\ -0.5 \cdot d_3 \cdot S\theta_A \\ 0.5 \cdot d_3 \cdot C\theta_A \end{Bmatrix}; \tilde{\mathbf{L}}_{EE}^{0} * \mathbf{u}_{E}^{0} = \begin{Bmatrix} P_{EE} \\ Q_{EE} \\ R_{EE} \end{Bmatrix} = \begin{Bmatrix} 0 \\ 0 \\ 0 \end{Bmatrix}$$

(3.205a)

– *Cycle C_F position vectors with respect to O and F:*

$$\mathbf{r}_A^0 = \begin{Bmatrix} 0 \\ 0 \\ 0 \end{Bmatrix}; \mathbf{r}_B^0 = \begin{Bmatrix} \lambda + 0.5 \cdot d_2 \\ 0.5 \cdot d_3 \cdot C\theta_A \\ 0.5 \cdot d_3 \cdot S\theta_A \end{Bmatrix}; \mathbf{r}_D^0 = \begin{Bmatrix} \lambda + d_2 \\ 0 \\ 0 \end{Bmatrix}; \mathbf{r}_F^0 = \begin{Bmatrix} \lambda + d_2 \\ 0.5 \cdot d_4 \cdot C\theta_A \\ 0.5 \cdot d_4 \cdot S\theta_A \end{Bmatrix};$$

$$\mathbf{L}_{FA}^0 = \mathbf{r}_A^0 - \mathbf{r}_E^0 = \begin{Bmatrix} -\lambda - d_2 \\ -0.5 \cdot d_4 \cdot C\theta_A \\ -0.5 \cdot d_4 \cdot S\theta_A \end{Bmatrix}; \mathbf{L}_{FB}^0 = \mathbf{r}_B^0 - \mathbf{r}_E^0 = \begin{Bmatrix} -0.5 \cdot d_2 \\ 0.5 \cdot (d_3 - d_4) \cdot C\theta_A \\ 0.5 \cdot (d_3 - d_4) \cdot S\theta_A \end{Bmatrix};$$

$$\mathbf{L}_{FD}^0 = \mathbf{r}_D^0 - \mathbf{r}_F^0 = \begin{Bmatrix} 0 \\ -0.5 \cdot d_4 \cdot C\theta_A \\ -0.5 \cdot d_4 \cdot S\theta_A \end{Bmatrix}; \mathbf{L}_{FF}^0 = \mathbf{r}_F^0 - \mathbf{r}_F^0 = \begin{Bmatrix} 0 \\ 0 \\ 0 \end{Bmatrix}$$

– *Cycle C_F moments of unit vectors with respect to F:*

$$\tilde{\mathbf{L}}_{FA}^0 * \mathbf{u}_A^0 = \begin{Bmatrix} P_{FA} \\ Q_{FA} \\ R_{FA} \end{Bmatrix} = \begin{Bmatrix} 0 \\ -0.5 \cdot d_4 \cdot S\theta_A \\ 0.5 \cdot d_4 \cdot C\theta_A \end{Bmatrix};$$

$$\tilde{\mathbf{L}}_{FB}^0 * \mathbf{u}_B^0 = \begin{Bmatrix} P_{FB} \\ Q_{FB} \\ R_{FB} \end{Bmatrix} = \begin{Bmatrix} 0 \\ +0.5 \cdot d_2 \cdot S\theta_A \\ -0.5 \cdot d_2 \cdot C\theta_A \end{Bmatrix};$$

$$\tilde{\mathbf{L}}_{FD}^0 * \mathbf{u}_D^0 = \begin{Bmatrix} P_{FD} \\ Q_{FD} \\ R_{FD} \end{Bmatrix} = \begin{Bmatrix} 0 \\ -0.5 \cdot d_4 \cdot S\theta_A \\ 0.5 \cdot d_4 \cdot C\theta_A \end{Bmatrix}; \tag{3.205b}$$

$$\tilde{\mathbf{L}}_{FF}^0 * \mathbf{u}_F^0 = \begin{Bmatrix} P_{FF} \\ Q_{FF} \\ R_{FF} \end{Bmatrix} = \begin{Bmatrix} 0 \\ 0 \\ 0 \end{Bmatrix}$$

3.9.3.2 Twist Velocity Matroidal Matrix

For single open cycle mechanisms in Chapter 2, Equation (2.251), the matrix $\mathbf{T}(\hat{\mathbf{t}})$ generates the kinematic equations.

- **Velocity matroidal matrix**
  The *velocity matroidal matrix* $\mathbf{C}(\hat{\mathbf{t}})$, Equation (3.206), generates the kinematic equations for closed multiple-cycle mechanisms by

addition of c columns to matrix $\mathbf{T}(\hat{\mathbf{t}})$. The extra c columns of matrix **U** represent the c edges in digraph, which are added to the tree to generate the c cycles:

$$[\mathbf{C}(\hat{\mathbf{t}})]=\begin{array}{c} \\ \end{array}\overset{\begin{array}{cccccc}\quad A & \quad B & \quad C & \quad D & \quad E & \quad F\end{array}}{\left[\begin{array}{cccc:cc} 1\cdot\hat{\mathbf{t}}_{E,A} & 1\cdot\hat{\mathbf{t}}_{E,B} & -1\cdot\hat{\mathbf{t}}_{E,C} & 0 & 1\cdot\hat{\mathbf{t}}_{E,E} & 0 \\ \hdashline 1\cdot\hat{\mathbf{t}}_{F,A} & 1\cdot\hat{\mathbf{t}}_{F,B} & 0 & -1\cdot\hat{\mathbf{t}}_{F,D} & 0 & 1\cdot\hat{\mathbf{t}}_{F,F} \end{array}\right]}\begin{array}{c} C_E \\ C_F \end{array} \tag{3.206}$$

Statement 3.12

The sum of twists on each row in Equation (3.206) is zero.

Proof: Equation (2.234) from section 2.17.17 is written for each cycle, where the twists on the right side are the twists for meshing joints E and F. In matrix form, for the two cycles and with vector entries, Equation (3.206) becomes

$$\underbrace{\overset{\begin{array}{cccccc}A & B & C & D & E & F\end{array}}{\left[\begin{array}{cccc:cc} \mathbf{u}_A^0 & \mathbf{u}_B^0 & -\mathbf{u}_C^0 & 0 & \mathbf{u}_E^0 & 0 \\ \tilde{\mathbf{L}}_{EA}^0 * \mathbf{u}_A^0 & \tilde{\mathbf{L}}_{EB}^0 * \mathbf{u}_B^0 & -\tilde{\mathbf{L}}_{EC}^0 * \mathbf{u}_C^0 & 0 & 0 & 0 \\ \hdashline \mathbf{u}_A^0 & \mathbf{u}_B^0 & 0 & -\mathbf{u}_D^0 & 0 & \mathbf{u}_F^0 \\ \tilde{\mathbf{L}}_{FA}^0 * \mathbf{u}_A^0 & \tilde{\mathbf{L}}_{FB}^0 * \mathbf{u}_B^0 & 0 & -\tilde{\mathbf{L}}_{FD}^0 * \mathbf{u}_D^0 & 0 & 0 \end{array}\right]}}_{[\mathbf{C}(\hat{\mathbf{u}})]}$$

$$* \underbrace{\left\{\begin{array}{c} \dot{\theta}_A \\ \dot{\theta}_B \\ \dot{\theta}_C \\ \dot{\theta}_D \\ \hdashline \dot{\theta}_E \\ \dot{\theta}_E \end{array}\right\}}_{\{\dot{\theta}_j\}} = \left\{\begin{array}{c} 0 \\ 0 \\ 0 \\ 0 \end{array}\right\} \tag{3.207}$$

where

$[\mathbf{C}(\hat{\mathbf{u}})]$ is the matroidal matrix of screws' vector components

$\{\dot{\theta}_j\}=\left\{\begin{array}{c}\dot{\theta}_{rev} \\ \dot{\theta}_{mesh}\end{array}\right\}$ is the joints' scalar angular velocity matrix

- Equations as functions of gear ratios
  Gear ratios:
  The gear ratio between nodes 2 and 3 in digraph, corresponding to meshing joint E, is defined as

$$i_E = \frac{d_3}{d_2} = \frac{N_3}{N_2} \tag{3.207a}$$

The gear ratio between nodes 2 and 4 in digraph, corresponding to meshing joint F, is defined as

$$i_F = \frac{d_4}{d_2} = \frac{N_4}{N_2} \tag{3.207b}$$

Cycle C_E equations for relative angular velocities, after dividing the fifth and sixth rows by d_2, are functions of gear ratio:

$$\begin{array}{cccccc} A & B & C & D & E & F \end{array}$$

$$\left[\begin{array}{cccc:cc} 1 & 0 & -1 & 0 & l_E & 0 \\ 0 & C\theta_A & 0 & 0 & m_E & 0 \\ 0 & S\theta_A & 0 & 0 & n_E & 0 \\ \hdashline 0 & 0 & 0 & 0 & 0 & 0 \\ -i_E & -1 & i_E & 0 & 0 & 0 \\ i_E & 1 & -i_E & 0 & 0 & 0 \end{array}\right] * \left\{\begin{array}{c} \dot{\theta}_A \\ \dot{\theta}_B \\ \dot{\theta}_C \\ \dot{\theta}_D \\ \hdashline \dot{\theta}_E \\ \dot{\theta}_F \end{array}\right\} = \left\{\begin{array}{c} 0 \\ 0 \\ 0 \\ 0 \\ \hdashline 0 \\ 0 \end{array}\right\} \tag{3.208}$$

From rows 1, 2, 3 and 5, the set of four independent equations is:

$$\left\{\begin{array}{c} l_E \\ m_E \\ n_E \end{array}\right\} \cdot \dot{\theta}_E = -\underbrace{\left\{\begin{array}{c} 1 \\ 0 \\ 0 \end{array}\right\}}_{\mathbf{u}_A^0} \cdot \dot{\theta}_A - \underbrace{\left\{\begin{array}{c} 0 \\ C\theta_A \\ S\theta_A \end{array}\right\}}_{\mathbf{u}_B^0} \cdot \dot{\theta}_B + \underbrace{\left\{\begin{array}{c} 1 \\ 0 \\ 0 \end{array}\right\}}_{\mathbf{u}_C^0} \cdot \dot{\theta}_C \tag{3.208a}$$

$$-i_E \cdot \dot{\theta}_A - \dot{\theta}_B + i_E \cdot \dot{\theta}_C = 0 \tag{3.208b}$$

Cycle C_F equations for relative angular velocities, after dividing the fifth and sixth rows by d_2, are functions of gear ratio:

$$\begin{array}{c} \begin{array}{cccccc} A & B & C & D & E & F \end{array} \\ \left[\begin{array}{cccc|cc} 1 & 0 & 0 & -1 & 0 & l_F \\ 0 & C\theta_A & 0 & 0 & 0 & m_F \\ 0 & S\theta_A & 0 & 0 & 0 & n_F \\ \hline 0 & 0 & 0 & 0 & 0 & 0 \\ -i_F & 1 & 0 & i_F & 0 & 0 \\ i_F & -1 & 0 & -i_F & 0 & 0 \end{array}\right] * \left\{\begin{array}{c} \dot{\theta}_A \\ \dot{\theta}_B \\ \dot{\theta}_C \\ \dot{\theta}_D \\ \hline \dot{\theta}_E \\ \dot{\theta}_F \end{array}\right\} = \left\{\begin{array}{c} 0 \\ 0 \\ 0 \\ 0 \\ \hline 0 \\ 0 \end{array}\right\} \end{array} \tag{3.209}$$

From rows 1, 2, 3 and 5, the set of four independent equations is:

$$\left\{\begin{array}{c} l_F \\ m_F \\ n_F \end{array}\right\} \cdot \dot{\theta}_F = -\underbrace{\left\{\begin{array}{c} 1 \\ 0 \\ 0 \end{array}\right\}}_{\mathbf{u}_A^0} \cdot \dot{\theta}_A - \underbrace{\left\{\begin{array}{c} 0 \\ C\theta_A \\ S\theta_A \end{array}\right\}}_{\mathbf{u}_B^0} \cdot \dot{\theta}_B + \underbrace{\left\{\begin{array}{c} 1 \\ 0 \\ 0 \end{array}\right\}}_{\mathbf{u}_D^0} \cdot \dot{\theta}_D \tag{3.209a}$$

$$-i_F \cdot \dot{\theta}_A + \dot{\theta}_B + i_F \cdot \dot{\theta}_D = 0 \tag{3.209b}$$

- The rank of velocity matroidal matrix
  Cycle C_E:
  Row 5 has the Q-unit screw components, whereas row 6 has the R-unit screw components. Dividing all entries in rows 5 and 6 by d_2, the entries are functions of gear ratio i_E. The matrix does not have the full rank 6. The rank of screws on cycle C_E drops to 4, since row 4 has all zero entries—therefore deleted, whereas rows 5 and 6 have proportional entries (contain same entries with opposite sign)—therefore only one row is independent. This drop-in rank is due to configuration of cycle's C_E screws; they make a fascicle of concurrent lines at point G. Therefore, the rank is

$$r(C_E) = 4 \tag{3.209c}$$

Cycle C_F:

Row 5 has the Q-unit screw components, whereas row 6 has the R-unit screw components. Dividing all entries in rows 5 and 6 by d_2, the entries are functions of fear ratio i_F.

The matrix does not have the full rank 6. The rank of screws on cycle C_F drops to 4, As previous cycle, row 4 has all zero entries—therefore deleted, whereas rows 5 and 6 have proportional entries (contain same entries with opposite sign)—therefore only one row is independent. This drop in rank is due to configuration of cycle's C_F screws; they make a fascicle of concurrent lines at point G. Therefore, the rank is

$$r(C_F) = 4 \tag{3.209d}$$

The rank of the velocity matroidal matrix is the sum of the previous two ranks and counts the number of independent equations:

$$r(C) = r(C_E) + r(C_F) = 8 \tag{3.209e}$$

- Number of equations and number of unknowns

  The number of equations is 8, coincidental to the rank. The unknowns are relative angular velocities (twist intensities) in joints A (x_0 component), B (x_0 component), E (x_0, y_0, z_0 components), and F (x_0, y_0, z_0 components).

- Solution

  The mechanism has M = 2; therefore, two gears are actuated. Let us consider actuation at C and D for gears 3 and 4, therefore $\dot{\theta}_C = \dot{\theta}_{0.3} = \omega_3$ and $\dot{\theta}_D = \dot{\theta}_{0.4} = \omega_4$ are provided.

  The system of two Equations (3.208b) and (3.209b):

$$\begin{cases} -i_E \cdot \dot{\theta}_A - \dot{\theta}_B + i_E \cdot \dot{\theta}_C = 0 \\ -i_F \cdot \dot{\theta}_A + \dot{\theta}_B + i_F \cdot \dot{\theta}_D = 0 \end{cases}$$

is solved for the unknown scalars: $\dot{\theta}_A$, $\dot{\theta}_B$:

$$\begin{cases} \dot{\theta}_A = \dfrac{1}{i_E + i_F}(i_E \cdot \omega_3 + i_F \cdot \omega_4) \\ \dot{\theta}_B = \dfrac{i_E \cdot i_F}{i_E + i_F}(\omega_3 - \omega_4) \end{cases} \tag{3.210}$$

The angular velocities in meshing joints E and F are determined from Equations (3.208a) and (3.209a):

$$\dot{\theta}_E = \begin{Bmatrix} -\dot{\theta}_A + \dot{\theta}_C \\ -C\theta_A \cdot \dot{\theta}_B \\ -S\theta_A \cdot \dot{\theta}_B \end{Bmatrix} = \begin{Bmatrix} \dfrac{i_F}{i_E + i_F}(\omega_3 - \omega_4) \\ -\dfrac{i_E \cdot i_F \cdot C\theta_A}{i_E + i_F}(\omega_3 - \omega_4) \\ -\dfrac{i_E \cdot i_F \cdot S\theta_A}{i_E + i_F}(\omega_3 - \omega_4) \end{Bmatrix};$$

$$\dot{\theta}_F = \begin{Bmatrix} -\dot{\theta}_A + \dot{\theta}_D \\ -C\theta_A \cdot \dot{\theta}_B \\ -S\theta_A \cdot \dot{\theta}_B \end{Bmatrix} = \begin{Bmatrix} -\dfrac{i_E}{i_E + i_F}(\omega_3 - \omega_4) \\ -\dfrac{i_E \cdot i_F \cdot C\theta_A}{i_E + i_F}(\omega_3 - \omega_4) \\ -\dfrac{i_E \cdot i_F \cdot S\theta_A}{i_E + i_F}(\omega_3 - \omega_4) \end{Bmatrix} \tag{3.211}$$

3.9.3.3 Absolute Angular Velocities of Gears, Planets, and Carriers

To determine the absolute angular velocities for gears, planets, and carriers, the following paths in the tree are considered:

- Path in the tree from node 0 to node 1:

$$\omega_1 = \mathbf{u}_A^0 \dot{\theta}_A = \begin{Bmatrix} \dfrac{1}{i_E + i_F}(i_E \cdot \omega_3 + i_F \cdot \omega_4) \\ 0 \\ 0 \end{Bmatrix}$$

- Path in the tree from node 0 to node 2:

$$\omega_2 = \mathbf{u}_A^0 \dot{\theta}_A + \mathbf{u}_B^0 \dot{\theta}_B = \begin{Bmatrix} \dfrac{1}{i_E + i_F}(i_E \cdot \omega_3 + i_F \cdot \omega_4) \\ \dfrac{i_E \cdot i_F \cdot C\theta_A}{i_E + i_F}(\omega_3 - \omega_4) \\ \dfrac{i_E \cdot i_F \cdot S\theta_A}{i_E + i_F}(\omega_3 - \omega_4) \end{Bmatrix} \tag{3.212}$$

- Path in the tree from node 0 to node 2:

$$\omega_3 = \mathbf{u}_C^0 \dot{\theta}_C = \begin{Bmatrix} \omega_3 \\ 0 \\ 0 \end{Bmatrix}$$

- Path in the tree from node 0 to node 4:

$$\omega_4 = \mathbf{u}_D^0 \dot{\theta}_D = \begin{Bmatrix} \omega_4 \\ 0 \\ 0 \end{Bmatrix}$$

Numerical example:

Data: $\omega_3 = \pi \cdot \text{rad/s} = 180 \cdot {}^\circ/\text{s}$; $\omega_4 = 2\pi \cdot \text{rad/s} = 360 \cdot {}^\circ/\text{s}$;

$d_2 = 3.2 \cdot \text{in.}$; $d_3 = d_4 = 9.8 \cdot \text{in.}$;

$$\lambda = 2.1 \cdot \text{in.};\ i_E = i_{E_{2,3}} = \frac{d_3}{d_2} = 3.0625;\ i_F = i_{E_{2,4}} = \frac{d_4}{d_2} = 3.0625$$

Results:

$$\omega_1 = \begin{Bmatrix} 270 \\ 0 \\ 0 \end{Bmatrix} {}^\circ/\text{s};\ |\omega_1| = 270{}^\circ/\text{s};\ \omega_2 = \begin{Bmatrix} 270 \\ -275.625 \cdot C\theta_A \\ -275.625 \cdot S\theta_A \end{Bmatrix} {}^\circ/\text{s};\ |\omega_2| = 385.8\ {}^\circ/\text{s};$$

$$\omega_3 = \begin{Bmatrix} 180 \\ 0 \\ 0 \end{Bmatrix} {}^\circ/\text{s};\ |\omega_3| = 180{}^\circ/\text{s};\ \omega_4 = \begin{Bmatrix} 360 \\ 0 \\ 0 \end{Bmatrix} {}^\circ/\text{s};\ |\omega_2| = 360\ {}^\circ/\text{s}$$

The SW simulated results are plotted in Figure 3.58: $|\omega_1| = 270\ {}^\circ/\text{s}$, $|\omega_2| = 385.8\ {}^\circ/\text{s}$, $|\omega_3| = 180\ {}^\circ/\text{s}$, $|\omega_4| = 360\ {}^\circ/\text{s}$, which are coincidental with the abovementioned calculated values.

3.9.3.4 Gears' Number of Teeth

From a preferred list of diametral pitches P, the values $P_E = 5$ teeth/in. and $P_F = 5$ teeth/in. are selected. Then, the gears' number of teeth is

$$N_2 = d_2 \cdot P_E = 16 \text{ teeth};\ N_3 = d_3 \cdot P_E = 49 \text{ teeth};\ N_4 = d_4 \cdot P_F = 49 \text{ teeth} \quad (3.213)$$

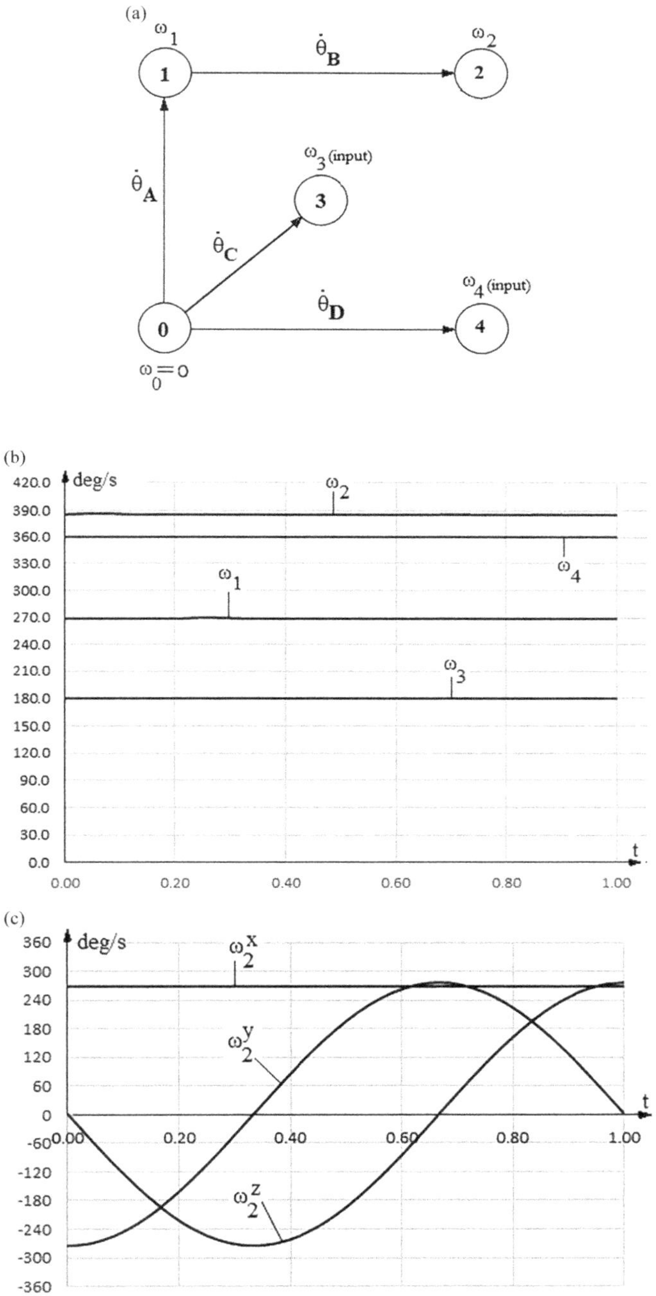

FIGURE 3.58
(a) Relative and absolute angular velocities on spanning tree; (b) SW simulated values for gears and carrier; (c) Planet angular velocity components with respect to x_0, y_0, z_0 axes.

3.9.3.5 Automatic Generation of BGT Equations

An automatic generation of coefficients in the kinematic equations is based on cycle matroid matrix **C** and the path matrix **Z** from Equations (3.190), (3.192), and (3.207) [9].

a. Equation (2.214) holds for the screws' upper vectors:

$$\underbrace{\left[\begin{array}{cccc|cc} A & B & C & D & E & F \\ 1\cdot \mathbf{u}_A^0 & \mathbf{u}_B^0 & -1\cdot \mathbf{u}_C^0 & 0 & 1\cdot \mathbf{u}_E^0 & 0 \\ \hline 1\cdot \mathbf{u}_A^0 & 1\cdot \mathbf{u}_B^0 & 0 & -1\cdot \mathbf{u}_D^0 & 0 & 1\cdot \mathbf{u}_F^0 \end{array}\right]}_{\left[\mathbf{C}(\mathbf{u}^0)\right]=\left[\mathbf{T}(\mathbf{u}^0)\,|\,\mathbf{U}(\mathbf{u}^0)\right]} * \underbrace{\begin{Bmatrix} \dot{\theta}_{rev} \\ \dot{\theta}_{mesh} \end{Bmatrix}}_{\dot{\theta}_j} = \begin{Bmatrix} 0 \\ 0 \end{Bmatrix} \tag{3.214}$$

where

$\left[\mathbf{C}(\mathbf{u}^0)\right]$ is the matroidal matrix for joint unit vectors

$\dot{\theta}_j$ is the joint scalar angular velocity matrix

Equation (3.215) determines the c meshing joint rates as functions of rates in the revolute joints:

$$\left\{\dot{\theta}_{mesh}\right\} = -\left[\mathbf{T}(\mathbf{u}^0)\right] * \left\{\dot{\theta}_{rev}\right\} \tag{3.215}$$

The solution is shown in Equations (3.208b) and (3.209b).

b. Equation (2.216) holds for the screws' lower vectors:

$$\underbrace{\left[\begin{array}{cccc|cc} A & B & C & D & E & F \\ 1\cdot Q_{E,A} & 1\cdot Q_{E,B} & -1\cdot Q_{E,C} & 0 & 1\cdot 0 & 0 \\ \hline 1\cdot Q_{F,A} & 1\cdot Q_{F,B} & 0 & -1\cdot Q_{F,D} & 0 & 1\cdot 0 \end{array}\right]}_{C(Q)=[T(Q)\,|\,0(Q)]} * \underbrace{\begin{Bmatrix} \dot{\theta}_{rev} \\ \dot{\theta}_{mesh} \end{Bmatrix}}_{\dot{\theta}_j} = \begin{Bmatrix} 0 \\ 0 \end{Bmatrix} \tag{3.216}$$

Equation (3.217) determines m − c revolute joints rates as functions of M actuating angular velocities in the revolute joints:

$$T(Q) * \dot{\theta}_{rev} = 0 \tag{3.217}$$

$$
\underbrace{\left[\begin{array}{cccc} A & B & C & D \\ \hline Q_{E,A} & Q_{E,B} & -Q_{E,C} & 0 \\ \hdashline Q_{F,A} & Q_{F,B} & 0 & -Q_{F,D} \end{array}\right]}_{T(Q)} * \underbrace{\left\{\begin{array}{c} \dot{\theta}_A \\ \dot{\theta}_B \\ \dot{\theta}_C\,(=\omega_3) \\ \dot{\theta}_D\,(=\omega_4) \end{array}\right\}}_{\dot{\theta}_{rev}} = \left\{\begin{array}{c} 0 \\ 0 \end{array}\right\}
$$

The solution is coincidental with that in Equation (3.210).

Equation (3.218) relates the gears' and carriers' absolute angular velocities to revolute joints' angular velocities. Each row is a path in the tree initiated at node 0 and ended at nodes 1, 2, 3, and 4:

$$
\{\omega_m\} = \left[\mathbf{Z}^T\left(\mathbf{u}^0\right)\right] * \left\{\dot{\theta}_{rev}\right\} \tag{3.218}
$$

$$
\underbrace{\left\{\begin{array}{c} \omega_1 \\ \omega_2 \\ \omega_3 \\ \omega_4 \end{array}\right\}}_{\omega_m} = \begin{array}{c} 1 \\ 2 \\ 3 \\ 4 \end{array} \underbrace{\left[\begin{array}{cccc} 1\cdot\mathbf{u}_A^0 & 0 & 0 & 0 \\ 1\cdot\mathbf{u}_A^0 & 1\cdot\mathbf{u}_B^0 & 0 & 0 \\ 0 & 0 & 1\cdot\mathbf{u}_C^0 & 0 \\ 0 & 0 & 0 & 1\cdot\mathbf{u}_D^0 \end{array}\right]}_{\mathbf{Z}^T\left(\mathbf{u}^0\right)}^{\begin{array}{cccc} A & B & C & D \end{array}} * \underbrace{\left\{\begin{array}{c} \dot{\theta}_A \\ \dot{\theta}_B \\ \dot{\theta}_C \\ \dot{\theta}_D \end{array}\right\}}_{\dot{\theta}_{rev}} = \left\{\begin{array}{c} \dot{\theta}_A \\ \dot{\theta}_A + \dot{\theta}_B \\ \dot{\theta}_C \\ \dot{\theta}_D \end{array}\right\}
$$

The solution is coincidental with that in Equation (3.212).

Problems

P3.1: The four-bar mechanism in Figure P3.1 (Animation P3.1) has the link lengths: $L_{AB} = 4''$, $L_{BD} = 24''$, $L_{DC} = 14''$, and $L_{CA} = 30''$.

For positional analysis, determine the following:

a. The mobility of mechanism
b. The digraph and number of cycles
c. Find the type of the four-bar mechanism, using Grashof conditions
d. The system of equations for positional analysis

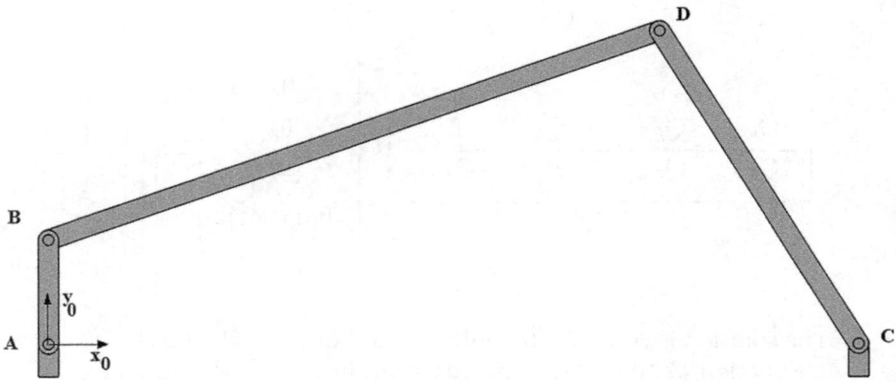

FIGURE P3.1
Four-bar mechanism.

e. The solution for the crank at 90°
f. The coordinates x and y for all joints, when the crank is at 90°
g. The transmission angle when the crank is at 90°
h. The results from positional analysis are saved as entries in the Latin matrix. Check that the sum of entries is zero on each row, and save the data for further velocity analysis

For velocity analysis, the crank rotates with a constant ccw angular velocity of 360 °/s.

Determine

i. The system of equations for velocity analysis
j. The system for the links' angular velocities when the crank is at 90°
k. The components x and y for linear velocities of all joints, when the crank is at 90°
l. The results from velocity analysis are saved as entries in the velocity matroidal matrix

Check that the sum of entries is zero on each row, and save the data for further acceleration analysis.

P3.2: The four-bar mechanism in Figure P3.1 has the link lengths: $L_{AB} = 4''$, $L_{BD} = 24''$, $L_{DC} = 14''$, and $L_{CA} = 30''$.

For positional analysis, determine

a. The system of equations for positional analysis as a function of crank angle $\theta_1(t) = 360 \cdot t$ [deg], considered increasing linearly. Consider the initial position at 0°
b. Create a spreadsheet with 25 rows for the x and y positional coordinates when the total time interval is 1 s

c. Create a spreadsheet with 25 rows for the x and y coordinates for each joint
d. Graph the y versus x coordinates to represent the path for each joint

P3.3: The double-crank four-bar mechanism in Figure P3.2 (Animation P3.2) has attached to the link BC a solar panel. The links' lengths are $L_{AB} = 30''$, $L_{BC} = 24''$, $L_{CD} = 14''$, and $L_{DA} = 4''$. The initial (home) position is considered when the crank AB makes 90° with the x-axis (the sunrise position for solar panel).

For positional analysis, determine

a. The mobility of mechanism
b. The digraph and number of cycles
c. Using Grashof conditions, show that the mechanism is a double crank
d. The system of equations for positional analysis
e. The solution when the crank AB makes 90° with the x-axis
f. The solution when the crank AB makes 234° with the x-axis

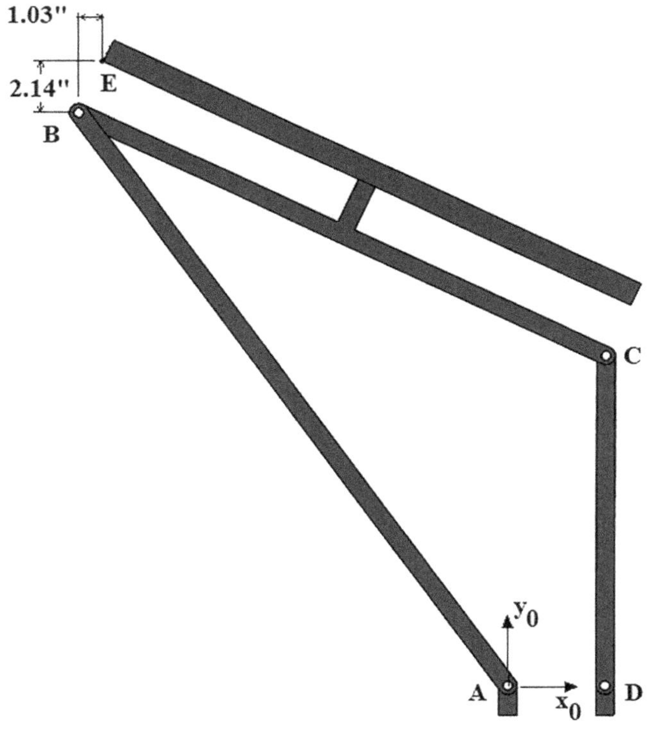

FIGURE P3.2
Four-bar mechanism with solar panel.

g. The coordinates x and y for all joints and the point of interest E when the crank is at 90° (sunrise) and 234° (sunset)

h. The transmission angle when the crank is at 90° and 234°

i. The results from positional analysis are saved as entries in the Latin matrix. Check that the sum of entries is zero on each row, and save the data for further velocity analysis

For velocity analysis, the crank AB rotates with 0.003 °/s ccw.

Determine

j. The system of equations for velocity analysis

k. The system for the links' angular velocities when the crank is at 90°

l. The components x and y for linear velocities of all joints, and the point of interest E when the crank is at 90°

m. The results from velocity analysis are saved as entries in the velocity matroidal matrix

Check that the sum of entries is zero on each row, and save the data for further acceleration analysis.

P3.4: The crank slider mechanism in Figure P3.3 (Animation P3.3) has the link lengths: L_{AB} = 4″ and L_{BD} = 24″. In the initial position (home position), the crank AB makes 0° with the horizontal, where the coordinates of contact point C_0 are (28″, 0″).

For positional analysis, determine

a. The mobility of mechanism

b. The digraph and number of cycles

c. The system of equations for positional analysis and the solution for the crank at 45°

d. The coordinates of all joints when the crank is at 45°

e. The maximum displacement d_C for the piston

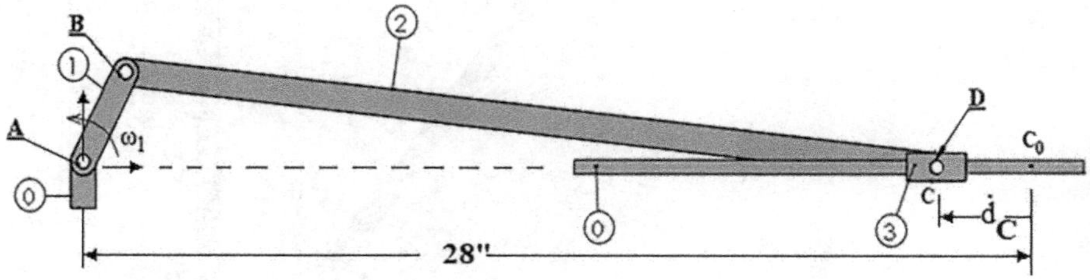

FIGURE P3.3
Crank slider planar mechanism.

f. The results from positional analysis are saved as entries in the Latin matrix. Check that the sum of entries is zero on each row, and save the data for further velocity analysis

For velocity analysis, the crank rotates with 360 °/s ccw.

Determine

g. The system of equations for velocity analysis
h. The system for the links' angular velocities when the crank is at 45°
i. The components x and y for linear velocities of all joints, and the point of interest E when the crank is at 45°
j. The results from velocity analysis are saved as entries in the velocity matroidal matrix

Check that the sum of entries is zero on each row, and save the data for further acceleration analysis

P3.5: The crank slider mechanism in Figure P3.3 has the link lengths: $L_{AB} = 4''$ and $L_{BD} = 24''$.

Determine

a. The system of equations for positional analysis as a function of crank angle $\theta_1(t) = 360 \cdot t$ [deg], considered increasing linearly. Consider the initial position at 0°
b. Create a spreadsheet with 25 rows for the piston displacement d_C when the total time interval is 1 s
c. Graph displacement d_C versus $\theta_1(t)$ to represent the input (crank angle)–output (piston displacement) relation
d. The system of equations for velocity analysis as a function of crank angle $\theta_1(t) = 360 \cdot t$ [deg]
e. Create a spreadsheet with 25 rows for the piston velocity $\dot{d}_C$, when the total time interval is 1 s
f. Graph piston velocity $\dot{d}_C$ versus θ_1 to represent the piston velocity as a function of crank angular displacement

P3.6: The planar mechanism in Figure P3.4 (Animation P3.4) has the link lengths: $L_{AB} = 2.85''$, $L_{BC} = 4.24''$, $L_{CD} = 5.62''$, $L_{BF} = 7.90''$, $L_{FG} = 7.87''$. The joints B, C, and F are on the same link, and BC is perpendicular to BF. In the initial position (piston's home position), the crank AB makes 0° with the horizontal, where the coordinates of contact point E_0 are (12.06″, −0.785″).

For positional analysis, determine

a. The mobility of mechanism
b. The digraph and number of cycles

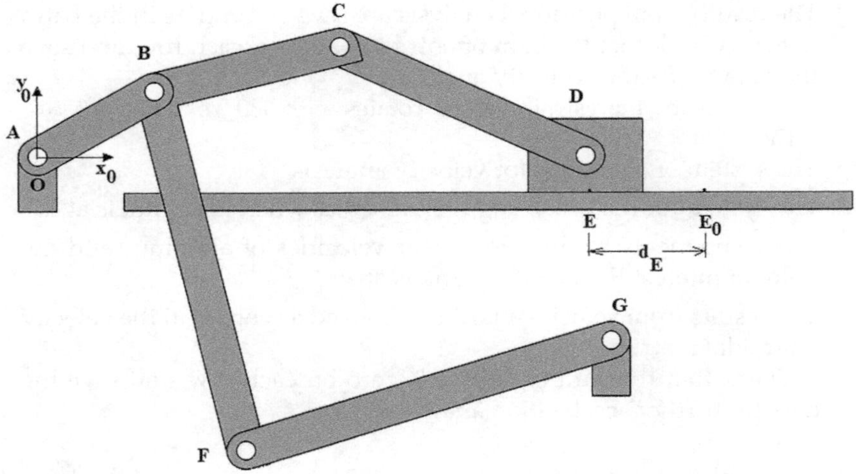

FIGURE P3.4
Planar mechanism.

c. The system of equations for positional analysis and the solution for the crank at 45°
d. The coordinates of all joints when the crank is at 45°
e. The maximum displacement d_E for the piston
f. The results from positional analysis are saved as entries in the Latin matrix. Check that the sum of entries is zero on each row, and save the data for further velocity analysis

   For velocity analysis, the crank rotates with 360 °/s ccw.
   Determine

g. The system of equations for velocities
h. Solve the system for the links' angular velocities and piston's linear velocity
i. The linear velocity magnitude for all joints

P3.7: The planar mechanism in Figure P3.5 (Animation P3.5) has the link lengths: L_{AB} = 2.78″, L_{BC} = 15.61″, L_{CD} = 5.72″, L_{DE} = 3.76″, L_{CF} = 10.94″, and L_{BF} = 11.14″. The joints B, F, and C are on the same link, and BF is perpendicular to BC. In the initial position (piston's home position), the crank AB makes 0° with the horizontal, where the coordinates of contact point G_0 are (13.2″, −4.725″).

For positional analysis
determine

- The same requirements a–f from P3.6

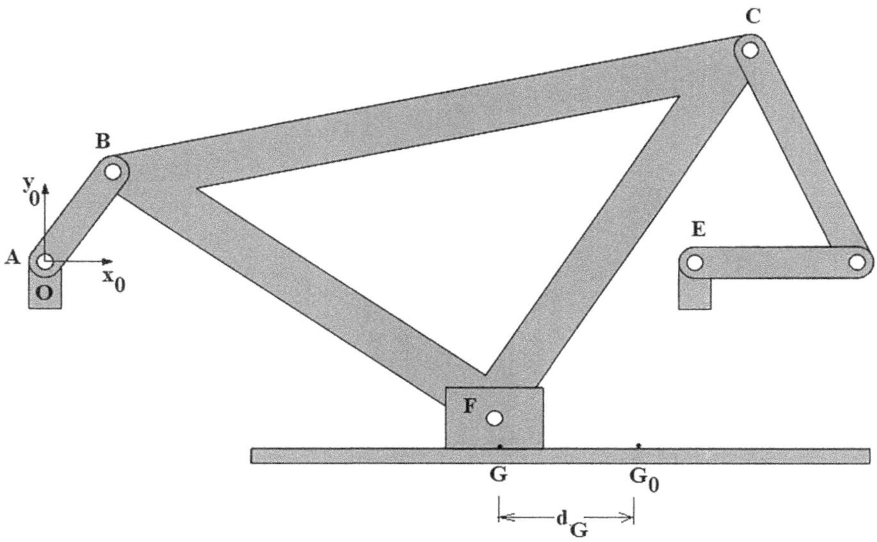

FIGURE P3.5
Planar mechanism with revolute and prismatic joints.

For velocity analysis, the crank rotates with 360 °/s ccw
Determine

- The same requirements g–i from P3.6

P3.8: The planar mechanism with revolute and prismatic joints in Figure P3.6 (Animation P3.6) has the link lengths: $L_{AB} = 5''$, $L_{AE} = 10''$, $x_{AF} = 7''$, $y_{AF} = 3.5''$, $L_{DH} = 6''$, and $L_{BG} = 16''$
For positional analysis, determine

a. The mobility of mechanism
b. The digraph and number of cycles
c. The system of equations for positional analysis and the solution for the crank at 72°
d. The coordinates of points of interest G and H when the crank is at 72°
e. The displacements d_C, d_D, and d_F
f. The results from positional analysis are saved as entries in the Latin matrix. Check that the sum of entries is zero on each row, and save the data for further velocity analysis
For velocity analysis, the crank rotates with 360 °/s ccw.
Determine
g. The system of equations for velocity analysis and the solution for the crank at 72°

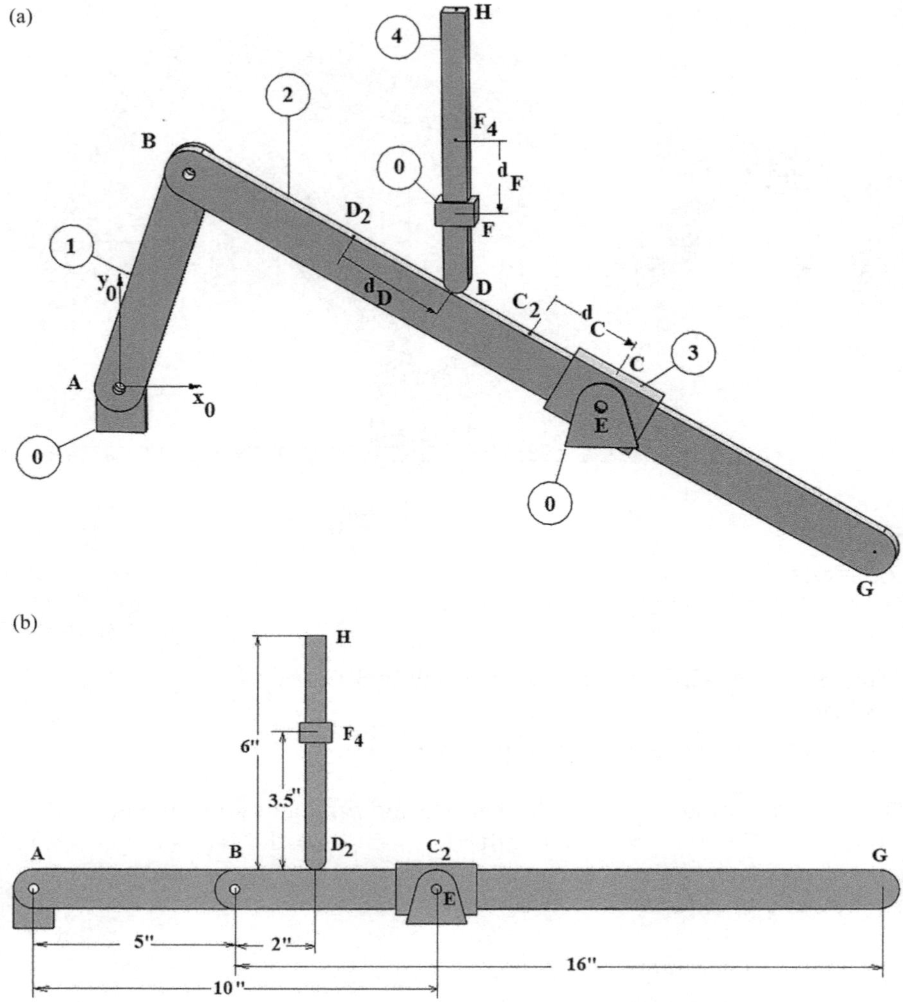

FIGURE P3.6
Planar mechanism with revolute, prismatic, and contact joints: (a) position for crank at 72°; (b) home position.

h. The x and y components of linear velocities for the points of interest G and H when the crank is at 72°

i. The results from velocity analysis are saved as entries in the velocity matroidal matrix. Check that the sum of entries is zero on each row, and save the data for further acceleration analysis

P3.9: The Jansen mechanism is illustrated in Animation P3.7. There are four identical mechanisms (legs) connected to frame 0, all actuated by the same crank AB. In Figure P3.7 (Animation P3.7a), one of them is illustrated, which has the link lengths: $L_{AB} = 17.23''$, $L_{BC} = 50.01''$, $L_{CE} = 56.01''$, $L_{DE} = 40''$, $L_{CD} = 41.78''$, $L_{AD} = 38.00''$, $L_{BG} = 61.05''$, $L_{FG} = 36.83''$, $L_{GH} = 49.04''$, and $L_{FH} = 65.75''$. The joints C, D, and E are on the same link 5. The joints F, G, and H are on the same link 7. The joint B connects links 1 and 2, whereas joint B′ connects links 1 and 3. Joint G connects links 3 and 4, whereas joint G′ connects links 3 and 7. In the initial position, the crank AB makes 0° with the x-axis. In contact with the ground, a point of interest H is considered. Due to the friction force at this point, the mechanism is able to move forward.

For positional analysis, determine

a. The mobility of mechanism
b. The digraph and number of cycles
c. The system of equations for positional analysis and the solution for the crank at 90°
d. The coordinates x and y of point of interest H when the crank is at 120°

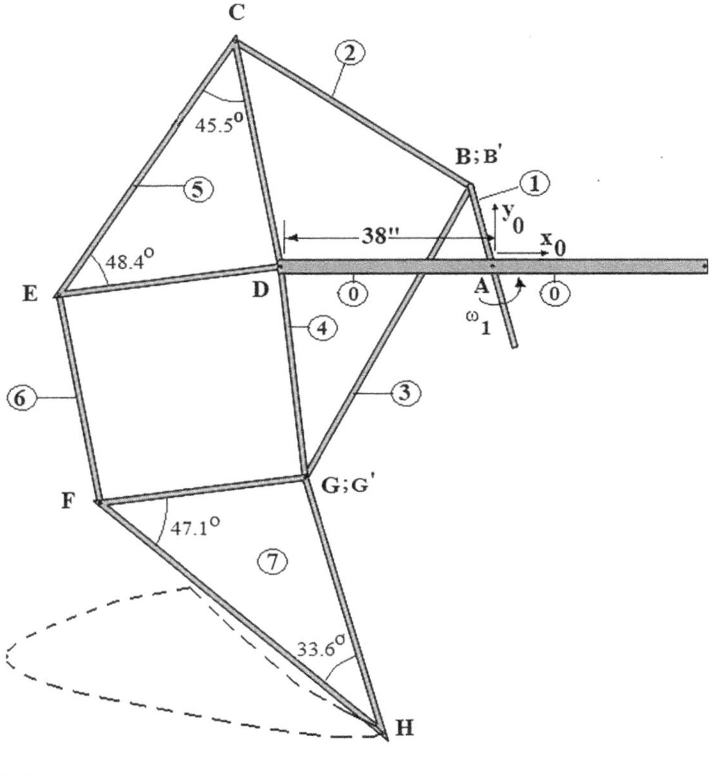

FIGURE P3.7
Planar Jansen mechanism.

e. The coordinates x and y of point of interest H when the crank is at 120°
f. For increments of 30° in crank angle, determine the coordinates x and y of point of interest H. Use an equation solver to solve the non-linear system of equations from positional analysis
g. Graph y_H versus x_H to represent the path of point H
   For velocity analysis, the crank rotates with 60 rpm (360 °/s) ccw.
   Determine
h. The system of equations for velocity analysis and the solution for the crank at 120°
i. The x and y velocity components for point H when the crank is at 120°
j. Create a spreadsheet with 12 rows (increments of 30° in crank angle) for the x and y velocity components and its magnitude v_H
k. Graph v_H versus the crank angle θ_1

P3.10: A cam with 2″ base radius is to be designed for a prescribed motion in an automated loader as follows: raise the follower 1″ within 1 s with *constant velocity*; lower the follower 1″ within 1 s with *constant velocity*; dwell for 2 s.
Determine

a. Time for one cycle
b. Required angular velocity of the cam
c. The cam rotation angle for each follower's interval $\Delta\theta_r$, $\Delta\theta_f$, $\Delta\theta_d$
d. The displacement equation for the follower rise from 0″ to $D = d_C^{max} = 1''$
e. The displacement equation for the follower fall from $D = d_C^{max} = 1''$ to 0″
f. The displacement equation for the cam's dwell
g. The coordinates x_B and y_B of contact cam–follower point B
h. Create spreadsheet, with x_B and y_B coordinates for one cam revolution. Graph the cam profile y_B versus x_B

P3.11: A cam with 2″ base radius is to be designed for a prescribed motion in an automated loader as follows: raise the follower 1″ within 1 s with *constant acceleration*; lower the follower 1″ within 1 s with *constant acceleration*; dwell for 2 s.
Determine
Same requirements from P3.10a–h

P3.12: A cam with 2″ base radius is to be designed for a prescribed motion in an automated loader as follows: raise the follower 1″ within 1 s with *constant velocity*; lower the follower 1″ within 1 s with *constant acceleration*; dwell for 2 s.
Determine
Same requirements from P3.10a–h

P3.13: The cam–follower opens and closes the valve from an auto distribution mechanism as illustrated in Figure P3.8 (Animation 3.8). The cam has 1″ base radius and is designed for a prescribed motion as follows: raise the follower 0.5″ in 0.33 s with *constant velocity*; lower the follower 0.5″ in 0.33 s with *constant velocity*; dwell for 1.34 s. The mechanism has 4 mobile links, two revolute joints at A and E, two prismatic joints at C and G, one cam joint at B, and two contact joints at F and D.

Determine

a. Draw the digraph and the cycles
b. Calculate the mechanism's mobility

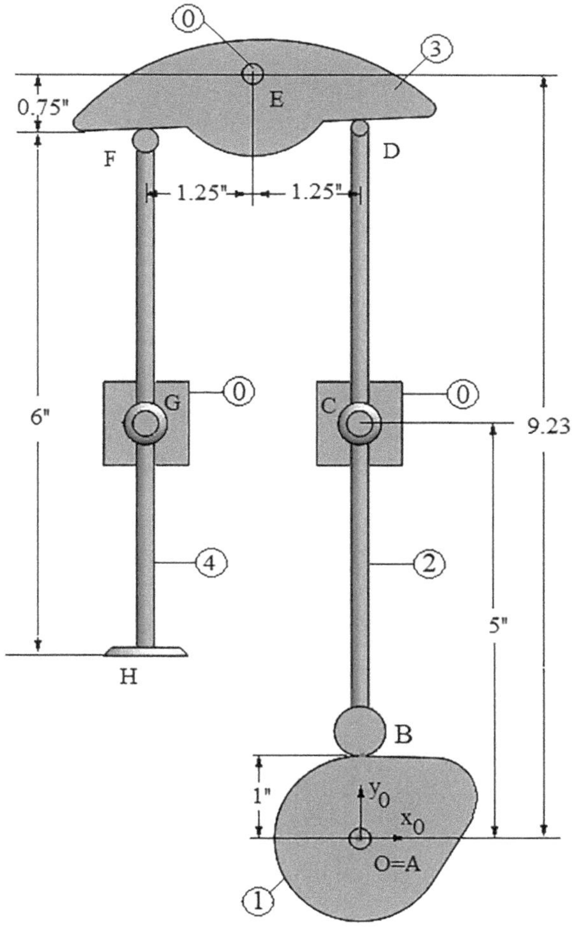

FIGURE P3.8
Auto distribution mechanism.

c. The angular velocity of the cam
d. Appendix 3.18 tabulates the position and velocity of follower. Use the data to determine the position and velocity of point H on the valve when the follower is at the end of rise
e. Determine the position and velocity of point H on the valve when the follower is at the end of fall
f. Determine the position and velocity of point H on the valve when the follower is at the dwell

P3.14: A fixed axes gear set (GT) is illustrated in Figure P3.9 (Animation P3.9). The two metric gears have the diameters $d_1 = 50$ mm and $d_2 = 120$ mm with a module $m_C = 2$ mm. The pinion 1 is in rotation with $n_1 = 60$ rpm.

Determine

a. Draw the digraph
b. The number of cycles and the mechanism's mobility

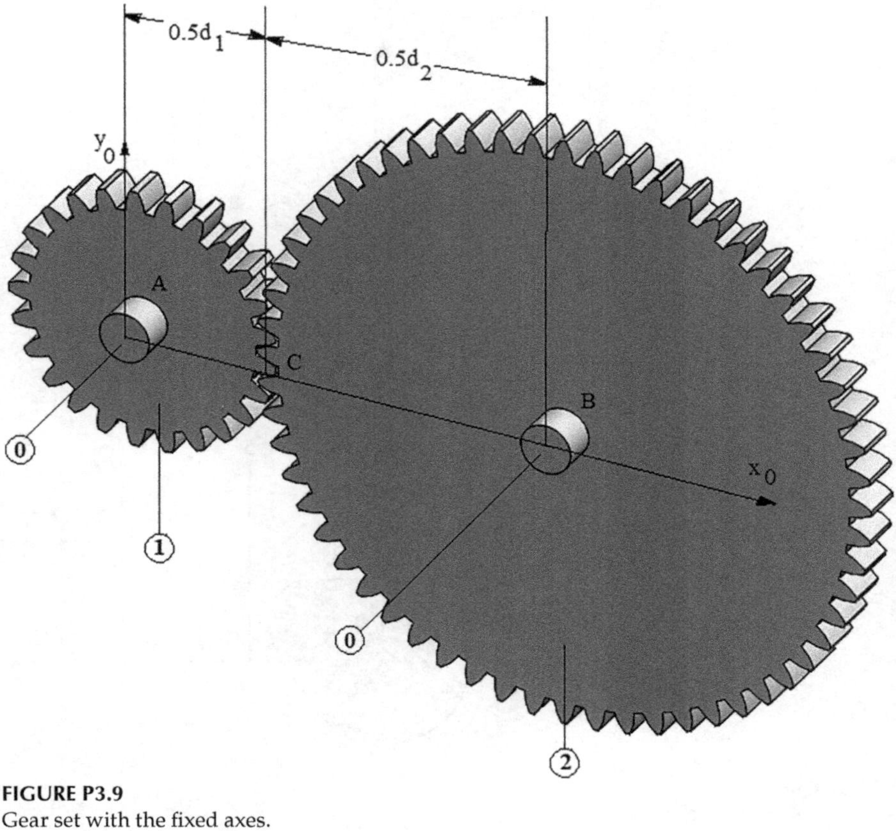

FIGURE P3.9
Gear set with the fixed axes.

c. The equation for angular velocities and solve for angular velocity of gear 2
d. The number of teeth for the two gears
e. The velocity ratio

P3.15: A planetary parallel axes gear set (EGT) is illustrated in Figure P3.10 (Animation P3.10). The two metric gears have the diameters $d_2 = 260$ mm, $d_3 = 46$ mm, and the module $m_D = 2$ mm. The carrier 1 is 107 mm long and rotates ccw with $\omega_1 = 360$ °/s. The gear 2 rotates ccw with $\omega_2 = 720$ °/s.

Determine

a. Draw the digraph
b. The number of cycles and the mechanism's mobility

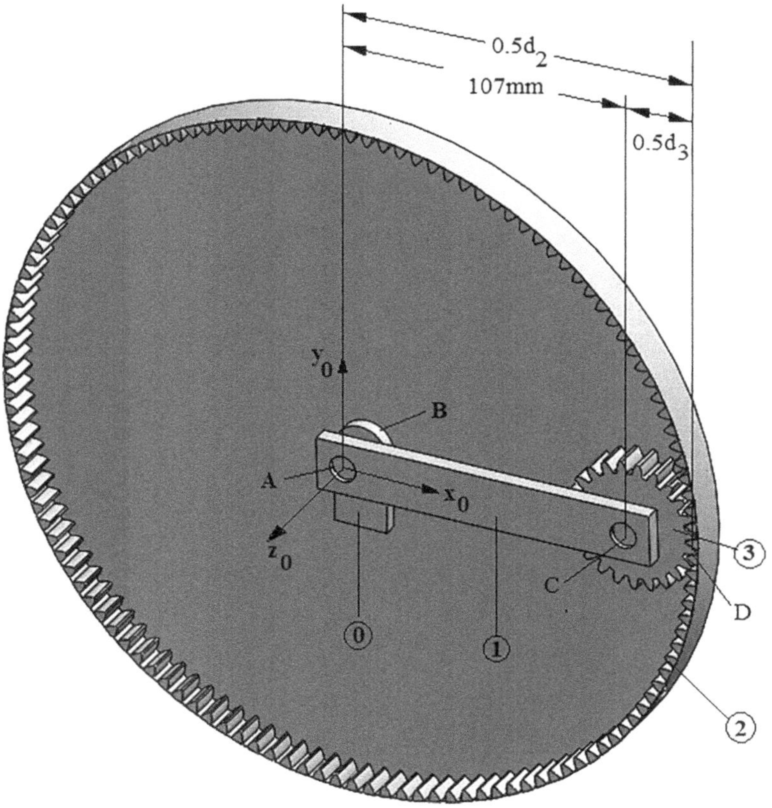

FIGURE P3.10
Planetary parallel axes mechanism.

c. The equation for angular velocities and solve it for angular velocity of planet 3
d. The number of teeth for gear 2 and planet 3

P3.16: The planetary parallel axes mechanism with three planet gears is illustrated in Figure P3.11 (Animation P3.11). The gear has the diameters 260 mm and the planet 46 mm. The module is 2 mm for gear and planets. The three branches of the carrier are all 107 mm long and rotates ccw with 360 °/s. The gear rotates ccw with 720 °/s.

a. Number the joints and links
b. Draw the digraph
c. Calculate the number of cycles
d. Calculate the mechanism's mobility
e. The equations for angular velocities and solve them for angular velocities of planets
f. Compare the results with those from P3.15. What advantage might be to have multiple planets in the design of EGT

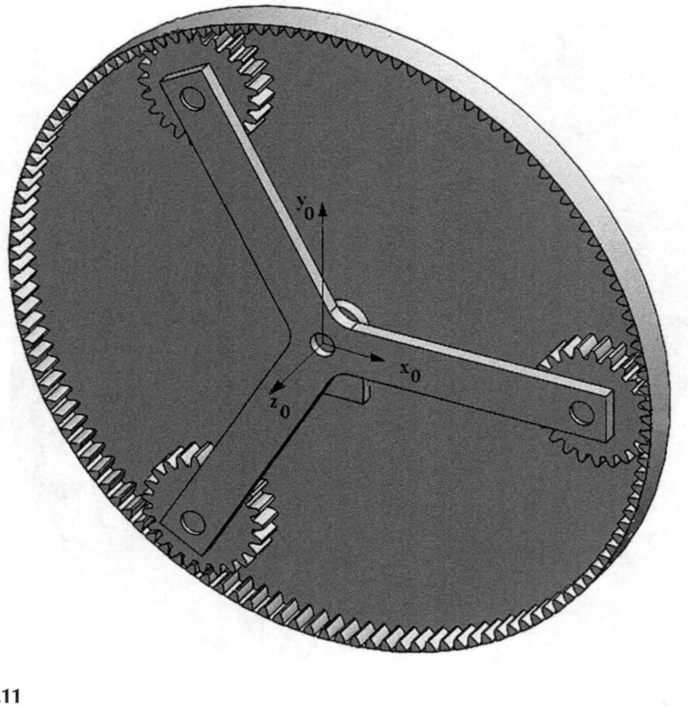

FIGURE P3.11
Planetary parallel EGT mechanism with three planet gears.

P3.17: The BGT illustrated in Figure P3.12 (Animation P3.12) is used as a robotic wrist mechanism. In Figures 1.6 and 1.10, the digraph and the cycles are illustrated. There are m = 6 mobile links, six revolute joints: A, B, C, D, E, F, and 3 meshing joints: G, H, and I. The gears' diameters are $d_2 = 9.8 \cdot \text{in.}$; $d_3 = 6.0 \cdot \text{in.}$; $d_4 = 4.5 \cdot \text{in.}$; $d_5 = 6.4 \cdot \text{in.}$; and $d_6 = 6.0 \cdot \text{in.}$, and the distance from carrier to gear 2 is $\lambda = 0.51$ in. The carrier 1 and gears 2 and 3 rotate ccw about x-axis with the angular velocities $\omega_1 = 360\ °/\text{s}$, $\omega_2 = 360\ °/\text{s}$, and $\omega_3 = 360\ °/\text{s}$, respectively.

a. Calculate the mechanism's mobility
b. Determine the equations for angular velocities and solve them for relative angular velocities in joints
c. Determine the absolute velocity components on x_0, y_0, z_0 for gears and carrier
d. Graph the absolute velocity components and magnitude for an interval of time 1 s

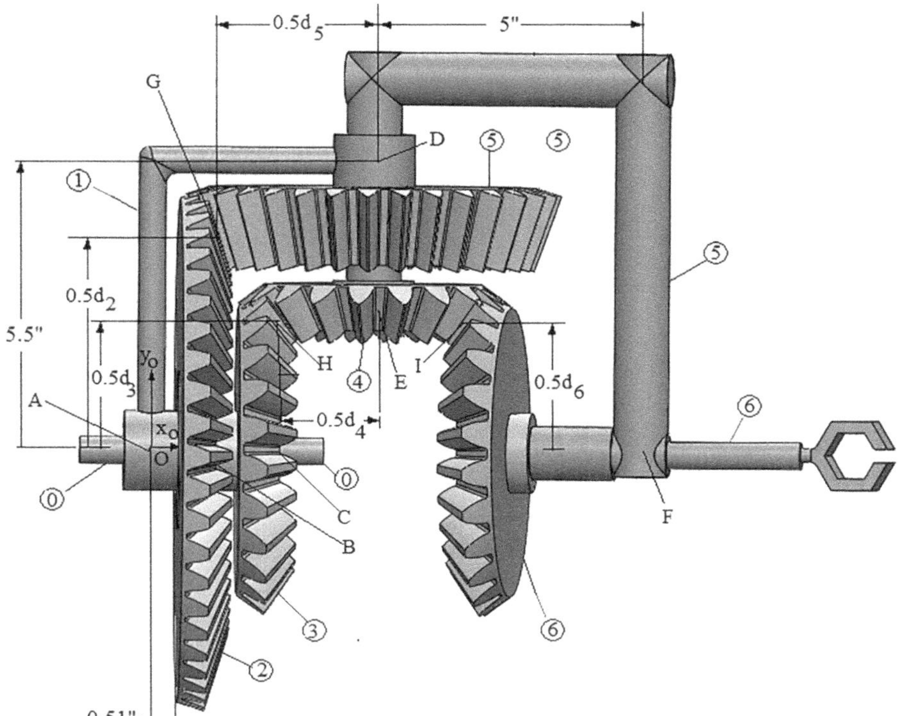

FIGURE P3.12
BGT robotic-wrist mechanism.

Appendix 3.1: Classification of Planar Mechanisms with M = 1

c (Cycles)	j_2	m (Mobile Links)	j_1	j (Joints)	Total number
1	0	3	4	4	**3**
1	1	2	2	3	
1	2	1	0	2	
2	0	5	7	7	4
2	1	4	5	6	
2	2	3	3	5	
2	3	2	1	4	
3	0	7	10	10	**6**
3	1	6	8	9	
3	2	5	6	8	
3	3	4	4	7	
3	4	3	2	6	
3	5	2	0	5	
4	0	9	13	13	**7**
4	1	8	11	12	
4	2	7	9	11	
4	3	6	7	10	
4	4	5	5	9	
4	5	4	3	8	
4	6	3	1	7	
5	0	11	16	16	**9**
5	1	10	14	15	
5	2	9	12	14	
5	3	8	10	13	
5	4	7	8	12	
5	5	6	6	11	
5	6	5	4	10	
5	7	4	2	9	
5	8	3	0	8	

Appendix 3.2: Classification of Planar Mechanisms with M = 2

c (Cycles)	j_2	m (Mobile links)	j_1	Total number
1	0	4	5	3
1	1	3	3	
1	2	2	1	
2	0	6	8	5
2	1	5	6	
2	2	4	4	
2	3	3	2	
2	4	2	0	
3	0	8	8	5
3	1	7	6	
3	2	6	4	
3	3	5	2	
3	4	4	0	
4	0	10	14	8
4	1	9	12	
4	2	8	10	
4	3	7	8	
4	4	6	6	
4	5	5	4	
4	6	4	2	
4	7	3	0	
5	0	12	17	9
5	1	11	15	
5	2	10	13	
5	3	9	11	
5	4	8	9	
5	5	7	7	
5	6	6	5	
5	7	5	3	
5	8	4	1	

Appendix 3.3: Classification of Planar Mechanisms with M = 3

c (Cycles)	j_2	m (Mobile links)	j_1	Total Number
1	0	2	6	3
1	1	5	4	
1	2	8	2	
2	0	4	9	5
2	1	7	7	
2	2	10	5	
2	3	13	3	
2	4	16	1	
3	0	6	12	7
3	1	9	10	
3	2	12	8	
3	3	15	6	
3	4	18	4	
3	5	21	2	
3	6	24	0	
4	0	8	15	8
4	1	11	13	
4	2	14	11	
4	3	17	9	
4	4	20	7	
4	5	23	5	
4	6	26	3	
4	7	29	1	
5	0	10	18	10
5	1	13	16	
5	2	16	14	
5	3	19	12	
5	4	22	10	
5	5	25	8	
5	6	28	6	
5	7	31	4	
5	8	34	2	
5	9	37	0	

Appendix 3.4: Classification of Parallel Axes Gear Trains with M = 1

c (Cycles)	j_2	**m (Mobile links)**	j_1	**Total Number**
1	1	2	2	1
2	2	3	3	1
3	3	4	4	1
4	4	5	5	1
5	5	6	6	1
6	6	7	7	1
7	7	8	8	1
8	8	9	9	1
9	9	10	10	1
10	10	11	11	1

Appendix 3.5: Classification of Parallel Axes Gear Trains with M = 2

c (Cycles)	j_2	m (Mobile links)	j_1	Total Number
1	1	3	3	1
2	2	4	4	1
3	3	5	5	1
4	4	6	6	1
5	5	7	7	1
6	6	8	8	1
7	7	9	9	1
8	8	10	10	1
9	9	11	11	1
10	10	12	12	1

Appendix 3.6: Classification of Parallel Axes Gear Trains with M = 3

c (Cycles)	j_2	m (Mobile links)	j_1	Total Number
1	1	4	4	1
2	2	5	5	1
3	3	6	6	1
4	4	7	7	1
5	5	8	8	1
6	6	9	9	1
7	7	10	10	1
8	8	11	11	1
9	9	12	12	1
10	10	13	13	1

Appendix 3.7: Latin Matrix Entries for Crank-Rocker Mechanism

θ_1 [deg]	x_{AB} [in]	y_{AB} [in]	x_{BD} [in]	y_{BD} [in]	x_{DC} [in]	y_{DC} [in]	x_{CA} [in]	y_{CA} [in]
0.0	4.000	0.000	20.308	12.791	5.692	-12.791	-30.000	0.000
15.0	3.864	1.035	20.830	11.920	5.306	-12.956	-30.000	0.000
30.0	3.464	2.000	21.332	10.997	5.204	-12.997	-30.000	0.000
45.0	2.828	2.828	21.776	10.090	5.396	-12.918	-30.000	0.000
60.0	2.000	3.464	22.145	9.253	5.855	-12.717	-30.000	0.000
75.0	1.035	3.864	22.436	8.521	6.528	-12.385	-30.000	0.000
90.0	-0.000	4.000	22.656	7.919	7.344	-11.919	-30.000	0.000
105.0	-1.035	3.864	22.810	7.465	8.226	-11.329	-30.000	0.000
120.0	-2.000	3.464	22.902	7.177	9.098	-10.641	-30.000	0.000
135.0	-2.828	2.828	22.933	7.075	9.895	-9.904	-30.000	0.000
150.0	-3.464	2.000	22.899	7.186	10.565	-9.186	-30.000	0.000
165.0	-3.864	1.035	22.789	7.529	11.075	-8.564	-30.000	0.000
180.0	-4.000	-0.000	22.588	8.110	11.412	-8.110	-30.000	0.000
195.0	-3.864	-1.035	22.286	8.907	11.578	-7.871	-30.000	0.000
210.0	-3.464	-2.000	21.880	9.862	11.584	-7.862	-30.000	0.000
225.0	-2.828	-2.828	21.385	10.894	11.443	-8.065	-30.000	0.000
240.0	-2.000	-3.464	20.836	11.911	11.164	-8.447	-30.000	0.000
255.0	-1.035	-3.864	20.283	12.830	10.752	-8.966	-30.000	0.000
270.0	0.000	-4.000	19.790	13.578	10.210	-9.578	-30.000	0.000
285.0	1.035	-3.864	19.418	14.104	9.546	-10.240	-30.000	0.000
300.0	2.000	-3.464	19.222	14.370	8.778	-10.906	-30.000	0.000
315.0	2.828	-2.828	19.231	14.359	7.941	-11.530	-30.000	0.000
330.0	3.464	-2.000	19.443	14.070	7.093	-12.070	-30.000	0.000
345.0	3.864	-1.035	19.822	13.531	6.314	-12.495	-30.000	0.000
360.0	4.000	0.000	20.308	12.791	5.692	-12.791	-30.000	0.000

Appendix 3.8: The Coordinates for All Joints for Crank-Rocker Mechanism

θ_1 [deg]	x_A [in]	y_A [in]	x_B [in]	y_B [in]	x_C [in]	y_C [in]	x_D [in]	y_D [in]
0.0	0.000	0.000	4.000	0.000	30.000	0.000	24.308	12.791
15.0	0.000	0.000	3.864	1.035	30.000	0.000	24.694	12.956
30.0	0.000	0.000	3.464	2.000	30.000	0.000	24.796	12.997
45.0	0.000	0.000	2.828	2.828	30.000	0.000	24.604	12.918
60.0	0.000	0.000	2.000	3.464	30.000	0.000	24.145	12.717
75.0	0.000	0.000	1.035	3.864	30.000	0.000	23.472	12.385
90.0	0.000	0.000	-0.000	4.000	30.000	0.000	22.656	11.919
105.0	0.000	0.000	-1.035	3.864	30.000	0.000	21.774	11.329
120.0	0.000	0.000	-2.000	3.464	30.000	0.000	20.902	10.641
135.0	0.000	0.000	-2.828	2.828	30.000	0.000	20.105	9.904
150.0	0.000	0.000	-3.464	2.000	30.000	0.000	19.435	9.186
165.0	0.000	0.000	-3.864	1.035	30.000	0.000	18.925	8.564
180.0	0.000	0.000	-4.000	-0.000	30.000	0.000	18.588	8.110
195.0	0.000	0.000	-3.864	-1.035	30.000	0.000	18.422	7.871
210.0	0.000	0.000	-3.464	-2.000	30.000	0.000	18.416	7.862
225.0	0.000	0.000	-2.828	-2.828	30.000	0.000	18.557	8.065
240.0	0.000	0.000	-2.000	-3.464	30.000	0.000	18.836	8.447
255.0	0.000	0.000	-1.035	-3.864	30.000	0.000	19.248	8.966
270.0	0.000	0.000	0.000	-4.000	30.000	0.000	19.790	9.578
285.0	0.000	0.000	1.035	-3.864	30.000	0.000	20.454	10.240
300.0	0.000	0.000	2.000	-3.464	30.000	0.000	21.222	10.906
315.0	0.000	0.000	2.828	-2.828	30.000	0.000	22.059	11.530
330.0	0.000	0.000	3.464	-2.000	30.000	0.000	22.907	12.070
345.0	0.000	0.000	3.864	-1.035	30.000	0.000	23.686	12.495
360.0	0.000	0.000	4.000	0.000	30.000	0.000	24.308	12.791

Appendix 3.9: Mobile Links' Orientation in Time for Crank-Rocker Mechanism. Calculated (EES) and Simulated (SW) Values

θ_1 [deg]	θ_{1sim} [deg]	θ_2 [deg]	θ_{2sim} [deg]	θ_3 [deg]	θ_{3sim} [deg]	τ [deg]	τ_{sim} [deg]
0.0	0.0	32.2	32.2	294.0	294.0	81.8	81.8
15.0	15.0	29.8	29.8	292.3	292.3	82.5	82.5
30.0	30.0	27.3	27.3	291.8	291.8	84.5	84.5
45.0	45.0	24.9	24.9	292.7	292.7	87.8	87.8
60.0	60.0	22.7	22.7	294.7	294.7	92.0	92.0
75.0	75.0	20.8	20.8	297.8	297.8	97.0	97.0
90.0	90.0	19.3	19.3	301.6	301.6	102.4	102.4
105.0	105.0	18.1	18.1	306.0	306.0	107.9	107.9
120.0	120.0	17.4	17.4	310.5	310.5	113.1	113.1
135.0	135.0	17.1	17.1	315.0	315.0	117.8	117.8
150.0	150.0	17.4	17.4	319.0	319.0	121.6	121.6
165.0	165.0	18.3	18.3	322.3	322.3	124.0	124.0
180.0	180.0	19.7	19.7	324.6	324.6	124.8	124.8
195.0	195.0	21.8	21.8	325.8	325.8	124.0	124.0
210.0	210.0	24.3	24.3	325.8	325.8	121.6	121.6
225.0	225.0	27.0	27.0	324.8	324.8	117.8	117.8
240.0	240.0	29.8	29.8	322.9	322.9	113.1	113.1
255.0	255.0	32.3	32.3	320.2	320.2	107.9	107.9
270.0	270.0	34.5	34.5	316.8	316.8	102.4	102.4
285.0	285.0	36.0	36.0	313.0	313.0	97.0	97.0
300.0	300.0	36.8	36.8	308.8	308.8	92.0	92.0
315.0	315.0	36.7	36.7	304.6	304.6	87.8	87.8
330.0	330.0	35.9	35.9	300.4	300.4	84.5	84.5
345.0	345.0	34.3	34.3	296.8	296.8	82.5	82.5
360.0	360.0	32.2	32.2	294.0	294.0	81.8	81.8

Appendix 3.10: Latin Matrix Entries Versus time for Crank Slider Mechanism

t (Time) [sec]	θ_1 [deg]	x_{CA} [in]	y_{CA} [in]	x_{AB} [in]	y_{AB} [in]	x_{BD} [in]	y_{BD} [in]	x_{DC} [in]	y_{DC} [in]
0.00	0.0	-27.979	-1.000	4.000	0.000	23.979	1.000	0.000	0.000
0.04	15.0	-27.864	-1.000	3.864	1.035	24.000	-0.035	0.000	0.000
0.08	30.0	-27.443	-1.000	3.464	2.000	23.979	-1.000	0.000	0.000
0.13	45.0	-26.759	-1.000	2.828	2.828	23.930	-1.828	0.000	0.000
0.17	60.0	-25.873	-1.000	2.000	3.464	23.873	-2.464	0.000	0.000
0.21	75.0	-24.864	-1.000	1.035	3.864	23.829	-2.864	0.000	0.000
0.25	90.0	-23.812	-1.000	-0.000	4.000	23.812	-3.000	0.000	0.000
0.29	105.0	-22.793	-1.000	-1.035	3.864	23.829	-2.864	0.000	0.000
0.33	120.0	-21.873	-1.000	-2.000	3.464	23.873	-2.464	0.000	0.000
0.38	135.0	-21.102	-1.000	-2.828	2.828	23.930	-1.828	0.000	0.000
0.42	150.0	-20.515	-1.000	-3.464	2.000	23.979	-1.000	0.000	0.000
0.46	165.0	-20.136	-1.000	-3.864	1.035	24.000	-0.035	0.000	0.000
0.50	180.0	-19.979	-1.000	-4.000	-0.000	23.979	1.000	0.000	0.000
0.54	195.0	-20.050	-1.000	-3.864	-1.035	23.914	2.035	0.000	0.000
0.58	210.0	-20.348	-1.000	-3.464	-2.000	23.812	3.000	0.000	0.000
0.63	225.0	-20.864	-1.000	-2.828	-2.828	23.693	3.828	0.000	0.000
0.67	240.0	-21.581	-1.000	-2.000	-3.464	23.581	4.464	0.000	0.000
0.71	255.0	-22.467	-1.000	-1.035	-3.864	23.502	4.864	0.000	0.000
0.75	270.0	-23.473	-1.000	0.000	-4.000	23.473	5.000	0.000	0.000
0.79	285.0	-24.537	-1.000	1.035	-3.864	23.502	4.864	0.000	0.000
0.83	300.0	-25.581	-1.000	2.000	-3.464	23.581	4.464	0.000	0.000
0.88	315.0	-26.521	-1.000	2.828	-2.828	23.693	3.828	0.000	0.000
0.92	330.0	-27.276	-1.000	3.464	-2.000	23.812	3.000	0.000	0.000
0.96	345.0	-27.777	-1.000	3.864	-1.035	23.914	2.035	0.000	0.000
1.00	360.0	-27.979	-1.000	4.000	0.000	23.979	1.000	0.000	0.000

Appendix 3.11: The Coordinates for All Joints for Crank-Rocker Mechanism

t (Time) [sec]	θ_1 [deg]	x_A [in]	y_A [in]	x_B [in]	y_B [in]	x_C [in]	y_C [in]	x_D [in]	y_D [in]	d_C [in]
0.00	0.0	0.000	0.000	4.000	0.000	27.979	1.000	27.979	1.000	-0.000
0.04	15.0	0.000	0.000	3.864	1.035	27.864	1.000	27.864	1.000	0.115
0.08	30.0	0.000	0.000	3.464	2.000	27.443	1.000	27.443	1.000	0.536
0.13	45.0	0.000	0.000	2.828	2.828	26.759	1.000	26.759	1.000	1.220
0.17	60.0	0.000	0.000	2.000	3.464	25.873	1.000	25.873	1.000	2.106
0.21	75.0	0.000	0.000	1.035	3.864	24.864	1.000	24.864	1.000	3.115
0.25	90.0	0.000	0.000	-0.000	4.000	23.812	1.000	23.812	1.000	4.167
0.29	105.0	0.000	0.000	-1.035	3.864	22.793	1.000	22.793	1.000	5.186
0.33	120.0	0.000	0.000	-2.000	3.464	21.873	1.000	21.873	1.000	6.106
0.38	135.0	0.000	0.000	-2.828	2.828	21.102	1.000	21.102	1.000	6.877
0.42	150.0	0.000	0.000	-3.464	2.000	20.515	1.000	20.515	1.000	7.464
0.46	165.0	0.000	0.000	-3.864	1.035	20.136	1.000	20.136	1.000	7.843
0.50	180.0	0.000	0.000	-4.000	-0.000	19.979	1.000	19.979	1.000	8.000
0.54	195.0	0.000	0.000	-3.864	-1.035	20.050	1.000	20.050	1.000	7.929
0.58	210.0	0.000	0.000	-3.464	-2.000	20.348	1.000	20.348	1.000	7.631
0.63	225.0	0.000	0.000	-2.828	-2.828	20.864	1.000	20.864	1.000	7.115
0.67	240.0	0.000	0.000	-2.000	-3.464	21.581	1.000	21.581	1.000	6.398
0.71	255.0	0.000	0.000	-1.035	-3.864	22.467	1.000	22.467	1.000	5.512
0.75	270.0	0.000	0.000	0.000	-4.000	23.473	1.000	23.473	1.000	4.506
0.79	285.0	0.000	0.000	1.035	-3.864	24.537	1.000	24.537	1.000	3.442
0.83	300.0	0.000	0.000	2.000	-3.464	25.581	1.000	25.581	1.000	2.398
0.88	315.0	0.000	0.000	2.828	-2.828	26.521	1.000	26.521	1.000	1.458
0.92	330.0	0.000	0.000	3.464	-2.000	27.276	1.000	27.276	1.000	0.703
0.96	345.0	0.000	0.000	3.864	-1.035	27.777	1.000	27.777	1.000	0.202
1.00	360.0	0.000	0.000	4.000	0.000	27.979	1.000	27.979	1.000	-0.000

Appendix 3.12: Mobile Links' Orientation in Time for Crank Slider Mechanism. Calculated (EES) and Simulated (SW) Values

(Time) [sec]	θ_1 [deg]	$\theta_{1,sim}$ [deg]	θ_2 [deg]	$\theta_{2,sim}$ [deg]	θ_3 [deg]	$\theta_{3,sim}$ [deg]	τ [deg]	τ_{sim} [deg]
0.00	0.0	0.0	2.4	2.4	0.0	0.0	87.6	87.6
0.04	15.0	15.0	359.9	359.9	0.0	0.0	89.9	89.9
0.08	30.0	30.0	357.6	357.6	0.0	0.0	87.6	87.6
0.13	45.0	45.0	355.6	355.6	0.0	0.0	85.6	85.6
0.17	60.0	60.0	354.1	354.1	0.0	0.0	84.1	84.1
0.21	75.0	75.0	353.1	353.1	0.0	0.0	83.1	83.1
0.25	90.0	90.0	352.8	352.8	0.0	0.0	82.8	82.8
0.29	105.0	105.0	353.1	353.1	0.0	0.0	83.1	83.1
0.33	120.0	120.0	354.1	354.1	0.0	0.0	84.1	84.1
0.38	135.0	135.0	355.6	355.6	0.0	0.0	85.6	85.6
0.42	150.0	150.0	357.6	357.6	0.0	0.0	87.6	87.6
0.46	165.0	165.0	359.9	359.9	0.0	0.0	89.9	89.9
0.50	180.0	180.0	2.4	2.4	0.0	0.0	87.6	87.6
0.54	195.0	195.0	4.9	4.9	0.0	0.0	85.1	85.1
0.58	210.0	210.0	7.2	7.2	0.0	0.0	82.8	82.8
0.63	225.0	225.0	9.2	9.2	0.0	0.0	80.8	80.8
0.67	240.0	240.0	10.7	10.7	0.0	0.0	79.3	79.3
0.71	255.0	255.0	11.7	11.7	0.0	0.0	78.3	78.3
0.75	270.0	270.0	12.0	12.0	0.0	0.0	78.0	78.0
0.79	285.0	285.0	11.7	11.7	0.0	0.0	78.3	78.3
0.83	300.0	300.0	10.7	10.7	0.0	0.0	79.3	79.3
0.88	315.0	315.0	9.2	9.2	0.0	0.0	80.8	80.8
0.92	330.0	330.0	7.2	7.2	0.0	0.0	82.8	82.8
0.96	345.0	345.0	4.9	4.9	0.0	0.0	85.1	85.1
1.00	360.0	360.0	2.4	2.4	0.0	0.0	87.6	87.6

Appendix 3.13: Links' Orientation Angles, Transmission Angle, and Slider Displacement for the Crank Slider Mechanism. Calculated (EES) and Simulated (SW) Values

Time (t) [sec]	θ_1 [deg]	$\theta_{1,sim}$ [deg]	θ_2 [deg]	$\theta_{2,sim}$ [deg]	θ_3 [deg]	$\theta_{3,sim}$ [deg]	τ [deg]	τ_{sim} [deg]	x_C [in]	x_{Csim} [in]	d_C [in]	d_{Csim} [in]
0.00	0.0	0.0	2.4	2.4	0.0	0.0	87.6	87.6	27.979	27.979	-0.000	0.000
0.04	15.0	15.0	359.9	359.9	0.0	0.0	89.9	89.9	27.864	27.864	0.115	0.116
0.08	30.0	30.0	357.6	357.6	0.0	0.0	87.6	87.6	27.443	27.443	0.536	0.536
0.13	45.0	45.0	355.6	355.6	0.0	0.0	85.6	85.6	26.759	26.759	1.220	1.221
0.17	60.0	60.0	354.1	354.1	0.0	0.0	84.1	84.1	25.873	25.873	2.106	2.106
0.21	75.0	75.0	353.1	353.1	0.0	0.0	83.1	83.1	24.864	24.864	3.115	3.115
0.25	90.0	90.0	352.8	352.8	0.0	0.0	82.8	82.8	23.812	23.812	4.167	4.167
0.29	105.0	105.0	353.1	353.1	0.0	0.0	83.1	83.1	22.793	22.793	5.186	5.186
0.33	120.0	120.0	354.1	354.1	0.0	0.0	84.1	84.1	21.873	21.873	6.106	6.106
0.38	135.0	135.0	355.6	355.6	0.0	0.0	85.6	85.6	21.102	21.102	6.877	6.877
0.42	150.0	150.0	357.6	357.6	0.0	0.0	87.6	87.6	20.515	20.515	7.464	7.464
0.46	165.0	165.0	359.9	359.9	0.0	0.0	89.9	89.9	20.136	20.136	7.843	7.843
0.50	180.0	180.0	2.4	2.4	0.0	0.0	87.6	87.6	19.979	19.979	8.000	8.000
0.54	195.0	195.0	4.9	4.9	0.0	0.0	85.1	85.1	20.050	20.050	7.929	7.929
0.58	210.0	210.0	7.2	7.2	0.0	0.0	82.8	82.8	20.348	20.348	7.631	7.631
0.63	225.0	225.0	9.2	9.2	0.0	0.0	80.8	80.8	20.864	20.864	7.115	7.115
0.67	240.0	240.0	10.7	10.7	0.0	0.0	79.3	79.3	21.581	21.581	6.398	6.398
0.71	255.0	255.0	11.7	11.7	0.0	0.0	78.3	78.3	22.467	22.467	5.512	5.512
0.75	270.0	270.0	12.0	12.0	0.0	0.0	78.0	78.0	23.473	23.473	4.506	4.506
0.79	285.0	285.0	11.7	11.7	0.0	0.0	78.3	78.3	24.537	24.537	3.442	3.442
0.83	300.0	300.0	10.7	10.7	0.0	0.0	79.3	79.3	25.581	25.581	2.398	2.398
0.88	315.0	315.0	9.2	9.2	0.0	0.0	80.8	80.8	26.521	26.521	1.458	1.458
0.92	330.0	330.0	7.2	7.2	0.0	0.0	82.8	82.8	27.276	27.276	0.703	0.703
0.96	345.0	345.0	4.9	4.9	0.0	0.0	85.1	85.1	27.777	27.777	0.202	0.202
1.00	360.0	360.0	2.4	2.4	0.0	0.0	87.6	87.6	27.979	27.979	-0.000	0.000

Appendix 3.14: Latin Matrix Entries Versus Time for RRTR Planar Mechanism

θ_1 [deg]	x_{CA} [in]	y_{CA} [in]	x_{AB} [in]	y_{AB} [in]	x_{BD} [in]	y_{BD} [in]	x_{DC} [in]	y_{DC} [in]
0.0	-14.250	0.000	4.250	0.000	9.264	2.611	0.736	-2.611
15.0	-14.250	0.000	4.105	1.100	9.370	2.200	0.775	-3.300
30.0	-14.250	0.000	3.681	2.125	9.279	2.557	1.290	-4.682
45.0	-14.250	0.000	3.005	3.005	9.087	3.174	2.158	-6.179
60.0	-14.250	0.000	2.125	3.681	8.814	3.867	3.311	-7.547
75.0	-14.250	0.000	1.100	4.105	8.471	4.569	4.679	-8.674
90.0	-14.250	0.000	-0.000	4.250	8.067	5.250	6.183	-9.500
105.0	-14.250	0.000	-1.100	4.105	7.611	5.892	7.739	-9.997
120.0	-14.250	0.000	-2.125	3.681	7.115	6.483	9.260	-10.163
135.0	-14.250	0.000	-3.005	3.005	6.591	7.015	10.665	-10.020
150.0	-14.250	0.000	-3.681	2.125	6.053	7.483	11.877	-9.608
165.0	-14.250	0.000	-4.105	1.100	5.520	7.885	12.836	-8.985
180.0	-14.250	0.000	-4.250	-0.000	5.008	8.220	13.492	-8.220
195.0	-14.250	0.000	-4.105	-1.100	4.538	8.488	13.817	-7.388
210.0	-14.250	0.000	-3.681	-2.125	4.137	8.691	13.794	-6.566
225.0	-14.250	0.000	-3.005	-3.005	3.831	8.830	13.424	-5.824
240.0	-14.250	0.000	-2.125	-3.681	3.656	8.904	12.719	-5.223
255.0	-14.250	0.000	-1.100	-4.105	3.654	8.905	11.696	-4.799
270.0	-14.250	0.000	0.000	-4.250	3.873	8.811	10.377	-4.561
285.0	-14.250	0.000	1.100	-4.105	4.367	8.577	8.783	-4.472
300.0	-14.250	0.000	2.125	-3.681	5.177	8.114	6.948	-4.433
315.0	-14.250	0.000	3.005	-3.005	6.292	7.284	4.953	-4.278
330.0	-14.250	0.000	3.681	-2.125	7.570	5.945	3.000	-3.820
345.0	-14.250	0.000	4.105	-1.100	8.681	4.157	1.464	-3.057
360.0	-14.250	0.000	4.250	0.000	9.264	2.611	0.736	-2.611

Appendix 3.15: The Coordinates for All Joints for RRTR Planar Mechanism

θ_1 [deg]	x_A [in]	y_A [in]	x_B [in]	y_B [in]	x_C [in]	y_C [in]	x_D [in]	y_D [in]
0.0	0.000	0.000	4.250	0.000	14.250	0.000	13.514	2.611
15.0	0.000	0.000	4.105	1.100	14.250	0.000	13.475	3.300
30.0	0.000	0.000	3.681	2.125	14.250	0.000	12.960	4.682
45.0	0.000	0.000	3.005	3.005	14.250	0.000	12.092	6.179
60.0	0.000	0.000	2.125	3.681	14.250	0.000	10.939	7.547
75.0	0.000	0.000	1.100	4.105	14.250	0.000	9.571	8.674
90.0	0.000	0.000	-0.000	4.250	14.250	0.000	8.067	9.500
105.0	0.000	0.000	-1.100	4.105	14.250	0.000	6.511	9.997
120.0	0.000	0.000	-2.125	3.681	14.250	0.000	4.990	10.163
135.0	0.000	0.000	-3.005	3.005	14.250	0.000	3.585	10.020
150.0	0.000	0.000	-3.681	2.125	14.250	0.000	2.373	9.608
165.0	0.000	0.000	-4.105	1.100	14.250	0.000	1.414	8.985
180.0	0.000	0.000	-4.250	-0.000	14.250	0.000	0.758	8.220
195.0	0.000	0.000	-4.105	-1.100	14.250	0.000	0.433	7.388
210.0	0.000	0.000	-3.681	-2.125	14.250	0.000	0.456	6.566
225.0	0.000	0.000	-3.005	-3.005	14.250	0.000	0.826	5.824
240.0	0.000	0.000	-2.125	-3.681	14.250	0.000	1.531	5.223
255.0	0.000	0.000	-1.100	-4.105	14.250	0.000	2.554	4.799
270.0	0.000	0.000	0.000	-4.250	14.250	0.000	3.873	4.561
285.0	0.000	0.000	1.100	-4.105	14.250	0.000	5.467	4.472
300.0	0.000	0.000	2.125	-3.681	14.250	0.000	7.302	4.433
315.0	0.000	0.000	3.005	-3.005	14.250	0.000	9.297	4.278
330.0	0.000	0.000	3.681	-2.125	14.250	0.000	11.250	3.820
345.0	0.000	0.000	4.105	-1.100	14.250	0.000	12.786	3.057
360.0	0.000	0.000	4.250	0.000	14.250	0.000	13.514	2.611

Appendix 3.16: Links' Orientation Angles, Slider Displacement, and Point of Interest P Coordinates for RRTR Planar Mechanism. Calculated (EES) and Simulated (SW) Values

t [sec]	θ_1 [deg]	θ_{1sim} [deg]	θ_2 [deg]	θ_{2sim} [in]	θ_3 [deg]	θ_{3sim} [deg]	d_D [in]	d_{Dsim} [in]	x_P [in]	x_{Psim} [in]	y_P [in]	y_{Psim} [in]	τ [deg]	τ_{sim} [deg]
0.00	0.0	0.0	15.7	15.7	285.7	285.7	0.000	0.000	9.299	9.299	17.566	17.566	90.0	90.0
0.04	15.0	15.0	13.2	13.2	283.2	283.2	-0.676	-0.676	10.079	10.079	17.767	17.767	90.0	90.0
0.08	30.0	30.0	15.4	15.4	285.4	285.4	-2.144	-2.144	9.401	9.401	17.594	17.594	90.0	90.0
0.13	45.0	45.0	19.3	19.3	289.3	289.3	-3.832	-3.832	8.232	8.232	17.229	17.229	90.0	90.0
0.17	60.0	60.0	23.7	23.7	293.7	293.7	-5.528	-5.529	6.919	6.919	16.713	16.713	90.0	90.0
0.21	75.0	75.0	28.3	28.3	298.3	298.3	-7.143	-7.143	5.586	5.586	16.062	16.062	90.0	90.0
0.25	90.0	90.0	33.1	33.1	303.1	303.1	-8.622	-8.622	4.295	4.295	15.296	15.296	90.0	90.0
0.29	105.0	105.0	37.7	37.7	307.7	307.7	-9.930	-9.930	3.078	3.078	14.431	14.431	90.0	90.0
0.33	120.0	120.0	42.3	42.3	312.3	312.3	-11.036	-11.037	1.958	1.958	13.490	13.490	90.0	90.0
0.38	135.0	135.0	46.8	46.8	316.8	316.8	-11.920	-11.920	0.949	0.949	12.496	12.496	90.0	90.0
0.42	150.0	150.0	51.0	51.0	321.0	321.0	-12.564	-12.564	0.061	0.061	11.478	11.478	90.0	90.0
0.46	165.0	165.0	55.0	55.0	325.0	325.0	-12.955	-12.955	-0.701	-0.701	10.466	10.466	90.0	90.0
0.50	180.0	180.0	58.6	58.6	328.6	328.6	-13.086	-13.086	-1.336	-1.336	9.495	9.495	90.0	90.0
0.54	195.0	195.0	61.9	61.9	331.9	331.9	-12.955	-12.955	-1.844	-1.844	8.605	8.605	90.0	90.0
0.58	210.0	210.0	64.5	64.5	334.5	334.5	-12.564	-12.564	-2.229	-2.229	7.844	7.844	90.0	90.0
0.63	225.0	225.0	66.5	66.5	336.5	336.5	-11.920	-11.920	-2.492	-2.492	7.264	7.264	90.0	90.0
0.67	240.0	240.0	67.7	67.7	337.7	337.7	-11.036	-11.037	-2.632	-2.632	6.933	6.933	90.0	90.0
0.71	255.0	255.0	67.7	67.7	337.7	337.7	-9.930	-9.930	-2.634	-2.634	6.928	6.928	90.0	90.0
0.75	270.0	270.0	66.3	66.3	336.3	336.3	-8.622	-8.622	-2.457	-2.457	7.344	7.344	90.0	90.0
0.79	285.0	285.0	63.0	63.0	333.0	333.0	-7.143	-7.143	-2.013	-2.013	8.281	8.281	90.0	90.0
0.83	300.0	300.0	57.5	57.5	327.5	327.5	-5.528	-5.529	-1.135	-1.135	9.817	9.817	90.0	90.0
0.88	315.0	315.0	49.2	49.2	319.2	319.2	-3.832	-3.832	0.439	0.439	11.930	11.930	90.0	90.0
0.92	330.0	330.0	38.1	38.1	308.1	308.1	-2.144	-2.144	2.978	2.978	14.353	14.353	90.0	90.0
0.96	345.0	345.0	25.6	25.6	295.6	295.6	-0.676	-0.676	6.368	6.368	16.460	16.460	90.0	90.0
1.00	360.0	360.0	15.7	15.7	285.7	285.7	0.000	0.000	9.299	9.299	17.566	17.566	90.0	90.0

Appendix 3.17: Links' Orientation and Slider Displacement for the Two-Cycle Planar Mechanism

t [sec]	θ_1 [deg]	θ_{1sim} [deg]	θ_2 [deg]	θ_{2sim} [in]	θ_3 [deg]	θ_{3sim} [deg]	θ_4 [deg]	θ_{4sim} [deg]	θ_5 [deg]	θ_{5sim} [deg]	d_C [in]	d_{Csim} [in]
0.00	0.0	0.0	32.2	32.2	294.0	294.0	327.8	327.8	0.0	0.0	0.000	0.000
0.04	15.0	15.0	29.8	29.8	292.3	292.3	327.3	327.3	0.0	0.0	-0.140	-0.141
0.08	30.0	30.0	27.3	27.3	291.8	291.8	327.2	327.2	0.0	0.0	-0.178	-0.179
0.13	45.0	45.0	24.9	24.9	292.7	292.7	327.4	327.4	0.0	0.0	-0.107	-0.108
0.17	60.0	60.0	22.7	22.7	294.7	294.7	328.0	328.0	0.0	0.0	0.059	0.058
0.21	75.0	75.0	20.8	20.8	297.8	297.8	328.9	328.9	0.0	0.0	0.293	0.293
0.25	90.0	90.0	19.3	19.3	301.6	301.6	330.2	330.2	0.0	0.0	0.564	0.564
0.29	105.0	105.0	18.1	18.1	306.0	306.0	331.8	331.8	0.0	0.0	0.842	0.842
0.33	120.0	120.0	17.4	17.4	310.5	310.5	333.7	333.7	0.0	0.0	1.101	1.101
0.38	135.0	135.0	17.1	17.1	315.0	315.0	335.6	335.6	0.0	0.0	1.325	1.325
0.42	150.0	150.0	17.4	17.4	319.0	319.0	337.5	337.5	0.0	0.0	1.504	1.504
0.46	165.0	165.0	18.3	18.3	322.3	322.3	339.1	339.1	0.0	0.0	1.636	1.635
0.50	180.0	180.0	19.7	19.7	324.6	324.6	340.3	340.3	0.0	0.0	1.720	1.719
0.54	195.0	195.0	21.8	21.8	325.8	325.8	340.9	340.9	0.0	0.0	1.761	1.760
0.58	210.0	210.0	24.3	24.3	325.8	325.8	340.9	340.9	0.0	0.0	1.762	1.762
0.63	225.0	225.0	27.0	27.0	324.8	324.8	340.4	340.4	0.0	0.0	1.728	1.727
0.67	240.0	240.0	29.8	29.8	322.9	322.9	339.4	339.4	0.0	0.0	1.658	1.658
0.71	255.0	255.0	32.3	32.3	320.2	320.2	338.1	338.1	0.0	0.0	1.553	1.553
0.75	270.0	270.0	34.5	34.5	316.8	316.8	336.5	336.5	0.0	0.0	1.410	1.410
0.79	285.0	285.0	36.0	36.0	313.0	313.0	334.7	334.7	0.0	0.0	1.228	1.228
0.83	300.0	300.0	36.8	36.8	308.8	308.8	333.0	333.0	0.0	0.0	1.008	1.007
0.88	315.0	315.0	36.7	36.7	304.6	304.6	331.3	331.3	0.0	0.0	0.754	0.754
0.92	330.0	330.0	35.9	35.9	300.4	300.4	329.8	329.8	0.0	0.0	0.483	0.482
0.96	345.0	345.0	34.3	34.3	296.8	296.8	328.6	328.6	0.0	0.0	0.220	0.219
1.00	360.0	360.0	32.2	32.2	294.0	294.0	327.8	327.8	0.0	0.0	0.000	0.000

Appendix 3.18: Cam for Constant Velocity Follower

t [sec]	θ_1 [deg]	d_C [in]	v_C [in/s]	a_C [in/s2]	x_B [in]	y_B [in]	t [sec]	θ_1 [deg]	d_C [in]	v_C [in/s]	a_C [in/s2]	x_B [in]	y_B [in]
0.00	0.0	0.000	1.515	0.000	0.000	1.000	1.04	187.5	0.000	0.000	0.000	-0.131	-0.991
0.04	7.5	0.063	1.515	0.000	0.139	1.054	1.08	195.0	0.000	0.000	0.000	-0.259	-0.966
0.08	15.0	0.125	1.515	0.000	0.291	1.087	1.13	202.5	0.000	0.000	0.000	-0.383	-0.924
0.13	22.5	0.188	1.515	0.000	0.455	1.098	1.17	210.0	0.000	0.000	0.000	-0.500	-0.866
0.17	30.0	0.250	1.515	0.000	0.625	1.083	1.21	217.5	0.000	0.000	0.000	-0.609	-0.793
0.21	37.5	0.312	1.515	0.000	0.799	1.041	1.25	225.0	0.000	0.000	0.000	-0.707	-0.707
0.25	45.0	0.375	1.515	0.000	0.972	0.972	1.29	232.5	0.000	0.000	0.000	-0.793	-0.609
0.29	52.5	0.438	1.515	0.000	1.141	0.875	1.33	240.0	0.000	0.000	0.000	-0.866	-0.500
0.33	60.0	0.500	-1.515	0.000	1.299	0.750	1.38	247.5	0.000	0.000	0.000	-0.924	-0.383
0.38	67.5	0.438	-1.515	0.000	1.329	0.550	1.42	255.0	0.000	0.000	0.000	-0.966	-0.259
0.42	75.0	0.375	-1.515	0.000	1.328	0.356	1.46	262.5	0.000	0.000	0.000	-0.991	-0.131
0.46	82.5	0.313	-1.515	0.000	1.302	0.171	1.50	270.0	0.000	0.000	0.000	-1.000	0.000
0.50	90.0	0.250	-1.515	0.000	1.250	-0.000	1.54	277.5	0.000	0.000	0.000	-0.991	0.131
0.54	97.5	0.187	-1.515	0.000	1.177	-0.155	1.58	285.0	0.000	0.000	0.000	-0.966	0.259
0.58	105.0	0.125	-1.515	0.000	1.087	-0.291	1.63	292.5	0.000	0.000	0.000	-0.924	0.383
0.63	112.5	0.063	-1.515	0.000	0.982	-0.407	1.67	300.0	0.000	0.000	0.000	-0.866	0.500
0.67	120.0	0.000	0.000	0.000	0.866	-0.500	1.71	307.5	0.000	0.000	0.000	-0.793	0.609
0.71	127.5	0.000	0.000	0.000	0.793	-0.609	1.75	315.0	0.000	0.000	0.000	-0.707	0.707
0.75	135.0	0.000	0.000	0.000	0.707	-0.707	1.79	322.5	0.000	0.000	0.000	-0.609	0.793
0.79	142.5	0.000	0.000	0.000	0.609	-0.793	1.83	330.0	0.000	0.000	0.000	-0.500	0.866
0.83	150.0	0.000	0.000	0.000	0.500	-0.866	1.88	337.5	0.000	0.000	0.000	-0.383	0.924
0.88	157.5	0.000	0.000	0.000	0.383	-0.924	1.92	345.0	0.000	0.000	0.000	-0.259	0.966
0.92	165.0	0.000	0.000	0.000	0.259	-0.966	1.96	352.5	0.000	0.000	0.000	-0.131	0.991
0.96	172.5	0.000	0.000	0.000	0.131	-0.991	2.00	360.0	0.000	0.000	0.000	0.000	1.000
1.00	180.0	0.000	0.000	0.000	-0.000	-1.000							

References

1. Talpasanu, I., Optimisation in kinematic and kinetostatic analysis of rigid systems with applications in machine design, *PhD Dissertation*, University Politehnica, Bucharest, 1991.
2. Talpasanu, I., Kinematics and dynamics of mechanical systems based on graph-matroid theory, *PhD Dissertation*, University of Texas at Arlington, Arlington, TX, 2004.
3. Cleghorn, W.L., and Dechev, N, *Mechanics of Machines*, Oxford University Press, New York, 2014.

4. Myszka, D., *Machines and Mechanisms*, Prentice Hall, Upper Saddle River, NJ, 2012.
5. Voinea, R., Atanasiu, M., and Talpasanu, I., Kinematics analysis of the articulated planar mechanisms through independent loop method, *The Fourth IFTOMM International Symposium on Theory and Practice of Mechanisms*, Bucharest, vol. 1, 1985.
6. Voinea, R., Atanasiu, M., Iordache, M., and Talpasanu, I., Matrix method for the kinetostatic analysis of planar mechanisms-VAIT Program, *The Fourth IFTOMM International Symposium on Theory and Practice of Mechanisms*, Bucharest, vol. 1, 1985.
7. Voinea, R, and Atanasiu, M., *New Analytical Methods in Theory of Mechanisms*, Edit. Tehnica: Bucharest, 1964.
8. Voinea, R., Atanasiu, M., Iordache, M., and Talpasanu, I., 1983, Determination of the kinematic parameters for a planar linkage mechanism by the independent loop method, *Proceedings of the 5th International Conference on Control Systems and Computer Science*, Bucharest, vol. 1, 1983.
9. Talpasanu, I., A general method for kinematic analysis of robotic wrist mechanisms, *ASME Journal of Mechanisms and Robotics*, 7(3), 2015.
10. Whitney, H., On the abstract properties of linear dependence, *American Journal of Mathematics*, 57, 1935.
11. Talpasanu, I., Yih, T.C., and Simionescu, P.A., Application of matroid method in kinematic analysis of parallel axes epicyclic gear trains, *ASME Journal of Mechanical Design*, 128, 2006.
12. Tsai, L.W., The kinematics of spatial robotic bevel-gear trains, *IEEE Journal of Robotics and Automation*, 4, 1988.
13. Norton, R.L., *Design of Machinery*, McGraw-Hill: New York, 2004.

4

Dynamic and Static Analysis of Mechanisms

4.1 Direct Angular Acceleration Analysis for Open Cycle Mechanisms

The topic covered in Chapter 2 was referring to angular and linear velocity vectors: relative with respect to local frames and absolute with respect to fixed frame 0. In this section, we refer to *relative acceleration* (angular or linear) vector when written with respect to the local frames and *absolute acceleration* (angular or linear) vector when written with respect to the fixed frame 0 (global).

Analog to the previous positional and velocity analysis, these vectors are attached to the nodes (links) and edges (joints) in a mechanism's digraph.

The concepts of force and angular momentum first require the determination of the linear and angular acceleration of a rigid body.

4.1.1 The Joint Relative Angular Acceleration

From solid rigid kinematics, at each joint, the relative acceleration vector has two components: *axial and complemental.*

The two components are implemented in a digraph edge, using the IT notation for displacements, introduced in Chapter 2:

- The *axial angular acceleration,* $\boldsymbol{\varepsilon}_{Z_{m-1,m}}$ (with scalar $\ddot{\theta}_{Z_{m-1,m}}$), along the axis of rotation, is due to the change in magnitude of the relative angular velocity $\boldsymbol{\omega}_{Z_{m-1,m}}$ (with scalar $\dot{\theta}_{Z_{m-1,m}}$).
- The *complemental angular acceleration,* $\boldsymbol{\varepsilon}^{com}_{Z_{m-1,m}}$, is perpendicular to the axis of rotation and is due to a change in the direction of vector $\boldsymbol{\omega}^{m}_{Z_{m-1,m}}$.

Notation: In calculations, the two subscripts for relative vectors are dropped:

$$\boldsymbol{\varepsilon}_{Z_{m-1,m}} = \boldsymbol{\varepsilon}_{Z} \text{ and } \boldsymbol{\omega}^{m}_{Z_{m-1,m}} = \boldsymbol{\omega}_{Z}.$$

4.1.2 The Joint Axial Angular Acceleration, ε_Z

The direction of the axial relative acceleration is that of the rotation axis of unit vector **u**, which is shown in Equation (4.1):

$$\varepsilon_Z^m = \ddot{\theta}_Z \cdot \mathbf{u}_Z^m \tag{4.1}$$

where

- $\ddot{\theta}_Z$ is the joint Z's axial angular acceleration magnitude, measured in [rad/s²] or [°/s²].
- $\mathbf{u}_Z^m$ is the unit vector of axis of rotation of joint Z, which has components with respect to the local frame *m*.

Considering Equation (2.39) to transfer the components from the local frame m to the fixed frame 0, then

$$\varepsilon_Z^0 = \mathbf{D}_m * \varepsilon_Z^m = (\mathbf{D}_m * \mathbf{u}_Z^m) \cdot \ddot{\theta}_Z \tag{4.2}$$

The symbol "*" in Equation (4.2) represents a multiplication between a matrix and a column vector.

This transfer is shown by a left-hand-side multiplication of the vector written in the local frame by its absolute rotation matrix.

Note:

1. Both vectors $\boldsymbol{\varepsilon}_Z$ and $\boldsymbol{\omega}_Z$ have the direction of axis of rotation with the unit vector $\mathbf{u}_Z$.
2. The scalar of relative angular acceleration $\ddot{\theta}_Z$ is an entry in the *joint relative* acceleration column matrix $\ddot{\mathbf{q}}_M(t)$. The double dot suggests that it is determined from the derivative with respect to time of the angular velocities, entry in $\dot{\mathbf{q}}_M(t)$ from Equation (2.178d).
3. The axial angular acceleration is zero for a constant angular velocity at the joint Z:

$$\text{If } \dot{\theta}_Z = \text{constant, then } \ddot{\theta}_Z = 0 \tag{4.3}$$

4. As shown in Figure 4.1, the joint's Z axial angular acceleration vector $\boldsymbol{\varepsilon}_Z$ is assigned to each edge Z in the digraph and written with respect to the local frame m − 1. An observer on the link m − 1 sees the link m in rotation at its scalar rate $\ddot{\theta}_Z$.

4.1.3 Constraint Equations for Axial Angular Acceleration

The axial relative acceleration for the prismatic joint is a zero vector:

$$\varepsilon_Z^0 = 0, \text{ for prismatic joint} \tag{4.4}$$

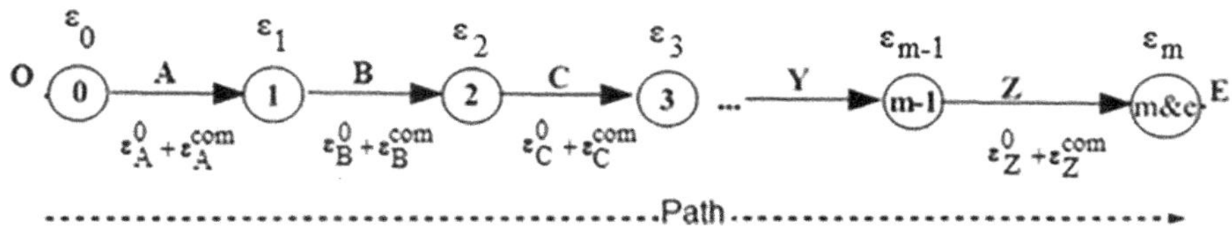

FIGURE 4.1
Relation between absolute and relative angular accelerations along the tree.

Its scalar (magnitude) is constrained at the prismatic joint ($\ddot{\theta}_Z = 0$), resulting in a zero vector.

The constraint in Equation (4.4) is the *constraint equation* for the null angular acceleration in the prismatic joint. It applies also between two points located on the same link, since all points on the same link rotate at the same rate.

4.1.4 The Joint Complemental Angular Acceleration, ε_Z^{com}

The *complemental angular acceleration* due to a change in the direction for vector $\boldsymbol{\omega}_Z^m$ is

$$\boldsymbol{\varepsilon}_Z^{com} = \boldsymbol{\omega}_{m-1} \times \boldsymbol{\omega}_Z^0 \tag{4.5}$$

The cross product of two vectors (×) is replaced by a matrix multiplication; the skew-symmetric matrix of the first vector, noted with a tilde "~," left-hand-side-multiplies the column matrix with components of the second vector, is shown in Equation (4.6):

$$\boldsymbol{\varepsilon}_Z^{com} = \tilde{\boldsymbol{\omega}}_{m-1} * \boldsymbol{\omega}_Z^0 = \begin{bmatrix} 0 & -\omega_{m-1}^z & \omega_{m-1}^y \\ \omega_{m-1}^z & 0 & -\omega_{m-1}^x \\ -\omega_{m-1}^y & \omega_{m-1}^x & 0 \end{bmatrix} * \begin{Bmatrix} \omega_Z^x \\ \omega_Z^y \\ \omega_Z^z \end{Bmatrix} \tag{4.6}$$

where

- ω_{m-1} is the absolute angular velocity for the link m − 1, with respect to fixed frame 0, which was determined using a velocity analysis.
- ω_Z^m is the relative angular velocity, which was determined using a velocity analysis. It has the direction of axis of rotation with the unit vector $\mathbf{u}_Z^m$. First, it is written with respect to the fixed frame m and then with respect to the fixed frame 0, as

$$\omega_Z^0 = \mathbf{D}_m * \omega_Z^m = (\mathbf{D}_m * \mathbf{u}_Z^m) \cdot \dot{\theta}_Z \tag{4.7}$$

Note: The complemental angular acceleration is zero for the prismatic joints.

The existing constraint $\dot{\theta}_Z = 0$ in the prismatic joint results in $\omega_Z^0 = 0$, therefore

$$\varepsilon_Z^{com} = 0 \quad \text{for the prismatic joint.} \tag{4.8}$$

4.1.5 Links' Absolute Angular Acceleration Matrix, ε_m

From the solid rigid kinematics, the vector difference between two absolute angular accelerations is the sum of two components: the axial angular acceleration plus the complemental angular acceleration, which is shown in Equation (4.9):

$$\varepsilon_m = \varepsilon_{m-1} + (\varepsilon_Z^0 + \varepsilon_Z^{com}) \tag{4.9}$$

where

- m is the number of mobile links.
- ε_{m-1} and ε_m are the *absolute angular accelerations* of the links m − 1 and m, with respect to the fixed frame, and their scalars are measured in [rad/s²] or [°/s²].
- ε_Z^0 is the axial angular acceleration, which is shown in Equation (4.2).
- ε_Z^{com} is the complemental angular acceleration, which is shown in Equation (4.6).

As shown in Figure 4.1, the absolute angular acceleration ε_m is assigned to each node m in the digraph and written with respect to the fixed frame. An observer on the ground sees the link m accelerating in rotation at its scalar rate ε_m.

Equation (4.9) is implemented to the digraph as follows:

- The absolute acceleration at node m is the result of addition to the previous node's m − 1 absolute angular acceleration by the sum of two relative components at the edge Z.
- A recursive procedure, m = 1, 2, 3, …, m, from Equation (4.9) leads to the following equations.

4.2 Governing Equation for Links' Absolute Angular Accelerations

Equation (4.9) is written by recursion along a path initiated at node 0 and ended at node m, which is shown in Equation (4.10).

Edge A: The sum of two components, axial and complemental relative (link 1 relative to link 0) accelerations:

$$\left(\boldsymbol{\varepsilon}_A^0 + \boldsymbol{\varepsilon}_A^{com}\right), \text{ with:}$$

$$\boldsymbol{\varepsilon}_A^0 = \mathbf{D}_1 * \boldsymbol{\varepsilon}_A^1 = \left(\mathbf{D}_1 * \mathbf{u}_A^1\right) \cdot \ddot{\theta}_A \tag{4.10}$$

$$\boldsymbol{\varepsilon}_A^{com} = \tilde{\boldsymbol{\omega}}_0 * \boldsymbol{\omega}_A^0$$

Node 1: Link 1's absolute acceleration is the result of previous link 0's absolute added to joint A's relative angular acceleration:

$$\boldsymbol{\varepsilon}_1 = \boldsymbol{\varepsilon}_0 + \left(\boldsymbol{\varepsilon}_A^0 + \boldsymbol{\varepsilon}_A^{com}\right) \tag{4.11}$$

Edge B: The sum of two components, axial and complemental relative (link 2 relative to link 1) accelerations:

$$\left(\boldsymbol{\varepsilon}_B^0 + \boldsymbol{\varepsilon}_B^{com}\right), \text{ with:}$$

$$\boldsymbol{\varepsilon}_B^0 = \left(\mathbf{D}_2 * \mathbf{u}_B^1\right) \cdot \ddot{\theta}_B \tag{4.12}$$

$$\boldsymbol{\varepsilon}_B^{com} = \tilde{\boldsymbol{\omega}}_1 * \boldsymbol{\omega}_B^0$$

Node 2: Link 2's absolute acceleration is the result of previous link 1's absolute added to joint's B relative angular acceleration:

$$\boldsymbol{\varepsilon}_2 = \boldsymbol{\varepsilon}_1 + \left(\boldsymbol{\varepsilon}_B^0 + \boldsymbol{\varepsilon}_B^{com}\right) \tag{4.13}$$

Edge C: The sum of two components, axial and complemental relative (link 3 relative to link 2) accelerations:

$$\left(\boldsymbol{\varepsilon}_C^0 + \boldsymbol{\varepsilon}_C^{com}\right), \text{ with}$$

$$\boldsymbol{\varepsilon}_C^0 = \left(\mathbf{D}_3 * \mathbf{u}_C^1\right) \cdot \ddot{\theta}_C \tag{4.14}$$

$$\boldsymbol{\varepsilon}_C^{com} = \tilde{\boldsymbol{\omega}}_2 * \boldsymbol{\omega}_C^0$$

Node 3: Link 3's absolute acceleration is the result of previous link 2's absolute added to joint's C relative angular acceleration:

$$\boldsymbol{\varepsilon}_3 = \boldsymbol{\varepsilon}_2 + \left(\boldsymbol{\varepsilon}_C^0 + \boldsymbol{\varepsilon}_C^{com}\right) \tag{4.15}$$

Edge Z: The sum of two components, axial and complemental relative (link m relative to link m – 1) accelerations:

$$\left(\boldsymbol{\varepsilon}_Z^0 + \boldsymbol{\varepsilon}_Z^{com}\right), \text{ with}$$

$$\boldsymbol{\varepsilon}_Z^{com} = \left(\mathbf{D}_m * \mathbf{u}_Z^m\right) \cdot \ddot{\theta}_Z \tag{4.16}$$

$$\boldsymbol{\varepsilon}_Z^{com} = \tilde{\boldsymbol{\omega}}_m * \boldsymbol{\omega}_Z^0$$

Node m: Link m's absolute acceleration is the result of previous link m – 1's absolute added to joint's Z relative angular acceleration:

$$\boldsymbol{\varepsilon}_m = \boldsymbol{\varepsilon}_{m-1} + \left(\boldsymbol{\varepsilon}_Z^0 + \boldsymbol{\varepsilon}_Z^{com}\right) \tag{4.17}$$

Therefore, along the path, Equation (4.18) holds.

Statement 4.1

The node's absolute acceleration is determined by the addition of all relative accelerations at edges located in the path before the node m, which is shown in Equation (4.18):

$$\boldsymbol{\varepsilon}_m = \left(\boldsymbol{\varepsilon}_A^0 + \boldsymbol{\varepsilon}_A^{com}\right)\Big|_A + \left(\boldsymbol{\varepsilon}_B^0 + \boldsymbol{\varepsilon}_B^{com}\right)\Big|_B + \left(\boldsymbol{\varepsilon}_C^0 + \boldsymbol{\varepsilon}_C^{com}\right)\Big|_C + \cdots + \left(\boldsymbol{\varepsilon}_Z^0 + \boldsymbol{\varepsilon}_Z^{com}\right)\Big|_Z \tag{4.18}$$

Equation (4.11) is the *governing equation for links' absolute angular accelerations* for spatial open cycle mechanisms.

Thus, to determine $\boldsymbol{\varepsilon}_1$, the term on the left-hand side of bracket $|_A$ is considered, whereas to determine $\boldsymbol{\varepsilon}_2$, all the terms on the left-hand side of bracket $|_B$ are considered.

Example 1: The spatial TRRT mechanism with four joints: translation (prismatic) T, revolute R, revolute R, and translation (prismatic) T has M = 4 degrees of freedom (DOF), the actuators' velocities are considered constant in time; therefore, the actuation accelerations—named as joint relative accelerations—are shown in Equation (4.19a):

$$\ddot{\mathbf{q}}_M(t) = \left\{ \begin{array}{cccc} \ddot{d}_A & \ddot{\theta}_B & \ddot{\theta}_C & \ddot{d}_D \end{array} \right\}^T = \left\{ \begin{array}{cccc} 0 & 0 & 0 & 0 \end{array} \right\}^T \tag{4.19a}$$

The links' absolute angular accelerations are determined using the digraph shown in Figure 4.2.

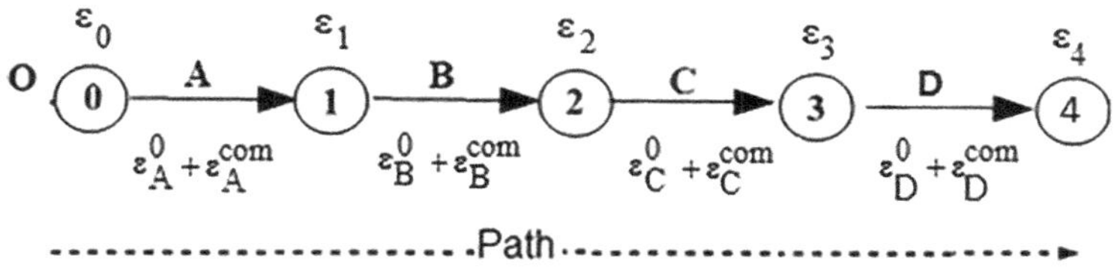

FIGURE 4.2
Relation between absolute and relative angular accelerations for TRRT mechanism.

- *Edge A* (prismatic joint): The sum of two components, axial and complemental relative accelerations, which is shown in Equation (4.10):

$$\left(\boldsymbol{\varepsilon}_A^0 + \boldsymbol{\varepsilon}_A^{com}\right) = (\mathbf{0} + \mathbf{0}) = 0 \tag{4.19b}$$

  where
  - *Axial relative*: $\varepsilon_A^0 = 0$, considering $\ddot{\theta}_A = 0$—the constraint equation at prismatic joint A, which is shown in Equation (4.4).
  - *Complemental relative*: $\boldsymbol{\varepsilon}_A^{com} = 0$, considering Equation (4.8).
- *Node 1*: The absolute angular acceleration of link 1 is determined using Equation (4.18), with consideration of the left-hand side of bracket $|_A$:

$$\boldsymbol{\varepsilon}_1 = \left(\boldsymbol{\varepsilon}_A^0 + \boldsymbol{\varepsilon}_A^{com}\right) = 0 \tag{4.19c}$$

- *Edge B* (revolute joint): The sum of two components, axial and complemental relative accelerations, which is shown in Equation (4.10):

$$\left(\boldsymbol{\varepsilon}_B^0 + \boldsymbol{\varepsilon}_B^{com}\right) = \left(\boldsymbol{\varepsilon}_B^0 + 0\right) = \begin{Bmatrix} 0 \\ \ddot{\theta}_B \\ 0 \end{Bmatrix} \tag{4.19d}$$

  where
  - *Axial relative*: $\boldsymbol{\varepsilon}_B^0 = \left(\mathbf{D}_2 * \mathbf{u}_B^2\right) \cdot \ddot{\theta}_B = \begin{Bmatrix} 0 \\ \ddot{\theta}_B \\ 0 \end{Bmatrix}$

$$\text{where } \mathbf{D}_2 * \mathbf{u}_B^2 = \begin{bmatrix} C\theta_B & 0 & S\theta_B \\ 0 & 1 & 0 \\ -S\theta_B & 0 & C\theta_B \end{bmatrix} * \begin{Bmatrix} 0 \\ 1 \\ 0 \end{Bmatrix} = \begin{Bmatrix} 0 \\ 1 \\ 0 \end{Bmatrix}$$

- *Complemental relative*: $\boldsymbol{\varepsilon}_B^{com} = \tilde{\boldsymbol{\omega}}_1 * \boldsymbol{\omega}_B^0 = 0_{3\times3} * \begin{Bmatrix} 0 \\ \dot{\theta}_B \\ 0 \end{Bmatrix} = \mathbf{0}$

The angular velocity $\omega_1 = \begin{Bmatrix} 0 \\ 0 \\ 0 \end{Bmatrix}$ is considered, with its skew-symmetric matrices $\tilde{\omega}_1 = \begin{bmatrix} 0 & -0 & 0 \\ 0 & 0 & -0 \\ -0 & 0 & 0 \end{bmatrix} = \mathbf{0}_{3\times3}$ and

$$\boldsymbol{\omega}_B^0 = \begin{Bmatrix} 0 \\ \dot{\theta}_B \\ 0 \end{Bmatrix}$$

- *Node 2*: The absolute angular acceleration of link 2 is determined using Equation (4.18), with consideration of the left-hand side of bracket $|_B$:

$$\boldsymbol{\varepsilon}_2 = \boldsymbol{\varepsilon}_1 + \left(\boldsymbol{\varepsilon}_B^0 + \boldsymbol{\varepsilon}_B^{com}\right) = 0 + \left(\boldsymbol{\varepsilon}_B^0 + 0\right) = \begin{Bmatrix} 0 \\ \ddot{\theta}_B \\ 0 \end{Bmatrix} \tag{4.19e}$$

- *Edge C* (revolute joint): The sum of two components, axial and complemental relative accelerations, which is shown in Equation (4.10):

$$\left(\boldsymbol{\varepsilon}_C^0 + \boldsymbol{\varepsilon}_C^{com}\right) = \begin{Bmatrix} \ddot{\theta}_C \cdot S\theta_B + \dot{\theta}_B \cdot \dot{\theta}_C \cdot C\theta_B \\ 0 \\ \ddot{\theta}_C C\theta_B - \dot{\theta}_B \cdot \dot{\theta}_C \cdot S\theta_B \end{Bmatrix} \tag{4.19f}$$

where

- *Axial relative*: $\boldsymbol{\varepsilon}_C^0 = \left(\mathbf{D}_3 * \mathbf{u}_C^3\right) \cdot \ddot{\theta}_C = \begin{Bmatrix} \ddot{\theta}_C \cdot S\theta_B \\ 0 \\ \ddot{\theta}_C \cdot C\theta_B \end{Bmatrix}$

where

$$\mathbf{D}_3 * \mathbf{u}_C^3 = \left[\begin{array}{c:c:c} C\theta_B \cdot C\theta_C & -C\theta_B \cdot S\theta_C & S\theta_B \\ S\theta_C & C\theta_C & 0 \\ -S\theta_B \cdot C\theta_C & S\theta_B \cdot S\theta_C & C\theta_B \end{array}\right] * \begin{Bmatrix} 0 \\ 0 \\ 1 \end{Bmatrix} = \begin{Bmatrix} S\theta_B \\ 0 \\ C\theta_B \end{Bmatrix}$$

- *Complemental relative*: $\boldsymbol{\varepsilon}_C^{com} = \tilde{\boldsymbol{\omega}}_2 * \boldsymbol{\omega}_C^0 = \begin{Bmatrix} \dot{\theta}_B \cdot \dot{\theta}_C \cdot C\theta_B \\ 0 \\ -\dot{\theta}_B \cdot \dot{\theta}_C \cdot S\theta_B \end{Bmatrix}$

  The angular velocity $\boldsymbol{\omega}_2 = \begin{Bmatrix} 0 \\ \dot{\theta}_B \\ 0 \end{Bmatrix}$, is considered with its skew-symmetric matrices $\tilde{\boldsymbol{\omega}}_2 = \begin{bmatrix} 0 & -0 & \dot{\theta}_B \\ 0 & 0 & -0 \\ -\dot{\theta}_B & 0 & 0 \end{bmatrix}$ and $\boldsymbol{\omega}_C^0 = \begin{Bmatrix} \dot{\theta}_C \cdot S\theta_B \\ 0 \\ \dot{\theta}_C \cdot C\theta_B \end{Bmatrix}$

- *Node 3*: The absolute angular acceleration of link 3 is determined using Equation (4.18), with consideration of the left-hand side of bracket $|_C$:

$$\boldsymbol{\varepsilon}_3 = \boldsymbol{\varepsilon}_2 + \left(\boldsymbol{\varepsilon}_C^0 + \boldsymbol{\varepsilon}_C^{com}\right) = \begin{Bmatrix} \ddot{\theta}_C \cdot S\theta_B + \dot{\theta}_B \cdot \dot{\theta}_C \cdot C\theta_B \\ \dot{\theta}_B \\ \ddot{\theta}_C \cdot C\theta_B - \dot{\theta}_B \cdot \dot{\theta}_C \cdot S\theta_B \end{Bmatrix} \tag{4.19g}$$

- *Edge D* (prismatic joint): The sum of two components, axial and complemental relative accelerations, which is shown in Equation (4.10):

$$\left(\boldsymbol{\varepsilon}_D^0 + \boldsymbol{\varepsilon}_D^{com}\right) = (\mathbf{0} + \mathbf{0}) = \mathbf{0} \tag{4.19h}$$

  where
  - *Axial relative*: $\boldsymbol{\varepsilon}_D^0$, considering $\ddot{\theta}_D = 0$, is the constraint equation at the prismatic joint D, which is shown in Equation (4.4)
  - *Complemental relative*: $\boldsymbol{\varepsilon}_D^{com}$, considering Equation (4.8)
- *Node 4*: The absolute angular acceleration of link 4 is determined using Equation (4.18), with consideration of the left-hand side of bracket $|_D$:

$$\boldsymbol{\varepsilon}_4 = \boldsymbol{\varepsilon}_3 + \left(\boldsymbol{\varepsilon}_D^0 + \boldsymbol{\varepsilon}_D^{com}\right) = \boldsymbol{\varepsilon}_3 + (\mathbf{0} + \mathbf{0}) = \begin{Bmatrix} \ddot{\theta}_C \cdot S\theta_B + \dot{\theta}_B \cdot \dot{\theta}_C \cdot C\theta_B \\ \ddot{\theta}_B \\ \ddot{\theta}_C C\theta_B - \dot{\theta}_B \cdot \dot{\theta}_C \cdot S\theta_B \end{Bmatrix} \tag{4.19i}$$

Note: For direct analysis, the angular accelerations for actuated revolute joints B and C are determined from the derivative with respect to time in Equation (4.178d).

4.3 Matroid Method for Inverse Angular Acceleration Analysis on Closed Cycle Mechanisms

The aforementioned direct analysis finds the end-effector angular acceleration based on angular accelerations at joints. The inverse analysis finds the actuating angular accelerations to move the end-effector in a desired angular acceleration task, which is required in the manipulator's control.

In the closed path, as shown in Figure 2.10a, b, the cut-edge is oriented by the same rule as all the other edges in digraph, that is, from the lower node number 0 to the higher node number m. The cut-edge for an open cycle is associated with a *fictitious joint* at E between node 0 and node m.

4.3.1 Cycle Basis Matrix Assigned to Relative Angular Accelerations

As shown in Figure 2.10a, a closed path (cycle) is generated when a c = 1 cut-edge is added to a given spanning tree. The direction of the cycle is assigned by the inverse cut-edge's direction, that is, starting from its tail. A superscript "–" sign is assigned to an edge, which is contrary to the direction of the cycle, and a superscript "+" sign is assigned to an edge if the edge is in the path's direction. The set of edges on the cycles is as follows: $C_E = \{A^+, B^+, C^+, \ldots, Y^+, Z^+, E^-\}$.

The cycle matroid is the basis incidence matrix, C, and was defined as a c × j matrix in which one row is assigned to a single cycle and each column is assigned to an edge, which is shown below:

$$\begin{array}{c} \text{Edges} \\ \begin{array}{ccccccc} A & B & C & & Y & Z & E \end{array} \end{array}$$

$$C = \begin{bmatrix} +1 & +1 & +1 & \ldots & +1 & +1 & -1 \end{bmatrix} \quad C_E \text{ Cycle}$$

4.3.2 The Relative Angular Acceleration Matrix, ε_j

$$\begin{array}{cccc} A & B & C & E \end{array}$$

$$\varepsilon_j = \left[+1\cdot\left(\varepsilon_A^0 + \varepsilon_A^{com}\right) + 1\cdot\left(\varepsilon_B^0 + \varepsilon_B^{com}\right) + 1\cdot\left(\varepsilon_C^0 + \varepsilon_C^{com}\right) \ldots - 1\cdot\left(\varepsilon_E^0 + \varepsilon_E^{com}\right)\right] \tag{4.20}$$

Equation (4.9) is written as follows:
$\boldsymbol{\varepsilon}_m - \boldsymbol{\varepsilon}_{m-1} = \boldsymbol{\varepsilon}_Z^0 + \boldsymbol{\varepsilon}_Z^{com}$, and applied for each entry in Equation (4.20).
Then, matrix $\boldsymbol{\varepsilon}_j$ has the entries shown in Equation (4.21):

$$\varepsilon_j = \left[\varepsilon_1 - \varepsilon_0 \quad \varepsilon_2 - \varepsilon_1 \quad \varepsilon_3 - \varepsilon_2 \quad \varepsilon_m - \varepsilon_{m-1} \quad \ldots \quad -(\varepsilon_m - \varepsilon_0) \right] \tag{4.21}$$

One could notice that each vector shows twice with opposite sign; therefore, the sum of entries is a zero vector.

Statement 4.2

The sum of entries on each row in the relative angular acceleration matrix $\boldsymbol{\varepsilon}_j$ is zero.

$$\sum c_Z \cdot \left(\boldsymbol{\varepsilon}_Z^0 + \boldsymbol{\varepsilon}_Z^{com} \right) = 0 \tag{4.22}$$

The scalar form of Equation (4.22) is equivalent to a system of three equations, solved for joint angular accelerations. For a desired task, the components in the absolute angular acceleration of end-effector are known.

$$\boldsymbol{\varepsilon}_E^{task} = \left\{ \boldsymbol{\varepsilon}_E^x \ \ \boldsymbol{\varepsilon}_E^y \ \ \boldsymbol{\varepsilon}_E^z \right\}^T \tag{4.23}$$

Example 1: Let us consider the TRRT spatial mechanism, Figure 2.1, where the task absolute angular acceleration is provided. From the inverse angular acceleration analysis, the relative angular accelerations at joints are determined, which are as follows:
Data:

- The task absolute angular velocity $\boldsymbol{\varepsilon}_E^{task} = \left\{ \boldsymbol{\varepsilon}_E^x \ \ \boldsymbol{\varepsilon}_E^y \ \ \boldsymbol{\varepsilon}_E^z \right\}^T$
- The constraint equations for the prismatic joints A and D: $\ddot{\theta}_A = \ddot{\theta}_D = 0$.

Solution:

- The relative angular velocities at joints are as follows:

$$\begin{aligned} &\varepsilon_1 - \varepsilon_0 = \begin{Bmatrix} 0 \\ 0 \\ 0 \end{Bmatrix}; \ \varepsilon_2 - \varepsilon_1 = \begin{Bmatrix} 0 \\ \ddot{\theta}_B \\ 0 \end{Bmatrix} \\ &\varepsilon_3 - \varepsilon_2 = \begin{Bmatrix} \ddot{\theta}_C \cdot S\theta_B + \dot{\theta}_B \cdot \dot{\theta}_C \cdot C\theta_B \\ 0 \\ \ddot{\theta}_C \cdot C\theta_B - \dot{\theta}_B \cdot \dot{\theta}_C \cdot S\theta_B \end{Bmatrix}; \ \varepsilon_4 - \varepsilon_3 = \begin{Bmatrix} 0 \\ 0 \\ 0 \end{Bmatrix} \end{aligned} \tag{4.24}$$

- The relative angular velocity matrix $\boldsymbol{\varepsilon}_j$ is as follows:

$$\begin{array}{ccccc} A & B & C & D & E \end{array}$$
$$\boldsymbol{\varepsilon}_j = \left[(\boldsymbol{\varepsilon}_1 - \boldsymbol{\varepsilon}_0) \quad (\boldsymbol{\varepsilon}_2 - \boldsymbol{\varepsilon}_1) \quad (\boldsymbol{\varepsilon}_3 - \boldsymbol{\varepsilon}_2) \quad (\boldsymbol{\varepsilon}_4 - \boldsymbol{\varepsilon}_3) \quad -(\boldsymbol{\varepsilon}_4 - \boldsymbol{\varepsilon}_0)\right]$$

- Equations from sum on entries being zero, written for each row, are as follows:

$$\begin{aligned} &\textit{Equation 1: } \ddot{\theta}_C \cdot S\theta_B + \dot{\theta}_B \cdot \dot{\theta}_C \cdot C\theta_B = \boldsymbol{\varepsilon}_E^x \\ &\textit{Equation 2: } \ddot{\theta}_B = \boldsymbol{\varepsilon}_E^y \\ &\textit{Equation 3: } \ddot{\theta}_C C\theta_B - \dot{\theta}_B \cdot \dot{\theta}_C \cdot S\theta_B = \boldsymbol{\varepsilon}_E^z \end{aligned} \tag{4.25}$$

The aforementioned set of three equations is added to the set of three equations from a linear acceleration analysis, which is discussed further in Section 4.4.3.

4.4 Governing Equation for Links' Absolute Linear Accelerations

- **The acceleration matroidal vector**

  This vector, noted $\ddot{\mathbf{L}}_{YZ}^0$, is defined as the difference between two frame origin's linear accelerations and noted with an upper double dot as

$$\ddot{\mathbf{L}}_{YZ}^0 = \mathbf{a}_{Z_m}^0 - \mathbf{a}_{Y_{m-1}}^0 \tag{4.26}$$

  where

  - m = 1, 2, … , is the mobile link mechanism.
  - $\mathbf{a}_{Z_m}^0$ and $\mathbf{a}_{Y_{m-1}}^0$ are the absolute linear accelerations of link m – 1's origin Y and link m's origin Z, measured in [ft/s²] or [m/s²].

The left-hand-side vector is assigned to an edge in digraph, Figure 4.3, and has two components according to the two vector equations from the solid rigid kinematics:

- The relation between accelerations of two points located on link m – 1 is as follows:

$$\mathbf{a}_{Z_{m-1}}^0 - \mathbf{a}_{Y_{m-1}}^0 = \mathbf{a}_{m-1}^\tau + \mathbf{a}_{m-1}^\nu \tag{4.27}$$

- The relation between velocities of two points located on different links m – 1 and m, joined at Z, noted *complemental linear acceleration*, is as follows:

$$\mathbf{a}^0_{Z_m} - \mathbf{a}^0_{Z_{m-1}} = \mathbf{a}^{com,0}_{Z} \tag{4.28}$$

By adding the terms on the left-hand and right-hand side in Equations (4.27) and (4.28), the acceleration matroidal vector is as follows:

$$\ddot{\mathbf{L}}^0_{YZ} = \mathbf{a}^0_{Z_m} - \mathbf{a}^0_{Y_{m-1}} = \mathbf{a}^{\tau}_{YZ} + \mathbf{a}^{\nu}_{YZ} + \mathbf{a}^{com,0}_{Z} \tag{4.29}$$

The three components in the right-hand side in matrix form are:

- The tangential acceleration is

$$\mathbf{a}^{\tau}_{YZ} = \boldsymbol{\varepsilon}^0_{m-1} \times \mathbf{L}^0_{YZ} = -\tilde{\mathbf{L}}^0_{YZ} * \boldsymbol{\varepsilon}^0_{m-1} \tag{4.30}$$

  where $\boldsymbol{\varepsilon}^0_{m-1}$ is the absolute angular velocity for link m – 1 expressed in the fixed frame 0, and measured in [rad/s²] or [°/s²]. The symbol "~" denotes a skew-symmetric matrix, and the symbol * represents a multiplication between a matrix and a column vector.

- The normal (centripetal) acceleration is shown in Equation (4.31):

$$\mathbf{a}^{\nu}_{YZ} = \boldsymbol{\omega}^0_{m-1} \times \boldsymbol{\omega}^0_{m-1} \times \mathbf{L}^0_{YZ} = [\tilde{\boldsymbol{\omega}}^0_{m-1}]^2 * \mathbf{L}^0_{YZ} \tag{4.31}$$

  It is measured in [ft/s²] or [m/s²] and is a vector collinear to $\mathbf{L}^0_{YZ}$ and oriented from Z toward Y.

  Its scalar components are determined, shown in Equation (4.31a):

$$\mathbf{a}^{\nu}_{YZ} = \left[\tilde{\omega}^0_{m-1}\right]^2 * \mathbf{L}^0_{YZ}$$

$$= \begin{bmatrix} -\left[\left(\omega^y_{m-1}\right)^2 + \left(\omega^z_{m-1}\right)^2\right] & \omega^x_{m-1}\cdot\omega^y_{m-1} & \omega^x_{m-1}\cdot\omega^z_{m-1} \\ \omega^y_{m-1}\cdot\omega^x_{m-1} & -\left[\left(\omega^x_{m-1}\right)^2 + \left(\omega^z_{m-1}\right)^2\right] & \omega^y_{m-1}\cdot\omega^z_{m-1} \\ \omega^z_{m-1}\cdot\omega^x_{m-1} & \omega^z_{m-1}\cdot\omega^y_{m-1} & -\left[\left(\omega^x_{m-1}\right)^2 + \left(\omega^y_{m-1}\right)^2\right] \end{bmatrix}$$

$$* \begin{Bmatrix} x^0_{YZ} \\ y^0_{YZ} \\ z^0_{YZ} \end{Bmatrix} \tag{4.31a}$$

- *The complemental linear acceleration* vector at the prismatic joint Z is determined, shown in Equation (4.32):

$$\mathbf{a}_Z^{com,0} = \ddot{\mathbf{d}}_Z^0 + 2 \cdot \boldsymbol{\omega}_{m-1}^0 \times \dot{\mathbf{d}}_Z^0 = \mathbf{u}_Z^0 \cdot \ddot{d}_Z - 2 \cdot \left(\tilde{\mathbf{u}}_Z^0 \dot{d}_Z^0\right) * \boldsymbol{\omega}_{m-1}^0 \tag{4.32}$$

The two components of complemental acceleration are as follows:

- The *relative linear acceleration at sliding joint Z*, for the link m sliding relative to the link m − 1, which is measured in [ft/s²] or [m/s²], is shown in Equation (4.33):

$$\ddot{\mathbf{d}}_Z^0 = \mathbf{u}_Z^0 \cdot \ddot{d}_Z \tag{4.33}$$

For the direct acceleration analysis, the scalar $\ddot{\mathbf{d}}_Z^0$ is the joint's rate of change in time of linear velocity. It is determined from the derivative with respect to time from linear velocity $\dot{d}_Z(t)$.

Note: In case the sliding direction is curvilinear, then the relative linear acceleration at the sliding joint Z has two components: the tangential component $\ddot{\mathbf{d}}_Z^{\tau} = \mathbf{u}_Z^0 \cdot \ddot{d}_Z$ and the normal component $\ddot{\mathbf{d}}_Z^{\upsilon} = \frac{n_Z^0 \cdot \dot{d}_Z^2}{\rho}$, where the vector $\mathbf{n}_Z^0$ is the unit vector of the normal direction and ρ is the radius of curvature.

- *Coriolis acceleration* at the prismatic joint Z is a vector perpendicular to the plane determined by the two vectors in the cross product. It is measured in [ft/s²] or [m/s²] and is shown in Equation (4.34):

$$\mathbf{a}_Z^{co} = 2 \cdot \boldsymbol{\omega}_{m-1}^0 \times \dot{\mathbf{d}}_Z^0 = -\tilde{\mathbf{u}}_Z^0 * \left(2 \cdot \dot{d}_Z^0 \cdot \boldsymbol{\omega}_{m-1}^0\right) \tag{4.34}$$

Its scalar components are determined, shown in Equation (4.34a):

$$\mathbf{a}_Z^{co} = -\begin{bmatrix} 0 & -u_Z^{z_0} & u_Z^{y_0} \\ u_Z^{z_0} & 0 & -u_Z^{x_0} \\ -u_Z^{y_0} & u_Z^{x_0} & 0 \end{bmatrix} * \begin{Bmatrix} \boldsymbol{\omega}_{m-1}^x \\ \boldsymbol{\omega}_{m-1}^y \\ \boldsymbol{\omega}_{m-1}^z \end{Bmatrix} \cdot 2 \cdot \dot{d}_Z^0$$

$$= \begin{Bmatrix} \boldsymbol{\omega}_{m-1}^y \cdot u_Z^z - \boldsymbol{\omega}_{m-1}^z \cdot u_Z^y \\ \boldsymbol{\omega}_{m-1}^z \cdot u_Z^x - \boldsymbol{\omega}_{m-1}^x \cdot u_Z^z \\ \boldsymbol{\omega}_{m-1}^x \cdot u_Z^y - \boldsymbol{\omega}_{m-1}^y \cdot u_Z^x \end{Bmatrix} \cdot 2 \cdot \dot{d}_Z^0 \tag{4.34a}$$

where

$\dot{\mathbf{d}}_Z^0 = \mathbf{u}_Z^0 \cdot \dot{d}_Z$ is the relative linear velocity at the joint Z, with respect to the fixed frame (4.34b)

$$\mathbf{u}_Z^0 = \mathbf{D}_m * \mathbf{u}_Z^m \quad \text{is the unit vector for relative linear velocity at the joint Z} \tag{4.34c}$$

$\mathbf{u}_Z^m$ is the unit vector for joint Z's axis of rotation. The relative velocity initially expressed in the local frame m and in the frame 0 by the left-hand-side multiplication with absolute rotation matrix

$\tilde{\mathbf{u}}_Z^0$ is the skew-symmetric matrix for the unit vector $\mathbf{u}_Z^0$.

Note:

- Since the complemental linear acceleration is assigned to an oriented edge in digraph, it shows with a "+" or a "−" for a direct or inverse direction to the path's orientation, respectively
- If a link is connected by the prismatic joint to the fixed link, the Coriolis acceleration is zero (since the angular velocity $\boldsymbol{\omega}_{m-1}^0$ is zero):

$$\mathbf{a}_Z^{co} = 0 \tag{4.34d}$$

- The complemental linear acceleration is zero for revolute joints (since the linear velocity $\dot{d}_Z^0$ is zero):

$$\mathbf{a}_Z^{com,0} = 0 \tag{4.34e}$$

Thus, for a link connected by two revolute joints at Y and Z, the Latin vector $\mathbf{L}_{YZ}^0$ has a constant magnitude, and the acceleration matroidal vector from Equation (4.29) becomes

$$\ddot{\mathbf{L}}_{YZ}^0 = \mathbf{a}_{Z_m}^0 - \mathbf{a}_{Y_{m-1}}^0 = \mathbf{a}_{YZ}^{\tau} + \mathbf{a}_{YZ}^{v} \tag{4.34f}$$

Notation:

Once defined on digraph for their relative meaning, the subscripts do not show in vector notations.

Thus, for $\mathbf{a}_{Y_{m-1}}^0$ and $\mathbf{a}_{Z_m}^0$, the subscripts m − 1 and m are dropped. Since they are absolute vectors, with respect to the fixed frame 0, the upper script is omitted.

4.4.1 Direct Linear Acceleration Analysis for Open Cycle Spatial Mechanisms

- **Joint Linear Accelerations**

  The link m frame origin's linear acceleration is the result of previous link m − 1 frame origin's linear acceleration added to the joint's Z acceleration matroidal vector, shown in the recursion Equation (4.35):

$$\mathbf{a}_Z = \mathbf{a}_Y + \ddot{\mathbf{L}}_{YZ}^0 \tag{4.35}$$

As previously developed in Chapter 2 for velocities, Equation (2.201), a recursive procedure for m = 1, 2, 3, …, m based on Equation (4.35) leads to the following conclusion.

Statement 4.3

The link m (node m-digraph) frame origin's linear acceleration is the result of addition of all acceleration matroidal vectors of edges located on the paths initiated at node 0 and which end at node m, as shown in Equation (4.36). For the complemental accelerations, the "+/–" edge signs are assigned:

$$\mathbf{a}_Z = \ddot{\mathbf{L}}^0_{OA}\Big|_A + \ddot{\mathbf{L}}^0_{AB}\Big|_B + \ddot{\mathbf{L}}^0_{BC}\Big|_C + \cdots + \ddot{\mathbf{L}}^0_{YZ}\Big|_Z \tag{4.36}$$

Thus, to determine $\mathbf{a}_A$, the term on the left-hand side of bracket $|_A$ is considered, whereas to determine $\mathbf{a}_B$, all the terms on the left-hand side of bracket $|_B$ are considered.

- **Center of Mass Acceleration**
  The center of mass (COM) acceleration for the link m − 1 is derived from Equation (4.35) for $\ddot{\mathbf{d}}_{G_{m-1}} = 0$, $\dot{\mathbf{d}}_{G_{m-1}} = 0$, since the points Y and G_{m-1} are located on the link m − 1 and connected by the constant magnitude vector $\mathbf{L}^0_{YG_{m-1}}$. Therefore:

$$\mathbf{a}_{G_{m-1}} = \mathbf{a}_Y + \ddot{\mathbf{L}}^0_{YG_{m-1}} \tag{4.37}$$

where

$$\ddot{\mathbf{L}}^0_{YG_{m-1}} = \mathbf{a}^{\tau}_{YG_{m-1}} + \mathbf{a}^{v}_{YG_{m-1}} = -\tilde{\mathbf{L}}^0_{YG_{m-1}} * \boldsymbol{\varepsilon}_{m-1} + \left[\tilde{\boldsymbol{\omega}}^0_{m-1}\right]^2 * \mathbf{L}^0_{YG_{m-1}} \tag{4.37a}$$

4.4.2 Example of Direct Linear Acceleration Analysis for Open Cycle TRRT Spatial Mechanism with 4 DOF

The acceleration analysis of spatial TRRT, Figure 2.1, mechanism is a continuation from velocity analysis developed in Chapter 2, Section 2.17.13. The M = 4 actuating joints are as follows: A (prism), B (rev), C (rev), and D (prism).

- *Joint constraints*: The linear relative accelerations for all joints (arcs a-digraph) are shown in Equation (4.38a):

$$\ddot{\mathbf{d}}_a = \left\{ \ddot{\mathbf{d}}_A \,\vdots\, \ddot{\mathbf{d}}_B \,\vdots\, \ddot{\mathbf{d}}_C \,\vdots\, \ddot{\mathbf{d}}_D \right\}^T = \left\{ \mathbf{u}^0_A \cdot \ddot{d}_A \,\vdots\, \mathbf{u}^0_B \cdot \ddot{d}_B \,\vdots\, \mathbf{u}^0_C \cdot \ddot{d}_C \,\vdots\, \mathbf{u}^0_D \cdot \ddot{d}_D \right\}^T$$

$$= \left\{ 0 \,\vdots\, 0 \,\vdots\, 0 \,\vdots\, 0 \right\}^T \tag{4.38a}$$

Since the joints B and C are revolute joints, they are constrained for linear displacement: $\ddot{d}_B = \ddot{d}_C = 0$ and for the joints A and D, they are constrained for linear velocity: $\ddot{d}_A = \ddot{d}_D = 0$.

The relative angular accelerations for all joints (arcs a-digraph) are shown in Equation (4.38b):

$$\ddot{\theta}_a = \left\{ \ddot{\theta}_A \;\vdots\; \ddot{\theta}_B \;\vdots\; \ddot{\theta}_C \;\vdots\; \ddot{\theta}_D \right\}^T = \left\{ \mathbf{u}_A^0 \cdot \ddot{\theta}_A \;\vdots\; \mathbf{u}_B^0 \cdot \ddot{\theta}_B \;\vdots\; \mathbf{u}_C^0 \cdot \ddot{\theta}_C \;\vdots\; \mathbf{u}_D^0 \cdot \ddot{\theta}_D \right\}^T$$

$$= \left\{ 0 \;\vdots\; 0 \;\vdots\; 0 \;\vdots\; 0 \right\}^T \tag{4.38b}$$

If the joints A and D are the prismatic joints, the angular displacement constraints are $\ddot{\theta}_A = \ddot{\theta}_D = 0$, and in this example, the angular velocities at the joints B and C are constant, then $\ddot{\theta}_B = \ddot{\theta}_C = 0$.

We consider the paths in Figure 4.3, where the acceleration matroidal vectors are shown at edges and the absolute linear accelerations are shown at nodes. Then, the absolute accelerations are determined as the sum of acceleration matroidal vectors in the joints situated within the path originated from the fixed link 0.

- The path matrix is shown in Equation (4.38c):

$$\mathbf{Z} = \begin{array}{c} \\ A \\ B \\ C \\ D \end{array} \begin{array}{c} \begin{array}{cccc} 1 & 2 & 3 & 4 \end{array} \\ \begin{bmatrix} +1 & +1 & +1 & +1 \\ 0 & +1 & +1 & +1 \\ 0 & 0 & +1 & +1 \\ 0 & 0 & 0 & +1 \end{bmatrix} \end{array}; \quad \mathbf{Z}^T = \begin{bmatrix} 1 & 0 & 0 & 0 \\ 1 & 1 & 0 & 0 \\ 1 & 1 & 1 & 0 \\ 1 & 1 & 1 & 1 \end{bmatrix} \tag{4.38c}$$

Equation (4.36) holds:

$$\left\{ \begin{array}{c} \mathbf{a}_A \\ \mathbf{a}_B \\ \mathbf{a}_C \\ \mathbf{a}_D \end{array} \right\} = \begin{array}{c} 1 \\ 2 \\ 3 \\ 4 \end{array} \begin{array}{c} \begin{array}{cccc} A & B & C & D \end{array} \\ \begin{bmatrix} 1 & 0 & 0 & 0 \\ 1 & 1 & 0 & 0 \\ 1 & 1 & 1 & 0 \\ 1 & 1 & 1 & 1 \end{bmatrix} \end{array} * \left\{ \begin{array}{c} \ddot{\mathbf{L}}_{OA} \\ \ddot{\mathbf{L}}_{AB} \\ \ddot{\mathbf{L}}_{BC} \\ \ddot{\mathbf{L}}_{CD} \end{array} \right\} \tag{4.38d}$$

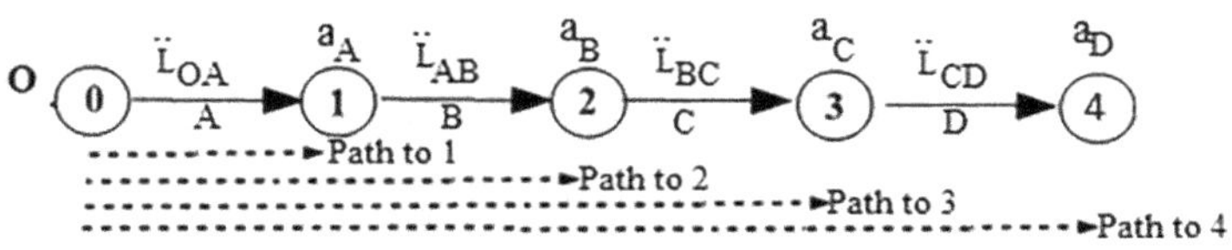

FIGURE 4.3
Acceleration matroidal vectors along the tree for TRRT mechanism.

The result of matrix multiplications in Equation (4.38d) is

$$\begin{Bmatrix} \mathbf{a}_A \\ \mathbf{a}_B \\ \mathbf{a}_C \\ \mathbf{a}_D \end{Bmatrix} = \begin{Bmatrix} \ddot{\mathbf{L}}_{OA} \\ \ddot{\mathbf{L}}_{OA} + \ddot{\mathbf{L}}_{AB} \\ \ddot{\mathbf{L}}_{OA} + \ddot{\mathbf{L}}_{AB} + \ddot{\mathbf{L}}_{BC} \\ \ddot{\mathbf{L}}_{OA} + \ddot{\mathbf{L}}_{AB} + \ddot{\mathbf{L}}_{BC} + \ddot{\mathbf{L}}_{CD} \end{Bmatrix} = \begin{Bmatrix} \mathbf{a}_O + \ddot{\mathbf{L}}_{OA} \\ \mathbf{a}_A + \ddot{\mathbf{L}}_{AB} \\ \mathbf{a}_B + \ddot{\mathbf{L}}_{BC} \\ \mathbf{a}_C + \ddot{\mathbf{L}}_{CD} \end{Bmatrix} \tag{4.38e}$$

Each row in Equation (4.38e) determines the links' absolute angular acceleration as a function of joints' relative accelerations in the joints situated within the path originated from the fixed link 0.

where

$$\begin{Bmatrix} \ddot{\mathbf{L}}_{OA} \\ \ddot{\mathbf{L}}_{AB} \\ \ddot{\mathbf{L}}_{BC} \\ \ddot{\mathbf{L}}_{CD} \end{Bmatrix} = \begin{Bmatrix} -\tilde{\mathbf{L}}^0_{OA} * \boldsymbol{\varepsilon}^0_0 + \left[\tilde{\boldsymbol{\omega}}^0_0\right]^2 * \mathbf{L}^0_{OA} + (\mathbf{u}^0_A \cdot \ddot{d}_A - 2\,\tilde{\mathbf{u}}^0_A \cdot \dot{d}^0_A * \boldsymbol{\omega}^0_0 \\ -\tilde{\mathbf{L}}^0_{AB} * \boldsymbol{\varepsilon}^0_1 + \left[\tilde{\boldsymbol{\omega}}^0_1\right]^2 * \mathbf{L}^0_{AB} \\ -\tilde{\mathbf{L}}^0_{BC} * \boldsymbol{\varepsilon}^0_2 + \left[\tilde{\boldsymbol{\omega}}^0_2\right]^2 * \mathbf{L}^0_{BC} \\ -\tilde{\mathbf{L}}^0_{CD} * \boldsymbol{\varepsilon}^0_3 + \left[\tilde{\boldsymbol{\omega}}^0_3\right]^2 * \mathbf{L}^0_{CD} + \left(\mathbf{u}^0_D \cdot \ddot{d}_D - 2 \cdot \tilde{\mathbf{u}}^0_D \cdot d^0_D * \boldsymbol{\omega}^0_3\right) \end{Bmatrix} \tag{4.38f}$$

- *The links' constraints*: $\varepsilon^0_0 = \varepsilon^0_1 = 0$;

The motion constraints for joints and the links are applied in Equation (4.38f), which become

$$\begin{Bmatrix} \ddot{\mathbf{L}}_{OA} \\ \ddot{\mathbf{L}}_{AB} \\ \ddot{\mathbf{L}}_{BC} \\ \ddot{\mathbf{L}}_{CD} \end{Bmatrix} = \begin{Bmatrix} 0 \\ 0 \\ \left[\tilde{\boldsymbol{\omega}}^0_2\right]^2 * \mathbf{L}^0_{BC} \\ -\tilde{\mathbf{L}}^0_{CD} * \boldsymbol{\varepsilon}^0_3 + \left[\tilde{\boldsymbol{\omega}}^0_3\right]^2 * \mathbf{L}^0_{CD} - 2 \cdot \left(\tilde{\mathbf{u}}^0_D \cdot \dot{d}^0_D\right) * \boldsymbol{\omega}^0_3 \end{Bmatrix} \tag{4.38g}$$

- *Joints' relative linear accelerations*:
  - The unit vectors for the actuated prismatic joints A and D are as follows:

$$\mathbf{u}^0_A = \begin{Bmatrix} 1 \\ 0 \\ 0 \end{Bmatrix}; \quad \mathbf{u}^0_D = \begin{Bmatrix} C\theta_B \cdot C\theta_C \\ S\theta_C \\ -S\theta_B \cdot C\theta_C \end{Bmatrix} \tag{4.38h}$$

  - Considering the frame origin's position vectors from Chapter 2, Equation (2.207), the Latin vectors are as follows:

$$\mathbf{L}_{BC}^{0} = \begin{Bmatrix} x_{BC} \\ y_{BC} \\ z_{BC} \end{Bmatrix} = \begin{Bmatrix} 0 \\ 25 \\ 0 \end{Bmatrix}; \quad \mathbf{L}_{CD}^{0} = \begin{Bmatrix} x_{CD} \\ y_{CD} \\ z_{CD} \end{Bmatrix}$$

$$= \begin{Bmatrix} (6.5+d_D)\cdot C\theta_B \cdot C\theta_C \\ (6.5+d_D)\cdot S\theta_C \\ -(6.5+d_D)\cdot S\theta_B \cdot C\theta_C \end{Bmatrix}; \quad \mathbf{L}_{DE}^{0} = \begin{Bmatrix} x_{DE} \\ y_{DE} \\ z_{DE} \end{Bmatrix} = \begin{Bmatrix} 38\cdot C\theta_B \cdot C\theta_C \\ 38\cdot S\theta_C \\ -38\cdot S\theta_B \cdot C\theta_C \end{Bmatrix} \tag{4.38i}$$

– Their skew-symmetric matrices are as follows:

$$\tilde{\mathbf{L}}_{BC}^{0} = \begin{bmatrix} 0 & 0 & 25 \\ 0 & 0 & 0 \\ -25 & 0 & 0 \end{bmatrix}; \quad \tilde{\mathbf{L}}_{CD}^{0} = \begin{bmatrix} 0 & -z_{CD} & y_{CD} \\ z_{CD} & 0 & -x_{CD} \\ -y_{CD} & x_{CD} & 0 \end{bmatrix};$$

$$\tilde{\mathbf{L}}_{DE}^{0} = \begin{bmatrix} 0 & -z_{DE} & y_{DE} \\ z_{DE} & 0 & -x_{DE} \\ -y_{DE} & x_{DE} & 0 \end{bmatrix} \tag{4.38j}$$

– Considering the absolute angular velocities from Chapter 2, Equation (2.207j):

$$\boldsymbol{\omega}_2 = \begin{Bmatrix} 0 \\ \dot{\theta}_B \\ 0 \end{Bmatrix}; \quad \boldsymbol{\omega}_3 = \begin{Bmatrix} S\theta_B \cdot \dot{\theta}_C \\ \dot{\theta}_B \\ C\theta_B \cdot \dot{\theta}_C \end{Bmatrix}; \quad \boldsymbol{\omega}_4 = \begin{Bmatrix} S\theta_B \cdot \dot{\theta}_C \\ \theta_B \\ C\theta_B \cdot \dot{\theta}_C \end{Bmatrix} \tag{4.38k}$$

– Considering the absolute angular accelerations from Equation (4.19):

$$\boldsymbol{\varepsilon}_2 = \begin{Bmatrix} 0 \\ \ddot{\theta}_B \\ 0 \end{Bmatrix} = \begin{Bmatrix} 0 \\ 0 \\ 0 \end{Bmatrix}; \quad \boldsymbol{\varepsilon}_3 = \begin{Bmatrix} \ddot{\theta}_C \cdot S\theta_B + \dot{\theta}_B \cdot \dot{\theta}_C \cdot C\theta_B \\ \ddot{\theta}_B \\ \ddot{\theta}_C \cdot C\theta_B - \dot{\theta}_B \cdot \dot{\theta}_C \cdot S\theta_B \end{Bmatrix}$$

$$= \begin{Bmatrix} \dot{\theta}_B \cdot \dot{\theta}_C \cdot C\theta_B \\ 0 \\ -\dot{\theta}_B \cdot \dot{\theta}_C \cdot S\theta_B \end{Bmatrix} \tag{4.38l}$$

- Joint's and end-effector's absolute linear accelerations
  - *Node 0*: The link 0 is fixed, and its absolute linear acceleration is zero:

$$\mathbf{a}_0 = 0 \tag{4.38m}$$

  - *Node 1*: The joint A linear acceleration from row 1 Equation (4.38e):

$$\mathbf{a}_A = \ddot{\mathbf{L}}_{OA} = \begin{Bmatrix} 0 \\ 0 \\ 0 \end{Bmatrix} \tag{4.38n}$$

  - *Node 2*: The joint B linear acceleration from row 2 Equation (4.38e):

$$\mathbf{a}_B = \mathbf{a}_A + \ddot{\mathbf{L}}_{AB}^0 = \begin{Bmatrix} 0 \\ 0 \\ 0 \end{Bmatrix} \tag{4.38o}$$

  - *Node 3*: The joint C linear acceleration from row 3 Equation (4.38e):

$$\mathbf{a}_C = \mathbf{a}_B + \ddot{\mathbf{L}}_{BC}^0 = \begin{Bmatrix} 0 \\ 0 \\ 0 \end{Bmatrix} \tag{4.38p}$$

  - *Node 4*: The joint D linear acceleration from row 4 Equation (4.38e):

$$\mathbf{a}_D = \mathbf{a}_C + \ddot{\mathbf{L}}_{CD}$$

$$= \mathbf{a}_C - \tilde{\mathbf{L}}_{CD}^0 * \boldsymbol{\varepsilon}_3^0 + \left[\tilde{\boldsymbol{\omega}}_3^0\right]^2 * \mathbf{L}_{CD}^0 - 2\cdot\left(\tilde{\mathbf{u}}_D^0 \cdot \dot{\mathbf{d}}_D^0\right) * \boldsymbol{\omega}_3^0 = \begin{Bmatrix} a_D^x \\ a_D^y \\ a_D^z \end{Bmatrix}$$

$$= \begin{Bmatrix} (6.5+d_D)\cdot\left[-C\theta_B\cdot C\theta_C\cdot\left((\dot{\theta}_B)^2+(\dot{\theta}_C)^2\right)+2\cdot S\theta_B\cdot S\theta_C\cdot\dot{\theta}_B\cdot\dot{\theta}_C \right. \\ \left. -2\cdot C\theta_B\cdot S\theta_C\cdot\dot{\theta}_C\cdot\dot{d}_D-2\cdot S\theta_B\cdot C\theta_C\cdot\dot{\theta}_B\cdot\dot{d}_D\right] \\ -(6.5+d_D)\cdot S\theta_C\cdot\left(\dot{\theta}_C\right)^2+2\cdot C\theta_C\cdot\dot{\theta}_C\cdot\dot{d}_D \\ (6.5+d_D)\cdot\left[S\theta_B\cdot C\theta_C\cdot\left((\dot{\theta}_B)^2+(\dot{\theta}_C)^2\right)+2\cdot C\theta_B\cdot S\theta_C\cdot\dot{\theta}_B\cdot\dot{\theta}_C \right. \\ \left. -2\cdot C\theta_B\cdot C\theta_C\cdot\dot{\theta}_B\cdot\dot{d}_D+2\cdot S\theta_B\cdot S\theta_C\cdot\dot{\theta}_C\cdot\dot{d}_D\right] \end{Bmatrix} \tag{4.38q}$$

- End-effector's E linear acceleration is determined from Equation (4.35), applied for the points D and E located on link 4:

$$\mathbf{a}_E = \mathbf{a}_D + \ddot{\mathbf{L}}_{DE}^{0} \tag{4.38r}$$

where

$$\ddot{\mathbf{L}}_{DE}^{0} = \mathbf{a}_4^{\tau} + \mathbf{a}_4^{\nu} = -\tilde{\mathbf{L}}_{DE}^{0} * \varepsilon_4 + \left[\tilde{\omega}_4^0\right]^2 * \mathbf{L}_{DE}^{0} \tag{4.38s}$$

The components of end-effector's linear acceleration are as follows:

$$\begin{Bmatrix} a_E^x \\ a_E^y \\ a_E^z \end{Bmatrix} = \begin{Bmatrix} (44.5+d_D)\cdot\left[-C\theta_B\cdot C\theta_C\cdot\left((\dot{\theta}_B)^2+(\dot{\theta}_C)^2\right)+2\cdot S\theta_B\cdot S\theta_C\cdot\dot{\theta}_B\cdot\dot{\theta}_C \right. \\ \left. -2\cdot C\theta_B\cdot S\theta_C\cdot\dot{\theta}_C\cdot\dot{d}_D-2\cdot S\theta_B\cdot C\theta_C\cdot\dot{\theta}_B\cdot\dot{d}_D\right] \\ -(44.5+d_D)\cdot S\theta_C\cdot\left(\dot{\theta}_C\right)^2+2\cdot C\theta_C\cdot\dot{\theta}_C\cdot\dot{d}_D \\ (44.5+d_D)\cdot\left[S\theta_B\cdot C\theta_C\cdot\left((\dot{\theta}_B)^2+(\dot{\theta}_C)^2\right)+2\cdot C\theta_B\cdot S\theta_C\cdot\dot{\theta}_B\cdot\dot{\theta}_C \right. \\ \left. -2\cdot C\theta_B\cdot C\theta_C\cdot\dot{\theta}_B\cdot\dot{d}_D+2\cdot S\theta_B\cdot S\theta_C\cdot\dot{\theta}_C\cdot\dot{d}_D\right] \end{Bmatrix} \tag{4.38t}$$

For the actuation shown in Figure 2.17, Figure 4.4, illustrates the frame's E acceleration components and its magnitude.

Numerical example: For the spatial TRRT mechanism, end-effector's acceleration components are determined.

Data: The position and velocities at t = 2s, from the previous positional and velocity analyses, are as follows:

$$\theta_B = 45°;\ \theta_C = 45°;\ \dot{\theta}_B = \pi/2\,[\text{rad/s}];\ \dot{\theta}_C = \pi/2\,[\text{rad/s}];$$
$$d_D = 20\,[\text{in.}];\ \dot{d}_D = 0\,[\text{in./s}]$$

Then, the components of acceleration calculated from Equation (4.38t) are as follows:

$$a_E^x = 0\left[\text{in./s}^2\right];\ a_E^y = -112.534\,[\text{in./s}^2];\ a_E^z = 318.295\,[\text{in./s}^2]$$

The simulated values SolidWorks (SW) are coincident with the calculated values:

$$a_E^{x,SW} = 0\,[\text{in./s}^2];\ a_E^{y,SW} = -112.534\,[\text{in./s}^2];\ a_E^{z,SW} = 318.295\,[\text{in./s}^2]$$

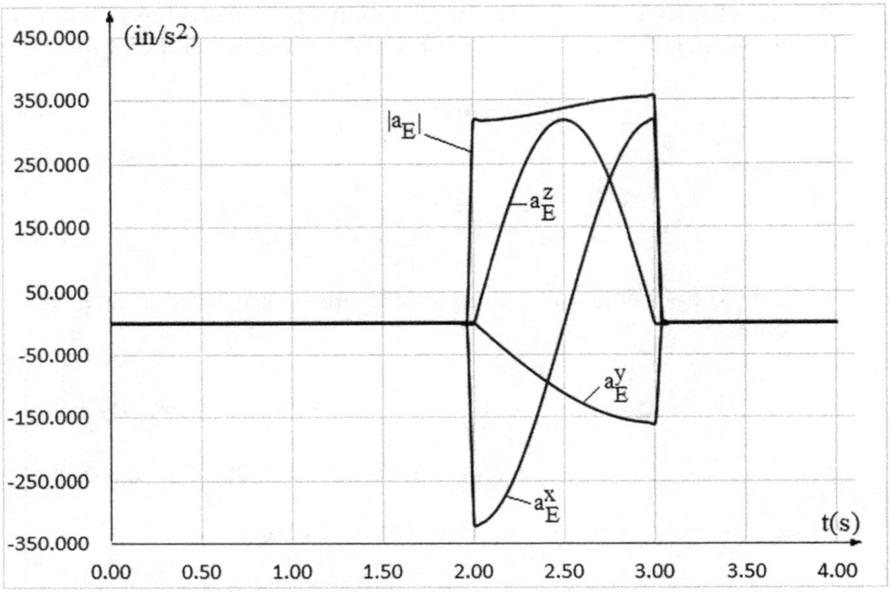

FIGURE 4.4
End-effector acceleration components and its magnitude.

4.4.3 The Matroid Method for Linear Acceleration Analysis of Single- and Multiple-Cycle Planar Mechanisms

- **The acceleration matroidal matrix**

  The previous equations for positional analysis were developed based on the property that the sum of entries in the Latin matrix **L** is zero. The equations for velocity analysis of closed cycle mechanisms were developed based on the property that the sum of entries in the velocity matroidal matrix $\dot{\mathbf{L}}$ is zero.

  The *acceleration matroidal matrix,* noted $\ddot{\mathbf{L}}$, is a c × n matrix with vector entries shown in Equation (4.39):

$$
\begin{array}{c} \qquad\qquad 0 \qquad\quad 1 \qquad \cdots \qquad m \\
\left[\ddot{\mathbf{L}}\right] = \left[\begin{array}{c|c|c|c} \ddot{\mathbf{L}}^{0}_{EA} & \ddot{\mathbf{L}}^{0}_{AB} & \ldots & \ddot{\mathbf{L}}^{0}_{ZE} \end{array}\right] \\
= \left[\begin{array}{c|c|c|c} \mathbf{a}_A - \mathbf{a}_E & \mathbf{a}_B - \mathbf{a}_A & \ldots & \mathbf{a}_E - \mathbf{a}_Z \end{array}\right]
\end{array} \tag{4.39}
$$

Statement 4.4

The sum of all acceleration matroidal vectors assigned to cycle edges is zero:

$$\ddot{\mathbf{L}}_{EA}^{0} + \ddot{\mathbf{L}}_{AB}^{0} + \ddot{\mathbf{L}}_{BC}^{0} + \cdots + \ddot{\mathbf{L}}_{YZ}^{0} + \ddot{\mathbf{L}}_{ZE}^{0} = \mathbf{0} \tag{4.40}$$

Proof: In the right-hand side of Equation (4.39), each vector acceleration shows twice and with opposite sign.

Equation (4.39) generates 2c independent scalar equations for the inverse acceleration analysis.

- *Acceleration matroidal vectors for planar mechanism:*
  For xy planar mechanisms, the components of acceleration matroidal vectors have a simplified form.
- If the link is connected by two revolute joints at Y and Z, then

$$\ddot{\mathbf{L}}_{YZ}^{0} = \mathbf{a}_{YZ}^{\tau} + \mathbf{a}_{YZ}^{\nu} \tag{4.41}$$

- If the link is connected by one revolute and one prismatic joint, then

$$\ddot{\mathbf{L}}_{YZ}^{0} = \mathbf{a}_{YZ}^{\tau} + \mathbf{a}_{YZ}^{\nu} + \mathbf{a}_{Z}^{com,0} \tag{4.42}$$

  where
- *The tangential acceleration*, Equation (4.30), is

$$\mathbf{a}_{YZ}^{\tau} = -\tilde{\mathbf{L}}_{YZ}^{0} * \boldsymbol{\varepsilon}_{m-1}^{0} = \begin{Bmatrix} -y_{YZ}^{0} \\ x_{YZ}^{0} \end{Bmatrix} \cdot \varepsilon_{m-1}^{0} \tag{4.43}$$

- *The normal acceleration*, Equation (4.31), is

$$\mathbf{a}_{YZ}^{\nu} = -\begin{Bmatrix} x_{YZ}^{0} \\ y_{YZ}^{0} \end{Bmatrix} \cdot \left(\omega_{m-1}^{z}\right)^{2} \tag{4.44}$$

- *The complemental (relative and Coriolis acceleration)*, Equation (4.34), for a prismatic joint Z is

$$\mathbf{a}_{Z}^{com} = \begin{Bmatrix} u_{Z}^{x} \\ u_{Z}^{y} \end{Bmatrix} \cdot \ddot{d}_{Z}^{0} - \begin{Bmatrix} -u_{Z}^{y} \\ u_{Z}^{x} \end{Bmatrix} \cdot \left(2 \cdot \dot{d}_{Z}^{0} \cdot \omega_{m-1}^{z}\right) \tag{4.45}$$

  The first term is the relative acceleration $\ddot{d}_{Z}^{0}$ allowed in a prismatic joint, whereas the second term is the Coriolis acceleration. The scalar of the Coriolis acceleration is $2 \cdot \dot{d}_{Z}^{0} \cdot \omega_{m-1}^{z}$, that is, the double product between the relative velocity $\dot{d}_{Z}^{0}$ at the joint Z and the absolute velocity of the link m − 1 (the tail of edge Z in digraph).

Equation (4.40) is written in matrix form as shown in Equation (4.46), in which motion constraints apply:

$$\left[\tilde{\mathbf{L}}^0\right]*\{\varepsilon_n\}+\left[\mathbf{C}(u^0)\right]*\{\ddot{d}_j\}=\left[\mathbf{L}^0\right]*\{(\omega_n)^2\}-\left[\mathbf{C}(\tilde{u}^0)\right]*\{2\cdot\dot{d}_j\cdot\omega\} \quad (4.46)$$

Equation (4.46) generates a linear system of 2c equations and solved for 2c accelerations.

Notice that the left-hand-side matrices $\tilde{\mathbf{L}}^0$ and $\mathbf{C}(u^0)$ are the same as in equations for inverse velocity analysis.

where

- n is the number of links (includes the fixed link), and j is the number of joints.
- $\left[\tilde{\mathbf{L}}^0\right]$ is a matrix with 2c rows and n columns. Its scalar entries are generated from the Latin matrix, interchanging the x components with the y components and then changing the sign for the y components.
- $\{\varepsilon_n\}$ is a column vector with scalar entries where all the links' angular accelerations are assigned to nodes in digraph.
- $\{\boldsymbol{\omega}_n\}$ is a column vector with scalar entries where all the links' angular velocities are assigned to nodes in digraph. The link constraints for motion are applied to these entries. The scalars are positive for a link in a counterclockwise rotation around z_0 and negative for a link in a clockwise rotation.
- $\{(\boldsymbol{\omega}_n)^2\}$ is a column vector with scalar entries where all the links' squared angular velocities are assigned to nodes in digraph.
- $\mathbf{C}_{c,j}=\begin{array}{c}\\ \left[\right.\end{array}\begin{array}{c|c|c|c} A & B & \cdots & E \\ c_A & c_B & \cdots & c_E \end{array}\begin{array}{c}\\ \left.\right]\end{array}$ is the cycle basis matroid **C**. For mechanisms with c-cycles, the matrix C is a $(c \times j)$ matrix.
- $\left[\mathbf{C}(u^0)\right]$ is a matrix with 2c rows and j columns. Its scalar entries are the components for joint's unit vectors multiplied by the corresponding entries c in matrix **C**.
- $\left[\mathbf{C}(\tilde{u}^0)\right]$ is the matrix of skew-symmetric components for unit vectors in joints, which are nonzero for prismatic joints.
- **Joint's linear accelerations**

  Starting from the joint A, the linear absolute accelerations for B, C, … , D are determined by equating the column entries in the left- and right-hand side of Equation (4.39):

$$\mathbf{a}_E=\mathbf{a}_A-\ddot{\mathbf{L}}_{EA};\ \mathbf{a}_B=\mathbf{a}_A+\ddot{\mathbf{L}}_{AB};\ldots\ \mathbf{a}_Z=\mathbf{a}_E-\ddot{\mathbf{L}}_{ZE} \quad (4.47)$$

4.4.4 Acceleration Analysis for Single-Cycle Planar Mechanisms with Revolute Joints

The following constraint equations are considered: $\dot{\mathbf{d}}_j = 0_j$, $\ddot{\mathbf{d}}_j = 0$ for null linear relative velocity and acceleration in the revolute joints and $\omega_0 = 0$, $\varepsilon_0 = 0$ for the fixed link. Therefore, Equation (4.46) holds a simplified form

$$\left[\tilde{L}^0\right] * \{\varepsilon_m\} = \left[L^0\right] * \left\{(\omega_m)^2\right\} \quad (4.48)$$

The unknowns are absolute angular accelerations assigned to nodes in digraph.

$$\begin{array}{c} \begin{matrix} \quad 1 & \quad 2 & & \quad m \end{matrix} \\ \left[\begin{array}{c|c|c|c} -y_{AB} & -y_{BC} & \cdots & -y_{ZE} \\ x_{AB} & x_{BC} & \cdots & x_{ZE} \end{array}\right] * \left\{\begin{array}{c} \varepsilon_1 \\ \varepsilon_2 \\ \vdots \\ \varepsilon_m \end{array}\right\} \end{array}$$

$$= \left[\begin{array}{c|c|c|c} x_{AB} & x_{BC} & \cdots & x_{ZE} \\ y_{AB} & y_{BC} & \cdots & y_{ZE} \end{array}\right] * \left\{\begin{array}{c} \omega_1^2 \\ \omega_2^2 \\ \vdots \\ \omega_m^2 \end{array}\right\} \quad (4.49)$$

Statement 4.5

For planar mechanisms with c-cycles and m mobile links, the system of 2c linear equations has (m – M) unknowns.

Proof: If the mechanism has m mobile links from which M of them are actuated, then m – M links have unknown angular accelerations. Let us consider the mobility equation: M = 3 m – 2 j, where m is the number of mobile links and j is the number of revolute joints. Then, m – M = 2j – 2m = 2c, which is the number of equations.

Example: The four-bar mechanism

The four-bar mechanism, as shown in Figure 4.5a, has the link lengths: $L_{AB} = 4''$, $L_{BD} = 24''$, $L_{DC} = 14''$, and $L_{CA} = 30''$.

The location of COM is as follows: $L_{AG_1}=2.120''$; $L_{BG_2}=11.863''$; $L_{DG_3}=6.722''$

The crank 1 rotates counterclockwise with the angular displacement $\theta_1 = 2\pi \cdot t\,[\text{rad}]$. Therefore, the crank's constant angular velocity is $\omega_1 = \dot{\theta}_1 = 2\pi\,[\text{rad/s}]$, and its angular acceleration is $\varepsilon_1 = \ddot{\theta}_1 = 0\,[\text{rad/s}^2]$. For the interval of t = 1 s, determine

a. The system of equations for acceleration analysis
b. Angular accelerations for links 2 and 3

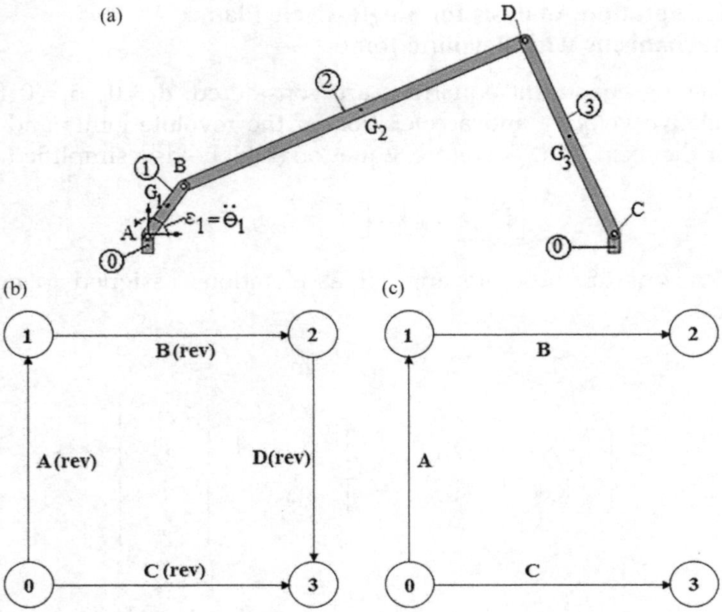

FIGURE 4.5
(a) Four-bar planar mechanism; (b) digraph; and (c) tree.

c. Linear accelerations of the joints A, B, C, and D
d. Linear accelerations of COM G_1, G_2, and G_3.

Solution:

a. The system of equations for accelerations analysis
The Latin matrix, Equation (3.56), and its skew-symmetric entries are

$$\begin{matrix} & 0 & 1 & 2 & 3 \end{matrix}$$

$$[L]=\left[\begin{array}{c|c|c|c} x_{CA} & x_{AB} & x_{BD} & x_{DC} \\ y_{CA} & y_{AB} & y_{BD} & y_{DC} \end{array}\right];$$

$$[\tilde{L}]=\left[\begin{array}{c|c|c|c} 0 & -y_{AB} & -y_{BD} & -y_{DC} \\ -30 & x_{AB} & x_{BD} & x_{DC} \end{array}\right] \tag{4.50a}$$

where $x_{CA}=-30;\ y_{CA}=0;\ x_{AB}=L_{AB}\cdot \mathrm{Cos}(\theta_1);\ y_{AB}=L_{AB}\cdot \mathrm{Sin}(\theta_1)$.

Equation (4.46) with the constraints (shown in round brackets) is

$$\left[\begin{array}{c:c:c:c} -y_{CA} & -y_{AB} & -y_{BD} & -y_{DC} \\ x_{CA} & x_{AB} & x_{BD} & x_{DC} \end{array}\right] * \left\{\begin{array}{c} \varepsilon_0(=0) \\ \varepsilon_1(=0) \\ \varepsilon_2 \\ \varepsilon_3 \end{array}\right\}$$

$$= \left[\begin{array}{c:c:c:c} x_{CA} & x_{AB} & x_{BD} & x_{DC} \\ y_{CA} & y_{AB} & y_{BD} & y_{DC} \end{array}\right] * \left\{\begin{array}{c} \omega_0^2(=0) \\ \omega_1^2(=4\pi^2) \\ \omega_2^2 \\ \omega_3^2 \end{array}\right\} \tag{4.50b}$$

Therefore,

$$\left[\begin{array}{c:c} -y_{BD} & -y_{DC} \\ x_{BD} & x_{DC} \end{array}\right] * \left\{\begin{array}{c} \varepsilon_2 \\ \varepsilon_3 \end{array}\right\} = \left\{\begin{array}{c} 157.914 \cdot Cos(\theta_1) + x_{BD} \cdot \omega_2^2 + x_{DC} \cdot \omega_3^2 \\ 157.914 \cdot Sin(\theta_1) + y_{BD} \cdot \omega_2^2 + y_{DC} \cdot \omega_3^2 \end{array}\right\} \tag{4.50c}$$

The 4c = 4 parameters x_{BD}, y_{BD}, x_{DC}, y_{DC} have been determined previously during positional analysis, as functions of crank angle θ_1. Also, the 2c = 2 angular velocities ω_2 and ω_3 have been determined previously at the velocity analysis as

$$\begin{aligned} \omega_2 &= \frac{4 \cdot \omega_1}{\Delta}\left[x_{DC} \cdot Sin(\theta_1) - y_{DC} \cdot Cos(\theta_1)\right]; \\ \omega_3 &= \frac{4 \cdot \omega_1}{\Delta}\left[y_{BD} \cdot Cos(\theta_1) - x_{BD} \cdot Sin(\theta_1)\right] \end{aligned} \tag{4.50d}$$

where $\Delta = x_{BD} \cdot y_{DC} - x_{DC} \cdot y_{BD}$

b. Angular accelerations for links 2 and 3, Figure 4.6a

The system of 2c = 2 linear equations is solved for m − M = 3 − 1 = 2 unknowns ε_2 and ε_3:

$$\left\{\begin{array}{l} \varepsilon_2 = \frac{1}{\Delta}\left[157.914 \cdot \left(x_{DC} \cdot Cos(\theta_1) + y_{DC} \cdot Sin(\theta_1)\right)\right. \\ \quad \left. + \left(x_{BD}x_{DC} + y_{BD}y_{DC}\right) \cdot \omega_2^2 + 196 \cdot \omega_3^2\right] \\ \varepsilon_3 = -\frac{1}{\Delta}\left[157.914 \cdot \left(x_{BD} \cdot Cos(\theta_1) + y_{BD} \cdot Sin(\theta_1)\right)\right. \\ \quad \left. + 576 \cdot \omega_2^2 + \cdot \left(x_{BD}x_{DC} + y_{BD}y_{DC}\right) \cdot \omega_3^2\right] \end{array}\right. \tag{4.50e}$$

Note:

The denominator Δ in Equations (4.50d) and (4.50e) is the determinant Δ of the matrix with the entries in columns 2 and 3 from the Equations (4.50a). If the determinant is zero, then the mechanism has a singularity for velocities and accelerations. For this position, the vectors L_{BD} and L_{DC} are collinear.

c. Linear acceleration of the joints A, B, C, and D

The matroidal acceleration matrix, Equation (4.39), is

$$\begin{matrix} & 0 & 1 & 2 & 3 \\ [\ddot{\mathbf{L}}] = \Big[& \ddot{\mathbf{L}}^0_{CA} & \ddot{\mathbf{L}}^0_{AB} & \ddot{\mathbf{L}}^0_{BD} & \ddot{\mathbf{L}}^0_{DC} \Big] \\ = \Big[& \mathbf{a}_A - \mathbf{a}_C & \mathbf{a}_B - \mathbf{a}_A & \mathbf{a}_D - \mathbf{a}_B & \mathbf{a}_C - \mathbf{a}_D \Big] \end{matrix} \tag{4.50f}$$

- The matroidal acceleration vectors are the columns in Equation (4.50f):

$$\begin{aligned} &\text{Column 0: } \ddot{\mathbf{L}}^0_{CA} = \begin{Bmatrix} -y_{CA} \\ x_{CA} \end{Bmatrix} \cdot \varepsilon_0 - \begin{Bmatrix} x_{CA} \\ y_{CA} \end{Bmatrix} \cdot \omega_0^2 = \begin{Bmatrix} 0 \\ 0 \end{Bmatrix}; \\ &\text{Column 1: } \ddot{\mathbf{L}}^0_{AB} = \begin{Bmatrix} -y_{AB} \\ x_{AB} \end{Bmatrix} \cdot \varepsilon_1 - \begin{Bmatrix} x_{AB} \\ y_{AB} \end{Bmatrix} \cdot \omega_1^2 = \begin{Bmatrix} -157.914 \cdot \mathrm{Cos}(\theta_1) \\ -157.914 \cdot \mathrm{Sin}(\theta_1) \end{Bmatrix}; \\ &\text{Column 2: } \ddot{\mathbf{L}}^0_{BD} = \begin{Bmatrix} -y_{BD} \\ x_{BD} \end{Bmatrix} \cdot \varepsilon_2 - \begin{Bmatrix} x_{BD} \\ y_{BD} \end{Bmatrix} \cdot \omega_2^2 = \begin{Bmatrix} -y_{BD} \cdot \varepsilon_2 - x_{BD} \cdot \omega_2^2 \\ x_{BD} \cdot \varepsilon_2 - y_{BD} \cdot \omega_2^2 \end{Bmatrix}; \\ &\text{Column 3: } \ddot{\mathbf{L}}^0_{DC} = \begin{Bmatrix} -y_{DC} \\ x_{DC} \end{Bmatrix} \cdot \varepsilon_3 - \begin{Bmatrix} x_{DC} \\ y_{DC} \end{Bmatrix} \cdot \omega_3^2 = \begin{Bmatrix} -y_{DC} \cdot \varepsilon_3 - x_{DC} \cdot \omega_3^2 \\ x_{DC} \cdot \varepsilon_3 - y_{DC} \cdot \omega_3^2 \end{Bmatrix} \end{aligned} \tag{4.50g}$$

- The joint's accelerations

  Starting from origin O coincidental with the fixed joint A ($\mathbf{a}_A = 0$), the linear absolute accelerations for B, C, and D are determined by equating the column entries in the left- and right-hand side of Equation (4.50f).

  Joint A (fixed joint):

$$\mathbf{a}_A = \begin{Bmatrix} 0 \\ 0 \end{Bmatrix}$$

Joint C (fixed joint):

$$\mathbf{a}_C = \mathbf{a}_A - \ddot{\mathbf{L}}^0_{CA} = \begin{Bmatrix} 0 \\ 0 \end{Bmatrix}$$

Joint B:

$$\mathbf{a}_B = \mathbf{a}_A + \ddot{\mathbf{L}}^0_{AB} = \begin{Bmatrix} -157.914 \cdot \mathrm{Cos}(\theta_1) \\ -157.914 \cdot \mathrm{Sin}(\theta_1) \end{Bmatrix} \tag{4.50h}$$

Joint D:

$$\mathbf{a}_D = \mathbf{a}_B + \ddot{\mathbf{L}}^0_{BD} = \begin{Bmatrix} -157.914 \cdot \mathrm{Cos}(\theta_1) - y_{BD} \cdot \varepsilon_2 - x_{BD} \cdot \omega_2^2 \\ -157.914 \cdot \mathrm{Sin}(\theta_1) + x_{BD} \cdot \varepsilon_2 - y_{BD} \cdot \omega_2^2 \end{Bmatrix}$$

Joint C (fixed joint): These are checkup equations

$$\mathbf{a}_C = \mathbf{a}_D + \ddot{\mathbf{L}}^0_{DC} = \begin{Bmatrix} -157.914 \cdot \mathrm{Cos}(\theta_1) - y_{BD} \cdot \varepsilon_2 - x_{BD} \cdot \omega_2^2 - y_{DC} \cdot \varepsilon_3 - x_{DC} \cdot \omega_3^2 \\ -157.914 \cdot \mathrm{Sin}(\theta_1) + x_{BD} \cdot \varepsilon_2 - y_{BD} \cdot \omega_2^2 + x_{DC} \cdot \varepsilon_3 - y_{DC} \cdot \omega_3^2 \end{Bmatrix}$$

$$= \begin{Bmatrix} 0 \\ 0 \end{Bmatrix}$$

Notice that Equation (4.50b) shows on the x and y entries for $\mathbf{a}_C$; therefore, both summations are zero.

d. Linear accelerations of com G_1, G_2, and G_3 for the provided location of COM:

$$L_{AG_1} = 2.120''; L_{BG_2} = 11.863''; L_{DG_3} = 6.722''$$

Center of mass link 1: From Equation (4.37a), the acceleration matroidal vector $\ddot{\mathbf{L}}^0_{AG_1}$ is

$$\ddot{\mathbf{L}}^0_{AG_1} = \mathbf{a}^{\tau}_{AG_1} + \mathbf{a}^{\nu}_{AG_1} = \begin{Bmatrix} -y_{AG_1} \\ x_{AG_1} \end{Bmatrix} \cdot \varepsilon_1 - \begin{Bmatrix} x_{AG_1} \\ y_{AG_1} \end{Bmatrix} \cdot \omega_1^2$$
$$= \begin{Bmatrix} -83.694 \cdot \mathrm{Cos}(\theta_1) \\ -83.694 \cdot \mathrm{Sin}(\theta_1) \end{Bmatrix} \tag{4.50i}$$

If A, G_1, and B are three collinear points on link 1, then the isometric Equation (4.50j) holds:

$$\begin{Bmatrix} x_{AG_1} \\ y_{AG_1} \end{Bmatrix} = \frac{L_{AG_1}}{L_{AB}} \begin{bmatrix} \mathrm{Cos}(0°) & -\mathrm{Sin}(0°) \\ \mathrm{Sin}(0°) & \mathrm{Cos}(0°) \end{bmatrix} \cdot \begin{Bmatrix} x_{AB} \\ y_{AB} \end{Bmatrix} = \begin{Bmatrix} 0.530 \cdot x_{AB} \\ 0.530 \cdot y_{AB} \end{Bmatrix} \tag{4.50j}$$

From Equation (4.37), the link 1's COM acceleration is

$$\mathbf{a}_{G_1} = \mathbf{a}_A + \ddot{\mathbf{L}}^0_{AG_1} = \begin{Bmatrix} -83.694 \cdot \mathrm{Cos}(\theta_1) \\ -83.694 \cdot \mathrm{Sin}(\theta_1) \end{Bmatrix} \tag{4.50k}$$

The magnitude of COM acceleration $\mathbf{a}_{G_1}$ is

$$|a_{G_1}| = \sqrt{\left(a^x_{G_1}\right)^2 + \left(a^y_{G_1}\right)^2} \tag{4.50l}$$

Center of mass link 2: From Equation (4.37a), the acceleration matroidal vector $\ddot{\mathbf{L}}^0_{BG_2}$ is

$$\ddot{\mathbf{L}}^0_{BG_2} = \mathbf{a}^{\tau}_{BG_2} + \mathbf{a}^{\nu}_{BG_2} = \begin{Bmatrix} -y_{BG_2} \\ x_{BG_2} \end{Bmatrix} \cdot \varepsilon_2 - \begin{Bmatrix} x_{BG_2} \\ y_{BG_2} \end{Bmatrix} \cdot \omega_2^2 = \begin{Bmatrix} -0.494 \cdot y_{BD} \cdot \varepsilon_2 - 0.494 \cdot x_{BD} \cdot \omega_2^2 \\ 0.494 \cdot x_{BD} \cdot \varepsilon_2 - 0.494 \cdot y_{BD} \cdot \omega_2^2 \end{Bmatrix} \tag{4.50m}$$

If B, G_2, and D are three collinear points on link 2, then the isometric Equation (4.50n) holds:

$$\begin{Bmatrix} x_{BG_2} \\ y_{BG_2} \end{Bmatrix} = \frac{L_{BG_2}}{L_{BD}} \begin{bmatrix} \mathrm{Cos}(0°) & -\mathrm{Sin}(0°) \\ \mathrm{Sin}(0°) & \mathrm{Cos}(0°) \end{bmatrix} \cdot \begin{Bmatrix} x_{BD} \\ y_{BD} \end{Bmatrix} = \begin{Bmatrix} 0.494 \cdot x_{BD} \\ 0.494 \cdot y_{BD} \end{Bmatrix} \tag{4.50n}$$

From Equation (4.37), the link 2's COM acceleration is

$$\mathbf{a}_{G_2} = \mathbf{a}_B + \ddot{\mathbf{L}}^0_{BG_2} = \begin{Bmatrix} -157.914 \cdot \mathrm{Cos}(\theta_1) - 0.494 \cdot y_{BD} \cdot \varepsilon_2 - 0.494 \cdot x_{BD} \cdot \omega_2^2 \\ -157.914 \cdot \mathrm{Sin}(\theta_1) + 0.494 \cdot x_{BD} \cdot \varepsilon_2 - 0.494 \cdot y_{BD} \cdot \omega_2^2 \end{Bmatrix} \tag{4.50o}$$

The magnitude of COM acceleration $\mathbf{a}_{G_2}$ is

$$|a_{G_2}| = \sqrt{(a_{G_2}^x)^2 + (a_{G_2}^y)^2} \tag{4.50p}$$

Center of mass link 3: From Equation (4.37a), the acceleration matroidal vector $\ddot{\mathbf{L}}_{DG_3}^0$ is

$$\ddot{\mathbf{L}}_{DG_3}^0 = \mathbf{a}_{DG_3}^\tau + \mathbf{a}_{DG_3}^v = \begin{Bmatrix} -y_{DG_3} \\ x_{DG_3} \end{Bmatrix} \cdot \varepsilon_3 - \begin{Bmatrix} x_{DG_3} \\ y_{DG_3} \end{Bmatrix} \cdot \omega_3^2$$
$$= \begin{Bmatrix} -0.480 \cdot y_{DC} \cdot \varepsilon_3 - 0.480 \cdot x_{DC} \cdot \omega_3^2 \\ 0.480 \cdot x_{DC} \cdot \varepsilon_3 - 0.480 \cdot y_{DC} \cdot \omega_3^2 \end{Bmatrix} \tag{4.50q}$$

If D, G_3, and C are three collinear points on link 3, then the isometric Equation (4.50r) holds:

$$\begin{Bmatrix} x_{DG_3} \\ y_{DG_3} \end{Bmatrix} = \frac{L_{DG_3}}{L_{DC}} \begin{bmatrix} \mathrm{Cos}(0°) & -\mathrm{Sin}(0°) \\ \mathrm{Sin}(0°) & \mathrm{Cos}(0°) \end{bmatrix} \cdot \begin{Bmatrix} x_{DC} \\ y_{DC} \end{Bmatrix}$$
$$= \begin{Bmatrix} 0.480 \cdot x_{DC} \\ 0.480 \cdot y_{DC} \end{Bmatrix} \tag{4.50r}$$

From Equation (4.37), the link 3's COM acceleration is

$$\mathbf{a}_{G_3} = \mathbf{a}_D + \ddot{\mathbf{L}}_{DG_3}^0$$
$$= \begin{Bmatrix} -157.914 \cdot \mathrm{Cos}(\theta_1) - y_{BD} \cdot \varepsilon_2 - x_{BD} \cdot \omega_2^2 - 0.480 \cdot y_{DC} \cdot \varepsilon_3 - 0.480 \cdot x_{DC} \cdot \omega_3^2 \\ -157.914 \cdot \mathrm{Sin}(\theta_1) + x_{BD} \cdot \varepsilon_2 - y_{BD} \cdot \omega_2^2 + 0.480 \cdot x_{DC} \cdot \varepsilon_3 - 0.480 \cdot y_{DC} \cdot \omega_3^2 \end{Bmatrix} \tag{4.50s}$$

The magnitude of COM acceleration $\mathbf{a}_{G_3}$ is (Figure 4.6b):

$$|a_{G_3}| = \sqrt{(a_{G_3}^x)^2 + (a_{G_3}^y)^2} \tag{4.50t}$$

4.4.5 Acceleration Analysis for Multiple-Cycle Planar Mechanisms

Example: The two-cycle planar mechanism

The mechanism in Figure 4.7a has given the link lengths: $L_{EA} = 30''$; $L_{AB} = 4''$; $L_{BF} = 24''$; $L_{FE} = 14''$; $L_{EG} = 7''$; $L_{GD} = 12''$; $L_{DC} = 0''$

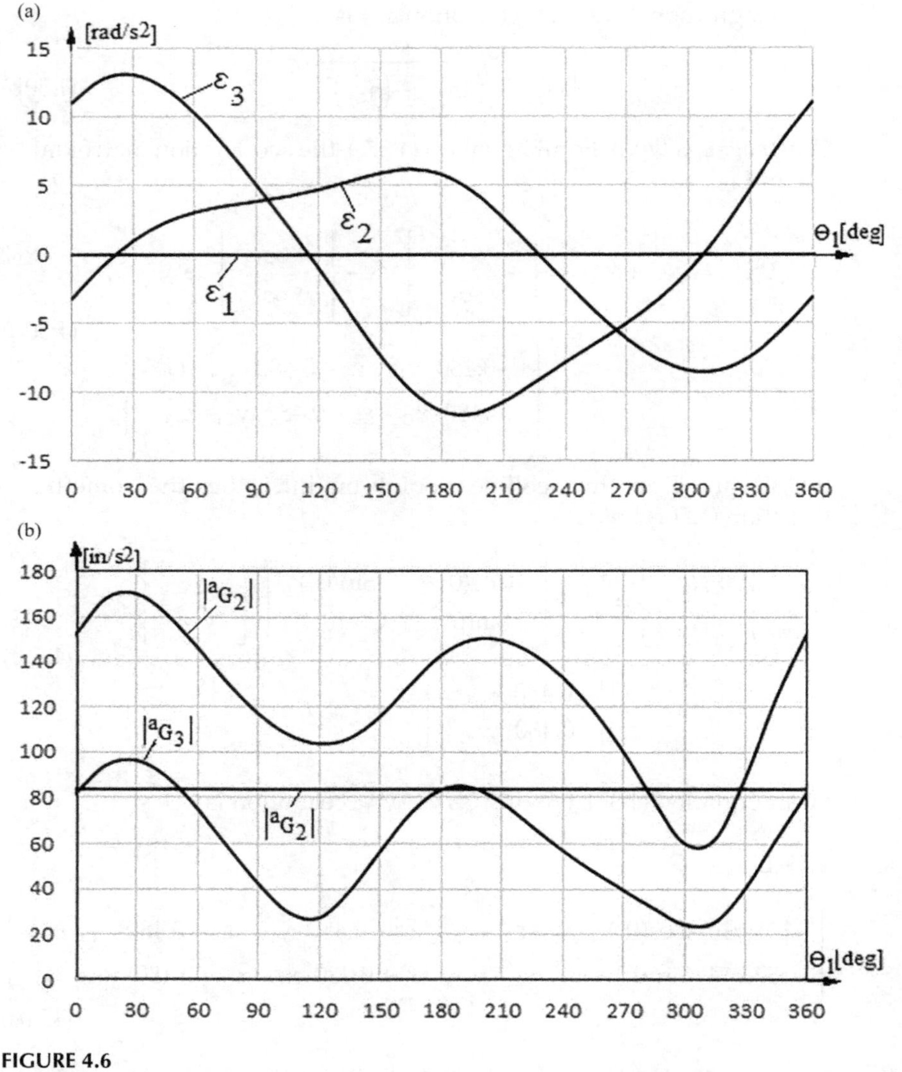

FIGURE 4.6
(a) The angular accelerations for links 2 and 3—four-bar mechanism and (b) the links' 1, 2, and 3 COM accelerations.

The crank's counterclockwise angular velocity is $\omega_1 = \dot{\theta}_1 = 2\pi$ [rad/s], and its angular acceleration is $\varepsilon_1 = \ddot{\theta}_1 = 0$ [rad/s^2]. For the interval of t = 1 s, determine

a. The system of equations for acceleration analysis
b. The angular accelerations for links 2, 3, 4, and the relative linear acceleration for link 5.

c. Linear accelerations of the joints B, C, D, E, F, and G
d. Linear accelerations of com G_1, G_2, G_3, G_4, and G_5 for the provided location of com:

$$L_{AG_1} = 2.330''; L_{BG_2} = 12.000''; L_{FG_3} = 6.658''; L_{GG_4} = 6.000''; L_{DG_5} = 0.000''$$

Solution:

a. The system of equations for acceleration analysis

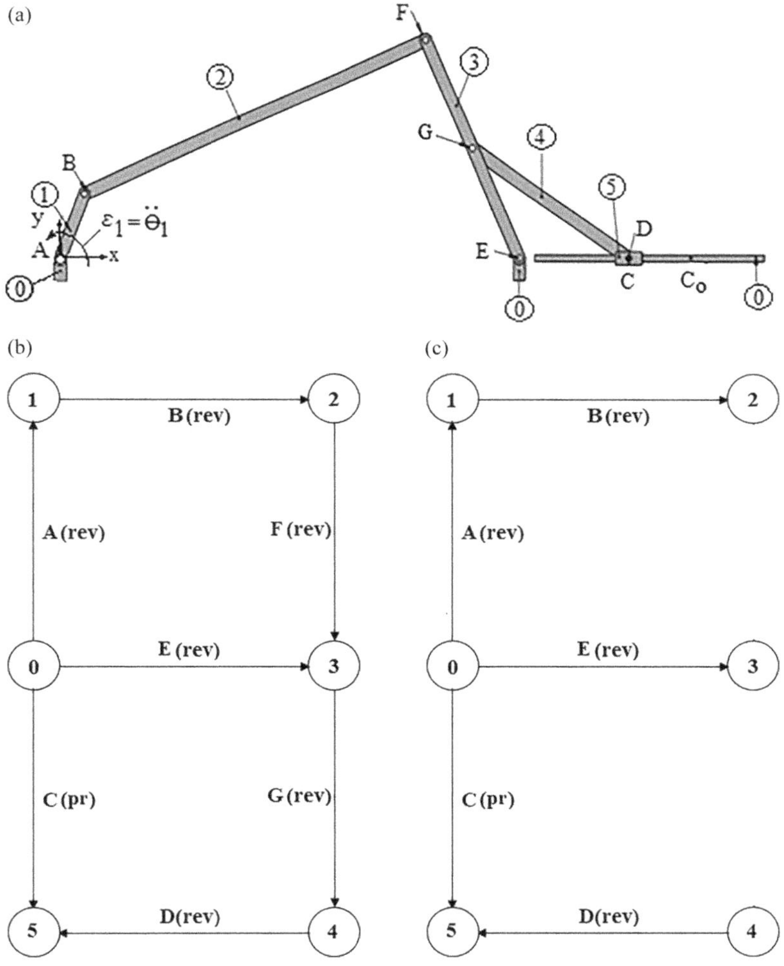

FIGURE 4.7
(a) Two-cycle mechanism; (b) digraph; (c) tree; and (d) cycles.

(Continued)

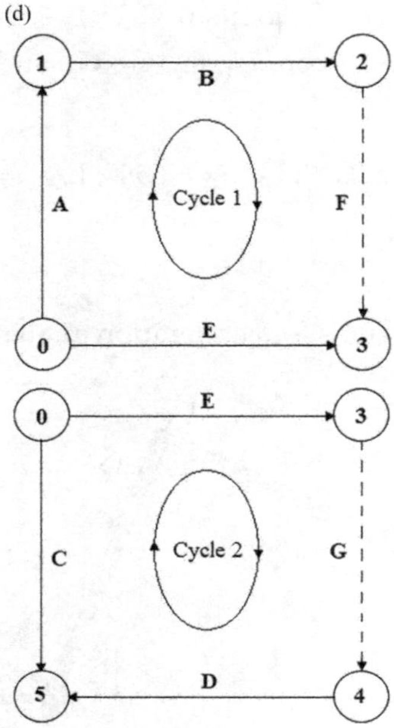

FIGURE 4.7 (CONTINUED)
(a) Two-cycle mechanism; (b) digraph; (c) tree; and (d) cycles.

Equation (4.46) becomes

$$[\tilde{L}]*\begin{Bmatrix} \varepsilon_0(=0) \\ \varepsilon_1(=0,\text{input}) \\ \varepsilon_2 \\ \varepsilon_3 \\ \varepsilon_4 \\ \varepsilon_5(=0) \end{Bmatrix} + [C(u)]*\begin{Bmatrix} \ddot{d}_A(=0) \\ \ddot{d}_B(=0) \\ \ddot{d}_C \\ \ddot{d}_D(=0) \\ \ddot{d}_E(=0) \\ \ddot{d}_F(=0) \\ \ddot{d}_G(=0) \end{Bmatrix}$$

$$= [L]*\begin{Bmatrix} (\omega_0)^2 \\ (\omega_1)^2 \\ (\omega_2)^2 \\ (\omega_3)^2 \\ (\omega_4)^2 \\ (\omega_5)^2 \end{Bmatrix} - [C(\tilde{u})]*\begin{Bmatrix} 2\cdot\dot{d}_A\cdot\omega_0(=0) \\ 2\cdot\dot{d}_B\cdot\omega_1(=0) \\ 2\cdot\dot{d}_C\cdot\omega_0(=0) \\ 2\cdot\dot{d}_D\cdot\omega_4(=0) \\ 2\cdot\dot{d}_E\cdot\omega_0(=0) \\ 2\cdot\dot{d}_F\cdot\omega_2(=0) \\ 2\cdot\dot{d}_G\cdot\omega_3(=0) \end{Bmatrix} \quad (4.51a)$$

where

- The terms within brackets illustrate the constraint velocities and accelerations.
- The *Latin matrix* **L**, from the positional analysis, Equation (3.141a), is a $2c \times n = 4 \times 6$ matrix with scalar entries defined around the nodes in the digraph and written in the direction of each cycle, as functions of the variable crank angle θ_1

$$[L] = \begin{array}{c} \begin{array}{cccccc} 0 & 1 & 2 & 3 & 4 & 5 \end{array} \\ \left[\begin{array}{c|c|c|c|c|c} x_{EA} & x_{AB} & x_{BF} & x_{FE} & 0 & 0 \\ y_{EA} & y_{AB} & y_{BF} & y_{FE} & 0 & 0 \\ \hline x_{CE} & 0 & 0 & x_{EG} & x_{GD} & x_{DC} \\ y_{CE} & 0 & 0 & y_{EG} & y_{GD} & y_{DC} \end{array}\right] \end{array}$$

$$= \left[\begin{array}{c|c|c|c|c|c} -30 & 4\cdot \mathrm{Cos}(\theta_1) & x_{BF} & x_{FE} & 0 & 0 \\ 0 & 4\cdot \mathrm{Sin}(\theta_1) & y_{BF} & y_{FE} & 0 & 0 \\ \hline d_C - 7.308 & 0 & 0 & -0.5x_{FE} & x_{GD} & 0 \\ 0 & 0 & 0 & -0.5y_{FE} & y_{GD} & 0 \end{array}\right] \quad (4.51b)$$

- The Latin matrix with skew-symmetric entries $\tilde{L}$ is

$$[\tilde{L}] = \begin{array}{c} \begin{array}{cccccc} 0 & 1 & 2 & 3 & 4 & 5 \end{array} \\ \left[\begin{array}{c|c|c|c|c|c} 0 & -4\cdot \mathrm{Sin}(\theta_1) & -y_{BF} & -y_{FE} & 0 & 0 \\ -30 & 4\cdot \mathrm{Cos}(\theta_1) & x_{BF} & x_{FE} & 0 & 0 \\ \hline 0 & 0 & 0 & 0.5y_{FE} & -y_{GD} & 0 \\ d_C - 7.308 & 0 & 0 & -0.5x_{FE} & x_{GD} & 0 \end{array}\right] \end{array} \quad (4.51c)$$

- The cycle basis matroid matrix, **C**

  Considering its definition (Equation (3.102)), the matrix **C** has two rows assigned to two cycles and each column assigned to a joint.

$$[C] = \begin{array}{c} \begin{array}{ccccccc} A & B & C & D & E & F & G \end{array} \\ \left[\begin{array}{ccccccc} 1 & 1 & 0 & 0 & -1 & 1 & 0 \\ 0 & 0 & -1 & 1 & 1 & 0 & 1 \end{array}\right] \end{array} \begin{array}{l} \text{Cycle 1} \\ \text{Cycle 2} \end{array} \quad (4.51d)$$

- The matrix $\mathbf{C}(u^0)$ for joint's unit vectors is multiplied each by the corresponding entry in matrix **C**.

The x and y components of joint's unit vectors are

$$\mathbf{u}_A = \mathbf{u}_B = \mathbf{u}_D = \mathbf{u}_E = \mathbf{u}_F = \mathbf{u}_G = \begin{Bmatrix} 0 \\ 0 \end{Bmatrix};$$

$$\mathbf{u}_C = \begin{Bmatrix} u_C^x \\ u_C^y \end{Bmatrix} = \begin{Bmatrix} u_C^x \\ u_C^y \end{Bmatrix} = \begin{Bmatrix} \cos(0°) \\ \sin(0°) \end{Bmatrix} = \begin{Bmatrix} 1 \\ 0 \end{Bmatrix} \quad (4.51e)$$

$$[\mathbf{C}(\mathbf{u})] = \overset{\begin{matrix} A & B & C & D & E & F & G \end{matrix}}{\begin{bmatrix} 1\cdot\mathbf{u}_A & 1\cdot\mathbf{u}_B & 0 & 0 & -1\cdot\mathbf{u}_E & 1\cdot\mathbf{u}_F & 0 \\ 0 & 0 & -1\cdot\mathbf{u}_C & 1\cdot\mathbf{u}_D & 1\cdot\mathbf{u}_E & 0 & 1\cdot\mathbf{u}_G \end{bmatrix}};$$

$$[\mathbf{C}(\mathbf{u})] = \overset{\begin{matrix} A & B & C & D & E & F & G \end{matrix}}{\left[\begin{array}{c:c:c:c:c:c:c} 0 & 0 & 0 & 0 & 0 & 0 & 0 \\ 0 & 0 & 0 & 0 & 0 & 0 & 0 \\ \hdashline 0 & 0 & -1 & 0 & 0 & 0 & 0 \\ 0 & 0 & 0 & 0 & 0 & 0 & 0 \end{array}\right]} \quad (4.51f)$$

- The matrix $[\mathbf{C}(\tilde{\mathbf{u}}^0)]$ with skew-symmetric entries is generated from Equation (4.51f) by interchanging the u^x with $(-u^y)$ components:

$$[\mathbf{C}(\tilde{\mathbf{u}})] = \overset{\begin{matrix} A & B & C & D & E & F & G \end{matrix}}{\begin{bmatrix} \tilde{\mathbf{u}}_A & \tilde{\mathbf{u}}_B & 0 & 0 & -\tilde{\mathbf{u}}_E & \tilde{\mathbf{u}}_F & 0 \\ 0 & 0 & -\tilde{\mathbf{u}}_C & \tilde{\mathbf{u}}_D & \tilde{\mathbf{u}}_E & 0 & \tilde{\mathbf{u}}_G \end{bmatrix}};$$

$$[\mathbf{C}(\tilde{\mathbf{u}})] = \overset{\begin{matrix} A & B & C & D & E & F & G \end{matrix}}{\left[\begin{array}{c:c:c:c:c:c:c} 0 & 0 & 0 & 0 & 0 & 0 & 0 \\ 0 & 0 & 0 & 0 & 0 & 0 & 0 \\ \hdashline 0 & 0 & 0 & 0 & 0 & 0 & 0 \\ 0 & 0 & -1 & 0 & 0 & 0 & 0 \end{array}\right]} \quad (4.51g)$$

The system of Equation (4.51a) is a system of $2c = 4$ scalar equations with $n - 1 - M = 6 - 1 - 1 = 4$ unknowns: $\varepsilon_2, \varepsilon_3, \varepsilon_4, \ddot{d}_C$

$$\begin{cases} -y_{BF}\cdot\varepsilon_2 - y_{FE}\cdot\varepsilon_3 = 16\pi^2\cdot\mathrm{Cos}(\theta_1) + x_{BF}\cdot(\omega_2)^2 + x_{FE}\cdot(\omega_3)^2 \\ x_{BF}\cdot\varepsilon_2 + x_{FE}\cdot\varepsilon_3 = 16\pi^2\cdot\mathrm{Sin}(\theta_1) + y_{BF}\cdot(\omega_2)^2 + y_{FE}\cdot(\omega_3)^2 \\ 0.5y_{FE}\cdot\varepsilon_3 - y_{GD}\cdot\varepsilon_4 - \ddot{d}_C = -0.5x_{FE}\cdot(\omega_3)^2 + x_{GD}\cdot(\omega_4)^2 \\ -0.5x_{FE}\cdot\varepsilon_3 + x_{GD}\cdot\varepsilon_4 = -0.5y_{FE}\cdot(\omega_3)^2 + y_{GD}\cdot(\omega_4)^2 \end{cases} \tag{4.51h}$$

b. The solution for the system of Equation (4.51h) is

$$\begin{cases} \varepsilon_2 = -\dfrac{1}{\Delta}\Big[16\pi^2\cdot\big(x_{FE}\mathrm{Cos}(\theta_1) + y_{FE}\cdot\mathrm{Sin}(\theta_1)\big) \\ \qquad +\big(x_{BF}\cdot x_{FE} + y_{BF}\cdot y_{FE}\big)\cdot(\omega_2)^2 + 196\cdot(\omega_3)^2\Big] \\ \varepsilon_3 = \dfrac{1}{\Delta}\Big[16\pi^2\cdot\big(x_{BF}\mathrm{Cos}(\theta_1) + y_{BF}\cdot\mathrm{Sin}(\theta_1)\big) \\ \qquad +576\cdot(\omega_2)^2 + \big(x_{BF}\cdot x_{FE} + y_{BF}\cdot y_{FE}\big)\cdot(\omega_3)^2\Big] \\ \varepsilon_4 = \dfrac{1}{x_{GD}}\big(0.5x_{FE}\cdot\varepsilon_3 - 0.5y_{FE}\cdot(\omega_3)^2 + y_{GD}\cdot(\omega_4)^2\big) \\ \ddot{d}_C = 0.5y_{FE}\cdot\varepsilon_3 - y_{GD}\cdot\varepsilon_4 + 0.5x_{FE}\cdot(\omega_3)^2 - x_{GD}\cdot(\omega_4)^2 \end{cases} \tag{4.51i}$$

Figure 4.8a and 4.8b illustrates the solution for one complete revolution of crank 1.

Note: The solution has the singularity factor $\Delta = x_{FE}\cdot y_{BF} - y_{FE}\cdot x_{BF}$ as denominator, the same as that from the velocity analysis. Although $\Delta \neq 0$ for this mechanism, the designer is aware that it might become zero for other link lengths. Therefore, Δ detects the potential singularities for velocities and accelerations.

c. Linear accelerations of the joints A, B, C, D, E, F, and G
   - The acceleration matroidal matrix is

$$\begin{matrix} & 0 & 1 & 2 & 3 & 4 & 5 \end{matrix}$$

$$[\ddot{\mathbf{L}}] = \left[\begin{array}{c:c:c:c:c:c} \ddot{\mathbf{L}}_{EA} & \ddot{\mathbf{L}}_{AB} & \ddot{\mathbf{L}}_{BF} & \ddot{\mathbf{L}}_{FE} & 0 & 0 \\ \ddot{\mathbf{L}}_{CE} & 0 & 0 & \ddot{\mathbf{L}}_{EG} & \ddot{\mathbf{L}}_{GD} & \ddot{\mathbf{L}}_{DC} \end{array}\right]$$

$$= \left[\begin{array}{c:c:c:c:c:c} \mathbf{a}_A - \mathbf{a}_E & \mathbf{a}_B - \mathbf{a}_A & \mathbf{a}_F - \mathbf{a}_B & \mathbf{a}_E - \mathbf{a}_F & 0 & 0 \\ \mathbf{a}_E - \mathbf{a}_C & 0 & 0 & \mathbf{a}_G - \mathbf{a}_E & \mathbf{a}_D - \mathbf{a}_G & \mathbf{a}_C - \mathbf{a}_D \end{array}\right] \tag{4.51j}$$

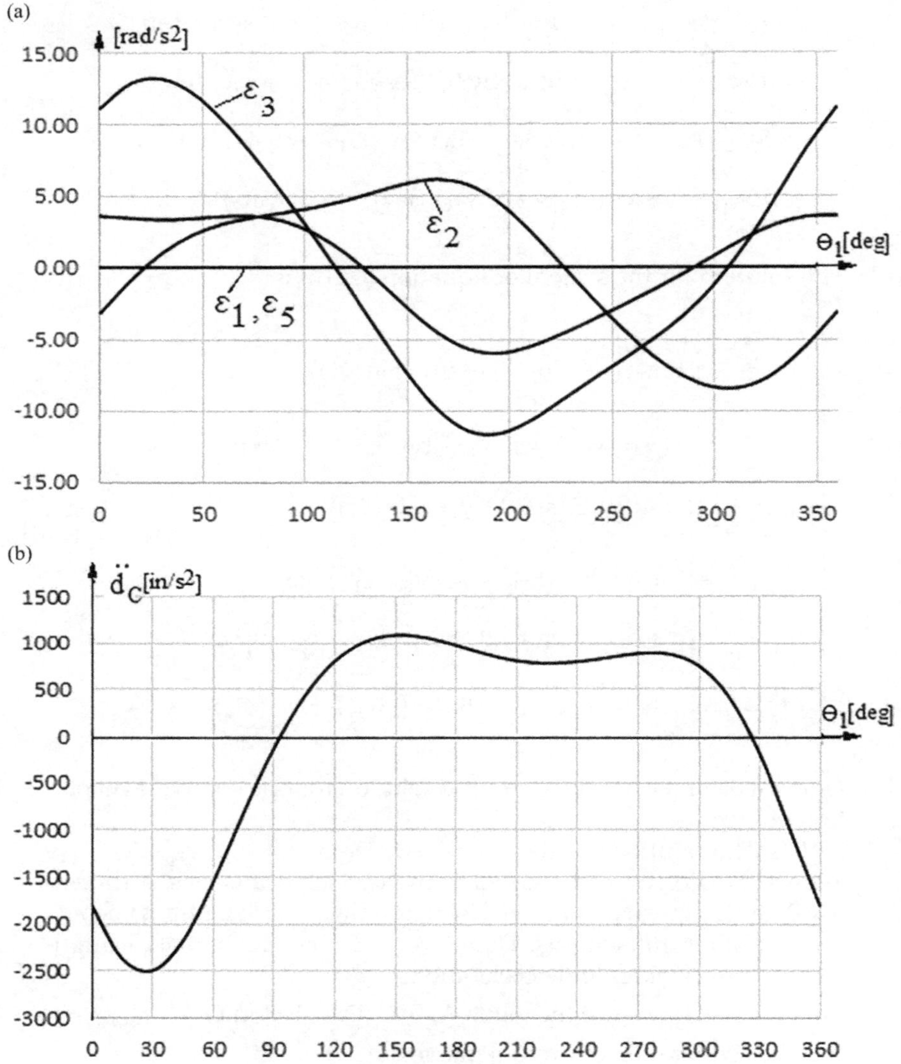

FIGURE 4.8
(a) The angular accelerations for links 1, 2, 3, 4, and 5 and (b) the link 5's linear acceleration.

- The matroidal acceleration vectors are the columns in Equation (4.51j):

  The coordinates of vectors $\ddot{\mathbf{L}}$ are expressed as functions of $x_{BF}, y_{BF}, x_{FE}, y_{FE}, x_{GD}, y_{GD}$ (from the positional equations), 2c = 4 velocities $\omega_2, \omega_3, \omega_4, \dot{d}_C$ (from the velocity equations), and 2c = 4 accelerations $\varepsilon_2, \varepsilon_3, \varepsilon_4, \ddot{d}_C$ (from the acceleration equations):

$$\text{Column 0: } \ddot{\mathbf{L}}^0_{EA} = \begin{Bmatrix} -y_{EA} \\ x_{EA} \end{Bmatrix} \cdot \varepsilon_0 - \begin{Bmatrix} x_{EA} \\ y_{EA} \end{Bmatrix} \cdot \omega_0^2 = \begin{Bmatrix} 0 \\ 0 \end{Bmatrix};$$

$$\ddot{\mathbf{L}}_{CE} = \begin{Bmatrix} -y_{CE} \\ x_{CE} \end{Bmatrix} \cdot \varepsilon_0 - \begin{Bmatrix} x_{CE} \\ y_{CE} \end{Bmatrix} \cdot \omega_0^2 - \mathbf{u}_C \cdot \ddot{d}_C + \tilde{\mathbf{u}}_C 2\dot{d}_C \omega_0$$

$$= -\mathbf{u}_C \cdot \ddot{d}_C = \begin{Bmatrix} -\ddot{d}_C \\ 0 \end{Bmatrix}$$

Note: Edge C is opposite to cycle 2's direction, then a minus is assigned to the complemental (relative and Coriolis) acceleration

$$\text{Column 1: } \ddot{\mathbf{L}}^0_{AB} = \begin{Bmatrix} -y_{AB} \\ x_{AB} \end{Bmatrix} \cdot \varepsilon_1 - \begin{Bmatrix} x_{AB} \\ y_{AB} \end{Bmatrix} \cdot \omega_1^2 = \begin{Bmatrix} -157.914 \cdot \mathrm{Cos}(\theta_1) \\ -157.914 \cdot \mathrm{Sin}(\theta_1) \end{Bmatrix}$$

$$\text{Column 2: } \ddot{\mathbf{L}}^0_{BF} = \begin{Bmatrix} -y_{BF} \\ x_{BF} \end{Bmatrix} \cdot \varepsilon_2 - \begin{Bmatrix} x_{BD} \\ y_{BD} \end{Bmatrix} \cdot \omega_2^2 = \begin{Bmatrix} -y_{BF} \cdot \varepsilon_2 - x_{BF} \cdot \omega_2^2 \\ x_{BF} \cdot \varepsilon_2 - y_{BF} \cdot \omega_2^2 \end{Bmatrix}$$

$$\text{Column 3: } \ddot{\mathbf{L}}^0_{FE} = \begin{Bmatrix} -y_{FE} \\ x_{FE} \end{Bmatrix} \cdot \varepsilon_3 - \begin{Bmatrix} x_{FE} \\ y_{FE} \end{Bmatrix} \cdot \omega_3^2 = \begin{Bmatrix} -y_{FE} \cdot \varepsilon_3 - x_{FE} \cdot \omega_3^2 \\ x_{FE} \cdot \varepsilon_3 - y_{FE} \cdot \omega_3^2 \end{Bmatrix};$$

$$\ddot{\mathbf{L}}^0_{EG} = \begin{Bmatrix} -y_{EG} \\ x_{EG} \end{Bmatrix} \cdot \varepsilon_3 - \begin{Bmatrix} x_{EG} \\ y_{EG} \end{Bmatrix} \cdot \omega_3^2 = \begin{Bmatrix} -y_{EG} \cdot \varepsilon_3 - x_{EG} \cdot \omega_3^2 \\ x_{EG} \cdot \varepsilon_3 - y_{EG} \cdot \omega_3^2 \end{Bmatrix}$$

$$\text{Column 4: } \ddot{\mathbf{L}}^0_{GD} = \begin{Bmatrix} -y_{GD} \\ x_{GD} \end{Bmatrix} \cdot \varepsilon_4 - \begin{Bmatrix} x_{GD} \\ y_{GD} \end{Bmatrix} \cdot \omega_4^2 = \begin{Bmatrix} -y_{GD} \cdot \varepsilon_4 - x_{GD} \cdot \omega_4^2 \\ x_{GD} \cdot \varepsilon_4 - y_{GD} \cdot \omega_4^2 \end{Bmatrix}$$

$$\text{Column 5: } \ddot{\mathbf{L}}^0_{DC} = \begin{Bmatrix} -y_{DC} \\ x_{DC} \end{Bmatrix} \cdot \varepsilon_5 - \begin{Bmatrix} x_{DC} \\ y_{DC} \end{Bmatrix} \cdot \omega_5^2 = \begin{Bmatrix} 0 \\ 0 \end{Bmatrix}$$

(4.51k)

- The joint's accelerations

  Starting from origin O coincidental with the fixed joint A ($\mathbf{a}_A = 0$), the linear accelerations for B C, D, E, F, and G are determined by equating the column entries in the left- and right-hand side of Equation (4.51j)

  Joints on cycle 1:

$$\text{Joint A(fixed joint): } \mathbf{a}_A = \begin{Bmatrix} 0 \\ 0 \end{Bmatrix}; \text{Joint E (fixed joint): } \mathbf{a}_E = \mathbf{a}_A - \ddot{\mathbf{L}}_{EA} = \begin{Bmatrix} 0 \\ 0 \end{Bmatrix}$$

$$\text{Joint B: } \mathbf{a}_B = \mathbf{a}_A + \ddot{\mathbf{L}}^0_{AB} = \begin{Bmatrix} -157.914 \cdot \mathrm{Cos}(\theta_1) \\ -157.914 \cdot \mathrm{Sin}(\theta_1) \end{Bmatrix}$$

$$\text{Joint F: } \mathbf{a}_F = \mathbf{a}_B + \ddot{\mathbf{L}}^0_{BF} = \left\{ \begin{array}{c} -157.914 \cdot \text{Cos}(\theta_1) - y_{BF} \cdot \varepsilon_2 - x_{BF} \cdot \omega_2^2 \\ -157.914 \cdot \text{Sin}(\theta_1) + x_{BF} \cdot \varepsilon_2 - y_{BF} \cdot \omega_2^2 \end{array} \right\}$$

Joint E(fixed joint): checkup equation

$$\mathbf{a}_E = \mathbf{a}_F + \ddot{\mathbf{L}}^0_{FE} = \left\{ \begin{array}{c} -157.914 \cdot \text{Cos}(\theta_1) - y_{BF} \cdot \varepsilon_2 - x_{BF} \cdot \omega_2^2 - y_{EG} \cdot \varepsilon_3 - x_{EG} \cdot \omega_3^2 \\ -157.914 \cdot \text{Sin}(\theta_1) + x_{BF} \cdot \varepsilon_2 - y_{BF} \cdot \omega_2^2 + x_{EG} \cdot \varepsilon_3 - y_{EG} \cdot \omega_3^2 \end{array} \right\}$$

$$= \left\{ \begin{array}{c} 0 \\ 0 \end{array} \right\} \tag{4.51l}$$

Notice that the x and y components in $\mathbf{a}_E$ are the first two Equations (4.51h), which are zero.

Joints on cycle 2:

Joint C:

$$\mathbf{a}_C = \mathbf{a}_E - \ddot{\mathbf{L}}^0_{CE} = \left\{ \begin{array}{c} \ddot{d}_C \\ 0 \end{array} \right\}; \text{ Joint G: } \mathbf{a}_G = \mathbf{a}_E + \ddot{\mathbf{L}}^0_{EG} = \left\{ \begin{array}{c} -y_{EG} \cdot \varepsilon_3 - x_{EG} \cdot \omega_3^2 \\ x_{EG} \cdot \varepsilon_3 - y_{EG} \cdot \omega_3^2 \end{array} \right\}$$

$$\text{Joint D: } \mathbf{a}_D = \mathbf{a}_G + \ddot{\mathbf{L}}^0_{DG} = \left\{ \begin{array}{c} -y_{EG} \cdot \varepsilon_3 - x_{EG} \cdot \omega_3^2 - y_{GD} \cdot \varepsilon_4 - x_{GD} \cdot \omega_4^2 \\ x_{EG} \cdot \varepsilon_3 - y_{EG} \cdot \omega_3^2 + x_{GD} \cdot \varepsilon_4 - y_{GD} \cdot \omega_4^2 \end{array} \right\}$$

d. Linear accelerations of com G_1, G_2, G_3, G_4, and G_3 for the provided location of COM:

$$L_{AG_1} = 2.330''; L_{BG_2} = 12.000''; L_{FG_3} = 6.658''; L_{GG_4} = 6.000''; L_{DG_5} = 0.000''$$

Center of mass link 1: From Equation (4.37a), the acceleration matroidal vector $\ddot{\mathbf{L}}^0_{AG_1}$ is

$$\ddot{\mathbf{L}}^0_{AG_1} = \mathbf{a}^{\tau}_{AG_1} + \mathbf{a}^{\nu}_{AG_1} = \left\{ \begin{array}{c} -y_{AG_1} \\ x_{AG_1} \end{array} \right\} \cdot \varepsilon_1 - \left\{ \begin{array}{c} x_{AG_1} \\ y_{AG_1} \end{array} \right\} \cdot \omega_1^2 = \left\{ \begin{array}{c} -92.064 \cdot \text{Cos}(\theta_1) \\ -92.064 \cdot \text{Sin}(\theta_1) \end{array} \right\} \tag{4.51m}$$

If A, G_1, and B are three collinear points on link 1, then

$$\left\{ \begin{array}{c} x_{AG_1} \\ y_{AG_1} \end{array} \right\} = \frac{L_{AG_1}}{L_{AB}} \left[\begin{array}{cc} \text{Cos}(0°) & -\text{Sin}(0°) \\ \text{Sin}(0°) & \text{Cos}(0°) \end{array} \right] \cdot \left\{ \begin{array}{c} x_{AB} \\ y_{AB} \end{array} \right\} = \left\{ \begin{array}{c} 0.583 \cdot x_{AB} \\ 0.583 \cdot y_{AB} \end{array} \right\}$$

From Equation (4.37), the link 1's COM acceleration is

$$\mathbf{a}_{G_1} = \mathbf{a}_A + \ddot{\mathbf{L}}^0_{AG_1}\,;\ \left|a_{G_1}\right| = \sqrt{\left(a^x_{G_1}\right)^2 + \left(a^y_{G_1}\right)^2}$$

Center of mass link 2: From Equation (4.37a), the acceleration matroidal vector $\ddot{\mathbf{L}}^0_{BG_2}$ is

$$\ddot{\mathbf{L}}^0_{BG_2} = \mathbf{a}^\tau_{BG_2} + \mathbf{a}^v_{BG_2} = \begin{Bmatrix} -y_{BG_2} \\ x_{BG_2} \end{Bmatrix} \cdot \varepsilon_2 - \begin{Bmatrix} x_{BG_2} \\ y_{BG_2} \end{Bmatrix} \cdot \omega_2^2$$

$$= \begin{Bmatrix} -0.500 \cdot y_{BF} \cdot \varepsilon_2 - 0.500 \cdot x_{BF} \cdot \omega_2^2 \\ 0.500 \cdot x_{BF} \cdot \varepsilon_2 - 0.500 \cdot y_{BF} \cdot \omega_2^2 \end{Bmatrix}$$

If B, G_2, and F are three collinear points on link 2, then

$$\begin{Bmatrix} x_{BG_2} \\ y_{BG_2} \end{Bmatrix} = \frac{L_{BG_2}}{L_{BF}} \begin{bmatrix} \mathrm{Cos}(0°) & -\mathrm{Sin}(0°) \\ \mathrm{Sin}(0°) & \mathrm{Cos}(0°) \end{bmatrix} \cdot \begin{Bmatrix} x_{BF} \\ y_{BF} \end{Bmatrix} = \begin{Bmatrix} 0.500 \cdot x_{BF} \\ 0.500 \cdot y_{BF} \end{Bmatrix}$$

From Equation (4.37), the link 2's COM acceleration is

$$\mathbf{a}_{G_2} = \mathbf{a}_B + \ddot{\mathbf{L}}^0_{BG_2}\,;\ \left|a_{G_2}\right| = \sqrt{\left(a^x_{G_2}\right)^2 + \left(a^y_{G_2}\right)^2}$$

Center of mass link 3: From Equation (4.37a), the acceleration matroidal vector $\ddot{\mathbf{L}}^0_{FG_3}$ is

$$\ddot{\mathbf{L}}^0_{FG_3} = \mathbf{a}^\tau_{FG_3} + \mathbf{a}^v_{FG_3} = \begin{Bmatrix} -y_{FG_3} \\ x_{FG_3} \end{Bmatrix} \cdot \varepsilon_3 - \begin{Bmatrix} x_{FG_3} \\ y_{FG_3} \end{Bmatrix} \cdot \omega_3^2$$

$$= \begin{Bmatrix} -0.476 \cdot y_{FE} \cdot \varepsilon_3 - 0.476 \cdot x_{FE} \cdot \omega_3^2 \\ 0.476 \cdot x_{FE} \cdot \varepsilon_3 - 0.476 \cdot y_{FE} \cdot \omega_3^2 \end{Bmatrix}$$

If F, G_3, and E are three collinear points on link 3, then

$$\begin{Bmatrix} x_{FG_3} \\ y_{FG_3} \end{Bmatrix} = \frac{L_{FG_3}}{L_{FE}} \begin{bmatrix} \mathrm{Cos}(0°) & -\mathrm{Sin}(0°) \\ \mathrm{Sin}(0°) & \mathrm{Cos}(0°) \end{bmatrix} \cdot \begin{Bmatrix} x_{FE} \\ y_{FE} \end{Bmatrix} = \begin{Bmatrix} 0.476 \cdot x_{FE} \\ 0.476 \cdot y_{FE} \end{Bmatrix}$$

From Equation (4.37), the link 3's COM acceleration is

$$\mathbf{a}_{G_3} = \mathbf{a}_F + \ddot{\mathbf{L}}^0_{FG_3}\,;\ \left|\mathbf{a}_{G_3}\right| = \sqrt{\left(a^x_{G_3}\right)^2 + \left(a^y_{G_3}\right)^2}$$

Center of mass link 4: From Equation (4.37a,) the acceleration matroidal vector $\ddot{\mathbf{L}}^0_{GG_4}$ is

$$\ddot{\mathbf{L}}^0_{GG_4} = \mathbf{a}^\tau_{GG_4} + \mathbf{a}^v_{GG_4} = \begin{Bmatrix} -y_{GG_4} \\ x_{GG_4} \end{Bmatrix} \cdot \varepsilon_4 - \begin{Bmatrix} x_{GG_4} \\ y_{GG_4} \end{Bmatrix} \cdot \omega_4^2$$

$$= \begin{Bmatrix} -0.500 \cdot y_{GD} \cdot \varepsilon_4 - 0.500 \cdot x_{GD} \cdot \omega_4^2 \\ 0.500 \cdot x_{GD} \cdot \varepsilon_4 - 0.500 \cdot y_{GD} \cdot \omega_4^2 \end{Bmatrix}$$

If G, G_4, and D are three collinear points on link 3, then

$$\begin{Bmatrix} x_{GG_4} \\ y_{GG_4} \end{Bmatrix} = \frac{L_{GG_4}}{L_{GD}} \begin{bmatrix} \mathrm{Cos}(0°) & -\mathrm{Sin}(0°) \\ \mathrm{Sin}(0°) & \mathrm{Cos}(0°) \end{bmatrix} \cdot \begin{Bmatrix} x_{GD} \\ y_{GD} \end{Bmatrix} = \begin{Bmatrix} 0.500 \cdot x_{GD} \\ 0.500 \cdot y_{GD} \end{Bmatrix}$$

From Equation (4.37), the link 3's COM acceleration is

$$\mathbf{a}_{G_3} = \mathbf{a}_F + \ddot{\mathbf{L}}^0_{FG_3}\,;\left|\mathbf{a}_{G_3}\right| = \sqrt{\left(a^x_{G_3}\right)^2 + \left(a^y_{G_3}\right)^2}$$

Center of mass link 5: From Equation (4.37a), the acceleration matroidal vector $\ddot{\mathbf{L}}^0_{DG_5}$ is

$$\ddot{\mathbf{L}}^0_{DG_5} = \mathbf{a}^\tau_{DG_5} + \mathbf{a}^v_{DG_5} = \begin{Bmatrix} -y_{DG_5} \\ x_{DG_5} \end{Bmatrix} \cdot \varepsilon_5 - \begin{Bmatrix} x_{DG_5} \\ y_{DG_5} \end{Bmatrix} \cdot \omega_5^2 = \begin{Bmatrix} 0 \\ 0 \end{Bmatrix}$$

If D and G_5 are coincidental, then the link 5's COM acceleration is

$$\mathbf{a}_{G_5} = \mathbf{a}_D + \ddot{\mathbf{L}}^0_{DG_5} = \mathbf{a}_D;\left|\mathbf{a}_{G_5}\right| = \sqrt{\left(a^x_{G_5}\right)^2 + \left(a^y_{G_5}\right)^2}$$

4.5 Governing Equations in Dynamics of Mechanisms

4.5.1 The Governing Force Equations for Open Cycle Mechanisms

The dynamics equations determine the dynamic reactions and the differential equations of motion.

For a mechanical system with m mobile links, let us consider three of them: m – 2, m – 1, and m. Link m – 1 is connected to two other links at the joints Y and Z (Figure 4.9). The graph and vector forms of dynamic equations are reported in [1]. The digraph approach and matrix form is developed further in this chapter for the modeling of dynamics equations [2,3].

Newton's second law for forces applies to node (link) m – 1, as shown in Equation (4.52):

$$m_{m-1} \cdot \mathbf{a}^0_{G_{m-1}} = +\mathbf{N}^0_Y - \mathbf{N}^0_Z + \mathbf{f}^{ext,0}_{m-1} \tag{4.52}$$

Equation (4.52) is adapted to the digraph in the form, as shown in Equation (4.53):

$$+\mathbf{N}^0_Y - \mathbf{N}^0_Z = \mathbf{f}^{in,0}_{m-1} - \mathbf{f}^{ext,0}_{m-1} \tag{4.53}$$

where

- m_{m-1} is the mass of link m – 1, attached to node m – 1 in digraph, a positive scalar measured in kg or slugs.
- $\mathbf{a}^0_{G_{m-1}}$ is the link m – 1's COM vector acceleration with respect to the fixed frame 0, which is shown in superscript. It is attached to the node m – 1 in the digraph.
- $\mathbf{f}^{in,0}_{m-1} = m_{m-1} \cdot \mathbf{a}^0_{G_{m-1}}$ is the link m – 1's *inertial force* with respect to the fixed frame 0.
- $\mathbf{f}^{ext,0}_{m-1}$ is the external resultant force acting on the link m – 1 which is written with respect to the fixed frame 0. It includes the force of gravity and all other forces acting on the link.
- $+\mathbf{N}^0_Y$ is the reaction force vector, for the link m – 2's (tail-arrow Y) action on the link m – 1(head-arrow Y), assigned to the joint Y (edge Y in digraph). The edge Y is incident (arrow enters) to the node m – 1, and the "+" is assigned to the reaction vector. The vector is written with respect to the fixed frame 0, which is shown in superscript.
- $-\mathbf{N}^0_Z$ is the reaction force vector, for the link m – 1's (tail-arrow Z) action on the link m (head-arrow Z). The edge Z is divergent (arrow leaves) from the node m – 1, and the "–" is assigned to the reaction vector. With a minus in front, the reaction force denotes the vector for the link m's action on the link m – 1. It is written with respect to the fixed frame 0, which is shown in superscript.

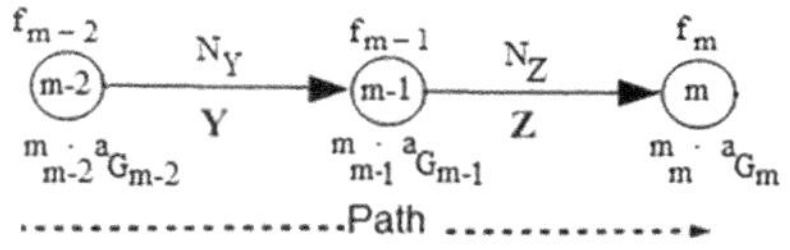

FIGURE 4.9
Reaction forces effective and external forces applied on link m – 1.

4.5.2 The Incidental and Transfer-IT Method on Dynamic Forces and Differential Equations for Open Cycle Mechanisms

Newton's second law is applied for each node (link) m on the tree, exemplified on the TRRT single open cycle spatial mechanism (Figure 4.10), for which we determine the governing equations for the dynamic reactions and the differential equations of motion. Further, the method is developed in matrix form for closed cycle mechanisms with single or multiple c-cycles [2,3].

Newton's second law is applied for each node (link), as shown in Equation (4.54):

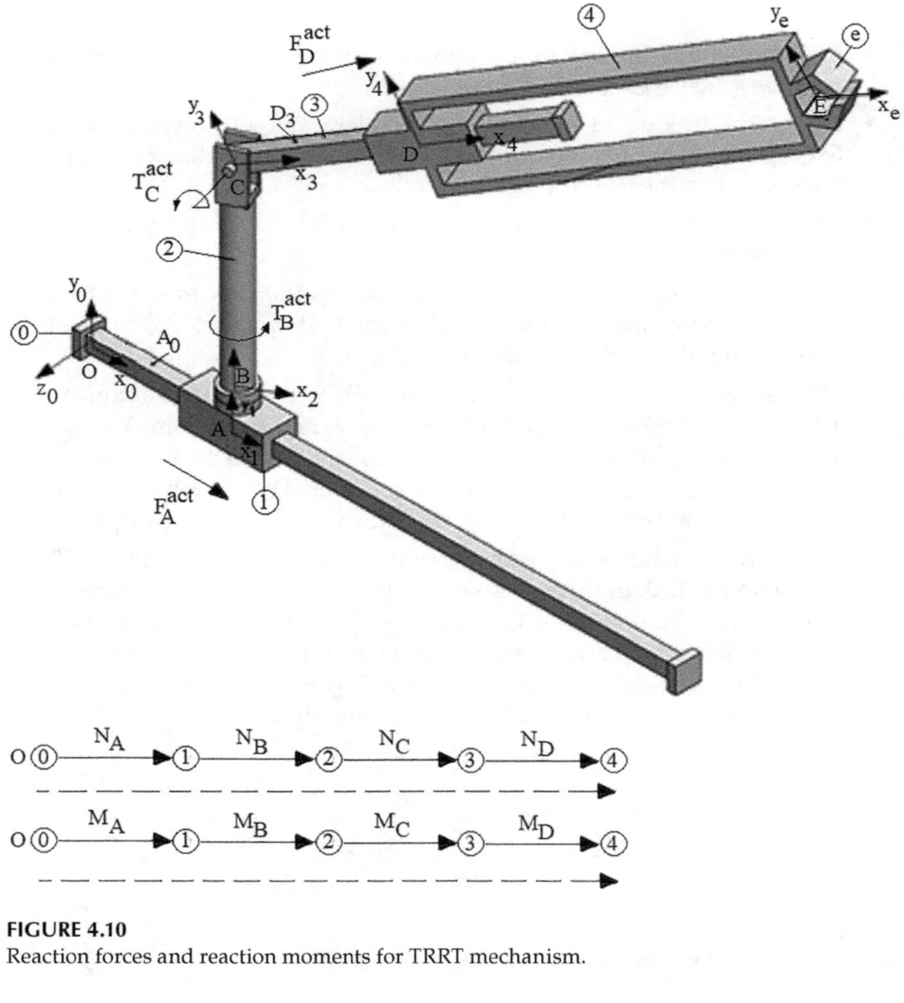

FIGURE 4.10
Reaction forces and reaction moments for TRRT mechanism.

$$\begin{aligned}
&\textit{Node } 1(\text{Link } 1): +\mathbf{N}_A^0 - \mathbf{N}_B^0 = \mathbf{f}_1^{in,0} - \mathbf{f}_1^{ext,0}\\
&\textit{Node } 2(\text{Link } 2): +\mathbf{N}_B^0 - \mathbf{N}_C^0 = \mathbf{f}_2^{in,0} - \mathbf{f}_2^{ext,0}\\
&\textit{Node } 3(\text{Link } 3): +\mathbf{N}_C^0 - \mathbf{N}_D^0 = \mathbf{f}_3^{in,0} - \mathbf{f}_3^{ext,0}\\
&\textit{Node } 4(\text{Link } 4): +\mathbf{N}_D^0 = \mathbf{f}_4^{in,0} - \mathbf{f}_4^{ext,0}
\end{aligned} \tag{4.54}$$

4.5.3 The Incidence Nodes-Edges Matrix

The digraph for an open cycle is a tree; the number of nodes in the tree is n = m + 1 (m mobile links plus the fixed link), and the number of edges (named arcs) is a. The property m = a holds for any tree.

The "Incidence Nodes-Edges Matrix" was introduced in Equation (1.11) and defines the arc–node connection in the tree. It is a n × a matrix in which each row is assigned to a node and each column is assigned to an arc. Its entries, noted $g_{n,a}$, are as follows:

- $g_{n,a} = -1$, if node n is the n_{tail}
- $g_{n,a} = +1$, if node n is the n_{head}
- $g_{n,a} = 0$, otherwise

Each column has two equal and opposite entries. Hence, the sum of entries in each column is zero. The entries for the digraph of a TRRT mechanism are as follows:

$$\underline{\mathbf{G}} = \begin{array}{c} \\ 0\\ 1\\ 2\\ 3\\ 4 \end{array}\begin{array}{c} \begin{array}{cccc} A & B & C & D \end{array}\\ \left[\begin{array}{cccc} -1 & 0 & 0 & 0\\ \hdashline +1 & -1 & 0 & 0\\ 0 & +1 & -1 & 0\\ 0 & 0 & +1 & -1\\ 0 & 0 & 0 & +1 \end{array}\right] \end{array} \tag{4.55}$$

The rows in matrix $\underline{\mathbf{G}}$ are not independent; this is proven by adding to the first row all the other rows and obtaining a row with all zero entries. By deleting the first row, corresponding to the fixed link 0, one can assure the independence of the other rows.

4.5.4 The Reduced Incidence Nodes-Edges Matrix, G

The "Reduced Incidence Nodes-Edges Matrix" is defined as a m × a square matrix by deleting the first row in matrix **G**, as shown in Equation (4.56)

$$
\mathbf{G} = \begin{array}{c} \\ 1 \\ 2 \\ 3 \\ 4 \end{array} \begin{array}{c} \begin{array}{cccc} A & B & C & D \end{array} \\ \begin{bmatrix} +1 & -1 & 0 & 0 \\ 0 & +1 & -1 & 0 \\ 0 & 0 & +1 & -1 \\ 0 & 0 & 0 & +1 \end{bmatrix} \end{array} \tag{4.56}
$$

Note: The coefficients in front of the dynamic reactions, **N**, in Equation (4.54) are the entries in the matrix **G**. Therefore, these equations in matrix form and vector entries are written as

$$
\begin{bmatrix} +1 & -1 & 0 & 0 \\ 0 & +1 & -1 & 0 \\ 0 & 0 & +1 & -1 \\ 0 & 0 & 0 & +1 \end{bmatrix} * \begin{Bmatrix} \mathbf{N}_A^0 \\ \mathbf{N}_B^0 \\ \mathbf{N}_C^0 \\ \mathbf{N}_D^0 \end{Bmatrix} = \begin{Bmatrix} m_1 \cdot \mathbf{a}_{G_1}^0 \\ m_2 \cdot \mathbf{a}_{G_2}^0 \\ m_3 \cdot \mathbf{a}_{G_3}^0 \\ m_4 \cdot \mathbf{a}_{G_4}^0 \end{Bmatrix} - \begin{Bmatrix} \mathbf{f}_1^{ext,0} \\ \mathbf{f}_2^{ext,0} \\ \mathbf{f}_3^{ext,0} \\ \mathbf{f}_4^{ext,0} \end{Bmatrix} = \begin{Bmatrix} \mathbf{P}_{G_1} \\ \mathbf{P}_{G_2} \\ \mathbf{P}_{G_3} \\ \mathbf{P}_{G_4} \end{Bmatrix} \tag{4.57}
$$

Notation: The resultant of inertial and external forces reduced to COM of each link is noted as $\mathbf{P}_{G_m}^0$:

$$
\mathbf{P}_{G_m}^0 = \mathbf{f}_m^{in,0} - \mathbf{f}_m^{ext,0} \tag{4.58}
$$

$$
\mathbf{P}_{G_m}^0 = \begin{Bmatrix} \mathbf{P}_{G_1} \\ \mathbf{P}_{G_2} \\ \mathbf{P}_{G_3} \\ \mathbf{P}_{G_4} \end{Bmatrix} = \begin{Bmatrix} m_1 \cdot \mathbf{a}_{G_1}^0 - \mathbf{f}_1^{ext,0} \\ m_2 \cdot \mathbf{a}_{G_2}^0 - \mathbf{f}_2^{ext,0} \\ m_3 \cdot \mathbf{a}_{G_3}^0 - \mathbf{f}_3^{ext,0} \\ m_4 \cdot \mathbf{a}_{G_4}^0 - \mathbf{f}_4^{ext,0} \end{Bmatrix} \tag{4.59}
$$

The column vectors in Equation (4.57) are noted as:

$$
\mathbf{N}_a^0 = \begin{Bmatrix} \mathbf{N}_A^0 \\ \mathbf{N}_B^0 \\ \mathbf{N}_C^0 \\ \mathbf{N}_D^0 \end{Bmatrix} \text{ the column of reaction forces at joints}
$$

Then, Equation (4.58) becomes

$$
\mathbf{G}_{m \times a} * \mathbf{N}_a^0 = \mathbf{P}_{G_m}^0 \tag{4.60}
$$

4.5.5 The Path Matrix, Z

This matrix was introduced in Equation (1.17), as a $a \times m$ square matrix in which each row is assigned to an arc in the tree and each column is assigned to an open path initiated at node 0 toward each node m. Its entries, noted $z_{a,m}$, are as follows:

- $z_{a,m} = +1$, if arc a is on the path and same direction as the path.
- $z_{a,m} = -1$, if arc a is on the path and opposite to the path.
- $z_{a,m} = 0$, if arc a is not on the path.

The entries for the digraph of TRRT mechanism is

$$\mathbf{Z} = \begin{array}{c} \\ A \\ B \\ C \\ D \end{array} \begin{array}{c} \begin{array}{cccc} 1 & 2 & 3 & 4 \end{array} \\ \begin{bmatrix} +1 & +1 & +1 & +1 \\ 0 & +1 & +1 & +1 \\ 0 & 0 & +1 & +1 \\ 0 & 0 & 0 & +1 \end{bmatrix} \end{array} = \begin{bmatrix} z_{A,1} & z_{A,2} & z_{A,3} & z_{A,4} \\ 0 & z_{B,2} & z_{B,3} & z_{B,4} \\ 0 & 0 & z_{C,3} & z_{C,4} \\ 0 & 0 & 0 & z_{D,4} \end{bmatrix} \tag{4.61}$$

The matrix Z for the TRRT mechanism is an upper triangular matrix, where all z entries are +1.

4.5.6 Relation for G and Z Matrices

The product between matrices **G** and **Z** is the unit matrix:

$$\mathbf{Z} * \mathbf{G} = \mathbf{G} * \mathbf{Z} = \mathbf{U}$$

Equation (4.57) is left-hand-side-multiplied by Z matrix, and considering Equation (4.61), then

$$\mathbf{N}_a^0 = \mathbf{Z}_{a \times m} * \mathbf{P}_{G_m}^0 \tag{4.62}$$

Notation: The right-hand-side term in Equation (4.62) is named *IT force, noted* $\mathbf{P}_a^0$:

$$\mathbf{P}_a^0 = \mathbf{Z}_{a \times m} * \mathbf{P}_{G_m}^0 \tag{4.63}$$

Its entries are determined next from:

$$\mathbf{P}_a^0 = \begin{Bmatrix} \mathbf{P}_A^0 \\ \mathbf{P}_B^0 \\ \mathbf{P}_C^0 \\ \mathbf{P}_D^0 \end{Bmatrix} = \begin{Bmatrix} z_{A,1} \cdot \mathbf{P}_{G1} + z_{A,2} \cdot \mathbf{P}_{G2} + z_{A,3} \cdot \mathbf{P}_{G3} + z_{A,4} \cdot \mathbf{P}_{G4} \\ z_{B,2} \cdot \mathbf{P}_{G2} + z_{B,3} \cdot \mathbf{P}_{G3} + z_{B,4} \cdot \mathbf{P}_{G4} \\ z_{C,3} \cdot \mathbf{P}_{G3} + z_{C,4} \cdot \mathbf{P}_{G4} \\ z_{D,4} \cdot \mathbf{P}_{G4} \end{Bmatrix} \tag{4.63a}$$

Conclusion:

- By multiplying by Z matrix, the resultant (inertial and external) forces are transferred from COM to tree's joints. Therefore, these forces are used in the calculation of joint reactions.
- The triangular form in Equation (4.63a) suggests a simple back substitution: the result from the last row is substituted in the row before the last, etc.

4.5.7 Equations for Reaction Forces on Open Cycle Mechanisms: The IT-Resistant Force

Considering Equations (4.62) and (4.63), then

$$\mathbf{N}_a^0 - \mathbf{P}_a^0 = 0 \tag{4.64}$$

The forces $\left(-\mathbf{P}_a^0\right)$, named *"IT-resistant forces,"* are assigned to joints (arcs tree).

Statement 4.6

The reaction forces and IT-resistant forces are a system of forces in fictitious equilibrium.

Equation (4.65) is the governing equation to determine the reaction forces in joints:

$$\mathbf{N}_a^0 = \mathbf{P}_a^0 \tag{4.65}$$

The characteristics of $\left(-\mathbf{P}_a^0\right)$ IT-resistant forces are as follows:

- They are dynamic forces, measured in [N] or [lbf].
- They are in a *fictitious static equilibrium* with the reaction forces in joints.
- The distinction between an IT-resistant force $\left(-\mathbf{P}_a^0\right)$ and the inertial force from Newton's law, $\mathbf{f}_m^{in,0} = -m_m \cdot \mathbf{a}_{Gm}^0$:

IT-resistant force $\left(-\mathbf{P}_a^0\right)$ is *incidental (assigned) to the joint a (edge tree), whereas*

$\mathbf{f}_m^{in,0}$ is *incidental (assigned) to the link m's (node tree)* COM

(4.66)

4.5.8 The Evaluation of IT Forces

The *IT forces* are the entries in a m × 1 column matrix, $\mathbf{P}_a^0$, *incidental* (assigned) to arcs in the tree. Its entries are vector forces, the difference between inertial and external forces, assigned to nodes. The multiplication by the path matrix **Z** *transfers* the above difference from COM to joints.

Then, the IT forces are determined from the tree using the recursion Equation (4.67):

$$\mathbf{P}_Z^0 = {}_{m=1}\big|z_{Z,1} \cdot \mathbf{P}_{G_1}^0 + {}_{m=2}\big|z_{Z,2} \cdot \mathbf{P}_{G_2}^0 + {}_{m=3}\big|z_{Z,3} \cdot \mathbf{P}_{G_3}^0 + {}_{m=4}\big|z_{Z,4} \cdot \mathbf{P}_{G_4}^0 \quad (4.67)$$

For the TRRT mechanism, in Equation (4.67), all entries in the matrix **Z** are +1.

Thus, for $\mathbf{P}_A^0$, all the terms on the right-hand side of $_{m=1}|$ are considered; for $\mathbf{P}_B^0$, all the terms on the right-hand side of $_{m=2}|$ are considered. This is equivalent to generating a set of paths (shown as dashed line); all paths are initiated at node 0 and ended in sequence at nodes 1, 2, 3, and 4.

- Joint A is in all paths initiated at node 0 and ended at nodes 1, 2, 3, and 4 (Figure 4.11):

$$\mathbf{P}_A^0 = z_{A,1} \cdot \left(\mathbf{f}_1^{in,0} - \mathbf{f}_1^{ext,0}\right) + z_{A,2} \cdot \left(\mathbf{f}_2^{in,0} - \mathbf{f}_2^{ext,0}\right) + z_{A,3} \cdot \left(\mathbf{f}_3^{in,0} - \mathbf{f}_3^{ext,0}\right) + z_{A,4} \cdot \left(\mathbf{f}_4^{in,0} - \mathbf{f}_4^{ext,0}\right) \quad (4.67a)$$

- Joint B is in the paths initiated at node 0 and ended at nodes 2, 3, and 4 (Figure 4.12):

$$\mathbf{P}_B^0 = z_{B,2} \cdot \left(\mathbf{f}_2^{in,0} - \mathbf{f}_2^{ext,0}\right) + z_{B,3} \cdot \left(\mathbf{f}_3^{in,0} - \mathbf{f}_3^{ext,0}\right) + z_{B,4} \cdot \left(\mathbf{f}_4^{in,0} - \mathbf{f}_4^{ext,0}\right) \quad (4.67b)$$

- Joint C is in the paths initiated at node 0 and ended at nodes 3 and 4 (Figure 4.13):

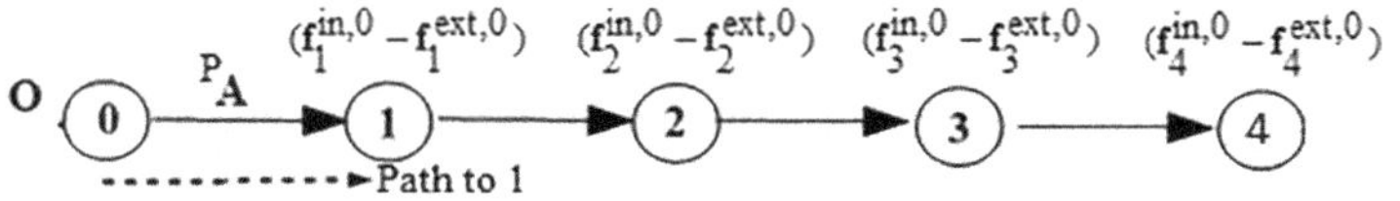

FIGURE 4.11
Edge A belongs to paths initiated at node 0 and ended at nodes 1, 2, 3, and 4.

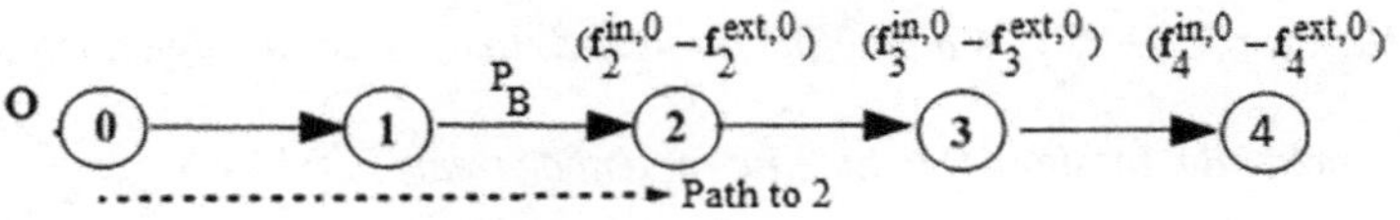

FIGURE 4.12
Edge B belongs to paths initiated at node 0 and ended at nodes 2, 3, and 4.

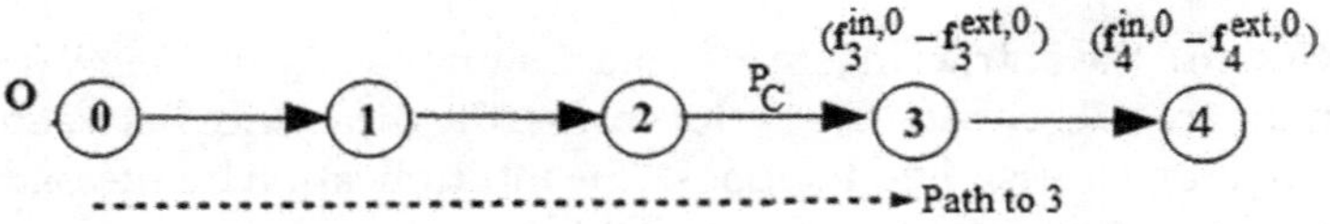

FIGURE 4.13
Edge C belongs to paths initiated at node 0 and ended at nodes 3 and 4.

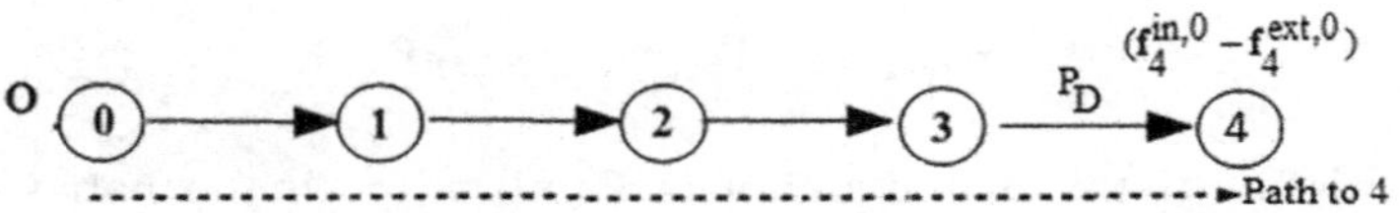

FIGURE 4.14
Edge D belongs to paths initiated at node 0 and ended at node 4.

$$P_C^0 = z_{C,3} \cdot \left(\mathbf{f}_3^{in,0} - \mathbf{f}_3^{ext,0}\right) + z_{C,4} \cdot \left(\mathbf{f}_4^{in,0} - \mathbf{f}_4^{ext,0}\right) \tag{4.67c}$$

- Joint D is in the paths initiated at node 0 and ended at node 4 (Figure 4.14):

$$\mathbf{P}_D^0 = +z_{D,4} \cdot \left(\mathbf{f}_4^{in,0} - \mathbf{f}_4^{ext,0}\right) \tag{4.67d}$$

4.6 The IT Method on Dynamic Moments and Differential Equations for Open Cycle Mechanisms

Newton's second law for moment of forces applies to the node (link) m – 1, as shown in Equation (4.68):

$$\dot{\mathbf{K}}_{G_{m-1}}^{in,0} = +\mathbf{M}_Y^0 - \mathbf{M}_Z^0 + \mu_{G_{m-1}}^{ext,0} + \left(\tilde{\mathbf{L}}_{G_{m-1}Y}^0 * \mathbf{N}_Y^0 - \tilde{\mathbf{L}}_{G_{m-1}Z}^0 * \mathbf{N}_Z^0\right) \tag{4.68}$$

It is adapted to the digraph in the form as shown in Equation (4.69) (Figure 4.15):

$$+\mathbf{M}_Y^0 - \mathbf{M}_Z^0 = \mathbf{Q}_{G_{m-1}}^0 - \left(\tilde{\mathbf{L}}_{G_{m-1}Y}^0 * \mathbf{N}_Y^0 - \tilde{\mathbf{L}}_{G_{m-1}Z}^0 * \mathbf{N}_Z^0\right) \tag{4.69}$$

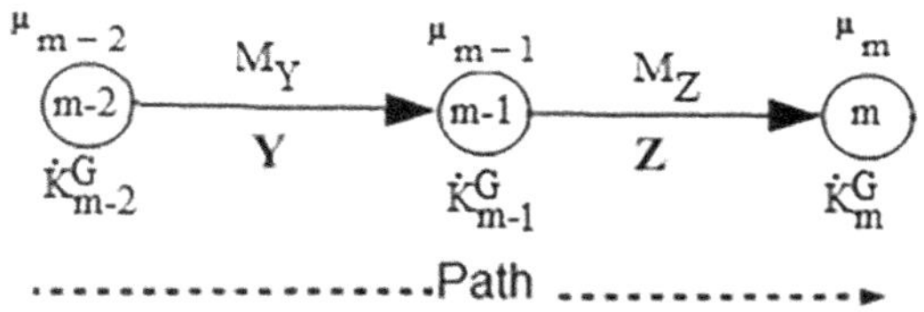

FIGURE 4.15
Moment reaction, rate of angular momentum, and moment of external forces.

Notation: The resultant moment of all inertial and external forces acting on links with respect to its COMs is

$$\mathbf{Q}^0_{G_{m-1}} = \dot{\mathbf{K}}^{in,0}_{G_{m-1}} - \boldsymbol{\mu}^{ext,0}_{G_{m-1}} \tag{4.70}$$

where

- $\mathbf{M}^0_Y$ and $\mathbf{M}^0_Z$ are the reaction couples in joints.
- $\boldsymbol{\mu}^{ext,0}_{G_{m-1}}$ is the resultant moment of all external forces acting on the link m − 1 with respect to its COM
- The resultant moment of reaction forces in the joints Y and Z, with respect to COM of link m − 1, viewed in the fixed frame, is shown in Equation (4.71):

$$+\tilde{\mathbf{L}}^0_{G_{m-1}Y} * \mathbf{N}^0_Y - \tilde{\mathbf{L}}^0_{G_{m-1}Z} * \mathbf{N}^0_Z \tag{4.71}$$

- $\dot{\mathbf{K}}^{in,0}_{G_{m-1}}$ is the rate of change of link m − 1's angular momentum, with respect to its COM at the fixed frame 0, and the link m − 1 is attached to the node m − 1 in diagraph. The rate of change of angular momentum is determined using Euler's formula, as a function of the mass moment of inertia, as shown in Equation (4.72):

$$\dot{\mathbf{K}}^{in,0}_{G_{m-1}} = \mathbf{I}_{G_{m-1}} * \boldsymbol{\varepsilon}_{m-1} + \tilde{\boldsymbol{\omega}}_{m-1} * \left(\mathbf{I}_{G_{m-1}} * \boldsymbol{\omega}_{m-1}\right) \tag{4.72}$$

- $\mathbf{I}_{G_{m-1}}$ is the moment and product of inertia 3 × 3 symmetric matrix, with respect to COM, as shown in Equation (4.73). The axial moments of inertia $I_{x_{m-1}}$, $I_{y_{m-1}}$, and $I_{z_{m-1}}$ are the entries on the diagonal. The products of inertias I_{xy}, I_{xz}, and I_{yz} become zero in the case where the local frame axes x, y, and z are the principal moments of inertia.

$$\mathbf{I}_{G_{m-1}} = \begin{bmatrix} I_{x_{m-1}} & -I_{x_{m-1}y_{m-1}} & -I_{x_{m-1}z_{m-1}} \\ -I_{x_{m-1}y_{m-1}} & I_{y_{m-1}} & -I_{y_{m-1}z_{m-1}} \\ -I_{x_{m-1}z_{m-1}} & -I_{y_{m-1}z_{m-1}} & I_{z_{m-1}} \end{bmatrix} \tag{4.73}$$

- ε_{m-1} is link m − 1's angular acceleration vector, with respect to the fixed frame 0.
- ω_{m-1} and $\tilde{\omega}_{m-1}$ are link m − 1's angular velocity vector and the 3 × 3 skew-symmetric matrix from angular velocity components, respectively, with respect to the fixed frame 0.
- $\tilde{\mathbf{L}}^0_{G_{m-1}Y}$ is the skew-symmetric matrix for vector $\mathbf{L}^0_{G_{m-1}Y}$

$$\tilde{\mathbf{L}}^0_{G_{m-1}Y} = \mathbf{r}_Y - \mathbf{r}_{G_{m-1}} \tag{4.74}$$

- $\tilde{\mathbf{L}}^0_{G_{m-1}Z}$ is the skew-symmetric matrix for vector $\mathbf{L}^0_{G_{m-1}Z}$, with

$$\tilde{\mathbf{L}}^0_{G_{m-1}Z} = \mathbf{r}_Z - \mathbf{r}_{G_{m-1}} \tag{4.75}$$

4.6.1 Equations for Reaction Moment on Open Cycle Mechanisms: The IT-Resistant Moment

The tree's m vector moment equations are reported in [2,3].

The method is exemplified on the TRRT single open cycle spatial mechanism for which we determine the governing equations for the dynamic reaction moments and the differential equations of motion. Further, the method is generalized for closed cycle mechanisms with single or multiple c-cycles.

Equation (4.35) is applied for each node (link), as shown in Equation (4.76):

$$\begin{aligned}
&\textit{Node } 1\,(\text{Link } 1){:} +\mathbf{M}^0_A - \mathbf{M}^0_B = \mathbf{Q}^0_{G_1} - \left(\tilde{\mathbf{L}}^0_{G_1A} * \mathbf{N}^0_A - \tilde{\mathbf{L}}^0_{G_1B} * \mathbf{N}^0_B\right)\\
&\textit{Node } 2\,(\text{Link } 2){:} +\mathbf{M}^0_B - \mathbf{M}^0_C = \mathbf{Q}^0_{G_2} - \left(\tilde{\mathbf{L}}^0_{G_2B} * \mathbf{N}^0_B - \tilde{\mathbf{L}}^0_{G_2C} * \mathbf{N}^0_C\right)\\
&\textit{Node } 3\,(\text{Link } 3){:}\ \mathbf{M}^0_C - \mathbf{M}^0_D = \mathbf{Q}^0_{G_3} - \left(\tilde{\mathbf{L}}^0_{G_3C} * \mathbf{N}^0_C - \tilde{\mathbf{L}}^0_{G_3D} * \mathbf{N}^0_D\right)\\
&\textit{Node } 4\,(\text{Link } 4){:} +\mathbf{M}^0_D = \mathbf{Q}^0_{G_4} - \left(\tilde{\mathbf{L}}^0_{G_4D} * \mathbf{N}^0_D\right)
\end{aligned} \tag{4.76}$$

4.6.2 The Position Vector Skew-Symmetric Matrix, $\mathbf{G}_{m\times a}(\tilde{\mathbf{L}})$ and $\mathbf{Z}_{a\times m}(\tilde{\mathbf{L}})$

- The matrix $\mathbf{G}_{m\times a}(\tilde{\mathbf{L}})$ is the m × a matrix whose entries are the *skew-symmetric matrices* for the position vectors originated from COM toward the tree's joints.

  It is generated from the **G** matrix entries, as shown in Equation (4.56), as:

$$\mathbf{G}_{m\times a} = \begin{array}{c} \\ 1 \\ 2 \\ 3 \\ 4 \end{array} \begin{array}{c} \begin{array}{cccc} A & B & C & D \end{array} \\ \left[\begin{array}{cccc} +1 & -1 & 0 & 0 \\ 0 & +1 & -1 & 0 \\ 0 & 0 & +1 & -1 \\ 0 & 0 & 0 & +1 \end{array}\right] \end{array} \Rightarrow \mathbf{G}_{m\times a}(\tilde{\mathbf{L}})$$

$$= \begin{bmatrix} +1\cdot\tilde{\mathbf{L}}^0_{G_1A} & -1\cdot\tilde{\mathbf{L}}^0_{G_1B} & 0 & 0 \\ 0 & +1\cdot\tilde{\mathbf{L}}^0_{G_2B} & -1\cdot\tilde{\mathbf{L}}^0_{G_2C} & 0 \\ 0 & 0 & +1\cdot\tilde{\mathbf{L}}^0_{G_3C} & -1\cdot\tilde{\mathbf{L}}^0_{G_3D} \\ 0 & 0 & 0 & +1\cdot\tilde{\mathbf{L}}^0_{G_4D} \end{bmatrix} \tag{4.77}$$

Note: The terms within brackets in Equation (4.76) are written as right-hand side multiplication of $\mathbf{G}_{m\times a}(\tilde{\mathbf{L}})$ by the column of reaction forces N_a

$$\underbrace{\begin{bmatrix} +1 & -1 & 0 & 0 \\ 0 & +1 & -1 & 0 \\ 0 & 0 & +1 & -1 \\ 0 & 0 & 0 & +1 \end{bmatrix}}_{\mathbf{G}_{m\times a}} * \underbrace{\begin{Bmatrix} \mathbf{M}^0_A \\ \mathbf{M}^0_B \\ \mathbf{M}^0_C \\ \mathbf{M}^0_D \end{Bmatrix}}_{\mathbf{M}^0_a} = \underbrace{\begin{Bmatrix} \mathbf{Q}^0_{G_1} \\ \mathbf{Q}^0_{G_2} \\ \mathbf{Q}^0_{G_3} \\ \mathbf{Q}^0_{G_4} \end{Bmatrix}}_{\mathbf{Q}^0_{G_m}}$$

$$- \underbrace{\begin{bmatrix} +1\cdot\tilde{\mathbf{L}}^0_{G_1A} & -1\cdot\tilde{\mathbf{L}}^0_{G_1B} & 0 & 0 \\ 0 & +1\cdot\tilde{\mathbf{L}}^0_{G_2B} & -1\cdot\tilde{\mathbf{L}}^0_{G_2C} & 0 \\ 0 & 0 & +1\cdot\tilde{\mathbf{L}}^0_{G_3C} & -1\cdot\tilde{\mathbf{L}}^0_{G_3D} \\ 0 & 0 & 0 & +1\cdot\tilde{\mathbf{L}}^0_{G_4D} \end{bmatrix}}_{\mathbf{G}_{m\times a}(\tilde{\mathbf{L}})} * \underbrace{\begin{Bmatrix} \mathbf{N}^0_A \\ \mathbf{N}^0_B \\ \mathbf{N}^0_C \\ \mathbf{N}^0_D \end{Bmatrix}}_{\mathbf{N}_a} \tag{4.78}$$

where

- Equation (4.62): $\mathbf{N}^0_a = \mathbf{Z}_{a\times m} * \mathbf{P}^0_{G_m}$ is written as:

$$\underbrace{\begin{Bmatrix} \mathbf{N}^0_A \\ \mathbf{N}^0_B \\ \mathbf{N}^0_C \\ \mathbf{N}^0_D \end{Bmatrix}}_{\mathbf{N}_a} = \underbrace{\begin{bmatrix} 1 & 1 & 1 & 1 \\ 0 & 1 & 1 & 1 \\ 0 & 0 & 1 & 1 \\ 0 & 0 & 0 & 1 \end{bmatrix}}_{\mathbf{Z}_{a\times m}} * \underbrace{\begin{Bmatrix} \mathbf{P}_{G_1} \\ \mathbf{P}_{G_2} \\ \mathbf{P}_{G_3} \\ \mathbf{P}_{G_4} \end{Bmatrix}}_{\mathbf{P}^0_{G_m}} \tag{4.79}$$

Therefore, Equation (4.78) in matrix form and vector entries is shown in equation (4.80):

$$\mathbf{G}_{m\times a} * \mathbf{M}_a^0 = \mathbf{Q}_{Gm}^0 - \mathbf{G}_{m\times a}(\tilde{\mathbf{L}}) * \mathbf{Z}_{a\times m} * \mathbf{P}_{Gm}^0 \tag{4.80}$$

Equation (4.80) is left-hand-side-multiplied by the path matrix **Z** matrix and considering the relation **Z** * **G** = **U**:

$$\mathbf{M}_a^0 = \mathbf{Z}_{a\times m} * \mathbf{Q}_{Gm}^0 + \underbrace{\left(-\mathbf{Z}_{a\times m} * \mathbf{G}_{m\times a}(\tilde{\mathbf{L}}) * \mathbf{Z}_{a\times m}\right)}_{\mathbf{Z}_{a\times m}(\tilde{\mathbf{L}})} * \mathbf{P}_{Gm}^0 \tag{4.80a}$$

- The matrix $\mathbf{Z}_{a\times m}(\tilde{\mathbf{L}})$ is the a × m matrix whose entries are the *skew-symmetric matrices* for the position vectors $\tilde{\mathbf{L}}$ originated from the tree's joints toward links' COM.

  It is generated from the **Z** matrix entries, as shown in Equation (4.80b):

$$\mathbf{Z} = \begin{array}{c} \\ A \\ B \\ C \\ D \end{array} \begin{array}{c} \begin{array}{cccc} 1 & 2 & 3 & 4 \end{array} \\ \begin{bmatrix} +1 & +1 & +1 & +1 \\ 0 & +1 & +1 & +1 \\ 0 & 0 & +1 & +1 \\ 0 & 0 & 0 & +1 \end{bmatrix} \end{array} \Rightarrow \mathbf{Z}_{m\times a}(\tilde{\mathbf{L}})$$

$$= \begin{bmatrix} +1\cdot\tilde{\mathbf{L}}_{AG_1}^0 & +1\cdot\tilde{\mathbf{L}}_{AG_2}^0 & +1\cdot\tilde{\mathbf{L}}_{AG_3}^0 & +1\cdot\tilde{\mathbf{L}}_{AG_4}^0 \\ 0 & +1\cdot\tilde{\mathbf{L}}_{BG_2}^0 & +1\cdot\tilde{\mathbf{L}}_{BG_3}^0 & +1\cdot\tilde{\mathbf{L}}_{BG_2}^0 \\ 0 & 0 & +1\cdot\tilde{\mathbf{L}}_{CG_3}^0 & +1\cdot\tilde{\mathbf{L}}_{CG_4}^0 \\ 0 & 0 & 0 & +1\cdot\tilde{\mathbf{L}}_{DG_4}^0 \end{bmatrix} \tag{4.80b}$$

An automatic generation is from Equation (4.80c):

$$\mathbf{Z}_{m\times a}(\tilde{\mathbf{L}}) = -\mathbf{Z}_{a\times m} * \mathbf{G}_{m\times a}(\tilde{\mathbf{L}}) * \mathbf{Z}_{a\times m} \tag{4.80c}$$

Equation (4.80a) becomes

$$\mathbf{M}_a^0 = \mathbf{Z}_{a\times m} * \mathbf{Q}_{Gm}^0 + \mathbf{Z}_{m\times a}(\tilde{\mathbf{L}}) * \mathbf{P}_{Gm}^0 \tag{4.81}$$

Notation:

- The right-hand-side term in Equation (4.80c) is named *IT moment*, $\mathbf{Q}_a^0$

$$\mathbf{Q}_a^0 = \mathbf{Z}_{a\times m} * \mathbf{Q}_{Gm}^0 + \mathbf{Z}_{m\times a}(\tilde{\mathbf{L}}) * \mathbf{P}_{Gm}^0 \tag{4.82}$$

Its entries are determined as a function of the path matrix **Z**, $\mathbf{P}^0_{Gm}$, and $\mathbf{Q}^0_{Gm}$

Equation (4.81) becomes

$$\mathbf{M}^0_a + \left(-\mathbf{Q}^0_a\right) = 0 \tag{4.83}$$

The moments $\left(-\mathbf{Q}^0_a\right)$, named *"IT-resistant moments,"* are assigned to joints (arcs tree).

Statement 4.7

The reaction moments in joint $\mathbf{M}^0_a$ and IT-resistant moments are a system of moments in fictitious equilibrium.

Equation (4.84) is the governing equation to determine the reaction moments in joints.

$$\mathbf{M}^0_a = \mathbf{Q}^0_a \tag{4.84}$$

The characteristics of IT-resistant moments, $\left(-\mathbf{Q}^0_a\right)$, are as follows:

- They are dynamic moments, measured in [N · m] or [lbf · ft].
- They are in a *fictitious equilibrium* with the reaction couples in the joint $\mathbf{M}^0_a$.
- Distinction between an IT-resistant moment $\left(-\mathbf{Q}^0_a\right)$ and the inertial rate of change in angular momentum, $\left(-\dot{\mathbf{K}}^{in,0}_{Gm}\right)$ determined using Euler's law (Equation (4.72)):
  - IT-resistant moment $\left(-\mathbf{Q}^0_a\right)$ is *incidental to the joint a (edge tree)*, whereas $\left(-\dot{\mathbf{K}}^{in,0}_{Gm}\right)$ is *incidental to the link m's* (node tree) *COM*
  - The IT-resistant moment includes the moments from the external forces that are involved in generating the dynamic reaction couple $\mathbf{M}^0_a$. The terms involved are evidenced by a nonzero +1 or –1 value of an entry $z_{a,m}$ in the matrix Z, whereas the terms not involved are evidenced by the $z_{a,m} = 0$ value.

4.6.3 The Evaluation of IT Moments

The IT moments are determined from the tree using Equation (4.82).

This is equivalent to generate a set of paths initiated at node 0 and ended in sequence at nodes 1, 2, 3, and 4. For the TRRT mechanism, all entries in the matrix **Z** are +1.

- Joint A is in all paths initiated at node 0 and ended at nodes 1, 2, 3, and 4:

$$\mathbf{Q}_A^0 = \left(\mathbf{Q}_{G_1}^0 + \tilde{\mathbf{L}}_{AG_1}^0 * \mathbf{P}_{G_1}^0\right) + \left(\mathbf{Q}_{G_2}^0 + \tilde{\mathbf{L}}_{AG_2}^0 * \mathbf{P}_{G_2}^0\right) + \left(\mathbf{Q}_{G_3}^0 + \tilde{\mathbf{L}}_{AG_3}^0 * \mathbf{P}_{G_3}^0\right) + \left(\mathbf{Q}_{G_4}^0 + \tilde{\mathbf{L}}_{AG_4}^0 * \mathbf{P}_{G_4}^0\right) \quad (4.85a)$$

- Joint B is in all paths initiated at node 0 and ended at nodes 2, 3, and 4:

$$\mathbf{Q}_B^0 = \left(\mathbf{Q}_{G_2}^0 + \tilde{\mathbf{L}}_{BG_2}^0 * \mathbf{P}_{G_2}^0\right) + \left(\mathbf{Q}_{G_3}^0 + \tilde{\mathbf{L}}_{BG_3}^0 * \mathbf{P}_{G_3}^0\right) + \left(\mathbf{Q}_{G_4}^0 + \tilde{\mathbf{L}}_{BG_4}^0 * \mathbf{P}_{G_4}^0\right) \quad (4.85b)$$

- Joint C is in all paths initiated at node 0 and ended at nodes 3 and 4:

$$\mathbf{Q}_C^0 = \left(\mathbf{Q}_{G_3}^0 + \tilde{\mathbf{L}}_{CG_3}^0 * \mathbf{P}_{G_3}^0\right) + \left(\mathbf{Q}_{G_4}^0 + \tilde{\mathbf{L}}_{CG_4}^0 * \mathbf{P}_{G_4}^0\right) \quad (4.85c)$$

- Joint D is in all paths initiated at node 0 and ended at node 4:

$$\mathbf{Q}_D^0 = \left(\mathbf{Q}_{G_4}^0 + \tilde{\mathbf{L}}_{DG_4}^0 * \mathbf{P}_{G_4}^0\right) \quad (4.85d)$$

Note:
The force and moment reactions are determined from Equations (4.65) and (4.84) by equating their components with the IT force's components. The prismatic actuated joint, where an actuating force is present, generates a differential equation of motion. The revolute joint, where an actuating torque is present, generates a differential equation.

4.6.4 Dynamic Force and Moment Reactions from Joint Constraints

Statement 4.8

- The components of the dynamic force reaction are defined from the components of joint's relative linear displacement.
- The components of the dynamic moment reaction are defined in local frames from joint's relative angular displacements.

Zero components in relative linear or angular velocities imply the existence of reaction forces and reaction moments along these directions. If we consider the scalar product (power) between vector relative linear velocity and reaction force and between relative angular velocity and moment, then the power of reaction forces and reaction moments is zero for ideal mechanisms (friction is neglected).

Since the relative velocities are expressed in local frames, the reactions are first expressed in local frames and then expressed in the fixed frame 0 (as shown in equations) by left-hand-side multiplication with the absolute rotation matrix.

Note:
For edges in a digraph, the subscript in the absolute rotation matrix $\mathbf{D}_m$ is arrow's $Z_{m-1,m}$ head for the node m. This is based on a previous notation that the reaction vectors $\mathbf{N}_Z$ and $\mathbf{M}_Z$ represent *link m − 1's (tail-node) action on the link m (head-node)*, where Z = A, B, …, (for j joints).
Example: The TRRT spatial mechanism, Figure 4.10

- *Joint A* (1 DOF, prismatic)
  - Reaction force at the joint A is defined from the linear velocity constraints in local frame 1:
  - The relative velocity along x_1 direction, $\dot{d}_A^1$, is allowed, which is also the direction of F_A^{act}—the actuating force from link 0 acting on link 1. The two zero components imply constraints along y_1 and z_1 directions; therefore, the reaction components Y_A^1 and Z_A^1 are introduced.

$$\dot{\mathbf{d}}_A^1 = \begin{Bmatrix} \dot{d}_A^1 \\ 0 \\ 0 \end{Bmatrix} \Rightarrow \mathbf{N}_A^1 = \begin{Bmatrix} F_A^{act} \\ Y_A^1 \\ Z_A^1 \end{Bmatrix} \tag{4.86a}$$

  - Reaction moment at the joint A is defined from the relative angular velocity constraints in local frame 1:
    The angular velocity $\dot{\boldsymbol{\theta}}_A^1$ is not allowed in the prismatic joint A. Instead, there are assigned reaction moment components for all three directions: $M_A^{x_1}$, $M_A^{y_1}$, and $M_A^{z_1}$:

$$\dot{\boldsymbol{\theta}}_A^1 = \begin{Bmatrix} 0 \\ 0 \\ 0 \end{Bmatrix} \Rightarrow \mathbf{M}_A^1 = \begin{Bmatrix} M_A^{x_1} \\ M_A^{y_1} \\ M_A^{z_1} \end{Bmatrix} \tag{4.86b}$$

- *Joint B* (1 DOF, revolute)
  - Reaction force at the joint B is defined from the linear velocity constraints in local frame 2:
    The linear velocity $\dot{\mathbf{d}}_B^2$ is not allowed in the revolute joint B. Instead, there are assigned reaction force components for all three directions: X_B^2, Y_B^2, and Z_B^2:

$$\dot{\mathbf{d}}_B^2 = \begin{Bmatrix} 0 \\ 0 \\ 0 \end{Bmatrix} \Rightarrow \mathbf{N}_B^2 = \begin{Bmatrix} X_B^2 \\ Y_B^2 \\ Z_B^2 \end{Bmatrix} \tag{4.86c}$$

- Reaction moment at the joint B is defined from the angular velocity constraints in local frame 2:

  An angular velocity $\dot{\theta}_B^2$ is allowed to rotate on the direction y_2, which is also the direction of T_B^{act}—the actuating torque from link 1 acting on link 2; then, it is constrained to rotate on the other directions x_2 and z_2; for these directions, the reaction components M_B^{x2} and M_B^{z2} are assigned.

$$\dot{\boldsymbol{\theta}}_B^2 = \begin{Bmatrix} 0 \\ \dot{\theta}_B^2 \\ 0 \end{Bmatrix} \Rightarrow \mathbf{M}_B^2 = \begin{Bmatrix} M_B^{x2} \\ T_B^{act} \\ M_B^{z2} \end{Bmatrix} \tag{4.86d}$$

- *Joint C* (1 DOF, revolute)
  - Reaction force at the joint C is defined from the linear velocity constraints in local frame 3:
    The linear velocity $\dot{\mathbf{d}}_C^3$ is not allowed in the revolute joint C, along x_3, y_3, or z_3 direction. Instead, the reaction force components for all three directions X_C^3, Y_C^3, and Z_C^3 are assigned.

$$\dot{\mathbf{d}}_C^3 = \begin{Bmatrix} 0 \\ 0 \\ 0 \end{Bmatrix} \Rightarrow \mathbf{N}_C^3 = \begin{Bmatrix} X_C^3 \\ Y_C^3 \\ Z_C^3 \end{Bmatrix} \tag{4.86e}$$

  - Reaction moment at the joint C is defined from the angular displacement constraints in local frame 3:

    An angular velocity $\dot{\theta}_C^3$ is allowed along z_3 direction, which is also the direction of $T_C^{act,3}$—the actuating torque from link 2 acting on link 3; then, it is constrained to rotate on the other directions x_3 and z_3; for these directions, the reaction components M_C^{x3} and M_C^{y3} are assigned.

$$\dot{\boldsymbol{\theta}}_C^3 = \begin{Bmatrix} 0 \\ 0 \\ \dot{\theta}_C^3 \end{Bmatrix} \Rightarrow \mathbf{M}_C^3 = \begin{Bmatrix} M_C^{x3} \\ M_C^{y3} \\ T_C^{act} \end{Bmatrix} \tag{4.86f}$$

- *Joint D* (1 DOF, prismatic)
  - Reaction force at the joint D is defined from the linear velocity constraints in local frame 4:

It allows the relative velocity $\dot{d}_D^4$ along x_4 direction, which is also the direction of F_D^{act}—the actuating force from link 3 acting on link 4. The two zero components imply constraints along y_4 and z_4 directions; therefore, the reaction components Y_D^4 and Z_A^4 are introduced.

$$\dot{\mathbf{d}}_D^4 = \begin{Bmatrix} \dot{d}_A^1 \\ 0 \\ 0 \end{Bmatrix} \Rightarrow \mathbf{N}_D^4 = \begin{Bmatrix} F_D^{act} \\ Y_D^4 \\ Z_D^4 \end{Bmatrix} \tag{4.86g}$$

- Reaction moment at the joint D is defined from the angular velocity constraints in local frame 4:

  The angular velocity $\dot{\boldsymbol{\theta}}_D^4$ is not allowed in the prismatic joint D. Instead, reaction moment components for all three directions M_D^{x4}, M_D^{y4}, and M_D^{z4} are assigned.

$$\dot{\boldsymbol{\theta}}_D^4 = \begin{Bmatrix} 0 \\ 0 \\ 0 \end{Bmatrix} \Rightarrow \mathbf{M}_D^4 = \begin{Bmatrix} M_D^{x4} \\ M_D^{y4} \\ M_D^{z4} \end{Bmatrix} \tag{4.86h}$$

The system of equations for reaction forces and moments in joints

The reaction forces and moments in dynamic Equations (4.65) and (4.84) are expressed in the fixed frame:

$$\begin{cases} \mathbf{N}_A^0 = \mathbf{P}_A^0 \\ \mathbf{N}_B^0 = \mathbf{P}_A^0 \\ \mathbf{N}_C^0 = \mathbf{P}_C^0 \\ \mathbf{N}_D^0 = \mathbf{P}_D^0 \end{cases}, \text{ and } \begin{cases} \mathbf{M}_A^0 = \mathbf{Q}_A^0 \\ \mathbf{M}_B^0 = \mathbf{Q}_A^0 \\ \mathbf{M}_C^0 = \mathbf{Q}_C^0 \\ \mathbf{M}_D^0 = \mathbf{Q}_D^0 \end{cases} \tag{4.87}$$

where $\mathbf{N}_Z^0 = \mathbf{D}_m * \mathbf{N}_Z^m$, for a = A, B, C, and D, and m = 1, 2, 3, and 4.

The solution for reaction forces and moments in joints

This solution does not require *the inverse matrix of coefficients*. Considering the property: $\mathbf{D}_m^T * \mathbf{D}_m = \mathbf{U}$, the left-hand-side multiplication by the transpose of each rotation matrix determines the solution, shown in Equation (4.88).

$$\begin{cases} \mathbf{N}_A^1 = \mathbf{D}_1^T * \mathbf{P}_A^0 \\ \mathbf{N}_B^2 = \mathbf{D}_2^T * \mathbf{P}_B^0 \\ \mathbf{N}_C^3 = \mathbf{D}_3^T * \mathbf{P}_C^0 \\ \mathbf{N}_D^4 = \mathbf{D}_4^T * \mathbf{P}_D^0 \end{cases}, \text{ and } \begin{cases} \mathbf{M}_A^1 = \mathbf{D}_1^T * \mathbf{Q}_A^0 \\ \mathbf{M}_B^2 = \mathbf{D}_2^T * \mathbf{Q}_B^0 \\ \mathbf{M}_C^3 = \mathbf{D}_3^T * \mathbf{Q}_C^0 \\ \mathbf{M}_D^4 = \mathbf{D}_4^T * \mathbf{Q}_D^0 \end{cases} \tag{4.88}$$

The solution for joint reactions is presented next based on the previous calculation of

- The reaction forces and moments from the joint constraints.
- The IT forces **P** and moments **Q** based on link masses and moments of inertia, links' COM linear and angular accelerations, and external forces and external moments reduced in links' COM.
- The absolute links' rotation matrices $\mathbf{D}_m$

- *Joint A* reactions:
  System of equations:

$$\text{Forces:}\begin{Bmatrix} F_A^{act} \\ Y_A^1 \\ Z_A^1 \end{Bmatrix} = \mathbf{D}_1^T * \begin{Bmatrix} P_A^{x_0} \\ P_A^{y_0} \\ P_A^{z_0} \end{Bmatrix};\ \text{moments:}\begin{Bmatrix} M_A^{x_1} \\ M_A^{y_1} \\ M_A^{z_1} \end{Bmatrix} = \mathbf{D}_1^T * \begin{Bmatrix} Q_A^{x_0} \\ Q_A^{y_0} \\ Q_A^{z_0} \end{Bmatrix},$$

$$\text{where } \mathbf{D}_1 = \begin{bmatrix} 1 & 0 & 0 \\ 0 & 1 & 0 \\ 0 & 0 & 1 \end{bmatrix}$$

Solution:

$$\text{Forces:}\begin{Bmatrix} F_A^{act} \\ Y_A^1 \\ Z_A^1 \end{Bmatrix} = \begin{Bmatrix} P_A^{x_0} \\ P_A^{y_0} \\ P_A^{z_0} \end{Bmatrix};\ \text{moments:}\begin{Bmatrix} M_A^{x_1} \\ M_A^{y_1} \\ M_A^{z_1} \end{Bmatrix} = \begin{Bmatrix} Q_A^{x_0} \\ Q_A^{y_0} \\ Q_A^{z_0} \end{Bmatrix} \quad (4.89a)$$

- *Joint B* reactions:
  System of equations:

$$\text{Forces:}\begin{Bmatrix} X_B^2 \\ Y_B^2 \\ Z_B^2 \end{Bmatrix} = \mathbf{D}_2^T * \begin{Bmatrix} P_B^{x_0} \\ P_B^{y_0} \\ P_B^{z_0} \end{Bmatrix};\ \text{moments:}\begin{Bmatrix} M_B^{x_2} \\ T_B^{act} \\ M_B^{z_2} \end{Bmatrix} = \mathbf{D}_2^T * \begin{Bmatrix} Q_B^{x_0} \\ Q_B^{y_0} \\ Q_B^{z_0} \end{Bmatrix},$$

$$\text{where } \mathbf{D}_2 = \begin{bmatrix} C\theta_B & 0 & S\theta_B \\ 0 & 1 & 0 \\ -S\theta_B & 0 & C\theta_B \end{bmatrix}$$

Solution:

$$\text{Forces:}\begin{Bmatrix} X_B^2 \\ Y_B^2 \\ Z_B^2 \end{Bmatrix} = \begin{Bmatrix} C\theta_B \cdot P_B^{x_0} - S\theta_B \cdot P_B^{z_0} \\ P_B^{y_0} \\ C\theta_B \cdot P_B^{z_0} + S\theta_B \cdot P_B^{x_0} \end{Bmatrix}; \text{moments:}\begin{Bmatrix} M_B^{x_2} \\ T_B^{act} \\ M_B^{z_2} \end{Bmatrix}$$

$$= \begin{Bmatrix} C\theta_B \cdot P_B^{x_0} - S\theta_B \cdot Q_B^{z_0} \\ Q_B^{y_0} \\ C\theta_B \cdot Q_B^{z_0} + S\theta_B \cdot Q_B^{x_0} \end{Bmatrix} \tag{4.89b}$$

- *Joint C* reactions:
  System of equations:

$$\text{Forces:}\begin{Bmatrix} X_C^3 \\ Y_C^3 \\ Z_C^3 \end{Bmatrix} = \mathbf{D}_3^T * \begin{Bmatrix} P_C^{x_0} \\ P_C^{y_0} \\ P_C^{z_0} \end{Bmatrix}; \text{moments:}\begin{Bmatrix} M_C^{x_3} \\ M_C^{y_3} \\ T_C^{act} \end{Bmatrix} = \mathbf{D}_3^T * \begin{Bmatrix} Q_C^{x_0} \\ Q_C^{y_0} \\ Q_C^{z_0} \end{Bmatrix},$$

$$\text{where } \mathbf{D}_3 = \begin{bmatrix} C\theta_B \cdot C\theta_C & -C\theta_B \cdot S\theta_C & S\theta_B \\ S\theta_C & C\theta_C & 0 \\ -S\theta_B \cdot C\theta_C & S\theta_B \cdot S\theta_C & C\theta_B \end{bmatrix}$$

Solution:

$$\text{Forces:}\begin{Bmatrix} X_C^3 \\ Y_C^3 \\ Z_C^3 \end{Bmatrix} = \begin{Bmatrix} C\theta_B \cdot C\theta_C \cdot P_C^{x_0} + S\theta_C \cdot P_C^{y_0} - S\theta_B \cdot C\theta_C \cdot P_C^{z_0} \\ -C\theta_B \cdot S\theta_C \cdot P_C^{x_0} + C\theta_C \cdot P_C^{y_0} + S\theta_B \cdot S\theta_C \cdot P_C^{z_0} \\ S\theta_B \cdot P_B^{x_0} + C\theta_B \cdot P_B^{z_0} \end{Bmatrix};$$

$$\text{moments:}\begin{Bmatrix} M_C^{x_3} \\ M_C^{y_3} \\ T_C^{act} \end{Bmatrix} = \begin{Bmatrix} C\theta_B \cdot C\theta_C \cdot Q_C^{x_0} + S\theta_C \cdot Q_C^{y_0} - S\theta_B \cdot C\theta_C \cdot Q_C^{z_0} \\ -C\theta_B \cdot S\theta_C \cdot Q_C^{x_0} + C\theta_C \cdot Q_C^{y_0} + S\theta_B \cdot S\theta_C \cdot Q_C^{z_0} \\ S\theta_B \cdot Q_B^{x_0} + C\theta_B \cdot Q_B^{z_0} \end{Bmatrix} \tag{4.89c}$$

- *Joint D* reactions:
  System of equations:

$$\begin{Bmatrix} F_D^{act} \\ Y_D^4 \\ Z_D^4 \end{Bmatrix} = \mathbf{D}_4^T * \begin{Bmatrix} P_D^{x_0} \\ P_D^{y_0} \\ P_D^{z_0} \end{Bmatrix}; \text{ moments: } \begin{Bmatrix} M_D^{x_4} \\ M_D^{y_4} \\ M_D^{z_4} \end{Bmatrix} = \mathbf{D}_4^T * \begin{Bmatrix} Q_D^{x_0} \\ Q_D^{y_0} \\ Q_D^{z_0} \end{Bmatrix},$$

$$\text{where } \mathbf{D}_4 = \begin{bmatrix} C\theta_B \cdot C\theta_C & -C\theta_B \cdot S\theta_C & S\theta_B \\ S\theta_C & C\theta_C & 0 \\ -S\theta_B \cdot C\theta_C & S\theta_B \cdot S\theta_C & C\theta_B \end{bmatrix}$$

Solution:

$$\text{Forces: } \begin{Bmatrix} F_D^{act} \\ Y_D^4 \\ Z_D^4 \end{Bmatrix} = \begin{Bmatrix} C\theta_B \cdot C\theta_C \cdot P_D^{x_0} + S\theta_C \cdot P_D^{y_0} - S\theta_B \cdot C\theta_C \cdot P_D^{z_0} \\ -C\theta_B \cdot S\theta_C \cdot P_D^{x_0} + C\theta_C \cdot P_D^{y_0} + S\theta_B \cdot S\theta_C \cdot P_D^{z_0} \\ S\theta_B \cdot P_D^{x_0} + C\theta_B \cdot P_D^{z_0} \end{Bmatrix};$$

$$\text{moments: } \begin{Bmatrix} M_D^{x_4} \\ M_D^{y_4} \\ M_D^{z_4} \end{Bmatrix} = \begin{Bmatrix} C\theta_B \cdot C\theta_C \cdot Q_D^{x_0} + S\theta_C \cdot Q_D^{y_0} - S\theta_B \cdot C\theta_C \cdot Q_D^{z_0} \\ -C\theta_B \cdot S\theta_C \cdot Q_D^{x_0} + C\theta_C \cdot Q_D^{y_0} + S\theta_B \cdot S\theta_C \cdot Q_D^{z_0} \\ S\theta_B \cdot Q_D^{x_0} + C\theta_B \cdot Q_D^{z_0} \end{Bmatrix} \quad (4.89d)$$

The TRRT mechanism has mobility M = 4; therefore, there are four actuating forces and torques:

$$F_A^{act}, T_B^{act}, T_C^{act}, F_D^{act}.$$

Differential equation of motion from actuating forces

Along the axes of linear motion, for example, joints A and D, the actuating forces F_A^{act} and F_D^{act} as a function of kinematic parameters $(d, \dot{d}, \ddot{d})$ are determined. Conversely, if the kinematic parameters are required for the provided actuation forces, then the two equations are integrated for the variable in time-kinematic parameters.

In the above solution, the force components along x_1 direction at A and along x_4 direction at D locate the differential equations:

$$\begin{aligned} F_A^{act} &= P_A^{X_0} \\ F_D^{act} &= C\theta_B \cdot C\theta_C \cdot P_D^{x_0} + S\theta_C \cdot P_D^{y_0} - S\theta_B \cdot C\theta_C \cdot P_D^{z_0} \end{aligned} \quad (4.90)$$

Differential equation of motion from actuating torques

Along the axes of relative rotation, for example, joints B and C, the actuating torques T_B^{act} and T_C^{act} as a function of kinematic parameters $(\theta, \dot{\theta}, \ddot{\theta})$ are determined. Conversely, if the kinematic parameters are required for the provided actuation torques, then the two equations are integrated for the variable in time-kinematic parameters.

In the above solution, the moment components along y_2 direction at B and along z_3 direction at C locate the differential equations:

$$\begin{aligned} T_B^{act} &= Q_B^{y_0} \\ T_C^{act} &= S\theta_B \cdot Q_B^{x_0} + C\theta_B \cdot Q_B^{z_0} \end{aligned} \tag{4.91}$$

The magnitudes of reaction forces

The magnitudes of reaction forces are invariant (the same value in the fixed frame as that in the local frame); therefore, they can be expressed from $\mathbf{P}_a^0$ in the fixed frame 0, shown in Equation (4.92):

$$\begin{aligned} &\left|N_A^1\right| = \left|N_A^0\right| = \sqrt{\left(P_A^{x_0}\right)^2 + \left(P_A^{y_0}\right)^2 + \left(P_A^{z_0}\right)^2}\,; \left|N_B^2\right| = \left|N_B^0\right| = \sqrt{\left(P_B^{x_0}\right)^2 + \left(P_B^{y_0}\right)^2 + \left(P_B^{z_0}\right)^2}\,; \\ &\left|N_C^3\right| = \left|N_C^0\right| = \sqrt{\left(P_C^{x_0}\right)^2 + \left(P_C^{y_0}\right)^2 + \left(P_C^{z_0}\right)^2}\,; \left|N_D^4\right| = \left|N_D^0\right| = \sqrt{\left(P_D^{x_0}\right)^2 + \left(P_D^{y_0}\right)^2 + \left(P_D^{z_0}\right)^2} \end{aligned} \tag{4.92}$$

The magnitudes of reaction moments

The magnitudes of reaction moments are invariants (the same value in the fixed frame as that in the local frame); therefore, they can be expressed from $\mathbf{Q}_a^0$ in the fixed frame 0, shown in Equation (4.93):

$$\begin{aligned} &\left|M_A^1\right| = \left|M_A^0\right| = \sqrt{\left(Q_A^{x_0}\right)^2 + \left(Q_A^{y_0}\right)^2 + \left(Q_A^{z_0}\right)^2}\,; \left|M_B^2\right| = \left|M_B^0\right| = \sqrt{\left(Q_B^{x_0}\right)^2 + \left(Q_B^{y_0}\right)^2 + \left(Q_B^{z_0}\right)^2}\,; \\ &\left|M_C^3\right| = \left|M_C^0\right| = \sqrt{\left(Q_C^{x_0}\right)^2 + \left(Q_C^{y_0}\right)^2 + \left(Q_C^{z_0}\right)^2}\,; \left|M_D^4\right| = \left|M_D^0\right| = \sqrt{\left(Q_D^{x_0}\right)^2 + \left(Q_D^{y_0}\right)^2 + \left(Q_D^{z_0}\right)^2} \end{aligned}$$

4.6.5 Review on Computation of IT Equations for Dynamics of Open Cycle Mechanisms

The computation of dynamic equations and differential equations of motion is performed as follows:

$$\text{Draw the digraph and spanning tree } \mathbf{Z}_{a\times m} \tag{4.93a}$$

$$\text{The matrix } \mathbf{Z}_{a\times m} \tag{4.93b}$$

$$\text{The matrix } \mathbf{Z}_{a\times m}(\tilde{\mathbf{L}}) \text{ written directly from } \mathbf{Z}_{a\times m} \text{ in step 2} \tag{4.93c}$$

- The force resultant and the moment resultant reduced to COM

$$\mathbf{P}_{G_m} = \mathbf{f}_m^{in,0} - \mathbf{f}_m^{ext,0} \tag{4.93d}$$

$$\mathbf{Q}_{G_m}^0 = \dot{\mathbf{K}}_m^{in,0} - \mu_m^{ext,0} \tag{4.93e}$$

- The IT force $\mathbf{P}_a^0$ and IT moment $\mathbf{Q}_a^0$ reduced to joint's tree. They are the coefficients on right-hand side of Equations (4.63) and (4.82):

$$\mathbf{P}_a^0 = \mathbf{Z}_{a\times m} * \mathbf{P}_{G_m} \tag{4.93f}$$

$$\mathbf{Q}_a^0 = \mathbf{Z}_{a\times m} * \mathbf{Q}_{G_m}^0 + \mathbf{Z}_{a\times m}(\tilde{\mathbf{L}}) * \mathbf{P}_{G_m}^0$$

Define the dynamic force and moment reactions from joint constraints (4.93g)

Moment equations:

- The IT moment equations

$$\mathbf{M}_a^0 = \mathbf{Q}_a^0 \tag{4.93h}$$

are solved for moment reactions in tree's joints

Force equations:

- The IT force equations

$$\mathbf{N}_a^0 = \mathbf{P}_a^0 \tag{4.93i}$$

are solved for joint reactions $\mathbf{N}_a^0$

The magnitude of force and moment reactions results from Equations (4.87) and their components on local axes from Equations (4.88).

Equations (4.93a) to (4.93i) are the *"IT equations for open cycle dynamic reactions and differential equations"* [2,3].

4.7 Closed Cycle Mechanisms: The IT Method for Dynamic Forces and Differential Equations

The digraph's tree m force equations are shown in this section. The method is exemplified on the TRRTC single-cycle spatial mechanism for which we determine the governing equations for the dynamic reactions and the differential equations of motion. Further, the method is generalized for mechanisms with c-cycles.

Example: A closed cycle TRRTC spatial mechanism (Figure 4.16a) with five joints: translation (prismatic) T, revolute R, revolute R, translation (prismatic) T, and contact joint C is obtained from the open cycle TRRT if the last link 4 is connected to the fixed link 0. This is the case when the end-effector will touch a fixed plane, assumed here to be the fixed (x_0, z_0) plane. A contact joint E allows five relative velocities between link 4 and link 0.

The number of cycles is determined by using Equation (1.7), from total number of joints j = 5 joints (edges digraph), n = m + 1 (nodes digraph), and m = 4 mobile links. The digraph in Figure 4.16b shows cut-edge E oriented from the lower node number 0 to the higher node number 4. The tree is illustrated in Figure 4.16c.

$$c = j - m = 5 - 4 = 1 \text{ cycle}$$

The mobility M can be predicted by Kutzbach formula, Equation (1.29), with

- m = 4 represents the mobile links 1, 2, 3, and 4.
- $j_1 = 4$ represents the prismatic and revolute joints A, B, C, and D.
- $j_5 = 1$ represents the contact joint at E.

$$M = 6 \cdot m - 5 \cdot j_1 - 4j_2 - 3 \cdot j_3 - 2j_4 - j_5 = 6 \cdot 4 - 5 \cdot 4 - 4 \cdot 0 - 3 \cdot 0 - 2 \cdot 0 - 1$$
$$= 3 \text{ DOF}$$

For a 3-DOF mechanism, three motors are required for actuation. In this example, A, B, and D are chosen as actuating joints (Figure 4.16a).

4.7.1 The Incidence Nodes-Edges Matrix: The Reduced and Row Reduced Matrix

- *The "Incidence Nodes-Edges Matrix,"* $\underline{\mathbf{G}}$

  For a mechanism with m mobile links and j joints, there is a matrix with m + 1 rows and j columns, noted $\underline{\mathbf{G}}_{m+1,j}$. If one deletes the first row—corresponding to link 0—shown as horizontal dashed line in the matrix, then one obtains the m × j reduced incidence matrix, **G**. Equation (4.94) shows the two matrices for the TRRTC mechanism with m = 4 mobile links and j = 5 joints.

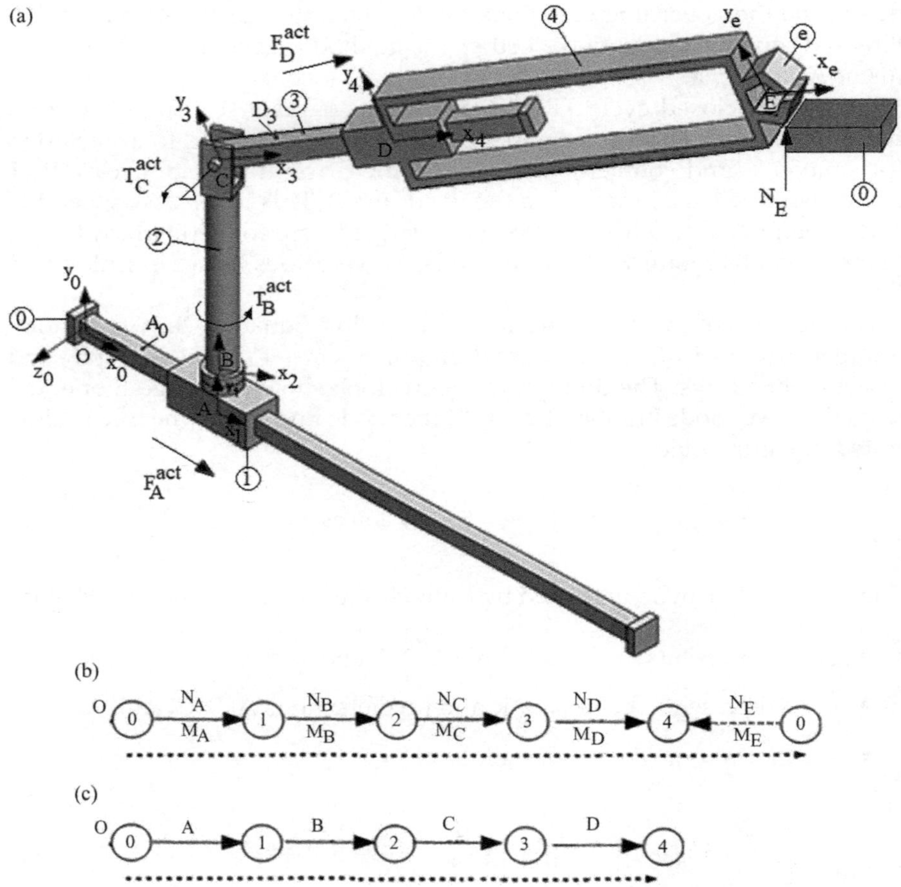

FIGURE 4.16
(a) TRRTC closed cycle spatial mechanism; (b) reaction forces and moments at cycle's joints; (c) tree.

$$
\underline{G} = \begin{array}{c} \\ 0 \\ 1 \\ 2 \\ 3 \\ 4 \end{array}
\begin{array}{c}
\begin{array}{ccccc} A & B & C & D & E \end{array} \\
\left[\begin{array}{cccc:c}
-1 & 0 & 0 & 0 & -1 \\ \hdashline
+1 & -1 & 0 & 0 & 0 \\
0 & +1 & -1 & 0 & 0 \\
0 & 0 & +1 & -1 & 0 \\
0 & 0 & 0 & +1 & 1
\end{array} \right]
\end{array};
$$

$$\mathbf{G}_{m\times j} = \begin{array}{c} \\ 1 \\ 2 \\ 3 \\ 4 \end{array} \begin{array}{c} \begin{array}{ccccc} A & B & C & D & E \end{array} \\ \left[\begin{array}{cccc:c} +1 & -1 & 0 & 0 & 0 \\ 0 & +1 & -1 & 0 & 0 \\ 0 & 0 & +1 & -1 & 0 \\ 0 & 0 & 0 & +1 & +1 \end{array}\right] \end{array} = \left[\begin{array}{c:c} \mathbf{G}_{m\times a} & \mathbf{G}^{cut}_{m\times c} \end{array}\right] \tag{4.94}$$

- *The partitions of reduced matrix*

  For a closed cycle mechanism, the reduced matrix **G** is partitioned into two sub-matrices, shown as a vertical dashed line:

  - **G** has the columns corresponding to the tree's edges. It is a square matrix since in the tree, m = a.
  - $\mathbf{G}^{cut}$ has the columns corresponding to cut-edges in the digraph. Its number of columns c is the same as the number of cycles in the mechanism.

As shown in Figure 4.16b, by cutting the edge E, the closed cycle becomes a tree, Figure 4.16c; therefore, $\mathbf{G}^{cut}$ is a one-column matrix, as shown in Equation (4.95):

$$\mathbf{G}_{m\times a} = \begin{array}{c} \\ 1 \\ 2 \\ 3 \\ 4 \end{array} \begin{array}{c} \begin{array}{cccc} A & B & C & D \end{array} \\ \left[\begin{array}{cccc} +1 & -1 & 0 & 0 \\ 0 & +1 & -1 & 0 \\ 0 & 0 & +1 & -1 \\ 0 & 0 & 0 & +1 \end{array}\right] \end{array} ; \mathbf{G}^{cut}_{m\times c} = \begin{array}{c} E \\ \left[\begin{array}{c} 0 \\ 0 \\ 0 \\ +1 \end{array}\right] \end{array} \tag{4.95}$$

Note: For mechanisms with c-cycles, matrix $\mathbf{G}^{cut}$ has c-columns corresponding to c-cut-edges

- *The path matrix,* **Z**.

$$\mathbf{Z}_{a\times m} = \begin{array}{c} \\ A \\ B \\ C \\ D \end{array} \begin{array}{c} \begin{array}{cccc} 1 & 2 & 3 & 4 \end{array} \\ \left[\begin{array}{cccc} +1 & +1 & +1 & +1 \\ 0 & +1 & +1 & +1 \\ 0 & 0 & +1 & +1 \\ 0 & 0 & 0 & +1 \end{array}\right] \end{array} = \left[\begin{array}{cccc} z_{A,1} & z_{A,2} & z_{A,3} & z_{A,4} \\ 0 & z_{B,2} & z_{B,3} & z_{B,4} \\ 0 & 0 & z_{C,3} & z_{C,4} \\ 0 & 0 & 0 & z_{D,4} \end{array}\right] \tag{4.95a}$$

- *The row-reduced incidence matrix,* $^{\otimes}\mathbf{G}$

Using row operations as in linear algebra, the matrix G becomes $^{\otimes}\mathbf{G}_{a\times j}$, as shown in Equation (4.96):

$$^{\otimes}\mathbf{G} = \left[\begin{array}{c:c} \mathbf{U}_{a\times a} & \mathbf{W}_{a\times c} \end{array}\right] \tag{4.96}$$

$$^{\otimes}\mathbf{G} = \begin{array}{c} \\ 1 \\ 2 \\ 3 \\ 4 \end{array} \begin{array}{c} \begin{array}{ccccc} A & B & C & D & E \end{array} \\ \left[\begin{array}{cccc:c} +1 & 0 & 0 & 0 & +1 \\ 0 & +1 & 0 & 0 & +1 \\ 0 & 0 & +1 & 0 & +1 \\ 0 & 0 & 0 & +1 & +1 \end{array}\right] \end{array}$$

$$= \left[\begin{array}{c:c} \mathbf{U}_{a\times a} & \mathbf{W}_{a\times c} \end{array}\right] \text{row reduced form of matrix } \mathbf{G}$$

A second approach in finding the row reduced is by left-hand-side multiplying in **G** with the path matrix **Z**:

$$^{\otimes}\mathbf{G} = \mathbf{Z}_{a\times m} * \left[\begin{array}{c:c} \mathbf{G}_{m\times a} & \mathbf{G}^{cut}_{m\times c} \end{array}\right] = \left[\begin{array}{c:c} \mathbf{Z}_{a\times m} * \mathbf{G}_{m\times a} & \mathbf{Z}_{a\times m} * \mathbf{G}^{cut}_{m\times c} \end{array}\right]$$
$$= \left[\begin{array}{c:c} \mathbf{U}_{a\times a} & \mathbf{W}_{a\times c} \end{array}\right] \tag{4.97a}$$

Considering the relation $\mathbf{Z} * \mathbf{G} = \mathbf{U}$, and also the notation $\mathbf{Z} * \mathbf{G}^{cut} = \mathbf{W}$, the two partitions of matrix $^{\otimes}\mathbf{G}$ hold, as shown in Equation (4.97b):

$$\mathbf{Z}_{a\times m} * \mathbf{G}_{m\times a} = \mathbf{U}_{a\times a} = \begin{array}{c} \\ 1 \\ 2 \\ 3 \\ 4 \end{array} \begin{array}{c} \begin{array}{cccc} A & B & C & D \end{array} \\ \left[\begin{array}{cccc} +1 & 0 & 0 & 0 \\ 0 & +1 & 0 & 0 \\ 0 & 0 & +1 & 0 \\ 0 & 0 & 0 & +1 \end{array}\right] \end{array} \tag{4.97b}$$

$$\mathbf{Z}_{a\times m} * \mathbf{G}^{cut}_{m\times c} = \mathbf{W}_{a\times c} = \begin{array}{c} \\ A \\ B \\ C \\ D \end{array} \begin{array}{c} E \\ \left[\begin{array}{c} +1 \\ +1 \\ +1 \\ +1 \end{array}\right] \end{array} = \left[\begin{array}{c} w_{AE} \\ w_{BE} \\ w_{CE} \\ w_{DE} \end{array}\right] \tag{4.97c}$$

4.7.2 The Weighting Matrix, W

This a × c matrix has entry w_{ac} as either +1, –1, or 0, which is used further in writing the dynamic and static equations for single- or multiple-cycle mechanisms.

$$\mathbf{W} = \mathbf{Z} * \mathbf{G}^{\text{cut}} \tag{4.98}$$

- *The weighting matrix* **W** *for multiple-cycle mechanisms*

  Next, examples of generated **W** matrices for mechanisms with two and three cycles commonly found in applications are shown. The weighting matrix can be generated for mechanisms with any number c of cycles. Examples of weighting matrix **W** for any mechanism with j joints, m mobile links, and c-cycles are shown, where

$$c = j - m; \; a = m \tag{4.99}$$

  - Two-cycle mechanisms have c = 2 cycles, j = 7 joints (A, B, C, D, E, F, and G), a = m = 5 arcs tree (A, B, C, D, and E), and c = j – m = 2 cut-edges (F, G)
  - Three-cycle mechanisms have c = 3 cycles, j = 9 joints (A, B, C, D, E, F, G, H, and I), a = m = 6 arcs tree (A, B, C, D, E, and F), and c = j – m = 3 cut-edges (G, H, and I)

$$\mathbf{W}_{a\times c} = \begin{matrix} \\ A \\ B \\ C \\ D \\ E \end{matrix} \begin{matrix} \begin{matrix} F & \quad G \end{matrix} \\ \begin{bmatrix} w_{AF} & w_{AG} \\ w_{BF} & w_{BG} \\ w_{CF} & w_{CG} \\ w_{DF} & w_{DG} \\ w_{EF} & w_{EG} \end{bmatrix} \end{matrix}; \quad \mathbf{W}_{a\times c} = \begin{matrix} \\ A \\ B \\ C \\ D \\ E \\ F \end{matrix} \begin{matrix} \begin{matrix} G & \quad H & \quad I \end{matrix} \\ \begin{bmatrix} w_{AG} & w_{AH} & w_{AI} \\ w_{BG} & w_{BH} & w_{BI} \\ w_{CG} & w_{CH} & w_{CI} \\ w_{DG} & w_{DH} & w_{DI} \\ w_{EG} & w_{EH} & w_{EI} \\ w_{FG} & w_{FH} & w_{FI} \end{bmatrix} \end{matrix} \tag{4.100}$$

4.7.3 The Cut-Set Matroid

The transposed matrix, ${}^{\otimes}\mathbf{G}^{T}$, from Equation (4.96) is named the *cut-set matroid*. It has m independent columns and j = a + c rows, shown in Equation (4.101):

$$
{}^{\otimes}\mathbf{G}^{\mathrm{T}}_{\mathrm{j\times m}} = \left[\begin{array}{c} \mathbf{U}_{\mathrm{a\times a}} \\ \hdashline \mathbf{W}^{\mathrm{T}}_{\mathrm{c\times a}} \end{array}\right] = \begin{array}{c} \\ \mathrm{A} \\ \mathrm{B} \\ \mathrm{C} \\ \mathrm{D} \\ \mathrm{E} \end{array} \begin{array}{c} \begin{array}{cccc} 1 & 2 & 3 & 4 \end{array} \\ \left[\begin{array}{cccc} +1 & 0 & 0 & 0 \\ 0 & +1 & 0 & 0 \\ 0 & 0 & +1 & 0 \\ 0 & 0 & 0 & +1 \\ \hdashline +1 & +1 & +1 & +1 \end{array}\right] \end{array} \tag{4.101}
$$

- *Digraph cut-sets*

  Let us draw curved lines (cuts) in the digraph, one at a time, between two points on the outside region of the digraph. Each cut intersects an arc in tree and the cut-edges. For the TRRTC digraph, the cut-sets are shown in Equation (4.102):

$$
\mathrm{G}_1 = \{\mathrm{A,E}\},\ \mathrm{G}_2 = \{\mathrm{B,E}\},\ \mathrm{G}_3 = \{\mathrm{C,E}\},\ \text{and } \mathrm{G}_4 = \{\mathrm{D,E}\} \tag{4.102}
$$

  Note: The entries from each cut-set correspond to nonzero entries on each column in Equation (4.101).

  The cut-set matroid in generating the equations for mechanisms' kinematics is reported in [4].

- *Orthogonality between cycle base matroid and cut-set matroid*

  The c × j cycle base matroid matrix, **C**, from the previous positional and velocity analysis, and the j × m cut-set matroid, ${}^{\otimes}\mathbf{G}^{\mathrm{T}}$, are orthogonal to each other; that is, their multiplication is the c × m zero matrix:

$$
\mathbf{C}_{\mathrm{c\times j}} * {}^{\otimes}\mathbf{G}^{\mathrm{T}}_{\mathrm{j\times m}} = 0_{\mathrm{c\times m}} \tag{4.103}
$$

4.7.4 Governing Dynamic Force Equations for Closed Cycle Mechanism

Newton's second law is applied for each node (link), as shown in Equation (4.104):

$$
\begin{aligned}
&\textit{Node } 1\,(\text{Link } 1){:} +\mathbf{N}^0_{\mathrm{A}} - \mathbf{N}^0_{\mathrm{B}} = \mathbf{f}^{\mathrm{in},0}_1 - \mathbf{f}^{\mathrm{ext},0}_1 \\
&\textit{Node } 2\,(\text{Link } 2){:} +\mathbf{N}^0_{\mathrm{B}} - \mathbf{N}^0_{\mathrm{C}} = \mathbf{f}^{\mathrm{in},0}_2 - \mathbf{f}^{\mathrm{ext},0}_2 \\
&\textit{Node } 3\,(\text{Link } 3){:} +\mathbf{N}^0_{\mathrm{C}} - \mathbf{N}^0_{\mathrm{D}} = \mathbf{f}^{\mathrm{in},0}_3 - \mathbf{f}^{\mathrm{ext},0}_3 \\
&\textit{Node } 4\,(\text{Link } 4){:} +\mathbf{N}^0_{\mathrm{D}} + \mathbf{N}^0_{\mathrm{E}} = \mathbf{f}^{\mathrm{in},0}_4 - \mathbf{f}^{\mathrm{ext},0}_4 \\
&\text{both reactions are incident (enter) the node 4, therefore taken with } +1
\end{aligned} \tag{4.104}
$$

One could notice that the coefficients in front of dynamic reactions: $\mathbf{N}_a^0$ for arcs and $\mathbf{N}_c^0$ for cut-edges, in Equation (4.105), are the entries in the matrices **G** and $\mathbf{G}^{cut}$. Therefore, these equations in the matrix form are written as:

$$\begin{bmatrix} +1 & -1 & 0 & 0 & \vdots & 0 \\ 0 & +1 & -1 & 0 & \vdots & 0 \\ 0 & 0 & +1 & -1 & \vdots & 0 \\ 0 & 0 & 0 & +1 & \vdots & +1 \end{bmatrix} * \begin{Bmatrix} \mathbf{N}_A^0 \\ \mathbf{N}_B^0 \\ \mathbf{N}_C^0 \\ \mathbf{N}_D^0 \\ \hdashline \mathbf{N}_E^0 \end{Bmatrix} = \begin{Bmatrix} \mathbf{P}_{G_1} \\ \mathbf{P}_{G_2} \\ \mathbf{P}_{G_3} \\ \mathbf{P}_{G_4} \end{Bmatrix} \tag{4.105}$$

The incidence matrix **G** is partitioned as follows:

$$\begin{bmatrix} +1 & -1 & 0 & 0 \\ 0 & +1 & -1 & 0 \\ 0 & 0 & +1 & -1 \\ 0 & 0 & 0 & +1 \end{bmatrix} * \begin{Bmatrix} \mathbf{N}_A^0 \\ \mathbf{N}_B^0 \\ \mathbf{N}_C^0 \\ \mathbf{N}_D^0 \end{Bmatrix} + \begin{bmatrix} 0 \\ 0 \\ 0 \\ +1 \end{bmatrix} \cdot \mathbf{N}_E^0 = \begin{Bmatrix} \mathbf{P}_{G_1} \\ \mathbf{P}_{G_2} \\ \mathbf{P}_{G_3} \\ \mathbf{P}_{G_4} \end{Bmatrix} \tag{4.106}$$

Then, Equation (4.106) is written as:

$$\mathbf{G}_{m\times a} * \mathbf{N}_a^0 + \mathbf{G}_{m\times c}^{cut} * \mathbf{N}_c^{cut,0} = \mathbf{P}_{G_m} \tag{4.107}$$

If Equation (4.107) is left-hand-side-multiplied by matrix **Z** and considering $\mathbf{P}_a^0 = \mathbf{Z} * \mathbf{P}_{G_m}$, then the equation for dynamic reactions becomes

$$\begin{bmatrix} \mathbf{U}_{a\times a} & \vdots & \mathbf{W}_{a\times c} \end{bmatrix} * \begin{Bmatrix} \mathbf{N}_a^0 \\ \hdashline \mathbf{N}_c^{cut,0} \end{Bmatrix} = \mathbf{P}_a^0 \tag{4.108}$$

Considering the row-reduced matrix ${}^{\otimes}\mathbf{G}$, then Equation (4.96), then:

$${}^{\otimes}\mathbf{G} * \mathbf{N}_j^0 = \mathbf{P}_a^0 \tag{4.109}$$

where

- The column vector with reaction forces in mechanism's j joints is

$$\mathbf{N}_j^0 = \begin{Bmatrix} \mathbf{N}_a^0 \\ \hdashline \mathbf{N}_c^{cut,0} \end{Bmatrix} \tag{4.110}$$

- $\mathbf{P}_{G_m} = \left(\mathbf{f}_m^{in,0} - \mathbf{f}_m^{ext,0}\right)$ is the resultant of inertial and external forces reduced to COM
- The column matrix *IT forces*, $\mathbf{P}_a^0$, in the tree's arcs:

$$\mathbf{P}_a^0 = \mathbf{Z}_{a\times m} * \mathbf{P}_{G_m} \tag{4.111}$$

The IT forces $\mathbf{P}_a^0$ are functions of COM's accelerations, determined previously using the acceleration analysis, and determined from the tree as:

Example for TRRTC mechanism, Figure 4.16:

$$\mathbf{P}_a^0 = \begin{Bmatrix} \mathbf{P}_A^0 \\ \mathbf{P}_B^0 \\ \mathbf{P}_C^0 \\ \mathbf{P}_D^0 \end{Bmatrix} = \begin{Bmatrix} \mathbf{P}_{G_1} + \mathbf{P}_{G_2} + \mathbf{P}_{G_3} + \mathbf{P}_{G_4} \\ \mathbf{P}_{G_2} + \mathbf{P}_{G_3} + \mathbf{P}_{G_4} \\ \mathbf{P}_{G_3} + \mathbf{P}_{G_4} \\ \mathbf{P}_{G_4} \end{Bmatrix} \tag{4.112}$$

4.7.5 The Cut-Set Reaction Forces in Joints

The *cut-set reaction* forces $\mathbf{N}_a^{cs,0}$ are entries in the column vector on the left-hand side of Equation (4.108):

$$\mathbf{N}_a^{cs,0} = {}^{\otimes}\mathbf{G} * \mathbf{N}_j^0 = \mathbf{N}_a^0 + \mathbf{W}_{a\times c} * \mathbf{N}_c^{cut,0} \tag{4.113}$$

Example for TRRTC mechanism, where all weighting w-entries are +1:

$$\begin{bmatrix} \mathbf{N}_A^{cs,0} \\ \mathbf{N}_B^{cs,0} \\ \mathbf{N}_C^{cs,0} \\ \mathbf{N}_D^{cs,0} \end{bmatrix} = \begin{bmatrix} \mathbf{N}_A^0 + w_{AE} \times \mathbf{N}_E^0 \\ \mathbf{N}_B^0 + w_{BE} \times \mathbf{N}_E^0 \\ \mathbf{N}_C^0 + w_{CE} \times \mathbf{N}_E^0 \\ \mathbf{N}_D^0 + w_{DE} \times \mathbf{N}_E^0 \end{bmatrix} \tag{4.113a}$$

Considering Equations (4.109) and (4.110), then

$$\mathbf{N}_a^{cs,0} = \mathbf{P}_a^0 \tag{4.114}$$

Equation (4.114) is the governing equation to determine the reaction forces in the joints.

Example for TRRTC mechanism:

$$\begin{Bmatrix} \mathbf{N}_A^0 + w_{AE} \times \mathbf{N}_E^0 \\ \mathbf{N}_B^0 + w_{BE} \times \mathbf{N}_E^0 \\ \mathbf{N}_C^0 + w_{CE} \times \mathbf{N}_E^0 \\ \mathbf{N}_D^0 + w_{DE} \times \mathbf{N}_E^0 \end{Bmatrix} = \begin{Bmatrix} \mathbf{P}_A^0 \\ \mathbf{P}_B^0 \\ \mathbf{P}_C^0 \\ \mathbf{P}_D^0 \end{Bmatrix} \tag{4.114a}$$

4.7.6 The IT-Resistant Force

Equation (4.114) is written as:

$$\mathbf{N}_{\mathrm{a}}^{\mathrm{cs},0} + \left(-\mathbf{P}_{\mathrm{a}}^{0}\right) = 0 \tag{4.115}$$

where the forces $\left(-\mathbf{P}_{\mathrm{a}}^{0}\right)$, named *"IT-resistant forces"* assigned to the joints (arcs tree), have been previously determined as:

$$\left(-\mathbf{P}_{\mathrm{a}}^{0}\right) = -\mathbf{Z}_{\mathrm{a}\times\mathrm{m}} * \mathbf{P}_{\mathrm{G}_{\mathrm{m}}} \tag{4.116}$$

Statement 4.9

For any closed cycle mechanism, the cut-set reaction forces and IT-resistant forces are a system of forces in fictitious equilibrium.

Note: This statement is equivalent to the d'Alembert principle in dynamics, developed in this text from a digraph approach. Equation (4.115) is applied for spatial and planar, single- or multiple-cycle mechanisms.

4.7.7 Reaction Forces in Arcs Tree Expressed from Reaction Forces in Cut-Edges

Considering the notations in Equations (4.113) and (4.114), then the reaction forces in a tree's joints are expressed as functions of the reaction forces in the cut-joints:

$$\mathbf{N}_{\mathrm{a}}^{0} = \mathbf{P}_{\mathrm{a}}^{0} - \mathbf{W}_{\mathrm{a}\times\mathrm{c}} * \mathbf{N}_{\mathrm{c}}^{\mathrm{cut},0} \tag{4.117}$$

Example: For single-cycle TRRTC mechanism, with:

$$\begin{bmatrix} \mathbf{N}_{\mathrm{A}}^{0} \\ \mathbf{N}_{\mathrm{B}}^{0} \\ \mathbf{N}_{\mathrm{C}}^{0} \\ \mathbf{N}_{\mathrm{D}}^{0} \end{bmatrix} = \begin{Bmatrix} \mathbf{P}_{\mathrm{A}}^{0} \\ \mathbf{P}_{\mathrm{B}}^{0} \\ \mathbf{P}_{\mathrm{C}}^{0} \\ \mathbf{P}_{\mathrm{D}}^{0} \end{Bmatrix} - \begin{bmatrix} w_{\mathrm{AE}} \cdot \mathbf{N}_{\mathrm{E}}^{0} \\ w_{\mathrm{BE}} \cdot \mathbf{N}_{\mathrm{E}}^{0} \\ w_{\mathrm{CE}} \cdot \mathbf{N}_{\mathrm{E}}^{0} \\ w_{\mathrm{DE}} \cdot \mathbf{N}_{\mathrm{E}}^{0} \end{bmatrix} \tag{4.117a}$$

- The system of scalar equations

  The unknown reactions were defined from constraints in the local frames. Therefore, Equation (4.117b) holds:

$$\begin{bmatrix} \mathbf{D}_{1} * \mathbf{N}_{\mathrm{A}}^{1} \\ \mathbf{D}_{2} * \mathbf{N}_{\mathrm{B}}^{2} \\ \mathbf{D}_{3} * \mathbf{N}_{\mathrm{C}}^{3} \\ \mathbf{D}_{4} * \mathbf{N}_{\mathrm{D}}^{4} \end{bmatrix} = \begin{Bmatrix} \mathbf{P}_{\mathrm{A}}^{0} \\ \mathbf{P}_{\mathrm{B}}^{0} \\ \mathbf{P}_{\mathrm{C}}^{0} \\ \mathbf{P}_{\mathrm{D}}^{0} \end{Bmatrix} - \begin{bmatrix} w_{\mathrm{AE}} \cdot \mathbf{N}_{\mathrm{E}}^{0} \\ w_{\mathrm{BE}} \cdot \mathbf{N}_{\mathrm{E}}^{0} \\ w_{\mathrm{CE}} \cdot \mathbf{N}_{\mathrm{E}}^{0} \\ w_{\mathrm{DE}} \cdot \mathbf{N}_{\mathrm{E}}^{0} \end{bmatrix} \tag{4.117b}$$

The system of scalar reaction components is obtained by equating the rows in Equation (4.117c).

$$\mathbf{D}_1 * \mathbf{N}_A^1 = \mathbf{P}_A^0 - \mathbf{N}_E^0; \mathbf{D}_2 * \mathbf{N}_B^2 = \mathbf{P}_B^0 - \mathbf{N}_E^0; \mathbf{D}_3 * \mathbf{N}_C^3 = \mathbf{P}_C^0 - \mathbf{N}_E^0; \mathbf{D}_4 * \mathbf{N}_D^4 = \mathbf{P}_D^0 - \mathbf{N}_E^0 \tag{4.117c}$$

4.7.8 Force Equations for Multiple-Cycle Mechanisms

- *Equations for two-cycle mechanisms*
  Considering Equation (4.100) for the weighting matrix **W**, then Equation (4.117) is

$$\begin{bmatrix} \mathbf{N}_A^0 \\ \mathbf{N}_B^0 \\ \mathbf{N}_C^0 \\ \mathbf{N}_D^0 \\ \mathbf{N}_E^0 \end{bmatrix} = \begin{Bmatrix} \mathbf{P}_A^0 \\ \mathbf{P}_B^0 \\ \mathbf{P}_C^0 \\ \mathbf{P}_D^0 \\ \mathbf{P}_E^0 \end{Bmatrix} - \begin{bmatrix} w_{AF} & w_{AG} \\ w_{BF} & w_{BG} \\ w_{CF} & w_{CG} \\ w_{DF} & w_{DG} \\ w_{EF} & w_{EG} \end{bmatrix} * \begin{bmatrix} \mathbf{N}_F^0 \\ \mathbf{N}_G^0 \end{bmatrix} \tag{4.118a}$$

Then, Equation (4.118b) holds:

$$\begin{aligned} &\mathbf{N}_A^0 = \mathbf{P}_A^0 - w_{AF} \cdot \mathbf{N}_F^0 - w_{AG} \cdot \mathbf{N}_G^0; \mathbf{N}_B^0 = \mathbf{P}_B^0 - w_{BF} \cdot \mathbf{N}_F^0 - w_{BG} \cdot \mathbf{N}_G^0; \\ &\mathbf{N}_C^0 = \mathbf{P}_C^0 - w_{CF} \cdot \mathbf{N}_F^0 - w_{CG} \cdot \mathbf{N}_G^0; \\ &\mathbf{N}_D^0 = \mathbf{P}_D^0 - w_{DF} \cdot \mathbf{N}_F^0 - w_{DG} \cdot \mathbf{N}_G^0; \mathbf{N}_E^0 = \mathbf{P}_E^0 - w_{EF} \cdot \mathbf{N}_F^0 - w_{EG} \cdot \mathbf{N}_G^0 \end{aligned} \tag{4.118b}$$

- The system of scalar equations is obtained from Equation (4.118b) substituting:

$$\mathbf{N}_z^0 = \mathbf{D}_m * \mathbf{N}_Z^m, \text{ for } Z = A, B, \ldots, E \tag{4.118c}$$

- *Equations for three-cycle mechanisms*
  Considering Equation (4.100) for the weighting matrix **W**, then Equation (4.117) is

$$\begin{bmatrix} \mathbf{N}_A^0 \\ \mathbf{N}_B^0 \\ \mathbf{N}_C^0 \\ \mathbf{N}_D^0 \\ \mathbf{N}_E^0 \\ \mathbf{N}_F^0 \end{bmatrix} = \begin{Bmatrix} \mathbf{P}_A^0 \\ \mathbf{P}_B^0 \\ \mathbf{P}_C^0 \\ \mathbf{P}_D^0 \\ \mathbf{P}_E^0 \\ \mathbf{P}_F^0 \end{Bmatrix} - \begin{bmatrix} w_{AG} & w_{AH} & w_{AI} \\ w_{BG} & w_{BH} & w_{BI} \\ w_{CG} & w_{CH} & w_{CI} \\ w_{DG} & w_{DH} & w_{DI} \\ w_{EG} & w_{EH} & w_{EI} \\ w_{FG} & w_{FH} & w_{FI} \end{bmatrix} * \begin{bmatrix} \mathbf{N}_G^0 \\ \mathbf{N}_H^0 \\ \mathbf{N}_I^0 \end{bmatrix} \tag{4.119a}$$

Then, Equation (4.119b) holds:

$$\begin{aligned} &\mathbf{N}_A^0 = \mathbf{P}_A^0 - w_{AG} \cdot \mathbf{N}_G^0 - w_{AH} \cdot \mathbf{N}_H^0 - w_{AI} \cdot \mathbf{N}_I^0; \\ &\mathbf{N}_B^0 = \mathbf{P}_B^0 - w_{BG} \cdot \mathbf{N}_G^0 - w_{BH} \cdot \mathbf{N}_H^0 - w_{BI} \cdot \mathbf{N}_I^0; \\ &\mathbf{N}_C^0 = \mathbf{P}_C^0 - w_{CG} \cdot \mathbf{N}_G^0 - w_{CH} \cdot \mathbf{N}_H^0 - w_{CI} \cdot \mathbf{N}_I^0; \\ &\mathbf{N}_D^0 = \mathbf{P}_D^0 - w_{DG} \cdot \mathbf{N}_G^0 - w_{DH} \cdot \mathbf{N}_H^0 - w_{DI} \cdot \mathbf{N}_I^0; \\ &\mathbf{N}_E^0 = \mathbf{P}_E^0 - w_{EG} \cdot \mathbf{N}_G^0 - w_{EH} \cdot \mathbf{N}_H^0 - w_{EI} \cdot \mathbf{N}_I^0; \\ &\mathbf{N}_F^0 = \mathbf{P}_F^0 - w_{FG} \cdot \mathbf{N}_G^0 - w_{FH} \cdot \mathbf{N}_H^0 - w_{FI} \cdot \mathbf{N}_I^0 \end{aligned} \tag{4.119b}$$

- The system of scalar equations is obtained from Equation (4.119b) substituting:

$$\mathbf{N}_z^0 = \mathbf{D}_m * \mathbf{N}_Z^m, \text{ for } Z = A, B, \ldots, F \tag{4.119c}$$

4.8 Closed Cycle Mechanisms: The IT Equations for Dynamic Moment Reactions and Differential Equations

Newton's second law is applied for each node (link), as shown in Equation (4.120).

$$\begin{aligned} &\textit{Node } 1 \,(\text{Link } 1){:} +\mathbf{M}_A^0 - \mathbf{M}_B^0 = \mathbf{Q}_{G_1}^0 - \left(\tilde{\mathbf{L}}_{G_1A}^0 * \mathbf{N}_A^0 - \tilde{\mathbf{L}}_{G_1B}^0 * \mathbf{N}_B^0\right) \\ &\textit{Node } 2 \,(\text{Link } 2){:} +\mathbf{M}_B^0 - \mathbf{M}_C^0 = \mathbf{Q}_{G_2}^0 - \left(\tilde{\mathbf{L}}_{G_2B}^0 * \mathbf{N}_B^0 - \tilde{\mathbf{L}}_{G_2C}^0 * \mathbf{N}_C^0\right) \\ &\textit{Node } 3 \,(\text{Link } 3){:}\ \mathbf{M}_C^0 - \mathbf{M}_D^0 = \mathbf{Q}_{G_3}^0 - \left(\tilde{\mathbf{L}}_{G_3C}^0 * \mathbf{N}_C^0 - \tilde{\mathbf{L}}_{G_3D}^0 * \mathbf{N}_D^0\right) \\ &\textit{Node } 4 \,(\text{Link } 4){:} +\mathbf{M}_D^0 - \mathbf{M}_E^0 = \mathbf{Q}_{G_4}^0 - \left(\tilde{\mathbf{L}}_{G_4D}^0 * \mathbf{N}_D^0 - \tilde{\mathbf{L}}_{G_4E}^0 * \mathbf{N}_E^0\right) \end{aligned} \tag{4.120}$$

The skew-symmetric position vectors originated from COM-toward all joints.
It is generated from the **G** matrix entries, shown in Equation (4.94):

$$\begin{array}{c} \\ 1 \\ 2 \\ 3 \\ 4 \end{array}\begin{array}{c} \begin{array}{ccccc} \quad\text{A} & \quad\quad\text{B} & \quad\quad\text{C} & \quad\quad\text{D} & \quad\quad\text{E} \end{array} \\ \left[\begin{array}{cccc|c} +1\cdot\tilde{\mathbf{L}}^0_{G_1A} & -1\cdot\tilde{\mathbf{L}}^0_{G_1B} & 0 & 0 & 0 \\ 0 & +1\cdot\tilde{\mathbf{L}}^0_{G_2B} & -1\cdot\tilde{\mathbf{L}}^0_{G_2C} & 0 & 0 \\ 0 & 0 & +1\cdot\tilde{\mathbf{L}}^0_{G_3C} & -1\cdot\tilde{\mathbf{L}}^0_{G_3D} & 0 \\ 0 & 0 & 0 & +1\cdot\tilde{\mathbf{L}}^0_{G_4D} & +1\cdot\tilde{\mathbf{L}}^0_{G_4E} \end{array}\right] \end{array} \tag{4.121}$$

$$= \left[\begin{array}{c|c} \mathbf{G}_{m\times a}(\tilde{\mathbf{L}}) & \mathbf{G}^{cut}_{m\times c}(\tilde{\mathbf{L}}) \end{array}\right]$$

The two partitions are as follows:

- $\mathbf{G}_{m\times a}(\tilde{\mathbf{L}})$ is the m × a matrix whose vector entries are the skew-symmetric position vectors originated from the COM-toward tree's joints.
- $\mathbf{G}^{cut}_{m\times c}(\tilde{\mathbf{L}})$ is the m × c matrix whose vector entries are the skew-symmetric position vectors originated from the COM-toward cut-joints, shown below.

$$\mathbf{G}_{m\times a}(\tilde{\mathbf{L}}) = \begin{bmatrix} +1\cdot\tilde{\mathbf{L}}^0_{G_1A} & -1\cdot\tilde{\mathbf{L}}^0_{G_1B} & 0 & 0 \\ 0 & +1\cdot\tilde{\mathbf{L}}^0_{G_2B} & -1\cdot\tilde{\mathbf{L}}^0_{G_2C} & 0 \\ 0 & 0 & +1\cdot\tilde{\mathbf{L}}^0_{G_3C} & -1\cdot\tilde{\mathbf{L}}^0_{G_3D} \\ 0 & 0 & 0 & +1\cdot\tilde{\mathbf{L}}^0_{G_4D} \end{bmatrix};$$

$$\mathbf{G}^{cut}_{m\times c}(\tilde{\mathbf{L}}) = \begin{bmatrix} 0 \\ 0 \\ 0 \\ +\tilde{\mathbf{L}}^0_{G_4E} \end{bmatrix} \tag{4.122}$$

The skew-symmetric position vectors originated from tree's joints toward the cut-joints.

- $\mathbf{W}_{a\times c}(\tilde{\mathbf{L}})$ is the a × c matrix with c-columns (equal to the number of cycles) whose vector entries are the *skew-symmetric position vectors.* It is generated from the $\mathbf{W}_{a\times c}$ matrix, shown in Equation (4.123):

1. Directly, by replacing the weighting coefficient w with the corresponding skew-symmetric matrix:

$$w_{a,c} \mapsto \tilde{\mathbf{L}}^{0}_{a,c} \tag{4.123}$$

Example: For TRRTC mechanism

$$\mathbf{W}_{a\times c} = \begin{matrix} \\ A \\ B \\ C \\ D \end{matrix} \begin{bmatrix} E \\ +1 \\ +1 \\ +1 \\ +1 \end{bmatrix} = \begin{bmatrix} w_{AE} \\ w_{BE} \\ w_{CE} \\ w_{DE} \end{bmatrix} \Rightarrow \mathbf{W}_{a\times c}(\tilde{\mathbf{L}}) = \begin{bmatrix} E \\ +1\cdot\tilde{\mathbf{L}}^{0}_{AE} \\ +1\cdot\tilde{\mathbf{L}}^{0}_{BE} \\ +1\cdot\tilde{\mathbf{L}}^{0}_{CE} \\ +1\cdot\tilde{\mathbf{L}}^{0}_{DE} \end{bmatrix} \tag{4.123a}$$

2. Automatic generation of its entries with matrix Equation (4.124), which is based on the partitions $\mathbf{G}$ and $\mathbf{G}^{cut}$ from Equation (4.122), as shown in [3].

$$\mathbf{W}_{a\times c}(\tilde{\mathbf{L}}) = \mathbf{Z}_{a\times m} * \left(\mathbf{G}^{cut}_{m\times c}(\tilde{\mathbf{L}}) - \mathbf{G}_{m\times j}(\tilde{\mathbf{L}}) * \mathbf{W}_{j\times c}\right) \tag{4.124}$$

Multiple-cycle mechanisms

- For two-cycle mechanisms, the position vector matrix $\mathbf{W}_{a\times c}(\tilde{\mathbf{L}})$ is generated directly from the weighting matrix $\mathbf{W}$, Equation (4.100), by replacing their entries, as shown in Equation (4.125).

$$\mathbf{W}_{a\times c}(\tilde{\mathbf{L}}) = \begin{matrix} \\ A \\ B \\ C \\ D \\ E \end{matrix} \begin{bmatrix} F & G \\ w_{AF}\cdot\tilde{\mathbf{L}}_{AF} & w_{AG}\cdot\tilde{\mathbf{L}}_{AG} \\ w_{BF}\cdot\tilde{\mathbf{L}}_{BF} & w_{BG}\cdot\tilde{\mathbf{L}}_{BG} \\ w_{CF}\cdot\tilde{\mathbf{L}}_{CF} & w_{CG}\cdot\tilde{\mathbf{L}}_{CG} \\ w_{DF}\cdot\tilde{\mathbf{L}}_{DF} & w_{DG}\cdot\tilde{\mathbf{L}}_{DG} \\ w_{EF}\cdot\tilde{\mathbf{L}}_{EF} & w_{EG}\cdot\tilde{\mathbf{L}}_{EG} \end{bmatrix} \tag{4.125}$$

Equation (4.120) in matrix form and vector entries is written as:

$$\begin{bmatrix} +1 & -1 & 0 & 0 & \vdots & 0 \\ 0 & +1 & -1 & 0 & \vdots & 0 \\ 0 & 0 & +1 & -1 & \vdots & 0 \\ 0 & 0 & 0 & +1 & \vdots & +1 \end{bmatrix} * \begin{Bmatrix} \mathbf{M}_A^0 \\ \mathbf{M}_B^0 \\ \mathbf{M}_C^0 \\ \mathbf{M}_D^0 \\ \hline \mathbf{M}_E^0 \end{Bmatrix} = \begin{Bmatrix} \mathbf{Q}_{G_1}^0 \\ \mathbf{Q}_{G_2}^0 \\ \mathbf{Q}_{G_3}^0 \\ \mathbf{Q}_{G_4}^0 \end{Bmatrix}$$

$$- \begin{bmatrix} +1 \cdot \tilde{\mathbf{L}}_{G_1A}^0 & -1 \cdot \tilde{\mathbf{L}}_{G_1B}^0 & 0 & 0 & \vdots & 0 \\ 0 & +1 \cdot \tilde{\mathbf{L}}_{G_2B}^0 & -1 \cdot \tilde{\mathbf{L}}_{G_2C}^0 & 0 & \vdots & 0 \\ 0 & 0 & +1 \cdot \tilde{\mathbf{L}}_{G_3C}^0 & -1 \cdot \tilde{\mathbf{L}}_{G_3D}^0 & \vdots & 0 \\ 0 & 0 & 0 & +1 \cdot \tilde{\mathbf{L}}_{G_4D}^0 & \vdots & +1 \cdot \tilde{\mathbf{L}}_{G_4E}^0 \end{bmatrix} * \begin{Bmatrix} \mathbf{N}_A^0 \\ \mathbf{N}_B^0 \\ \mathbf{N}_C^0 \\ \mathbf{N}_D^0 \\ \hline \mathbf{N}_E^0 \end{Bmatrix} \tag{4.126}$$

4.8.1 The Governing Equations to Evaluate the Reactions in Cut-Joints

To decrease the number of unknown force reactions in the moment equations, the joint reactions from Equation (4.117) are substituted in the right-hand side of Equation (4.126). Equation (4.126) is left-hand-side multiplied by path matrix **Z**:

$$\begin{aligned} &\underbrace{\left(\mathbf{Z}_{a\times m} * \mathbf{G}_{m\times a}\right)}_{\mathbf{U}_{a\times a}} * \mathbf{M}_a^0 + \underbrace{\left(\mathbf{Z}_{a\times m} * \mathbf{G}_{m\times c}^{cut}\right)}_{\mathbf{W}_{a\times c}} * \mathbf{M}_c^{cut,0} \\ &+ \underbrace{\mathbf{Z}_{a\times m} * \left(\mathbf{G}_{m\times c}^{cut}(\tilde{\mathbf{L}}) - \mathbf{G}_{m\times a}(\tilde{\mathbf{L}}) * \mathbf{W}_{j\times c}\right)}_{\mathbf{W}_{a\times c}(\tilde{\mathbf{L}})} * \mathbf{N}_{c\times 1}^{cut,0} \\ &= \mathbf{Z}_{a\times m} * \mathbf{Q}_{G_m}^0 + \underbrace{\left(-\mathbf{Z}_{a\times m} * \mathbf{G}_{m\times a}(\tilde{\mathbf{L}}) * \mathbf{Z}_{a\times m}\right)}_{\mathbf{Z}_{m\times a}(\tilde{\mathbf{L}})} * \mathbf{P}_{G_m}^0 \end{aligned} \tag{4.127}$$

Notice that the following terms are grouping

- Equation (4.124) for matrix $\mathbf{W}_{a\times c}(\tilde{\mathbf{L}})$
- Equation (4.80c) for matrix $\mathbf{Z}_{m\times a}(\tilde{\mathbf{L}})$:

$$\mathbf{Z}_{m\times a}(\tilde{\mathbf{L}}) = -\mathbf{Z}_{a\times m} * \mathbf{G}_{m\times a}(\tilde{\mathbf{L}}) * \mathbf{Z}_{a\times m}$$

- Equation (4.82) for the IT moment $\mathbf{Q}_a^0$ in the tree's joints:

$$\mathbf{Q}_a^0 = \mathbf{Z}_{a\times m} * \mathbf{Q}_{Gm}^0 + \mathbf{Z}_{a\times m}(\tilde{\mathbf{L}}) * \mathbf{P}_{Gm}^0 \quad (4.128)$$

Then, Equation (4.127) becomes

$$\mathbf{M}_a^0 + \mathbf{W}_{a\times c} * \mathbf{M}_c^{cut,0} + \mathbf{W}_{a\times c}(\tilde{\mathbf{L}}) * \mathbf{N}_c^{cut,0} = \mathbf{Q}_a^0 \quad (4.129)$$

Equation (4.129) is the IT dynamic moment equation, with a low number of variables $\mathbf{N}_c^{cut,0}$ and $\mathbf{M}_c^{cut,0}$. These equations have been generated from cut-set matroid entries.

4.8.2 The Cut-Set Reaction Moment in Tree's Joints: The Cut-Set Matroid Method

The *cut-set reaction* moment $\mathbf{M}_a^{cs,0}$ is the column vector on the left-hand side of Equation (4.129).

$$\mathbf{M}_a^{cs,0} = \mathbf{M}_a^0 + \mathbf{W}_{a\times c} * \mathbf{M}_c^{cut,0} + \mathbf{W}_{a\times c}(\tilde{\mathbf{L}}) * \mathbf{N}_c^{cut,0} \quad (4.130)$$

Considering Equations (4.129) and (4.130), then

$$\mathbf{M}_a^{cs,0} = \mathbf{Q}_a^0 \quad (4.131)$$

4.8.3 Reaction Moment Equations for Closed Cycle Mechanisms: The Resistant Moment

Equation (4.131) is written as:

$$\mathbf{M}_a^{cs,0} + \left(-\mathbf{Q}_a^0\right) = 0 \quad (4.132)$$

where $\left(-\mathbf{Q}_a^0\right)$ are the *"IT-resistant moments,"* assigned to joints (arcs tree).

Statement 4.10

For any closed cycle mechanism, the cut-set reaction moments and IT-resistant moments are a system in fictitious equilibrium.

Note: This statement is equivalent to the d'Alembert principle in dynamics, developed in this text from a digraph approach. Equation (4.132) is applied for spatial and planar, single- or multiple-cycle mechanisms.

4.8.4 Review on Computation of IT Equations for Dynamics of Closed Cycle Mechanisms

The computation of dynamic equations and differential equations of motion is performed as follows:

$$\text{Draw the digraph and spanning tree} \tag{4.133a}$$

$$\text{The matrices } \mathbf{Z}_{a\times m},\ \mathbf{G}^{cut}_{m\times c},\ \text{and } \mathbf{W}_{a\times c} = \mathbf{Z} * \mathbf{G}^{cut} \tag{4.133b}$$

$$\begin{gathered}\text{The matrices } \mathbf{Z}_{a\times m}(\tilde{\mathbf{L}}) \text{ and } \mathbf{W}_{a\times c}(\tilde{\mathbf{L}}) \\ \text{written directly from } \mathbf{Z}_{a\times m} \text{ and } \mathbf{W}_{a\times c}\end{gathered} \tag{4.133c}$$

The force resultant and the moment resultant reduced to COM:

$$\mathbf{P}_{G_m} = \mathbf{f}^{in,0}_m - \mathbf{f}^{ext,0}_m \tag{4.133d}$$

$$\mathbf{Q}^0_{G_m} = \dot{\mathbf{K}}^{in,0}_m - \boldsymbol{\mu}^{ext,0}_m \tag{4.133e}$$

- The IT force $\mathbf{P}^0_a$ and IT moment $\mathbf{Q}^0_a$ reduced to joint's tree. They are the coefficients on the right-hand side of Equations (4.111) and (4.128):

$$\mathbf{P}^0_a = \mathbf{Z}_{a\times m} * \mathbf{P}_{G_m} \tag{4.133f}$$

$$\mathbf{Q}^0_a = \mathbf{Z}_{a\times m} * \mathbf{Q}^0_{G_m} + \mathbf{Z}_{a\times m}(\tilde{\mathbf{L}}) * \mathbf{P}^0_{G_m} \tag{4.133g}$$

Moment equations:

- The IT moment equations

$$\mathbf{M}^0_a + \mathbf{W}_{a\times c} * \mathbf{M}^{cut,0}_c + \mathbf{W}_{a\times c}(\tilde{\mathbf{L}}) * \mathbf{N}^{cut,0}_c = \mathbf{Q}^0_a \tag{4.133h}$$

are solved for reactions in cut-joints $\mathbf{N}^{cut,0}_c$ and $\mathbf{M}^{cut,0}_c$

Force equations:

- The IT force equations

$$\mathbf{N}^0_a + \mathbf{W}_{a\times c} * \mathbf{N}^{cut,0}_c = \mathbf{P}^0_a \tag{4.133i}$$

are solved for joint reactions $\mathbf{N}^0_a$

Equations (4.133a) to (4.133i) are the *"IT equations for dynamic reactions and differential equations,"* [2,3]

Note:

- The moment Equation (4.133h) is solved for a minimal number of cut-set unknowns: $\mathbf{M}^0_a$, $\mathbf{M}^{cut,0}_c$, and $\mathbf{N}^{cut,0}_c$.
- Then, the reaction forces in the joint's tree are determined from Equation (4.133i) as functions of the previously determined $\mathbf{N}^{cut,0}_c$ (cut-set reaction forces).

- Applications for single- or multiple-cycle, spatial and planar mechanisms are presented next.
  1. **Single-cycle IT dynamic equations**
     For single-cycle mechanisms with c = 1 cycle, m = 4 mobile links, j = 5 joints (A, B, C, D, and E), and c = 1 cut-joint (E), the equations are as follows:
     Moment equations:

$$\begin{cases} \mathbf{M}_A^0 + w_{AE} \cdot \mathbf{M}_E^0 + \tilde{\mathbf{L}}_{AE}^0 * \mathbf{N}_E^0 = \mathbf{Q}_A^0 \\ \mathbf{M}_B^0 + w_{BE} \cdot \mathbf{M}_E^0 + \tilde{\mathbf{L}}_{BE}^0 * \mathbf{N}_E^0 = \mathbf{Q}_B^0 \\ \mathbf{M}_C^0 + w_{CE} \cdot \mathbf{M}_E^0 + \tilde{\mathbf{L}}_{CE}^0 * \mathbf{N}_E^0 = \mathbf{Q}_C^0 \\ \mathbf{M}_D^0 + w_{DE} \cdot \mathbf{M}_E^0 + \tilde{\mathbf{L}}_{DE}^0 * \mathbf{N}_E^0 = \mathbf{Q}_D^0 \end{cases} \tag{4.134a}$$

     Note: The number of moment vector equations is m = 4 (the number of mobile links).
     Force equations:

$$\begin{cases} \mathbf{N}_A^0 + w_{AE} \cdot \mathbf{N}_E^0 = \mathbf{P}_A^0 \\ \mathbf{N}_B^0 + w_{BE} \cdot \mathbf{N}_E^0 = \mathbf{P}_B^0 \\ \mathbf{N}_C^0 + w_{CE} \cdot \mathbf{N}_E^0 = \mathbf{P}_C^0 \\ \mathbf{N}_D^0 + w_{DE} \cdot \mathbf{N}_E^0 = \mathbf{P}_D^0 \end{cases} \tag{4.134b}$$

     Note: The number of force equations is a = j − c = 4 (the number of joints in the tree).
  2. **Two-cycle IT dynamic equations**
     For two-cycle mechanisms with c = 2 cycles, m = 5 mobile links, j = 7 joints (A, B, C, D, E, F, and G), and c = 2 cut-joints (F and G), the equations are as follows:
     Moment equations:

$$\begin{cases} \mathbf{M}_A^0 + w_{AF} \cdot \mathbf{M}_F^0 + w_{AG} \cdot \mathbf{M}_G^0 + \tilde{\mathbf{L}}_{AF}^0 * \mathbf{N}_F^0 + \tilde{\mathbf{L}}_{AG}^0 * \mathbf{N}_G^0 = \mathbf{Q}_A^0 \\ \mathbf{M}_B^0 + w_{BF} \cdot \mathbf{M}_F^0 + w_{BG} \cdot \mathbf{M}_G^0 + \tilde{\mathbf{L}}_{BF}^0 * \mathbf{N}_F^0 + \tilde{\mathbf{L}}_{BG}^0 * \mathbf{N}_G^0 = \mathbf{Q}_B^0 \\ \mathbf{M}_C^0 + w_{CF} \cdot \mathbf{M}_F^0 + w_{CG} \cdot \mathbf{M}_G^0 + \tilde{\mathbf{L}}_{CF}^0 * \mathbf{N}_F^0 + \tilde{\mathbf{L}}_{CG}^0 * \mathbf{N}_G^0 = \mathbf{Q}_C^0 \\ \mathbf{M}_D^0 + w_{DF} \cdot \mathbf{M}_F^0 + w_{DG} \cdot \mathbf{M}_G^0 + \tilde{\mathbf{L}}_{DF}^0 * \mathbf{N}_F^0 + \tilde{\mathbf{L}}_{DG}^0 * \mathbf{N}_G^0 = \mathbf{Q}_D^0 \\ \mathbf{M}_E^0 + w_{EF} \cdot \mathbf{M}_F^0 + w_{EG} \cdot \mathbf{M}_G^0 + \tilde{\mathbf{L}}_{EF}^0 * \mathbf{N}_F^0 + \tilde{\mathbf{L}}_{EG}^0 * \mathbf{N}_G^0 = \mathbf{Q}_E^0 \end{cases} \tag{4.135a}$$

     Note: The number of moment equations is m = 5 (the number of mobile links).

Force equations:

$$\begin{cases} \mathbf{N}_A^0 = \mathbf{P}_A^0 - w_{AF} \cdot \mathbf{N}_F^0 - w_{AG} \cdot \mathbf{N}_G^0 \\ \mathbf{N}_B^0 = \mathbf{P}_B^0 - w_{BF} \cdot \mathbf{N}_F^0 - w_{BG} \cdot \mathbf{N}_G^0 \\ \mathbf{N}_C^0 = \mathbf{P}_C^0 - w_{CF} \cdot \mathbf{N}_F^0 - w_{CG} \cdot \mathbf{N}_G^0 \\ \mathbf{N}_D^0 = \mathbf{P}_D^0 - w_{DF} \cdot \mathbf{N}_F^0 - w_{DG} \cdot \mathbf{N}_G^0 \\ \mathbf{N}_E^0 = \mathbf{P}_E^0 - w_{EF} \cdot \mathbf{N}_F^0 - w_{EG} \cdot \mathbf{N}_G^0 \end{cases} \tag{4.135b}$$

Note: The number of force equations is a = j − c = 5 (the number of joints in the tree).

3. **Three-cycle IT dynamic equations**

For three-cycle mechanisms with c = 3 cycles, m = 6 mobile links, j = 9 joints (A, B, C, D, E, F, G, H, and I), and c = 3 cut-joints (G, H, and I), the equations are as follows:

Moment equations:

$$\begin{cases} \mathbf{M}_A^0 + w_{AG} \cdot \mathbf{M}_G^0 + w_{AH} \cdot \mathbf{M}_H^0 + w_{AI} \cdot \mathbf{M}_I^0 + \tilde{\mathbf{L}}_{AG}^0 * \mathbf{N}_G^0 + \tilde{\mathbf{L}}_{AH}^0 * \mathbf{N}_H^0 + \tilde{\mathbf{L}}_{AI}^0 * \mathbf{N}_I^0 = \mathbf{Q}_A^0 \\ \mathbf{M}_B^0 + w_{BG} \cdot \mathbf{M}_G^0 + w_{BH} \cdot \mathbf{M}_H^0 + w_{BI} \cdot \mathbf{M}_I^0 + \tilde{\mathbf{L}}_{BG}^0 * \mathbf{N}_G^0 + \tilde{\mathbf{L}}_{BH}^0 * \mathbf{N}_H^0 + \tilde{\mathbf{L}}_{BI}^0 * \mathbf{N}_I^0 = \mathbf{Q}_B^0 \\ \mathbf{M}_C^0 + w_{CG} \cdot \mathbf{M}_G^0 + w_{CH} \cdot \mathbf{M}_H^0 + w_{CI} \cdot \mathbf{M}_I^0 + \tilde{\mathbf{L}}_{CG}^0 * \mathbf{N}_G^0 + \tilde{\mathbf{L}}_{CH}^0 * \mathbf{N}_H^0 + \tilde{\mathbf{L}}_{CI}^0 * \mathbf{N}_I^0 = \mathbf{Q}_C^0 \\ \mathbf{M}_D^0 + w_{DG} \cdot \mathbf{M}_G^0 + w_{DH} \cdot \mathbf{M}_H^0 + w_{DI} \cdot \mathbf{M}_I^0 + \tilde{\mathbf{L}}_{DG}^0 * \mathbf{N}_G^0 + \tilde{\mathbf{L}}_{DH}^0 * \mathbf{N}_H^0 + \tilde{\mathbf{L}}_{DI}^0 * \mathbf{N}_I^0 = \mathbf{Q}_D^0 \\ \mathbf{M}_E^0 + w_{EG} \cdot \mathbf{M}_G^0 + w_{EH} \cdot \mathbf{M}_H^0 + w_{EI} \cdot \mathbf{M}_I^0 + \tilde{\mathbf{L}}_{EG}^0 * \mathbf{N}_G^0 + \tilde{\mathbf{L}}_{EH}^0 * \mathbf{N}_H^0 + \tilde{\mathbf{L}}_{EI}^0 * \mathbf{N}_I^0 = \mathbf{Q}_E^0 \\ \mathbf{M}_F^0 + w_{FG} \cdot \mathbf{M}_G^0 + w_{FH} \cdot \mathbf{M}_H^0 + w_{FI} \cdot \mathbf{M}_I^0 + \tilde{\mathbf{L}}_{FG}^0 * \mathbf{N}_G^0 + \tilde{\mathbf{L}}_{FH}^0 * \mathbf{N}_H^0 + \tilde{\mathbf{L}}_{FI}^0 * \mathbf{N}_I^0 = \mathbf{Q}_F^0 \end{cases} \tag{4.136a}$$

Note: The number of moment equations is m = 6 (the number of mobile links).

Force equations:

$$\begin{cases} \mathbf{N}_A^0 = \mathbf{P}_A^0 - w_{AG} \cdot \mathbf{N}_G^0 - w_{AH} \cdot \mathbf{N}_H^0 - w_{AI} \cdot \mathbf{N}_I^0 \\ \mathbf{N}_B^0 = \mathbf{P}_B^0 - w_{BG} \cdot \mathbf{N}_G^0 - w_{BH} \cdot \mathbf{N}_H^0 - w_{BI} \cdot \mathbf{N}_I^0 \\ \mathbf{N}_C^0 = \mathbf{P}_C^0 - w_{CG} \cdot \mathbf{N}_G^0 - w_{CH} \cdot \mathbf{N}_H^0 - w_{CI} \cdot \mathbf{N}_I^0 \\ \mathbf{N}_D^0 = \mathbf{P}_D^0 - w_{DG} \cdot \mathbf{N}_G^0 - w_{DH} \cdot \mathbf{N}_H^0 - w_{DI} \cdot \mathbf{N}_I^0 \\ \mathbf{N}_E^0 = \mathbf{P}_E^0 - w_{EG} \cdot \mathbf{N}_G^0 - w_{EH} \cdot \mathbf{N}_H^0 - w_{EI} \cdot \mathbf{N}_I^0 \\ \mathbf{N}_F^0 = \mathbf{P}_F^0 - w_{FG} \cdot \mathbf{N}_G^0 - w_{FH} \cdot \mathbf{N}_H^0 - w_{FI} \cdot \mathbf{N}_I^0 \end{cases} \tag{4.136b}$$

Note: The number of force equations is a = j − c = 6 (the number of joints in the tree).

Although the mechanisms with four or more than four cycles are rare, the equations are written by *permutation of joint labels* and by adding rows in moment equations (coincidental with m mobile links) and by adding rows in force equations (coincidental with a = j − c joints in the tree).

Thus, the quadruple-cycle IT equations are generated by adding one more row to the triple-cycle moment equations and one more row to the force equations, as shown next.

4. **Four-cycle IT dynamic equations**

For four-cycle mechanisms with c = 4 cycles, m = 7 mobile links, j = 11 joints (A, B, C, D, E, F, G, H, I, J, and K), and c = 4 cut-joints (H, I, J, and K), the equations are as follows:

Moment equations:

$$\begin{cases} \mathbf{M}_A^0 + w_{AH}\cdot\mathbf{M}_H^0 + w_{AI}\cdot\mathbf{M}_I^0 + w_{AJ}\cdot\mathbf{M}_J^0 + w_{AK}\cdot\mathbf{M}_K^0 + \tilde{\mathbf{L}}_{AH}^0 * \mathbf{N}_H^0 + \tilde{\mathbf{L}}_{AI}^0 * \mathbf{N}_I^0 \\ \quad + \tilde{\mathbf{L}}_{AJ}^0 * \mathbf{N}_J^0 + \tilde{\mathbf{L}}_{AK}^0 * \mathbf{N}_K^0 = \mathbf{Q}_A^0 \\ \mathbf{M}_B^0 + w_{BH}\cdot\mathbf{M}_H^0 + w_{BI}\cdot\mathbf{M}_I^0 + w_{BJ}\cdot\mathbf{M}_J^0 + w_{BK}\cdot\mathbf{M}_K^0 + \tilde{\mathbf{L}}_{BH}^0 * \mathbf{N}_H^0 + \tilde{\mathbf{L}}_{BI}^0 * \mathbf{N}_I^0 \\ \quad + \tilde{\mathbf{L}}_{BJ}^0 * \mathbf{N}_J^0 + \tilde{\mathbf{L}}_{BK}^0 * \mathbf{N}_K^0 = \mathbf{Q}_B^0 \\ \mathbf{M}_C^0 + w_{CH}\cdot\mathbf{M}_H^0 + w_{CI}\cdot\mathbf{M}_I^0 + w_{CJ}\cdot\mathbf{M}_J^0 + w_{CK}\cdot\mathbf{M}_K^0 + \tilde{\mathbf{L}}_{CH}^0 * \mathbf{N}_H^0 + \tilde{\mathbf{L}}_{CI}^0 * \mathbf{N}_I^0 \\ \quad + \tilde{\mathbf{L}}_{CJ}^0 * \mathbf{N}_J^0 + \tilde{\mathbf{L}}_{CK}^0 * \mathbf{N}_K^0 = \mathbf{Q}_C^0 \\ \mathbf{M}_D^0 + w_{DH}\cdot\mathbf{M}_H^0 + w_{DI}\cdot\mathbf{M}_I^0 + w_{DJ}\cdot\mathbf{M}_J^0 + w_{DK}\cdot\mathbf{M}_K^0 + \tilde{\mathbf{L}}_{DH}^0 * \mathbf{N}_H^0 + \tilde{\mathbf{L}}_{DI}^0 * \mathbf{N}_I^0 \\ \quad + \tilde{\mathbf{L}}_{DJ}^0 * \mathbf{N}_J^0 + \tilde{\mathbf{L}}_{DK}^0 * \mathbf{N}_K^0 = \mathbf{Q}_D^0 \\ \mathbf{M}_E^0 + w_{EH}\cdot\mathbf{M}_H^0 + w_{EI}\cdot\mathbf{M}_I^0 + w_{EJ}\cdot\mathbf{M}_J^0 + w_{EK}\cdot\mathbf{M}_K^0 + \tilde{\mathbf{L}}_{EH}^0 * \mathbf{N}_H^0 + \tilde{\mathbf{L}}_{EI}^0 * \mathbf{N}_I^0 \\ \quad + \tilde{\mathbf{L}}_{EJ}^0 * \mathbf{N}_J^0 + \tilde{\mathbf{L}}_{EK}^0 * \mathbf{N}_K^0 = \mathbf{Q}_E^0 \\ \mathbf{M}_F^0 + w_{FH}\cdot\mathbf{M}_H^0 + w_{FI}\cdot\mathbf{M}_I^0 + w_{FJ}\cdot\mathbf{M}_J^0 + w_{FK}\cdot\mathbf{M}_K^0 + \tilde{\mathbf{L}}_{FH}^0 * \mathbf{N}_H^0 + \tilde{\mathbf{L}}_{FI}^0 * \mathbf{N}_I^0 \\ \quad + \tilde{\mathbf{L}}_{FJ}^0 * \mathbf{N}_J^0 + \tilde{\mathbf{L}}_{FK}^0 * \mathbf{N}_K^0 = \mathbf{Q}_F^0 \\ \mathbf{M}_G^0 + w_{GH}\cdot\mathbf{M}_H^0 + w_{GI}\cdot\mathbf{M}_I^0 + w_{GJ}\cdot\mathbf{M}_J^0 + w_{GK}\cdot\mathbf{M}_K^0 + \tilde{\mathbf{L}}_{GH}^0 * \mathbf{N}_H^0 + \tilde{\mathbf{L}}_{GI}^0 * \mathbf{N}_I^0 \\ \quad + \tilde{\mathbf{L}}_{GJ}^0 * \mathbf{N}_J^0 + \tilde{\mathbf{L}}_{GK}^0 * \mathbf{N}_K^0 = \mathbf{Q}_G^0 \end{cases} \tag{4.137a}$$

Note: The number of moment equations is m = 7 (the number of mobile links).

Force equations:

$$\begin{cases} \mathbf{N}_A^0 = \mathbf{P}_A^0 - w_{AH} \cdot \mathbf{N}_H^0 - w_{AI} \cdot \mathbf{N}_I^0 - w_{AJ} \cdot \mathbf{N}_J^0 - w_{AK} \cdot \mathbf{N}_K^0 \\ \mathbf{N}_B^0 = \mathbf{P}_B^0 - w_{BH} \cdot \mathbf{N}_H^0 - w_{BI} \cdot \mathbf{N}_I^0 - w_{BJ} \cdot \mathbf{N}_J^0 - w_{BK} \cdot \mathbf{N}_K^0 \\ \mathbf{N}_C^0 = \mathbf{P}_C^0 - w_{CH} \cdot \mathbf{N}_H^0 - w_{CI} \cdot \mathbf{N}_I^0 - w_{CJ} \cdot \mathbf{N}_J^0 - w_{CK} \cdot \mathbf{N}_K^0 \\ \mathbf{N}_D^0 = \mathbf{P}_D^0 - w_{DH} \cdot \mathbf{N}_H^0 - w_{DI} \cdot \mathbf{N}_I^0 - w_{DJ} \cdot \mathbf{N}_J^0 - w_{DK} \cdot \mathbf{N}_K^0 \\ \mathbf{N}_E^0 = \mathbf{P}_E^0 - w_{EH} \cdot \mathbf{N}_H^0 - w_{EI} \cdot \mathbf{N}_I^0 - w_{EJ} \cdot \mathbf{N}_J^0 - w_{EK} \cdot \mathbf{N}_K^0 \\ \mathbf{N}_F^0 = \mathbf{P}_F^0 - w_{FH} \cdot \mathbf{N}_H^0 - w_{FI} \cdot \mathbf{N}_I^0 - w_{FJ} \cdot \mathbf{N}_J^0 - w_{FK} \cdot \mathbf{N}_K^0 \\ \mathbf{N}_G^0 = \mathbf{P}_G^0 - w_{GH} \cdot \mathbf{N}_H^0 - w_{GI} \cdot \mathbf{N}_I^0 - w_{GJ} \cdot \mathbf{N}_J^0 - w_{GK} \cdot \mathbf{N}_K^0 \end{cases} \quad (4.137b)$$

Note: The number of force equations is a = j − c = 7 (the number of joints in the tree).

The solution for reaction forces and moments in joints

The reactions for spatial mechanisms are defined in the local frames as $\mathbf{N}_a^m$, $\mathbf{M}_a^m$ considering the relative velocity constraints. Then, by a left-hand-side multiplication with the rotation matrix, the reactions are expressed in the fixed frame 0 as:

$$\mathbf{N}_a^0 = \mathbf{D}_m * \mathbf{N}_a^m, \text{ and } \mathbf{M}_a^0 = \mathbf{D}_m * \mathbf{M}_a^m \quad (4.138)$$

The system of linear equations in scalar form is obtained from the left-hand-side multiplication in moment and force equations by the transpose of the rotation matrix and considering the property:

$$\mathbf{D}_m^T * \mathbf{D}_m = \mathbf{U} \quad (4.139)$$

Then, Equation (4.138) becomes

$$\mathbf{N}_a^m = \mathbf{D}_m^T * \mathbf{N}_a^0, \text{ and } \mathbf{M}_a^m = \mathbf{D}_m^T * \mathbf{M}_a^0 \quad (4.140)$$

where a = A, B, C, D, …, are the arcs (edges tree), and m = 1, 2, 3, 4,…, are the mobile links.

4.8.5 Examples of Mechanisms with Single and Multiple Cycles: Singularity Coefficient

Example 1: The spatial TRRTC mechanism, Figure 4.16, with c = 1 cycle, j = 5, m = 4, and M = 3 DOF. The common steps for all mechanisms in finding the dynamic equations are as follows:

The matrices $\mathbf{W}$ *and* $\mathbf{W}(\tilde{\mathbf{L}})$ *were determined using Equation (4.97c) and reproduced here*:

$$\mathbf{W}_{a\times c} = \begin{matrix} \\ A \\ B \\ C \\ D \end{matrix} \begin{matrix} E \\ \begin{bmatrix} +1 \\ +1 \\ +1 \\ +1 \end{bmatrix} \end{matrix} = \begin{bmatrix} w_{AE} \\ w_{BE} \\ w_{CE} \\ w_{DE} \end{bmatrix} \Rightarrow \mathbf{W}_{a\times c}(\tilde{\mathbf{L}}) = \begin{bmatrix} +1\cdot\tilde{\mathbf{L}}^0_{AE} \\ +1\cdot\tilde{\mathbf{L}}^0_{BE} \\ +1\cdot\tilde{\mathbf{L}}^0_{CE} \\ +1\cdot\tilde{\mathbf{L}}^0_{DE} \end{bmatrix} \quad (4.141a)$$

- **The single-cycle IT dynamic moment equations**

With all weighting w entries, as +1, the moment IT Equation (4.134a) is

$$\begin{cases} \mathbf{M}^0_A + \mathbf{M}^0_E + \tilde{\mathbf{L}}^0_{AE} * \mathbf{N}^0_E = \mathbf{Q}^0_A \\ \mathbf{M}^0_B + \mathbf{M}^0_E + \tilde{\mathbf{L}}^0_{BE} * \mathbf{N}^0_E = \mathbf{Q}^0_B \\ \mathbf{M}^0_C + \mathbf{M}^0_E + \tilde{\mathbf{L}}^0_{CE} * \mathbf{N}^0_E = \mathbf{Q}^0_C \\ \mathbf{M}^0_D + \mathbf{M}^0_E + \tilde{\mathbf{L}}^0_{DE} * \mathbf{N}^0_E = \mathbf{Q}^0_D \end{cases} \quad (4.141b)$$

Equation (4.141a) are left-hand-side-multiplied by the transpose matrix, for each joint (arc):

$$\begin{cases} \mathbf{M}^1_A + \mathbf{D}^T_1 * \left(\mathbf{M}^0_E + \tilde{\mathbf{L}}^0_{AE} * \mathbf{N}^0_E\right) = \mathbf{D}^T_1 * \mathbf{Q}^0_A \\ \mathbf{M}^1_B + \mathbf{D}^T_2 * \left(\mathbf{M}^0_E + \tilde{\mathbf{L}}^0_{BE} * \mathbf{N}^0_E\right) = \mathbf{D}^T_2 * \mathbf{Q}^0_B \\ \mathbf{M}^1_C + \mathbf{D}^T_3 * \left(\mathbf{M}^0_E + \tilde{\mathbf{L}}^0_{CE} * \mathbf{N}^0_E\right) = \mathbf{D}^T_3 * \mathbf{Q}^0_C \\ \mathbf{M}^1_D + \mathbf{D}^T_4 * \left(\mathbf{M}^0_E + \tilde{\mathbf{L}}^0_{DE} * \mathbf{N}^0_E\right) = \mathbf{D}^T_4 * \mathbf{Q}^0_D \end{cases} \quad (4.141c)$$

If $\mathbf{M}^0_E = 0$ (no reaction moment at a contact joint E), the system of Equation (4.141c) becomes

$$\begin{cases} \mathbf{M}^1_A + \mathbf{D}^T_1 * \left(\tilde{\mathbf{L}}^0_{AE} * \mathbf{N}^0_E\right) = \mathbf{D}^T_1 * \mathbf{Q}^0_A \\ \mathbf{M}^1_B + \mathbf{D}^T_2 * \left(\tilde{\mathbf{L}}^0_{BE} * \mathbf{N}^0_E\right) = \mathbf{D}^T_2 * \mathbf{Q}^0_B \\ \mathbf{M}^1_C + \mathbf{D}^T_3 * \left(\tilde{\mathbf{L}}^0_{CE} * \mathbf{N}^0_E\right) = \mathbf{D}^T_3 * \mathbf{Q}^0_C \\ \mathbf{M}^1_D + \mathbf{D}^T_4 * \left(\tilde{\mathbf{L}}^0_{DE} * \mathbf{N}^0_E\right) = \mathbf{D}^T_4 * \mathbf{Q}^0_D \end{cases} \quad (4.141d)$$

Defining the joint reactions from constraint displacements

The TRRTC mechanism has M = 3 DOF. The joints A, B, and D are considered actuated by F_A^{act}, T_B^{act}, and F_D^{act}. The joint reactions are defined from the relative linear and angular velocity allowed by the joints: joint A-prismatic along x_1 direction; joint B-revolute along y_2 direction; joint C-revolute along z_3 direction; joint D-prismatic along x_4 direction; and joint E-the normal to the contact along y_0 direction. The procedure was presented in the previous section for open cycle TRRT mechanism, which is adapted next to the closed cycle with three actuated joints.

- *Joint A* (1 DOF, prismatic, actuated):
  - Reaction force at the joint A is defined from the linear velocity constraints in local frame 1:

  The relative velocity along x_1 direction $\dot{d}_A^1$ is allowed, which is also the direction of F_A^{act}—the actuating force from link 0 acting on link 1. The two zero components imply constraints along y_1 and z_1 directions; therefore, the reaction components Y_A^1 and Z_A^1 are introduced.

$$\dot{\mathbf{d}}_A^1 = \begin{Bmatrix} \dot{d}_A^1 \\ 0 \\ 0 \end{Bmatrix} \Rightarrow \mathbf{N}_A^1 = \begin{Bmatrix} F_A^{act} \\ Y_A^1 \\ Z_A^1 \end{Bmatrix} \tag{4.141e}$$

  - Reaction moment at the joint A is defined from the angular velocity constraints in local frame 1:

  The angular velocity $\dot{\boldsymbol{\theta}}_A^1$ is not allowed in the prismatic joint A; then, the reaction moment components for all three directions $M_A^{x_1}$, $M_A^{y_1}$, and $M_A^{z_1}$ are assigned.

$$\dot{\boldsymbol{\theta}}_A^1 = \begin{Bmatrix} 0 \\ 0 \\ 0 \end{Bmatrix} \Rightarrow \mathbf{M}_A^1 = \begin{Bmatrix} M_A^{x_1} \\ M_A^{y_1} \\ M_A^{z_1} \end{Bmatrix} \tag{4.141f}$$

- *Joint B* (1 DOF, revolute, actuated):
  - Reaction force at the joint B is defined from the linear velocity constraints in local frame 2:

  The linear velocity $\dot{\mathbf{d}}_B^2$ is not allowed in the revolute joint B; then, the reaction force components for all three directions X_B^2, Y_B^2, and Z_B^2 are assigned.

$$\dot{\mathbf{d}}_B^2 = \begin{Bmatrix} 0 \\ 0 \\ 0 \end{Bmatrix} \Rightarrow \mathbf{N}_B^2 = \begin{Bmatrix} X_B^2 \\ Y_B^2 \\ Z_B^2 \end{Bmatrix} \tag{4.141g}$$

- Reaction moment at the joint B is defined from the angular velocity constraints in local frame 2:

An angular velocity $\dot{\theta}_B^2$ is allowed along y_2 direction, which is also the direction of T_B^{act}—the actuating torque from link 1 acting on link 2; then, it is constrained to rotate on the other directions x_2 and z_2; the reaction components M_B^{x2} and M_B^{z2} are assigned for these directions.

$$\dot{\theta}_B^2 = \begin{Bmatrix} 0 \\ \dot{\theta}_B^2 \\ 0 \end{Bmatrix} \Rightarrow \mathbf{M}_B^2 = \begin{Bmatrix} M_B^{x2} \\ T_B^{act} \\ M_B^{z2} \end{Bmatrix} \tag{4.141h}$$

- *Joint C* (1 DOF, revolute, non-actuated):
  - Reaction force at the joint C is defined from the linear velocity constraints in local frame 3:

The linear velocity $\dot{\mathbf{d}}_C^3$ is not allowed in the revolute joint C, along x_3, y_3, or z_3 direction; then, the reaction force components for all three directions X_C^3, Y_C^3, and Z_C^3 are assigned.

$$\dot{\mathbf{d}}_C^3 = \begin{Bmatrix} 0 \\ 0 \\ 0 \end{Bmatrix} \Rightarrow \mathbf{N}_C^3 = \begin{Bmatrix} X_C^3 \\ Y_C^3 \\ z_C^3 \end{Bmatrix} \tag{4.141i}$$

- Reaction moment at the joint C is defined from the angular velocity constraints in local frame 3:

An angular velocity $\dot{\theta}_C^3$ is allowed along z_3 direction and is constrained to rotate on the other directions x_3 and z_3; the reaction components M_C^{x3} and M_C^{y3} are assigned for these directions.

$$\dot{\theta}_C^3 = \begin{Bmatrix} 0 \\ 0 \\ \dot{\theta}_C^3 \end{Bmatrix} \Rightarrow \mathbf{M}_C^3 = \begin{Bmatrix} M_C^{x3} \\ M_C^{y3} \\ 0 \end{Bmatrix} \tag{4.141j}$$

- *Joint D* (1 DOF, prismatic, actuated):
  - Reaction force at the joint D is defined from the linear velocity constraints in local frame 4:

  It allows the relative velocity $\dot{d}_D^4$ along x_4 direction, which is also the direction of F_D^{act}—the actuating force from link 3 acting on link 4. The two zero components imply constraints along y_4 and z_4 directions; therefore, the reaction components Y_D^4 and Z_A^4 are introduced.

$$\dot{\mathbf{d}}_D^4 = \begin{Bmatrix} \dot{d}_A^1 \\ 0 \\ 0 \end{Bmatrix} \Rightarrow \mathbf{N}_D^4 = \begin{Bmatrix} F_D^{act} \\ Y_D^4 \\ z_D^4 \end{Bmatrix} \tag{4.141k}$$

  - Reaction moment at the joint D is defined from the angular velocity constraints in local frame 4:

The angular velocity $\dot{\boldsymbol{\theta}}_D^4$ is not allowed in the prismatic joint D; then, reaction moment components for all three directions M_D^{x4}, M_D^{y4}, and M_D^{z4} are assigned.

$$\dot{\boldsymbol{\theta}}_D^4 = \begin{Bmatrix} 0 \\ 0 \\ 0 \end{Bmatrix} \Rightarrow \mathbf{M}_D^4 = \begin{Bmatrix} M_D^{x4} \\ M_D^{y4} \\ M_D^{z4} \end{Bmatrix} \tag{4.141l}$$

- *Joint E* (5 DOF, contact, non-actuated)

  Contact joint's E reactions are expressed in the fixed frame 0.

  There is no linear velocity on y_0 direction; therefore, a reaction force Y_E^0 is assigned.

  Angular velocities are allowed about all three joint axes; thus, there are no reaction moments.

$$\begin{Bmatrix} \dot{d}_D^{x0} \\ 0 \\ \dot{d}_D^{z0} \end{Bmatrix} \Rightarrow \mathbf{N}_E^0 = \begin{Bmatrix} 0 \\ Y_E^0 \\ 0 \end{Bmatrix}; \mathbf{M}_E^0 = \begin{Bmatrix} 0 \\ 0 \\ 0 \end{Bmatrix} \tag{4.141m}$$

  - The cross products in Equation (4.141c) are determined by left-hand-side multiplication of the reactions by the skew-symmetric matrices:

$$\tilde{\mathbf{L}}_{AE}^{0} * \mathbf{N}_{E}^{0} = \begin{bmatrix} 0 & -z_{AE} & y_{AE} \\ z_{AE} & 0 & -x_{AE} \\ -y_{AE} & x_{AE} & 0 \end{bmatrix} * \begin{Bmatrix} 0 \\ Y_{E}^{0} \\ 0 \end{Bmatrix} = \begin{Bmatrix} -z_{AE} \cdot Y_{E}^{0} \\ 0 \\ x_{AE} \cdot Y_{E}^{0} \end{Bmatrix}$$

$$\tilde{\mathbf{L}}_{BE}^{0} * \mathbf{N}_{E}^{0} = \begin{Bmatrix} -z_{BE} \cdot Y_{E}^{0} \\ 0 \\ x_{BE} \cdot Y_{E}^{0} \end{Bmatrix}; \ \tilde{\mathbf{L}}_{CE}^{0} * \mathbf{N}_{E0,4}^{0} = \begin{Bmatrix} -z_{CE} \cdot Y_{E}^{0} \\ 0 \\ x_{CE} \cdot Y_{E}^{0} \end{Bmatrix}$$

$$\tilde{\mathbf{L}}_{DE}^{0} * \mathbf{N}_{E}^{0} = \begin{Bmatrix} -z_{DE} \cdot Y_{E}^{0} \\ 0 \\ x_{DE} \cdot Y_{E}^{0} \end{Bmatrix} \tag{4.141n}$$

The system of Equation (4.141d) has 12 scalar equations:
Joint A:

$$\mathbf{M}_{A}^{1} + \mathbf{D}_{1}^{T} * \left(\tilde{\mathbf{L}}_{AE}^{0} * \mathbf{N}_{E}^{0} \right) = \mathbf{D}_{1}^{T} * \mathbf{Q}_{A}^{0}$$

$$\begin{Bmatrix} M_{A}^{x_1} - z_{AE} \cdot Y_{E}^{0} \\ M_{A}^{y_1} \\ M_{A}^{z_1} - x_{AE} \cdot Y_{E}^{0} \end{Bmatrix} = \begin{Bmatrix} Q_{A}^{x_0} \\ Q_{A}^{y_0} \\ Q_{A}^{z_0} \end{Bmatrix}; \text{With } \mathbf{D}_1 = \begin{bmatrix} 1 & 0 & 0 \\ 0 & 1 & 0 \\ 0 & 0 & 1 \end{bmatrix} \tag{4.141o}$$

Joint B:

$$\mathbf{M}_{B}^{2} + \mathbf{D}_{2}^{T} * \left(\tilde{\mathbf{L}}_{BE}^{0} * \mathbf{N}_{E}^{0} \right) = \mathbf{D}_{2}^{T} * \mathbf{Q}_{B}^{0}$$

$$\begin{Bmatrix} M_{B}^{x_2} - \left(z_{BE} \cdot C\theta_B + x_{BE} \cdot S\theta_B \right) \cdot Y_{E}^{0} \\ T_{B}^{act} \\ M_{B}^{z_2} + \left(-z_{BE} \cdot S\theta_B + x_{BE} \cdot C\theta_B \right) \cdot Y_{E}^{0} \end{Bmatrix} = \begin{Bmatrix} C\theta_B \cdot Q_{B}^{x_0} + S\theta_B \cdot Q_{B}^{z_0} \\ Q_{B}^{y_0} \\ S\theta_B \cdot Q_{B}^{x_0} + C\theta_B \cdot Q_{B}^{z_0} \end{Bmatrix};$$

$$\text{With } \mathbf{D}_2 = \begin{bmatrix} C\theta_B & 0 & S\theta_B \\ 0 & 1 & 0 \\ -S\theta_B & 0 & C\theta_B \end{bmatrix} \tag{4.141p}$$

Joint C:

$$\mathbf{M}_C^3 + \mathbf{D}_3^T * \left(\tilde{\mathbf{L}}_{CE}^0 * \mathbf{N}_E^0\right) = \mathbf{D}_3^T * \mathbf{Q}_C^0$$

$$\left\{\begin{array}{c} M_C^{x3} - (C\theta_B \cdot C\theta_C \cdot z_{CE} + S\theta_B \cdot C\theta_C \cdot x_{CE}) \cdot Y_E^0 \\ M_C^{y3} + (C\theta_B \cdot S\theta_C \cdot z_{CE} + S\theta_B \cdot S\theta_C \cdot x_{CE}) \cdot Y_E^0 \\ (-S\theta_B \cdot z_{CE} + C\theta_B \cdot x_{CE}) \cdot Y_E^0 \end{array}\right\}$$

$$= \left\{\begin{array}{c} C\theta_B \cdot C\theta_C \cdot Q_C^{x0} + S\theta_C \cdot Q_C^{y0} - S\theta_B \cdot C\theta_C \cdot Q_C^{z0} \\ -C\theta_B \cdot S\theta_C \cdot Q_C^{x0} + C\theta_C \cdot Q_C^{y0} + S\theta_B \cdot S\theta_C \cdot Q_C^{z0} \\ S\theta_C \cdot Q_C^{x0} + C\theta_B \cdot Q_C^{z0} \end{array}\right\}; \qquad (4.141q)$$

$$\text{With } \mathbf{D}_3^0 = \left[\begin{array}{c|c|c} C\theta_B \cdot C\theta_C & -C\theta_B \cdot S\theta_C & S\theta_B \\ S\theta_C & C\theta_C & 0 \\ -S\theta_B \cdot C\theta_C & S\theta_B \cdot S\theta_C & C\theta_B \end{array}\right]$$

Joint D:

$$\mathbf{M}_D^4 + \mathbf{D}_4^T * \left(\tilde{\mathbf{L}}_{DE}^0 * \mathbf{N}_E^0\right) = \mathbf{D}_4^T * \mathbf{Q}_D^0$$

$$\left\{\begin{array}{c} M_D^{x4} - (C\theta_B \cdot C\theta_C \cdot z_{CE} + S\theta_B \cdot C\theta_C \cdot x_{CE}) \cdot Y_E^0 \\ M_D^{y4} + (C\theta_B \cdot S\theta_C \cdot z_{CE} + S\theta_B \cdot S\theta_C \cdot x_{CE}) \cdot Y_E^0 \\ M_D^{z4} + (-S\theta_B \cdot z_{CE} + C\theta_B \cdot x_{CE}) \cdot Y_E^0 \end{array}\right\}$$

$$= \left\{\begin{array}{c} C\theta_B \cdot C\theta_C \cdot Q_D^{x0} + S\theta_C \cdot Q_D^{y0} - S\theta_B \cdot C\theta_C \cdot Q_D^{z0} \\ -C\theta_B \cdot S\theta_C \cdot Q_D^{x0} + C\theta_C \cdot Q_D^{y0} + S\theta_B \cdot S\theta_C \cdot Q_D^{z0} \\ S\theta_C \cdot Q_D^{x0} + C\theta_B \cdot Q_D^{z0} \end{array}\right\}; \qquad (4.141r)$$

$$\text{With } \mathbf{D}_4^0 = \left[\begin{array}{c|c|c} C\theta_B \cdot C\theta_C & -C\theta_B \cdot S\theta_C & S\theta_B \\ S\theta_C & C\theta_C & 0 \\ -S\theta_B \cdot C\theta_C & S\theta_B \cdot S\theta_C & C\theta_B \end{array}\right]$$

- **The solution for single-cycle IT dynamic equations**
  - The linear system of 12 equations is solved symbolically for 12 unknown reaction components:

$$M_A^{x1},\ M_A^{y1},\ M_A^{z1},\ M_B^{x2},\ M_B^{z2},\ M_C^{x3},\ M_C^{y3},\ M_D^{x4},\ M_D^{y4},\ M_D^{z4},\ Y_E^0,\ T_B^{act}$$

Firstly, the third Equation (4.141q) has one unknown and is solved for the cut-set reaction Y_E^0.

$$Y_E^0 = \frac{S\theta_C}{\Delta} \cdot Q_C^{x0} + \frac{C\theta_B}{\Delta} \cdot Q_C^{z0} \qquad (4.141s)$$

- **Singularity coefficient**: The denominator in Y_E^0 is named *singularity coefficient*, as shown in Equation (4.141s):

$$\Delta = -S\theta_B \cdot z_{CE} + C\theta_B \cdot x_{CE}$$

If $\Delta = 0$, all of the reactions' *magnitudes become very large*. The mechanism designer should be aware and avoid this mechanism's position.

- **Solution for the single-cycle IT dynamic moment equations**

$$\begin{aligned}
&M_A^{x1} = Q_A^{x0} + z_{AE} \cdot Y_E^0;\ M_A^{y1} = Q_A^{y0};\ M_A^{z1} = Q_A^{z0} + x_{AE} \cdot Y_E^0 \\
&M_B^{x2} = C\theta_B \cdot Q_B^{x0} - S\theta_B \cdot Q_B^{z0} + (z_{BE} \cdot C\theta_B + x_{BE} \cdot S\theta_B) \cdot Y_E^0;\ T_B^{act} = Q_B^{y0} \\
&M_B^{z2} = S\theta_B \cdot Q_B^{x0} + C\theta_B \cdot Q_B^{z0} - (-z_{BE} \cdot S\theta_B + x_{BE} \cdot C\theta_B) \cdot Y_E^0 \\
&M_C^{x3} = C\theta_B \cdot C\theta_C \cdot Q_C^{x0} + S\theta_C \cdot Q_C^{y0} - S\theta_B \cdot C\theta_C \cdot Q_C^{z0} \\
&\qquad + (C\theta_B \cdot C\theta_C \cdot z_{CE} + S\theta_B \cdot C\theta_C \cdot x_{CE}) \cdot Y_E^0 \\
&M_C^{y3} = -C\theta_B \cdot S\theta_C \cdot Q_C^{x0} + C\theta_C \cdot Q_C^{y0} + S\theta_B \cdot S\theta_C \cdot Q_C^{z0} \\
&\qquad - (C\theta_B \cdot S\theta_C \cdot z_{CE} + S\theta_B \cdot S\theta_C \cdot x_{CE}) \cdot Y_E^0 \\
&M_D^{x4} = C\theta_B \cdot C\theta_C \cdot Q_D^{x0} + S\theta_C \cdot Q_D^{y0} - S\theta_B \cdot C\theta_C \cdot Q_D^{z0} \\
&\qquad + (C\theta_B \cdot C\theta_C \cdot z_{CE} + S\theta_B \cdot C\theta_C \cdot x_{CE}) \cdot Y_E^0 \\
&M_D^{y4} = -C\theta_B \cdot S\theta_C \cdot Q_D^{x0} + C\theta_C \cdot Q_D^{y0} + S\theta_B \cdot S\theta_C \cdot Q_D^{z0} \\
&\qquad - (C\theta_B \cdot S\theta_C \cdot z_{CE} + S\theta_B \cdot S\theta_C \cdot x_{CE}) \cdot Y_E^0 \\
&M_D^{z4} = S\theta_C \cdot Q_D^{x0} + C\theta_B \cdot Q_D^{z0} - (-S\theta_B \cdot z_{CE} + C\theta_B \cdot x_{CE}) \cdot Y_E^0
\end{aligned} \qquad (4.141t)$$

- **Solution for the single-cycle IT dynamic force equations**

With all weighting w entries, as +1, the moment IT Equation (4.134b) are

$$\begin{cases} \mathbf{N}_A^0 = \mathbf{P}_A^0 - \mathbf{N}_E^0 \\ \mathbf{N}_B^0 = \mathbf{P}_B^0 - \mathbf{N}_E^0 \\ \mathbf{N}_C^0 = \mathbf{P}_C^0 - \mathbf{N}_E^0 \\ \mathbf{N}_D^0 = \mathbf{P}_D^0 - \mathbf{N}_E^0 \end{cases} \tag{4.141u}$$

Joint A:

$$\mathbf{N}_A^1 = \mathbf{D}_1^T * \left(\mathbf{P}_A^0 - \mathbf{N}_E^0\right) \Rightarrow \begin{Bmatrix} F_A^{act} \\ Y_A^1 \\ Z_A^1 \end{Bmatrix} = \begin{Bmatrix} P_A^{x_0} \\ P_A^{y_0} - Y_E^0 \\ P_A^{z_0} \end{Bmatrix} \tag{4.141v}$$

Joint B:

$$\mathbf{N}_B^2 = \mathbf{D}_2^T * \left(\mathbf{P}_B^0 - \mathbf{N}_E^0\right) \Rightarrow \begin{Bmatrix} X_B^2 \\ Y_B^2 \\ Z_B^2 \end{Bmatrix} = \begin{Bmatrix} C\theta_B \cdot P_B^{x_0} - S\theta_B \cdot P_B^{z_0} \\ P_B^{y_0} - Y_E^0 \\ S\theta_B \cdot P_B^{x_0} - C\theta_B \cdot P_B^{z_0} \end{Bmatrix} \tag{4.141w}$$

Joint C:

$$\mathbf{N}_C^3 = \mathbf{D}_3^T * \left(\mathbf{P}_C^0 - \mathbf{N}_E^0\right) \Rightarrow \begin{Bmatrix} X_C^3 \\ Y_C^3 \\ Z_C^3 \end{Bmatrix}$$

$$= \begin{Bmatrix} C\theta_B \cdot C\theta_C \cdot P_C^{x_0} + S\theta_C \cdot \left(P_C^{y_0} - Y_E^0\right) - S\theta_B \cdot C\theta_C \cdot P_C^{z_0} \\ -C\theta_B \cdot S\theta_C \cdot P_C^{x_0} + C\theta_C \cdot \left(P_C^{y_0} - Y_E^0\right) - S\theta_B \cdot S\theta_C \cdot P_C^{z_0} \\ S\theta_C \cdot P_C^{x_0} - C\theta_B \cdot P_C^{z_0} \end{Bmatrix} \tag{4.141x}$$

Joint D:

$$\mathbf{N}_D^4 = \mathbf{D}_4^T * \left(\mathbf{P}_D^0 - \mathbf{N}_E^0\right) \Rightarrow \begin{Bmatrix} F_D^{act} \\ Y_D^4 \\ Z_D^4 \end{Bmatrix}$$

$$= \begin{Bmatrix} C\theta_B \cdot C\theta_C \cdot P_D^{x_0} + S\theta_C \cdot \left(P_D^{y_0} - Y_E^0\right) - S\theta_B \cdot C\theta_C \cdot P_D^{z_0} \\ -C\theta_B \cdot S\theta_C \cdot P_D^{x_0} + C\theta_C \cdot \left(P_D^{y_0} - Y_E^0\right) - S\theta_B \cdot S\theta_C \cdot P_D^{z_0} \\ S\theta_C \cdot P_D^{x_0} - C\theta_B \cdot P_D^{z_0} \end{Bmatrix} \tag{4.141y}$$

- **The differential equations of motion**

$$F_A^{act} = P_A^{x_0}$$

$$T_B^{act} = Q_B^{y_0}$$

$$F_D^{act} = C\theta_B \cdot C\theta_C \cdot P_D^{x_0} + S\theta_C \cdot \left(P_D^{y_0} - \frac{S\theta_C}{\Delta} \cdot Q_C^{x_0} - \frac{C\theta_B}{\Delta} \cdot Q_C^{z_0} \right) - S\theta_B \cdot C\theta_C \cdot P_D^{z_0} \tag{4.141z}$$

Example 2: The four-bar planar mechanism (Figure 4.5a), with its digraph shown in Figure 4.5b and its tree shown in Figure 4.5c, has m = 3 mobile links (labeled 1, 2, and 3) and j = 4 revolute joints (labeled A, B, C, and D). The c = 1 cut-joint is the joint D. The mechanism has c = j − m = 1 cycle and M = 3m − 2j = 1 DOF. Link 1 (crank) is actuated by the torque $T_A^{act,0}$. The link masses and moments of inertia relative to links COM are as follows: $m_1 = 0.6$ [lbm]; $I_{G_1} = 1.05$ [lbm · in^2]; $m_2 = 3.94$ [lbm]; $I_{G_2} = 197.74$ [lbm · in^2]; $m_3 = 2.39$ [lbm]; $I_{G_3} = 40.89$ [lbm · in^2].

The dynamic reactions at the joints and the actuating torque $T_A^{act,0}$ are determined next.

Solution:

- The digraph matrices **G** and **Z** are

$$[\underline{\mathbf{G}}] = \begin{array}{c} \\ 0 \\ 1 \\ 2 \\ 3 \end{array} \begin{array}{c} \begin{array}{cccc} A & B & C & D \end{array} \\ \left[\begin{array}{ccc:c} -1 & 0 & -1 & 0 \\ \hdashline 1 & -1 & 0 & 0 \\ 0 & 1 & 0 & -1 \\ 0 & 0 & 1 & 1 \end{array}\right] \end{array} ; \mathbf{G} = \begin{array}{c} \\ 1 \\ 2 \\ 3 \end{array} \begin{array}{c} \begin{array}{ccc} A & B & C \end{array} \\ \begin{bmatrix} 1 & -1 & 0 \\ 0 & 1 & 0 \\ 0 & 0 & 1 \end{bmatrix} \end{array};$$

$$\mathbf{G}^{cut} = \begin{array}{c} D \\ \begin{bmatrix} 0 \\ -1 \\ 1 \end{bmatrix} \end{array} ; \mathbf{Z} = \begin{array}{c} \\ A \\ B \\ C \end{array} \begin{array}{c} \begin{array}{ccc} 1 & 2 & 3 \end{array} \\ \begin{bmatrix} +1 & +1 & 0 \\ 0 & +1 & 0 \\ 0 & 0 & +1 \end{bmatrix} \end{array} \tag{4.142a}$$

The matrices $\mathbf{Z}(\tilde{\mathbf{L}})$ and $\mathbf{W}(\tilde{\mathbf{L}})$ are

$$\mathbf{Z} \Rightarrow \mathbf{Z}(\tilde{\mathbf{L}}) = \begin{matrix} \\ A \\ B \\ C \end{matrix} \begin{bmatrix} 1 & 2 & 3 \\ +1\cdot\tilde{\mathbf{L}}^0_{AG_1} & +1\cdot\tilde{\mathbf{L}}^0_{AG_2} & 0 \\ 0 & +1\cdot\tilde{\mathbf{L}}^0_{BG_2} & 0 \\ 0 & 0 & +1\cdot\tilde{\mathbf{L}}^0_{CG_3} \end{bmatrix};$$

$$\mathbf{W} = \mathbf{Z} * \mathbf{G}^{cut} = \begin{matrix} \\ A \\ B \\ C \end{matrix} \begin{bmatrix} D \\ -1 \\ -1 \\ 1 \end{bmatrix} \Rightarrow \mathbf{W}(\tilde{\mathbf{L}}) = \begin{bmatrix} -1\cdot\tilde{\mathbf{L}}^0_{AD} \\ -\cdot\tilde{\mathbf{L}}^0_{BD} \\ 1\cdot\tilde{\mathbf{L}}^0_{CD} \end{bmatrix} \tag{4.142b}$$

- The resultant of inertial and external forces reduced to COM of each link

$$\mathbf{P}^0_{G_m} = \begin{Bmatrix} \mathbf{P}^0_{G_1} \\ \mathbf{P}^0_{G_2} \\ \mathbf{P}^0_{G_3} \end{Bmatrix} = \begin{Bmatrix} \mathbf{f}^{in,0}_{G_1} - \mathbf{f}^{ext,0}_{G_1} \\ \mathbf{f}^{in,0}_{G_2} - \mathbf{f}^{ext,0}_{G_2} \\ \mathbf{f}^{in,0}_{G_3} - \mathbf{f}^{ext,0}_{G_3} \end{Bmatrix}$$

Therefore, the scalar components in [lbf] are

$$\begin{aligned} \mathbf{P}^0_{G_1} &= \left\{ m_1\cdot a^x_{G_1}; \quad m_1\left(a^y_{G_1}+g\right); \quad 0 \right\}^T; \\ \mathbf{P}^0_{G_2} &= \left\{ m_2\cdot a^x_{G_2}; \quad m_2\left(a^y_{G_2}+g\right); \quad 0 \right\}^T; \\ \mathbf{P}^0_{G_3} &= \left\{ m_3\cdot a^x_{G_3}; \quad m_3\left(a^y_{G_3}+g\right); \quad 0 \right\}^T \end{aligned} \tag{4.142c}$$

- The IT forces are calculated from the matrix equation with vector entries, $\mathbf{P}^0_a = \mathbf{Z} * \mathbf{P}^0_{Gm}$:

$$\begin{Bmatrix} \mathbf{P}^0_A \\ \mathbf{P}^0_B \\ \mathbf{P}^0_C \end{Bmatrix} = \begin{bmatrix} +1 & +1 & 0 \\ 0 & +1 & 0 \\ 0 & 0 & +1 \end{bmatrix} * \begin{Bmatrix} \mathbf{P}^0_{G_1} \\ \mathbf{P}^0_{G_2} \\ \mathbf{P}^0_{G_3} \end{Bmatrix}$$

Therefore, the scalar components in [lbf] are

$$\mathbf{P}_A^0 = \mathbf{P}_{G1}^0 + \mathbf{P}_{G2}^0 = \left\{ m_1 \cdot a_{G1}^x + m_2 \cdot a_{G2}^x; \quad m_1\left(a_{G1}^y + g\right) + m_2\left(a_{G2}^y + g\right); \quad 0 \right\}^T$$

$$\mathbf{P}_B^0 = \mathbf{P}_{G2}^0 = \left\{ m_2 \cdot a_{G2}^x; \quad m_2\left(a_{G2}^y + g\right); \quad 0 \right\}^T;$$

$$\mathbf{P}_C^0 = \mathbf{P}_{G3}^0 = \left\{ m_3 \cdot a_{G3}^x; \quad m_3\left(a_{G3}^y + g\right); \quad 0 \right\}^T \tag{4.142d}$$

The resultant of inertial and external moments reduced to com of each link

$$\mathbf{Q}_{Gm}^0 = \begin{Bmatrix} \mathbf{Q}_{G1}^0 \\ \mathbf{Q}_{G2}^0 \\ \mathbf{Q}_{G3}^0 \end{Bmatrix} = \begin{Bmatrix} \dot{\mathbf{K}}_{G1}^{acc,0} - \mu_{G1}^{ext,0} \\ \dot{\mathbf{K}}_{G2}^{acc,0} - \mu_{G2}^{ext,0} \\ \dot{\mathbf{K}}_{G2}^{acc,0} - \mu_{G2}^{ext,0} \end{Bmatrix} = \begin{Bmatrix} \mathbf{I}_{G1} * \varepsilon_1 \\ \mathbf{I}_{G2} * \varepsilon_2 \\ \mathbf{I}_{G3} * \varepsilon_3 \end{Bmatrix} \tag{4.142e}$$

If the external forces are the links' weights applied in COM, then $\mu_{Gm}^{ext,0} = 0$ for m = 1, 2, and 3, and for a planar mechanism:

$$\dot{\mathbf{K}}_{Gm}^{acc,0} = \begin{Bmatrix} \dot{\mathbf{K}}_{G1}^{acc,0} \\ \dot{\mathbf{K}}_{G2}^{acc,0} \\ \dot{\mathbf{K}}_{G3}^{acc,0} \end{Bmatrix} = \begin{Bmatrix} \mathbf{I}_{G1} * \varepsilon_1 + \tilde{\omega}_1 * (\mathbf{I}_{G1} * \omega_1) \\ \mathbf{I}_{G1} * \varepsilon_1 + \tilde{\omega}_1 * (\mathbf{I}_{G1} * \omega_1) \\ \mathbf{I}_{G1} * \varepsilon_1 + \tilde{\omega}_1 * (\mathbf{I}_{G1} * \omega_1) \end{Bmatrix} = \begin{Bmatrix} \mathbf{I}_{G1} * \varepsilon_1 \\ \mathbf{I}_{G2} * \varepsilon_2 \\ \mathbf{I}_{G3} * \varepsilon_3 \end{Bmatrix}$$

- The IT moments are calculated from the matrix equation with vector entries:

$$\mathbf{Q}_a^0 = \mathbf{Z} * \mathbf{Q}_{Gm}^0 + \mathbf{Z}(\tilde{\mathbf{L}}) * \mathbf{P}_{Gm}^0$$

$$\begin{Bmatrix} \mathbf{Q}_A^0 \\ \mathbf{Q}_B^0 \\ \mathbf{Q}_C^0 \end{Bmatrix} = \begin{matrix} A \\ B \\ C \end{matrix} \begin{bmatrix} +1 & +1 & 0 \\ 0 & +1 & 0 \\ 0 & 0 & +1 \end{bmatrix} * \begin{Bmatrix} \mathbf{Q}_{G1}^0 \\ \mathbf{Q}_{G2}^0 \\ \mathbf{Q}_{G3}^0 \end{Bmatrix}$$

$$+ \begin{bmatrix} +1 \cdot \tilde{\mathbf{L}}_{AG1}^0 & +1 \cdot \tilde{\mathbf{L}}_{AG2}^0 & 0 \\ 0 & +1 \cdot \tilde{\mathbf{L}}_{BG2}^0 & 0 \\ 0 & 0 & +1 \cdot \tilde{\mathbf{L}}_{CG3}^0 \end{bmatrix} * \begin{Bmatrix} \mathbf{P}_{G1}^0 \\ \mathbf{P}_{G2}^0 \\ \mathbf{P}_{G3}^0 \end{Bmatrix}$$

Therefore, the scalar components in [lbf · in] are

$$\mathbf{Q}_A^0 = \mathbf{Q}_{G_1}^0 + \mathbf{Q}_{G_2}^0 + \tilde{\mathbf{L}}_{AG_1}^0 * \mathbf{P}_{G_1}^0 + \tilde{\mathbf{L}}_{AG_2}^0 * \mathbf{P}_{G_2}^0$$

$$= \{0;\ 0;\quad I_{G_1} \cdot \varepsilon_1 + I_{G_2} \cdot \varepsilon_2 - y_{AG_1}^0 \cdot P_{G_1}^x + x_{AG_1}^0 \cdot P_{G_1}^y$$

$$- y_{AG_2}^0 \cdot P_{G_2}^x + x_{AG_2}^0 \cdot P_{G_2}^y \}^T$$

$$\mathbf{Q}_B^0 = \mathbf{Q}_{G_2}^0 + \tilde{\mathbf{L}}_{BG_2}^0 * \mathbf{P}_{G_2}^0 = \left\{ 0;\quad 0;\quad I_{G_2} \cdot \varepsilon_2 - y_{BG_2}^0 \cdot P_{G_2}^x + x_{BG_2}^0 \cdot P_{G_2}^y \right\}^T$$

$$\mathbf{Q}_C^0 = \mathbf{Q}_{G_3}^0 + \tilde{\mathbf{L}}_{CG_3}^0 * \mathbf{P}_{G_3}^0 = \left\{ 0;\quad 0;\quad I_{G_3} \cdot \varepsilon_3 - y_{CG_3}^0 \cdot P_{G_3}^x + x_{CG_3}^0 \cdot P_{G_3}^y \right\}^T$$

(4.142f)

- Single-cycle IT dynamic equations
  The IT moment Equation (4.134a) become

$$\left\{ \begin{matrix} \mathbf{M}_A^0 \left(= \mathbf{T}_A^{act} \right) \\ \mathbf{M}_B^0 (= 0) \\ \mathbf{M}_C^0 (= 0) \end{matrix} \right\} + \begin{bmatrix} -1 \\ -1 \\ 1 \end{bmatrix} * \{\mathbf{M}_D (= 0)\} + \begin{bmatrix} -\tilde{\mathbf{L}}_{AD}^0 \\ -\tilde{\mathbf{L}}_{BD}^0 \\ \tilde{\mathbf{L}}_{CD}^0 \end{bmatrix} * \{\mathbf{N}_D\} = \left\{ \begin{matrix} \mathbf{Q}_A^0 \\ \mathbf{Q}_B^0 \\ \mathbf{Q}_C^0 \end{matrix} \right\}$$

where the actuating torque at the revolute joint A and no reaction moments in the joints B, C, and D are shown within the brackets. Therefore

$$\left\{ \begin{matrix} \mathbf{T}_A^{act} - \tilde{\mathbf{L}}_{AD}^0 * \mathbf{N}_D^0 = \mathbf{Q}_A^0 \\ -\tilde{\mathbf{L}}_{BD}^0 * \mathbf{N}_D^0 = \mathbf{Q}_B^0 \\ \tilde{\mathbf{L}}_{CD}^0 * \mathbf{N}_D^0 = \mathbf{Q}_C^0 \end{matrix} \right. \tag{4.142g}$$

- *The joint reactions based on constraint equation.*
  For a planar (x_0, y_0) mechanism with the revolute joints, the revolute axes are all parallel to z_0. The linear velocity is not allowed in the revolute joints A, B, C, and D; therefore, there are reaction components on the x and y directions.

$$\mathbf{N}_A^0 = \begin{Bmatrix} X_A^0 \\ Y_A^0 \\ 0 \end{Bmatrix}; \mathbf{T}_A^{act} = \begin{Bmatrix} 0 \\ 0 \\ T_A^{act} \end{Bmatrix}; \mathbf{N}_B^0 = \begin{Bmatrix} X_B^0 \\ Y_B^0 \\ 0 \end{Bmatrix}; \mathbf{M}_B^0 = \begin{Bmatrix} 0 \\ 0 \\ 0 \end{Bmatrix};$$

$$\mathbf{N}_C^0 = \begin{Bmatrix} X_C^0 \\ Y_C^0 \\ 0 \end{Bmatrix}; \mathbf{M}_C^0 = \begin{Bmatrix} 0 \\ 0 \\ 0 \end{Bmatrix}; \mathbf{N}_D^0 = \begin{Bmatrix} X_D^0 \\ Y_D^0 \\ 0 \end{Bmatrix}; \mathbf{M}_D^0 = \begin{Bmatrix} 0 \\ 0 \\ 0 \end{Bmatrix} \tag{4.142h}$$

Magnitude at the joint A: $|\mathbf{N}_A^0| = \sqrt{\left(X_A^0\right)^2 + \left(Y_A^0\right)^2}$, analog for other joints.

The cross products in Equation (4.142g) are determined by the left-hand-side multiplication of reactions by the skew-symmetric matrices:

$$\tilde{\mathbf{L}}_{AD}^0 * \mathbf{N}_D^0 = \begin{bmatrix} 0 & 0 & y_{AD} \\ 0 & 0 & -x_{AD} \\ -y_{AD} & x_{AD} & 0 \end{bmatrix} * \begin{Bmatrix} X_D^0 \\ Y_D^0 \\ 0 \end{Bmatrix} = \begin{Bmatrix} 0 \\ 0 \\ -y_{AD} \cdot X_D^0 + x_{AD} \cdot Y_D^0 \end{Bmatrix};$$

$$\tilde{\mathbf{L}}_{BD}^0 * \mathbf{N}_D^0 = \begin{Bmatrix} 0 \\ 0 \\ -y_{BD} \cdot X_D^0 + x_{BD} \cdot Y_D^0 \end{Bmatrix}; \tag{4.142i}$$

$$\tilde{\mathbf{L}}_{CD}^0 * \mathbf{N}_D^0 = \begin{Bmatrix} 0 \\ 0 \\ -y_{CD} \cdot X_D^0 + x_{CD} \cdot Y_D^0 \end{Bmatrix}$$

- *The system of Equation (4.142g) has* $2c + M = 2 + 1 = 3$ *scalar equation*

$$\begin{Bmatrix} T_A^{act} + y_{AD} \cdot X_D^0 - x_{AD} \cdot Y_D^0 \\ y_{BD} \cdot X_D^0 - x_{BD} \cdot Y_D^0 \\ -y_{CD} \cdot X_D^0 + x_{CD} \cdot Y_D^0 \end{Bmatrix} = \begin{Bmatrix} Q_A^{z_0} \\ Q_B^{z_0} \\ Q_C^{z_0} \end{Bmatrix} \tag{4.142j}$$

- *Solution*:

  The system of 2c + M = 3 scalar equations is solved for 3 unknowns:

  - 2c = 2 cut-set reaction forces X_D^0, and Y_D^0 from the second and third equations
  - The actuating torque T_A^{act}

  Thus, the cut-joint reaction components are

$$X_D^0 = \frac{1}{\Delta} \cdot \left(x_{CD} \cdot Q_B^{z0} + x_{BD} \cdot Q_C^{z0} \right)$$
$$Y_D^0 = \frac{1}{\Delta} \cdot \left(y_{CD} \cdot Q_B^{z0} + y_{BD} \cdot Q_C^{z0} \right) \quad (4.142k)$$

- *Singularity coefficient*:

$$\Delta = x_{CD} \cdot y_{BD} - x_{BD} \cdot y_{CD}$$

  This position corresponds to links BD and CD in extension. The magnitude of the reactions is very large, and the designer should be aware of such a singularity position.

- *The differential equation of motion*:

  - M = 1 actuating torque at A, T_A^{act}. The first equation for T_A^{act} represents the required crank's torque for moving the mechanism at the desired crank's angular velocity and acceleration.

  By substituting the solution in Equation (4.142k) in the first equation, the differential equation of motion is generated.

$$T_A^{act} = Q_A^{z0} + \frac{x_{AD} \cdot y_{CD} - x_{CD} \cdot y_{AD}}{\Delta} \cdot Q_B^{z0} + \frac{x_{AD} \cdot y_{BD} - x_{BD} \cdot y_{AD}}{\Delta} \cdot Q_C^{z0} \quad (4.142l)$$

- The IT dynamic force Equation (4.134b) are

$$\begin{Bmatrix} \mathbf{N}_A^0 \\ \mathbf{N}_B^0 \\ \mathbf{N}_C^0 \end{Bmatrix} = \begin{Bmatrix} \mathbf{P}_A^0 \\ \mathbf{P}_B^0 \\ \mathbf{P}_C^0 \end{Bmatrix} - \begin{bmatrix} -1 \\ -1 \\ 1 \end{bmatrix} * \mathbf{N}_D^0$$

The reactions in the tree's joints A, B, and C are (Figure 4.17):

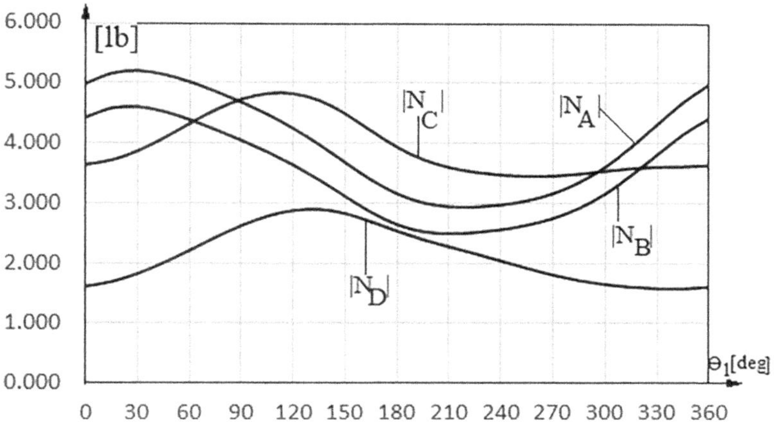

FIGURE 4.17
Magnitude of dynamic reaction forces in the joints A, B, C, and D.

$$\mathbf{N}_A^0 = \mathbf{P}_A^0 + \mathbf{N}_D^0 = \left\{ P_A^{x_0} + \frac{1}{\Delta}\cdot\left(x_{CD}\cdot Q_B^{z_0} + x_{BD}\cdot Q_C^{z_0}\right); \quad P_A^{y_0} + \frac{1}{\Delta}\cdot\left(y_{CD}\cdot Q_B^{z_0} + y_{BD}\cdot Q_C^{z_0}\right); \quad 0 \right\}^T$$

$$\mathbf{N}_B^0 = \mathbf{P}_B^0 + \mathbf{N}_D^0 = \left\{ P_B^{x_0} + \frac{1}{\Delta}\cdot\left(x_{CD}\cdot Q_B^{z_0} + x_{BD}\cdot Q_C^{z_0}\right); \quad P_B^{y_0} + \frac{1}{\Delta}\cdot\left(y_{CD}\cdot Q_B^{z_0} + y_{BD}\cdot Q_C^{z_0}\right); \quad 0 \right\}^T \tag{4.142m}$$

$$\mathbf{N}_C^0 = \mathbf{P}_C^0 + \mathbf{N}_D^0 = \left\{ P_C^{x_0} + \frac{1}{\Delta}\cdot\left(x_{CD}\cdot Q_B^{z_0} + x_{BD}\cdot Q_C^{z_0}\right); \quad P_C^{y_0} + \frac{1}{\Delta}\cdot\left(y_{CD}\cdot Q_B^{z_0} + y_{BD}\cdot Q_C^{z_0}\right); \quad 0 \right\}^T$$

Example 3: The two-cycle planar mechanism with revolute and prismatic joints is shown in Figure 4.7a, for which the reaction forces at the joints and the differential equation of motion are determined.

The calculations are symbolically developed.

Solution:

The links are labeled with numbers, starting with 0 which is assigned to the fixed link. There are five mobile links, m = 5, which have been assigned the numbers 1, 2, 3, 4, and 5. There are seven joints labeled with capital letters: A, B, C, D, E, F, and G. All of the mechanism's joints are 1 DOF each: revolute at A, B, D, E, F, and G and one prismatic at C; therefore, $j_1 = 7$. There are no joints with 2 DOF; therefore, $j_2 = 0$. The total number of joints is $j = j_1 + j_2 = 7 + 0 = 7$ joints.

The digraph attached to the mechanism is illustrated in Figure 4.7b, with its nodes corresponding to the mechanism's links and the edges corresponding to the mechanism's joints. The number of cycles is determined from Equation (1.7): c = j − m = 7 − 5 = 2 cycles. The tree is shown in Figure 4.7b, and the cycles are illustrated in Figure 4.7d. The mobility of the four-bar planar mechanism is determined from Equation (1.28):

$$M = 3 \cdot m - 2 \cdot j_1 - j_2 = 3 \cdot 5 - 2 \cdot 7 - 0 = 1 \text{ DOF}$$

- The digraph matrices **G** and **Z** are

$$[\underline{\mathbf{G}}] = \begin{array}{c} \\ 0 \\ 1 \\ 2 \\ 3 \\ 4 \\ 5 \end{array} \begin{array}{c} \begin{array}{ccccc:cc} A & B & C & D & E & F & G \end{array} \\ \left[\begin{array}{ccccc:cc} -1 & 0 & -1 & 0 & -1 & 0 & 0 \\ \hdashline 1 & -1 & 0 & 0 & 0 & 0 & 0 \\ 0 & 1 & 0 & 0 & 0 & -1 & 0 \\ 0 & 0 & 0 & 0 & 1 & 1 & -1 \\ 0 & 0 & 0 & -1 & 0 & 0 & 1 \\ 0 & 0 & 1 & 1 & 0 & 0 & 0 \end{array}\right] \end{array};$$

$$[\mathbf{G}] = \begin{array}{c} \\ 1 \\ 2 \\ 3 \\ 4 \\ 5 \end{array} \begin{array}{c} \begin{array}{ccccc} A & B & C & D & E \end{array} \\ \left[\begin{array}{ccccc} 1 & -1 & 0 & 0 & 0 \\ 0 & 1 & 0 & 0 & 0 \\ 0 & 0 & 0 & 0 & 1 \\ 0 & 0 & 0 & -1 & 0 \\ 0 & 0 & 1 & 1 & 0 \end{array}\right] \end{array};$$

$$[\mathbf{G}^{cut}] = \begin{array}{c} \begin{array}{cc} F & G \end{array} \\ \left[\begin{array}{cc} 0 & 0 \\ -1 & 0 \\ 1 & -1 \\ 0 & 1 \\ 0 & 0 \end{array}\right] \end{array}; \mathbf{Z} = \begin{array}{c} \\ A \\ B \\ C \\ D \\ E \end{array} \begin{array}{c} \begin{array}{ccccc} 1 & 2 & 3 & 4 & 5 \end{array} \\ \left[\begin{array}{ccccc} +1 & +1 & 0 & 0 & 0 \\ 0 & +1 & 0 & 0 & 0 \\ 0 & 0 & 0 & +1 & +1 \\ 0 & 0 & 0 & -1 & 0 \\ 0 & 0 & +1 & 0 & 0 \end{array}\right] \end{array} \quad (4.143a)$$

- The matrices $\mathbf{Z}(\tilde{\mathbf{L}})$ and $\mathbf{W}(\tilde{\mathbf{L}})$ are

$$\mathbf{Z} \Rightarrow \mathbf{Z}(\tilde{\mathbf{L}}) = \begin{bmatrix} +1\cdot\tilde{\mathbf{L}}^0_{AG_1} & +1\cdot\tilde{\mathbf{L}}^0_{AG_2} & 0 & 0 & 0 \\ 0 & +1\cdot\tilde{\mathbf{L}}^0_{BG_2} & 0 & 0 & 0 \\ 0 & 0 & 0 & +1\cdot\tilde{\mathbf{L}}^0_{CG_4} & +1\cdot\tilde{\mathbf{L}}^0_{CG_5} \\ 0 & 0 & 0 & -1\cdot\tilde{\mathbf{L}}^0_{DG_4} & 0 \\ 0 & 0 & +1\cdot\tilde{\mathbf{L}}^0_{EG_3} & 0 & 0 \end{bmatrix};$$

$$\mathbf{W} = \mathbf{Z} * \mathbf{G}^{cut} = \begin{matrix} \\ A \\ B \\ C \\ D \\ E \end{matrix} \begin{matrix} \begin{matrix} F & \quad G \end{matrix} \\ \begin{bmatrix} -1 & 0 \\ -1 & 0 \\ 0 & 1 \\ 0 & -1 \\ 1 & -1 \end{bmatrix} \end{matrix} \Rightarrow \mathbf{W}(\tilde{\mathbf{L}}) = \begin{bmatrix} -1\cdot\tilde{\mathbf{L}}^0_{AF} & 0 \\ -1\cdot\tilde{\mathbf{L}}^0_{BF} & 0 \\ 0 & +1\cdot\tilde{\mathbf{L}}^0_{CG} \\ 0 & -1\cdot\tilde{\mathbf{L}}^0_{DG} \\ +1\cdot\tilde{\mathbf{L}}^0_{EF} & -1\cdot\tilde{\mathbf{L}}^0_{EG} \end{bmatrix} \tag{4.143b}$$

- The resultant of inertial and external forces reduced to com of each link:

$$\mathbf{P}^0_{G_m} = \begin{Bmatrix} \mathbf{P}^0_{G_1} \\ \mathbf{P}^0_{G_2} \\ \mathbf{P}^0_{G_3} \\ \mathbf{P}^0_{G_4} \\ \mathbf{P}^0_{G_5} \end{Bmatrix} = \begin{Bmatrix} \mathbf{f}_1^{in,0} - \mathbf{f}_1^{ext,0} \\ \mathbf{f}_2^{in,0} - \mathbf{f}_2^{ext,0} \\ \mathbf{f}_3^{in,0} - \mathbf{f}_3^{ext,0} \\ \mathbf{f}_4^{in,0} - \mathbf{f}_4^{ext,0} \\ \mathbf{f}_5^{in,0} - \mathbf{f}_5^{ext,0} \end{Bmatrix}$$

Therefore, the scalar components in [lbf] are

$$\begin{aligned} \mathbf{P}^0_{G_1} &= \left\{ m_1 \cdot a^x_{G_1};\quad m_1\left(a^y_{G_1} + g\right);\quad 0 \right\}^T; \\ \mathbf{P}^0_{G_2} &= \left\{ m_2 \cdot a^x_{G_2};\quad m_2\left(a^y_{G_2} + g\right);\quad 0 \right\}^T; \\ \mathbf{P}^0_{G_3} &= \left\{ m_3 \cdot a^x_{G_3};\quad m_3\left(a^y_{G_3} + g\right);\quad 0 \right\}^T; \\ \mathbf{P}^0_{G_4} &= \left\{ m_4 \cdot a^x_{G_4};\quad m_4\left(a^y_{G_4} + g\right);\quad 0 \right\}^T; \\ \mathbf{P}^0_{G_5} &= \left\{ m_5 \cdot a^x_{G_5};\quad m_3\left(a^y_{G_5} + g\right);\quad 0 \right\}^T \end{aligned} \tag{4.143c}$$

- The IT forces are calculated from the matrix equation with vector entries, $\mathbf{P}_a^0 = \mathbf{Z} * \mathbf{P}_{G_m}^0$:

$$\begin{Bmatrix} \mathbf{P}_A^0 \\ \mathbf{P}_B^0 \\ \mathbf{P}_C^0 \\ \mathbf{P}_D^0 \\ \mathbf{P}_E^0 \end{Bmatrix} = \begin{bmatrix} +1 & +1 & 0 & 0 & 0 \\ 0 & +1 & 0 & 0 & 0 \\ 0 & 0 & 0 & +1 & +1 \\ 0 & 0 & 0 & -1 & 0 \\ 0 & 0 & +1 & 0 & 0 \end{bmatrix} * \begin{Bmatrix} \mathbf{P}_{G_1}^0 \\ \mathbf{P}_{G_2}^0 \\ \mathbf{P}_{G_3}^0 \\ \mathbf{P}_{G_4}^0 \\ \mathbf{P}_{G_5}^0 \end{Bmatrix}$$

Therefore, the scalar components in [lbf] are

$$\mathbf{P}_A^0 = \mathbf{P}_{G_1}^0 + \mathbf{P}_{G_2}^0 = \left\{ m_1 \cdot a_{G_1}^x + m_2 \cdot a_{G_2}^x; \quad m_1\left(a_{G_1}^y + g\right) + m_2\left(a_{G_2}^y + g\right); \quad 0 \right\}^T$$

$$\mathbf{P}_B^0 = \mathbf{P}_{G_2}^0 = \left\{ m_2 \cdot a_{G_2}^x; \quad m_2\left(a_{G_2}^y + g\right); \quad 0 \right\}^T;$$

$$\mathbf{P}_C^0 = \mathbf{P}_{G_4}^0 + \mathbf{P}_{G_5}^0 = \left\{ m_4 \cdot a_{G_4}^x + m_5 \cdot a_{G_5}^x; \quad m_4\left(a_{G_4}^y + g\right) + m_5\left(a_{G_5}^y + g\right); \quad 0 \right\}^T;$$

$$\mathbf{P}_D^0 = -\mathbf{P}_{G_4}^0 = \left\{ -m_4 \cdot a_{G_4}^x; \quad -m_4\left(a_{G_4}^y + g\right); \quad 0 \right\}^T; \tag{4.143d}$$

$$\mathbf{P}_E^0 = \mathbf{P}_{G_3}^0 = \left\{ m_3 \cdot a_{G_3}^x; \quad m_3\left(a_{G_3}^y + g\right); \quad 0 \right\}^T$$

- The resultant of inertial and external moments reduced to COM of each link:

$$\mathbf{Q}_{G_m}^0 = \begin{Bmatrix} \mathbf{Q}_{G_1}^0 \\ \mathbf{Q}_{G_2}^0 \\ \mathbf{Q}_{G_3}^0 \\ \mathbf{Q}_{G_4}^0 \\ \mathbf{Q}_{G_5}^0 \end{Bmatrix} = \begin{Bmatrix} \dot{\mathbf{K}}_{G_1}^{acc,0} - \boldsymbol{\mu}_{G_1}^{ext,0} \\ \dot{\mathbf{K}}_{G_2}^{acc,0} - \boldsymbol{\mu}_{G_2}^{ext,0} \\ \dot{\mathbf{K}}_{G_3}^{acc,0} - \boldsymbol{\mu}_{G_3}^{ext,0} \\ \dot{\mathbf{K}}_{G_4}^{acc,0} - \boldsymbol{\mu}_{G_4}^{ext,0} \\ \dot{\mathbf{K}}_{G_5}^{acc,0} - \boldsymbol{\mu}_{G_5}^{ext,0} \end{Bmatrix} = \begin{Bmatrix} \mathbf{I}_{G_1} * \boldsymbol{\varepsilon}_1 \\ \mathbf{I}_{G_2} * \boldsymbol{\varepsilon}_2 \\ \mathbf{I}_{G_3} * \boldsymbol{\varepsilon}_3 \\ \mathbf{I}_{G_4} * \boldsymbol{\varepsilon}_4 \\ 0 \end{Bmatrix}$$

If the external forces are the links' weights applied in COM, then $\mu_{G_m}^{ext,0} = 0$ for m = 1, 2, 3, 4, and 5. Link 5 is in translation with $\varepsilon_5 = 0$.

- The IT moments are calculated from the matrix equation with vector entries:

$$\mathbf{Q}_a^0 = \mathbf{Z} * \mathbf{Q}_{G_m}^0 + \mathbf{Z}(\tilde{\mathbf{L}}) * \mathbf{P}_{G_m}^0$$

$$\begin{Bmatrix} \mathbf{Q}_A^0 \\ \mathbf{Q}_B^0 \\ \mathbf{Q}_C^0 \\ \mathbf{Q}_D^0 \\ \mathbf{Q}_E^0 \end{Bmatrix} = \begin{bmatrix} +1 & +1 & 0 & 0 & 0 \\ 0 & +1 & 0 & 0 & 0 \\ 0 & 0 & 0 & +1 & +1 \\ 0 & 0 & 0 & -1 & 0 \\ 0 & 0 & +1 & 0 & 0 \end{bmatrix} * \begin{Bmatrix} \mathbf{Q}_{G_1}^0 \\ \mathbf{Q}_{G_2}^0 \\ \mathbf{Q}_{G_3}^0 \\ \mathbf{Q}_{G_4}^0 \\ \mathbf{Q}_{G_5}^0 \end{Bmatrix}$$

$$+ \begin{bmatrix} +1\cdot\tilde{\mathbf{L}}_{AG_1}^0 & +1\cdot\tilde{\mathbf{L}}_{AG_2}^0 & 0 & 0 & 0 \\ 0 & +1\cdot\tilde{\mathbf{L}}_{BG_2}^0 & 0 & 0 & 0 \\ 0 & 0 & 0 & +1\cdot\tilde{\mathbf{L}}_{CG_4}^0 & +1\cdot\tilde{\mathbf{L}}_{CG_5}^0 \\ 0 & 0 & 0 & -1\cdot\tilde{\mathbf{L}}_{DG_4}^0 & 0 \\ 0 & 0 & +1\cdot\tilde{\mathbf{L}}_{EG_3}^0 & 0 & 0 \end{bmatrix} * \begin{Bmatrix} \mathbf{P}_{G_1}^0 \\ \mathbf{P}_{G_2}^0 \\ \mathbf{P}_{G_3}^0 \\ \mathbf{P}_{G_4}^0 \\ \mathbf{P}_{G_5}^0 \end{Bmatrix}$$

Therefore, the scalar components in [lbf · in] are

$$\begin{aligned} \mathbf{Q}_A^0 &= \mathbf{Q}_{G_1}^0 + \mathbf{Q}_{G_2}^0 + \tilde{\mathbf{L}}_{AG_1}^0 * \mathbf{P}_{G_1}^0 + \tilde{\mathbf{L}}_{AG_2}^0 * \mathbf{P}_{G_2}^0 \\ &= \{0;\ 0;\ \ I_{G_1}\cdot\varepsilon_1 + I_{G_2}\cdot\varepsilon_2 - y_{AG_1}^0\cdot P_{G_1}^x \\ &\quad + x_{AG_1}^0\cdot P_{G_1}^y - y_{AG_2}^0\cdot P_{G_2}^x + x_{AG_2}^0\cdot P_{G_2}^y\}^T; \end{aligned}$$

$$\mathbf{Q}_B^0 = \mathbf{Q}_{G_2}^0 + \tilde{\mathbf{L}}_{BG_2}^0 * \mathbf{P}_{G_2}^0 = \{\ 0;\quad 0;\quad I_{G_2}\cdot\varepsilon_2 - y_{BG_2}^0\cdot P_{G_2}^x + x_{BG_2}^0\cdot P_{G_2}^y\ \}^T;$$

$$\begin{aligned} \mathbf{Q}_C^0 &= \mathbf{Q}_{G_4}^0 + \mathbf{Q}_{G_5}^0 + \tilde{\mathbf{L}}_{CG_4}^0 * \mathbf{P}_{G_4}^0 + \tilde{\mathbf{L}}_{CG_5}^0 * \mathbf{P}_{G_5}^0 \\ &= \{\ 0;\quad 0;\quad I_{G_4}\cdot\varepsilon_4 - y_{CG_4}^0\cdot P_{G_4}^x + x_{CG_4}^0\cdot P_{G_4}^y - y_{CG_5}^0\cdot P_{G_5}^x + x_{CG_5}^0\cdot P_{G_5}^y\ \}^T; \end{aligned}$$

$$\mathbf{Q}_D^0 = -\mathbf{Q}_{G_4}^0 - \tilde{\mathbf{L}}_{DG_4}^0 * \mathbf{P}_{G_4}^0 = \{\ 0;\quad 0;\quad -I_{G_4}\cdot\varepsilon_4 + y_{DG_4}^0\cdot P_{G_4}^x - x_{DG_4}^0\cdot P_{G_4}^y\ \}^T;$$

$$\mathbf{Q}_E^0 = \mathbf{Q}_{G_3}^0 + \tilde{\mathbf{L}}_{EG_3}^0 * \mathbf{P}_{G_3}^0 = \{\ 0;\quad 0;\quad I_{G_3}\cdot\varepsilon_3 - y_{EG_3}^0\cdot P_{G_3}^x + x_{EG_3}^0\cdot P_{G_3}^y\ \}^T \tag{4.143e}$$

- The two-cycle IT dynamic equations

  The IT dynamic moment Equation (4.135a) is

$$\{\mathbf{M}_a^0\} + [\mathbf{W}] * \{\mathbf{M}^{cut}\} + [\mathbf{W}(\tilde{\mathbf{L}})] * \{\mathbf{N}^{cut}\} = \{\mathbf{Q}_a^0\}$$

$$\begin{Bmatrix} \mathbf{M}_A^0 (= \mathbf{T}_A^{act}) \\ \mathbf{M}_B^0 (= 0) \\ \mathbf{M}_C^0 (= 0) \\ \mathbf{M}_D^0 (= 0) \\ \mathbf{M}_E^0 (= 0) \end{Bmatrix} + \begin{bmatrix} -1 & 0 \\ -1 & 0 \\ 0 & +1 \\ 0 & -1 \\ +1 & -1 \end{bmatrix} * \begin{Bmatrix} \mathbf{M}_F^0 (= 0) \\ \mathbf{M}_G^0 (= 0) \end{Bmatrix} + \begin{bmatrix} -1 \cdot \tilde{\mathbf{L}}_{AF}^0 & 0 \\ -1 \cdot \tilde{\mathbf{L}}_{BF}^0 & 0 \\ 0 & +1 \cdot \tilde{\mathbf{L}}_{CG}^0 \\ 0 & -1 \cdot \tilde{\mathbf{L}}_{DG}^0 \\ +1 \cdot \tilde{\mathbf{L}}_{EF}^0 & -1 \cdot \tilde{\mathbf{L}}_{EG}^0 \end{bmatrix}$$

$$* \begin{Bmatrix} \mathbf{N}_F^0 \\ \mathbf{N}_G^0 \end{Bmatrix} = \begin{Bmatrix} \mathbf{Q}_A^0 \\ \mathbf{Q}_B^0 \\ \mathbf{Q}_C^0 \\ \mathbf{Q}_D^0 \\ \mathbf{Q}_E^0 \end{Bmatrix} \tag{4.143f}$$

where the actuating torque at the revolute joint A and no reaction moments in the joints B, C, D, E, and F are shown within the brackets. Therefore

$$\begin{cases} \mathbf{T}_A^{act} - \tilde{\mathbf{L}}_{AF} * \mathbf{N}_F^0 = \mathbf{Q}_a^0 \\ -\tilde{\mathbf{L}}_{BF} * \mathbf{N}_F^0 = \mathbf{Q}_B^0 \\ \tilde{\mathbf{L}}_{CG} * \mathbf{N}_G^0 = \mathbf{Q}_C^0 \\ -\tilde{\mathbf{L}}_{DG} * \mathbf{N}_G^0 = \mathbf{Q}_D^0 \\ \tilde{\mathbf{L}}_{EF} * \mathbf{N}_F^0 - \tilde{\mathbf{L}}_{EG} * \mathbf{N}_G^0 = \mathbf{Q}_E^0 \end{cases} \tag{4.143g}$$

- The joint reactions based on constraint equation

  For a planar (x_0, y_0) mechanism, the revolute axes are all parallel to z_0. The prismatic joint C allows relative linear velocity in the x direction; therefore, reaction $\mathbf{N}_C^0$ has y component only.

$$\mathbf{N}_A^0 = \begin{Bmatrix} X_A^0 \\ Y_A^0 \\ 0 \end{Bmatrix}; \mathbf{M}_A^0 = \begin{Bmatrix} 0 \\ 0 \\ T_A^{act} \end{Bmatrix}; \mathbf{N}_B^0 = \begin{Bmatrix} X_B^0 \\ Y_B^0 \\ 0 \end{Bmatrix}; \mathbf{M}_B^0 = \begin{Bmatrix} 0 \\ 0 \\ 0 \end{Bmatrix};$$

$$\mathbf{N}_C^0 = \begin{Bmatrix} 0 \\ Y_C^0 \\ 0 \end{Bmatrix}; \mathbf{M}_C^0 = \begin{Bmatrix} 0 \\ 0 \\ 0 \end{Bmatrix}; \mathbf{N}_D^0 = \begin{Bmatrix} X_D^0 \\ Y_D^0 \\ 0 \end{Bmatrix}; \mathbf{M}_D^0 = \begin{Bmatrix} 0 \\ 0 \\ 0 \end{Bmatrix};$$

$$\mathbf{N}_E^0 = \begin{Bmatrix} X_E^0 \\ Y_E^0 \\ 0 \end{Bmatrix}; \mathbf{M}_E^0 = \begin{Bmatrix} 0 \\ 0 \\ 0 \end{Bmatrix}; \mathbf{N}_F^0 = \begin{Bmatrix} X_F^0 \\ Y_F^0 \\ 0 \end{Bmatrix}; \mathbf{M}_F^0 = \begin{Bmatrix} 0 \\ 0 \\ 0 \end{Bmatrix};$$

$$\mathbf{N}_G^0 = \begin{Bmatrix} X_G^0 \\ Y_G^0 \\ 0 \end{Bmatrix}; \mathbf{M}_G^0 = \begin{Bmatrix} 0 \\ 0 \\ 0 \end{Bmatrix} \tag{4.143h}$$

The cross products in Equation (4.143g) are determined by the left-hand-side multiplication of reactions by the skew-symmetric matrices:

$$\tilde{\mathbf{L}}_{AF}^0 * \mathbf{N}_F^0 = \begin{Bmatrix} 0 \\ 0 \\ -y_{AF} \cdot X_F^0 + x_{AF} \cdot Y_F^0 \end{Bmatrix}; \tilde{\mathbf{L}}_{BF}^0 * \mathbf{N}_F^0 = \begin{Bmatrix} 0 \\ 0 \\ -y_{BF} \cdot X_F^0 + x_{BF} \cdot Y_F^0 \end{Bmatrix};$$

$$\tilde{\mathbf{L}}_{CG}^0 * \mathbf{N}_G^0 = \begin{Bmatrix} 0 \\ 0 \\ -y_{CG} \cdot X_G^0 + x_{CG} \cdot Y_G^0 \end{Bmatrix};$$

$$\tilde{\mathbf{L}}_{DG}^0 * \mathbf{N}_G^0 = \begin{Bmatrix} 0 \\ 0 \\ -y_{DG} \cdot X_G^0 + x_{DG} \cdot Y_G^0 \end{Bmatrix}; \tilde{\mathbf{L}}_{EF}^0 * \mathbf{N}_F^0 = \begin{Bmatrix} 0 \\ 0 \\ -y_{EF} \cdot X_F^0 + x_{EF} \cdot Y_F^0 \end{Bmatrix};$$

$$\tilde{\mathbf{L}}_{EG}^0 * \mathbf{N}_G^0 = \begin{Bmatrix} 0 \\ 0 \\ -y_{EG} \cdot X_G^0 + x_{EG} \cdot Y_G^0 \end{Bmatrix} \tag{4.143i}$$

The system of Equation (4.143g) has 2c + M = 5 *scalar equations*

$$\left\{\begin{array}{c} T_A^{act} + y_{AF} \cdot X_F^0 - x_{AF} \cdot Y_F^0 \\ y_{BF} \cdot X_F^0 - x_{BF} \cdot Y_F^0 \\ -y_{CG} \cdot X_G^0 + x_{CG} \cdot Y_G^0 \\ y_{DG} \cdot X_G^0 - x_{DG} \cdot Y_G^0 \\ -y_{EF} \cdot X_F^0 + x_{EF} \cdot Y_F^0 + y_{EG} \cdot X_G^0 - x_{EG} \cdot Y_G^0 \end{array}\right\} = \left\{\begin{array}{c} Q_A^{z_0} \\ Q_B^{z_0} \\ Q_C^{z_0} \\ Q_D^{z_0} \\ Q_E^{z_0} \end{array}\right\} \quad (4.143j)$$

- *Solution*:

  The system of 2c + M = 5 scalar equations is solved for five 5 unknowns: the 2c = 4 cut-set reaction forces X_F^0, Y_F^0, X_G^0, Y_G^0 (from the last four equations) and

  - Solution for c = 2 components of the reaction force in cut-joint G, solving the third and fourth equations

$$X_G^0 = -\frac{1}{\Delta_G}\left(x_{DG} \cdot Q_C^{z_0} + x_{CG} \cdot Q_D^{z_0}\right)$$
$$Y_G^0 = -\frac{1}{\Delta_G}\left(y_{DG} \cdot Q_C^{z_0} + y_{CG} \cdot Q_D^{z_0}\right) \quad (4.143k)$$

- *Singularity factor*: $\Delta_G = x_{DG} \cdot y_{CG} - x_{CG} \cdot y_{DG}$

  If $\Delta_G = 0$, the mechanism has C, G, and D collinear and corresponds to the links DG and CG in extension.

  The magnitude of the reaction force in the cut-joint G becomes very large.

  - Solution for c = 2 components of the reaction force in the cut-joint F, solving the second and fifth equations:

$$X_F^0 = -\frac{1}{\Delta_F}\left(x_{EF} \cdot Q_B^{z_0} + x_{BF} \cdot Q_E'^{z_0}\right)$$
$$Y_F^0 = -\frac{1}{\Delta_F}\left(y_{EF} \cdot Q_B^{z_0} + y_{BF} \cdot Q_E'^{z_0}\right) \quad (4.143l)$$

  where

$$Q_E'^{z_0} = Q_E^{z_0} + \frac{1}{\Delta_G}\left[\left(x_{DG} \cdot y_{EG} - x_{EG} \cdot y_{DG}\right)Q_C^{z_0} + \left(x_{CG} \cdot y_{EG} - x_{EG} \cdot y_{CG}\right)Q_D^{z_0}\right]$$

- *Singularity factor*: $\Delta_F = x_{EF} \cdot y_{BF} - x_{BF} \cdot y_{EF}$

  If $\Delta_F = 0$, the mechanism has E, F, and B collinear and corresponds to the links EF and BF in extension.

The magnitude of the reaction force in the cut-joint F becomes very large.

Note: The set of 2c = 4 equations for the calculation of c = 2 cut-joint reactions is split into two sets of two equations, which brings a simplification in the calculation of dynamic reactions. If the system of four equations with four unknowns is considered, then the determinant Δ for the matrix of coefficients is the product of the above-mentioned singularity factors:

$$\Delta = \Delta_F \cdot \Delta_G$$

- The differential equation of motion

  The M = 1 actuating torque T_A^{act} substituting the previous cut-joint reactions in the first equation. This equation is the differential equation of motion:

$$T_A^{act} = Q_C^{z0} - y_{AF} \cdot X_F^0 + x_{AF} \cdot Y_F^0 \tag{4.143m}$$

- The IT dynamic force Equation (4.135b) is

$$\{\mathbf{N}_a^0\} = \{\mathbf{P}_a^0\} - [\mathbf{W}] * \{\mathbf{N}^{cut}\}$$

$$\begin{Bmatrix} \mathbf{N}_A^0 \\ \mathbf{N}_B^0 \\ \mathbf{N}_C^0 \\ \mathbf{N}_D^0 \\ \mathbf{N}_E^0 \end{Bmatrix} = \begin{Bmatrix} \mathbf{P}_A^0 \\ \mathbf{P}_B^0 \\ \mathbf{P}_C^0 \\ \mathbf{P}_D^0 \\ \mathbf{P}_E^0 \end{Bmatrix} - \begin{bmatrix} -1 & 0 \\ -1 & 0 \\ 0 & +1 \\ 0 & -1 \\ +1 & -1 \end{bmatrix} * \begin{Bmatrix} \mathbf{N}_F^0 \\ \mathbf{N}_G^0 \end{Bmatrix}$$

The scalar components for reactions in the tree's joints A, B, C, D, and E are (Figure 4.18):

$$\begin{aligned}
\mathbf{N}_A^0 &= \mathbf{P}_A^0 + \mathbf{N}_F^0 = \{\; P_A^{x0} + X_F^0; \quad P_A^{y0} + Y_F^0; \quad 0 \;\}^T \\
\mathbf{N}_B^0 &= \mathbf{P}_B^0 + \mathbf{N}_F^0 = \{\; P_B^{x0} + X_F^0; \quad P_B^{y0} + Y_F^0; \quad 0 \;\}^T \\
\mathbf{N}_C^0 &= \mathbf{P}_C^0 - \mathbf{N}_G^0 = \{\; P_C^{x0} - X_G^0; \quad P_C^{y0} - Y_G^0; \quad 0 \;\}^T \\
\mathbf{N}_D^0 &= \mathbf{P}_D^0 + \mathbf{N}_G^0 = \{\; P_D^{x0} + X_G^0; \quad P_D^{y0} + Y_G^0; \quad 0 \;\}^T \\
\mathbf{N}_E^0 &= \mathbf{P}_E^0 - \mathbf{N}_F^0 + \mathbf{N}_G^0 = \{\; P_E^{x0} - X_F^0 + X_G^0; \quad P_E^{y0} - Y_F^0 + Y_G^0; \quad 0 \;\}^T
\end{aligned} \tag{4.143n}$$

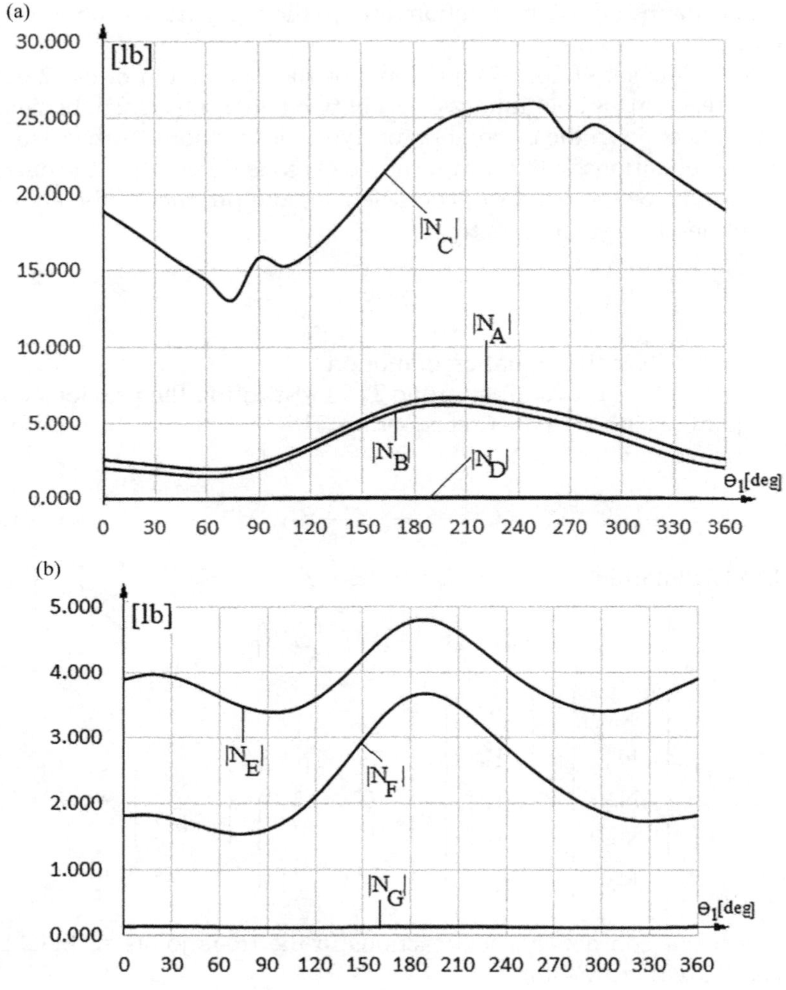

FIGURE 4.18
Magnitude of dynamic reaction forces in joints: (a) joints A, B, C, and D and (b) joints E, F, and G.

4.8.6 Example: Dynamic Reactions and Differential Equation for a Mechanism with Gears

The gear set of the single-cycle planar parallel axes is illustrated in Figure 4.19a, for which the reaction forces at the joints and the differential equation of motion are determined.

Solution: The links are labeled with numbers, starting with 0 which is assigned to the fixed link. There are two mobile links, m = 2, for which the numbers 1 and 2 have been assigned. There are three joints labeled with capital letters: A, B, and C. The mechanism has two revolute joints with 1 DOF each at A and B; therefore, $j_1 = 2$. There is one gear joint with 2 DOF at C; therefore, $j_2 = 1$. This gear joint is labeled with the last letter C. The total number of joints from Equation (1.27) is $j = j_1 + j_2 = 2 + 1 = 3$ joints. The digraph attached to the mechanism and the tree is illustrated in Figure 4.19b,c, with the nodes corresponding to the mechanism's links and the edges corresponding to the mechanism's joints.

The number of cycles is determined using Equation (1.7): $c = j - m = 3 - 2 = 1$ cycle. The mobility of the planar mechanism is determined using Equation (1.28):

$$M = 3 \cdot m - 2 \cdot j_1 - j_2 = 3 \cdot 2 - 2 \cdot 2 - 1 = 1$$

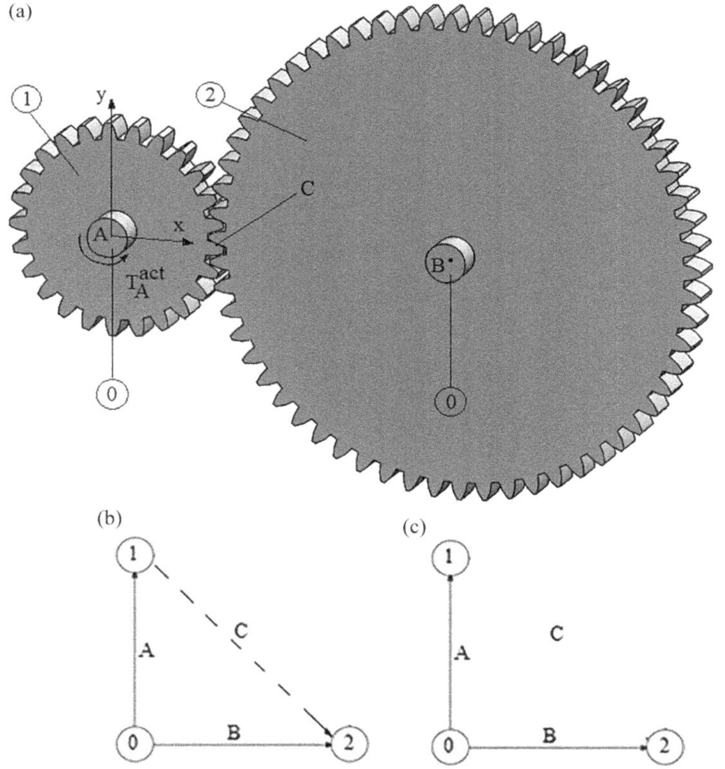

FIGURE 4.19
(a) Gear set; (b) digraph; and (c) tree.

Solution:

- The digraph matrices **G** and **Z** are

$$[\underline{\mathbf{G}}] = \begin{array}{c} \\ 0 \\ 1 \\ 2 \end{array} \begin{array}{c} \begin{array}{ccc} A & B & C \end{array} \\ \left[\begin{array}{cc|c} -1 & -1 & 0 \\ \hline 1 & 0 & -1 \\ 0 & 1 & 1 \end{array}\right] \end{array}; [\mathbf{G}] = \begin{array}{c} \\ 1 \\ 2 \end{array} \begin{array}{c} \begin{array}{cc} A & B \end{array} \\ \left[\begin{array}{cc} 1 & 0 \\ 0 & 1 \end{array}\right] \end{array};$$

$$[\mathbf{G}^{cut}] = \begin{array}{c} D \\ \left[\begin{array}{c} -1 \\ 1 \end{array}\right] \end{array}; \mathbf{Z} = \begin{array}{c} \\ A \\ B \end{array} \begin{array}{c} \begin{array}{cc} 1 & 2 \end{array} \\ \left[\begin{array}{cc} +1 & 0 \\ 0 & +1 \end{array}\right] \end{array} \tag{4.144a}$$

- The matrices $\mathbf{Z(\tilde{L})}$ and $\mathbf{W(\tilde{L})}$ are

$$\mathbf{Z} \Rightarrow \mathbf{Z(\tilde{L})} = \begin{array}{c} \\ A \\ B \end{array} \begin{array}{c} \begin{array}{cc} 1 & 2 \end{array} \\ \left[\begin{array}{cc} +1 \cdot \tilde{\mathbf{L}}^0_{AG_1} & 0 \\ 0 & +1 \cdot \tilde{\mathbf{L}}^0_{BG_2} \end{array}\right] \end{array};$$

$$\mathbf{W} = \mathbf{Z} * \mathbf{G}^{cut} = \begin{array}{c} \\ A \\ B \end{array} \begin{array}{c} C \\ \left[\begin{array}{c} -1 \\ +1 \end{array}\right] \end{array} \Rightarrow \mathbf{W(\tilde{L})} = \left[\begin{array}{c} -1 \cdot \tilde{\mathbf{L}}^0_{AC} \\ +1 \cdot \tilde{\mathbf{L}}^0_{BC} \end{array}\right] \tag{4.144b}$$

- The resultant of inertial and external forces reduced to COM of each link:

$$\mathbf{P}^0_{G_m} = \left\{ \begin{array}{c} \mathbf{P}^0_{G_1} \\ \mathbf{P}^0_{G_2} \end{array} \right\} = \left\{ \begin{array}{c} \mathbf{f}_1^{in,0} - \mathbf{f}_1^{ext,0} \\ \mathbf{f}_2^{in,0} - \mathbf{f}_2^{ext,0} \end{array} \right\} = \left\{ \begin{array}{c} -\mathbf{f}_1^{ext,0} \\ -\mathbf{f}_2^{ext,0} \end{array} \right\} \tag{4.144c}$$

If each gear's COM is located on the fixed axes of rotation, then their linear accelerations are zero. Therefore, $\mathbf{f}_1^{in,0} = \mathbf{f}_2^{in,0} = 0$. The external forces are the weights of the two gears. Therefore, the scalar components in [lbf] are

$$\mathbf{P}_{G_1}^0 = -\mathbf{f}_1^{ext,0} = -\begin{Bmatrix} 0 \\ -m_1 g \\ 0 \end{Bmatrix} = \begin{Bmatrix} 0 \\ m_1 g \\ 0 \end{Bmatrix};$$

$$\mathbf{P}_{G_2}^0 = -\mathbf{f}_2^{ext,0} = -\begin{Bmatrix} 0 \\ -m_2 g \\ 0 \end{Bmatrix} = \begin{Bmatrix} 0 \\ m_2 g \\ 0 \end{Bmatrix}$$

- The IT forces are calculated from the matrix equation with vector entries, $\mathbf{P}_a^0 = \mathbf{Z} * \mathbf{P}_{G_m}^0$:

$$\begin{Bmatrix} \mathbf{P}_A^0 \\ \mathbf{P}_B^0 \end{Bmatrix} = \begin{bmatrix} +1 & 0 \\ 0 & +1 \end{bmatrix} * \begin{Bmatrix} \mathbf{P}_{G_1}^0 \\ \mathbf{P}_{G_2}^0 \end{Bmatrix}$$

Therefore, the scalar components in [lbf] are

$$\mathbf{P}_A^0 = \mathbf{P}_{G_1}^0 = \begin{Bmatrix} 0 \\ m_1 g \\ 0 \end{Bmatrix}; \mathbf{P}_B^0 = \mathbf{P}_{G_2}^0 = \begin{Bmatrix} 0 \\ m_2 g \\ 0 \end{Bmatrix} \tag{4.144d}$$

- The resultant of inertial and external moments reduced to COM of each link:

$$\mathbf{Q}_{G_m}^0 = \begin{Bmatrix} \mathbf{Q}_{G_1}^0 \\ \mathbf{Q}_{G_2}^0 \end{Bmatrix} = \begin{Bmatrix} \dot{\mathbf{K}}_{G_1}^{acc,0} - \boldsymbol{\mu}_{G_1}^{ext,0} \\ \dot{\mathbf{K}}_{G_2}^{acc,0} - \boldsymbol{\mu}_{G_2}^{ext,0} \end{Bmatrix} = \begin{Bmatrix} \mathbf{I}_{G_1} * \boldsymbol{\varepsilon}_1 \\ \mathbf{I}_{G_2} * \boldsymbol{\varepsilon}_2 \end{Bmatrix} \tag{4.144e}$$

If the external forces are the links' weights applied in COM, then $\boldsymbol{\mu}_{G_m}^{ext,0} = 0$ for m = 1, 2, and for a planar mechanism:

$$\dot{\mathbf{K}}_{G_m}^{acc,0} = \begin{Bmatrix} \dot{\mathbf{K}}_{G_1}^{acc,0} \\ \dot{\mathbf{K}}_{G_2}^{acc,0} \end{Bmatrix} = \begin{Bmatrix} \mathbf{I}_{G_1} * \boldsymbol{\varepsilon}_1 + \tilde{\boldsymbol{\omega}}_1 * (\mathbf{I}_{G_1} * \boldsymbol{\omega}_1) \\ \mathbf{I}_{G_1} * \boldsymbol{\varepsilon}_1 + \tilde{\boldsymbol{\omega}}_1 * (\mathbf{I}_{G_1} * \boldsymbol{\omega}_1) \end{Bmatrix} = \begin{Bmatrix} \mathbf{I}_{G_1} * \boldsymbol{\varepsilon}_1 \\ \mathbf{I}_{G_2} * \boldsymbol{\varepsilon}_2 \end{Bmatrix}$$

- The IT moments are calculated from the matrix equation with vector entries:

$$\mathbf{Q}_a^0 = \mathbf{Z} * \mathbf{Q}_{G_m}^0 + \mathbf{Z}(\tilde{\mathbf{L}}) * \mathbf{P}_{G_m}^0$$

$$\left\{\begin{matrix}\mathbf{Q}_A^0\\ \mathbf{Q}_B^0\end{matrix}\right\} = \begin{matrix}A\\ B\end{matrix}\begin{bmatrix}+1 & 0\\ 0 & +1\end{bmatrix} * \left\{\begin{matrix}\mathbf{Q}_{G_1}^0\\ \mathbf{Q}_{G_2}^0\end{matrix}\right\}$$

$$+\begin{bmatrix}+1\cdot\tilde{\mathbf{L}}_{AG_1}^0 & 0\\ 0 & +1\cdot\tilde{\mathbf{L}}_{BG_2}^0\end{bmatrix} * \left\{\begin{matrix}\mathbf{P}_{G_1}^0\\ \mathbf{P}_{G_2}^0\end{matrix}\right\}$$

Therefore, the scalar components in [lbf· in] are

$$\mathbf{Q}_A^0 = \mathbf{Q}_{G_1}^0 + \tilde{\mathbf{L}}_{AG_1}^0 * \mathbf{P}_{G_1}^0 = \left\{\begin{matrix}0\\ 0\\ I_{G_1}\cdot\varepsilon_1 - y_{AG_1}^0\cdot P_{G_1}^x + x_{AG_1}^0\cdot P_{G_1}^y\end{matrix}\right\} = \left\{\begin{matrix}0\\ 0\\ I_{G_1}\cdot\varepsilon_1\end{matrix}\right\}$$

$$\mathbf{Q}_B^0 = \mathbf{Q}_{G_2}^0 + \tilde{\mathbf{L}}_{BG_2}^0 * \mathbf{P}_{G_2}^0 = \left\{\begin{matrix}0\\ 0\\ I_{G_2}\cdot\varepsilon_2 - y_{BG_2}^0\cdot P_{G_2}^x + x_{BG_2}^0\cdot P_{G_2}^y\end{matrix}\right\} = \left\{\begin{matrix}0\\ 0\\ I_{G_2}\cdot\varepsilon_2\end{matrix}\right\}$$

(4.144f)

If the COM are located at A and B, then $x_{AG_1}^0 = y_{AG_1}^0 = 0; x_{BG_2}^0 = y_{BG_2}^0 = 0$

The moments of inertia about axes of rotation are $I_{G_1} = \frac{m_1\cdot d_1^2}{8}$; $I_{G_2} = \frac{m_2\cdot d_2^2}{8}$ [kg m²]

- The single-cycle IT dynamic equations

$$\{\mathbf{M}_a^0\} + [\mathbf{W}] * \{\mathbf{M}^{cut}\} + [\mathbf{W}(\tilde{\mathbf{L}})] * \{\mathbf{N}^{cut}\} = \{\mathbf{Q}_a^0\}$$

The IT moment equation (4.134a) becomes

$$\left\{\begin{matrix}\mathbf{M}_A^0(=\mathbf{T}_A^{act})\\ \mathbf{M}_B^0(=0)\end{matrix}\right\} + \begin{bmatrix}-1\\ 1\end{bmatrix} * \{\mathbf{M}_C(=0)\} + \begin{bmatrix}-1\cdot\tilde{\mathbf{L}}_{AC}^0\\ +1\cdot\tilde{\mathbf{L}}_{BC}^0\end{bmatrix} * \{N_C\} = \left\{\begin{matrix}\mathbf{Q}_A^0\\ \mathbf{Q}_B^0\end{matrix}\right\}$$

where the actuating torque at the revolute joint A and no reaction moments in the joints B and C are shown within the brackets.

- The single-cycle IT dynamic equations

$$\left\{\begin{matrix}\mathbf{T}_A^{act} - \tilde{\mathbf{L}}_{AC}^0 * \mathbf{N}_C^0 = \mathbf{Q}_A^0\\ \tilde{\mathbf{L}}_{BC}^0 * \mathbf{N}_C^0 = \mathbf{Q}_B^0\end{matrix}\right. \quad (4.144g)$$

- The joint reactions based on constraint equation

  Joint A (revolute, actuated):

  For a planar (x_0, y_0) mechanism the revolute axes are all parallel to z_0. The linear velocity is not allowed in the revolute joint A; therefore, there are reaction components in the x and y directions.

$$\mathbf{N}_A^0 = \begin{Bmatrix} X_A^0 \\ Y_A^0 \\ 0 \end{Bmatrix}; \mathbf{T}_A^{act} = \begin{Bmatrix} 0 \\ 0 \\ T_A^{act} \end{Bmatrix} \tag{4.144h}$$

Joint B (revolute):

There are reaction force components in the x and y directions:

$$\mathbf{N}_B^0 = \begin{Bmatrix} X_B^0 \\ Y_B^0 \\ 0 \end{Bmatrix}; \mathbf{M}_B^0 = \begin{Bmatrix} 0 \\ 0 \\ 0 \end{Bmatrix} \tag{4.144i}$$

Joint C (meshing):

- Tangent-normal directions for reaction forces in meshing joints

  In case of planar mechanisms with meshing joints, the allowed relative motion is along the common tangent direction at the contact point, noted x^τ; the normal direction, noted y^ν, defines the direction of a reaction force expressed in the additional local frame (x^τ, y^ν). The contact point is located at the contact between the two pitch circles with diameters d_1 and d_2. The additional frame on gear m at the contact point is rotated relative to the local frame (x^m, y^m) by an angle Φ. The angle measured between x^m and x^τ is the *pressure angle*. The reaction is defined for convenience in the additional frame. Then, by the left-hand-side multiplication with the rotation matrix of angle Φ, the reaction is expressed in the fixed frame 0. For example, a mechanism has a meshing joint C with x^τ rotated about z^m.

  The joint C allows the relative displacement along x^τ direction therefore, there is a constraint along the normal direction y^ν.

$$\begin{Bmatrix} \dot{d}_C^\nu \\ 0 \\ 0 \end{Bmatrix} \Rightarrow \mathbf{N}_C^\nu = \begin{Bmatrix} 0 \\ Y_C^\nu \\ 0 \end{Bmatrix} \tag{4.144j}$$

With respect to local frame 2, rotated by an angle Φ then

$$\mathbf{N}_C^0 = \mathbf{D}(\Phi) * \mathbf{N}_C^v = \begin{bmatrix} C\Phi & -S\Phi & 0 \\ S\Phi & C\Phi & 0 \\ 0 & 0 & 1 \end{bmatrix} * \begin{Bmatrix} 0 \\ Y_C^v \\ 0 \end{Bmatrix} = \begin{Bmatrix} -S\Phi \cdot Y_C^v \\ C\Phi \cdot Y_C^v \\ 0 \end{Bmatrix} \tag{4.144k}$$

where $\mathbf{D}(\Phi) = \begin{bmatrix} C\Phi & -S\Phi & 0 \\ S\Phi & C\Phi & 0 \\ 0 & 0 & 1 \end{bmatrix}$ is the rotation about z matrix of the frame (x^τ, y^v) by an angle Φ.

An angular velocity about z_2 of gear 2 relative to gear 1, $\dot{\theta}_C^2$, is allowed at the meshing joint C; therefore, $M_C^{z2} = 0$ (no reaction moment about direction z_2). If this is a planar gear set, then the reaction moments about directions x_2 and y_2 are null.

$$\dot{\theta}_C^2 = \begin{Bmatrix} 0 \\ 0 \\ \dot{\theta}_C^3 \end{Bmatrix} \Rightarrow \mathbf{M}_C^0 = \begin{Bmatrix} 0 \\ 0 \\ 0 \end{Bmatrix} \tag{4.144l}$$

The cross products in Equation (4.144g) are determined by left-hand-side multiplication of reactions by the skew-symmetric matrices:

$$\tilde{\mathbf{L}}_{AC}^0 * \mathbf{N}_C^0 = \begin{bmatrix} 0 & 0 & 0 \\ 0 & 0 & -x_{AC} \\ 0 & x_{AC} & 0 \end{bmatrix} * \begin{Bmatrix} -S\Phi \cdot Y_C^v \\ C\Phi \cdot Y_C^v \\ 0 \end{Bmatrix}$$

$$= \begin{Bmatrix} 0 \\ 0 \\ x_{AC} \cdot C\Phi \cdot Y_C^v \end{Bmatrix} = \begin{Bmatrix} 0 \\ 0 \\ 0.5d_1 \cdot C\Phi \cdot Y_C^v \end{Bmatrix}$$

(4.144m)

$$\tilde{\mathbf{L}}_{BC}^0 * \mathbf{N}_C^0 = \begin{bmatrix} 0 & 0 & 0 \\ 0 & 0 & -x_{BC} \\ 0 & x_{BC} & 0 \end{bmatrix} * \begin{Bmatrix} -S\Phi \cdot Y_C^v \\ C\Phi \cdot Y_C^v \\ 0 \end{Bmatrix}$$

$$= \begin{Bmatrix} 0 \\ 0 \\ x_{BC} \cdot C\Phi \cdot Y_C^v \end{Bmatrix} = \begin{Bmatrix} 0 \\ 0 \\ -0.5d_2 \cdot C\Phi \cdot Y_C^v \end{Bmatrix}$$

- The system Equation (4.144g) has m = 2 scalar equations:

$$\left\{\begin{array}{c} T_A^{act} - 0.5d_1 \cdot C\Phi \cdot Y_C^v \\ -0.5d_2 \cdot C\Phi \cdot Y_C^v \end{array}\right\} = \left\{\begin{array}{c} I_{G_1} \cdot \varepsilon_1 \\ I_{G_2} \cdot \varepsilon_2 \end{array}\right\} \quad (4.144n)$$

The system of two scalar equations is solved for two unknowns: Y_C^v and $T_A^{act,0}$.

- *Solution*:
  - The cut-set reaction force at the meshing joint C:

$$Y_C^v = -\frac{I_{G_2} \cdot \varepsilon_2}{0.5d_2 \cdot C\Phi}$$

  - The differential equation of motion:

$$T_A^{act} = I_{G_1}\varepsilon_1 + 0.5d_1 \cdot C\Phi \cdot Y_C^v = I_{G_1} \cdot \varepsilon_1 - \left(\frac{d_1}{d_2}\right) \cdot I_{G_2}\varepsilon_2 = \left[I_{G_1} + \left(\frac{N_1}{N_2}\right)^2 I_{G_2}\right] \cdot \varepsilon_1 \quad (4.144o)$$

Notation:

$I_A^{red} = \left[I_{G_1} + \left(\frac{N_1}{N_2}\right)^2 I_{G_2}\right]$ represents the moment of inertia for the overall gear set—reduced to the joint A.

$\varepsilon_1 = \ddot{\theta}_A$ is the angular acceleration for actuated joint A, [rad/s^2].

$\frac{d_1}{d_2} = \frac{N_1}{N_2}$ is the ratio of the number of teeth for the two gears.

$\varepsilon_2 = -\frac{d_1}{d_2}\varepsilon_1 = -\frac{N_1}{N_2}\varepsilon_1$—from the parallel axes gear-set acceleration analysis.

Then, for the differential equation of motion, Equation (4.144p) holds:

$$T_A^{act} = I_A^{red} \cdot \ddot{\theta}_A \quad (4.144p)$$

For the inverse dynamics, the actuation $T_A^{act,0}$ is provided and the angular displacement $\theta_A(t)$ is required. Two integrations of the differential equation are required when solving for angular displacement.

- The IT dynamic force Equation (4.135b) is

$$\{\mathbf{N}_a\} = \{\mathbf{P}_a\} - [\mathbf{W}] * \{\mathbf{N}^{cut}\}$$

$$\begin{Bmatrix} \mathbf{N}_A^0 \\ \mathbf{N}_B^0 \end{Bmatrix} = \begin{Bmatrix} \mathbf{P}_A^0 \\ \mathbf{P}_B^0 \end{Bmatrix} - \begin{bmatrix} -1 \\ 1 \end{bmatrix} * \mathbf{N}_C^0 \tag{4.144q}$$

$$\begin{cases} \mathbf{N}_A^0 = \mathbf{P}_A^0 + \mathbf{N}_C^0 \\ \mathbf{N}_B^0 = \mathbf{P}_b^0 - \mathbf{N}_C^0 \end{cases}$$

The scalar components for reactions in the tree's joints A and B are

$$\mathbf{N}_A^0 = \mathbf{P}_A^0 + \mathbf{N}_C^0 \Rightarrow \begin{Bmatrix} X_A^0 \\ Y_A^0 \\ 0 \end{Bmatrix} = \begin{Bmatrix} 0 \\ m_1 \cdot g \\ 0 \end{Bmatrix} + \begin{Bmatrix} -S\Phi \cdot Y_C^v \\ C\Phi \cdot Y_C^v \\ 0 \end{Bmatrix}$$

$$= \begin{Bmatrix} \dfrac{I_{G_2} \cdot \varepsilon_2 \cdot \operatorname{Tan}\Phi}{0.5d_2} & \vdots & m_1 \cdot g - \dfrac{I_{G_2} \cdot \varepsilon_2}{0.5d_2} & \vdots & 0 \end{Bmatrix}^T$$

$$\tag{4.144r}$$

$$\mathbf{N}_B^0 = \mathbf{P}_B^0 - \mathbf{N}_C^0 \Rightarrow \begin{Bmatrix} X_B^0 \\ Y_B^0 \\ 0 \end{Bmatrix} = \begin{Bmatrix} 0 \\ m_2 \cdot g \\ 0 \end{Bmatrix} - \begin{Bmatrix} -S\Phi \cdot Y_C^v \\ C\Phi \cdot Y_C^v \\ 0 \end{Bmatrix}$$

$$= \begin{Bmatrix} -\dfrac{I_{G_2} \cdot \varepsilon_2 \cdot \operatorname{Tan}\Phi}{0.5d_2} & \vdots & m_2 \cdot g + \dfrac{I_{G_2} \cdot \varepsilon_2}{0.5d_2} & \vdots & 0 \end{Bmatrix}^T$$

- Interpretation of results:
  - The reaction forces at A and B have x components (radial) and y components (tangential) transmitted to the shafts at the revolute joints A and B.
  - The reaction Y_C^v at the meshing joint C is assigned in the digraph to the cut-edge C, as $Y_{C_{1,2}}^v$.

In the IT notation, we stated that this is the force from gear 1 that is acting on gear 2 (edge C in digraph leaves node 1 and enters node 2).

4.9 Statics of Mechanisms and Machines

4.9.1 Background

The equations for static reaction forces and static reaction moments are developed from those presented in Sections 4.5–4.8, where the substitution for all kinematic velocities (linear and angular) and accelerations (linear and angular) as zero vectors is made. Therefore, the inertial vector force and the rate of change for angular momentum are zero.

$$\begin{aligned} &\mathbf{a}_{G_m}^{0} = 0 \Rightarrow \mathbf{f}_{m}^{in,0} = 0 \\ &\omega_m = 0;\ \varepsilon_m = 0 \Rightarrow \dot{\mathbf{K}}_{G_m}^{in,0} = 0 \end{aligned} \tag{4.145}$$

The break force $\mathbf{f}_m^{br,0}$ and break torque $\mathbf{T}_{G_m}^{br,0}$ are introduced as external forces and moments reduced to link COM.

The output from static analysis is the calculation of reaction forces and moments and external break forces and break torques. If the mobility of a mechanical system with rigid links is $M > 0$, it is a mechanism in equilibrium, and if $M \leq 0$, it is a *frame*. The count of equations and unknowns is performed, to evaluate whether the mechanism or frame is hyper-static. Hyper-static systems, although beyond the scope of this text, require additional equations, leaving the assumption of rigid link.

4.9.2 The Governing Equations for Closed Cycle Mechanisms

The computation of equations is performed as follows:

$$\text{Draw the digraph and spanning tree} \tag{4.146a}$$

$$\text{The matrices } \mathbf{Z}_{a\times m},\ \mathbf{G}_{m\times c}^{cut}, \text{ and } \mathbf{W}_{a\times c} = \mathbf{Z} * \mathbf{G}^{cut} \tag{4.146b}$$

$$\text{The matrices } \mathbf{Z}_{a\times m}(\tilde{\mathbf{L}}) \text{ and } \mathbf{W}_{a\times c}(\tilde{\mathbf{L}}) \text{ written directly from } \mathbf{Z}_{a\times m} \text{ and } \mathbf{W}_{a\times c} \tag{4.146c}$$

- The static force resultant and the static moment resultant reduced to COM. They include the break force and moment

$$\begin{aligned} &\mathbf{P}_{G_m}^{0} = -\mathbf{f}_m^{ext,0} = -\mathbf{f}_m^{br,0} \\ &\mathbf{Q}_{G_m}^{0} = -\boldsymbol{\mu}_{G_m}^{ext,0} = -\mathbf{T}_{G_m}^{br,0} \end{aligned} \tag{4.146d}$$

- The IT force $\mathbf{P}_a^0$ and IT moment $\mathbf{Q}_a^0$ reduced to joint's tree. They are the coefficients on right-hand side of Equations (4.111) and (4.128):

$$\mathbf{P}_a^0 = \mathbf{Z}_{a\times m} * \mathbf{P}_{Gm} \tag{4.146e}$$

$$\mathbf{Q}_a^0 = \mathbf{Z}_{a\times m} * \mathbf{Q}_{Gm}^0 + \mathbf{Z}_{a\times m}(\tilde{\mathbf{L}}) * \mathbf{P}_{Gm}^0 \tag{4.146f}$$

Moment equations:

- The IT moment equations

$$\mathbf{M}_a^0 + \mathbf{W}_{a\times c} * \mathbf{M}_c^{cut,0} + \mathbf{W}_{a\times c}(\tilde{\mathbf{L}}) * \mathbf{N}_c^{cut,0} = \mathbf{Q}_a^0 \tag{4.146g}$$

are solved for static reactions in cut-joints $\mathbf{N}_c^{cut,0}$ and $\mathbf{M}_c^{cut,0}$, break force $\mathbf{f}_m^{br,0}$, and torque $\mathbf{T}_m^{br,0}$

Force equations:

- The IT force equations

$$\mathbf{N}_a^0 + \mathbf{W}_{a\times c} * \mathbf{N}_c^{cut,0} = \mathbf{P}_a^0 \tag{4.146h}$$

are solved for the joint reactions $\mathbf{N}_a^0$

Equations (4.146a) to (4.146h) are the "*IT equations for static reactions, force, and break torque*" [2,3]

Note:

- The IT moment equations generate a linear system with significantly reduced number of equations and unknowns. Thus, for planar systems (frames, mechanisms), there are 2c equations only that are solved for 2c cut-forces and moments, instead of 3m equations of equilibrium written for the m links. For example, let us assume a frame with the revolute joints only:
- if c = 1 cycle and m = 3 links, then there are solved 2c = 2 equations instead of 3m = 9.
- if c = 2 cycles and m = 4 links, then there are solved 2c = 4 equations instead of 3m = 12.

4.9.3 Example: Static Reactions and Break Torque Calculation for a Mechanism with Gears

The gear set for single-cycle planar parallel axes is illustrated in Figure 4.19a, for which the break torque acting on gear 2 that holds in equilibrium, the actuating torque $\mathbf{T}_A^{act}$, and the static reactions in joints are determined.

Solution:

- From Equation (4.144a), the digraph matrices **G** and **Z** are

$$[\underline{\mathbf{G}}] = \begin{matrix} \\ 0 \\ 1 \\ 2 \end{matrix}\begin{matrix} A & B & C \\ \left[\begin{matrix} -1 & -1 \\ 1 & 0 \\ 0 & 1 \end{matrix}\right. & \left.\begin{matrix} 0 \\ -1 \\ 1 \end{matrix}\right] \end{matrix}; [\mathbf{G}] = \begin{matrix} & A & B \\ 1 & 1 & 0 \\ 2 & 0 & 1 \end{matrix};$$

$$[\mathbf{G}^{cut}] = \begin{matrix} D \\ \begin{bmatrix} -1 \\ 1 \end{bmatrix} \end{matrix}; \mathbf{Z} = \begin{matrix} & 1 & 2 \\ A & +1 & 0 \\ B & 0 & +1 \end{matrix} \tag{4.147a}$$

- From Equation (4.144b), the matrices $\mathbf{Z}(\tilde{\mathbf{L}})$ and $\mathbf{W}(\tilde{\mathbf{L}})$ are

$$\mathbf{Z} \Rightarrow \mathbf{Z}(\tilde{\mathbf{L}}) = \begin{matrix} & 1 & 2 \\ A & +1 \cdot \tilde{\mathbf{L}}^0_{AG_1} & 0 \\ B & 0 & +1 \cdot \tilde{\mathbf{L}}^0_{BG_2} \end{matrix};$$

$$\mathbf{W} = \mathbf{Z} * \mathbf{G}^{cut} = \begin{matrix} & C \\ A & -1 \\ B & +1 \end{matrix} \Rightarrow \mathbf{W}(\tilde{\mathbf{L}}) = \begin{bmatrix} -1 \cdot \tilde{\mathbf{L}}^0_{AC} \\ +1 \cdot \tilde{\mathbf{L}}^0_{BC} \end{bmatrix} \tag{4.147b}$$

- The resultant external forces reduced to COM of each link:

$$\mathbf{P}^0_{G_m} = \begin{Bmatrix} \mathbf{P}^0_{G_1} \\ \mathbf{P}^0_{G_2} \end{Bmatrix} = \begin{Bmatrix} -\mathbf{f}_1^{ext,0} \\ -\mathbf{f}_2^{ext,0} \end{Bmatrix} \tag{4.147c}$$

If the external forces are the weights of the two gears, then:

$$\mathbf{P}^0_{G_1} = -\mathbf{f}_1^{ext,0} = -\begin{Bmatrix} 0 \\ -m_1 g \\ 0 \end{Bmatrix} = \begin{Bmatrix} 0 \\ m_1 g \\ 0 \end{Bmatrix};$$

$$\mathbf{P}^0_{G_2} = -\mathbf{f}_2^{ext,0} = -\begin{Bmatrix} 0 \\ -m_2 g \\ 0 \end{Bmatrix} = \begin{Bmatrix} 0 \\ m_2 g \\ 0 \end{Bmatrix}$$

- The static IT forces are calculated from the matrix equation with vector entries, $\mathbf{P}_a^0 = \mathbf{Z} * \mathbf{P}_{G_m}^0$:

$$\left\{ \begin{array}{c} \mathbf{P}_A^0 \\ \mathbf{P}_B^0 \end{array} \right\} = \left[\begin{array}{cc} +1 & 0 \\ 0 & +1 \end{array} \right] * \left\{ \begin{array}{c} \mathbf{P}_{G_1}^0 \\ \mathbf{P}_{G_2}^0 \end{array} \right\}$$. Therefore, the scalar components in [lbf] are

$$\mathbf{P}_A^0 = \mathbf{P}_{G_1}^0 = \left\{ \begin{array}{c} 0 \\ m_1 g \\ 0 \end{array} \right\}; \mathbf{P}_B^0 = \mathbf{P}_{G_2}^0 = \left\{ \begin{array}{c} 0 \\ m_2 g \\ 0 \end{array} \right\} \tag{4.147d}$$

- The resultant of external and break moments reduced to COM of each link:

$$\mathbf{Q}_{G_m}^0 = \left\{ \begin{array}{c} \mathbf{Q}_{G_1}^0 \\ \mathbf{Q}_{G_2}^0 \end{array} \right\} = \left\{ \begin{array}{c} 0 \\ -\mathbf{T}_{G_2}^{br,0} \end{array} \right\}$$; therefore the scalar components are

$$\mathbf{Q}_{G_1}^0 = \left\{ \begin{array}{c} 0 \\ 0 \\ 0 \end{array} \right\}; \mathbf{Q}_{G_2}^0 = -\mathbf{T}_{G_2}^{br,0} = -\left\{ \begin{array}{c} 0 \\ 0 \\ T_{G_2}^{br,0} \end{array} \right\} = \left\{ \begin{array}{c} 0 \\ 0 \\ -T_{G_2}^{br,0} \end{array} \right\} \tag{4.147e}$$

- The IT moments are calculated from the matrix equation with vector entries:

$$\mathbf{Q}_a^0 = \mathbf{Z} * \mathbf{Q}_{G_m}^0 + \mathbf{Z}(\tilde{\mathbf{L}}) * \mathbf{P}_{G_m}^0$$

$$\left\{ \begin{array}{c} \mathbf{Q}_A^0 \\ \mathbf{Q}_B^0 \end{array} \right\} = \begin{array}{c} A \\ B \end{array} \left[\begin{array}{cc} +1 & 0 \\ 0 & +1 \end{array} \right] * \left\{ \begin{array}{c} \mathbf{Q}_{G_1}^0 \\ \mathbf{Q}_{G_2}^0 \end{array} \right\}$$

$$+ \left[\begin{array}{cc} +1 \cdot \tilde{\mathbf{L}}_{AG_1}^0 & 0 \\ 0 & +1 \cdot \tilde{\mathbf{L}}_{BG_2}^0 \end{array} \right] * \left\{ \begin{array}{c} \mathbf{P}_{G_1}^0 \\ \mathbf{P}_{G_2}^0 \end{array} \right\}$$

Therefore, the scalar components in [lbf· in] are

$$\mathbf{Q}_A^0 = \mathbf{Q}_{G_1}^0 + \tilde{\mathbf{L}}_{AG_1}^0 * \mathbf{P}_{G_1}^0 = \left\{ \begin{array}{c} 0 \\ 0 \\ -y_{AG_1}^0 \cdot P_{G_1}^x + x_{AG_1}^0 \cdot P_{G_1}^y \end{array} \right\} = \left\{ \begin{array}{c} 0 \\ 0 \\ 0 \end{array} \right\}$$

$$\mathbf{Q}_B^0 = \mathbf{Q}_{G_2}^0 + \tilde{\mathbf{L}}_{BG_2}^0 * \mathbf{P}_{G_2}^0 = \left\{ \begin{array}{c} 0 \\ 0 \\ -T_{G_2}^{br,0} - y_{BG_2}^0 \cdot P_{G_2}^x + x_{BG_2}^0 \cdot P_{G_2}^y \end{array} \right\} = \left\{ \begin{array}{c} 0 \\ 0 \\ -T_{G_2}^{br,0} \end{array} \right\} \tag{4.147f}$$

If the COM are located at A and B, then $x_{AG_1}^0 = y_{AG_1}^0 = x_{BG_2}^0 = y_{BG_2}^0 = 0$

- The single-cycle IT dynamic equations

$$\left\{ \begin{array}{c} \mathbf{M}_A^0 (= \mathbf{T}_1^{act,0}) \\ \mathbf{M}_B^0 (= 0) \end{array} \right\} + \left[\begin{array}{c} -1 \\ 1 \end{array} \right] * \{\mathbf{M}_C (= 0)\} + \left[\begin{array}{c} -1 \cdot \tilde{\mathbf{L}}_{AC}^0 \\ +1 \cdot \tilde{\mathbf{L}}_{BC}^0 \end{array} \right] * \{N_C\} = \left\{ \begin{array}{c} \mathbf{Q}_A^0 \\ \mathbf{Q}_B^0 \end{array} \right\}$$

where the actuating torque at the revolute joint A and no reaction moments in the joints B and C are shown within the brackets.

$$\left\{ \begin{array}{c} \mathbf{T}_1^{act,0} - \tilde{\mathbf{L}}_{AC}^0 * \mathbf{N}_C^0 = \mathbf{Q}_A^0 \\ \tilde{\mathbf{L}}_{BC}^0 * \mathbf{N}_C^0 = \mathbf{Q}_B^0 \end{array} \right. \tag{4.147g}$$

- The joint reactions based on constraint equation

Joint A (revolute, actuated):

$$\mathbf{N}_A^0 = \left\{ \begin{array}{c} X_A^0 \\ Y_A^0 \\ 0 \end{array} \right\}; \mathbf{T}_A^{act} = \left\{ \begin{array}{c} 0 \\ 0 \\ T_A^{act} \end{array} \right\} \tag{4.147h}$$

Joint B (revolute):

$$\mathbf{N}_B^0 = \left\{ \begin{array}{c} X_B^0 \\ Y_B^0 \\ 0 \end{array} \right\}; \mathbf{M}_B^0 = \left\{ \begin{array}{c} 0 \\ 0 \\ 0 \end{array} \right\} \tag{4.147i}$$

Joint C (meshing):

$$\mathbf{N}_C^0 = \left\{ \begin{array}{c} -S\Phi \cdot Y_C^v \\ C\Phi \cdot Y_C^v \\ 0 \end{array} \right\}; \mathbf{M}_C^0 = \left\{ \begin{array}{c} 0 \\ 0 \\ 0 \end{array} \right\} \tag{4.147j}$$

From Equation (4.144g), the cross products are

$$\tilde{\mathbf{L}}_{AC}^{0} * \mathbf{N}_{C}^{0} = \begin{Bmatrix} 0 \\ 0 \\ 0.5d_1 \cdot C\Phi \cdot Y_C^v \end{Bmatrix}; \tilde{\mathbf{L}}_{BC}^{0} * \mathbf{N}_{C}^{0} = \begin{Bmatrix} 0 \\ 0 \\ -0.5d_2 \cdot C\Phi \cdot Y_C^v \end{Bmatrix} \tag{4.147k}$$

- The system Equation (4.147g) has m = 2 scalar equations:

$$\begin{Bmatrix} T_A^{act} - 0.5d_1 \cdot C\Phi \cdot Y_C^v \\ -0.5d_2 \cdot C\Phi \cdot Y_C^v \end{Bmatrix} = \begin{Bmatrix} 0 \\ -T_{G_2}^{br} \end{Bmatrix} \tag{4.147l}$$

The system of two scalar equations is solved for two unknowns: Y_C^v and $T_{G_2}^{br,0}$

- *Solution*:
  The cut-set reaction force at the meshing joint C and the break torque on gear 2:

$$Y_C^v = \frac{T_A^{act}}{0.5d_1 \cdot C\Phi}$$

$$T_{G_2}^{br,0} = T_A^{act} \cdot \frac{d_2}{d_1} = T_A^{act} \cdot \frac{N_2}{N_1}$$

- The IT dynamic force Equation (4.135b) is

$$\{\mathbf{N}_a\} = \{\mathbf{P}_a\} - [\mathbf{W}] * \{\mathbf{N}^{cut}\}$$

$$\begin{Bmatrix} \mathbf{N}_A^0 \\ \mathbf{N}_B^0 \end{Bmatrix} = \begin{Bmatrix} \mathbf{P}_A^0 \\ \mathbf{P}_B^0 \end{Bmatrix} - \begin{bmatrix} -1 \\ 1 \end{bmatrix} * \mathbf{N}_C^0 \tag{4.147m}$$

$$\begin{cases} \mathbf{N}_A^0 = \mathbf{P}_A^0 + \mathbf{N}_C^0 \\ \mathbf{N}_B^0 = \mathbf{P}_B^0 - \mathbf{N}_C^0 \end{cases}$$

The scalar components for reactions in the tree's joints A and B are

$$\mathbf{N}_A^0 = \mathbf{P}_A^0 + \mathbf{N}_C^0 \Rightarrow \begin{Bmatrix} X_A^0 \\ Y_A^0 \\ 0 \end{Bmatrix} = \begin{Bmatrix} 0 \\ m_1 \cdot g \\ 0 \end{Bmatrix} + \begin{Bmatrix} -S\Phi \cdot Y_C^v \\ C\Phi \cdot Y_C^v \\ 0 \end{Bmatrix}$$

$$= \left\{ \frac{-T_A^{act} \cdot \text{Tan}\,\Phi}{0.5d_1} \;\middle|\; m_1 \cdot g + \frac{T_A^{act}}{0.5d_1} \;\middle|\; 0 \right\}^T$$

$$\mathbf{N}_B^0 = \mathbf{P}_B^0 - \mathbf{N}_C^0 \Rightarrow \begin{Bmatrix} X_B^0 \\ Y_B^0 \\ 0 \end{Bmatrix} = \begin{Bmatrix} 0 \\ m_2 \cdot g \\ 0 \end{Bmatrix} - \begin{Bmatrix} -S\Phi \cdot Y_C^v \\ C\Phi \cdot Y_C^v \\ 0 \end{Bmatrix} \tag{4.147n}$$

$$= \left\{ \frac{T_A^{act} \cdot \text{Tan}\,\Phi}{0.5d_1} \;\middle|\; m_2 \cdot g - \frac{T_A^{act}}{0.5d_1} \;\middle|\; 0 \right\}^T$$

4.10 Conclusions

- The matroid method is defined on two matroids; in kinematics, the cycle basis matroid, and in dynamics, the cut-set matroid. A digraph attached to a mechanism in addition to the two matroids proves to be a promising approach in the development of IT dynamic equations (developed by Ilie Talpasanu) for differential equations and dynamic reactions. The text includes the procedures not found in other books.

  The calculated results coincide with those from simulations. Thus, this approach involves only matrix multiplications and is ideal for programming to validate the design of a mechanism. Since the mechanism's velocities and accelerations are included in the IT force, $\mathbf{P}_a^0$, and the IT moment, $\mathbf{Q}_a^0$, the method for IT dynamic equations does not require the development of the derivatives from positional parameters.
- Compared with Lagrange equations, where for each type of mechanism the kinetic and potential energy functions and then their derivatives are required, the IT matroidal method is advantageous since there are many mechanisms, which share the same digraph. For mechanisms with the same digraph, the equations are the same.

The matroid method can be extended to any number of cycles and links by adding rows and columns in the cycle matroid.

- Compared with Kane equations, where the derivatives of kinematic parameters are developed for each particular mechanism, the IT equations have the coefficients generated from the digraph matrices, ideal for the development of efficient computer implementations of algorithms.

Problems

P4.1: A double pendulum used in manipulating objects located at E (end-effector point) is illustrated in Figure P4.1a. The links' length are $\mathbf{L}_{AB} = 10''$, $\mathbf{L}_{BP} = 5''$, and the location of COM on links is $L_{AG_1} = 4.612''$; $L_{BG_2} = 2.855''$. The links' mass and moments of inertia are $m_1 = 1.050$ [lbm]; $I_{G_1} = 8.370$ [lbm · in.^{2}]; $m_2 = 0.703$ [lbm]; $I_{G_2} = 1.467$ [lbm · in.^{2}]. The relative angular displacements for actuated joints A and B are $\theta_A(t) = 360 \cdot t^2$ [°] and $\theta_B(t) = 270 \cdot t^2$ [°]. The home position (initial position) is defined as the mechanism in the position with both angular actuator displacements set to zero (Figure P4.1b).

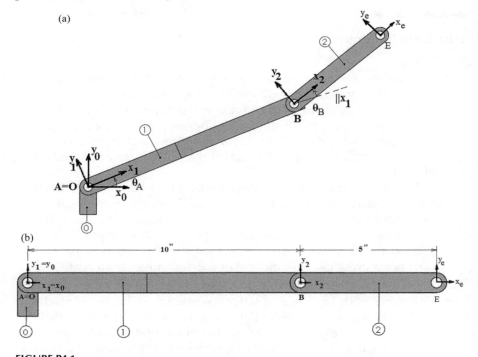

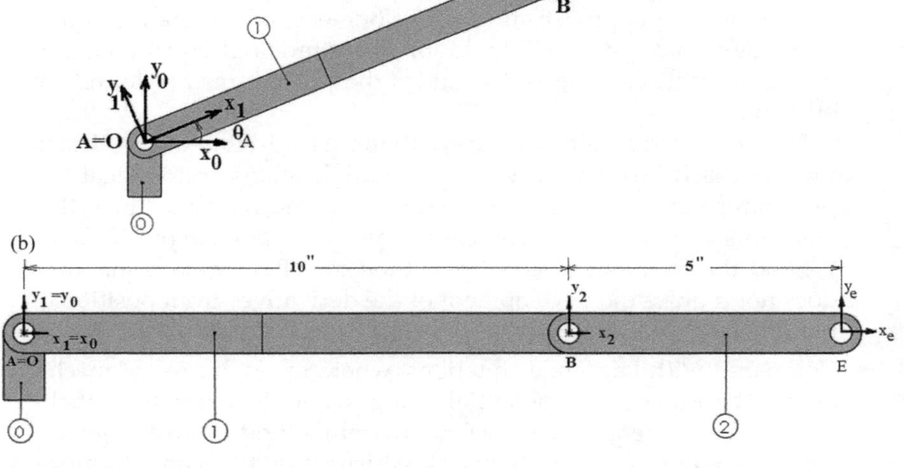

FIGURE P4.1
(a) Double pendulum and (b) home position.

Determine

1. For positional analysis
   a. The mobility of mechanism
   b. The digraph and its matrices **G** and **Z**
   c. Create a spreadsheet with 25 rows for the x and y positional coordinates of joints A, B and COM G_1, G_2 when the total time interval is 1 sec.
2. For velocity analysis, the joints' relative angular velocities are $\dot{\theta}_A(t) = 720 \cdot t\,[°/s]$ and $\dot{\theta}_B(t) = 540 \cdot t\,[°/s]$, respectively.
   a. Create a spreadsheet with 25 rows for links' 1 and 2 absolute angular velocities.
3. For the acceleration analysis, the joints' relative angular accelerations are $\ddot{\theta}_A(t) = 720\,[°/s^2]$ and $\ddot{\theta}_B(t) = 540 \cdot [°/s^2]$.
   a. Create a spreadsheet with 25 rows for links' 1 and 2 absolute angular accelerations
   b. Create a spreadsheet with 25 rows for the x and y components of the linear acceleration for the joints A, B and COM G_1, G_2 when the total time interval is 1 sec
4. For dynamic analysis
   a. Write the IT equations for open cycle mechanism
   b. The differential equations of motion
   c. Create a spreadsheet with 25 rows for actuating torques T_A^{act} and T_B^{act}. Graph their variation when the total time interval is 1 sec
   d. Create a spreadsheet with 25 rows for the x and y components of dynamic reaction forces in the joints A and B. Graph the magnitude of reactions when the total time interval is 1 sec

P4.2: The four-bar mechanism in Figure P4.2 has the link lengths: $L_{AB} = 4''$, $L_{BD} = 24''$, $L_{DC} = 14''$, $L_{CA} = 30''$. The location of COM is as follows: $L_{AG_1} = 2.120''$; $L_{BG_2} = 11.863''$; $L_{DG_3} = 6.722''$.

The crank 1 rotates counterclockwise with the angular displacement $\theta_1 = 2\pi \cdot t\,[\text{rad}]$, angular velocity $\omega_1 = \dot{\theta}_1 = 2\pi\,[\text{rad/s}]$, and angular acceleration $\varepsilon_1 = \ddot{\theta}_1 = 0\,[\text{rad/s}^2]$.

The mass and moment of inertia relative to each link's COM are as follows: $m_1 = 0.596\,[\text{lbm}]$; $I_{G_1} = 1.047\,[\text{lbm} \cdot \text{in.}^2]$; $m_2 = 3.939\,[\text{lbm}]$; $I_{G_2} = 197.744\,[\text{lbm} \cdot \text{in.}^2]$; $m_3 = 2.389\,[\text{lbm}]$; $I_{G_3} = 40.887\,[\text{lbm} \cdot \text{in.}^2]$. At $t = 0.25\,\text{s}$, the crank is in vertical position $\theta_1 = \pi/2\,[\text{rad}]$. For this position, determine

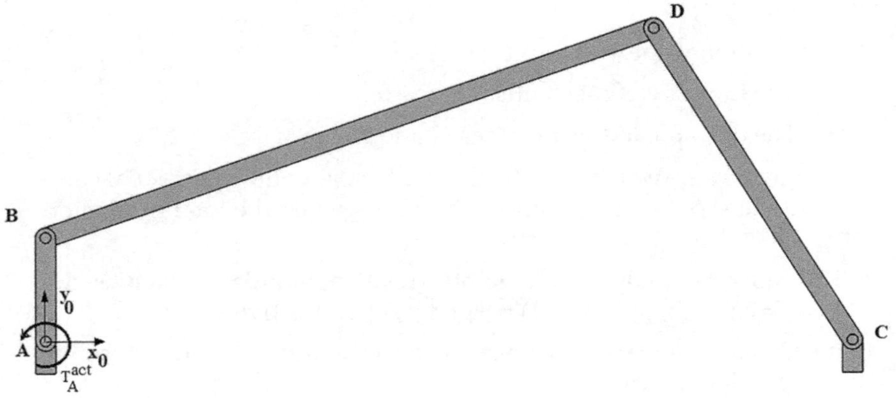

FIGURE P4.2
Four-bar planar mechanism.

a. x and y coordinates for the joints A, B, C, D and com G_1, G_2, G_3
b. Absolute angular velocities for links 2 and 3
c. Absolute angular accelerations for links 2 and 3
d. The x and y components of the linear acceleration for the joints A, B, C, and D
e. The x and y components of the linear acceleration of com G_1, G_2, and G_3
f. Write the IT equations for the single-cycle mechanism
g. The actuating torque T_A^{act}
h. The dynamic reactions in the joints A, B, C, and D
i. The break torque T_3^{br} applied at the link 3's COM to hold the mechanism in equilibrium in the position $\theta_1 = \pi/2$ [rad].

P4.3: The crank-slider offset mechanism shown in Figure P4.3 has the link lengths: $L_{AB} = 4''$, $L_{BD} = 24''$, $L_{DC} = 0.19''$. The location of COM is as follows: $L_{AG_1} = 2.330''$; $L_{BG_2} = 12''$; $L_{DG_3} = 0''$. In the initial position (home position),

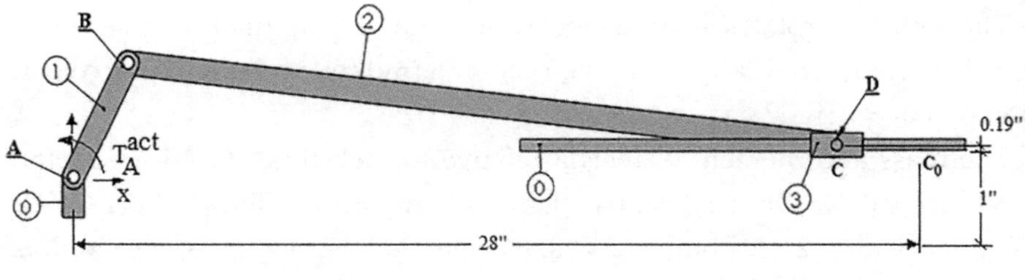

FIGURE P4.3
The crank-slider offset mechanism.

the crank AB makes 0° with the horizontal, where the coordinates of contact point C_0 are (28″, 1″). The crank 1 rotates counterclockwise with the angular displacement $\theta_1 = 2\pi \cdot t$ [rad], angular velocity $\omega_1 = \dot{\theta}_1 = 2\pi$ [rad/s], and angular acceleration $\varepsilon_1 = \ddot{\theta}_1 = 0$ [rad/s^2].

The link masses and moments of inertia relative to links COM are as follows: $m_1 = 0.542$ [lbm]; $I_{G_1} = 0.780$ [lbm · in.^{2}]; $m_2 = 3.9$ [lbm]; $I_{G_2} = 192.299$ [lbm · in.^{2}]; $m_3 = 0.184$ [lbm].

At $t = 0.125$ s, the crank makes with x-axis $\theta_1 = \pi/4$ [rad]. For this position, determine

a. x and y coordinates for the joints A, B, C, D and COM G_1, G_2, G_3
b. Absolute angular velocity for links 2 and 3
c. Absolute angular acceleration for links 2 and 3
d. Linear accelerations of the joints A, B, C, and D
e. Linear accelerations of COM G_1, G_2, and G_3
f. Write the IT equations for the single-cycle mechanism
g. The actuating torque T_A^{act}
h. The dynamic reactions in the joints A, B, C, and D
i. The break force f_3^{br} applied in the x direction on link 3 to hold the mechanism in equilibrium in the position $\theta_1 = \pi/4$ [rad].

P4.4: The two-cycle mechanism shown in Figure P4.4 has given the following link lengths: $L_{EA} = 30''$; $L_{AB} = 4''$; $L_{BF} = 24''$; $L_{FE} = 14''$; $L_{EG} = 7''$; $L_{GD} = 12''$; $L_{DC} = 0.19''$. The location of COM is as follows: $L_{AG_1} = 2.330''$; $L_{BG_2} = 12.000''$; $L_{FG_3} = 6.658''$; $L_{GG_4} = 6.000''$; $L_{DG_5} = 0.000''$.

In the initial position (home position), the crank AB makes 0° with the horizontal, where the coordinates of contact point C_0 are (37.31″, −0.19″). The crank's counterclockwise angular velocity is $\omega_1 = \dot{\theta}_1 = 2\pi$ [rad/s], and its angular acceleration is $\varepsilon_1 = \ddot{\theta}_1 = 0$ [rad/s^2].

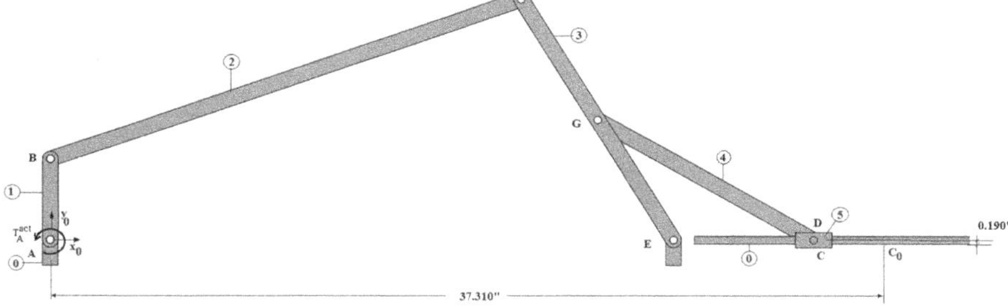

FIGURE P4.4
Two-cycle planar mechanism.

The mass and moment of inertia relative to each link's COM are as follows: $m_1 = 0.542$ [lbm]; $I_{G_1} = 0.780$ [lbm · in.^{2}]; $m_2 = 3.900$ [lbm]; $I_{G_2} = 192.299$ [lbm · in.^{2}]; $m_3 = 2.158$ [lbm]; $I_{G_3} = 34.273$ [lbm · in.^{2}]; $m_4 = 0.250$ [lbm]; $I_{G_4} = 3.177$ [lbm · in.^{2}]; $m_5 = 0.026$ [lbm]. At t = 0.25 s, the crank is in vertical position $\theta_1 = \pi/2$ [rad]. For this position, determine

a. x and y coordinates for all joints
b. The absolute angular velocities for links 2, 3, and 4
c. The absolute angular accelerations for links 2, 3, and 4
d. Linear accelerations of the joints B, C, D, E, F, and G
e. Linear accelerations of COM G_1, G_2, G_3, G_4, and G_5
f. Write the IT equations for the two-cycle mechanism
g. The actuating torque T_A^{act}
h. The dynamic reactions in the joints A, B, C, D, E, F, and G
i. The break force f_5^{br} applied in the x direction on link 5 to hold the mechanism in equilibrium in the position with the crank 1 at $\theta_1 = \pi/2$ [rad].

P4.5: The two-cycle gear train (GT) with fixed parallel axes (Figure P4.5). Figure 3.56c illustrates its associated graph. The GT has metric gears with $d_1 = 50$mm; $d_2 = 120$mm; $d_3 = 125$mm; $d_2' = 60$mm. The modules are selected as $m_D = 2$ mm and $m_E = 2.5$ mm. Then, the gears' number of teeth is as follows: $N_1 = 25$ teeth; $N_2 = 60$ teeth; $N_3 = 50$ teeth; $N_2' = 24$ teeth, where link 2 is a compound gear with two sets of teeth: N_2 and N_2', rotating as a whole. The actuated gear 1 rotates with counterclockwise (ccw) ω_1. Equation (3.183) illustrates the solution for output angular velocities of gears 2 and 3:

$\omega_2 = -\frac{1}{i_D}\omega_1;\ \omega_3 = -\frac{1}{i_D \cdot i_E}\omega_1$, where the constant ratios are defined as

$i_D = \frac{d_2}{d_1} = 2.4;\ i_E = \frac{d_3}{d_2'} = 2.083.$

If there are constant ratios, then the gears' angular accelerations are

$\varepsilon_2 = -\frac{1}{i_D}\varepsilon_1;\ \varepsilon_3 = -\frac{1}{i_D \cdot i_E}\varepsilon_1.$

Appendix 4.1 tabulates the joints' and links' COM position, and Appendix 4.2 tabulates the gear mass and central moment of inertia for each gear.

a. Write the IT equations for the two-cycle mechanism
b. The differential equation of motion for actuating torque T_A^{act}
c. The dynamic reactions in the joints A, B, and C

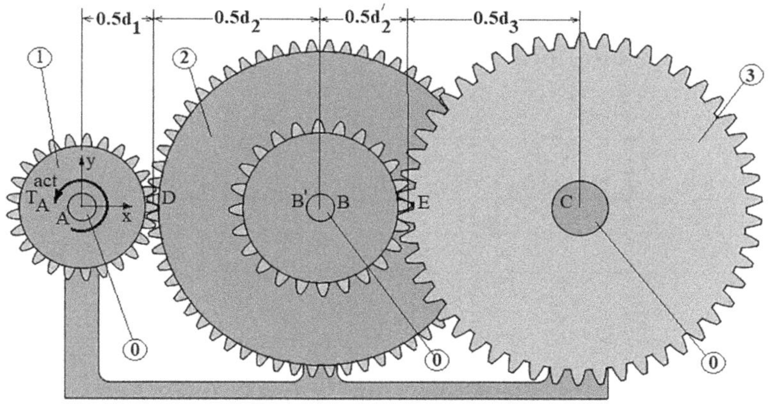

FIGURE P4.5
GT with fixed parallel axes.

d. The reactions in the meshing joints D and E
e. The break force T_3^{br} applied on gear 3 to stop the GT's rotation.

P4.6: Figure P4.6 illustrates a planar frame. Appendix 4.3 tabulates the joints' and links' COM position, and Appendix 4.4 tabulates the external forces and moments reduced to link's COM.

a. Determine the digraph, number of cycles, and the tree
b. Evaluate whether the system is either a frame or a mechanism
c. The static IT moment and static IT force equations
d. Determine the joints' static reaction forces

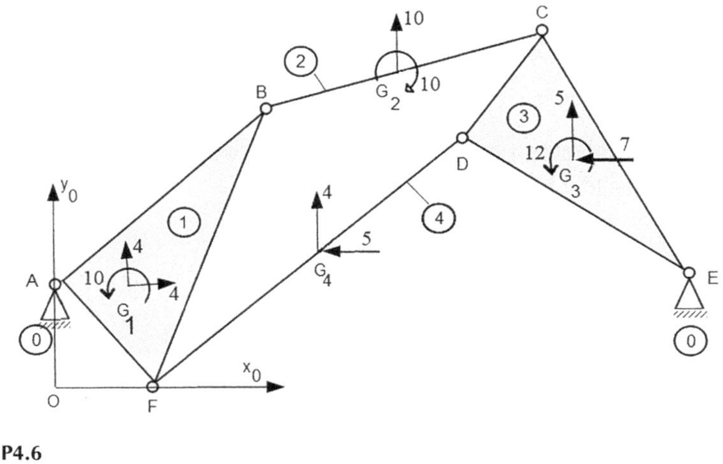

FIGURE P4.6
Planar frame of rigid links.

Appendix 4.1: The Joints and Links' COM Position for Problem P4.5.

	A	B	B'	C	D	E	G_1	G_2	G'_2	G_3
x [mm]	0	85	85	177.5	25	115	0	85	85	177.5
y [mm]	0	0	0	0	0	0	0	0	0	0
z [mm]	0	0	24	24	0	24	0	0	24	24

Appendix 4.2: The Gear Mass and Central Moment of Inertia for Problem P4.5.

	1	2	2'	3
m[grams]	14.666	88.931	21.355	94.427
I_G^z [grams·mm²]	2687.076	81976.347	5032.556	94421.518

Appendix 4.3: The Joints and Links' COM Position for Problem P4.6.

	A	B	C	D	E	F	G_1	G_2	G_3	G_4
x[in]	0	3	18	10	11	5	2	13	12	7
y[in]	2	7	11	7	2	0	3	9.67	5	2.8

Appendix 4.4: The External Forces and Moments Reduced to Link's COM for Problem P4.6.

m →	1	2	3	4
$f_m^{ext,x}$ [lbf]	4	0	-7	-5
$f_m^{ext,y}$ [lbf]	4	10	5	4
$\mu_m^{ext,z}$ [lbf in]	10	-10	12	0

References

1. Wittenburg, J., *Dynamics of Systems of Rigid Bodies*, B.G. Teubner: Stuttgart, 1977.
2. Talpasanu, I., Optimisation in kinematic and kinetostatic analysis of rigid systems with applications in machine design, *PhD Dissertation*, University Politehnica, Bucharest, 1991.
3. Talpasanu, I., Kinematics and dynamics of mechanical systems based on graph-matroid theory, *PhD Dissertation*, University of Texas at Arlington, Arlington, TX, 2004.
4. Talpasanu, I., and Murty, P., Open chain systems based on oriented graph-matroid theory, *SAE International Journal of Passenger Cars-Mechanical Systems*, 1, 2009.

Index

Page numbers followed by *f* indicate figures.

J

K

L